The Legacy of Kurt Schütte

Kurt Schütte, 1909–1998
With kind permission of the family of Kurt Schütte (© by the family of Kurt Schütte)

Reinhard Kahle · Michael Rathjen
Editors

The Legacy of Kurt Schütte

Springer

Editors
Reinhard Kahle
Departamento de Matemática
Universidade Nova de Lisboa
Caparica, Portugal

Michael Rathjen
School of Mathematics
University of Leeds
Leeds, UK

ISBN 978-3-030-49423-0 ISBN 978-3-030-49424-7 (eBook)
https://doi.org/10.1007/978-3-030-49424-7

Mathematics Subject Classification: 03F03, 03F05, 03F15

Cover illustration: Kurt Schütte 1994, photography by Michael Rathjen

This Springer imprint is published by the registered company Springer Nature Switzerland AG
The registered company address is: Gewerbestrasse 11, 6330 Cham, Switzerland

Zu Kurt Schüttes 111. Geburtstag

Kurt Schütte, 1994

Photography by Michael Rathjen

Preface

This book is dedicated to Kurt Schütte, one of the greatest proof theorists of the 20th century. He was born in 1909, in the same year as Gerhard Gentzen, and died in 1998. Schütte studied in Berlin and Göttingen. He was David Hilbert's last doctoral student with a dissertation on the decision problem in mathematical logic, in 1933. He subsequently had to spend the war years up to 1945 as a meteorologist. After the war, gradually reestablishing himself in the academic world, Schütte became a pioneer in infinitary proof theory where one considers proof systems—dubbed "halbformale Systeme" by him—accommodating inference rules with infinitely many premises. The complexity of these derivations is measured via several ordinal magnitudes, and the cost of their proof-theoretic transformations, such as cut elimination, is calibrated in terms of ordinal functions, notably ones developed by G. Cantor and O. Veblen. In the 1950s the proof theory of systems with the ω-rule, a rule which had already been proposed by Hilbert [4], flourished in the hands of Schütte. He extended his approach to systems of ramified analysis, bringing this technique to perfection in his 1960 monograph "Beweistheorie" [11].

In the 1950s Schütte also greatly advanced techniques for proving completeness of proof systems. In his 1956 paper [9] (submitted 1954) he introduced the method of search trees (Suchbäume) where one associates a canonical tree (aka the universal tree), $\mathbb{B}_F$, with a formula F. With it he not only gave a very transparent proof of Gödel's completeness theorem for first order logic but also for infinitary ω-logic.[1] Moreover, in his 1956 paper Schütte also proved that infinitary ω-logic is already complete for cut-free computable derivations in which the premises of the ω-rule are always given by a recursive function (see [9, Theorem 6]).[2]

Schütte was a great minimalist. He saw deeply into Gentzen's calculus, discerning its potential for concentration, and developed the concepts of negative and positive

[1] The technique of search trees is related to Beth's semantic tableaux method [1] and Hintikka's sets of formulas method (nowadays called Hintikka sets) [5, 6] that were developed independently roughly at the same time.

[2] The latter result is sometimes referred to as Shoenfield's completeness theorem as Shoenfield established the completeness of the recursive ω-rule in 1959 in [17] via a different method (no canonical tree); but Schütte clearly has the priority here.

part that generalize Gentzen's of antecedent and succedent.[3] Schütte was rather fond of his calculus, but to his great regret even his former doctoral students later abandoned it in favor of sequent-type calculi. In a manuscript from 1991 (not meant for publication), Schütte recast the ordinal analysis of the theory *KPM* of [8] in his favorite framework, making several technical changes and stressing that the detour via a Tait-style calculus in [8] was superfluous as the essential distinctions are much more transparently captured via positive and negative parts: "*Der überflüssige Umweg über ein Tait-artiges System wurde vermieden, da sich ja mit den äußerst elementaren Begriffen der Positiv- und Negativteile in einfachster Weise genau das ausdrücken läßt, worauf es in der Beweistheorie ankommt.*" [16].

At the end of the 1950s, Schütte made an important contribution to the solution of Takeuti's Fundamental Conjecture (TFC), a problem that was at the center of attention of proof theorists at the time. TFC asserts the eliminability of all cuts in the simple theory of types. Some special cases studied by Takeuti himself via syntactic methods were encouraging. Schütte's main contribution to TFC was a reformulation in equivalent semantic terms, namely, that suitable partial valuations could be extended to total ones [10]. The Fundamental Conjecture was solved eventually along the lines of Schütte's reformulation, first by Bill Tait for second-order type theory and later by Dag Prawitz [7] and Moto-o Takahashi [19] in full and independently.

Perhaps Schütte's most famous contribution to logic is the determination of the limit of predicativity in the guise of the ordinal Γ_0 that he and Solomon Feferman achieved via different methods [2, 12, 13]. However, Schütte's proof-theoretic work of the 1950s and in his book Beweistheorie was crucial for the treatment of predicative analysis in both cases. Γ_0 is known as the Feferman-Schütte ordinal.

The type of predicativity delineated by this ordinal is one described in terms of autonomous progressions of theories, the autonomy condition being due to Kreisel who combined ideas of Poincaré and Russell on predicativity. Without any autonomy conditions, progressions of theories had been studied under the name of ordinal logics by Alan Turing in his 1939 Princeton doctoral thesis [20].

Schütte also worked on non-classical logics. Shortly after Kripke had introduced his semantics for modal and intuitionistic first-order logics, Schütte presented compactness, completeness and soundness proofs for these logic in his usual elegant way in a short monograph [14].[4]

In the year 1977, Springer published Schütte's third monograph with the title "Proof Theory" [15]. Originally it was planed as a translation of his "Beweistheorie" into English but it evolved into a completely different book. It is a masterly economical book that nobody can blame for being too talkative. As a student, the second editor of the present volume was intrigued by it and studied it from cover to cover. Despite having previously read monographs on model theory, recursion theory and set theory, nothing struck him as mysterious and difficult as "Proof Theory". It

[3] These parts constitute the fulcrum of the proof system in [9] and also of the proof system of the 1960 monograph [11] and everything that came after that. The related notion of signed formula was subsequently used in [18], but without reference to Schütte. See also p. 221, p. 223 this volume.

[4] For more on this book's significance see p. 235 this volume.

certainly lived up to Schütte's maxim: "The master builder removes the scaffolding when the building is completed".

The year 1988 saw the publication of "Proof Theory of Impredicative Subsystems of Analysis" by Wilfried Buchholz and Kurt Schütte. The ordinal analyses of theories in this book are based on Buchholz' Ω_σ-rules and the framework of Schütte's proof calculus of positive and negative forms. It is a difficult yet fascinating book that furnishes an amazingly compact and complete treatment of many subsystems of analysis within the span of just 119 pages.

After his retirement Schütte remained very active, keenly following the latest developments in ordinal analysis. By that time he was almost blind. When the second editor visited Schütte in his flat in the middle of the 1990s he helped him lighting his cigarettes as he could only feel their tip. It is quite a miraculous feat how under these circumstances Schütte could not only penetrate the technically intricate and most advanced ordinal analyses of *KPM* and *KP* + Π_3-Reflection but was able to recast them in his favorite calculus, modifying and simplifying the treatment in a very elegant way and thereby furnishing alternative approaches. The last part of this book makes some of these very late papers available for the first time. Hermann Weyl, in one of his last papers [21][5] wrote about aging as a mathematician, recalling a well-known passage from Hardy's *A Mathematician's Apology* [3]:

> *"The mood which Hardy's words reflect with such obvious sincerity is not alien to me who long ago passed sixty, and I agree wholeheartedly with him that 'mathematics is a young man's game.' "*

Fortunately, Schütte gave us a proof that this need not be one's fate.

This book is divided into four parts. The first, titled *History and Memories*, is devoted to Schütte's ways as a proof theorist and as a person. It also contains two articles on the history of logic and proof theory that Schütte (one jointly with Helmut Schwichtenberg) wrote in German. The second part, *Proof Theory at Work*, is concerned with current developments in proof theory that are very close to Schütte's own work in proof theory. The third part, *Further Legacy*, contains contributions by authors who were either in close contact with him, collaborated with him or look at ordinal analysis and constructivity from a different viewpoint. The last part, *Kurt Schüttes Spätwerk*, is comprised of four papers (three of which were previously unpublished) that Schütte wrote in the last decade of his life.

During the preparation of this volume two great logicians whose work was closely intertwined with Schütte's died: Solomon Feferman (1928–2016) and Gaisi Takeuti (1926–2017).

Feferman's contribution to this volume consists of a talk he gave at the Schütte memorial colloquium in Munich on 14 November 1999. It contains his own recollections of Schütte together with a list of challenges for proof theory on the eve of the year 2000. We are happy to include this paper in Chapter 2 of this volume.

Takeuti remembered Schütte in his "Memoirs of a Proof Theorist" [22], and we like to cite the corresponding passages:

---

[5] Written after 1953; Weyl died in 1955 one month after his 70th birthday.

The reader may not believe it, but I think that hardly anybody in the world except my students, Schütte and his disciples read my articles on this subject seriously. Notably, Schütte's interest in my work was the result of Gödel's influence, as I shall explain. Gödel thought that, for the progress of my fundamental conjecture, it would be useful to put Schütte and myself together, and so he invited Schütte to the Institute for Advanced Study. When I dropped by at the Institute one late summer day, a stranger came directly to me, and asked "Would you know Takeuti?" I replied that I was the person himself and then he introduced himself as Schütte and told me: "I have just talked with Gödel, and found that Gödel is interested in your fundamental conjecture" and so on and so forth. I imagine Gödel told him that there might be some kind of relationship between what Schütte had worked on and my fundamental conjecture, and he suggested some research directions to him.

Schütte was a person other than Gödel who became interested in my work. I think Schütte's interest was aroused by Gödel. Looking back, we can say that Gödel's foresight was correct, considering that the results on my fundamental conjecture by Motoo Takahashi and Prawitz were based on Schütte's work done at the Institute at that time.

Thanks to Gödel, during the two years of my stay at the Institute, many logicians such as Bernays, Schütte, and Feferman were there. Smullyan and Putnam were at the University as well. They held a logic seminar every week, and the logic group was very lively. In particular, there were two proof-theorists in the rare Gentzen style together (Schütte and Takeuti), and so we were high-spirited. Smullyan would make me laugh by referring to us in a joking manner: "Is your name TakéSchütte?"

Acknowledgements The editors are grateful to Natalie Clarius who helped with the typesetting of the papers. During the preparation of the volume, the first editor was supported by the Portuguese Science Foundation, FCT, through the project *Hilbert's 24th Problem*, PTDC/MHC-FIL/2583/2014, and UID/MAT/00297/2013 (*Centro de Matemática e Aplicações*) and by the Udo-Keller-Stiftung. The second editor was supported by a grant from the John Templeton Foundation ("A new dawn of intuitionism: mathematical and philosophical advances," ID 60842). He is also very grateful to the Hausdorff Mathematical Research Center in Bonn which allowed him to advance the book in 2018 during the Trimester *Types, Sets and Constructions*.

Tübingen and Leeds, *Reinhard Kahle*
June 2020 *Michael Rathjen*

References

1. Evert Willem Beth. Semantic entailment and formal derivability. *Mededelingen van de Koninklijke Nederlandse Akademie van Wetenschappen*, 18(13):309–342, 1955.
2. Solomon Feferman. Systems of predicative analysis. *Journal of Symbolic Logic*, 29:1–30, 1964.
3. G. H. Hardy. *A Mathematician's Apology*. New York: Cambridge University Press, 1940.
4. David Hilbert. Die Grundlegung der elementaren Zahlenlehre. *Mathematische Annalen*, 104:485–494, 1931.
5. K. Jaako Hintikka. Form and content in quantification theory. Two papers on symbolic logic. *Acta philosophica Fennica*, 8:7–55, 1995.
6. K. Jaako Hintikka. Notes on quantification theory. *Commentationes physico-mathematicae*, 17:13, 1995.
7. Dag Prawitz. Hauptsatz for higher order logic. *Journal of Symbolic Logic*, 33:452–457, 1968.
8. Michael Rathjen. Proof-theoretic analysis of KPM. *Archive for Mathematical Logic*, 30:377–403, 1991.

9. Kurt Schütte. Ein System des verknüpfenden Schließens. *Archiv für Mathematische Logik und Grundlagenforschung*, 2:55–67, 1956.
10. Kurt Schütte. Syntactical and semantical properties of simple type theory. *Journal of Symbolic Logic*, 25:305–326, 1960a.
11. Kurt Schütte. *Beweistheorie*. Springer, Berlin, 1964.
12. Kurt Schütte. Eine Grenze für die Beweisbarkeit der transfiniten Induktion in der verzweigten Typenlogik. *Archiv für Mathematische Logik und Grundlagenforschung*, 67:45–60, 1964.
13. Kurt Schütte. Predicative well-orderings. In Dummet Crossley, editor, *Formal systems and recursive functions*, pages 176–184. North Holland, Amsterdam, 1965.
14. Kurt Schütte. *Vollständige Systeme Modaler und Intuitionistischer Logik*. Ergebnisse der Mathematik und ihrer Grenzgebiete, Band 42. Springer Verlag, Berlin, 1968.
15. Kurt Schütte. *Proof Theory*, volume 225 of *Grundlehren der mathematischen Wissenschaften*. Springer, 1977.
16. Kurt Schütte. Beweistheoretische Analyse von KPM. Type-written manuscript, München, 30.10.1991, 25 pages, 1991.
17. J. R. Shoenfield. On a restricted ω-rule. *L'Academie Polonaise des Sciences, Bulletin, Série des sciences Mathématiques, Astronomiques et Physiques*, 7:405–407, 1959.
18. Raymond M. Smullyan. *First-Order Logic*. Ergebnisse der Mathematik und ihrer Grenzgebiete. Springer-Verlag, Berlin, Heidelberg, New York, 1971.
19. Moto-o Takahashi. A proof of cut-elimination in simple type theory. *Journal of the Mathematical Society of Japan*, 19:399–410, 1967.
20. Alan M. Turing. Systems of logic based on ordinals. *Proceedings of the London Mathematical Society*, 2:161–228, 1939.
21. Hermann Weyl. Axiomatic versus constructive procedures in mathematics. *The Mathematical Intelligencer*, 7:10–38, 1985.
22. Mariko Yasugi and Nicholas Passell, editors. *Memoirs of a Proof Theorist*. World Scientific, 2003. English translation of a collection of essays written by Gaisi Takeuti.

Contents

Part IV Kurt Schüttes Spätwerk

List of Contributors

Toshiyasu Arai
Graduate School of Mathematical Sciences, The University of Tokyo, 3-8-1 Komaba, Meguro-ku, Tokyo 153-8914, Japan, e-mail: tosarai@ms.u-tokyo.ac.jp

Michael Bärtschi
Institut für Informatik, Universität Bern, Neubrückstrasse 10, CH-3012 Bern, Switzerland, e-mail: baertsch@inf.unibe.ch

Wolfgang Bibel
Darmstadt University of Technology, Darmstadt, Germany, e-mail: bibel@gmx.net

Wilfried Buchholz
Mathematisches Institut, LMU München, Germany, e-mail: buchholz@mathematik.uni-muenchen.de

Curt Christian (†)
Institut für Logistik der Universität Wien, Austria

John N. Crossley
Faculty of Information Technology, Monash University, Australia, e-mail: John.Crossley@monash.edu

Solomon Feferman (†)
Stanford University, Stanford, USA

Gerhard Jäger
Institut für Informatik, Universität Bern, Neubrückstrasse 10, CH-3012 Bern, Switzerland, e-mail: jaeger@inf.unibe.ch ·

Reinhard Kahle
Theorie und Geschichte der Wissenschaften, Universität Tübingen, Keplerstr. 2, 72074 Tübingen, Germany and CMA, FCT, Universidade Nova de Lisboa, 2829-516 Caparica, Portugal, e-mail: kahle@mat.uc.pt

Grigori Mints (†)
Stanford University, Stanford, USA

Horst Osswald
Mathematische Institut, LMU München, Germany, e-mail: osswald@mathematik.uni-muenchen.de

Jens Otten
University of Oslo, Oslo, Norway, e-mail: jeotten@ifi.uio.no

Wolfram Pohlers
Institut für Mathematische Logik und Grundlagenforschung, Einsteinstraße 62, 48149 Münster, Germany, e-mail: pohlers@uni-muenster.de

Michael Rathjen
Department of Pure Mathematics, University of Leeds Leeds, LS2 9JT, UK, e-mail: M.Rathjen@leeds.ac.uk

Diana Schmidt
Hochschule Heilbronn, Max-Planck-Str. 39, 74081 Heilbronn, Germany, e-mail: diana.schmidt@hs-heilbronn.de

Kurt Schütte (†)
Mathematisches Institut, LMU München, Germany

Helmut Schwichtenberg
Mathematisches Institut, LMU München, Germany, e-mail: schwicht@mathematik.uni-muenchen.de

Anton Setzer
Department of Computer Science, Swansea University Bay Campus, Fabian Way, Swansea SA1 8EN, UK, e-mail: a.g.setzer@swansea.ac.uk

Stephen G. Simpson
Vanderbilt University, Nashville TN 34720, USA and Pennsylvania State University, University Park, PA 16802, USA, e-mail: sgslogic@gmail.com

Ian Alexander Thomson
School of Mathematics, University of Leeds, Leeds, LS2 9JT, UK, e-mail: i.alecthomson@gmail.com

Sergei Tupailo,
School of Mathematics, University of Leeds, Leeds, LS2 9JT, UK, e-mail: sergei@cs.ioc.ee

Jeroen van der Meeren
Gent Universitet, Vakgroep Wiskunde, Krijgslaan 281, Gebouw S22, 9000 Ghent, Belgium, e-mail: jeroen.vandermeeren@ugent.be

Andreas Weiermann
Gent Universitet, Vakgroep Wiskunde, Krijgslaan 281, Gebouw S22, 9000 Ghent, Belgium, e-mail: andreas.weiermann@ugent.be

Gunnar Wilken
Structural Cellular Biology Unit, Okinawa Institute of Science and Technology, 1919-1 Tancha, Onna-son, 904-0495 Okinawa, Japan. e-mail: wilken@oist.jp

Part I
History and Memories

Chapter 1
"Sehr geehrter Herr Professor!"
Proof Theory in 1949 in a Letter from Schütte to Bernays

Reinhard Kahle

Abstract We present a letter which Kurt Schütte sent in 1949 to his former *de-facto* PhD supervisor Paul Bernays. This letter contains an outline of the proof-theoretic methods which became standard in infinitary proof theory.

1.1 Hilbert's Programme after Gödel and Gentzen

In the 1920s, David Hilbert had conceived a foundational programme in Mathematical Logic, which aimed to provide formal consistency proofs, carried out by "weak means", of formalized mathematics. One rationale behind it was to rebut the criticisms leveled against classical mathematics by intuitionism, and to beat Brouwer at his own game: the metamathematical tools were supposed to be acceptable from an intuitionistic point of view, and if one could prove the consistency of stronger theories by such tools, intuitionists would have to accept them. This was somehow the mathematical strategy of Hilbert, even if the full story was rather more involved, and, at some point, overshadowed by personal quarrels between Hilbert and Brouwer.

Initially, Hilbert had proposed *finitist mathematics* as the metamathematical framework to carry out the intended consistency proofs. Even without a clear specification of finitist mathematics, it is an immediate consequence of Gödel's results that Hilbert's original ideas cannot be carried out. According to testimonies of Bernays and Ackermann, Hilbert immediately adopted a "new meta-mathematical standpoint";[1] today this shift manifests itself by the replacement of finitist mathematics with *constructive mathematics*, as a framework which should still be acceptable

R. Kahle
Theorie und Geschichte der Wissenschaften, Universität Tübingen, Keplerstr. 2, 72074 Tübingen, Germany · CMA, FCT, Universidade Nova de Lisboa, 2829-516 Caparica, Portugal, e-mail: kahle@mat.uc.pt

[1] "Besonders interessiert hat mich der neue meta-mathematische Standpunkt, den Sie jetzt einnehmen und der durch die Gödelsche Arbeit veranlaßt worden ist." [1, p.1f].

R. Kahle and M. Rathjen (eds.), *The Legacy of Kurt Schütte*,
https://doi.org/10.1007/978-3-030-49424-7_1

from an intuitionistic point of view. Soon after Gödel, Gentzen was able to provide a consistency proof of Arithmetic in terms of transfinite induction up to ε_0. While obviously no longer finitist, it is a consistency proof in terms of Hilbert's new methodological standpoint.[2] As much as this result has to be appreciated, one may observe that the development of proof theory in terms of consistency proofs for stronger systems than Arithmetic got somehow stuck. There were several important methodological advances in proof theory and, most notably, Bernays compiled the state-of-the-art of proof theory in the seminal two volume monograph *Grundlagen der Mathematik*, published together with Hilbert [11, 12]. But Gentzen was well aware of the fact that the consistency proof for Arithmetic could, at best, only be a first step towards a consistency proof for Analysis. He wrote in 1938, [6, p. 235f.]:

> Indeed, it seems not entirely unreasonable to me to suppose that *contradictions* might possibly be concealed even in classical *analysis*.
>
> ... the most important [consistency] proof of all in practice, that for *analysis*, is still outstanding.

It is reported that Gentzen worked, up to the end of his life in 1945 in a prison in Prague, on a consistency proof for Analysis.[3] But no concrete result was published.

Schütte was originally attracted to mathematical logic by reading Oskar Becker's *Mathematische Existenz* [2] while studying in Berlin in 1930, [23, p. 93 in this volume]. Upon his return to Göttingen, he found the perfect environment for studying mathematical logic. Eventually, he finished his PhD under the *de-facto* supervision of Paul Bernays in 1933 with a thesis on the decision problem, [16]. David Hilbert was only his formal supervisor, and Schütte actually met him only twice personally, [23, p. 95 in this volume]. One may note that the topic of his thesis was not really about proof theory in the line of Hilbert's Programme. Apparently, by that time he was not particularly involved in Hilbert's Programme, neither in the early, finitist version, nor in the revised, constructive one. Concerning Gentzen—who finished his PhD only a couple of months after him, but already under the formal supervision of Hermann Weyl—he said that they had no personal contact, but only met occasionally in the hallways of the Mathematical Institute in Göttingen.[4] In any case, after his PhD Schütte first pursued the qualifications needed to be a secondary school teacher and switched, in 1937, to meteorology. He, thus, served during World War II as meteorologist in the German army.

[2] For a more detailed discussion of Gentzen's results in this context, see [13].

[3] See, in particular, the last paragraph on page 3 of Schütte's letter, below.

[4] Personal communication by H. Schwichtenberg.

1.2 Schütte's Return to Logic

After the war, Schütte returned to Göttingen and finalized his school teacher education. Having been, for more than a decade, outside of University research, he met Arnold Schmidt by chance in Göttingen. Arnold Schmidt reports about this encounter to Paul Bernays in a letter from May 23rd, 1948:[5]

> Probably it might interest you that Mr. Schütte surfaced here again. During the Nazi time he was at the weather service and returned to school teaching after the end of the war. I met him him—after I had lost sight of him since the time I left Göttingen—some months ago here in Göttingen; and when I asked how his relation to logic had developed, he reacted entirely negatively. He kept himself busy with mathematics, in particular with algebra, once in a while in his leisure hours; but mathematical logic and foundational research doesn't interest him at all any longer; after all, there would be "absolutely nothing to do any longer"; it is totally thought through to the end and in a state of stagnation. I strongly pointed out to him that this would not be correct; and I told him that, if he would come to me, I could tell him immediately a dozen of interesting problems. Several months later he recently arrived; he had just finished his teacher exam and was keen on filling the ensuing compulsory leisure time with mathematics. I informed him about the state of the art of logic and foundational research and tried to give him a number of suggestions; to my joy he dedicates himself with increasing enthusiasm to the recommended reading and suggested lines of thought; maybe in a while it will lead to something interesting.

Bernays's replied, on July 31st, 1948, that he was very much interested and happy to hear about Schütte and that Schmidt should convey his regards.[6]

---

[5] German original:

> Übrigens wird es Sie vielleicht interessieren, dass Herr Schütte hier wieder aufgetaucht ist. Er war während der Nazizeit im Wetterdienst und ist nach Kriegsende zum Schuldienst zurückgekehrt. Ich traf ihn, nachdem ich ihn seit meinem damaligen Weggang von Göttingen ganz aus den Augen verloren hatte, vor einigen Monaten zufällig hier in Göttingen, und als ich fragte, wie sein Verhältnis zur Logik sich entwickelt habe, verhielt er sich vollkommen ablehnend. Er habe sich zwar mit Mathematik, vor allem mit Algebra, noch ab und zu in seinen Mussestunden etwas befasst, aber die mathematische Logik und Grundlagenforschung interessiere ihn garnicht mehr, da sei doch "garnichts mehr zu machen", sie sei doch völlig ausgedacht und stagniert. Ich habe ihn sehr eindringlich darauf hingewiesen, dass das nicht richtig sei, und habe ihm gesagt, wenn er zu mir komme, könne ich ihm gleich ein Dutzend interessanter Probleme sagen. Nach mehreren Monaten kam er vor kurzem an; er hatte sein Referendarexamen gerade gemacht und wollte die nun eintretende erzwungene Mussezeit mathematisch ausfüllen. Ich habe ihn über den Stand der Logik und Grundlagenforschung informiert und ihm eine Reihe von Anregungen zu geben versucht, und ich sehe zu meiner Freude, dass er sich mit wachsendem Eifer den von mir vorgeschlagenen Lektüren und Überlegungen widmet; möglicherweise kommt da in einiger Zeit etwas dabei heraus.

[6] "Was Sie mir über Herrn Schütte schreiben, hat mich lebhaft interessiert; es war mir sehr erfreulich, nach langer Zeit wieder Nachricht über ihn zu haben. Wenn Sie ihn wieder sprechen, möchten Sie ihn von mir grüssen."

On August 9th, 1948, Schmidt can report the first "fruit" of Schütte's renewed interest in Logic:[7]

> I will give your regards to Mr. Schütte. In my last letter I described to you, how he was initially totally opposed to mathematical logic and foundational research and as I succeeded very slowly to interest him again in the important problems. These efforts yielded in the meantime an extraordinary fruit. Mr. Schütte concerned himself with shortenings of Gentzen's proof, however, I could point out to him that these were defective. As a result, he fully committed himself with all his working power to the consistency problem; and after he had completely mastered the topic, as I could clearly see, he extended the method of Gentzen to a consistency proof of analysis, using the formalism of analysis which you presented at the end of the last appendix of the second volume of your book. After a first, rough perusal of the proof, it seems to me that this time the matter is fully correct; I hope to be able to carry out the more detailed checking in the coming week[.]
>
> For the time being, the crucial snag thereby is that Schütte's extension of Gentzen's procedure seems to be formalizable in analysis itself; Mr Schütte is intensely searching for the cause of the ensuing contradiction with Gödel's theorems, whereby he considers a mistake in his proof to be the least likely outcome; he rather believes that hidden side conditions of Gödel's theorem or his own procedure might be the culprit. Therefore, the issue is apparently in a state where one should not talk about it, however, I wanted to let especially you know, very informally, about the embryonic state of developments that take place here.

[7] German original:

> Herrn Schütte werde ich Ihre Grüsse ausrichten. Ich schilderte Ihnen in meinem letzten Brief, wie er zunächst ganz gegen die mathematische Logik und Grundlagenforschung eingestellt war und wie es mir dann ganz allmählich gelang, ihn wieder für die wichtigen Probleme zu interessieren. Dieses Bemühen hat nun inzwischen eine ganz erstaunliche Frucht getragen. Herr Schütte befasste sich zunächst mit Abkürzungen des Gentzen-Beweises, die ich ihm allerdings als fehlerhaft nachweisen konnte. Daraufhin hat er sich nun wirklich mit vollem Ernst und ganzer Arbeitskraft auf das Widerspruchsfreiheitsproblem gestürzt, und nachdem er, wie ich deutlich feststellen konnte, ganz in der Sache firm war, hat er die Gentzensche Methode zu einem Widerspruchsfreiheitsbeweis der Analysis erweitert, wobei jener Formalismus der Analysis zugrundegelegt ist, den Sie im letzten Anhang des 2. Bandes Ihres Buches als letzten angaben. Bei erster, grober Durchsicht des Beweises scheint mir die Sache diesmal ganz intakt zu sein; die genauere Prüfung hoffe ich in der kommenden Woche vornehmen zu können[.]
>
> Der entscheidende Haken dabei ist vorläufig der, dass die von Schütte vorgenommene Erweiterung des Gentzenschen Verfahrens in der Analysis formalisierbar zu sein scheint; Herr Schütte ist eifrig dabei, die Ursache des hieraus entfliessenden scheinbaren Widerspruchs zum Gödelschen Satze nachzuspüren, wobei er einen Fehler in seinem Beweis für am wenigsten wahrscheinlich hält; eher glaubt er an versteckte Nebenvoraussetzungen des Gödelschen Satzes oder seines eigenen Verfahrens. Insofern ist die Angelegenheit offenbar noch in einem Zustande, in dem man nicht von ihr reden sollte, immerhin wollte ich gerade Ihnen ganz inoffiziell doch eben schon einmal von diesen neuen hiesigen embryonalen Entwicklungen Mitteilung machen.

1.3 Schütte to Bernays, August 26th, 1949

In Bernays's *Nachlass* one can find two letters of Schütte to him from 1934. Afterwards the correspondence apparently stopped and only on September 16th, 1948 Schütte resumed the correspondence. In this letter he gives his former supervisor a brief report about his life since 1935 (without even mentioning his new family) and then turning straight to logical matters in line with the style of Arnold Schmidt's letters to him. A letter of November 4th, 1948 reports on some further progress in his work, enclosing some manuscripts (which are not preserved). On April 16th, 1949 he sends Bernays a copy of what became his publication [17]. Four months later, on August 26th, 1949 he sends him a letter which we will reproduce in the following. It contains an outline of Schütte's strategy for the ordinal analysis of (weak systems of) analysis by use of the ω-rule, a strategy which one can consider to be the blueprint for *Schütte-style proof theory* as it is popular in proof theory up to this day.

Fig. 1.1 Photography from the meeting *Kolloquium zur Logistik und der mathematischen Grundlagenforschung „unter der geistigen Leitung von Paul Bernays"* in Oberwolfach in autumn 1949.[9] From left to the right: Irmgard Süß, Hans-Heinrich Ostmann, Paul Bernays, Gisbert Hasenjäger, Arnold Schmidt, Herbert von Kaven, Kurt Schütte.
Sources: Archives of the Mathematisches Forschungsinstitut Oberwolfach and Universitätsarchiv Freiburg. Reprinted with permission.

---

[9] Documents concerning this meeting can be found in the digital archive of the *Mathematisches Forschungsinstitut, Oberwolfach*: https://oda.mfo.de/handle/mfo/1998.

(204) Göttingen, 26. August 1949
Hospitalstr. 4b.

Sehr geehrter Herr Professor!

Ich möchte mir erlauben, Ihnen von einer Abänderung meines Widerspruchsfreiheitsbeweises für die gestufte Analysis zu berichten. Gegenüber dem ursprünglichen Beweis, den ich Ihnen zugesandt hatte, ziehe ich jetzt nämlich eine etwas andere Beweisführung vor, die zu denselben und teilweise noch umfassenderen Ergebnissen führt, aber wesentlich durchsichtiger und kürzer ist. Ich glaube, hiermit eine Form gefunden zu haben, in welcher die Zusammenhänge am klarsten hervortreten. Verwertet werden

1. die von Gentzen entdeckte Möglichkeit, einen logischen Kalkul mit „umweglosen" Herleitungen zu entwickeln (Gentzenscher Hauptsatz),

2. die von Lorenzen entdeckte Möglichkeit, den Gentzenschen Hauptsatz auch unter Einbeziehung der vollständigen Induktion zu retten, und zwar durch Heranziehung von Schlüssen mit unendlich vielen Prämissen,

3. die von Gentzen vorgenommene Zuordnung von Ordnungszahlen zu den Herleitungen und die metamathematisch angewandte transfinite Induktion über ein Anfangsstück der II. Zahlklasse.

Diese Grundgedanken wende ich an:

1. mit meinem aufbauenden logischen Kalkul (ohne Sequenzen und mit logischen Grundzeichen für „oder", „nicht" und „alle"),

2. mit dem Schluß der „unendlichen Induktion"

$$\frac{\mathfrak{A}(z) \vee \mathfrak{B} \quad \text{für alle Ziffern } z}{(x)\,\mathfrak{A}(x) \vee \mathfrak{B}}$$

wie er entsprechend bereits von Hilbert empfohlen wurde,

3. mit einer Zuordnung von Ordnungszahlen zu den Herleitungsformeln, die nur folgenden Bedingungen unterliegt:

a) Bei einem umformenden Schluß haben Ober- und Unterformel gleiche Ordnung,

b) Bei einem aufbauenden Schluß und bei einem Schnitt hat die Unterformel eine größere Ordnung als jede Oberformel.

Dabei sind die Herleitungen finit beschreibbare unendliche Figuren, die noch gewissen Beschränkungen unterliegen. Der Widerspruchsfreiheitsbeweis wird in Form eines Beweises der Schnitt-Eliminierbarkeit geführt. Dieser Beweisgedanke entspricht vollkommen dem von Lorenzen angewandten, die Durchführung ist eine etwas andere. Ich glaube, daß meine Untersuchungen neben denen von Lorenzen deshalb nicht überflüssig sind, weil die

Fig. 1.2 Page 1 of Schütte's letter to Bernays.

Göttingen, August 26th, 1949
Hospitalstr. 4b

Dear Professor!

I would like to take the liberty to report to you on a modification of my consistency proof of stratified analysis.[10] Namely, in contrast to the original proof which I sent to you, I now prefer a slightly different proof strategy, which leads to the same but partly even more comprehensive results, however, which is substantially more transparent and shorter. I think that hereby I found a form in which the connections attain their clearest expression. Herein the following are utilized:

1. the possibility, discovered by Gentzen, of developing a logical calculus with "detour-free" derivations (Gentzen's *Hauptsatz*),
2. the possibility, discovered by Lorenzen, of preserving Gentzen's *Hauptsatz* also under inclusion of formal induction, namely by means of inferences with an infinite number of premises,
3. Gentzen's assignment of ordinals to derivations combined with transfinite induction applied metamathematically over an initial segment of the second number class.

The key ideas I apply together:

1. with my buildup logical calculus (without sequences, using the logical symbols for "or", "not", and "all"),
2. with the inference of "infinite induction"

$$\frac{\mathfrak{A}(\mathfrak{z}) \vee \mathfrak{N} \qquad \text{for all numerals } \mathfrak{z}}{(x)\mathfrak{A}(x) \vee \mathfrak{N}}$$

   as it was accordingly already recommended by Hilbert,
3. with an assignment of ordinals to the formulas of the derivation which is only subject to the following conditions:
   a) in the case of a structural inference, premise and conclusion have the same order,
   b) in case of a buildup inference as well as a cut the conclusion has a higher order than each of the premises.

In this process, the derivations are finitistically describable infinite figures which are subject to certain constraints. This proof idea is entirely in keeping with the one applied by Lorenzen, but the implementation is a slightly different one. I think, that my investigations are not redundant when put next to those of Lorenzen because the

---

[10] "Geschichtete Analysis", as Schütte calls it, is here translated as "stratified analysis". In his 1952 paper, however, he calls it "verzweigte Analysis". The word "verzweigt" also occurs in this letter and we translate it as "ramified".

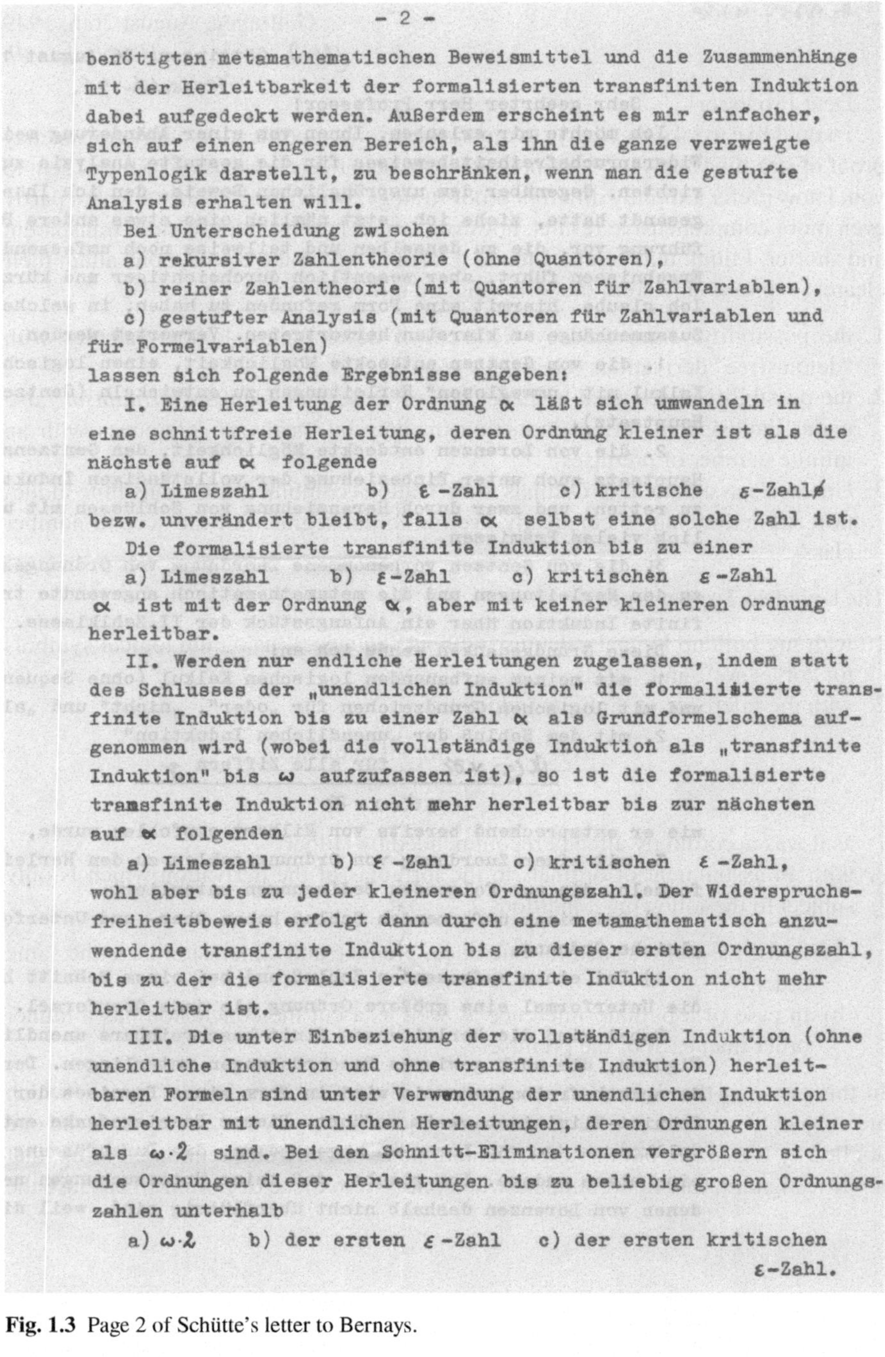

\- 2 -

benötigten metamathematischen Beweismittel und die Zusammenhänge mit der Herleitbarkeit der formalisierten transfiniten Induktion dabei aufgedeckt werden. Außerdem erscheint es mir einfacher, sich auf einen engeren Bereich, als ihn die ganze verzweigte Typenlogik darstellt, zu beschränken, wenn man die gestufte Analysis erhalten will.

Bei Unterscheidung zwischen

a) rekursiver Zahlentheorie (ohne Quantoren),

b) reiner Zahlentheorie (mit Quantoren für Zahlvariablen),

c) gestufter Analysis (mit Quantoren für Zahlvariablen und für Formelvariablen)

lassen sich folgende Ergebnisse angeben:

I. Eine Herleitung der Ordnung α läßt sich umwandeln in eine schnittfreie Herleitung, deren Ordnung kleiner ist als die nächste auf α folgende

a) Limeszahl b) ε-Zahl c) kritische ε-Zahl

bezw. unverändert bleibt, falls α selbst eine solche Zahl ist.

Die formalisierte transfinite Induktion bis zu einer

a) Limeszahl b) ε-Zahl c) kritischen ε-Zahl

α ist mit der Ordnung α, aber mit keiner kleineren Ordnung herleitbar.

II. Werden nur endliche Herleitungen zugelassen, indem statt des Schlusses der „unendlichen Induktion" die formalisierte transfinite Induktion bis zu einer Zahl α als Grundformelschema aufgenommen wird (wobei die vollständige Induktion als „transfinite Induktion" bis ω aufzufassen ist), so ist die formalisierte tramsfinite Induktion nicht mehr herleitbar bis zur nächsten auf α folgenden

a) Limeszahl b) ε-Zahl c) kritischen ε-Zahl,

wohl aber bis zu jeder kleineren Ordnungszahl. Der Widerspruchsfreiheitsbeweis erfolgt dann durch eine metamathematisch anzuwendende transfinite Induktion bis zu dieser ersten Ordnungszahl, bis zu der die formalisierte transfinite Induktion nicht mehr herleitbar ist.

III. Die unter Einbeziehung der vollständigen Induktion (ohne unendliche Induktion und ohne transfinite Induktion) herleitbaren Formeln sind unter Verwendung der unendlichen Induktion herleitbar mit unendlichen Herleitungen, deren Ordnungen kleiner als $\omega \cdot 2$ sind. Bei den Schnitt-Eliminationen vergrößern sich die Ordnungen dieser Herleitungen bis zu beliebig großen Ordnungszahlen unterhalb

a) $\omega \cdot 2$ b) der ersten ε-Zahl c) der ersten kritischen ε-Zahl.

Fig. 1.3 Page 2 of Schütte's letter to Bernays.

necessary metamathematical means of proof and the connections with the deducibility of the formalized transfinite induction are revealed in this way. Moreover, it seems to me easier to restrict oneself to a narrower domain than presented by the system of full ramified type if one wants to preserve stratified analysis (gestufte Analysis).

By distinguishing between

a) recursive number theory (without quantifiers),
b) pure number theory (with quantifiers for number variables)
c) stratified analysis (with quantifiers for number variables and for formula variables),

one can state the following results:

I. A derivation of order α can be transformed into a cut-free derivation whose order is smaller than, respectively, the next
   a) limit number b) ε number c) criticial ε number
   after α, or remains unaltered, if α happens to such a number.
   The formalized transfinite induction up to a, respectively,
   a) limit number b) ε number c) criticial ε number
   α ist derivable with order α, but not with a smaller order.
II. If one permits only finite derivations in that instead of the inference of "infinite induction" one adopts only formalized transfinite induction up to the number α (where ordinary formal induction has to be conceived as "transfinite induction" up to ω), then the formalized transfinite induction is no longer derivable for the next
   a) limit number b) ε number c) criticial ε number, respectively,
   that comes after α, yet is derivable for every smaller ordinal. Thus the consistency proof follows then via a metamathematically applied transfinite induction up to this first ordinal for which the formalized transfinite induction is no longer derivable.
III. The formulae derivable under inclusion of ordinary formal induction (but without the inference of infinite induction and also without transfinite induction) become derivable by means of infinite induction, yielding infinite derivations whose orders are smaller than $\omega \cdot 2$. Subjected to cut elimination, the orders of these derivations increase up to arbitrarily large ordinals below, respectively,
   a) $\omega \cdot 2$ b) the first ε number c) the first criticial ε number.

\- 3 -

Der Widerspruchsfreiheitsbeweis erfolgt durch eine matamathematische transfinite Induktion bis zu dieser Stelle.

Bis zur ersten kritischen ε-Zahl habe ich die transfinite Induktion in der Weise finit dargestellt, daß die Ordnungszahlen als rekursiv eingeführte Zahlzeichen erscheinen.

Bei diesen Widerspruchsfreiheitsbeweisen wird der Gentzensche Höhenbegriff entbehrlich. Die Ordnungszahlen spiegeln in der natürlichsten Weise die Kompliziertheit der Herleitungen wieder. Man sieht, wie durch die Schnitt-Eliminationen die Herleitungen komplizierter und damit die Ordnungszahlen größer werden. So erkennt man, weshalb die metamathematische transfinite Induktion so weit hinauf geführt werden muß.

Die Ergebnisse dürften mit denen von Lorenzen in der Weise zusammenhängen, daß die von Lorenzen angewandte Satzinduktion von dem gleichen Charakter einer transfiniten Induktion der II.Zahlklasse ist wie die über die Ordnungszahlen der Herleitungen erstreckte transfinite Induktion. Bei unbeschränkter Zulassung der Schlüsse mit unendlich vielen Prämissen, wie dies bei Lorenzen der Fall ist, erstreckt sich die entsprechende transfinite Induktion sogar beliebig weit über die II.Zahlklasse.

Meine Untersuchungen haben wir im letzten Semester teilweise im Seminar behandelt, und ich habe auch einen Vortrag darüber vor der Mathematischen Gesellschaft gehalten. Die Arbeiten sollen nun nacheinander in 3 Teilen in den Mathematischen Annalen erscheinen, von denen der 1.Teil die Logik behandelt („Schlußweisenkalküle der Prädikatenlogik"), der 2.Teil die Zahlentheorie (mit der „unendlichen Induktion") und der 3.Teil die gestufte Analysis (ebenfalls mit der „unendlichen Induktion").

Ich habe jetzt auch Aufzeichnungen aus dem Nachlaß von Gentzen gesehen, die allerdings in ihrer knappen Form nicht recht zu verstehen sind. Es scheint mir jedoch folgendes daraus hervorzugehen: 1.Gentzen hat sich dabei anscheinend einen Widerspruchsfreiheitsbeweis für die volle ungestufte Analysis vorgenommen. 2.Die Beweismittel sollten anscheinend von dem Umfang einer gestuften Analysis sein. 3.Die Aufzeichnungen scheinen sich nur auf eine Untersuchung von verschiedenen möglichen Ansätzen zu beziehen, ohne daß darin schon ein zum Ziele führender Weg angegeben ist. Falls Gentzen ihn gehabt hat, wird er verloren gegangen sein.

Fig. 1.4 Page 3 of Schütte's letter to Bernays.

> The consistency proof follows by a metamathematical transfinite induction up to this place.

Up to the first critical ε number, I have provided a finitistic representation of the corresponding transfinite induction in which the ordinals make their appearance as certain recursively introduced numerals.

In these consistency proofs, Gentzen's notion of height can be dispensed with. The ordinals mirror in the most natural way the complexity of the derivations. One sees how cut elimination renders derivations more complex and how thereby the assigned ordinals grow larger. In this way one comprehends why the metamathematical transfinite induction must be performed thus far upwards.

The results should relate to those of Lorenzen in that the sentence induction [Satzinduktion] employed by Lorenzen is of the same character of a transfinite induction over the second number class as the transfinite induction ranging over the orders of derivations. If one allows arbitrary inferences with infinitely many premises, as is the case with Lorenzen, the corresponding transfinite induction will reach arbitrarily high levels of the second number class.

We have studied part of my investigations in a seminar in the last semester, and I also gave a talk about them at the *Mathematische Gesellschaft*. The work should now successively be published in three parts in the *Mathematische Annalen*, of which the first part covers the logic ("Inference calculi of predicate logic"), the second part number theory (with "infinite induction"), and the third part stratified Analysis (also using "infinite induction").

I have now also seen notes from the Nachlass of Gentzen, which due to their condensed form are hardly graspable. But it seems to me that the following can be extracted from them: 1. Gentzen apparently set out to provide a consistency proof for full unstratified Analysis. 2. The methods of proof should apparently lie within the scope of stratified analysis. 3. The notes seem to address only possible approaches to attacking the problem without already indicating a path that leads to its resolution. If Gentzen saw such a path, we will have lost it.

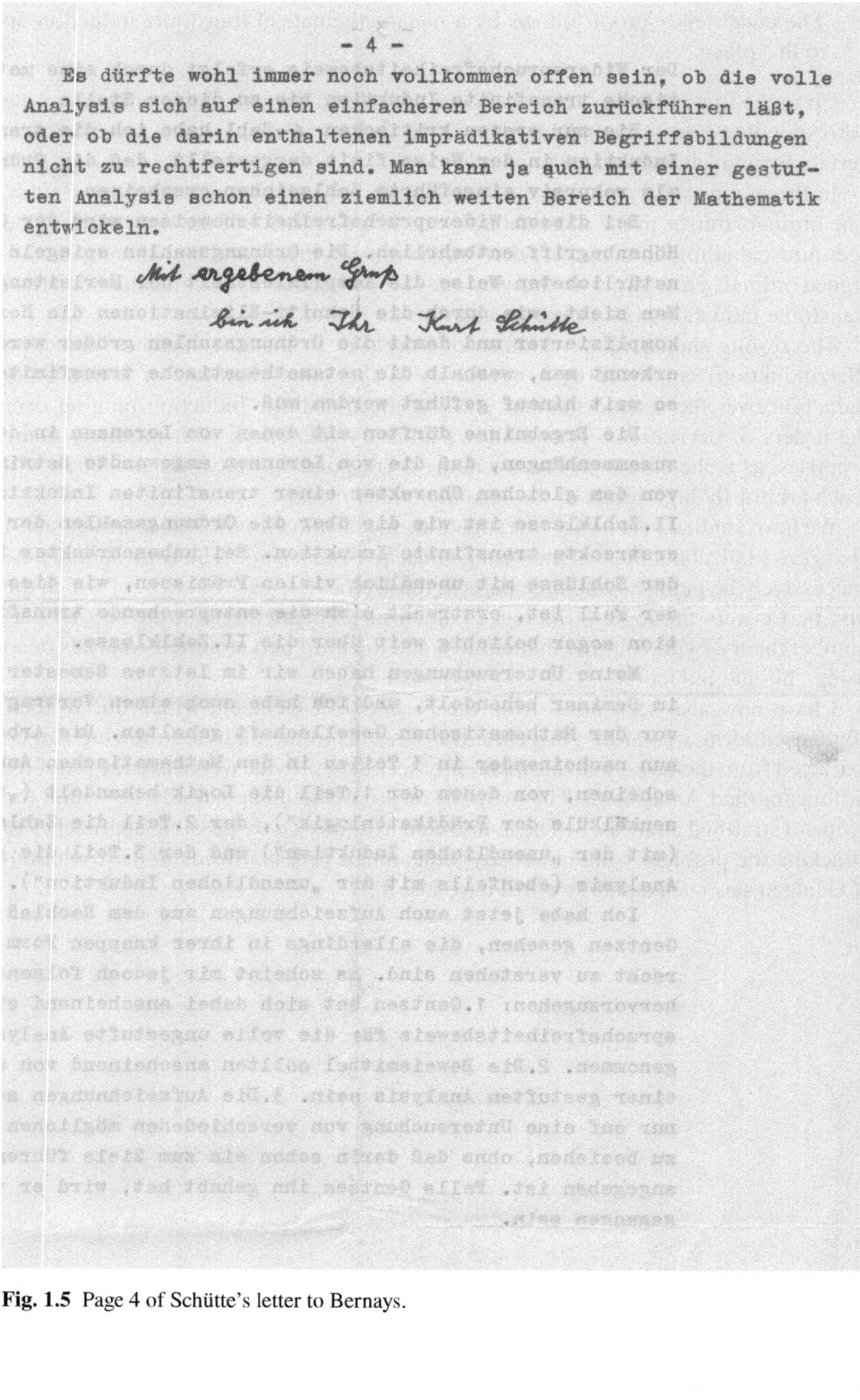

\- 4 -

Es dürfte wohl immer noch vollkommen offen sein, ob die volle Analysis sich auf einen einfacheren Bereich zurückführen läßt, oder ob die darin enthaltenen imprädikativen Begriffsbildungen nicht zu rechtfertigen sind. Man kann ja auch mit einer gestuften Analysis schon einen ziemlich weiten Bereich der Mathematik entwickeln.

Mit ergebenem Gruß

bin ich Ihr Kurt Schütte

Fig. 1.5 Page 4 of Schütte's letter to Bernays.

It is probably still completely open whether full analysis allows for a reduction to a simpler realm, or whether the impredicative concept constructions that it embodies cannot be justified at all. In the end, one can already develop a fairly large chunk of mathematics within stratified analysis.

With most devoted greetings
Yours truly, Kurt Schütte

Of course, this letter tells its own story. We like to comment only on a couple of points.

The central issue, the ω-rule and the frequent references to Lorenzen will be examined in the following section.

On page 1, Schütte refers to a consistency proof for ramified analysis which he had sent earlier to Bernays. We have no further evidence about this proof. When he speaks on page 3 of three parts to be published in the *Mathematische Annalen*, he refers to the publications [17, 18, 19]. The first of it was sent to Bernays with a previous letter on August 26th, 1949, and even already formally submitted to the journal, as the printed version bears the submission date July 31st, 1949. It would not be surprising if Bernays would have been involved in the referee process not only of this, but of all three papers. As they contain all technical details of the results communicated to Bernays, there is no need to discuss these results here in more detail.

In the last paragraph on page 3, Schütte refers to notes of Gentzen which apparently made its way from Prague to Göttingen. The evaluation is clear: "If Gentzen had [a consisteny proof of full unramified Analysis], it will have been lost." It is not clear what happened to these particular notes. It appears to be unlikely that they are related to the notes which were published recently by von Plato [26], as the latter ones don't relate to Analysis.

The final paragraph is of interest with respect to Schütte's skepticism concerning the hope for a *reductive* proof-theoretic analysis of full Analysis. It even contemplates the usefulness of a restricted fragment of Analysis for Mathematics. We will take up this question in the last section.

1.4 The ω-rule and Paul Lorenzen

Paul Lorenzen published, in 1951, a paper with the title "Algebraische und logistische Untersuchungen über freie Verbände" [15]. It contains "a constructive proof of consistency for ramified type theory without the axiom of reducibility" [4, p. 1] and, most notably, uses some form of ω-rule to reach this result. The paper was actually in preparation since the 1940s—see the discussion in [4]—and Schütte will have received a copy of it through Bernays or Arnold Schmidt at the time when he started his work described in the letter to Bernays.[11] Obviously, the various references to Lorenzen in this letter are to this article. Thus, Schütte seems to be historically correct to credit Lorenzen for the introduction of the ω-rule into proof theory. However, there is a specific twist, as Schütte remarks that this idea was "already recommended by Hilbert". Of course, Schütte refers to obscure remarks of Hilbert in his last two papers [9, 8]—one of them based on a talk given by Hilbert on July 17, 1931 in Göttingen, which Schütte might well have attended. But there are two remarks in order: first, to our knowledge, Lorenzen never referred to Hilbert; his motivation for the ω-rule is, in fact, completely independent of Hilbert's considerations and is based on a lattice-theoretic perspective. Thus, we take it for granted that Lorenzen invented the ω-rule independently of Hilbert, and that the link to Hilbert was only observed by Schütte. Secondly, an inspection of Hilbert's wording in [9, 8] shows that he did not propose an ω-rule in a formal calculus.[12] He discussed a "meta-rule" which would allow to introduce a new *axiom* ("Anfangsformel" in German) $\forall n.A(n)$ when one has proofs for all instances of natural numbers $A(0), A(1), \ldots$, (in addition, A is even restricted to be quantifier-free).[13] As close as Hilbert might have been to the ω-rule as we know it today, there is a bold conceptional step necessary to incorporate such a rule at the object level. And while Lorenzen deserves the credit to have seen the potential of the ω-rule for proof theory, it was Schütte who actually worked out the full potential of using it in what we call today *semi-formal systems*, see [21, 22].[14]

1.5 The Legacy of Kurt Schütte

Paul Bernays visited Göttingen for the first time after the war in 1947, but Schütte missed the opportunity to meet him at that time. Shortly after the date of the letter given above, however, both met again in person on the occasion of one of the first

---

[11] Schütte apparently had no personal contact with Lorenzen before 1950 when he wrote a letter to Lorenzen, asking explicitly for permission to cite his still unpublished paper, [20]. Lorenzen studied in Göttingen from 1936 to 1939, at a time when Schütte had already left the University. In 1962, Lorenzen, then Professor in Kiel, wrote a positive letter of reference for Schütte who got his first professorship there in 1963, [27, fn. 20, p. 82]. It is known that both over time fell out with each other, having also followed different paths, Schütte in Mathematics, Lorenzen in Philosophy.

[12] In fact, in [10, p. 370], he had mocked infinitary proofs:

> . . . some stress the stipulation, as a kind of restrictive condition, that, if mathematics is to be rigorous, only a finite number of inferences is admissible in a proof—as if anyone had ever succeeded in carrying out an infinite number of them!

[13] For the proper infinitary rule, Feferman [5, p. 213] writes: "Apparently, the first to consider such a frankly infinitary rule was Tarski in a lecture he gave in 1927, although his first publication in which this was stated is [[25]]." He also refers to Carnap who formulated such a rule in 1934 and 1935, so that Rosser, in 1937, used the designation "Carnap's rule".

[14] A more detailed account to the relation between Lorenzen and Schütte concerning the ω-rule can be found in [14].

international meetings at the *Mathematisches Forschungsinstitut* in Oberwolfach in September 1949 (see the photography on page 7).

Schütte pursued his proof-theoretic investigations, first as an assistent of Arnold Schmidt in Marburg, with international visits to the USA and Switzerland, from 1963 as Professor in Kiel, and, finally, from 1966 onwards as full Professor in Munich where he was able to attract a group of students and to build the *Schütte school*.

In the second half of the 20th century, Kurt Schütte was one of the leading figures shaping proof theory as we know it today. As remarked above, Hilbert's Programme got somehow stuck after Gentzen's consistency proof for Arithmetic. While Lorenzen had the ingenious idea to use an ω-rule for an algebraic analysis of formal derivations, it was Schütte who turned this rule into a handy mathematical tool to give fine-grained proof-theoretic analyses of mathematical theories.

It is interesting to see that Schütte starts his book on *Beweistheorie* [21] with a staunch intuitionistic rejection of the *tertium-non-datur*; and the argument is kept literally the same in the 1977 English edition [22, p. 1f]: "There is no sense in accepting a mathematical truth which is not possibly provable. ... Therefore there is no immediate sense in saying that the principle of *Tertium non datur* ... be accepted as satisfactory for mathematical propositions." He then goes on to argue that the *Tertium non datur* should be justified in just the way Hilbert had proposed. But this is a risky trail. It would imply, in the last instance, that without proper consistency proofs, parts of mathematics may remain "unjustified". Let us recall a sentence of the last paragraph of Schütte's letter to Bernays where he uses the heavily philosophically burdened notion of *justification* (*Rechtfertigung*): "It is probably still completely open whether full analysis allows for a reduction to a simpler realm, or whether the impredicative concept constructions that it embodies cannot be justified at all."

Later, Schütte made it clear, that he sees Hilbert's revised programme at an end at Predicative Analysis (which has as proof-theoretic ordinal Γ_0): "With the limit of predicativity one reached already the limit for the feasibility of the revised Hilbert Programme." [24, p. 78 in this volume]. But—contrary to some philosophically minded logicians, as Lorenzen—this was no reason for Schütte to stop here; neither by rejecting any form of impredicative mathematics, nor by rejecting proof-theoretic studies beyond Γ_0. To the contrary, he promoted proof theory independently of any philosophical motivation, concentrating on the mathematical fruitfulness of the proof-theoretic concepts, tools, and results. Schütte finished his book *Beweistheorie* still with a semi-philosophical reflection on "the different possibilities for the foundation of Analysis" [21, § 36]; in the English edition this is gone and the final paragraph is a technical one on *Proofs of Well-Ordering* [22, § 29]. More generally, "since the beginning of the 1950s Schütte developed proof theory into a structure theory of mathematical proof possibility, independent of the problems of foundations to a large extent." [3, p. 653][15]. In a similar manner, Takeuti [28, p. 97] writes: "One characteristic of the Schütte School is that they shelve the troublesome problem of

[15] In German: "[Schütte] entwickelte seit Anfang der 1950er Jahre die Beweistheorie zu einer von der Grundlagenproblematik weitgehend unabhängigen Strukturtheorie mathematischer Beweismöglichkeiten."

Hilbert's Program, which is entangled with Gödel's theorem, and confine themselves to the theme of determining the proof-theoretic ordinals of formal systems." In our view, this is the true legacy of Kurt Schütte: to have liberated proof theory from any philosophical constraints. And concerning the tension between intuitionism and classical mathematics, Schütte declared in his last public talk[16] unambiguously:

Klassische Mathematik hat sich durchgesetzt!

Acknowledgements We are grateful to the *Hochschularchiv der ETH Zürich* for the permission to reprint the letter of Schütte to Bernays (Hs. 975: 4230). I'm also thankful to Michael Rathjen who revised the English translation of the German texts. Research supported by the Portuguese Science Foundation, FCT, through the project *Hilbert's 24th Problem*, PTDC/MHC-FIL/2583/2014, and UID/MAT/00297/2013 (*Centro de Matemática e Aplicações*) and by the Udo-Keller-Stiftung.

References

1. Wilhelm Ackermann. Letter to David Hilbert, August 23rd, 1933, Niedersächsische Staats- und Universitätsbibliothek Göttingen, Cod. Ms. D. Hilbert 1.
2. Oskar Becker. *Mathematische Existenz*. M. Niemeyer, 1927.
3. Wilfried Buchholz. Schütte, Kurt Wilhelm. In *Neue Deutsche Biographie (NDB)*, volume 23, pages 653–654. Duncker & Humblot, Berlin, 2007.
4. Thierry Coquand and Stefan Neuwirth. An introduction to Lorenzen's "Algebraic and logistic investigations on free lattices" (1951). arXiv: 1711.06139v1 [math.LO], 2017.
5. Solomon Feferman. Introductory note to *[Gödel 1931c]*. In S. Feferman et al., editors, *Kurt Gödel: Collected works, I: Publications 1929–1936*, pages 208–213. Oxford University Press, 1986.
6. Gerhard Gentzen. *Collected Works*. North-Holland, 1969. Edited by M. E. Szabo.
7. David Hilbert. Über das Unendliche. *Mathematische Annalen*, 95:161–190, 1926.
8. David Hilbert. Beweis des Tertium non datur. *Nachrichten von der Gesellschaft der Wissenschaften zu Göttingen, Mathematisch-Physikalische Klasse*, pages 120–125, 1931. Talk given on July 17, 1931 in Göttingen.
9. David Hilbert. Die Grundlegung der elementaren Zahlenlehre. *Mathematische Annalen*, 104(1):485–494, 1931. Talk given in December 1930 in Hamburg.
10. David Hilbert. On the infinite. In Jean van Heijenoort, editor, *From Frege to Gödel*, pages 367–392. Harvard University Press, 1967. English translation of [7].
11. David Hilbert and Paul Bernays. *Grundlagen der Mathematik I*. Die Grundlehren der mathematischen Wissenschaften in Einzeldarstellungen, 40. Springer, 1934. 2nd edition 1968.
12. David Hilbert and Paul Bernays. *Grundlagen der Mathematik II*. Die Grundlehren der mathematischen Wissenschaften in Einzeldarstellungen, 50. Springer, 1939. 2nd edition 1970.
13. Reinhard Kahle. Gentzen's theorem in context. In Reinhard Kahle and Michael Rathjen, editors, *Gentzen's Centenary: The quest for consistency*, pages 3–24. Springer, 2015.
14. Reinhard Kahle and Isabel Oitavem. Lorenzen between Gentzen and Schütte. In Gerhard Heinzmann and Gereon Wolters, editors, *Paul Lorenzen—Mathematician and Logician*. Springer. To appear.
15. Paul Lorenzen. Algebraische und logistische Untersuchungen über freie Verbände. *Journal of Symbolic Logic*, 16(81–106), 1951.

[16] The author of this paper attended this talk; as Schütte was already blind by that time, he didn't read the manuscript as reprinted in this volume [23], but spoke freely from memory.

16. Kurt Schütte. Untersuchungen zum Entscheidungsproblem der mathematischen Logik. *Mathematische Annalen*, 109:572–603, 1934.
17. Kurt Schütte. Schlußweisen-Kalküle der Prädikatenlogik. *Mathematische Annalen*, 122:47–65, 1950.
18. Kurt Schütte. Beweistheoretische Erfassung der unendlichen Induktion in der Zahlentheorie. *Mathematische Annalen*, 122:369–389, 1951.
19. Kurt Schütte. Beweistheoretische Untersuchung der verzweigten Analysis. *Mathematische Annalen*, 124:123–147, 1952.
20. Kurt Schütte. Letter to Paul Lorenzen, Göttingen, May 1st, 1950. Lorenzen-Nachlass, Philosophisches Archiv der Universität Konstanz, PL 1-1-45.
21. Kurt Schütte. *Beweistheorie*, volume 103 of *Grundlehren der Mathematischen Wissenschaften*. Springer, 1960.
22. Kurt Schütte. *Proof theory*, volume 225 of *Grundlehren der Mathematischen Wissenschaften*. Springer, 1977.
23. Kurt Schütte. Bemerkungen zur Hilbertschen Beweistheorie. *Acta Borussica*, 5:241–244, 1995. Reprinted and translated into English in Chapter 8 of this volume.
24. Kurt Schütte and Helmut Schwichtenberg. Mathematische Logik. In Winfried Scharlau, editor, *Ein Jahrhundert Mathematik 1890–1990*, volume 6 of *Dokumente zur Geschichte der Mathematik*, pages 717–740. Vieweg, 1990. Reprinted in Chapter 7 of this volume.
25. Alfred Tarski. Einige Betrachtungen über die Begriffe der ω-Widerspruchsfreiheit und der ω-Vollständigkeit. *Monatshefte für Mathematik und Physik*, 40:97–112, 1933.
26. Jan von Plato. *Saved from the Cellar*. Sources and Studies in the History of Mathematics and Physical Sciences. Springer, 2017.
27. Matthias Wille. Zwischen Algebra und Erlanger Schule. Paul Lorenzens Beiträge zur Beweistheorie. In Ralf Krömer and Gregor Nickel, editors, *Siegener Beiträge zur Geschichte und Philosophie der Mathematik*, volume 1, pages 79–108. universi, 2013.
28. Mariko Yasugi and Nicholas Passell, editors. *Memoirs of a Proof Theorist*. World Scientific, 2003. English translation of a collection of essays written by Gaisi Takeuti.

Chapter 2
Kurt Schütte's Way

Solomon Feferman

Abstract This is the text of an oral contribution to the *Schütte Kolloquium* held at the Mathematical Institute, Munich, Nov. 14, 1999.

2.1 Introduction

I'm very appreciative of the invitation that Wilfried Buchholz and Helmut Schwichtenberg extended to me to open this colloquium in memory of Kurt Schütte. This means a great deal to me both personally and because of his great significance for our field.

I first met Prof. Schütte just forty years ago, in 1959, when we each began a year's visit at the Institute for Advanced Study in Princeton. We last met just a little over two years ago here at the Institute in Munich. During the intervening years we did not have frequent personal contact, but each occasion had special meaning for me. One of the first came in 1965; I had been spending a year's sabbatical in Paris and Amsterdam, and he kindly invited me for a visit to Kiel, where he held a position for a few years before going to Munich. I was invited to stay in his home and was immediately made to feel very comfortable by the warm personal hospitality of Prof. and Mrs. Schütte. Among later memorable occasions were the celebration of his 65th birthday in 1974 at a symposium on proof theory that took place in Kiel, and once more of his 75th birthday here in Munich ten years later. In between, in the latter part of 1980, my wife and I enjoyed a couple of visits to Munich of two weeks each arranged by Helmut Schwichtenberg. That trip was from Oxford, where I was once again on sabbatical. We arrived in a Volkswagen Rabbit and left in a BMW 320i, which I still have and which still runs well, though it is no longer our number one vehicle.

S. Feferman (†)
Stanford University, USA

R. Kahle and M. Rathjen (eds.), *The Legacy of Kurt Schütte*,
https://doi.org/10.1007/978-3-030-49424-7_2

The occasion of my last meeting with Prof. Schütte was in connection with my participation in a Workshop on Reduction, here in Munich, organized by Godehard Link for the Gesellschaft für Analytische Philosophie, and Helmut used the opportunity to invite me to give a lecture for the logic seminar, which I happily accepted. I knew that Prof. Schütte was not in good health, and thus was very surprised to learn that he was planning to attend. His granddaughter brought him about a half hour prior to my lecture and we were able to have a heartwarming conversation, during which we shared many memories. I knew that he had continued to work into his late years despite his blindness, and asked about what he had been up to. He said that he was no longer working but that not long before that he had returned to, and at long last settled, a problem in graph theory connected with a special case of the decision problem for the predicate calculus that had occupied his attention in 1934—that's when he was just starting out after his doctoral work, as Hilbert's last student. That is an example of his remarkable tenacity and clarity of mind to the end. For my logic seminar, Helmut suggested that for the benefit of Prof. Schütte that I speak in a very loud, slow voice, repeating everything that was on my transparencies. He was very attentive throughout, and though he did not ask any questions, I felt that he appreciated the substance of the talk.

From the time of our first meeting to the last, he addressed me as "Professor Feferman", and I addressed him as "Professor Schütte", and though the exchange was overtly formal, and though our contact was relatively infrequent, I always felt a strong personal connection, due as much as anything to his fine, gentle, upstanding and unpretentious personality. His character was mirrored in his features and his bearing, which seemed hardly to change over the years. The image of him that I carry is that of the photograph used as the frontispiece for the 1974 Kiel Symposion volume: the broad forehead, the expressive and welcoming eyebrows and eyes, the firm nose and, finally, the straight mouth with just a touch of humor at the corners.

I plan to spend a fair amount of time today recalling various moments from the 1950s onwards in the development of Kurt Schütte's work and of the school in proof theory that he founded, of its significance for our field and for me personally, and how it related to other approaches to proof theory over the same period, especially in the United States. This has been a period of remarkable progress, dominated by certain paradigms, of which one of the most important was that of ordinal analysis in the style of the Munich school. It is of course coincidental that we are memorializing Prof. Schütte just as we are turning our eyes to the year 2000 and the opening up of a new century. But one can't help but feel that we are witnessing the end of an era, and that this is an appropriate time to begin to assess what has been accomplished and to see where we ought to go from here. I have tried to initiate such an assessment via a list of ten questions that I circulated a few months ago to a number of you and to other workers in the field, and—in the last part of my talk today—I want to tell you something about what we have learned from the responses to that, and of my own views on the subject.

2.2 Beweistheorie

It is an example of Kurt Schütte's perseverance that after having to spend the years 1936–1945 as a meteorologist and the five following years making his way back into the academic world, he was able to return to logic as Arnold Schmidt's assistant in Marburg in 1950 and complete his Habilitation in 1952 at the age of 43. Beginning in 1951 he published a series of fundamental papers in proof theory: on the ordinal analysis of number theory and of ramified analysis, both via the use of semi-formal systems employing infinitary rules of inference, and on systems of ordinal functions and effective systems of ordinal notations used for that purpose but potentially going far beyond those applications. Further contributions concerned a type-free system extending one introduced by Ackermann, and novel formulations of rules of inference for proof-theoretical purposes. All of these things, and more, made their appearance in Schütte's *Beweistheorie*, which was published in 1960. It is by no means my intention to survey his work in or outside of proof theory here, but I think it is worth noting that he also did much work on geometry in the 1950s.

It's my recollection that Schütte was doing the final proof reading of his *Beweistheorie* when he came to Princeton in 1959. That was an exciting year for logic at the Institute. Another distinguished senior visitor was Paul Bernays. The three of us represented three distinct generations. Schütte was there without his family, and Bernays was unmarried. Both gentle and friendly, they would often visit us in our Institute flat on Einstein Drive, and Bernays would play the piano for our young daughters. Other visitors were Gaisi Takeuti (with whom I shared an office), Roger Lyndon and Anne Davis Morel, who had studied with Alfred Tarski a few years before me. Lurking in the wings was Paul Cohen, who was going around asking everybody, and the logicians in particular, what the most important problem would be to solve in their field; we know the outcome of that. The main attraction at the Institute for the logicians was of course Kurt Gödel, who was of some availability via discussions in his office to the senior visitors, as well as, but to a lesser extent to junior visitors like me; he made no public appearances. We visitors in logic at the Institute had a regular seminar with the logicians at Princeton University, led by Alonzo Church; Hilary Putnam was there as a young faculty member, and Martin Davis would come down regularly from New York to join us. Finally, Raymond Smullyan, who was finishing his Ph.D. under Church, was also an active participant. At the time I was immersed in my paper on transfinite progressions of axiomatic systems, which reworked and extended Turing's work on ordinal logics that had been carried out under Church's direction in 1939; curiously, Church had no comment to make about the historical connections when I presented my results in our joint seminar. Though the idea of autonomous progressions was already in the air, I did not start working on them until 1961. I did not know in 1959–60 how important Schütte's work in his papers of the 1950s and in his book would turn out to be to me for the treatment of predicative analysis via autonomous ramified progressions. Moreover, I did not yet consider myself to be one of the proof theorists in any usual sense of the word, and paid only modest attention at that time to what *they* were concentrating on. That year, at the Institute, it was Takeuti's Fundamental Conjecture as to the eliminability of cuts in the simple theory of types that held the center of their attention. Takeuti himself had obtained various special cases by syntactic arguments. One of Schütte's main results in that period was a reformulation of the Fundamental Conjecture in the semantic terms of the extendibility of suitable partial valuations to total ones. That may seem uncharacteristic, given his primary identification with syntactic methods and with the generalized Hilbert program. But he did not shy away from semantic notions, and that plus his exploitation of infinitary methods shows his flexibility of mind, methodologically speaking. The Fundamental Conjecture was solved in the latter part of the 1960s in terms of Schütte's semantical reformulation, first by Bill Tait for second-order type theory and later in full and independently by Dag Prawitz and M. Takahashi. But that only established cut-eliminability or normalizability, not the termination of prima-facie normalization steps. The latter was finally obtained in 1970 independently by Jean-Yves Girard, Per Martin-Löf and Dag Prawitz, by using Girard's method of *candidats de réductibilité* in connection with his functional interpretation of simple type theory. Despite the beauty of the method and the satisfaction of the final result, I think it is fair to say that that was a dead end, since one does not in general have a useful subformula property for normal derivations in type theory, and since there was no associated ordinal analysis for type theory over the natural numbers or other useful information forthcoming.

Now I have to bring Georg Kreisel into the picture. Since the early 1950s he had been evolving his own program in proof theory, that I have described in my article, "Kreisel's unwinding program" for the volume *Kreiseliana*, edited by P. Odifreddi. Kreisel aimed to replace Hilbert's consistency program—one that was in his view philosophically untenable—by a program of extracting mathematically useful information from proofs. Methodologically, he concentrated on functional interpretations, first via generalized Herbrand no-counter-example interpretations extracted from the epsilon-substitution work of Ackermann and later via Gödel's *Dialectica* interpretation. Kreisel first came into my life in 1956 during a visit to Berkeley when I was finishing my work on consistency statements and interpretability of theories, and I became increasingly under his influence. After I moved to Stanford in September of that year he became a regular visitor to Stanford, eventually joining the faculty on a part-time basis. Bill Tait was brought in as an assistant professor of philosophy in 1959 and for a considerable period following that devoted himself totally to the logical and foundational problems raised by Kreisel.

The first point of the connection of Kreisel with Schütte was Kreisel's heavily critical seven page review of *Beweistheorie* in the September 1960 issue of *The Journal of Symbolic Logic*. The criticisms ranged from points of detail to central aims; in addition, the review attacked the book as much for what it *didn't* do as for what it *did* do, an underhanded tactic that I only began to recognize later as typically Kreiselian. I think it was when I had just returned to Stanford from the Institute that he showed me an advance copy of the review, and I was stunned by its unrelenting negativity. I imagined how Schütte might react, and encouraged Kreisel to mollify the criticism somewhat. This came out as follows: "Excellent and permanent as many of the sections of this book obviously are, the title *Beweistheorie* is an exaggeration ... [but] even if it is not a comprehensive account of all *Beweistheorie*, it is a marvellous contribution: one would have hardly believed possible such a combination of completeness, elegance, and conciseness in sustained, purely syntactic reasoning." At least in this final assessment, Kreisel did summarize well the characteristic virtues of Schütte's work. But I imagine he was still badly stung, if not devastated by the review, though I never heard of one word of complaint in response. Nevertheless, as we shall see later, he did take a number of points under consideration when it came to preparing the second, considerably revised, edition.

2.3 Predicativity

One of Kreisel's criticisms of what Schütte did *not* do in *Beweistheorie* concerned the suitability of ramified analysis as a system in which to formalize predicative mathematics. He pointed out that when a prenex statement is proved in ramified analysis, the effect of existential quantifiers is stronger, while that of universal quantifiers is weaker than in the matching statement of unramified analysis. So, he said, "if both kinds of quantifiers occur . . . anything may happen", i.e. it is not clear what its mathematical significance will be in the usual understanding of analysis; these *defects*, he continued, *are gratuitous*. Kreisel said that Schütte should have considered instead unramified systems like that based on Δ^1_1-CA, which have a prima-facie predicative interpretation. What lay behind this comment were the following three results: (1) Kleene's theorem matching sets definable in the ramified hierarchy at level α with sets definable in the hyperarithmetic hierarchy at level $\omega(1 + \alpha)$; (2) Kleene's characterization of the hyperarithmetic sets as exactly those definable in Δ^1_1 form; and (3) Kreisel's result that the hyperarithmetic sets form an ω-model of the scheme Σ^1_1-AC, and hence of Δ^1_1-AC.

The larger background for all this was the increasing synergy in the United States between the fields of higher recursion theory, set theory, model theory and proof theory, that had its first flowering at a remarkable five-week long institute for symbolic logic held at Cornell University in July 1957. This brought together leading representatives of all four fields, together with many of their students. Among the speakers there was Kleene's student Clifford Spector, and one of his two lectures was devoted to predicative set theory in a sense that had been suggested by Hao Wang. Spector proposed to identify the predicative ordinals with the recursive ordinals, and the predicative sets with the ramified analytic sets up to ω_1^{CK}. He considered this proposal to be bolstered by his theorem which showed that the hyperarithmetic well orderings do not go beyond the recursive well-orderings in order type. This can be rephrased as an autonomy condition on definability, but we had to wait until 1960 for Kreisel to propose an autonomy condition on provability as the way to characterize the predicatively provable ordinals. It was only in 1961 that Schütte and I began independently to tackle what that would come to, and that led to our publications in 1964. The results are well known and I have also given an informal account about the way things went in my appendix to the second edition of Takeuti's *Proof Theory*, so I don't need to repeat the details here. Suffice it to say, as I did there, that the upper bound, Γ_0, came essentially ready made from Schütte's ordinal analysis of the ramified systems in his book and that he had obtained in the early 1950s, and that the more difficult part for both of us was to show that this bound was best possible, which we did by rather different methods. Incidentally, I have no recollection as to what led me to choose the letter 'Γ' for this purpose; I don't believe that that had been used previously in Schütte's writings on systems of ordinal notation. Also incidentally, one criticism that *I* had concerning *Beweistheorie*, was that it was retrograde with respect to his own work on such systems. In his 1954 paper "Kennzeichnung von Ordnungszahlen durch rekursiv erklärte Funktionen", the interpretation of the klammersymbols in terms of critical hierarchies of ordinal functions going back to Veblen 1902 was made quite clear, but this was entirely missing from Schütte's book, perhaps to make the treatment look completely constructive. If one had not seen and understood the earlier work, the treatment in the book would look entirely mysterious.

The rest of the story concerning predicativity was quite different for Schütte and me. He did no more work on the subject after that point, whereas that was something I kept coming back to over the following years, first to find predicatively reducible unramified systems of analysis, then to see what parts of mathematics could be formalized directly in such systems, and finally to view predicativity as a special cases of general notions of the reflective closure or unfolding of schematic systems and in particular of predicativity as the unfolding of basic schematic non-finitist arithmetic. When I visited Schütte at Kiel in 1965 I don't believe we spoke at all about predicativity. Returning as a footnote to Kreisel's criticisms, it was not until 1967 that the picture was completed with Harvey Friedman's result that the system Σ^1_1-AC is proof-theoretically equivalent to $(\Pi^0_1\text{-CA})_{<\varepsilon_0}$ and hence has proof-theoretic ordinal $\varphi(\varepsilon_0, 0)$, making it much smaller than Γ_0.

2.4 Breaching the Impredicative Barrier

In the summer of 1963 Kreisel led a seminar on the foundations of analysis at Stanford. Among the participants were Bill Howard, Bill Tait and myself. The primary focus, which at that time was quite distant from my concerns, was Spector's 1962 extension of Gödel's *Dialectica* interpretation of arithmetic in the primitive recursive functionals of finite type to a functional interpretation of full 2nd order analysis by the class of bar recursive functionals, and the question to what extent that could be considered constructive. Since the bar recursive functionals are continuous, Kreisel's thought was that the only way in which constructivity could be established would be via the inductive generation of the representing functions of such functionals, or to be more precise of the indices of such. This led him to consider theories of iterated generalized (i.e. arithmetical) inductive definitions. But then he saw that these would not go beyond something like Δ^1_2-CA + BI, and hence would fall far short of securing full analysis. His conclusion was that in answer to the question, "Can bar recursion of finite type be constructively justified?" the answer "is negative by a wide margin, since not even bar recursion of type 2 can be proved consistent". Despite this disappointment, the seminar opened up a new series of proof-theoretical problems, concerning subsystems of full 2nd order AC and systems of iterated inductive definitions. Howard quickly made significant contributions concerning Spector's proof, simplifying it and making it more perspicuous, and also by investigating special cases of Bar Induction BI, in particular one which turned out to be equivalent in strength to the theory ID_1 of one generalized inductive definition.

Now, Prof. Schütte had spent the year 1962–63 visiting Penn State, where Howard was a student, and they began to have significant contact at that point. Howard has written briefly about this in his article for *Kreiseliana*; I wish I knew more about their contact at that time and later. I believe they had considerable correspondence, and Howard should be encouraged to tell more about how that went. At any rate, it was Howard who established the second genuine connection of the U.S. proof-theorists with Schütte's ordinal analysis program when he determined the proof-theoretic ordinal of a system of Bar Induction equivalent to ID_1, via its functional interpretation. This put a new ordinal star on the map, the Bachmann-Howard ordinal $\varphi(\varepsilon_{\Omega+1}, 0)$ and with it began the study of systems of notation for this and much larger ordinals based on Bachmann's approach first by Harvey Gerber, Hilbert Levitz, and David Isles, and later by Helmut Pfeiffer and others. At the important Buffalo, New York, 1968 conference on proof theory and intuitionism, Gerber gave a recursive ordering for the Bachmann-Howard ordinal, Levitz showed how certain of Bachmann's ordinals matched up with Schütte's systems $\Sigma(n)$ on the one hand and Takeuti's ordinal diagrams $O(n)$ on the other. Isles extended the Bachmann hierarchies and assignment of fundamental sequences to ordinals up to the first inaccessible cardinal. Thus higher set theory made a prima-facie integral appearance in the construction of systems of ordinal notation.

For me, one of the big steps at the Buffalo conference was Harvey Friedman's reduction of the system Σ^1_2-AC to Π^1_1-CA iterated through ε_0. I then showed how α times and $<\alpha$ times iterated Π^1_1-CA could be matched up with suitable iterations of classical theories of generalized inductive definitions of recursive well-founded trees. This suggested that an extension of the constructive ordinal analysis of ID_1 to ID_α would lead to one for Σ^1_2-AC and other interesting subsystems of analysis, and I started plugging that as a route to follow. I began already at Buffalo to look at a functional interpretation of a theory OR_1 of ordinals suitable to interpret ID_1, but its natural extensions to OR_α proved intractable. I then got one of my students at that time—Jeffery Zucker—working on the problem, and though he obtained a number of interesting results relating theories of inductive definitions, theories of ordinals and subsystems of analysis in his 1971 dissertation, the problem of obtaining an ordinal analysis did not make any progress. It was only in 1973 that a new connection with the Schütte school led to major breakthroughs. In that year Wolfram Pohlers attended lectures I gave on that problem at a workshop on inductive definitions in Tübingen, and within the next couple of years he was able to bring to bear the skills that he had mastered in Munich in order to obtain the long sought for results. Shortly after, Wilfried Buchholz was able to obtain the same results using another approach via his $\Omega_{\mu+1}$ rules. And the related problem of reducing classical theories of iterated generalized inductive definitions to corresponding intuitionistic theories of accessibility inductive definitions was solved by them and, independently in part by Wilfried Sieg, following an earlier line suggested by Bill Tait. The full story is told in my introduction, "How we got from there to here", to the 1980 Lecture Notes volume coauthored with Buchholz, Pohlers and Sieg, and I shall not try to elaborate any further on that here.

2.5 Proof Theory, 2nd Edition

The second revised edition of Prof. Schütte's book on proof theory, in excellent English translation by John Crossley, appeared in 1977, and I reviewed it for the Bulletin of the American Mathematical Society a little over a year later. As Schütte himself said in the preface, it was originally intended to be a second edition in the usual sense but in fact it was completely rewritten. Much of the beginning material was carried over from *Beweistheorie*, but following that there was an impressive amount of fresh material in the book, some of it implicitly responding to Kreisel's criticisms of the first edition. For example, included are an exposition of Gödel's functional interpretation of arithmetic, the determination of Γ_0 as the least non-autonomous ordinal of ramified analysis, ordinal analyses of the systems of Δ^1_1-CA and Π^1_1-CA, connections with iterated inductive definitions, and an exposition of Buchholz' $\bar{\theta}$ ordinal notation system which had emerged in the early 70s from a suggestion of mine through work of Peter Aczel and Jane Bridge. All this was very much to the good, but I could not help but voice new criticisms of my own. One concerned the continued use of the treatment of rules of inference in terms of positive and negative parts of formulas instead of more explicit sequential form. I gather from the title of Wolfram Pohlers' talk to follow mine today that there are some genuine advantages, after all, of Schütte's approach, but I could not see any at the time. A second criticism I made in the review was that he omitted the original ordinal analysis of arithmetic via the ω-rule, and instead exposited Howard's assignment of ordinals $<\varepsilon_0$ to the terms for primitive recursive functionals in Gödel's interpretation. It would have seemed more in keeping with the spirit of his earlier more perspicuous approach to at least use Tait's treatment of with the spirit of his earlier more perspicuous approach to at least use Tait's treatment of that via infinitely long terms. And when he came to the treatment of Π^1_1-CA, he followed Pohlers' original treatment assigning ordinals to Takeuti's reduction procedures, instead of one of the subsequent treatments via ordinal analysis of iterated inductive definitions. One final criticism I made concerned the stated general aims of proof theory, but on that I want to delay discussion.

What seemed to me to be particularly retrograde was the assignment of ordinals to reduction procedures for finite derivations à la Gentzen and Takeuti which seemed to have no intrinsic motivation, in contrast to the mathematically clear-cut assignment of ordinals to infinite derivations that Schütte had pioneered and that had changed the face of proof-theoretical ordinal analysis. Wilfried Buchholz' work in recent years showing how intimately the two approaches can be closely connected, after all, for a variety of important formal systems, now undercuts that criticism, though of course there was not an inkling at the time that one could establish anything like such tight relationships.

In the beginning of my review of the second edition I complimented the author on his "meticulous attention to details, careful definitions, complete but compact proofs, and assimilable organization into units and subunits." These are the marks of Kurt Schütte's "ways", and the mark of his great success as a teacher is that he has put that distinctive stamp on the work of his students and his students' students. One general stylistic complaint that I made, though, was that he was too sparing with informal explanations and motivations; happily, that is one respect in which his logical descendants have not followed him. Of course, with the critical aspects of my review, I was now in a position where I wondered how it would affect Prof. Schütte, but there was never any indication that he took these amiss or that it affected our relationship in any way. In general, his equanimity was one of the most admirable aspects of his character.

2.6 Evolution of the Munich School

The contact with Prof. Schütte's students, Wolfram Pohlers and Wilfried Buchholz, that was forged in the mid-1970s flowered not only through my interaction with them as I have already indicated, but also through their students and even their students' students. I could go through a dozen names of those individuals from the Munich school and its extensions to Bern and Münster among other places, with whom I have had useful contact and, in a number of cases, even extended visits at Stanford. Most importantly for me, this has included significant collaboration with Gerhard Jäger since 1979, and more recently with his student, Thomas Strahm. What I have seen in the process is an evolution of what, for simplicity, I shall simply call the Munich school, of approaches to proof theory, which continues to combine the absolute technical mastery so characteristic of the Munich school with a full use of knowledge and motivation gained from higher recursion theory and set theory, and in addition is widening its horizons considerably, especially through applications to computer science. As much as anything, this is the legacy that Kurt Schütte has bequeathed us.

2.7 Where Do We Stand, and Where Do We Go from Here?

A natural reaction to the original ten questions I posed is to ask, why send out such a survey at all? Is reading it going to affect what anyone is doing? Doesn't the field just self-propel in all its different directions without having to think about where we're heading? I understand such doubts and don't have a convincing answer. But I think that all along proof theory has been more introspective than most other fields of research, and that has to do with its origins in Hilbert's program and the necessary steady retreat that has been made from that. We can point to a remarkable record of accomplishment, but in many cases we have difficulty in saying to the outside world what the nature of that accomplishment is. I think of this as a search for our authentic identity, and that some exercises like this in self-assessment can help us find that.

To turn now to the questions themselves, like several other respondents I have found it easier to deal with the questions more or less as a whole rather than one by one, though of course taking them as a point of departure. In doing so, I mainly want to concentrate on open problems/neglected areas, which are relevant to all the questions.

Conceived of most broadly, proof theory should aim to study proofs in the widest sense possible to make it a subject for systematic metamathematical treatment. Mathematical proofs are the richest and most closely examined source of material for such a study, and it is natural to give greatest prominence to them. In a catholic view of proof theory one should not be so confined, and I do so only because that's what I know the best. Even with this restriction, it is not clear that our theory of proofs does justice to the subject matter. I take it for granted that there seems to be no way that various properties of proofs of interest to mathematicians—such as being intelligible, beautiful, clever, novel, surprising, etc.—can be part of a systematic metamathematical study. Relatedly, I don't expect our theory could explain what makes one proof *more* intelligible, beautiful, etc. than another. On the other hand, there is a basic relation between proofs, of interest to mathematicians, which we would expect to be accounted for in our theory and which is not, namely their identity:

(Q1) When are two proofs (essentially) the same?

It's a bit of a scandal that we have no convincing theoretical criterion for that.

Let's step back for a moment. Granted that we have to take an idealized model of proofs for a metamathematical study, in what sense and to what extent can we support the thesis that every proof can be formalized? Of course we know that the issue of the problematic role of diagrammatic and other heuristic arguments in geometric proofs goes back to ancient times, but they seemed to be essential for discovery and communication. Officially, they can be dispensed with, but in practice they can't. Coming up to the present day, one leading mathematician has told me that there are proofs in topology which can't be formalized, because they involve elements of visualization in some essential way. And some logicians in recent years, such as Barwise and Etchemendy, have been investigating what they call heterogeneous reasoning, e.g. proofs that are manipulated on computer screens that involve diagrammatic and iconic elements in some significant way. It's not clear to me if they are making the case that these are non-formalizable in anything like the usual sense. So this leads to the following question:

(Q2) Can all mathematical proofs be formalized?

An answer to this question would require something like Church's Thesis:

(Q3) What is the most general concept of formal proof?

The common view is that the notion of formal proof is relative to the concept of formal axiomatic system. But mathematical proofs don't wear their underlying assumptions (*axioms*) on their sleeves, and the usual process of representing proofs formally does so only with reference to more or less standard background systems. Is there a sense in which proofs can be considered in isolation, or must they always be referenced to a background system? Supposing one had an answer to (Q3). Then one could hope to engage in case studies of challenges to a positive answer to (Q2). But I think it would be profitable to pursue such case studies even without an answer to (Q3).

Here's a related matter that I find puzzling. It's common experience that the standard sort of formal representation of informal statements captures their structure and content quite well, and that there is little uncertainty about how to pass back and forth by hand between the informal and formal statements. Students take to this readily, and we can quickly and convincingly identify errors in mis-representation for them. Note that none of this seems to require prior determination of a background formal language in which these are to be represented. Moreover, once such a language has been specified, it is easy to check mechanically whether a proposed formal representation of a statement is indeed well-formed according to the language. [An aside: it's my impression that there is no generally accepted standard representation of natural language statements at large. What is it about mathematical statements that makes them special in this respect?] But now, when we come to proofs, there is no generally agreed upon standard representation, and for the various kinds of representations (e.g. Frege-Hilbert style, Gentzen-Prawitz natural deduction style, Gentzen sequent style, etc.) which have been developed for proof theory, there is no immediate passage from the informal proof to its formal representation (or even vice-versa) and no generally accepted means of checking such proofs for well-formedness (*correctness*) mechanically, let alone by hand. The proof theory that evolved from Hilbert's program didn't concern itself with such questions. Rather, once one was convinced that a given body of proofs can be represented in principle in a certain way, the only concern was whether a contradiction might be derivable from such a body of proofs.

(Q4) Why isn't there an easier transition from informal but in-principle formalizable proofs to actually formalized, mechanically checkable, proofs?

Even if one has doubts, as I do, about the value of having proofs checked mechanically (because, it seems to me, if you can be clear enough about a proof to set it up for checking, you can already convince yourself of its correctness), there should be value in analyzing what's needed to deal with (Q4), since the conceptual problems that will need to be handled along the way are of independent interest. Perhaps we've been trying to do much in one go—from informal proof to proof on the machine—and instead one should break it up in a series of steps as has been done for the transition from informal algorithms to their implementations.

All of the preceding belongs, I guess, to structural proof theory, a subject which has certainly flourished in recent years through the study of substructural logics, especially linear logic and its kin. Being completely out of that area, I'm in no position to assess its current state of health and prospects, though the fact that there continues to be high activity is evidence that both are good. Another area of considerable recent development is the application of notions of complexity to propositional proof systems, which likewise seems to be being pursued energetically, which is all to the good. Since some of the basic general problems of complexity theory such as $\mathsf{P} \stackrel{?}{=} \mathsf{NP}$ are equivalent to statements about such systems, there is evidently hope that the answer will come through an application of proof-theoretical methods. Time will tell; I'm optimistic that the problem itself will be settled within the next decade, and then we can see what method(s) were the successful ones. Of course, it would be a feather in the cap of proof-theory if the answer came from our community rather than, say, the computer science community or the finite-model theory field.

I turn now to the part of proof theory with which I am most familiar, and which concerns formal systems for various parts of mathematics, ranging from fragments of arithmetic through number theory and analysis up to systems of set theory, with way stations in algebra along the way. The main directions in this are the *reductive* line of development stemming from Hilbert's program, and what I've called the *extractive* line emphasized by Kreisel, but of course stemming from a variety of sources (characterization of the provably recursive functions, Kleene realizability, Gödel functional interpretation, etc.). The reductive line has really split into two related lines, that for lack of better descriptions I will call here the *strength* line, dominated by the project of ordinal analysis as well as other measures of strength, and the *interrelation* line concerned with relations of interpretation, (partial) conservation, etc., between formal systems. Looking at the developments in this part of our subject over the last 50 years, I think the progress we have made is remarkable, and we can truly speak of having mapped out a substantial proof-theoretical landscape, under steady expansion. Moreover, the proof-theorists following these lines of work have a great variety of tools at their disposal and interesting questions of comparative methodology have emerged. Regrettably, the achievements that have been made have been insufficiently appreciated by logicians working in other areas, let alone by the wider mathematical public. I think one of the reasons for this is the welter of formal systems that have been studied, and it is here that some introductory texts for general consumption could serve a useful purpose. They might concentrate on some few principal, *typical* systems for examples and then explain how and why one is led to consider refinements (e.g. via restrictions of comprehension, choice and/or induction principles). By comparison, the promoters of reverse mathematics have been able to reach a wider audience by fixing on their five or six systems that every reasonably literate person can understand, even if the reasons for fixing on those particular systems are not prima-facie compelling in themselves. Of course, every interesting subject has its welter of examples and data (viz., degree theory or large cardinal theory) but the entry to such other subjects is not prohibitive. If students are to be encouraged to pursue these lines of proof theory, and I think they should, more expository work will be needed to bring out the main features of the proof-theoretical landscape.

Now for some questions, beginning with the extractive line. The very familiar but still basic question asked by Kreisel is:

(Q5) What more than its truth do we know of a statement when it has been proved by restricted means?

It is hard to know how this might be answered in such generality. In special cases, for example Π_2^0 statements or Π_1^1 statements, we can say something, such as that we have somehow characterized the associated provably recursive function or the associated ordinal (of unsecured sequences), resp. And for Σ_1^0 statements we would like to say that a witness can be extracted from a proof. But none of these kinds of answers is responsive to:

(Q6) What more than its truth do we know of a Π_1^0 statement when it has been proved by restricted means?

This could be considered a question for the reductive line, where consistency statements are to be established by restricted means, e.g. by quantifier-free transfinite induction on a "natural" well-ordering; but what does the least ordinal of such an ordering tell us about the statement? Perhaps (as suggested to me by Peter Aczel) something simpler can be said, namely as to where these statements sit in their ordering by implication (over some weak system). Note that one can cook up incomparable consistency statements, so the order is not linear. Perhaps there are even natural examples of this sort, given by consistency statements canonically associated with finitely presented (not necessarily finitely axiomatized) formal systems, and that leads us to:

(Q7) What is the ordering relation between canonical consistency statements?

Large-cardinal theorists point to the fact of the *well-ordering* of the consistency statements associated with large cardinals, the motivations for which came from all sorts of different directions, as evidence of some inner harmony among them (and perhaps for their justification). But that's just an empirical fact about one class of consistency statements. It would be remarkable if that were true of canonical consistency statements in general.

To get back on the extractive track, one slogan that has motivated work in this area is that if one can prove a Π_2^0 statement constructively (and, indirectly, even classically) then one can extract a program from the proof. And that accomplishes a double purpose of providing one with a proof of correctness. So far as I know, no new programs of interest have been obtained in this way. In general, if some procedure is proved to exist by usual mathematical reasoning, the associated function is not feasibly computable. For example, it is trivial to prove constructively that any list can be sorted (relative to a given ordering of its items). But completely separate algorithmic analyses have been necessary to find optimal sorting procedures. That shifts the burden to the computer scientist. One could try to get around this by restricting to fragments of arithmetic (or associated fragments of analysis) whose provably recursive functions are feasible in principle. But I think even in these cases, one would not come out with optimal solutions. To be sure, there is a value in developing systematic procedures for extracting such programs even if they don't give optimal solutions, and even if they are for already known problems. What I would hope to see, though, is an answer to:

(Q8) Can we extract previously unknown constructive information such as in-principle algorithms or bounds from constructive or classical proofs?

Once one has such information, one could then subject the situation to ad hoc non-proof-theoretical analysis to see to what extent the information can be optimized. Relatedly, one would hope to obtain more applications of proof theory to mathematics. I researched the literature on this for my article in the *Kreiseliana* volume, and found that the existing applications are few and far between, despite the high hopes (and a certain amount of hype) that originally pushed the "unwinding" program. What is needed is a sustained effort and the development of general methods like those which led to many mathematical applications of model theory, especially to algebra.

(Q9) Can proof-theoretical methods be applied to obtain results of clear mathematical interest comparable to those obtained by model-theory?

Let's come finally to the reductive direction of work. Here a basic problem is that the line of development from the Hilbert program has become so attenuated that one can no longer say in any clear-cut way what is accomplished. The leading work in that direction is that of Arai and Rathjen on the subsystem of analysis based on Π^1_2 comprehension. The Gentzen-Schütte-Takeuti aim was to prove consistency by constructive means. However, if the means employed are so complicated—as they are in the cited work—one cannot assert that one's confidence in the consistency of this system is at all increased thereby. Setting aside the putative value of demonstrating consistency, one could try to see if some reduction to a prima facie constructive system results from that work, in the form of the interrelation line that I suggested in my article on Hilbert's program relativized. As far as I know, no such system is on offer. The other aim that has been pursued was to give up foundational goals in favor of technical results, namely to give an ordinal analysis of the system in question. But we have no criterion for what constitutes an ordinal analysis in general, in terms of which we can see what the cited work achieves. I am in the process of preparing a paper, "Does reductive proof theory have a viable rationale?", which goes further into these matters. I don't currently have an answer to the question raised in that title and which include the work of Arai and Rathjen, but I hope strongly that an answer will emerge that preserves the philosophical significance of reductive proof theory. At any rate, to be specific:

(Q10) What has the proof-theoretical work on Π^1_2 comprehension accomplished? At the very least, how can the work be made more understandable?

My belief is that this will require some new concepts, whether what is accomplished is foundational or is simply an ordinal analysis. In the latter case, it comes back to the old question:

(Q11) What is a natural system of ordinal representation?

Finally, the ordinal analysis program leads us to questions of methodology. As it has been carried out in its most extensive development, this has required the employment of concepts from higher recursion theory and, through that, of higher set theory, involving large cardinal notions, their admissible analogues and their formal use in ordinal representation systems. The phenomenon is even more widespread, and there must be a good, unifying reason as to why it works as well as it does:

(Q12) What explains the success of large cardinal analogues in admissible set theory, explicit mathematics, constructive set theory, constructive type theory, and ordinal representation systems?

In very recent years another method has emerged at the hands of Jäger and his co-workers to deal with impredicative systems such as ID_1 and beyond, that uses only techniques from predicative proof theory (namely, partial cut-elimination and asymmetric interpretation).

(Q13) What explains the success of metapredicative methods? What is their limit, if any? Does metapredicativity have foundational significance?

It may be that metapredicativity stands to locally predicative methods (used for prevailing ordinal analysis) as the use of ordinal representation systems built via transfinite types over the countable ordinals stands to those built over higher number classes.

Still another interesting approach to ordinal analysis is the use of model-theoretic methods recently published by Avigad and Sommer. This has worked successfully on a variety of predicative systems but it is not clear yet if it can break into impredicative territory. If it should be able to compete successfully with ordinal analysis at higher levels, that in my view would constitute the final detachment from the Gentzen-Schütte-Takeuti line, and then ordinal analysis would have to stand on its own as a contribution to the strength line.

To conclude, I have to admit that my feeling about the foundationally reductive line of proof theory had been rather one of discouragement, and that we may be at a dead end, or at least at a loss where to go from here. But, instead, thinking further about the ten questions I had posed and reading the many useful responses I received has made me feel *very* encouraged about the current state and future of our subject as a whole. The further questions I have raised here are my ideas of general directions in which I think it would be profitable for proof theory to move in the coming years. Other workers in the field have in some cases quite different ideas, in other cases ideas that overlap or complement the above. If one thought that all the interesting questions had already been answered or could be answered without any essentially new ideas, we would be in a bad state. I'm convinced just the opposite is true. The fact that there are so many good questions and such good people in the field to deal with them ensures its continuing vitality into the next century.

To be sure, seeing the way through for a foundationally and technically informative proof theory of strong subsystems of analysis will be a great challenge, but if the past is any predictor, we can be confident that people will rise to that challenge. I thus look forward with great optimism to the developments in the coming years.

Chapter 3
. . . and so on: Schütte on Naming Ordinals

John N. Crossley

Abstract Kurt Schütte prized elegance in his work. Ever since the work of Gerhard Gentzen, ordinal numbers have been used as a measure in developing consistency proofs for ever more powerful theories. Here I discuss Schütte's development of his elegant systems of recursive ordinal notations, which went through a small number of phases, but the central idea for his systems of ordinal notation originated from a 1908 paper of Oswald Veblen. However Schütte did not immediately reveal how he had arrived at his quite complicated notations; he simply presented his *Klammersymbolen* as a *fait accompli*. This chapter aims at clarifying the development. Further, when Veblen's technique is viewed as a *process*, the systems can be extended much further. Successive processes seem 'natural' but the quest to characterise 'natural well-orderings' (or even their extent) seems illusory, but Schütte only needed a limited range of ordinals, which he treated elegantly.

3.1 Introduction

Kurt Schütte, the sartorially elegant Munich professor, albeit always with a cigarette in his hand, also prized elegance in his works on logic. In this chapter I shall discuss his development of elegant systems of ordinal notations, which process went through a small number of phases. Many people subsequently both developed his systems and also related them to those of other logicians, notably the ordinal diagrams of Gaisi Takeuti (see [44]). One should also mention the parallels and connections with the work of Sol Feferman (*inter alia* his [16]). The roots of Schütte's work can surely be traced back to his graduate student days in Göttingen though I have not seen records of this. It is well known that Schütte was the last PhD student of David Hilbert, though Paul Bernays was largely his *de facto* supervisor. Schütte graduated in 1934, just

J. N. Crossley
Faculty of Information Technology, Monash University, Australia, e-mail: John.Crossley@monash.edu

R. Kahle and M. Rathjen (eds.), *The Legacy of Kurt Schütte*,
https://doi.org/10.1007/978-3-030-49424-7_3

one year after Gerhard Gentzen. Gentzen had moved to Göttingen in 1929 and wrote his thesis *Untersuchungen über das logische Schliessen* (Investigations of logical deduction), under the supervision of Paul Bernays, which was accepted in June 1933 and published in two parts in 1934–35 ([36], p. 667, [20], translated in [33])). For his degree Schütte had been working on the decision problem, which was a favourite problem of mathematical logic in the 1930s (and subsequently!).

He went into schoolteaching in 1935 and then worked in meteorology from 1936 and throughout World War II. When the war ended he went back to school teaching but was also involved in the renewed but small circle of mathematical logicians in Göttingen. Then in 1950 he became an assistant under the formidable H. Arnold Schmidt in Marburg. In these years he wrote his first book, *Beweistheorie* [41], which I, and others, believe is his most beautiful and elegant work.[1] This contained his first presentation in a book of a system of ordinal notations. 'Schütte . . . indicated that the central idea [for his ordinal notations in [41]] came from a paper written in 1908 by O. Veblen ([46])' (see [11], p. 58).[2] O. Veblen was clearly interested in logic since Gödel talked about his double negation translation of classical into intuitionist arithmetic ([24]) 'in the Menger colloquium, to wit, in the presence of O. Veblen who shortly afterwards went to Göttingen.' ([36], p. 671). Veblen had been in Göttingen in 1932, when Schütte and Gentzen were students there. It seems plausible, therefore, that, being aware of Veblen's interests, Schütte learnt either then or later of Veblen's 1908 paper [46].[3]

Gentzen's work was obviously the major stimulus for Schütte who was one of a 'chosen few' who read it.[4] As von Plato puts it, 'Despite Bernays'[s] efforts, the methods of Gentzen's 1943 paper, or the earlier 1938 paper [[21, 22]], did not become widely used: Both works remained by and large unread in the 1940s and early 1950s' ([36], p. 712). 'Gentzen's last paper [[22]] marked the beginning of ordinal proof theory' ([36], p.668). In the 1950s Schütte was writing his Habilitationschrift on ramified analysis, which probably motivated him to use Veblen's work as this features large in his first edition of the *Beweistheorie* book ([41]).

3.2 A Few Technicalities

As is well known, Gentzen had shown that transfinite induction up to ε_0 was necessary and sufficient to prove the consistency of arithmetic. He also showed that for ordinals, α less than ε_0, that could be 'represented' in arithmetic, transfinite induction up to α was provable in arithmetic. In this way Gentzen demonstrated a true but unprovable

---

[1] It has some deficiencies, which will be mentioned below.

[2] The bibliographic reference numbers in quoted passages have been changed to fit the references listed at the end.

[3] Schütte would have also met Veblen in Princeton when the former was a Visiting Professor in the Institute for Advanced Study in Princeton, but that was in 1959/60 so the book would have been written by then (see [7], p. 653).

[4] Von Plato mentions Schütte, Kreisel, and Takeuti, [36], p. 712.

sentence of arithmetic, since an ordering of type ε_0 can be represented in arithmetic, something he regarded as an advance on Gödel's result. 'As Gentzen points out, transfinite induction is a mathematical principle, whereas Gödel's original proof of incompleteness had produced a "metamathematical" principle that was coded into arithmetic through Gödel numbering' ([36], p. 712).

It is not difficult to show that one can construct well-orderings of type $\leq \varepsilon_0$ representable in arithmetic. It should be remembered that it was not much earlier that Gödel had introduced the notion of representing predicates in formal arithmetic and that such representations were involved for recursive predicates. One way to construct such orderings is by looking at a system that has *finite* presentations for the ordinals less than some ordinal α. Then, just as Gödel did with his Gödel numbering one can get an ordering of a (recursive) set of natural numbers by using the prime numbers.

The first step is to get the representations of the ordinals. Schütte worked from Veblen's *sequences* of ordinals (for a non-successor ordinal α).Veblen had been motivated by Hardy's attempts (see [25]) to find a (unique, necessarily uncountable) sequence of ordinals right through to the ordinal of the continuum. Veblen wrote, 'One of the most interesting problems in the theory of transfinite numbers arises in connection with HARDY's [[25]] scheme for obtaining a subset of the continuum of type Ω.' ([46], p. 280.) 'His main aim though was to extend the initial segment of the countable limit ordinals for which fundamental sequences can be given effectively. Veblen's main tools are the operations of derivation and transfinite iteration applied to continuous increasing functions of ordinals.' ([11], p. 60.)

For ε_0 this was easy: essentially one exploited the Cantor normal form for such ordinals (see, e.g. [8]). To go further Veblen, recognising that ε_0 was a fixed point of the function $\lambda\alpha.\omega^\alpha$, looked at fixed points of other continuous increasing functions of the ordinals, indeed the simplest ones starting with 'the smallest ε-number equal to its subscript' ([46], p. 290). Extending this he defined a function E and noted, 'Of course this function E(x) may be made the basis of a new set T' and the expression of ordinals in terms of smaller ordinals continued indefinitely' ([46], p. 291). This apparently inexhaustible and perennial source of further extensions has not been characterised mathematically and I shall return to it below in section 3.4.

Having defined such functions it is then just a matter of coding the notations for the ordinals using prime factorisation of natural numbers in a simple way. This is enough to guarantee the ordering relation is recursive.

Schütte did not spell this out in his *Beweistheorie* book. Feferman in his review of the second edition wrote: 'What is particularly welcome in the exposition here is that the ordinal-theoretic base is frankly explained and accepted as the source of the notation systems, rather than suppressed as was the case in [41]' ([17], p. 226).

Schütte went through several versions of the ordinal notations becoming more explicit in print about the encoding before he published *Beweistheorie*. It was only after the first edition of *Beweistheorie* was published that he moved to this greater clarity and there is a noteworthy stage in his lectures at Pennsylvania State University when he was visiting there in 1962–3 (see [42], vol. II) where he first gives a standard introduction to ordinal numbers (chapters I and II), before he introduces his notation system in chapter III. Cantor normal form is then expressed in terms of his representation of ordinals by natural numbers ([42], vol. II, p. 58, Chapter III, Theorem 5.5). Observe, however, that the Pennsylvania State Notes [42] were written *after* Schütte's first edition of *Beweistheorie* [41] was published.[5]

3.3 The *Klammersymbol* Revelation

The approach in [42] may be regarded as a condensed version of the *Klammersymbole* (bracket symbols) that he introduced in1954 in [40]. These *Klammersymbole* gave a clear identification of the iterated derived functions of Veblen [46] (especially Theorem 6).

That was not my reaction when I first encountered them! However, like G. H. Hardy in the well-known, and perhaps apocryphal story,[6] I now think Schütte was right when he said:

> Offenbar werden hierdurch die Werte φA in rekursiver Weise eindeutig bestimmt.[7]

Formally the definition proceeds as follows:
A *Klammersymbol* is an expression of the form

$$\begin{pmatrix} a_0 & a_1 & \dots & a_n \\ \alpha_0 & \alpha_1 & \dots & \alpha_n \end{pmatrix}$$

where $a_0, a_1, \dots, a_n, \alpha_0, \alpha_1, \dots, \alpha_n$ are ordinals and $0 \leq \alpha_0 < \alpha < \dots < \alpha_n$. He makes the important definition of equality for *Klammersymbole* almost *sotto voce*:

> Es soll dann und nur dann $A = B$ sein, wenn die Klammersymbole A, B durch Einschaltungen bzw. Streichungen von Spalten $\begin{pmatrix} 0 \\ \alpha_i \end{pmatrix}$ auseinander hervorgehen.[8]

Given a continuous increasing function φ on the ordinals, i.e. one that satisfies $\alpha < \beta \rightarrow \varphi(\alpha) < \varphi(\beta)$ and $\varphi(\lim(\alpha_\beta)) = \lim(\varphi(\alpha_\beta))$, Schütte sets

$$\varphi\begin{pmatrix} a \\ 0 \end{pmatrix} = \varphi(a)$$

$\varphi\begin{pmatrix} a_0 & a_1 & \dots & a_n \\ 0 & \alpha_1 & \dots & \alpha_n \end{pmatrix}$ (with $a_1 \neq 0$) is the a_0-th common solution x of all equations

---

[5] Feferman's recollection is that Schütte was doing the final proof reading of [41] when he was at Princeton in 1959 (see his chapter in this volume).

[6] Hardy is alleged to have said, 'That's obvious', then thought about it, unconvinced. He then walked round the quad at Trinity, came back and said, 'Yes, it *is* obvious.'

[7] [40], p. 17: Obviously the values φA are hereby uniquely determined in a recursive way.

[8] $A = B$ if, and only if, the a and b differ only by the *insertion or deletion* of columns of the form $\begin{pmatrix} 0 \\ \alpha_i \end{pmatrix}$.

$$\varphi \begin{pmatrix} x & a_1^* & a_2 & \dots & a_n \\ \alpha_1^* & \alpha_1 & \alpha_2 & \dots & \alpha_n \end{pmatrix} = x$$

$$\text{with } \alpha_1^* < \alpha_1 \text{ and } a_1^* < a_1.$$

We may consider the αs in $\varphi \begin{pmatrix} a_0 & a_1 & \dots & a_n \\ 0 & \alpha_1 & \dots & \alpha_n \end{pmatrix}$ as indicating the (iterated) successively higher 'derivatives' and the as as the a-th fixed points of those derivatives. The definition clearly gives a new increasing function and taking the sup ensures the continuity. Finally, because he is taking the common solution, the new function builds on the earlier ones and guarantees the determination of an ordinal bigger than all previous ones.

In encoding these *Klammersymbole* into the natural numbers, he first regards a natural number α as its prime factorisation $p_0^{\alpha_0} \cdot p_1^{\alpha_1} \cdot \dots \cdot p_n^{\alpha_n}$ before further decomposing the α_i according to their own prime factorisations.[9] It is this version that is found in *Beweistheorie* [41], pp. 106 ff. The use of prime numbers manages to obscure the (relative) simplicity of the procedure of taking derivatives. In the second edition (pp. 86ff) the definition of the representations of ordinals was spelt out in much more direct detail as was gratefully noted above in Feferman's review ([17], p. 226).

The Veblen approach was quite sufficient even for the ordinal, now known as the Feferman-Schütte ordinal, Γ_0, for predicative analysis (see Feferman's chapter in this volume). The problem with Veblen's approach, in terms of providing (constructive) well-orderings of that size in the natural numbers, was that one could clearly consider the smallest fixed point of *all* the Veblen functions–and start all over again–and again. In later work Schütte did indeed extend the system and this time in a startling new way. I remember being amazed and finding it incomprehensible when Schütte told me of this new approach–it must have been around 1970. On reflection one knows there are lots of unused ordinals and that, as far as a notation system is concerned, the names of such ordinals will just be formal symbols (that can be represented by finite numbers, provided there are not too many). Moreover one can 'collapse' the orderings so that there are no gaps.[10] The technique of using uncountable ordinals in finding sequences of ordinals began with the work of Bachmann [3] and was developed in particular by Pfeiffer [35] and Isles [26, 27]. Subsequently these were developed by Sol Feferman [16] and [18], p. 462, and my former students Peter Aczel [1, 2] and Jane Bridge Kister [28, 29] and then constructively characterised by Buchholz [6]. (See [10], section 6, pp. 67–69 and [11], p. 68 for more detail, and also [43].)

An alternative approach to that employed by Veblen is simply to regard the whole Veblen technique as a *process*. If we consider the process, let us call it $\mathcal{P}$, that Veblen had used for generating all ordinals less than a certain ordinal, say $V_0 = |\mathcal{P}(0)|$. If we now add a notation for V_0, say Υ_0 then we can apply the process again to get

[9] There are bizarrely amusing examples on pp. 107 and 109 of [41] for 8,000,000 and then showing 34560 < 32805.

[10] This collapsing idea may have had its roots in the work of Mostowski, [34], on collapsing models of set theory, see his Theorem 3, p. 147.

$\mathcal{P}(\mathcal{P}(0) \cup \Upsilon_0) = \mathcal{P}^1(0)$ and so on. Eventually we have $\mathcal{P}^\omega(0) = \bigcup \mathcal{P}^n(0)$, which is clearly closed under $\mathcal{P}$ since all the Veblen derivatives have been taken into account. Again, however, we can add a notation for the ordinal of this well-ordering obtained and continue. Intuitively the procedure seems clear, but no-one seems to have been able to define it precisely. This kind of idea, and the never-ending sequence of possibilities for taking higher and higher order processes seems to have built on the popularity of what was then (in the 1960s), quite simple set theory and led me and many others (see e.g. [13] but also [10]) to search for a notion, or to try to find a characterisation of 'natural well-orderings'.

3.4 Natural Well-orderings

What can be seen in the development of ordinal notations is a combination of curiosity–the persistent attraction of infinite numbers, pragmatic adaptation of what is (or was) available and inventiveness. However the culture in which the notion and notations were developed have played a pivotal, but generally uncommented, role.

The developments subsequent to Veblen's work and prior to Schütte's were largely shaped by pure curiosity (specifically in the desire to provide a constructive theory of ordinals) and the need for infinite well-orderings that had constructive properties (in the development of proof theory) (see the preface to [43], p. 2).

With the advent of Intuitionism and constructive approaches to mathematics in general, attempts were made to create a constructive theory of ordinals. The strict intuitionistic approach as pursued by Brouwer (see, e.g., Kreisel's review [31] and Brouwer [5]) led to a complicated and turgid approach that has not attracted much direct attention since as a correlate to Cantor's classical theory.[11] On the other hand, with the definition of recursive (computable) functions, various studies were undertaken considering representations of infinite ordinals by natural numbers (or, what is equivalent in a rather wide sense, notation systems with finite alphabets).

Kleene's 1938 attempt [30] certainly provided notations for a significant class of ordinals, Kleene's system $\mathcal{O}$, namely up to Church-Kleene ω_1^{CK}– an ordinal not otherwise computed. Unfortunately, for most infinite ordinals it provided sequences up to such ordinals that were not 'constructively comparable'. (See e.g. [14] for the case of ω^2.)

In retrospect it seems that the search for 'natural well-orderings' was reminiscent of the paradox of Achilles and the Tortoise, except that we did not catch the tortoise.[12] There have been many discussions of the 'problem' of 'natural' well-orderings.[13] It

[11] One should not, however, forget all the work done on the Brouwer-Kleene ordinals using Kleene's system $\mathcal{O}$, see below.

[12] [11] summed up the situation in the late 1980s and gives much more detail on the players in the period from around 1960 to 1986, while [38], see below, is the most up to date consideration I have found.

[13] Curiously the problem was recently described in [45] as 'Feferman's *natural well-ordering problem*', although the authors referred to [11] and its discussion.

is not clear who originated the phrase, though perhaps it was Kreisel,[14] and the more one looks into it the less well-defined it becomes (see the end of [11]).

More recently Rathjen has sensibly questioned the notion of 'natural well-ordering' ([38], p. 225):

> ... in [32], Kreisel seems to seek naturalness in algebraic characterizations of ordered structures. Feferman, in [18], discerns the properties of completeness, repleteness,[15] relative categoricity and preservation of these under iteration of the critical process as significant features of systems of natural representation. Girard [23] appears to propose dilators to capture the abstract notion of a notation system for ordinals. However, in my opinion, similar attempts in the Philosophy of Science of defining 'natural properties' and the complete failure of these attempts show that it is futile to look for a formal definition of 'natural well-ordering' that will exclude every pathological example.[16]

There are obviously various possible definitions of 'constructively comparable' and the most obvious is that there be a recursive isomorphisms between two such sequences, or to put it another way, recursive isomorphisms between two representations of a given ordinal. Following Dekker and Myhill's approach to constructive cardinals (see [15]) the theory of constructive order types ([9, 12] provided an approach to classification but did not show how to construct notations for larger and larger ordinals.

It was from the 1960s that interest in 'natural well-orderings' seems to have flourished most, though researchers used earlier results from classical set theory, the most notable source in addition to Veblen's work [46] being the constructions of Heinz Bachmann. (See [3, 4] for the original work and [1, 2, 29] for the revisiting of Bachmann's work). However 'the subject of natural well-orderings may be said to have been born about the same time as Professor Schütte!' [who was born in 1909]. ([11], p. 58.)

A Personal Note: One Last Song

When we started the Logic Seminar in Oxford in the early 1960s Hao Wang, who was then Reader in the Philosophy of Mathematics, proposed we study Schütte's (then) new book *Beweistheorie* [41]. We did so and, like Feferman (see above Section 3.2), we were perplexed by the recursive well-ordering. However Michael Dummett and I invited Schütte to speak at the 1963 Logic Colloquium in Oxford. Subsequently he gave me a copy of his Penn State notes [42], which made life easier. He later told me that his Munich seminar had studied my thesis (mostly published in [9]), which I still regard as a great honour. He also invited me to Munich for the first of numerous visits. He was always generous, supportive and, like his wife, very kind. I have benefitted greatly from these visits, especially working with Martin Wirsing (whom I first met when he was Schütte's graduate student) and our students and

[14] He mentions it in his survey [32], but the term was already in use in the early 1960s.

[15] I.e. no ordinals are missed out.

[16] See also his article [39].

colleagues writing many papers and a book [37]. I have developed a great affection for Munich and Müncheners over forty years. I am eternally grateful to Herr Schütte who made all that possible.

Acknowledgements Apologies to the later Sol Feferman re the title, see [19]. Many thanks to Wilfried Buchholz for helpful comments on, and an important correction to, an earlier draft.

References

1. Peter Henry George Aczel. A new approach to the Bachmann method for describing countable ordinals. Preliminary summary. Unpublished., 1969.
2. Peter Henry George Aczel. Describing ordinals using functionals of transfinite type. *Journal of Symbolic Logic*, 37:35–47, 1972.
3. Heinz Bachmann. Die Normalfunktionen und das Problem der ausgezeichneten Folgen von Ordinalzahlen. *Vierteljahresschrift der Naturforschenden Gesellschaft in Zürich*, 95:115–147, 1950.
4. Heinz Bachmann. *Transfinite Zahlen*. Springer, Berlin, 1955.
5. Luitzen Egbertus Jan Brouwer. Begründung der Mengenlehre unabhängig vom logischen Satz vom ausgeschlossen Dritten, Erster Teil: Allgemeine Mengenlehre. In Arend Heyting, editor, *L. E. J. Brouwer, Collected Works. volume 1, Philosophy and Foundations of Mathematics*, pages 150–190. North Holland Publishing Company, Amsterdam, Oxford, New York, 1975. Previously published in KNAW Verhandelingen 1^e sectie 12 no. 5, 43 pp.
6. Wilfried Buchholz. *Rekursive Bezeichnungssysteme für Ordinalzahlen auf der Grundlage der Feferman-Aczelschen Normalfunktionen* Θ_α. PhD thesis, Ludwig-Maximilians-Universität, Munich, 1974.
7. Wilfried Buchholz. Schütte, Kurt Wilhelm. In Bayerische Akademie der Wissenschaften. Historische Kommission, editor, *Neue Deutsche Biographie, volume 23*, pages 653–654. Duncker & Humblot, Berlin, 2007.
8. Georg Cantor. Beiträge zur Begründung der transfiniten Mengenlehre. *Mathematische Annalen*, 49(2):207–246, 1897.
9. John Newsome Crossley. Constructive Order Types, I. In John Newsome Crossley and Michael Anthony Eardley Dummett, editors, *Formal Systems and Recursive Functions*, pages 189–264. North Holland Pub. Co., Amsterdam, 1965.
10. John Newsome Crossley and Jane Bridge Kister. Natural well-orderings. *Archiv für mathematische Logik und Grundlagenforschung*, 26(1):57–76, 1987.
11. John Newsome Crossley and Jane Elizabeth Kister. Natural well-orderings. *Archiv für mathematische Logik und Grundlagenforschung*, 26:57–76, 1986/87.
12. John Newsome Crossley and Anil Nerode. *Combinatorial functors*, volume 81 of *Ergebnisse der Mathematik und ihrer Grenzgebiete*. Springer, New York, 1979.
13. John Newsome Crossley and Rohit J. Parikh. On isomorphisms of recursive well-orderings (abstract). *Journal of Symbolic Logic*, 28:308, 1963. See correction in *J.S.L.* 31:525–538, 1966.
14. John Newsome Crossley and Kurt Schütte. Non-uniqueness at ω^2 in Kleene's O. *Arch. f. Math. Logik Grundlagenforsch.*, 9:95–101, 1966.
15. Jacob C. E. Dekker and John Myhill. Recursive equivalence types. *University of California Publications in Mathematics (NS)*, 3:67–214, 1960.
16. Solomon Feferman. Systems of Predicative Analysis, II, Representations of ordinals. *Journal of Symbolic Logic*, 33:193–220, 1968.
17. Solomon Feferman. Review of *proof theory*, by Kurt Schütte. *Bulletin of the American Mathematical Society*, 1(1):224–228, 1979.

18. Solomon Feferman. Proof Theory: a personal report. pages 447–485. 1987. In [44].
19. Solomon Feferman. And so on . . . : Reasoning with infinite diagrams. *Synthese*, 186(1):71–386, 2012.
20. Gerhard Gentzen. Untersuchungen über das logische Schließen. *Mathematische Zeitschrift*, pages 176–210, 405–431, 1934–35.
21. Gerhard Gentzen. Neue Fassung des Widerspruchsfreiheitsbeweises für die reine Zahlentheorie. *Forschungen zur Logik und zur Grundlegung der exakten Wissenschaften*, pages 19–44, 1938.
22. Gerhard Gentzen. Beweisbarkeit und Unbeweisbarkeit von Anfangsfällen der transfiniten Induktion in der reinen Zahlentheorie. *Mathematische Annalen*, pages 140–161, 1943.
23. Jean-Yves Girard. Π^1_2-logic. Part I: Dilators. *Annals of Mathematical Logic*, 21, 1981.
24. Kurt Gödel. Zur intuitionistischen Arithmetik und Zahlentheorie. *Ergebnisse eines mathematischen Kolloquiums*, 4:34–38, 1933.
25. G. H. Hardy. A theorem concerning the infinite cardinal numbers. *Quarterly Journal of Mathematics*, 35:87–94, 1904.
26. David Isles. Regular ordinals and normal forms. In Akiko Kino, John Myhill, and Richard E. Vesley, editors, *Intuitionism and proof theory; Proceedings of the Summer Conference at Buffalo, NY, 1968*, pages 339–361, Amsterdam, 1970. North Holland Publishing Company.
27. David Isles. Natural well-orderings. *Journal of Symbolic Logic*, 36, 1971.
28. Jane Elizabeth Bridge [Kister]. *Some problems in mathematical logic. Systems of ordinal functions and ordinal notations*. Dphil thesis, Oxford University, 1972.
29. Jane Elizabeth Bridge [Kister]. A simplification of the Bachmann method for generating large countable ordinals. *Journal of Symbolic Logic*, 40:171–185, 1975.
30. Stephen C. Kleene. On notation for ordinal numbers. *The Journal of Symbolic Logic*, 3(4):150–155, 1938.
31. Georg Kreisel. Review of Crossley [9]. *Zentralblatt für Mathematik und ihrer Grenzgebiete*, 145:10–12, 1968.
32. Georg Kreisel. A survey of proof theory. *Journal of Symbolic Logic*, 3:321–388, 1968.
33. editor Michael E. Szabo, editor. *The Collected Papers of Gerhard Gentzen*. North-Holland Pub. Co., Amsterdam, 1969.
34. Andrzej Mostowski. An undecidable arithmetical statement. *Fundamenta Mathematicae*, XXXVI:143–164, 1949.
35. Hans Pfeiffer. *Ausgezeichnete Folgen für gewisse Abschnitte der zweiten und weiterer Zahlklassen*. Phd thesis, University of Hannover, 1964.
36. Jan von Platon. Gentzen's logic. In Dov M. Gabbay and John Hayden Woods, editors, *Logic from Russell to Church*, volume 5 of *Handbook of the History of Logic*. 2009.
37. Iman Hafiz Poernomo, John Newsome Crossley, and Martin Wirsing. *Adapting proofs-as-programs*. Springer, New York, 2005.
38. Michael Rathjen. The realm of ordinal analysis. In S. Cooper and J. Truss, editors, *Sets and Proofs*, pages 219–279. Cambridge University Press, 1999.
39. Michael Rathjen. Theories and ordinals in proof theory. *Synthese*, 148(3):719–743, 2006.
40. Kurt Schütte. Kennzeichnung von Ordnungszahlen durch rekursiv erklärte Funktionen. *Mathematische Annalen*, pages 15–32, 1954.
41. Kurt Schütte. *Beweistheorie*, volume 103 of *Grundlehren der mathematischen Wissenschaften in Einzeldarstellungen mit besonderer Berücksichtigung der Anwendungsgebiete*. Springer Verlag, Berlin, 1960.
42. Kurt Schütte. *Lecture Notes in Mathematical Logic*. Department of Mathematics, The Pennsylvania State University, University Park, Pennsylvania, 1962/3. Vol.I (Fall 1962), Vol.II (Winter 1963).
43. Kurt Schütte. *Proof theory*, volume 225 of *Grundlehren der mathematischen Wissenschaften*. Springer Verlag, Berlin, New York, 1977. Tr. by J. N. Crossley from the German. (This is a very heavily revised version of the first edition: [41], 1960.).
44. Gaisi Takeuti. *Proof theory*. North-Holland Pub. Co., Amsterdam, second edition, 1987.

45. Jeroen van der Meeren and Andreas Weiermann. How to compare Buchholz-style ordinal notation systems with Gordeev-style notation systems. In A. Beckmann, V. Mitrana, and M. Soskova, editors, *CiE (Computability in Europe), 2015, volume 9136 of LNCS*, pages 353–362, Berlin, 2015. Springer.
46. Oswald Veblen. Continuous increasing functions of finite and transfinite ordinals. *Transactions of the American Mathematical Society*, 9:280–292, 1908.

Kapitel 4
Kurt Schütte

Curt Christian

Zusammenfassung Dies ist ein Nachruf auf Kurt Schütte, den Curt Christian für die Österreichische Akademie der Wissenschaften verfaßt hatte und der zuerst im *Almanach der Österreichischen Akademie der Wissenschaften, 149. Jahrgang (1999), S. 447–453* veröffentlicht wurde.

Am 18. August 1998 verstarb das korrespondierende Mitglied im Ausland der Österreichischen Akademie der Wissenschaften Kurt Wilhelm Schütte, der international hochangesehene emeritierte o. Professor für Mathematik an der Universität München mit dem Hauptarbeitsgebiet der Mathematischen Logik und Grundlagentheorie und dem Spezialgebiet der Beweistheorie.

Sein Wirken fällt in die Zeit nach Bewältigung der durch Russells Paradoxie verursachten Grundlagenkrise der Mathematik und ist gekennzeichnet durch die kritisch-schöpferische Auseinandersetzung mit den Grundlagenpositionen des Logizismus, des Intuitionismus, des Platonismus, der Hilbertschen Beweistheorie („Formalismus"), die Auseinandersetzung mit den Gödelschen Resultaten (Vollständigkeitssatz, 1. und 2. Unvollständigkeitssatz) und dem Gödel-Cohenschen Resultat der Unentscheidbarkeit von Auswahlaxiom und Continuumshypothese in der Basismengenlehre, ferner die Auseinandersetzung mit den verschiedenen Systemen der Mengentheorie, mit der Modelltheorie als—dem Vollständigkeitssatz gemäßes—Korrelat der als Consistenztheorie gefaßten Beweistheorie, endlich durch die Auseinandersetzung mit der die theoretische Grundlage der Informatik bildenden Rekursionstheorie. Für Schüttes Entwicklungen der Beweistheorie war Ausgang das Gödelsche Resultat, daß sich die Widerspruchsfreiheit der formalen Zahlentheorie mit den Mitteln derselben auf Metaebene allein nicht beweisen läßt. Da die finiten Methoden in der formalen Zahlentheorie aber enthalten sind, so genügen diese nicht, die Widerspruchsfreiheit der Zahlentheorie zu beweisen. Hilberts Erwartung eines

C. Christian (†)
Institut für Logistik der Universität Wien, Austria

R. Kahle and M. Rathjen (eds.), *The Legacy of Kurt Schütte*,
https://doi.org/10.1007/978-3-030-49424-7_4

zahlentheoretischen Widerspruchsfreiheitsbeweises in finiter Weise konnte nicht bestätigt werden.

Die Widerspruchsfreiheitsbeweise der Zahlentheorie von Gentzen und Ackermann verwenden Transfinite Induktion bis ε_0.

Auch alle anderen Widerspruchsfreiheitsbeweise für formale mathematische Teilsysteme von Feferman, Schütte, Takeuti, H. Friedman, Simpson u. a. verwenden höhere Mittel, als den Teilsystemen auf Metaebene entsprechen würde.

Bei diesen Entwicklungen spielt der Begriff der beweistheoretischen Ordinalzahl eines formalen Teilsystems der Mathematik sowie der Begriff der Prädikativität eine tragende Rolle.

Im Anschluß an anfangs der sechziger Jahre gleichzeitig und unabhängig voneinander gemachten Untersuchungen von Schütte und Feferman konnte letzterer den prädikativ interpretierbaren Teil der Analysis genau charakterisieren (durch Einschränkung der Comprehension auf sog. Δ_1^1-Formeln, in denen Mengenquantoren nur in sehr schwacher Weise auftreten).

Den ersten Widerspruchsfreiheitsbeweis für ein wesentlich imprädikatives Teilsystem der Analysis, die sog. Π_1^1-Analysis, hat in Weiterentwicklung der Gentzenschen Methoden G. Takeuti 1967 gegeben. Die beweistheoretische Ordinalzahl dieses und anderer Systeme wurde von den Schütte-Mitarbeitern Buchholz, Jäger, Pohlers bestimmt.

Was nun den Lebensweg Kurt Wilhelm Schüttes betrifft, so wurde er am 14. Oktober 1909 in Salzwedel als Sohn von August Schütte und Martha geb. Schröder geboren. Von 1916 bis 1919 besuchte er die Vorschule des humanistischen Gymnasiums Franziszeum in Zerbst i. Anhalt, anschließend eine Klasse des Gymnasiums selbst, von Herbst 1920 bis Ostern 1928 besuchte er das König Wilhelms-Gymnasium in Magdeburg, wo er am 6. März 1928 die Reifeprüfung ablegte. Er studierte Mathematik, Physik, Chemie und Philosophie, und zwar SS 1928 bis WS 1929 an der Universität Göttingen, SS 1930 an der Universität Berlin, dann WS 1930/31 bis WS 1934/35 wieder an der Universität Göttingen.

Nach Abschluß seiner Studien promovierte er am 10. Mai 1933 in Göttingen zum Dr. phil. in Mathematik und den Nebenfächern Theoretische Physik und Philosophie. Sein Doktorvater war David Hilbert, seine Dissertation lautete: „Untersuchungen zum Entscheidungsproblem der mathematischen Logik". Am 21. Juni 1935 legte er in Göttingen die wissenschaftliche Prüfung für das Lehramt an höheren Schulen mit den Hauptfächern Mathematik, Physik und dem Nebenfach Chemie ab.

Vom Herbst 1935 bis Herbst 1936 war er Studienreferendar am Gymnasium in Neuhaldensleben. Vom Herbst 1936 bis Sommer 1937 praktizierte er an der Wetterwarte in Magdeburg und machte am 25. Mai 1937 die Meteorologische Ergänzungsprüfung an der Universität Berlin. Vom Sommer 1937 bis 1945 war er Meteorologe im Reichswetterdienst, zunächst auf dem Fliegerhorst in Delmenhorst, dann während des Krieges bei verschiedenen Luftwaffen-Dienstellen in Deutschland, Rußland, Griechenland und Ungarn.

Am 28. Dezember 1937 heiratete er Hanna Lechte, die ihm zwei Töchter schenkte: Sigrid (geb. 1939) und Gisela (geb. 1942).

Nach Kriegsende war er vom Herbst 1947 bis Ostern 1948 Studienreferendar in Göttingen, ebenda legte er am 11. März die pädagogische Prüfung für das Lehramt an höheren Schulen ab und wurde am 1. November 1948 zum Studienassessor ernannt.

Er war zeitweise als Lehrer an der Felix Klein-Oberschule, zeitweise als wissenschaftliche Hilfskraft am Mathematischen Institut der Universität Göttingen tätig.

Seit 1. November 1950 war er wissenschaftlicher Assistent am Mathematischen Seminar der Universität Marburg a. d. Lahn, wo er sich am 21. Dezember 1951 für Mathematik habilitierte; seiner Vorlesungstätigkeit als Dozent folgte ab 28. Juli 1958 die als außerplanmäßiger Professor.

Im akademischen Jahr 1959/60 war er Member of the Institute for Advanced Study in Princeton, USA; danach war er wieder an der Universität Marburg tätig. Im SS 1961 und WS 1961/62 war er Gastprofessor an der Eidgenössischen Technischen Hochschule in Zürich. Im akademischen Jahr 1962/63 war er Visiting Professor an der Pennsylvania State University.

Am 1. April 1963 wurde er zum o. Professor für Logik und Wissenschaftslehre an der Universität Kiel ernannt.

Am 1. April 1966 erfolgte die Ernennung zum o. Professor für Mathematik an der Universität München, wo er als Forscher und Hochschullehrer am Mathematischen Institut bis zu seiner Emeritierung am 1. April 1977 tätig war.

Von 1958 bis 1979 war er Consulting Editor of the Journal of Symbolic Logic.

Seit 1968 war er Mitherausgeber des Archivs für mathematische Logik und Grundlagenforschung.

Er war seit 1973 o. Mitglied der Bayerischen Akademie der Wissenschaften, seit 15. Mai 1984 korrespondierendes Mitglied im Ausland der Österreichischen Akademie der Wissenschaften.

Schüttes letzte Lebensjahre waren durch eine fortschreitende schwere, in Blindheit endende Sehbehinderung gekennzeichnet. Konnte er zunächst noch mit einem Lesegerät arbeiten, so mußte er sich später eines Tonbandgerätes bedienen. Die Arbeit[1] über „Bemerkungen zur Hilbertschen Beweistheorie. Acta Borussica. 5 (München 1995)“ ist mit Hilfe eines solchen Tonbandgerätes entstanden, ebenso ein in „Elemente der Mathematik“ im Erscheinen begriffener Beitrag[2] zu einem Problem von gerichteten Graphen, der mit einer Kurzbiographie, verfaßt von seinem ehemaligen Schüler Pohlers, versehen sein wird.

Nach dem Tode seiner Gattin (1990) war Schütte auf die Hilfe seiner beiden Töchter und dreier Enkelinnen sowie zweier pensionierter ehemaligen Sekretärinnen des Mathematischen Instituts angewiesen.

Trotz dieser Behinderungen war sein Geist bis zu seinem durch einen Herzinfarkt herbeigeführten Tod wach, ideenreich, diskussionsfreudig und voll innerer Harmonie.

Die Mathematik—insbesondere die Mathematische Logik—verliert mit Kurt Schütte eine ihrer profiliertesten Gestalten.

---

[1] Nachgedruckt in Kapitel 8 dieses Buches.

[2] Kurt Schütte. Ein Beitrag zum Problem von gerichteten Graphen. *Elemente der Mathematik*, 55:63–66, 2000.

Chapter 4
Kurt Schütte

Curt Christian

Abstract This is an obituary for Kurt Schütte, written by Curt Christian for the Austrian Academy of Science and published in the *Almanach der Österreichischen Akademie der Wissenschaften, 149. Jahrgang (1999), pp. 447–453.*[1]

Kurt Schütte, a corresponding foreign member of the Austrian Academy of Sciences, passed away on 18 August 1998. He was an internationally highly regarded researcher and a professor emeritus of mathematics at the University of Munich, working in mathematical logic and the foundations of mathematics with a specialisation in proof theory.

His work took place in the period after the foundational crisis of mathematics caused by Russell's paradox had been settled, and is characterised by the critical-creative debate with the foundational positions of logicism, intuitionism, platonism, Hilbert's proof theory ("formalism"), the debate regarding the results by Gödel (completeness theorem, first and second Incompleteness Theorem) and the results by Gödel and Cohen that the axiom of choice and the continuum hypothesis cannot be decided in basic set theory, furthermore the debate regarding the various systems of set theory, with model theory as a – according to the completeness theorem – correlate of proof theory considered as a theory of consistency, finally by the interaction with recursion theory, which was about to form the theoretical foundation of computer science. Schütte's development of proof theory originated from the result by Gödel, which showed that the consistency of formal number theory cannot be proved meta theoretically in number theory itself. Since finite methods are contained in formal number theory, they are not sufficient to prove the consistency of number theory. Hilbert's expectation of a consistency proof of number theory using finite methods could not be satisfied.

C. Christian (†)
Institut für Logistik der Universität Wien, Austria

[1] Translated by Anton Setzer, Swansea.

R. Kahle and M. Rathjen (eds.), *The Legacy of Kurt Schütte*,
https://doi.org/10.1007/978-3-030-49424-7_4

The consistency proofs of number theory by Gentzen and Ackermann used transfinite induction up to ε_0.

All other consistency proofs for formal mathematical subsystems by Feferman, Schütte, Takeuti, H. Friedman, Simpson and others use as well methods which are stronger then what corresponds to these subsystems meta theoretically.

In these developments, the notion of a proof theoretic ordinal of a formal subsystem of mathematics as well as the notion of predicativity plays a foundational rôle.

Following parallel and independent investigations by Schütte and Feferman in the beginning of the sixties, the latter was able to characterise precisely the predicatively interpretable part of analysis (by restriction of comprehension to so called Δ_1^1-formulas, in which quantifiers over sets occur only in a very weak form).

The first consistency proof for an essentially impredicative subsystem of analysis, the so called Π_1^1-analysis,j was given by G. Takeuti 1967 by further developing the methods by Gentzen. The proof theoretic ordinal of this and further systems was determined by Buchholz, Jäger and Pohlers, who were collaborators of Schütte.

Let us now review the personal life of Kurt Wilhelm Schütte. He was born on 14 October 1909 in Salzwedel as the son of August Schütte and Martha nee Schröder. From 1916 to 1919 he attended the preschool of the humanistic gymnasium Franziszeum in Zerbst in Anhalt. He then attended one year at the gymnasium itself. From autumn 1920 until Easter 1928 he attended the König Wilhelms-Gymnasium in Magdeburg, where he passed the school-leaving examination on 6 March 1928. He studied mathematics, physics, chemistry, and philosophy, from summer semester 1928 until winter semester 1929 at the University of Göttingen, during the summer semester 1930 at the University of Berlin, then again from winter semester 1930/31 until winter semester 1934/35 at the University of Göttingen.

After completing his studies he took the doctoral degree on 10 May 1933 in Göttingen with the title of a Dr. phil. in mathematics and the secondary subjects theoretical physics and philosophy. His doctoral supervisor was David Hilbert, and the title of his dissertation was "Untersuchungen zum Entscheidungsproblem der mathematischen Logik" ("Investigations of the decision problem of mathematical logic"). On 21 June 1935 he passed the scientific exam for becoming a teacher at high schools with the main subjects mathematics, physics, and the secondary subject chemistry.

From autumn 1935 until autumn 1936 he was trainee teacher at the Gymnasium in Neuhaldensleben. From autumn 1936 until summer 1937 he practised at the meteorological observatory in Magdeburg and passed on 25 May 1937 the meteorological supplementary examination at the University of Berlin. From Summer 1937 until 1945 he was meteorologist at the meteorological service of the German Reich, first at the airbase in Delmenhorst, then during the war at various offices of the air force in Germany, Russia, Greece, and Hungary.

On 28 December 1937 he married Hanna Lechte with whom he had two daughters: Sigrid (born 1939) and Gisela (born 1942).

After the end of the war he was from autumn 1947 until Easter 1948 teacher trainee in Göttingen, where he passed on 11 March 1948 the pedagogical exam for teaching at high schools, and was appointed on 1 November 1948 "Studienassessor" (probationary teacher).

Temporarily he was teacher at the Felix Klein-High school, and scientific assistant at the mathematical institute of the University of Göttingen.

Since 1 November 1950 he was scientific assistant at the Mathematical Seminar of the University of Marburg a. d. Lahn, where he passed the habilitation in mathematics on 21 December 1951; after lecturing as a docent he became extraordinary professor on 28 July 1958.

During the academic year 1959/60 he was member of the Institute for Advanced Studies in Princeton, USA; afterwards he returned to his position at the University of Marburg. In the summer semester 1961 and winter semester 1961/62 he was visiting professor at the Swiss Federal Institute of Technology in Zurich (ETH Zurich). In the academic year 1962/63 he was visiting professor at the Pennsylvania State University.

On 1 April 1963 he was appointed ordinary professor for logic and theory of science at the University of Kiel.

Subsequently, he was appointed on 1 April 1966 ordinary professor for mathematics at the University of Munich, where he stayed as researcher and university lecturer at the Mathematical Institute until his retirement on 1 April 1977 when he became professor emeritus.

From 1958 to 1979 he was consulting editor of the Journal of Symbolic Logic.

Since 1968 he was co-editor of the Archive for Mathematical Logic and Foundational Research.

He was since 1973 ordinary member of the Bavarian Academy of Sciences, since 15 May 1984 corresponding member abroad of the Austrian Academy of Sciences.

The last years in the life of Schütte were affected by his increasingly strong visual impairment, which ended in blindness. Initially he could still work with a reading device, but later he needed to use an audio tape recording system. The paper[2] on "Remarks on Hilbert's proof theory. Acta Borrussica 5 (Munich 1995)" was created with the help of such an audio tape recording system, as well as an article[3] which is about to appear in "Elements of Mathematics" on a problem of directed graphs, which will contain as an attachment a short biography of Schütte, written by a former student of Pohlers.

After the death of his wife (1990) Schütte was dependent on the help of his two daughters and three grand daughters as well as two retired former secretaries of the Mathematical Institute.

Despite this disability his spirit stayed awake, full of ideas, open to discussions and full of inner harmony, until his death due to a heart attack.

Mathematics—especially Mathematical Logic—will loose with Kurt Schütte one of his most outstanding representatives.

---

[2] Reprinted and translated into English in Chapter 8 of this volume.

[3] Kurt Schütte. Ein Beitrag zum Problem von gerichteten Graphen. *Elemente der Mathematik*, 55:63–66, 2000.

Chapter 5
Memories of Kurt Schütte and the logic group in Munich: A personal report

Wolfram Pohlers

Abstract After a brief curriculum vitae of Kurt Schütte I describe some episodes of the emergence of the Schütte School of Proof Theory as I remember it.

Kurt Wilhelm Schütte was born on October 14, 1909 in Salzwedel, Altmark. He studied mathematics in Berlin and Göttingen, where he was the last doctoral student of David Hilbert in 1933 with a dissertation on the decision problem of mathematical logic [3]. According to his own statement he didn't see Hilbert all too often (twice, if I remember correctly) but worked primarily with Paul Bernays, to whom he kept contact until Paul Bernays' death. After the war, during which he served as a meteorologist in the "Reichswetterdienst", he resumed a career as a high school teacher (Gymnasiallehrer), a career for which he had laid the cornerstones already before the war and which was absolutely common for mathematicians at that time. In addition to his academic degree he had also passed the "Staatsexamen", examinations which were mandatory for high school teachers. In 1947 he started to work as a "Studienreferendar" (the degree for beginners) and was promoted to "Studienassessor" after having passed the "Assessor Examen" in 1948. However, during this time he by chance met Arnold Schmidt, another former doctoral student of Hilbert, who succeeded in reviving Schütte's interest in Mathematical Logic. Schütte then simultaneously started also to work as teaching assistant (wissenschaftliche Hilfskraft) at the Mathematical Institute of the University in Göttingen. In 1950 he followed Arnold Schmidt as his assistant to Marburg, where he habilitated in 1952.

Beginning in the late 50s, after having also worked on the foundations of geometry and packing problems—together with B. L. van der Waerden he succeeded in 1953 to give the first exact proof of the kissing number problem in three dimensions—he concentrated mainly on proof theory. In 1960 his monograph "Beweistheorie" appeared in the "Grundlehren der Mathematik" in which he, among other things,

W. Pohlers
Institut für Mathematische Logik und Grundlagenforschung, Einsteinstraße 62, 48149 Münster, Germany, e-mail: pohlers@uni-muenster.de

R. Kahle and M. Rathjen (eds.), *The Legacy of Kurt Schütte*,
https://doi.org/10.1007/978-3-030-49424-7_5

pursued and perfected Gerhard Gentzen's approach to a proof–theoretic analysis of axiom system by means of Cantors ordinal numbers. Parts of the years 1959/1960 he spent as a visiting professor at the Institute for Advanced Studies in Princeton where he was invited by Kurt Gödel who was interested in Schütte's work on the decision problem. During the years 1960/1961 he was visiting the ETH in Zurich, probably at Bernays' invitation and in 1962/1963 he worked as a guest professor at the Pennsylvania State University. In 1963 he accepted an offer by the University of Kiel where he worked as full professor until 1966 when the University of Munich offered him a chair for Mathematical Logic. There he worked until his retirement in 1977.

Schütte's outstanding contribution to the further development of Gentzen's program of proof-theoretic analyses was the systematic use of semi-formal systems. These systems extend the Gentzen calculi by allowing inferences with infinitely many premises. Semi-formal proofs are infinite well-founded trees whose heights are canonically measured by transfinite ordinals. In this way the cryptic ordinal assignments in Gentzen's consistency proof got a natural explanation by the heights of the semi-formal proofs emerging during the proof-theoretic analysis.

Another concept that Schütte considered very important and significant were his positive and negative parts of formulae. Roughly speaking these are propositional sub-formulae of a formula whose truth (positive part) resp. falsity (negative part) already draw the truth of the entire formula. Schütte's aim was to use this concept to merge the advantages of the Hilbert calculus, which derives formulae, with those of Gentzen's sequent calculus, deriving sequents of formulae. His idea was to replace inferences with main formula in the antecedent by inferences whose main formula is a negative part and those with main formula in the succedent by inferences whose main formula is a positive part. To check the usefulness of this concept he tried to use it in a completeness proof for first order predicate logic. On that occasion he invented his method of search trees (quasi deduction trees as he used to call them). A method which turned out to be very powerful.

During his stay at Princeton he got in contact with Gaisi Takeuti (who, according to Schütte, was Gentzen's best student, although Gentzen and Takeuti never met). Takeuti had just discovered that the generalization of Gentzen's Hauptsatz to simple type theory would imply the consistency of full second order arithmetic. A constructive proof of cut elimination for simple type theory, similar to Gentzen's proof for first order logic, would therefore entail a constructive consistency proof for second order arithmetic. It was, however, wide open whether the Hauptsatz holds for simple type theory at all, let alone if there is a constructive proof à la Gentzen. Here Schütte could use his search trees to show that there is a semantical equivalent to Gentzen's Hauptsatz [4]. Later Tait, Takahashi and independently Prawitz used this equivalence to give (non constructive) proofs of the Hauptsatz, first for second order logic (Tait in [6]) and then for full simple type theory (Takahasi in [7] and Prawitz in [2]).

One of the most spectacular results of Schütte's proof theoretic work was the exact determination of the ordinal that limits predicativity. A result which was independently obtained by Solomon Feferman. Schütte reported about this result in an invited one hour key note lecture at the International Congress of Mathematicians 1966 in Moscow.

Schütte became a member of the Bavarian Academy of Sciences and Humanities in 1973 and a corresponding member of the Austrian Academy of Sciences in 1984. He cofounded the "Archiv für mathematische Logik und Grundlagenforschung" (now Archive for Mathematical Logic) and was a founding member of the DVMLG (German Association for Mathematical Logic and the Foundations of the Exact Sciences).

I met Schütte for the first time in the summer term 1968 when I attended his lectures on Mathematical Logic. I was fascinated by the clear and complete presentation of the material and fell in love with logic immediately. It was a series of four hours lecture per week, spread over two days of two hours each. In the middle of the two hours there was a break of fifteen minutes. I used this break to ask questions which Schütte willingly answered and, little by little, it became a habit for Schütte and myself to spend the intermissions together wandering up and down and discussing logical questions. That was probably the reason why Schütte offered me a teaching assistant position although I had not yet passed my Vordiplom. So I became a kind of "junior member" of the logic group. Until the move of the Mathematical Institute to the new building at Theresienstraße 39 during the year 1972, Schütte's logic group was housed on one floor of a former residential building in Schellingstraße 10 in the immediate vicinity of the "Dreierinstitut" which housed besides mathematics two further institutes. Since the floor heights of both buildings did not fit together there was only a transition to the Dreierinstitut on some special floors. If I remember that correctly one of the transitions was on the first floor while we were located at the second. I got a small room, apparently the former servant's chamber of the flat as an "office", and keys for the flat and the transition door. The members of the then logic group were Justus Diller, Klaus Vitzthum and Wolfgang Bibel. I hardly remember Klaus Vitzthum who left the group before I was really settled in but remember well his successor Johannes Czermak. Also Wolfgang Bibel left the group pretty soon and was replaced by Werner Carstengerdes.

My task was first and foremost to work out Schüttes lecture notes and seminar notes. This was not hard work since Schütte usually had completely worked out handwritten manuscripts which were typed by our secretary Anne Ernst on a wax matrix onto which I had to carve the mathematical symbols that were not available on the typewriter. Very rarely I detected a mistake in Schütte's manuscripts.

I remember well Schütte's course on set theory, held in the winter term 1969/70 which I had to "work out" and which appeared as a mimeographed booklet. Starting with chapter I "Grundlagen einer axiomatischen Mengenlehre" it sets up a formal logical frame in which in fact all future proofs are to be formalized. This perfectly reflects Schütte's view of mathematics, which can also be seen in his research papers. Nevertheless the booklet contains a brief "informal" introduction about "naive set theory". This introduction includes a proof of the Cantor Bernstein Theorem. In the period of 1968 to the early seventies the university was in turmoil. The uproar was less pronounced in the faculties for mathematics and sciences than at more philosophically and social science oriented faculties. But there were also groups among the math students that wanted to see more emphasis on the social responsibility of mathematics and mathematics teaching. I remember a pamphlet in which my "informal" proof of the Cantor Bernstein Theorem (which was certainly in principle Schütte's proof) was taken as an example for over-formalized mathematics as it should not be taught. Unfortunately the alternative given in the pamphlet was erroneous.

Another consequence of this manuscript was that Schütte once asked me to visit Wolfgang Stegmüller. Stegmüller, an analytical philosopher, was very fond of formal systems and therefore took great pleasure in the booklet. He got, however, confused that I had marked inferences as "modus ponens" when in fact they were applications of cut, which was not among the basic inferences of our formal system. So it became my task to assist Stegmüller in reading the text and I wandered more or less regularly down to Kaulbachstraße to visit Stegmüller in his institute. Decades later, when I met Stegmüller at the occasion of a talk given by him in the philosophy department in Münster, he still recognized me and remembered this episode. Anyway, despite of its over-formalization this little booklet contains a chapter on ordinal arithmetic which in this detail can rarely be found in any textbook on set theory. Today, however, I would hardly use it as an introduction to set theory.

In 1970 Carstengerdes quit his academic career for an employment in industry and Justus Diller was promoted to "Wissenschaftlicher Rat and Professor". Until the winter term 1971/72 the Munich logic group only included Schütte, as full professor, Diller, as "wissenschaftlicher Rat", Czermak as Scientific Assistant and myself as teaching assistant. During 1971 the group was reinforced by Horst Osswald, coming from Hannover, who had already earned his doctorate, and I was promoted to "Verwalter der Dienstgeschäfte eines wissenschaftlichen Assistenten" as soon as I had finished my diploma thesis and passed the examination in spring 1971. In 1972 Czermak left Munich for a position in Austria, his homeland, and was replaced by Wilfried Buchholz. Justus Diller left Munich 1973 for a chair at the University of Münster and Horst Osswald was promoted to Wissenschaftlicher Rat und Professor in 1974. In the same year Schütte engaged Wolfgang Mass as assistant and the nucleus of the "Munich School of Proof Theory" was completed with already an offshoot in Münster. Important to mention is also Gerhard Jäger who joined us in the eighties on a position financed by the German Research Foundation. As far as I know he became Schütte's last doctoral student (and my first one as a co-supervisor).

Schütte was a friendly, distinguished person. He never became loud and demanding, but nevertheless enjoyed absolute respect and influence among his colleagues. I don't remember decisions ever being made that were contrary to his declared will. However, he was also willing to compromise on many things.

He had a broad knowledge in general mathematics. When I was preparing my doctoral examination he had no problem to give me offhand the definition and the essentials of a Riemannian surface in high precision. He regularly attended the general mathematical colloquium and the social meeting afterwards. I hardly remember that he took part in the official discussion after the talk but I remember well occasional intensive discussions during the subsequent social meetings. He was also astonishingly well informed about Riemannian geometry and general relativity.

His elder daughter told me that even short before his death he tried to explain general relativity to the doctors in the hospital.

Nevertheless, the matter close to his heart was mathematical logic, especially proof theory and here especially proof theory based on ordinals, ordinal analysis as we call it today. As a supervisor, however, he gave us all the freedom we liked.

There were no concise research programs at that time. When I was asking for a topic for a dissertation he suggested me to continue his dissertation and to work on the decidability of prenex formulae with suffix $\exists \ldots \exists\forall\forall\exists \ldots \exists$ in the first order predicate calculus with identity. In [3] he could show that this class of formulae without identity is decidable. A result which had already been proved by Gödel [1] but Schütte—not knowing this paper—could additionally show that formulae of this type are finitely satisfiable. It was his conjecture that the same holds true for formulae in the language of predicate logic with identity (which was in accordance with a conjecture by Gödel). After some reading I found out that this was not the part of mathematical logic that fascinated me and I decided to try to work on Takeuti's analysis of an axiom system for Π^1_1-comprehension. He accepted my decision without objections and further comments. As far as I remember, the original problem was later proposed to Martin Wirsing. Of course he could not solve the problem—Warren Goldfarb showed in 1984 that this class is in fact undecidable—but obviously the subject possessed sufficient substance for a successful dissertation. I remember a talk of Schütte in Oberwolfach, in the late seventies or early eighties (I could not check it offhand), in which he addressed the decision problem for this class and accurately circumscribed which stumbling blocks would have been to overcome in order to obtain finite satisfiability. But I cannot remember any details of his presentation.

Usually he let us work in a completely independent way and hardly interfered with the progress of the work. As an exception I only remember the dissertation of Wilfried Buchholz, who worked on extensions of ordinal notation systems. Schütte was so fascinated by the topic, which was based on an idea by Sol Feferman and partly worked out in a dissertation by Jane Bridge under the supervision of Peter Aczel, that he himself, almost in competition with Wilfried, was working on it. However, Wilfried could cope with the pressure and today he is famous for his pioneering work in ordinal notations.

The planned second edition of his monograph Beweistheorie in English developed into a completely new work during the preparations. Schütte tried to involve as much of the most recent developments as possible and often consulted Wilfried Buchholz and myself for advice. Here we occasionally tried to convince him to involve a bit more motivation but even when he agreed in the beginning he felt uncomfortable with the motivational text and—true to his maxim: "The master builder also removes the scaffolding when the building is finished"—erased it in the end. This maxim becomes also very visible in all his original papers.

The book appeared in 1977 and was reviewed by Solomon Feferman in 1979. Schütte was not happy with the review. In fact it was the first time that I saw him upset. However, Sol's review was in no way meant to diminish the achievements of the book. He was "historically amused" that simple type theory was treated as "pure logic", despite the original (historical) intention to justify mathematics via simple type theory and Takeuti's results on the consequences of cut-elimination in simple type theory for the consistency of (parts of) mathematics. But this was not really a criticism. The only real objections were that Schütte was *"too sparing with informal explanations"* and that the reviewer complained about the use of positive and negative parts which *"kept finding* [him]*self translating the set-up into more familiar terms"*, by which he certainly meant sequents. Schütte could live with the criticism not to have given too much informal information as long all the mathematical details were present. This was in accordance with his maxim. What really offended him was the disdain that was shown for his positive and negative parts. However, here I have to confess that today even I have difficulties to read my own old papers which are based on positive and negative parts.[1] I don't think that Schütte's anger lasted too long. He soon was back to good terms with Feferman.

Schütte retired in 1977. Too early for us, since his original plan was not to retire before the age of 70. For that reason three of us, Wilfried Buchholz, Wolfgang Maass and myself, all not yet habilitated, sped up in writing down our theses and all three of us were habilitated during the year 1978. Schütte was succeeded by Helmut Schwichtenberg who cautiously lead the institute into a new era.

I left Munich in 1985 for a chair in Mathematical Logic in Münster but never lost contact with Schütte. I was present on the funeral of his wife, Hanna Schütte, who died in 1990. After her death he moved to a flat in the Belgradstaße (close to my former abode). There it was much easier for his relatives to take care of him than at the house in "Am Brombeerschlag" located at the diametric end of Munich.

Schütte had always had problems with his eyes. He was a heavy smoker which did not help to alleviate this condition. At that time he was practically blind. So his daughters and granddaughters, who took care of him and visited him regularly, were very concerned that he could put away a burning cigarette and not find it again. Therefore he was given strict advice to smoke only on the balcony. He obeyed this advice strictly. Also our former secretary, Mrs. Anne Ernst (to whom I still am in contact) and the secretary of a former colleague, Mrs. Eberle, visited him regularly, read to him or went out with him for walks. Despite the tremendous limitations that this disease meant for him and his work, I never heard him complaining.

Whenever I was in Munich I tried to visit him in his flat. Mostly standing on the balcony while he was smoking. In the evening we used to have dinner, occasionally accompanied by my wife, at a wine restaurant close to his apartment called Rolandseck, which was also well-known to my wife and myself. There he was a regular guest respectfully addressed as "Herr Professor".

Schütte showed an intense interest in the work of our institute in Münster. That this was more than benevolent interest follows from the fact that he transferred papers of Michael Rathjen to his "negative– and positive–parts calculus" trying to include technical improvements. These versions have been distributed in mimeographed form. In a preamble to these distributions he emphasizes that these papers are entirely based on the work of Michael Rathjen. As he assured me by telephone and

[1] However, in preparing my (scientific) contribution to this volume I realized that using Schütte's positive and negative parts calculus would have made the convention in the beginning of Section 10.6 superfluous.

then also in writing it was not his intention to compete with us and he therefore did not plan to publish these papers. Nevertheless, I believe that these papers can and should now be published and therefore welcome the decision of the editors to include these papers into the volume.

Since he was practically blind he used a dictaphone for his texts, including mathematical texts. I do not know who typed the aforementioned papers. But I helped to publish his last paper "Ein Beitrag zum Problem von gerichteten Graphen" [5] which has been typed by my secretary Martina Pfeifer using Schütte's dictaphone which he then donated to me. It is still in my possession. The paper was published posthumously in 2000. Schütte died on August 18, 1998 and it was my sad duty to give the farewell speech at his funeral.

Acknowledgements I am indebted to Wilfried Buchholz and Michael Rathjen for proofreading and supplementing a first version of theses memories.

References

1. Kurt Gödel. Ein Spezialfall des Entscheidungsproblems der theoretischen Logik. *Ergebnisse eines Mathematischen Kolloquiums*, 2:27–28, 1932.
2. Dag Prawitz. Hauptsatz for higher order logic. *Journal of Symbolic Logic*, 33:452–457, 1968.
3. Kurt Schütte. Untersuchungen zum Entscheidungsproblem der mathematischen Logik. *Mathematische Annalen*, 109:572–603, 1934.
4. Kurt Schütte. Syntactical and semantical properties of simple type theory. *Journal of Symbolic Logic*, 25:305–326, 1960.
5. Kurt Schütte. Ein Beitrag zum Problem von gerichteten Graphen. *Elemente der Mathematik*, 55:63–66, 2000.
6. William W. Tait. A non constructive proof of Gentzen's Hauptsatz for second order predicate logic. *Bulletin of the American Mathematical Society*, 72:980–983, 1966.
7. Moto-o Takahashi. A proof of cut-elimination in simple type theory. *Journal of the Mathematical Society of Japan*, 19:399–410, 1967.

P.S.: In proofreading this note I received the sad news that Schütte's former secretary Mrs. Anne Ernst passed away on August 31, 2019 at the age of 96.

Chapter 6
Reminiscences of Kurt Schütte

Wolfgang Bibel

Abstract The article reflects some of the reminiscences of Kurt Schütte by his first PhD student based on the numerous interactions during the years of collaboration.

6.1 Introduction

Kurt Schütte (1909–1998) supervised my PhD work, and I was a member of his research group in Munich in its initial three years, a period covered also in my memoirs [1], there viewed from a more personal perspective. During that time we maintained a close communicational exchange both in the context of the daily work as well as in social encounters. From these discussions I had the opportunity to learn about many aspects of his career, work and life. In this note I would like to share some of this knowledge with the community.

6.2 Fragments of Schütte's Professional Career

Schütte was the last PhD student of David Hilbert (1862–1943) at the University of Göttingen and received his doctorate in 1934 with a thesis entitled "*Untersuchungen zum Entscheidungsproblem der mathematischen Logik*". But he kept telling us that it was actually Paul Bernays (1888–1977) who supervised his thesis work in all its details. Hilbert was the "Ordentlicher Professor" (full professor) while Bernays at that time was just an "Außerordentlicher Professor" (associate professor) who anyway closely collaborated with Hilbert on the latter's programme for the foundation of Mathematics. Under these circumstances it was clear that Hilbert, although over 70 already, took the responsibility as the main advisor.

W. Bibel
Darmstadt University of Technology, Darmstadt, Germany, e-mail: bibel@gmx.net

R. Kahle and M. Rathjen (eds.), *The Legacy of Kurt Schütte*,
https://doi.org/10.1007/978-3-030-49424-7_6

Since Bernays lost his position by way of the takeover by the National Socialists and since Hilbert was too old for taking new initiatives, the chances for an academic career at the time had vanished for Schütte. Therefore he tried to enter the career of a high-school teacher and passed the respective examination (2. Staatsexamen) in 1935. Unfortunately, teaching positions were particularly hard to obtain in those years. The state rather built up the army including the air force which led to the opening of new meteorology positions well suited for a trained mathematician. For this reason Schütte, in 1936, took the opportunity to enter the career as a meteorologist for the years to follow. After the end of the war the demand for army meteorologists evaporated. In order to make a living, he re-entered the career as a high-school teacher passing another obligatory examination (Assessorexamen) to this aim in 1948.

At the same time he resumed his research in Logic. He reactivated the contacts with the logician Hermann Arnold Schmidt (1902–1967) whom he knew from his years in Göttingen where Schmidt also worked with Hilbert as an assistant (until 1936) and also got his PhD there in 1934 under Hilbert's supervision. Schmidt in 1946 became "Außerplanmäßiger Professor" in Göttingen and in 1950 Ordinarius (full professor) in Marburg. There he agreed to engage the seven years younger, but already 40 years old Schütte as his assistant (Wissenschaftlicher Assistent) in the same year 1950, an obviously uneasy constellation to say the least.

Apparently, after the war Schütte became determined to make every attempt for a turn into an academic career and started to work extremely hard to make up for the lost time since the completion of his PhD. His wife described the situation of those years before 1950 as a truly miserable one. The family of four with two little daughters was cramped into a single living room and Schütte would work in a corner of it writing on wallpaper and filling plenty of it with formulas. Whenever his wife tried to address him, he used to be so concentrated in his work that he would not even notice the approach. Thus it was in those years that he laid the foundation for his outstanding insights into Proof Theory, resulting in three first and remarkable journal publications already in 1950/1951 including [6].

The inflexible German university system nevertheless required the regular career steps also in his rather extraordinary case, with the Habilitation—based on the thesis "*Beweistheoretische Untersuchungen der verzweigten Analysis*"—in 1952 and the consequential promotion from Wissenschaftlicher Assistent to Privatdozent in 1955 so that finally in 1958 he became an Außerplanmäßiger Professor at the age of by now 49, all steps done in Marburg under a dominating Ordinarius Schmidt.

By that time he had earned already an international reputation and was invited by Gödel (1906–1978) to spend the academic year 1959/1960 as a visiting professor at the prestigeous Institute for Advanced Study in Princeton. This was his chance to escape the narrow-minded intellectual environment in Marburg for sure. In Princeton he completed the first edition of his book on Proof Theory [12]. It happened that Gaisi Takeuti (1926–2017) also spent the same year at Princeton. Not least through the interactions with him Schütte was stimulated to the research which led to his 1960 paper in the Journal of Symbolic Logic on Type Theory [7]. Occasionally, he would tell us from his encounters with Gödel; for instance, he mentioned the peculiarity that Gödel during a lecture would keep facing exclusively the blackboard and practically never the audience.

Schütte's PhD mentor Bernays invited him to spend the subsequent academic year 1961/1962 at the ETH Zürich. In coordination with Bernays' former student, Richard Büchi (1924–1984), then at Purdue in the US, they together managed to arrange for a visiting professorship for Schütte during the subsequent academic year 1962/1963 at Pennsylvania State University. Thereafter, in 1963, he finally got an offer for a full professorship at the University of Kiel where he stayed for three years until he was offered an esteemed, well remunerated and well equipped professorship at the Ludwig-Maximilians-Universität (LMU) in München, at the time aged 56 already. Only now he was in a position to build up a research group and supervise students in his field of expertise.

6.3 The Beginnings of Schütte's Logic Research Group

I had received my diploma from the LMU in 1964 under the supervision of Karl Stein (1913–2000) and had suffered thereafter several failures in attempts to do research towards a PhD thesis. In a psychologically rather desperate situation I by chance met my school and university mate Gerd Fischer, later a mathematics professor, in front of the Mathematical Institute in Munich. He told me that a newly appointed professor, named Schütte, was looking for assistants, an opportunity which I tried out immediately. Since I could exemplify my interest in logic by pointing to some earlier study of quantum logic, he agreed to engage me as a PhD student and assistent (formally as "Verwalter einer wissenschaftlichen Assistentenstelle") in his group which started its work in May 1966. The other members initially were Justus Diller as a postdoc, another PhD student Klaus Vitzthum (working on some Geometry topic) and the secretary Anne Ernst. Our offices were located in a former flat in the second floor of the building Schellingstraße 10, next to the Math building.

It is notable that all three initial group members later became professors (as did many if not most of his later PhD students). Diller was already an established scientist who had received his PhD in 1963 with a thesis in some Geometry topic under the supervision of Friedrich Bachmann in Kiel. Each member could pursue his individual work in full concentration, but in a friendly and colloquial atmosphere. Schütte could be approached by us at any point in time, quite in contrast to the usual attitude of German professors of the time. We often went out together for lunch in one of the restaurants around the area. At those occasions Schütte used to drink one or even two beers (of half a liter each) with—to our surprise—no noticeable effects at all. And he used to smoke one cigarette after another at those occasions (and during most of the rest of the day). When addressed because of this heavy smoking habit, he would choke by some reply like: "*It's not that much since during sleep I do not smoke at all.*"

Occasionally, we met also in private settings, at Schütte's rented house Am Brombeerschlag 34 in Munich, in Diller's rented flat in Gauting or in my rented apartment in the Trainsjochstraße 2. A particularly memorable evening of this kind took place at the occasion of a visit by Gaisi Takeuti in Munich who was not only a famous logician but also excelled by drinking enormous amounts of beer at this occasion in my apartment. It is telling for the unusually good atmosphere that Schütte even lent me his family's piano which was of no use anymore for them. Not unusual for that time, we all in the group nevertheless addressed each other by surnames and the formal "Sie" (rather than the more colloquial "Du").

6.4 Aspects of Schütte's Professional Work

Under a formal point of view Schütte gave excellent lectures, in the first year attended by only a dozen or so students. Usually in Mathematics, such lectures would consist of the presented formalism and the accompanying "baggage" helpful for the human understanding. Schütte reduced the "baggage" rigorously to a minimum. He held the conviction that everything could be found in the theorems so that no "baggage" was in fact needed. For someone like me who was new in this area this was a real challenge. One of our diploma students, Dietrich Fischer, joined the group as a research assistant in order to write up Schütte's lecture with some additional "baggage" helpful for the students. In this combination Schütte's lectures soon became more popular attracting top-ranking students from the department. Typically, Schütte would not just present in those lectures the material from his published book or papers. Rather he reworked the material each time anew and in a further optimized form.

Although Schütte rarely told us anything about his own research work, we could observe that during those years he worked hard on new original publications, on his 1968 book about non-classical logics [11], or on invited talks. These observations could be made by us because after weekends he would bring his handwritten manuscripts to Mrs. Ernst who would carefully typewrite them into the machine. Remarkably, I have never seen a single page of those numerous handwritten sheets of paper with any correction made by him in the text. Schütte must have forced himself to rewriting a whole page rather than allowing himself the correction of even a slight typo in it.

He engaged in several activities involving logical topics. Already in the first year 1966 he gave an invited talk to the International Congress of Mathematicians which took place in Moscow and which established a good relationship with Russian logicians. In the same year the "Kolloquium über Logik und Grundlagen der Mathematik" was held in Hannover attracting celebrities in Logic and Philosophy from all parts of the world. Most impressive for me at this occasion was the performance of the great Alfred Tarski (1901–1983) who after the talk of an Italian countess on supposed mistakes in Gödel stood up and spontaneously taught the poor lady (and the amused audience) a brilliant lesson about Gödel's incompleteness proof. Schütte was involved in the event's program committee and before the conference had discussed with all three of us the quality of the submissions in his share of reviewing. It was obvious from those discussions that he entertained a rather low appreciation of any logical work lacking mathematical rigor.

His particular professional engagement was in the activities of the "Deutsche Vereinigung für Mathematische Logik und für Grundlagenforschung der Exakten Wissenschaften" (DVMLG) which—along with Arnold Schmidt—he co-founded in 1962. He also co-founded the journal "Archiv für Mathematische Logik und Grundlagen der Wissenschaften". During the "Arbeitstagung zur mathematischen Logik" which was held at the Mathematisches Forschungsinstitut Oberwolfach in April 1967 the leading German logicians had intensive and rather difficult discussions on how to reorganize the DVMLG chaired from the beginning by Schmidt. When Schmidt died a few months later, Diller mentioned to Schütte with a sigh: "*How much effort could we have saved, had we foreseen his forthcoming death.*"

At this workshop I met for the first time leading logicians like Bernays, Hermes, Specker and many others. Schütte presented a talk entitled "*Zur Berechenbarkeit von Funktionen höherer Typen*" [10]. Incidentally, since Oberwolfach was so hard to reach by public transportation and since I was the only one in the group owning a car, Schütte and Diller travelled there with me in that year and again to the analogue event in the following year, at which I then also presented a talk on my thesis results.

Schütte was the prototype of a totally committed scientist with rather limited interest in anything going on outside of his close environment. Even as a mathematician he was focussed on a rather narrow range of subjects. These include besides Logic and Proof Theory also Number Theory, where he made a contribution to the "Kußzahlenproblem" in a 1953 publication jointly with van der Waerden, as well as (foundations of) Geometry. His interest in the latter according to him had the organisational reason that as a professor he was expected to offer more courses than just those on Logic and Proof Theory in order to fulfill the teaching requirements and also to attract students. Although he regularly attended the weekly seminars in the department, I cannot recall him participating in the discussions following the talks except if the subject of the presentation was close to his own areas of interest. Every year he, along with us, also attended the more socially motivated "Bayerisches Mathematiker Kolloquium" held at various places in Bavaria such as Eichstätt in 1968 and Memmingen in 1969.

Influenced by his strict concentration just on research we took little notice even of the roaring student demonstrations taking place not far from the institute in Munich around Easter 1968 whereby two people even came to death. Rather we tended to live in a separate world which, however, began to attract researchers from around the world for visiting stays of differing durations. One of them was Johannes Czermak, later a professor at the University of Salzburg. Another one was Mr. Gudjonsen from Iceland. And, of course, more excellent PhD students soon began to join the group contributing to the establishment of what later became known as the Munich School for Proof Theory.

6.5 Schütte's Role as My Supervisor

Concerning my thesis work Schütte had discovered in the mid-sixties what is known as the Schütte-Feferman limit ordinal κ_0 (also termed Γ_0) characterizing the predicative part of analysis [9, 8, 5]. He speculated that there might be a syntactically defined part of Type Theory in which cut elimination could be proved by induction right up to this ordinal κ_0. To find out whether such a part could be found was the task he suggested to me. Unfortunately, I could not identify such a part (nor did anyone else in the meantime up to this day to the best of my knowledge). However, in the course of this search I came up with an optimized syntactic proof of Gentzen's Hauptsatz about cut elimination for a restricted class of formulas in Type Theory. It involved a more constructive denotation system for ordinals due to Schütte. Altogether this formed the core of my PhD thesis completed in 1968. The resulting paper with its main results was published as [2].

Apparently, Schütte was somewhat disappointed about my thesis results so that he graded it with a disillusioning "cum laude". In consequence, we both agreed that my career would preferably turn into a direction better suited to my interests which have reached out beyond just Proof Theory. He recommended me for a position in the newly founded Informatics department at the Technical University München where I started working in May 1969. In November that year he attended the opening celebration of this department at which Richard Büchi was invited to present a keynote address. As we three logicians sat together in the casual gathering afterwards, Friedrich L. Bauer shocked us by claiming in passing: "*Logic has diminished its relevance for Informatics to a peripheral degree.*" It was not amusing for me to see the two internationally renowned logicians swallowing this ignorant and offending remark without even the slightest attempt for a rebuttal. By the way, the encounter with Büchi at that occasion happened to become instrumental for my later appointment as assistant professor at Wayne State University in Detroit in the academic year 1970/1971.

Schütte was once again involved in an important process in my career, namely as a referee for my Habilitation thesis. I became one of the rare candidates in Germany who failed to pass this postdoctoral procedure required in Germany as a qualification for teaching. Schütte's report dated 11 March 1976 was not helpful at all in preventing this failure although it ended with the words: "*Whether certain parts of the work might be of interest for Informatics, I cannot judge.*" Seen from his own narrow area of research at the time of writing the report might be understood as a fair one.

Judged from today's perspective it is truly hard to understand why he seemed to have totally failed to recognize the contributions offered in this thesis entitled "*Programmieren in der Sprache der Prädikatenlogik*" (programming in the language of first-order logic). For one, the paper, completed in 1974, was one of the few pioneering papers—written independently of those by Colmerauer and Kowalsky of the same year—which eventually led to the prospering area of Logic Programming. Note that without the programming language PROLOG resulting from that area the programs presented in the paper by Bibel and Otten within this volume could not have been written. Second, the paper included the publication [4] and thus the core ideas of the approach to Theorem Proving presented in this same paper just mentioned. One might understand why—besides his peculiar motivations in the matter—a physicist and numerical mathematician like Bauer, the premier referee for the Habilitation, at that time had no feel whatsoever for both subjects. But what about a logician whose formal system formed the basis of the work? But Schütte's report is merely beckmessering on issues of secondary importance and missing the main contributions altogether.

Schütte was in good company in this respect though, since also prominent logicians like Dana Scott in those years failed to appreciate the relevance of our work in Automated Deduction, judging in a personal communication the development of a formal system like GS_3 presented in the same mentioned paper as irrelevant. Similarly, Donald Knuth in his role as the decisive referee recommended rejectance of the paper [3] when this was first submitted to Acta Informatica in 1971 (with two rounds of refereeing lasting for more than two years).

In the many years thereafter I had rare encounters with Schütte although our offices in Munich were only a few blocks apart. As one of the few instances, I attended the celebrations of his 80th birthday and, sad enough, his funeral.

References

1. L. Wolfgang Bibel. *Reflexionen vor Reflexen – Memoiren eines Forschers.* Cuvillier Verlag, Göttingen, 2017.
2. Wolfgang Bibel. Schnittelimination in einem Teilsystem der einfachen Typenlogik. *Archiv für Mathematische Logik*, 12:159–178, 1969.
3. Wolfgang Bibel. An approach to a systematic theorem proving procedure in first-order logic. *Computing*, 12:43–55, 1974. First presented to the GI Annual Conference in 1971; also available as Bericht Nr. 7207, Technische Universität München, Abteilung Mathematik (1972).
4. Wolfgang Bibel and Joachim Schreiber. Proof search in a Gentzen-like system of first-order logic. In E. Gelenbe and D. Poitier, editors, *Proceedings of the International Computing Symposium*, pages 205–212, Amsterdam, 1975. North–Holland.
5. Solomon Feferman. Systems of predicative analysis. *Journal for Symbolic Logic*, 29:1–30, 1964.
6. Kurt Schütte. Schlußweisen-Kalküle der Prädikatenlogik. *Mathematische Annalen*, 122:47–65, 1950.
7. Kurt Schütte. Syntactical and semantical properties of simple type theory. *The Journal of Symbolic Logic*, 25:305–326, 1960.
8. Kurt Schütte. Eine Grenze für die Beweisbarkeit der Transfiniten Induktion in der verzweigten Typenlogik. *Archiv für mathematische Logik und Grundlagenforschung*, 7:45–60, 1965.
9. Kurt Schütte. Predicative well-orderings. In J. Crossley and M. Dummett, editors, *Formal Systems and Recursive Functions*, pages 279–302. North-Holland, Amsterdam, 1965.
10. Kurt Schütte. Zur Berechenbarkeit von Funktionen höherer Typen. Mathematisches Forschungsinstitut Oberwolfach, http://oda.mfo.de/view/bsz/325107009/DEFAULT/110/, April 1967.
11. Kurt Schütte. *Vollständige Systeme Modaler und Intuitionistischer Logik.* Ergebnisse der Mathematik und ihrer Grenzgebiete, Band 42. Springer Verlag, Berlin, 1968.
12. Kurt Schütte. *Proof Theory.* Grundlehren der mathematischen Wissenschaften 225. Springer-Verlag, Berlin, 1977. English and revised version of: Beweistheorie (1960).

Kapitel 7
Mathematische Logik

Kurt Schütte und Helmut Schwichtenberg

Abstract This is a survey article about the development of mathematical logic in Germany, written for the centenary volume of the German Mathematical Society in 1990. It is reprinted from *Winfried Scharlau, editor. Ein Jahrhundert Mathematik 1890–1990, volume 6 of Dokumente zur Geschichte der Mathematik, pages 717–740. Vieweg, 1990.*

Kurz vor der Gründung der DMV waren die „Grundlagen der Arithmetik" (1884) von Frege und „Was sind und was sollen die Zahlen?" (1888) von Dedekind erschienen. Bald darauf folgten die „Beiträge zur Begründung der transfiniten Mengenlehre" (1895–97) von Cantor. Hiermit begann eine systematische Grundlagenforschung der Mathematik und zugleich die Entwicklung der transfiniten Mengenlehre. Voraussetzung für die mathematischen Grundlagenuntersuchungen war die hauptsächlich von Frege und von Peano geschaffene Systematik einer exakten mathematischen Logik.

Wir skizzieren in diesem Artikel zunächst die Grundlegung der modernen mathematischen Logik und beschreiben dann die verschiedenen Richtungen (Logizismus, Beweistheorie und Intuitionismus), die in den ersten Jahrzehnten dieses Jahrhunderts zur Begründung der Mathematik eingeschlagen wurden. Anschließend kehren wir zur Mengenlehre zurück, die zwar schon vor der Jahrhundertwende begründet, aber erst später in völlig formalisierter Weise mit den Methoden der mathematischen Logik behandelt wurde. Weiterhin berichten wir hauptsächlich über Ergebnisse, die von deutschen Logikern auf anderen Teilgebieten der mathematischen Logik erzielt wurden, insbesondere in der Rekursionstheorie, in der Modelltheorie, in Anwendungen auf die Algebra, in der Nichtstandard-Analysis und im Zusammenhang mit der Informatik. Dieser kurze Bericht kann natürlich keinen Anspruch auf Vollständigkeit erheben.

Für ihre Hilfe bei der Abfassung dieses Berichts möchten wir H.-D. Donder, H. Osswald, A. Prestel und M. Ziegler herzlich danken.

K. Schütte (†) · H. Schwichtenberg
Mathematisches Institut, LMU München, Germany, E-mail: schwicht@mathematik.uni-muenchen.de

R. Kahle and M. Rathjen (eds.), *The Legacy of Kurt Schütte*,
https://doi.org/10.1007/978-3-030-49424-7_7

7.1 Die Grundlegung der modernen mathematischen Logik

Die formale Logik, die bereits Aristoteles in exakter Weise begründet hatte, war zwar von der Scholastik in verschiedenen Richtungen weiterentwickelt worden, wurde aber erst in der Mitte des vorigen Jahrhunderts in mathematischer Weise völlig neu geschaffen, und zwar zunächst in Analogie zur Algebra ohne jegliche Bezugnahme auf die vorhergegangenen Behandlungsweisen der Logik. Die mathematische Logik begann mit der Entwicklung einer Booleschen Algebra, die eine Formalisierung der Aussagenlogik und verwandter mathematischer Strukturen liefert, aber noch nicht die volle Ausdrucksfähigkeit der modernen mathematischen Logik besitzt.

Die weiterführenden Begriffsbildungen, die der heutigen mathematischen Logik zugrunde liegen, wurden in Deutschland hauptsächlich von Gottlob Frege (1848–1925) eingeführt und in klarer Weise zur Anwendung gebracht. In seiner „Begriffsschrift" (1879) hat Frege erstmalig eine lange Reihe logischer Formeln in streng formaler Weise aus wenigen Axiomen abgeleitet. Hierbei tritt im Unterschied zu den Formalisierungen, die in Analogie zur Algebra durchgeführt wurden, die fundamentale Bedeutung der Implikation klar in Erscheinung. Die Fregeschen Axiomenschemata

$$A \to (B \to A) \quad \text{und} \quad (A \to (B \to C)) \to ((A \to B) \to (A \to C))$$

der Implikation liefern nämlich zusammen mit der Schlußregel des modus ponens ein fundamentales System der Logik, das sich durch Hinzunahme von sehr einfachen und naheliegenden Axiomen zur vollen klassischen Aussagenlogik erweitern läßt.

Die Werke von Frege enthalten eine Fülle von damals völlig neuartigen Einsichten, insbesondere bezüglich des Unterschiedes von Variablen und Konstanten, des Begriffs der logischen Funktion und des Quantors sowie der Einführung von Kennzeichnungstermen. Hierzu hatte Frege eine ausdrucksvolle logische Systematik geschaffen, mit der er eine Grundlegung der gesamten Mathematik in Angriff nahm. Seine Grundlagenuntersuchungen der Mathematik gehören dem von ihm begründeten Logizismus an, den wir im nächsten Abschnitt besprechen werden.

Die von Frege verwendete Begriffsschrift, die zwar in klarer Weise die logischen Beziehungen veranschaulicht, konnte sich jedoch nicht durchsetzen, weil sie im Unterschied zu allen üblichen Schreibweisen zweidimensional ist. Daher schloß sich die später in der mathematischen Logik verwendete Symbolik nicht an Frege, sondern im wesentlichen an die von G. Peano eingeführte Symbolik an.

Es ist erstaunlich, daß die bahnbrechenden Arbeiten von Frege zunächst überhaupt nicht beachtet wurden. Der Hauptgrund war wohl, daß die von Frege verwendete zweidimensionale Symbolik allgemein abgelehnt und häufig mißverstanden wurde. Es hat seit dem Erscheinen der Begriffsschrift zwanzig Jahre gedauert, bis man auf Frege aufmerksam wurde, und es hat weitere zwanzig Jahre gedauert, bis die strenge Exaktheit der Fregeschen Arbeiten in der mathematischen Logik wieder erreicht wurde. Es war Bertrand Russell, der zuerst die Bedeutung Freges erkannte und ausdrücklich erklärte, daß ihm die Ideen von Frege sehr wertvoll seien.

Ein weiterer erfolgreicher Logiker dieser Zeit war in Deutschland Leopold Löwenheim (1878–1957). Er verfaßte mehrere Arbeiten über Eliminations- und Reduktionsverfahren im Rahmen der Algebra der Logik von Ernst Schröder. 1915 bewies Löwenheim als erster den als Satz von Löwenheim und Skolem bekannten Fundamentalsatz der Logik, der besagt, daß jede Theorie in der Sprache der Prädikatenlogik erster Stufe, die ein Modell hat, bereits ein abzählbares Modell hat. Löwenheim bewies diesen Satz unter Bezugnahme auf eine höhere Prädikatenlogik, während Th. Skolem denselben Satz 1920 allein im Rahmen der Prädikatenlogik erster Stufe bewies.

7.2 Der Logizismus

Die Fortschritte, die gegen Ende des vorigen Jahrhunderts in der mathematischen Logik erzielt waren, reizten dazu, nun eine Grundlegung der Mathematik mit den Methoden der mathematischen Logik zu unternehmen. Der von Frege begründete Logizismus war der Ansicht, daß sich die gesamte Mathematik allein in der reinen Logik entwicklen läßt. Die Tatsache, daß das Unendlichkeitsaxiom in der Mathematik unentbehrlich ist, war für den Logizismus kein Einwand gegen seine These, weil der Übergang von der Denkbarkeit eines unbeschränkten Iterationsprozesses zur Denkbarkeit einer unendlichen Gesamtheit als rein logisch empfunden wurde, wie es ja auch in der Schrift „Was sind und was sollen die Zahlen?“ von Dedekind zum Ausdruck kommt.

Frege hat in seinen Büchern „Die Grundlagen der Arithmetik“ (1884) und „Grundgesetze der Arithmetik“ (1893–1904) die Mathematik in voller Ausführlichkeit mit reiner Logik zu begründen versucht. Sein Aufbau der Mathematik ist jedoch, da er völlig typenfrei ist, den logischen Antinomien ausgesetzt, insbesondere der Russellschen Antinomie der Menge aller Mengen, die nicht sich selbst als Element enthalten. Frege hat, nachdem dies bemerkt wurde, selbst eingesehen, daß sein Unternehmen gescheitert ist. Immerhin sind seine Untersuchungen in mancherlei Hinsicht sehr instruktiv gewesen.

Ein zweiter Versuch zur Begründung der Mathematik in der reinen Logik wurde in dem umfangreichen Werk „Principia Mathematica“ (1910–1913) von Bertrand Russell und Alfred North Whitehead unternommen. Hier wurde, um die logischen Antinomien auszuschließen, eine verzweigte Typenlogik zugrunde gelegt. Das heißt, daß hier jedes Prädikat, in dem gewisse Quantoren auftreten, einem höheren Typ zugerechnet wird als die Typen der in ihm autretenden Quantoren, so daß ein Objekt, das unter Bezugnahme auf eine gewisse Gesamtheit definiert ist, nicht selbst dieser Gesamtheit zugerechnet wird. Ein derartiges System ist völlig konstruktiv. Es vermeidet nicht nur die bekannten logischen Antinomien, sondern ist sogar nachweisbar widerspruchsfrei.

In dieser verzweigten Typenlogik läßt sich jedoch nicht die klassische Mathematik entwickeln. Man erhält hier anstatt des klassischen Begriffs der reellen Zahl eine Hierarchie von reellen Zahlen verschiedener Ordnungen. Daher konnte in den

Principia Mathematica die klassische Mathematik nur durch Hinzunahme eines Reduzibilitätsaxioms gewonnen werden. Dieses Axiom besagt im wesentlichen, daß jedes Prädikat, dessen Bedeutungsbereich eine Menge von Grundobjekten ist, extensional äquivalent einem entsprechenden Prädikat kleinsten Typs ist. Hierdurch wird jedoch der konstruktive Ansatz der verzweigten Typenlogik aufgehoben, so daß nach Hinzunahme des Reduzibilitätsaxioms keine Garantie mehr für die Widerspruchsfreiheit des Systems besteht.

Daraufhin haben verschiedene Logiker, insbesondere Chwistek und Ramsey bemerkt, daß es einfacher und völlig ausreichend sei, anstatt der verzweigten Typenlogik eine einfache Typenlogik, die kein Reduzibilitätsaxiom benötigt, zugrunde zu legen. In der einfachen Typenlogik wird nur zwischen Grundobjekten, Mengen von Grundobjekten, Mengen von Mengen usw. unterschieden, aber keine Typenverzweigung zwischen verschiedenen Mengen von Grundobjekten vorgenommen. Man kann zwar bisher nicht beweisen, daß eine Formalisierung der klassischen Mathematik im Rahmen der einfachen Typenlogik widerspruchsfrei ist, vermeidet aber bereits in der einfachen Typenlogik die bekannten logischen Antinomien, so daß die Benutzung der einfachen Typenlogik nicht problematischer als der in den Principia Mathematica eingeschlagene Weg ist.

Die vom Logizismus unternommenen Versuche, die Mathematik in der reinen Logik zu begründen, konnten jedenfalls ihr Ziel nicht erreichen, haben aber die Weiterentwicklung der mathematischen Logik wesentlich gefördert. So sind die Principia Mathematica für mehrere Jahrzehnte das wichtigste Lehrbuch der mathematischen Logik geworden. Es hatte ja sonst keine so vollständige und aufschlußreiche Systematik der mathematischen Logik gegeben, wie sie in den Principia Mathematica zur Darstellung gekommen ist.

7.3 Die Grundlagenkrise der Mathematik

Die Besinnung auf die Grundlagen der Mathematik, die vom Logizismus ausgegangen war, hatte darauf aufmerksam gemacht, daß die bisherige Mathematik durchaus nicht so unproblematisch ist, wie es bis dahin im allgemeinen angenommen wurde. Schon viel früher hatten bereits Kronecker und Poincaré Einwände gegen die indirekten Beweisführungen in der Mathematik erhoben. Nun war es vor allem L. E. J. Brouwer, der das den indirekten Beweisen zugrunde liegende *tertium non datur* entschieden ablehnte.

Ein zweiter Einwand richtete sich gegen die imprädikative Verwendung des Begriffs der reellen Zahl. Man erhält ja die Vollständigkeitseigenschaft der reellen Zahlen nur dadurch, daß man die obere Grenze einer beliebigen nach oben beschränkten Menge reeller Zahlen unter Bezugnahme auf die Gesamtheit der reellen Zahlen, also in imprädikativer Weise definiert. Auf die Problematik dieses Verfahrens hat Hermann Weyl in seinem Buch „Das Kontinuum" (1918) hingewiesen mit der Erklärung, daß man die Theorie der reellen Zahlen unter Verlust der Vollständigkeitseigenschaft erheblich einschränken müsse, um ihre Exaktheit sicherzustellen.

Aufgrund dieser gegen die bisherige Mathematik erhobenen Einwände sprach man in den zwanziger und dreißiger Jahren von einer Grundlagenkrise der Mathematik, nachdem 1921 in der Mathematischen Zeitschrift ein Aufsatz von Hermann Weyl erschienen war mit dem Titel „Über die neue Grundlagenkrise der Mathematik".

Es wurden nun zur Neubegründung der Mathematik zwei verschiedene Richtungen eingeschlagen, nämlich die des Intuitionismus von L. E. J.Brouwer und die der Beweistheorie von David Hilbert, die damals als Formalismus bezeichnet wurde.

Hilbert hatte es jedoch stets abgelehnt, als Formalist bezeichnet zu werden, denn es kam ihm ja nicht darauf an, bedeutunglose Formalismen zu entwickeln, sondern die klassische Mathematik in streng formaler Weise als widerspruchsfrei und somit als mathematisch bedeutungsvoll sicherzustellen. Wir wollen uns, wie er sagte, nicht aus dem Paradies vertreiben lassen, das uns Cantor beschert hat.

7.4 Die Hilbertsche Beweistheorie

Hilberts Auffassung von der Mathematik, die ihm schon lange vor der sogenannten Grundlagenkrise der Mathematik völlig klar war, kommt in einem Brief, den er 1899 an Gottlob Frege schrieb, deutlich zum Ausdruck. In diesem Brief heißt es:

> Sie schreiben: Aus der Wahrheit der Axiome folgt, daß sie einander nicht widersprechen. Es hat mich sehr interessiert, gerade diesen Satz bei Ihnen zu lesen, da ich nämlich, solange ich über solche Dinge denke, schreibe und vortrage, immer gerade umgekehrt sage: Wenn sich die willkürlich gesetzten Axiome nicht widersprechen, mit sämtlichen Folgen, so sind sie wahr, so existieren die durch die Axiome definierten Dinge.

Man könnte für diese Auffassung von Hilbert den Vollständigkeitssatz der Prädikatenlogik anführen, der allerdings damals noch nicht vorlag. Nach diesem Satz hat ja jede Theorie im Rahmen der Prädikatenlogik erster Stufe, wenn sie widerspruchsfrei ist, ein Modell, so daß die Objekte, die durch die Axiome der betreffenden Theorie definiert sind, tatsächlich existieren.

Es war nun das Ziel der Hilbertschen Beweistheorie, die einzelnen Teile der klassischen Mathematik in vollformalisierter Gestalt als widerspruchsfrei nachzuweisen. Hierzu hat man nicht nur die Objekte, sondern auch die Aussagemöglichkeiten, Axiome und Schlußregeln der mathematischen Theorie formal zu fixieren, um in einer metamathematischen Untersuchung ihre Widerspruchsfreiheit nachweisen zu können. Die Metamathematik oder Beweistheorie, deren Untersuchungsobjekt die als widerspruchsfrei nachzuweisende formalisierte mathematische Theorie ist, sollte sich nach dem ursprünglichen Plan von Hilbert auf eine sehr elementare Untersuchungsmethode beschränken, die Hilbert als finit bezeichnete, nämlich auf die kombinatorische Behandlung von zwar beliebig umfangreichen, aber stets als endlich anzusehenden Zeichenfiguren.

In der Hilbertschen Metamathematik wird nicht wie im Logizismus der Versuch unternommen, die natürlichen Zahlen in der reinen Logik zu entwickeln, weil dies viel zu aufwendig und durchaus nicht elementar wäre. Es wird vielmehr ein sehr elementarer Teil der Arithmetik bereits der Metamathemtik zugrunde gelegt. Man kann ja die nichtnegativen ganzen Zahlen in einer sehr einfachen Weise durch sogenannte Ziffern repräsentieren, nämlich nach folgender induktiven Definition:

(z1) Das Symbol 0 ist eine Ziffer.
(z2) Ist z eine Ziffer, so ist auch z' (der Nachfolger von z) eine Ziffer.

Bei derartigen induktiven Definitionen ist es vielfach üblich eine weitere Definitionsregel hinzuzufügen, die besagt, daß die Menge der Ziffern die kleinste Menge mit den Eigenschaften (z1) und (z2) ist. Die Hinzunahme einer solchen Regel wäre hier nicht nur überflüssig, sondern sogar irreführend, da sie auf den allgemeinen Mengenbegriff Bezug nimmt, von dem in der Metamathematik überhaupt keine Rede ist. Ziffern sind nach ihrer induktiven Definition genau diejenigen Zeichen oder Zeichenreihen, die sich durch Anwendungen der Regeln (z1) und (z2) ergeben, wobei in jedem Fall natürlich nur endlich viele Anwendungen dieser Regeln in Betracht kommen. Für diese Ziffern gilt definitionsgemäß das Prinzip der vollständigen Induktion. Man braucht hierfür in der Metamathematik kein Axiom. Erst bei der Formalisierung eines weniger elementaren Teils der Mathematik wird ein Axiom oder Axiomenschema der vollständigen Induktion benötigt.

Die ersten Widerspruchsfreiheitsbeweise für Teile der Mathematik wurden nach einer Idee von Hilbert in den Jahren 1924–25 von Wilhelm Ackermann durchgeführt, und zwar für quantorenfreie Formalisierungen von mathematischen Theorien, die anstelle der Axiome und Schlußregeln für Quantoren nur das Axiomenschema $A(t) \longrightarrow A(\varepsilon_x A(x))$ für beliebige Terme t haben.

Hierbei bezeichnet $\varepsilon_x A(x)$ ein Objekt mit der Eigenschaft A, falls es ein solches gibt, andernfalls ein beliebiges Objekt. Eine derartig formalisierte Theorie erweist sich als widerspruchsfrei, wenn sich in jeder formalen Herleitung einer numerischen Formel alle ε-Terme unter Erhaltung des Herleitungszusammenhanges eliminieren lassen, weil in einer Herleitung, die keine ε-Terme enthält, keine falschen numerischen Formeln auftreten können.

Ackermann gelang es, ein allgemeines Verfahren zur Elimination der ε-Terme zu entwickeln. Hiermit konnte man jedoch nicht einmal die reine Zahlentheorie als widerspruchsfrei nachweisen, da man hierfür nicht beweisen konnte, daß das Ackermannsche Eliminationsverfahren stets nach endlich vielen Schritten abbricht. Ein solcher Beweis war, wie sich nach einigen Jahren herausstellte, allerdings in finiter Weise überhaupt nicht möglich.

Nachdem in Wien Kurt Gödel 1930 in seiner Dissertation erstmalig den Vollständigkeitssatz der Prädikatenlogik bewiesen hatte, bewies er 1931 in seiner Habilitationsschrift „Über formal unentscheidbare Sätze der Principia Mathematica und verwandter Systeme“ zwei Sätze von sensationeller Bedeutung. Der erste Satz besagt, daß jedes hinreichend starke formale System grundsätzlich unvollständig ist, das heißt, Sätze zu formulieren gestattet, die in dem System weder beweisbar noch widerlegbar sind, sofern das System nicht widerspruchsvoll ist. Der zweite Satz besagt, daß die Widerspruchsfreiheit eines formalen Systems, das die rekursive Zahlentheorie und gewisse natürliche Ableitungsregeln enthält, nicht mit den Mitteln dieses Systems beweisbar ist. Aus diesem zweiten Unvollständigkeitssatz folgt insbesondere, daß die Widerspruchsfreiheit der reinen Zahlentheorie nicht in finiter Weise beweisbar ist, weil die finiten Methoden in der reinen Zahlentheorie enthalten sind.

Weder Gödel noch Hilbert haben diese neuen Erkenntnisse als eine Widerlegung des Hilbertschen Programms aufgefaßt. Beide haben hieraus nur die Konsequenz gezogen, daß der von Hilbert ursprünglich eingenommene streng finite Standpunkt zu einem allgemeineren konstruktiven Standpunkt erweitert werden müsse, um die Widerspruchsfreiheit von mathematisch bedeutungsvollen Theorien beweisen zu können.

So hat Hilbert bald nach dem Erscheinen der Gödelschen Unvollständigkeitssätze in einem Göttinger Kolloquiumsvortrag vorgeschlagen, die formalisierte vollständige Induktion, deren beweistheoretische Behandlung besonders schwierig ist, durch die konstruktive Anwendung einer Schlußregel mit unendlich vielen Prämissen zu ersetzen. Diese von Hilbert als *unendliche Induktion* bezeichnete Schlußregel besagt, daß auf die Allformel $(x)A(x)$ geschlossen werden, darf, wenn in konstruktiver Weise $A(z)$ für jede Ziffer z hergeleitet werden kann. Diese Idee von Hilbert wurde zunächst nicht beachtet, sondern erst nach fast zwanzig Jahren aufgegriffen und ist heute für beweistheoretische Untersuchungen unentbehrlich geworden.

In Göttingen hat dann Gerhard Gentzen zunächst in seiner Dissertation „Untersuchungen über das logische Schließen" (1934–35) mit seinem System des natürlichen Schließens und mit seinem Sequenzenkalkul eine sehr durchsichtige neue Systematik der Prädikatenlogik entwickelt, die sich als äußerst fruchtbar erwiesen hat und zum Allgemeingut aller Logiker geworden ist. Anschließend gelang es ihm 1936, die Widerspruchsfreiheit der reinen Zahlentheorie zu beweisen, und zwar mittels transfiniter Induktion bis zur kleinsten ε-Zahl ε_0. Diese transfinite Induktion überschreitet zwar den finiten Standpunkt, wie es nach Gödel erforderlich ist, verläuft aber in einer völlig konstruktiven Weise, so daß der Beweis von Gentzen als ein Evidenznachweis für die reine Zahlentheorie im Sinne des erweiterten Hilbertschen Programms anzusehen ist.

Durch transfinite Induktion bis ε_0 konnte nun auch Wilhelm Ackermann 1940 beweisen, daß sein Verfahren zur Elimination der ε-Terme im formalen System der reinen Zahlentheorie stets nach endlich vielen Schritten zum Abschluß kommt, womit sich ein weiterer Widerspruchsfreiheitsbeweis für die reine Zahlentheorie ergab.

Durch die Untersuchungen von Gentzen wurde erstmalig erkannt, daß sich die Beweisstärke eines formalen Systems durch eine Ordinalzahl kennzeichnen läßt. Man bezeichnet heute als die beweistheoretische Ordinalzahl eines formalen Systems der Mathematik die kleinste Ordinalzahl α mit der Eigenschaft, daß sich keine rekursive Wohlordnung vom Ordnungstyp α in dem betreffenden formalen System als wohlgeordnet beweisen läßt. Wie Gentzen 1943 in seiner Habilitationsschrift bewies, ist ε_0 die beweistheoretische Ordinalzahl der reinen Zahlentheorie.

Gerhard Gentzen war zuletzt Dozent an der Universität Prag und ist am Ende des Krieges in tschechischer Gefangenschaft umgekommen, nachdem sich sowohl amerikanische als auch russische Mathematiker vergeblich um seine Freilassung bemüht hatten.

Nach dem Kriege wurden in Anknüpfung an die Methoden von Gentzen auch gewisse Teilsysteme der klassischen Analysis als widerspruchsfrei nachgewiesen, und zwar im Rahmen einer Prädikatenlogik zweiter Stufe, in der sowohl über die Grundobjekte als auch über Mengen von Grundobjekten quantifiziert wird. Die reellen Zahlen werden hier durch Mengen von Grundobjekten (natürlichen oder rationalen Zahlen) repräsentiert. Zur Mengenbildung dient das Komprehensionsaxiom, das bezüglich einer Formel $A(x)$ die Menge $\{x : A(x)\}$ aller Grundobjekte mit der Eigenschaft A zu bilden gestattet. Bei uneingeschränkter Verwendung dieses Komprehensionsaxioms erhält man die volle klassische Theorie der reellen Zahlen, von der nicht zu erwarten ist, daß sie sich durch einen Widerspruchsfreiheitsbeweis auf grundsätzlich einfachere Beziehungen zurückführen läßt. Durch Einschränkungen des Komprehensionsaxioms lassen sich jedoch Teilsysteme der Analysis abgrenzen, für die konstruktive Widerspruchsfreiheitsbeweise erbracht werden konnten. Auf andere Weise hat auch Paul Lorenzen gewisse Teile der klassischen Analysis konstruktiv begründet.

Unter Verwendung der von Hilbert vorgeschlagenen unendlichen Induktion konnte auch der Gentzensche Widerspruchsfreiheitsbeweis etwas einfacher und durchsichtiger gestaltet werden, weil sich mit Hilfe dieser Schlußregel alle Quantoren in den Herleitungen numerischer Formeln eliminieren lassen, was in dem von Gentzen behandelten formalen System nicht allgemein möglich ist.

Durch die beweistheoretischen Untersuchungen konnte auch eine Grenze für die Prädikativität ermittelt werden. Zu Beginn der sechziger Jahre wurde nämlich von S. Feferman und K. Schütte gleichzeitig und unabhängig die kleinste Ordinalzahl Γ_0 bestimmt, welche die Eigenschaft hat, daß keine Wohlordnung vom Ordnungstyp Γ_0 in prädikativer Weise als wohlgeordnet beweisbar ist. Hiermit konnte Feferman den prädikativ interpretierbaren Teil der klassischen Analysis genau charakterisieren, nämlich durch Einschränkung der Komprehension auf sogenannte Δ^1_1-Formeln, in denen die Mengenquantoren nur in einer besonders schwachen Weise auftreten.

Mit der Grenze der Prädikativität ist bereits die Grenze für die Durchführbarkeit des erweiterten Hilbertschen Programms erreicht. Man hat für jedes prädikativ interpretierbare Teilsystem der Analysis einen Widerspruchsfreiheitsbeweis durch transfinite Induktion bis zu einer Ordinalzahl $< \Gamma_0$, der den Anforderungen des erweiterten Hilbertschen Programms genügt. Ein Teilsystem der Analysis, das nicht prädikativ interpretierbar ist, läßt sich aber nicht in prädikativer Weise und somit nicht im Sinne des erweiterten Hilbertschen Programms als widerspruchsfrei nachweisen.

Immerhin hat sich gezeigt, daß der prädikativ interpretierbare Teil der klassischen Analysis wesentlich umfangreicher ist als der auf arithmetische Komprehension beschränkte Teil, auf den Hermann Weyl in seinem Buch „Das Kontinuum" die Analysis einzuschränken gedachte. Fast alle bisher bewiesenen fundamentalen Sätze der klassischen Analysis gehören dem prädikativ interpretierbaren Teil an.

Nachdem sich nun herausgestellt hatte, daß der klassischen Mathematik eine viel stärkere Abstraktion zugrunde liegt, als Hilbert erwartet hatte, wurden auch für wesentlich imprädikative Teile der Mathematik Widerspruchsfreiheitsbeweise durchgeführt. Diese Beweise sind allerdings nicht als Evidenznachweise im Sinne des erweiterten Hilbertschen Programms anzusehen, geben aber Aufschluß über die Beweisstärken der betreffenden formalen Systeme.

Die Grundlagenkrise der Mathematik ist inzwischen überwunden. Schon 1930 hatte Paul Bernays, der Mitarbeiter Hilberts, in einer seiner philosophischen Abhandlungen erklärt:

> Es erscheint als eine unberechtigte Zumutung von seiten der Philosophie an die Mathematik, daß sie ihre einfachere und leistungsfähigere Methode zugunsten einer beschwerlicheren und an Systematik zurückstehenden Methode aufgeben solle, ohne durch eine innere Nötigung dazu veranlaßt zu sein.

Diese Auffassung hat sich durchgesetzt. Es besteht mit vereinzelten Ausnahmen kein Zweifel mehr an der Exaktheit der klassischen Mathematik.

Den ersten Widerspruchsfreiheitsbeweis für ein wesentlich imprädikatives Teilsystem der Analysis hat in Weiterentwicklung der Gentzenschen Methoden 1967 G. Takeuti erbracht, und zwar für die sogenannte Π_1^1-Analysis, die äquivalent ist mit demjenigen Teilsystem der Analysis, in dem die Komprehension auf beliebige Formeln mit höchstens einem Mengenquantor eingeschränkt ist. Für dieses und einige bedeutend stärkere Teilsysteme der Analysis sowie für entsprechende Teilsysteme der Mengenlehre haben in München W. Buchholz, G. Jäger und W. Pohlers durch Widerspruchsfreiheitsbeweise die beweistheoretischen Ordinalzahlen bestimmt.

Unter Bezugnahme auf diese beweistheoretischen Ordinalzahlen konnten H. Friedman und S.G. Simpson die Beweisbarkeit oder Unbeweisbarkeit gewisser kombinatorischer Sätze in verschiedenen Teilsystemen der Analysis nachweisen.

7.5 Der Intuitionismus

Der zur Zeit der Grundlagenkrise der Mathematik von L.E.J. Brouwer begründete Intuitionismus hatte nicht nur das *tertium non datur*, auf dem die indirekten Beweisführungen beruhen, sondern überhaupt das axiomatisch-formale Verfahren der klassischen Mathematik abgelehnt. Die Brouwersche Mathematik weicht daher sehr stark von der klassischen Mathematik ab. An die Stelle des Mengenbegriffs tritt hier der Begriff der Wahlfolge, womit sich ein völlig anderes Kontinuum als in der klassischen Mathematik ergibt. Jede Funktion reeller Zahlen ist hier stetig, und jede Operation auf einem Funktional natürlicher Zahlen ist hier bereits eindeutig bestimmt durch die Werte dieses Funktionals in einem hinreichend großen Abschnitt.

Die intuitionistische Mathematik hat sich nicht durchgesetzt, aber in mancherlei Hinsicht zu neuen Erkenntnissen geführt. Wesentliche Teile der intuitionistischen Mathematik wurden von S. C. Kleene, J. Myhill, G. Kreisel und A. S. Troelstra und anderen in einer Weise interpretiert, die auch für die klassische Mathematik bedeutungsvoll ist.

Insbesondere hat die von A. Heyting 1930 durchgeführte Formalisierung der intuitionistischen Prädikatenlogik eine fundamentale Bedeutung bekommen. Diese intuitionistische Prädikatenlogik, in der das *tertium non datur* nicht ableitbar ist, kann sowohl als ein Teilsystem als auch als eine Erweiterung der klassischen Prädikatenlogik aufgefaßt werden. Sie ist ein Teilsystem insofern, als die in ihr herleitbaren Formeln eine echte Teilmenge der in der klassischen Prädikatenlogik herleitbaren Formeln bilden. Anderseits kann die gesamte klassische Prädikatenlogik in einer Weise ausgedrückt werden, die auch in der intuitionistischen Prädikatenlogik gültig ist. Dazu hat man die Disjunktion und Existenzquantifizierung als definiert anzusehen (durch Negation und Konjunktion beziehungsweise durch Negation und den Allquantor), und man muß ferner für jedes Prädikatensymbol P das Prinzip des indirekten Beweisens

$$\forall x_1 \dots x_n(\neg\neg P x_1 \dots x_n \to P x_1 \dots x_n)$$

(auch Stabilität von P genannt) annehmen. Auf diese Weise hat man dann die klassische Prädikatenlogik mit dem in ihr gültigen tertium non datur in der intuitionistischen Prädikatenlogik interpretiert, die außerdem noch eine in der klassischen Prädikatenlogik nicht definierbare Disjunktion, Implikation und Existenzquantifizierung enthält. Derartige Zurückführungen von formalen Systemen der klassischen Mathematik auf entsprechende Systeme der intuitionistischen Mathematik wurden vielfach vorgenommen, insbesondere von Kurt Gödel bei der Einführung seiner Funktionale endlicher Typen (1958) und von Clifford Spector in einer Arbeit, die nach seinem frühen Tode von Georg Kreisel 1962 herausgegeben wurde unter dem Titel: „Provably recursive functionals of analysis: a consistency proof of analysis by an extension of principles formulated in current intuitionistic mathematics.“ In dieser Arbeit führt Spector die Widerspruchsfreiheit der klassischen Analysis zurück auf die Existenz von gewissen Funktionalen, die im Rahmen der intuitionistischen Mathematik durch Bar-Rekursion definiert sind. Es handelt sich hierbei um eine Rekursion, die es in der klassischen Mathematik nicht gibt, sondern nur in der intuitionistischen Mathematik aufgrund der Voraussetzung, daß jede Operation auf einem Funktional natürlicher Zahlen bereits durch die Werte dieses Funktionals in einem hinreichend großen Abschnitt bestimmt ist.

Die Bar-Rekursion höherer Typen, die von Spector gebraucht wird, bedarf jedoch einer besonderen Begründung. Eine solche Begründung wurde zuerst von B. Scarpellini 1970 durch ein topologisches Modell für die Bar-Rekursion höherer Typen gegeben. Horst Luckhardt hat dann 1973 in den Springer Lecture Notes 306 in ausführlicher Systematik eine Begründung der Gödelschen Funktional-Interpretation und der Bar-Rekursion höherer Typen entwickelt, die im Zusammenhang mit der Arbeit von Spector zwar keine Begründung der Analysis im Sinne des erweiterten Hilbertschen Programms liefert, aber als eine Begründung der klassischen Analysis im Sinne der intuitionistischen Mathematik angesehen werden kann.

7.6 Die Mengenlehre

Nachdem Richard Dedekind (1831-1916) in seiner Schrift „Was sind und was sollen die Zahlen?" 1888 die Theorie der reellen Zahlen schon in mengentheoretischer Weise begründet hatte, wurde die allgemeine Mengenlehre bereits vor der Jahrhundertwende von Georg Cantor (1845-1918) geschaffen. Cantor setzte sowohl den Begriff der Kardinalzahl als auch den der Ordinalzahl ins Transfinite fort, wobei die Ordinalzahlen, die im Finiten den Kardinalzahlen entsprechen, eine viel feinere Zahlenstruktur als die Kardinalzahlen liefern. Hiermit wird die vollständige Induktion der Arithmetik zu dem viel stärkeren Beweisverfahren einer transfiniten Induktion erweitert. Die allgemeine Anwendbarkeit der transfiniten Induktion folgt aus dem Wohlordnungssatz, den Ernst Zermelo 1904 bewies. Der Beweis dieses Satzes beruht auf dem Auswahlaxiom, dessen Annahme zwar recht naheliegend ist, das aber zu so starken völlig unkonstruktiven Ergebnissen führt, daß seine Gültigkeit nicht unproblematisch zu sein scheint. Nach dem Wohlordnungssatz läßt sich ja unter anderen auch die Menge der reellen Zahlen wohlordnen, während man sich überhaupt nicht vorzustellen vermag, wie eine solche Wohlordnung hergestellt werden kann. Das Auswahlaxiom hat daher gewisse Kritik hervorgerufen, wird aber im allgemeinen der mathematischen Forschung zugrunde gelegt. So ist zum Beispiel das Zornsche Lemma, das nur mit dem Auswahlaxiom beweisbar ist, ein fundamentales Beweismittel der Algebra geworden.

Es wurde sehr bald erkannt, daß der zunächst äußerst allgemein aufgefaßte Mengenbegriff einer gewissen Einschränkung bedarf, da zum Beispiel der Begriff der Menge aller Mengen, ja schon der Begriff der Menge aller Kardinalzahlen oder der Menge aller Ordinalzahlen zu Widersprüchen führt. Man ist deshalb dazu übergegangen, beliebige Gesamtheiten nicht als Mengen, sondern als Klassen zu bezeichnen, wobei die Mengen spezielle Klassen sind und nicht jede Klasse eine Menge ist.

Daraufhin wurden Axiomatisierungen der Mengenlehre hauptsächlich von E. Zermelo 1908, A. Fraenkel 1925–28 und J. v. Neumann 1925 entwickelt.

Um die Ordinalzahlen, die sich als Äquivalenzklassen wohlgeordneter Mengen ergeben, in einem axiomatischen System der Mengenlehre nicht als echte Klassen, sondern als Mengen zu erhalten, hat man jede Ordinalzahl durch ein spezielles Element ihrer Äquivalenzklasse zu repräsentieren. Dies geschieht nach J.v. Neumann am einfachsten dadurch, daß man die Ordinalzahlen durch diejenigen Mengen repräsentiert, die bezüglich der $\in$-Relation wohlgeordnet sind. Die Ordinalzahlen erscheinen dann als die Mengen ihrer Vorgänger, also die Ordinalzahl 0 als die leere Menge, die Ordinalzahl 1 als die Menge, deren einziges Element die leere Menge ist, usw. Diese Charakterisierung der Ordinalzahlen als die Mengen ihrer Vorgänger ist allerdings nur für den allgemeinen Ordinalzahlenbegriff im Rahmen eines axiomatischen Systems der Mengenlehre sinnvoll. Man kann ja gewisse Ordinalzahlenabschnitte auch durch rekursiv definierte Wohlordnungsrelationen auf der Menge der natürlichen Zahlen repräsentieren. Für die in dieser Weise bezeichneten Ordinalzahlen kann natürlich keine Rede davon sein, daß sie die Mengen ihrer Vorgänger sind.

Ausführliche systematische Untersuchungen einer im Rahmen der Prädikatenlogik erster Stufe axiomatisierten Mengenlehre wurden zuerst hauptsächlich von

Paul Bernays durchgeführt und im Journal of Symbolic Logic 1937–38 publiziert, später auch in dem gemeinsam mit A. Fraenkel herausgegebenen Buch „Axiomatic Set Theory“ (1958).

Einen bedeutenden Fortschritt in der axiomatischen Mengenlehre hat Kurt Gödel mit einem relativen Widerspruchsfreiheitsbeweis für das Auswahlaxiom und die verallgemeinerte Kontinuumhypothese erzielt. Diese Hypothese besagt, daß die Kardinalität der Potenzmenge einer jeden unendlichen Menge genau um 1 größer ist als die Kardinalität der betrachteten Ausgangsmenge. Gödel hat nach dem Erscheinen seiner berühmten Habilitationsschrift noch in Wien eine Vorlesung über seinen relativen Widerspruchsfreiheitsbeweis gehalten, der dann 1938 in Princeton publiziert wurde.

Gödel bewies, daß aus der Widerspruchsfreiheit eines formalen Systems der Prädikatenlogik erster Stufe, das die üblichen Axiome der Mengenlehre, aber weder das Auswahlaxiom noch die Kontinuumhypothese enthält, die Widerspruchsfreiheit eines entsprechenden formalen Systems mit Auswahlaxiom und verallgemeinerter Kontinuumhypothese folgt. Er hat seinen Beweis der Anschaulichkeit halber modelltheoretisch formuliert, indem er von einem Modell des gegebenen formalen Systems ausging und hieraus durch Einschränkung der Mengen auf sogenannte konstruktible Mengen ein Modell bildete, in dem außer den vorgegebenen mengentheoretischen Axiomen auch das Auswahlaxiom und die verallgemeinerte Kontinuumhypothese gilt. Der Beweis ist aber im Grunde von rein syntaktischer Natur, da hier allein aus der Widerspruchsfreiheit des gegebenen formalen Systems in konstruktiver Weise auf die Widerspruchsfreiheit des Systems der konstruktiblen Mengen geschlossen wird.

Die nächste Frage war nun, ob das Auswahlaxiom und die Kontinuumhypothese abhängig oder unabhängig von den übrigen Axiomen der Mengenlehre sind. Gödel hat, wie aus seinen späteren Äußerungen hervorgeht, auch in dieser Hinsicht Untersuchungen unternommen und ist wahrscheinlich der Lösung dieser Frage bereits sehr nahe gekommen, hat aber hierüber nichts veröffentlicht.

Es war dann Paul Cohen, der 1963 als erster die Unabhängigkeit des Auswahlaxioms und der Kontinuumhypothese bewies. Seine Beweise verwenden eine neue Erzwingungsmethode, mit der in einer gewissen syntaktischen Weise Modelle mit den erwünschten Eigenschaften konstruiert werden. Bald darauf wurde von D. Scott und R. Solovay eine weitere Methode für Unabhängigkeitsbeweise eingeführt, bei der den Formeln anstelle von Wahrheitswerten Werte in einer Booleschen Algebra zugeordnet werden. Diese Methode ist rein modelltheoretischer Natur und setzt die Existenz eines sehr starken Universums voraus.

Hiermit setzte eine intensive mengentheoretische Forschung ein, die eine Fülle neuer Unabhängigkeitsbeweise erbrachte und auch die Konsequenzen von stärkeren mengentheoretischen Axiomen untersucht.

Ronald Jensen, der in Bonn bei Hasenjaeger studiert hatte, hat sehr viel zur frühen Verbreitung der Erzwingungmethode in Deutschland beigetragen. Der Höhepunkt seiner frühen Arbeit auf diesem Gebiet war die Konstruktion eines Modells der Mengenlehre, in dem die Souslinhypothese und die allgemeine Kontinuumshypothese gelten. Die Souslinhypothese besagt, daß jede nichtleere vollständig und dicht geordnete Menge ohne erstes und letztes Element, in der jede disjunkte Menge von offenen Intervallen höchstens abzählbar ist, isomorph zu den reellen Zahlen ist.

Diese Hypothese war auch ein Ausgangspunkt für die erste Hauptrichtung der Forschung von Jensen, nämlich der Untersuchung des konstruktiblen Universums L. Jensen zeigte 1968, daß in L die Souslinhypothese falsch ist. Die Beweisanalyse führte zu dem kombinatorischen Prinzip $\diamond$, einer Verstärkung der Kontinuumshypothese, das inzwischen mannigfaltige Anwendungen in der allgemeinen Topologie, Algebra und kombinatorischen Mengenlehre gefunden hat. In gewisser Weise benötigt der Beweis von $\diamond$ in L nur „grobe" Eigenschaften dieses Universums, die im wesentlichen schon Gödel bekannt waren.

Schon in seiner Habilitationsschrift aus dem Jahre 1967 begann Jensen jedoch eine neue detaillierte Untersuchung des konstruktiblen Universums. Hier wird genau die „kleinste" Stelle in dem hierarchischen Aufbau von L untersucht, wo sich bestimmte Dinge ereignen. Dies wird heute als Feinstrukturtheorie von L bezeichnet. Mit ihrer Hilfe hat Jensen eine fast vollständige Analyse der kombinatorischen Struktur von L durchgeführt. Die dabei destillierten Prinzipien, z. B. $\square$ und Moräste, fanden wie $\diamond$ vielfältige Anwendungen.

Im Jahre 1974 kehrte Jensen nach längerer Abwesenheit aus Deutschland für einige Jahre nach Bonn zurück. Gleich am Anfang dieser Periode bewies er zwei tiefe und überraschende Struktursätze. Aus seiner Beschäftigung mit dem singulären Kardinalzahlproblem entwickelte er den Überdeckungssatz für L. Grob gesprochen besagt dieser Satz, daß jedes innere Modell nahe bei L liegt oder sehr viel größer als L ist. Sein Kodierungssatz besagt, daß jedes abzählbare transitive Modell der Mengenlehre eine klassengenerische Erweiterung besitzt, in der alle Mengen aus einer festen reellen Zahl konstruktibel sind.

Der Überdeckungssatz für L führte zu einer neuen Entwicklung in der Theorie der inneren Modelle. Zusammen mit Tony Dodd führte Jensen das Kernmodell K ein, welches die Lücke zwischen L und den von Kunen untersuchten natürlichen inneren Modellen für meßbare Kardinalzahlen füllte. Dies führte zu weiteren wichtigen Überdeckungssätzen. Außerdem zeigte sich, daß die Feinstrukturtheorie grundlegend für die Untersuchung des „vollen" Universums V ist.

In gewisser Weise ist das Kernmodell K eine natürliche Erweiterung von L, das heißt die Elemente von K sind fast konstruktibel. In den letzten Jahren hat sich Jensen neben anderen sehr bemüht, noch bessere „konstruktible" Approximationen an das Universum V zu entwickeln. Dieser grundlegende Zweig der mengentheoretischen Forschung ist noch nicht am Ziel angelangt und überschneidet sich mit der Suche nach natürlichen inneren Modellen für große Kardinalzahlen. Schon jetzt ist aber deutlich, daß eine verallgemeinerte Feinstrukturtheorie grundlegend für diese Untersuchungen ist.

In Deutschland haben außerdem noch H.-D. Donder und B. Koppelberg wertvolle Beiträge zur Mengenlehre geliefert.

7.7 Die Rekursionstheorie

Das wohl einfachste Rekursionsschema ist das der primitiven Rekursion: es definiert eine zahlentheoretische Funktion f aus gegebenen Funktionen g und h durch die Gleichungen

$$f(0, x_2, \ldots, x_n) = g(x_2, \ldots, x_n),$$
$$f(k+1, x_2, \ldots, x_n) = h(k, f(k, x_2, \ldots, x_m), x_2, \ldots, x_n).$$

Dieses zunächst recht speziell aussehende Definitionsschema ist erstaunlich stark. Besonders Skolem hat gezeigt, daß die darauf aufgebaute sogenannte primitiv rekursive Zahlentheorie zur Formalisierung der elementaren Zahlentheorie völlig ausreicht. Hilbert hat dann die Frage gestellt, ob es überhaupt berechenbare zahlentheoretische Funktionen gibt, die nicht mehr primitiv rekursiv sind. Dieses Problem wurde 1928 von Ackermann gelöst: Definiert man

$$f_1(x, y) = x + y, \quad f_2(x, y) = xy, \quad f_3(x, y) = x^y$$

und allgemein $f_n(x, y)$ durch die entsprechende Fortsetzung, so ist die durch $g(n, x, y) := f_n(x, y)$ definierte Funktion offenbar berechenbar, aber aus Wachstumsgründen nicht mehr primitiv rekursiv.

Die Ausgangsfrage der Rekursionstheorie im eigentlichen Sinne ist die nach dem allgemeinen Begriff eines Algorithmus, also eines Berechnungsverfahrens, bei dem die Argumente (Eingabedaten) und die Werte (Ausgabedaten) konkrete Objekte (etwa Zeichenreihen) sind. Schon immer hat man sich in der Mathematik darum bemüht, Algorithmen zur Lösung von Problemen anzugeben. Ein frühes Beispiel ist etwa der Euklidische Algorithmus zur Bestimmung des größten gemeinsamen Teilers zweier natürlichen Zahlen.

Es hat jedoch in der Mathematik auch immer Fragestellungen gegeben, die sich einer algorithmischen Lösung widersetzt haben. Ein Beispiel ist das zehnte Problem aus der berühmten Liste, die Hilbert in seinem 1900 gehaltenen Vortrag vor dem Internationalen Mathematikerkongress in Paris angegeben hat.

> Eine diophantische Gleichung mit irgendwelchen Unbekannten und mit ganzen rationalen Zahlenkoeffizienten sei vorgelegt. *Man soll ein Verfahren angeben, nach welchem sich mittels einer endlichen Anzahl von Operationen entscheiden läßt, ob die Gleichung in den ganzen rationalen Zahlen lösbar ist.*

Hier haben viele erfolglose Lösungsversuche schließlich zu der Vermutung geführt, daß es überhaupt keinen Algorithmus gibt, der über die Lösbarkeit einer vorgelegten diophantischen Gleichung entscheidet; wir werden weiter unten darauf zurückkommen. Damit wird jedoch die Fragestellung auf eine andere Ebene gehoben. Es geht jetzt nicht mehr darum, einen konkreten Algorithmus anzugeben, sondern man muß sich über die allgemeine Natur von Algorithmen soweit klar werden, daß man einen Beweis für die Nichtexistenz eines Algorithmus der verlangten Art führen kann.

Ein erster Schritt in diese Richtung wurde von Herbrand getan. In einem Brief vom 7. April 1931 an Gödel möchte Herbrand eine Axiomatisierung der Arithmetik genau angeben. Er schreibt

> ... Außerdem haben wir in der Arithmetik andere Funktionen, zum Beispiel durch Rekursion definierte Funktionen, die ich mit folgenden Axiomen definieren werde. Nehmen wir an, daß wir alle die Funktionen $f_n(x_1, x_2, \ldots, x_{p_n})$ einer gewissen endlichen oder unendlichen Menge F definieren wollen. Jedes $f_n(x_1, \ldots)$ wird gewisse Definitionsaxiome haben. .... Diese Axiome werden folgenden Bedingungen genügen.
>
> 1. Die Definitionsaxiome von f_n enthalten, außer dem f_n, nur Funktionen von kleinerem Index.
> 2. Diese Axiome enthalten nur freie Variable und Konstanten.
> 3. Man muß mit intuitionistischen Beweisen zeigen können, daß es möglich ist, den Wert der Funktionen für jedes bestimmte Wertsystem ihrer Argumente mit diesen Axiomen eindeutig zu berechnen.
>
> Zum Beispiel haben wir folgende Beispiele:
>
> a) Die Funktionen $x + y$ und $x \cdot y$, die die Menge E_1 machen, die mit folgenden Axiomem definiert sind:
>
> $$x + 0 = x \qquad x + (y + 1) = (x + y) + 1$$
> $$x \cdot 0 = 0 \qquad x(y + 1) = xy + x$$

Herbrand gibt dann unter b) die Axiome der primitiven Rekursion an und erklärt unter c), daß nach seinem allgemeinen Schema noch viele andere Funktionen definiert sein können, wobei er als Beispiel die Ackermannsche Funktion erwähnt.

Es handelt sich hier um den ersten schriftlich vorliegenden Versuch, den Begriff der rekursiven, also durch einen Algorithmus gegebenen Funktion zu definieren.

Gödel hat 1934 in einer Vorlesung in Princeton als „suggested by Herbrand" die folgende Definition einer rekursiven Funktion angegeben:

> If φ denotes an unknown function, and $\psi_1, \ldots, \psi_k$ are known functions, and if the ψ's and φ are substituted in one another in the most general fashions and certain pairs of the resulting equations are equated, then, if the resulting set of functional equations has one and only one solution for φ, φ is a recursive function.

Gödel hat diesen Definitionsversuch Herbrand irrtümlich zugeschrieben. Denn aus dem oben auszugsweise zitierten Brief Herbrands an Gödel und aus Gödels Antwort vom 25. Juli 31 (die Herbrand, der am 27. Juli 31 in den Alpen tödlich verunglückt ist, nicht mehr erreicht hat) geht hervor, daß dieser Briefwechsel die einzige Kontaktaufnahme zwischen beiden war. Ferner ergibt sich aus einem Briefwechsel von Gödel mit J. van Heijenoort, daß Gödel 1934 bei der Vorbereitung der Vorlesungen in Princeton der Brief Herbrands nicht mehr vorlag und er sich auf sein Gedächtnis verlassen mußte.

Kalmar hat 1956 gezeigt, daß diese Definition nicht den intendierten Begriff trifft: Er hat eine algorithmisch nicht berechenbare Funktion angegeben, die sich als eindeutige Lösung eines Gleichungssystems darstellen läßt. Später haben Grzegorczyk, Mostowski und Ryll–Nardzewski diejenigen Funktionen charakterisiert, die als eindeutige Lösungen von Gleichungssystemen auftreten. Es sind dies genau die hyperarithmetischen Funktionen, von denen weiter unten noch die Rede sein wird.

Gödel selbst hat schon in der Vorlesung von 1934 eine wichtige Modifikation an diesem Definitionsversuch vorgenommen. Er verlangt, daß für jedes Argumententupel der Funktionswert durch gewisse einfache formale Regeln (im wesentlichen eine Einsetzungs- und eine Ersetzungsregel) ausrechenbar sein muß. Diese Definition ist 1936 (in leicht veränderter Form) von Kleene zur Grundlage einer ersten Ausarbeitung der Theorie der rekursiven (und der partiell-rekursiven) Funktionen gemacht worden. Kleene bewies dort unter wesentlicher Verwendung der 1931 von Gödel eingeführten Technik der Numerierung von Formeln einige inzwischen klassischen Resultate: den Substitutionssatz (auch S-m-n Theorem genannt), den Rekursionssatz, ferner — mittels des Kleeneschen Normalformensatzes — die Übereinstimmung der rekursiven Funktionen mit den sogenannten μ-rekursiven Funktionen, die aus einfachen Ausgangsfunktionen durch Anwendung der Definitionsschemata der Einsetzung, der primitiven Rekursion und des folgenden μ-Schemas erzeugt werden: Ist eine Funktion g bereits erzeugt, und hat g die Eigenschaft, daß es zu jedem $x_1, \ldots, x_n$ ein y gibt mit $g(x_1, \ldots, x_n, y) = 0$, so darf man übergehen zu der durch $f(x_1, \ldots, x_n) := \mu y(g(x_1, \ldots, x_n, y) = 0) :=$ das kleinste y mit $g(x_1, \ldots, x_n, y) = 0$ definierten Funktion.

Daraufhin hat A. Church 1936 die heute nach ihm benannte These aufgestellt, daß der oben diskutierte Begriff der allgemeinen rekursiven Funktion und der von ihm eingeführte äquivalente Begriff der λ-definierbaren Funktion mit dem der „effektiv berechenbaren“ Funktion (in einem etwas vagen intuitiven Sinn) zusammenfällt. Diese These ist heute allgemein akzeptiert. Eine wesentliche Stützung hat sie erfahren durch die von A. Turing 1937 durchgeführte Analyse der Operationen, die ein menschlicher Rechner ausführt. (Eine ähnliche Überlegung wurde gleichzeitig und unabhängig von E. Post angestellt.) Diese Analyse führte Turing auf ein abstraktes Maschinenmodell, das heute nach ihm benannt wird. Es zeigte sich, daß auch die durch solche Turing-Maschinen berechenbaren Funktionen mit den allgemein rekursiven Funktionen zusammenfallen.

Als äquivalent mit dem Begriff der allgemeinen rekursiven Funktionen erwies sich dann auch der Begriff der kombinatorisch definierbaren Funktion, der sich im Rahmen der von H.B. Curry 1930 in seiner Göttinger Dissertation (bei Hilbert) begründeten kombinatorischen Logik ergibt. Der λ-Kalkül von Church ist äquivalent mit der viel früher von Curry entwickelten kombinatorischen Logik. Die Formeln des λ-Kalküls sind übersichtlicher als die der kombinatorischen Logik, während die kombinatorische Logik den Vorteil hat, im Unterschied zum λ-Kalkül völlig ohne gebundene Variablen auszukommen. Besonders der λ-Kalkül, aber teilweise auch die kombinatorische Logik haben Anwendungen in der Informatik gefunden.

Auch von den 1954 eingeführten Markov-Algorithmen hat sich gezeigt, daß die hiermit definierbaren Funktionen mit den allgemeinen rekursiven Funktionen zusammenfallen.

Bereits 1931 hat schon Gödel mit dem Begriff der in einem formalen System S entscheidungsdefiniten Relation eine Definition formuliert, von der sich herausgestellt hat, daß sie mit dem Begriff der rekursiven Relation (d.h. einer Relation, deren charakteristische Funktion rekursiv ist) zusammenfällt. Dies hat Gödel selbst 1946 bemerkt, nachdem er schon vorher eine gewisse Absolutheit (d.h. Unabhängigkeit von dem zugrunde liegenden System) dieses Begriffs festgestellt hatte.

Nachdem so der Begriff des Algorithmus eine befriedigende und mathematisch präzise Klärung gefunden hatte, war erstmals die Möglichkeit eröffnet, von gewissen mathematischen Problemen zu zeigen, daß sie algorithmisch *nicht* lösbar sind. Innerhalb der Theorie der rekursiven Funktionen lassen sich solche Probleme leicht angeben: Kleene konstruierte 1936 ein Beispiel einer primitiv rekursiven Relation R derart, daß $\{ n : \forall x R(x, n) \}$ nicht rekursiv ist. Nun hatte Gödel bereits 1931 ein Verfahren angegeben, das zu jedem primitiv rekursiven R eine prädikatenlogische Formel erster Stufe liefert, deren Erfüllbarkeit mit der Richtigkeit von $\forall x R(x, n)$ äquivalent ist. Es folgt also, daß die Menge der erfüllbaren prädikatenlogischen Formeln erster Stufe nicht rekursiv ist, das heißt nicht algorithmisch charakterisiert werden kann. Dieses Resultat wurde explizit zuerst von Church formuliert.

Später sind dann vielfältige Verschärfungen hiervon gefunden worden: für viele syntaktisch abgegrenzte Klassen $\mathcal{F}$ von Formeln der Prädikatenlogik erster Stufe konnte man zeigen, daß die Frage nach der Erfüllbarkeit von Formeln $F \in \mathcal{F}$ unentscheidbar bzw. entscheidbar ist. An diesen Untersuchungen waren besonders Herbrand, Gödel, Hao Wang, Börger und Goldfarb beteiligt. Als besonders nützlich für Unentscheidbarkeitsbeweise hat es sich herausgestellt, den Begriff der Rekursivität durch sogenannte Registermaschinen (anstelle von Turingmaschinen) zu charakterisieren. Eine Übersicht über den heutigen Stand der Kenntnisse auf diesem Gebiet findet man in dem 1985 erschienenen Buch „Berechenbarkeit, Komplexität, Logik" von E. Börger.

Auch für viele in der Mathematik aufgetretenen Fragestellungen konnte die algorithmische Unlösbarkeit nachgewiesen werden: Aufbauend im wesentlichen auf Vorarbeiten von Julia Robinson, die sich auf Lösungen der Pellschen Gleichung beziehen, zeigte Matijasevitch 1970 die Unentscheidbarkeit des zehnten Hilbertschen Problems. Boone und Novikov bewiesen in den fünfziger Jahren die Unentscheidbarkeit des Wortproblems der Gruppentheorie.

Der Begriff der Rekursivität läßt sich offenbar relativieren: Eine Menge A natürlicher Zahlen heißt rekursiv in einer Menge B (Schreibweise: $A \leq_T B$), wenn es eine Turingmaschine M gibt, die die Zugehörigkeit einer vorgelegten Zahl zu der Menge A entscheidet, wobei M im Verlauf der Rechnung B als „Orakel" benutzen kann, das heißt beliebig oft Anfragen stellen kann, ob ein Zwischenergebnis zur Menge B gehört oder nicht. Die Äquivalenzklassen bezüglich der von $\leq_T$ erzeugten Äquivalenzrelation nennt man rekursive Grade; Untersuchungen über ihre Struktur sind sehr schwierig und machen ein Hauptgebiet der heutigen Rekursionstheorie aus. Von besonderem Interesse sind rekursiv aufzählbare Grade, das heißt die Grade mit rekursiv aufzählbaren Repräsentanten. Eine Menge heißt rekursiv aufzählbar, wenn sie sich in der Form $\{ n : \exists x R(x, n) \}$ mit rekursiven R definieren läßt. Post hat 1944 das Problem formuliert, ob es überhaupt rekursiv aufzählbare Grade gibt, die von $\mathbf{0}$ (dem Grad der rekursiven Mengen) und dem maximalen rerursiv aufzählbaren Grad $\mathbf{0}'$ (dessen Existenz man sich leicht überlegen kann) verschieden sind. Diese Frage wurde 1956 unabhängig voneinander von Friedberg und Muchnik positiv beantwortet. Beide Autoren verwendeten zum Beweis eine neuartige Methode, die sogenannte Prioritätsmethode, die seitdem in vielerlei Hinsicht erweitert worden ist und die heute noch diesen Teil der Rekursionstheorie charakterisiert. In Deutschland hat hierzu besonders Ambos-Spieß wichtige Beiträge geliefert.

Bisher war von Rekursionstheorie nur über den natürlichen Zahlen die Rede gewesen. Schon früh hat man jedoch versucht, den Begriff der Rekursivität auch über anderen Bereichen zu studieren. Hier stellt sich sofort die Frage, welche Bereiche denn angemessen sind. Eine Antwort darauf haben Kripke und Platek gegeben: Sie konnten zeigen, daß die sogenannten zulässigen Ordinalzahlen oder etwas allgemeiner die Abschnitte L_α von Gödels Hierarchie der konstruktiblen Mengen mit einer zulässigen Ordinalzahl α geeignete Bereiche für Rekursionstheorien sind. Die kleinste zulässige Ordinalzahl oberhalb von ω ist ω_1^{CK} (CK steht für Church und Kleene), das heißt die kleinste Ordinalzahl, die nicht mehr durch eine rekursive Wohlordnung dargestellt werden kann. In der Rekursiontheorie über zulässigen Ordinalzahlen, der sogenannten α-Rekursionstheorie, ist es von großer Wichtigkeit, den Begriff der Endlichkeit geeignet zu verallgemeinern; darauf hat zuerst Kreisel hingewiesen. Im Fall der Rekursionstheorie über ω_1^{CK} haben Kreisel und Sacks den Endlichkeitsbegriff zum Begriff einer hyperarithmetischen, also Δ_1^1-Teilmenge der natürlichen Zahlen verallgemeinert. Im allgemeinen Fall der α-Rekursiontheorie für zulässiges α ist der passende Endlichkeitsbegriff der eines α-rekursiven Bildes einer Ordinalzahl $< \alpha$.

Später haben insbesondere S.Friedman und Sacks versucht, Rekursionstheorie auch auf L_β mit beliebigen Limeszahlen β zu verallgemeinern. In dieser sogenannten β-Rekursionstheorie lassen sich die Begriffe der β-rekursiv aufzählbaren Menge und der β-rekursiven Funktion in natürlicher Weise definieren. Weniger naheliegend war die folgende Definition eines angemessenen Endlichkeitsbegriffs, die von W. Maaß gefunden wurde: eine Teilmenge von L_β heißt invariant endlich, wenn sie ein Element der kleinsten rekursiv invarianten Klasse von β-rekursiven beschränkten Teilmengen von L_β ist.

Schließlich ist noch über eine Verallgemeinerung der Rekursionstheorie zu berichten, in der die betrachteten Objekte nicht mehr nur Funktionen über den natürlichen Zahlen sind, sondern Funktionale oder von noch höherem Typ. In einer viel beachteten Arbeit „Über eine noch nicht benützte Erweiterung des finiten Standpunkts" von 1958 hat Gödel die eingangs betrachteten primitiv rekursiven Funktionen auf höhere Typen erweitert. Die zugehörige Theorie T hat dann die volle Stärke der Arithmetik.

Die volle Rekursiontheorie hat Kleene auf höhere Typen erweitert. Er hat dabei grundlegende Zusammenhänge mit der Theorie der hyperarithmetischen Mengen aufgedeckt: eine Menge natürlicher Zahlen ist hyperarithmetisch genau dann, wenn sie rekursiv ist in dem Typ-2-Objekt E, das 0 oder 1 ist je nachdem ob sein Funktionsargument eine Nullstelle besitzt oder nicht. Kleenes Rekursionstheorie höherer Typen hat als Grundannahme, daß alle betrachteten Objekte total, das heißt überall definiert sein müssen. Das führt dazu, daß man unendliche Rechnungen zulassen muß. Denn wenn ein Typ-2-Objekt wie etwa das obige E als Argument in einer Rechnung auftritt, dann kann dieses Argument ja nur so benutzt werden, daß man es anwendet auf eine (nach Kleenes Grundanahme totale) Funktion, das heißt auf ein unendliches Objekt, das vorher durch die Rechnung bereitgestellt werden muß. Es ist deshalb nicht verwunderlich, daß die weitere Entwicklung dieser Art von Rekursionstheorie auf Fragestellungen führt, bei denen die Antwort von den zugrunde gelegten mengentheoretischen Axiomen (wie etwa $V = L$) abhängt. Genaueres hierüber findet man etwa in dem 1978 erschienen Buch „Recursion-Theoretic Hierarchies" von P. Hinman.

Eine Rekursionstheorie höherer Typen, in der alle Rechnungen endlich sind und die deshalb von vornherein partiell definierte und stetige Objekte zugrunde legt, ist (nach Vorarbeiten von Platek) von D. Scott und Yu. Ersov entwickelt worden. Die Grundidee ist hier, die unendlichen Objekte als Limiten (genauer: Ideale) endlicher Approximationen darzustellen. Ein Funktional heißt dann rekursiv, wenn das Ideal seiner endlichen Approximationen rekursiv aufzählbar ist. Diese sogenannte Bereichstheorie (domain theory) ist in der Informatik von grundlegender Bedeutung, insbesondere für die Semantik von Programmsprachen.

7.8 Die Modelltheorie

Neben der Beweistheorie entwickelte sich seit Beginn der zwanziger Jahre die „Modelltheorie". In ihr steht nicht die Frage nach der Widerspruchsfreiheit eines (formalen) Axiomensystems für die Mathematik oder Teilgebiete davon im Vordergrund — diese wird, im Gegenteil, als gegeben angenommen — sondern die Untersuchung der Modelle eines solchen Axiomensystems. Ihre Wurzeln hat die Modelltheorie einerseits in der schon in Abschnitt 1 erwähnten Arbeit von Löwenheim (1915) und in einer Arbeit von Skolem aus dem Jahr 1934, andererseits in Gödels Dissertation (1930). In der Arbeit von Skolem wird mit einer Konstruktion, die im wesentlichen die Ultraproduktkonstruktion darstellt, unter anderem gezeigt, daß es kein (formales) Axiomensystem geben kann, das die natürlichen Zahlen bis auf Isomorphie charakterisiert. Diese Methode der Konstruktion von sogenannten Nichtstandard-Modellen eines Axiomensystems stand Pate bei der Entwicklung der Nichtstandard-Analysis durch Abraham Robinson in den sechziger Jahren. In seiner Arbeit von 1930 zeigte Gödel, daß die auf Frege, Whitehead, Russell und Hilbert zurückgehende Formalisierung der Mathematik alle mathematischen Schlußweisen vollständig erfaßt, daß also eine Unbeweisbarkeit notwendig in einem „Gegenbeispiel" begründet sein muß. Die Konstruktion dieses Gegenbeispiels stellt eine andere grundlegende Methode der heutigen Modelltheorie dar.

Die beiden eben angesprochenen Methoden wurden in den dreißiger Jahren von A. Tarski durch weitere ergänzt und systematisiert. Tarski begann auch das Studium konkreter algebraischer Theorien unter modelltheoretischen Gesichtspunkten. Seine Untersuchungen erbrachten die Entscheidbarkeit einiger wesentlicher Teiltheorien der Mathematik, so zum Beispiel der Theorie der komplexen Zahlen, der Theorie der reellen Zahlen und der Theorie der euklidischen Ebene. Tarskis geometrische Untersuchungen wurden in den sechziger Jahren in Deutschland von W. Schwabhäuser fortgeführt. Er zeigte unter anderem die Entscheidbarkeit der ebenen hyperbolischen Geometrie.

In den sechziger Jahren erhielt die Modelltheorie wesentliche Impulse einerseits durch den tiefliegenden Satz von Morley, andererseits durch die aufsehenerregende Anwendung modelltheoretischer Methoden durch J. Ax und S. Kochen in dem Problemkreis diophantischer Gleichungen.

Mit den Arbeiten von Ax und Kochen eröffnete sich der Modelltheorie ein weites Feld der Anwendung — Fragen der Entscheidbarkeit und Axiomatisierbarkeit in der Algebra, speziell der Körpertheorie, bis hin zur Behandlung rein algebraischer Fragestellungen. In Deutschland lieferten insbesondere A. Prestel, V. Weisspfenning und M. Ziegler hierzu zahlreiche Beiträge.

Der Satz von Morley besagt, daß für eine Theorie je zwei Modelle gleicher, aber beliebiger überabzählbarer Mächtigkeit isomorph sind, falls dies schon für mindestens eine überabzählbare Mächtigkeit richtig ist. Der Beweis dieses Satzes brachte für die Modelltheorie wiederum neue Konstruktionsmethoden, die sich für die weitere Entwicklung als grundlegend erwiesen haben. Insbesondere wurde hierdurch die sogenannte Stabilitätstheorie und damit letzlich S. Shelahs Klassifikationsprogramm für Teiltheorien der Mathematik inspiriert. Theorien, deren Modellklasse eine Strukturtheorie ähnlich der Theorie algebraisch abgeschlossener Körper erlaubt, gelten dabei als besonders einfach (sie sind insbesondere stabil), während Theorien von Strukturen mit einer linearen Ordnung, wie zum Beispiel die reellen Zahlen, als besonders kompliziert gelten (sie sind insbesondere instabil). Beiträge hierzu lieferten in Deutschland unter anderen U. Felgner und J. Reineke zur Theorie der Gruppen und Ringe und M. Ziegler zur Theorie der Moduln.

Wie schon oben erwähnt, entwickelte in den sechziger Jahren A.Robinson seine Nichtstandard-Analysis, die eine späte Rechtfertigung des Rechnens mit infinitesimalen Zahlen von Leibniz brachte. In gewisser Weise stellt eine Arbeit von Schmieden und Laugwitz von 1958 einen Vorläufer davon dar. Mit modelltheoretischen Überlegungen (Saturiertheitsprinzip) wurde gezeigt, daß man, ohne auf Widersprüche zu stoßen, die Existenz von unendlich großen und unendlich kleinen „reellen“ Zahlen annehmen kann. Dieser neue Zahlenbereich bildet wieder einen angeordneten Körper, der dieselben formalen Eigenschaften wie die üblichen reellen Zahlen hat (Transferprinzip). Das Vollständigkeitsaxiom gilt nur für die internen Mengen (Prinzip der definitorischen Abgeschlossenheit der internen Mengen). Diese drei Prinzipien der Nichtstandard-Analysis wurden auf die vielen Arten topologischer Strukturen und auf die natürlichen Zahlen angewandt. So entwickelte sich die Nichtstandard-Analysis, obwohl logischen Ursprungs, zu einem eigenständigen Teilgebiet der Mathematik. Einen besonderen Aufschwung erlebte die Nichtstandard-Analysis durch Arbeiten von Loeb, Anderson und Keisler in den siebziger Jahren, in denen gezeigt wurde, daß man alle straffen Wahrscheinlichkeitsmaße als Zählmaße auffassen und deshalb stochastische Integrale (bis auf unendlich kleine Fehler) pfadweise definieren kann. In dem Buch mit dem Titel „Nonstandard Methods in Stochastic Analysis and Mathematical Physics“ (Academic Press, 1986), das S. Albeverio (Bochum) zusammen mit drei Norwegern geschrieben hat, findet man einen Überblick auch über Teile der in Deutschland entstandenen Anwendungen der Nichtstandard-Analysis auf den Gebieten Mathematische Logik (Wahrscheinlichkeitslogik), Funktionalanalysis, Stochastik und Mathematische Physik.

Der topologische Übertragungsprozeß, der der Nichtstandard-Analysis zugrunde liegt, initiierte die Entwicklung der sogenannten topologischen Modelltheorie, deren Entwicklung und Anwendung in Deutschland besonders Flum, Prestel und Ziegler vorantrieben. Robinson hatte 1972 die Frage gestellt, welche Logik für topologische Strukturen dieselbe Rolle spielen könnte wie die Logik erster Stufe für algebraische Strukturen. Die zu betrachtenden Strukturen haben offenbar die Form $(\mathcal{M}, \tau)$, wobei τ eine Topologie auf der Trägermenge der erststufigen Struktur $\mathcal{M}$ ist. Beispiele sind etwa topologische Räume ($\mathcal{M}$ ist hier nur eine Menge), topologische Gruppen und topologische Vektorräume. Es hat mehrere Versuche gegeben, die Frage von Robinson angemessen zu beantworten, und zwar durch folgende formale Systeme:

1. $\mathcal{L}^{mon}$, eine Version der monadischen Logik zweiter Stufe, in der die Mengenquantoren über offene Mengen laufen.
2. $\mathcal{L}^t$; dies ist eine Einschränkung von $\mathcal{L}^{mon}$, in der Mengenquantoren (die also wieder über offene Mengen laufen) nur erlaubt sind im Kontext

   $$\exists X(t \in X \wedge \varphi)$$

   wobei X nur negativ in φ vorkommt, beziehungsweise im dualen Kontext

   $$\forall X(t \in X \rightarrow \varphi)$$

   mit X nur positiv in φ.
3. $\mathcal{L}^{I^n}$. Hier hat man die übliche Logik erster Stufe um einen n-stelligen Operator zur Bildung des Inneren erweitert.

Man kann zeigen, daß die Ausdrucksfähigkeit dieser Sprachen echt abnimmt, also $\mathcal{L}^{mon} > \mathcal{L}^t > \mathcal{L}^{I^n}$.

Es stellt sich die Frage, welche dieser Sprachen sich für eine topologische Modelltheorie eignet. Eine überzeugende Antwort ist im wesentlichen durch Untersuchungen von Ziegler gegeben worden: $\mathcal{L}^t$ (im Gegensatz zu $\mathcal{L}^{mon}$) erfüllt den Kompaktheitssatz (das heißt: eine Menge von $\mathcal{L}^t$-Sätzen hat ein topologisches Modell genau dann wenn jede endliche Teilmenge eines besitzt) und den Satz von Löwenheim–Skolem. Es gilt sogar eine Form des Lindström-Satzes für $\mathcal{L}^t$, daß nämlich $\mathcal{L}^t$ eine maximale Logik mit diesen beiden Eigenschaften ist. Es scheint also, daß mit $\mathcal{L}^t$ eine Antwort auf die eingangs gestellte Frage von Robinson gefunden ist.

Für weitere wichtige Eigenschaften von $\mathcal{L}^t$ (etwa rekursive Axiomatisierbarkeit, Interpolationssatz) müssen wir auf die Literatur verweisen, etwa auf das Buch „Topological Model Theory" von J. Flum und M. Ziegler (Springer Lecture Notes 769). Dort werden auch Anwendungen behandelt, etwa die (erbliche) Unentscheidbarkeit der Theorie der T_2-Räume und eine vollständige Axiomatisierung der Theorie des topologischen Körpers der komplexen Zahlen.

Kapitel 8
Bemerkungen zur Hilbertschen Beweistheorie

Kurt Schütte

Zusammenfassung Vortrag am 13. Mai 1993 im Gedenkkolloquium der Arbeitsgemeinschaft für Ost- und Westpreußische Landeskunde an der Universität München zum 50-jährigen Todestag von David Hilbert.[1]

Als ich 1930 in meinem 5. Semester in Berlin studierte, fiel mir im dortigen Philosophischen Seminar eine Abhandlung des Bonner Philosophen Oskar Becker über Mathematische Existenz in die Hände. Hierdurch erfuhr ich von dem damaligen Grundlagenstreit der Mathematik zwischen L. E. J. Brouwer und David Hilbert.

Brouwer verwarf die von Dedekind und Cauchy begründete klassische Mathematik, insbesondere die allgemeine Verwendung des tertium non datur und hiermit indirekte Beweisführungen. Statt dessen setzte er sich für eine von ihm als intuitionistisch bezeichnete Mathematik ein, die wesentlich von der klassischen Mathematik verschieden ist. So ist z.B. das intuitionistische Kontinuum bedeutend schwächer als das klassische Kontinuum der reellen Zahlen.

Demgegenüber verfolgte Hilbert das Ziel, die gesamte klassische Mathematik mit finiten, das heißt mit einfachsten kombinatorischen Methoden als widerspruchsfrei nachzuweisen, um auf diese Weise ihre mathematische Exaktheit sicherzustellen. Die mathematische Theorie, welche diese Widerspruchsfreiheitsbeweise im Rahmen der mathematischen Logik durchführen sollte, wurde von Hilbert Beweistheorie genannt.

Oskar Becker hatte in seiner Abhandlung beide Auffassungen der Mathematik eingehend diskutiert und sich für den Intuitionismus entschieden. Trotzdem schien mir die von Hilbert erstrebte Sicherstellung der klassischen Mathematik sinnvoller als die inutitionistische Einschränkung der Mathematik zu sein. Daraufhin kehrte ich im folgenden Semester nach Göttingen zurück, um mich mit der von Hilbert verwendeten mathematischen Logik vertraut zu machen.

K. Schütte (†)
Mathematisches Institut, LMU München, Germany

[1] Zuerst abgedruckt in *Acta Borussica, 5:241–244, 1995.*

R. Kahle and M. Rathjen (eds.), *The Legacy of Kurt Schütte*,
https://doi.org/10.1007/978-3-030-49424-7_8

Bereits 1931 stellte sich jedoch durch die Gödelschen Unvollständigkeitssätze heraus, daß nicht einmal die reine Zahlentheorie mit streng finiten Methoden als widerspruchsfrei beweisbar ist. Hierdurch sah sich Hilbert veranlaßt, seinen zunächst eingenommenen streng finiten Standpunkt zu einem allgemeineren konstruktiven Standpunkt zu erweitern.

1936 konnte nun Gerhard Gentzen die Widerspruchsfreiheit der reinen Zahlentheorie beweisen, und zwar durch transfinite Induktion bis zur kleinsten ε-Zahl ε_0. Diese transfinite Induktion ist noch vollständig konstruktiv, so daß der Gentzensche Widerspruchsfreiheitsbeweis den Anforderungen des erweiterten Hilbertschen Programms genügt.

Hilbert hat selbst keine Widerspruchsfreiheitsbeweise durchgeführt, sondern nur die Anregung dazu gegeben. Er hat aber bald nach dem Erscheinen der Gödelschen Unvollständigkeitssätze und noch vor dem Gentzenschen Widerspruchsfreiheitsbeweis in einem Göttinger Kolloquiumsvortrag vorgeschlagen, in Erweiterung des finiten Standpunktes das Axiomenschema der vollständigen Induktion durch die konstruktive Verwendung einer Schlußregel mit unendlich vielen Prämissen zu ersetzen. Diese von Hilbert vorgeschlagene Methode wurde zunächst überhaupt nicht beachtet, sondern erst fast 20 Jahre später aufgegriffen und ist heute für beweistheoretische Untersuchungen von starken Teilsystemen der Analysis unentbehrlich geworden.

Gegen die klassische Analysis waren wegen ihres imprädikativen Charakters besonders starke Einwände erhoben worden. Man bezeichnet einen Begriff als imprädikativ, wenn er unter Bezugnahme auf eine gewisse Gesamtheit definiert ist und selbst dieser Gesamtheit hinzugerechnet wird. Man hat, um die Vollständigkeitseigenschaft der reellen Zahlen zu gewinnen, allerdings zuzulassen, daß reelle Zahlen unter Bezugnahme auf die Gesamtheit der reellen Zahlen definiert werden. Der klassische Begriff der reellen Zahl ist also imprädikativ. Daher konnte man in den fünfziger Jahren nur sehr schwache Teilsysteme der Analysis in konstruktiver Weise als widerspruchsfrei nachweisen.

Zu Beginn der sechziger Jahre ließ sich nun die kleinste Ordinalzahl α bestimmen, welche die Eigenschaft hat, daß es keine rekursive Wohlordnung vom Ordnungstyp α gibt, die in prädikativer Weise als wohlgeordnet beweisbar ist. Als diese Ordinalzahl ergab sich die kleinste stark kritische Ordinalzahl Γ_0. Unter Bezugnahme auf Γ_0 konnten nun diejenigen Teilsysteme der Analysis charakterisiert werden, die prädikativ interpretierbar sind. Es handelt sich hierbei um schwache Teilsysteme der Analysis, in denen aber schon die wichtigsten Fundamentalsätze der klassischen Analysis beweisbar sind.

Mit dieser Grenze der Prädikativität ist bereits die Grenze für die Durchführbarkeit des erweiterten Hilbertschen Programms erreicht, denn ein Teilsystem der Analysis ist genau dann in konstruktiver Weise als widerspruchsfrei beweisbar, wenn es prädikativ interpretierbar ist.

Hiermit haben die beweistheoretischen Untersuchungen gezeigt, daß die klassische Analysis auf einer viel stärkeren Abstraktion beruht, als Hilbert vermutet hatte. Sie beruht auf einer sehr starken, aber doch sehr natürlichen Abstraktion, die sich jedenfalls als äußerst fruchtbar erwiesen hat. Man hat daher eingesehen, daß durchaus kein Grund besteht, die klassische Mathematik etwa deshalb abzulehnen, weil sie sich grundsätzlich nicht auf eine elementare Systematik zurückführen läßt.

Nun wurden auch für wesentlich imprädikative Teilsysteme der Analysis und der Mengenlehre Widerspruchsfreiheitsbeweise durchgeführt, und zwar durch transfinite Induktionen über Γ_0 hinaus. Hierbei legte man Wert darauf, zugleich mit diesen Widerspruchsfreiheitsbeweisen die beweistheoretische Ordinalzahlen der betrachteten formalen Systeme zu ermitteln. Man versteht unter der beweistheoretischen Ordinalzahl eines formalen mathematischen Systems — kurz gesagt — die kleinste Ordinalzahl α mit der Eigenschaft, daß die Widerspruchsfreiheit des betreffenden Systems durch transfinite Induktion bis α beweisbar ist.

Für eine umfangreiche Folge von immer stärkeren Teilsystemen der Analysis und der Mengenlehre konnten zunächst in München und in den letzten Jahren weitergehend auch in Münster die beweistheoretischen Ordinalzahlen ermittelt werden. Diese Ordinalzahlen charakterisieren die Beweisstärken der betreffenden Systeme und geben Aufschluß darüber, ob gewisse mathematische Sätze in diesen Systemen beweisbar oder unbeweisbar sind.

Zusammenfassend läßt sich sagen:

1. Die klassische Mathematik, für die sich Hilbert intensiv eingesetzt hatte, hat sich gegenüber dem Intuitionismus durchgesetzt.
2. Die von Hilbert geschaffene Beweistheorie ist ein wichtiges Gebiet der mathematischen Logik geworden, das heute außer in Deutschland auch in mehreren anderen europäischen Ländern, in den USA und in Japan gepflegt wird.

Hilberts Vorlesungen und Vorträge, die ich gehört habe, zeichneten sich ebenso wie seine Publikationen durch besondere Klarheit und leichte Verständlichkeit aus.

Ich selbst hatte nur mit Hilberts Mitarbeiter Paul Bernays, der bis 1933 außerordentlicher Professor in Göttingen war, einen engen wissenschaftlichen Kontakt gewonnen. Mit Hilbert bin ich persönlich nur zweimal zusammengekommen, nämlich erstens bei meinem Besuch in seiner Wohnung vor meiner Doktorprüfung und zweitens bei der mündlichen Prüfung vor genau 60 Jahren als letzter Doktorand von Hilbert, wobei ich beide Male seine menschliche Güte erfahren habe.

Schließlich möchte ich noch erwähnen, daß sich Hilbert seiner ostpreußischen Heimat stets eng verbunden gefühlt hat. Er hat von Göttingen aus immer wieder Königsberg besucht und seinen ostpreußischen Tonfall bis ins hohe Alter beibehalten.

Chapter 8
Remarks on Hilbert's Proof Theory

Kurt Schütte

Abstract Lecture given on 13 May 1993 at the memorial colloquium of the Working Group for East and West Prussian Regional Studies at the University of Munich on occasion of the 50th anniversary of the death of David Hilbert.[1]

In 1930, when I was studying mathematics in Berlin in my 5th semester, at the Philosophical Seminar of Berlin I came across an exposition of the philosopher Oskar Becker from Bonn on the mathematical existence. From this I learned about the dispute at that time between L. E. J. Brouwer and David Hilbert on the foundations of mathematics.

Brouwer rejected classical mathematics as founded by Dedekind and Cauchy, especially the general use of the tertium non datur, and therefore as well the use of indirect proofs. Instead, he supported a version of mathematics which he called intuitionistic, which substantially differs from classical mathematics. For instance, the intuitionistic continuum is substantially weaker then the classical continuum of real numbers.

By contrast, Hilbert pursued the goal of proving the consistency of the entirety of classical mathematics using finitary methods, that is by means of the simplest combinatorial methods, in order to guarantee its mathematical accuracy. Hilbert called the mathematical discipline, which should carry out such consistency proofs in the context of mathematical logic, proof theory.

Oskar Becker discussed in his exposition both stances in depth and decided to be in favour of intuitionism. However, to me it makes much more sense to secure classical mathematics as pursued by Hilbert than to intuitionistically restrict mathematics. In the following semester I returned to Göttingen, in order to familiarise myself with mathematical logic as used by Hilbert.

K. Schütte (†)
Mathematisches Institut, LMU München, Germany.

[1] Translated by Anton Setzer, Swansea.

R. Kahle and M. Rathjen (eds.), *The Legacy of Kurt Schütte*,
https://doi.org/10.1007/978-3-030-49424-7_8

However, already in 1931 it was discovered, as a consequence of the incompleteness theorems of Gödel, that not even pure number theory could be proved to be consistent using strictly finitistic methods. Therefore Hilbert felt compelled to extend his original strictly finitistic point of view to a more general constructive point of view.

In 1936 Gerhard Gentzen was able to prove the consistency of pure number theory by using transfinite induction up to the smallest ε number ε_0. This transfinite induction is still completely constructive, so that Gentzen's consistency proof satisfies the requirements of the extended Hilbertian programme.

Hilbert didn't carry out any consistency proofs himself but only motivated others to carry them out. However, soon after the publication of Gödel's incompleteness theorems and even before Gentzen's consistency proof he proposed in a lecture given in the colloquium in Göttingen, as part of an extension of the finite point of view, to replace the axiom scheme of complete induction by the constructive use of an inference rule with infinitely many premises. This method proposed by Hilbert was initially not recognised at all, but was taken up almost 20 years later and has nowadays become indispensable for the proof theoretic investigations of strong subsystems of analysis.

Against classical analysis rather strong objections have been raised because of its impredicative character. One calls a notion impredicative, if it is defined by referring to a totality of which this notion is itself a member. In order to obtain the completeness property of the real numbers, one has however to allow that the real numbers are defined referring to the totality of all real numbers. The classical notion of a real number is therefore impredicative. Therefore in the fifties only very weak subsystems of analysis could be shown to be consistent.

At the beginning of the sixties one could determine the smallest ordinal α, which has the property that there is no recursive well ordering of order type α, which can be shown to be well ordered in a predicative way. It turned out that this ordinal is the smallest strongly critical ordinal Γ_0. Using Γ_0 one could now characterise those subsystems of analysis, which can be interpreted predicatively. These are weak subsystems of analysis, in which however already the most important fundamental theorems of classical analysis can be shown.

With this limit of predicativity one has already reached the limit of what can be obtained by the extended Hilbertian programme, since a subsystem of analysis can be shown constructively to be consistent if and only if it can be interpreted predicatively.

Therefore the proof-theoretic investigations have shown that classical analysis is based on a much stronger abstraction than what was conjectured by Hilbert. It is based on a very strong, but nevertheless very natural abstraction, which at any rate has turned out to be very fruitful. One has therefore recognised, that there is no real reason, to reject classical mathematics only because it cannot in principal be reduced to an elementary systematic approach.

Now consistency proofs even for genuinely impredicative subsystems of analysis and set theory have been carried out, by using transfinite induction beyond Γ_0. Here one puts emphasis on determining together with the consistency proof the proof theoretic ordinal of the formal systems under consideration. Here the proof theoretic ordinal of a formal mathematical system is — roughly speaking — the smallest ordinal α with the property, that the consistency of the system under consideration can be shown by transfinite induction up to α.

For a considerable sequence of ever stronger subsystems of analysis and set theory one was able to determine their proof-theoretic ordinals, initially in Munich and then in the last years mainly in Münster. These ordinals characterise the proof strength of the corresponding systems and determine whether certain mathematical theorems are provable or unprovable in those systems.

In summary one can say:

1. Classical mathematics, intensively fought for by Hilbert, has prevailed over intuitionism.
2. Proof theory as created by Hilbert has become an important area of mathematical logic, which nowadays is pursued not only in Germany but also in several European countries, in the USA, and in Japan.

The lectures and talks of Hilbert I have listened to, as well his publications, are distinguished by their outstanding clarity and ready comprehensibility.

I myself was only in close scientific contact with Hilbert's collaborator Paul Bernays, who until 1933 held the position of extraordinary professor in Göttingen. With Hilbert I had only two personal meetings, namely the first one when I visited him at his home before my doctoral exam and a second time at my oral exam exactly 60 years ago as the last doctoral student of Hilbert, whereby in both cases I have experienced his human kindness.

Finally I wanted to mention that Hilbert felt always closely connected to his home region in East Prussia. Whilst living in Göttingen, he regularly visited Königsberg and maintained his East Prussian intonation into old age.

Part II
Proof Theory at Work

Chapter 9
Having a Look Again at Some Theories of Proof-Theoretic Strengths around Γ_0

Michael Bärtschi and Gerhard Jäger

Abstract In the first part of this article we introduce generalizations of ATR_0 and FP_0: arithmetical properties are replaced by Δ^1_1 properties, and the positivity requirement of the definition clauses of FP_0 is liberalized to monotonicity. Afterwards we turn to systems of admissible sets above the natural numbers as urelements. In this context we look, in particular, at principles of Σ and Π reduction.

Key words: Subsystems of second order arithmetic and set theory, proof-theoretic strength

Mathematics subject classification: 03E30, 03F15, 03F35

9.1 Introduction

One of the big challenges in proof theory of the sixties of the previous century was the determination of the precise ordinal bound for predicative mathematics. This problem was solved independently by Feferman [5] and Schütte [15, 16]. Feferman and Schütte showed that the least non-predicatively provable ordinal is the ordinal Γ_0 that can be characterized in terms of the Veblen functions $(\varphi_\alpha : \alpha \in On)$ as the least α such that $\varphi_\alpha(0) = \alpha$. For establishing this ordinal bound, Feferman and Schütte made use of an intricate interplay between systems of ramified analysis, well-ordering proofs, cut elimination and a bootstrapping process. See the original articles mentioned above or Schütte's text book [17] for further details.

In parallel to the formal characterization of predicative mathematics an issue to be dealt with was the question of how much mathematics could be developed pred-

M. Bärtschi · Gerhard Jäger
Institut für Informatik, Universität Bern, Neubrückstrasse 10, CH-3012 Bern, Switzerland, e-mail: {michael.baertschi,gerhard.jaeger}@inf.unibe.ch

R. Kahle and M. Rathjen (eds.), *The Legacy of Kurt Schütte*,
https://doi.org/10.1007/978-3-030-49424-7_9

icatively. This was pursued theoretically by identifying equivalent formal systems and alternative approaches to predicativity and by means of detailed case studies.

Fresh blood was brought into the whole discussion in connection with the Friedman-Simpson program of reverse mathematics. In this program five subsystems of second order arithmetic play a prominent role, and the fourth of these Big Five is the theory ATR_0 of arithmetical transfinite recursion. The proof-theoretic ordinal of ATR_0 is the Feferman-Schütte ordinal Γ_0, and ATR_0 is equivalent, for example, to the fixed point theory FP_0 and Σ^1_1 separation Σ^1_1-Sep_0; cf. Avigad [1] and Simpson [18].

The first part of this article introduces generalizations of ATR_0 and FP_0: arithmetical properties are replaced by Δ^1_1 properties, and the positivity requirement of the definition clauses of FP_0 is liberalized to monotonicity. We show that the corresponding theories are equivalent to ATR_0, thus answering a question of Feferman. The proof of these equivalences goes via a system for weak transfinite dependent choice, which is interesting on its own.

Afterwards we turn to systems of admissible sets above the natural numbers as urelements. In this context we look, in particular, at principles of Σ and Π reduction. For establishing their upper proof-theoretic bounds we make us of an adaptation of Simpson's suitable trees. For dealing with the system BS^0 of basic set theory without $\in$-induction and complete induction restricted to sets, we need a preparatory step consisting of partial cut elimination and asymmetric interpretations.

9.2 Subsystems of Second Order Arithmetic

As mentioned in the introduction, the subsystems of second order arithmetic that interest us most in this article are variants of the theory ATR_0 of arithmetical transfinite recursion and of the fixed point theory FP_0. We begin this section with describing the general context of these systems. Afterwards we introduce the principles (Δ^1_1-TR), ($\mathsf{M}\Delta^1_1$-FP), and (weak-Σ^1_1-TDC) and prove their equivalence over ACA_0.

9.2.1 The General Context

Let $\mathcal{L}_1$ be a standard first order language with (i) a countably infinite supply of first order variables, (ii) a numeral $\overline{n}$ for every natural number n, (iii) function and relation symbols for all primitive recursive functions and relations. All theories considered in this section are formulated in the second order language $\mathcal{L}_2$ that extends $\mathcal{L}_1$ by adding countably many second order variables and the relation symbol $\in$ for elementhood between first and second order objects. In $\mathcal{L}_2$ the first order objects are, of course, supposed to range over natural numbers and the second order objects to range over sets of natural numbers. The number terms and formulas of $\mathcal{L}_2$ are built up as usual; in general, we simply write n for the numeral $\overline{n}$.

We use the following categories of letters (possibly with subscripts) as metavariables:

- $a, b, c, u, v, w, x, y, z$ for first order variables;
- $R, S, T, U, V, W, X, Y, Z$ for second order variables;
- r, s, t for number terms;
- A, B, C for formulas.

A formula without bound set variables is called *arithmetical*. The Σ^1_1 and Π^1_1 formulas are those of the form $\exists X A$ and $\forall X A$, respectively, where A is arithmetical. A formula is a Σ^1 formula iff its negation normal form does not contain a universal set quantifier, and a Π^1 formula iff its negation normal form does not contain an existential set quantifier.

Equality is only taken as basic symbol between numbers. The subset relation on sets and equality of sets are defined by

$$\begin{aligned} S \subseteq T &:= \forall x(x \in S \to x \in T), \\ S = T &:= S \subseteq T \wedge T \subseteq S. \end{aligned}$$

In the following, we will make use of the standard primitive recursive coding machinery in $\mathcal{L}_2$: (a, b) stands for the primitive recursively formed ordered pair of a, b. For the corresponding projection functions we write $(.)_0$ and $(.)_1$ such that $a = ((a)_0, (a)_1)$ iff a codes an ordered pair.

Sets of natural numbers can be considered as codes of binary relations via this pairing function. For every set R we write

$$\begin{aligned} a \preceq_R b &:= (a, b) \in R, \\ a \prec_R b &:= (a, b) \in R \wedge (b, a) \notin R. \end{aligned}$$

We say that a belongs to the *field* of R—in symbols $Fd[R, a]$—iff there exists a b such that $a \preceq_R b$ or $b \preceq_R a$, and we write $LO[R]$ to express that R is a *linear ordering* of its field, i.e., R is a set of ordered pairs satisfying

(LO.1) $a \preceq_R b \wedge b \preceq_R c \to a \preceq_R c$,
(LO.2) $a \preceq_R b \wedge b \preceq_R a \to a = b$,
(LO.3) $a \preceq_R b \vee b \preceq_R a$

for all a, b, c from the field of R. We also make use of the following auxiliary notations:

$$\begin{aligned} b \in (S)_a &:= (a, b) \in S, \\ b \in (S)_{Ra} &:= b = ((b)_0, (b)_1) \wedge (b)_0 \prec_R a \wedge b \in S, \\ WO[R] &:= LO[R] \wedge \forall X(X \neq \emptyset \to (\exists x \in X)\forall y(y \prec_R x \to y \notin X)), \\ \mathcal{H}_A[R, S] &:= LO[R] \wedge \forall x(Fd[R, x] \to \forall y(y \in (S)_x \leftrightarrow A[(S)_{Rx}, y])). \end{aligned}$$

where $A[X, y]$ is an arithmetical formula with distinguished free variables X and y; it may contain other free variables than those displayed. The formula $\mathcal{H}_A[R, S]$ says that S describes the hierarchy obtained by iterating comprehension with respect to $A[X,y]$ along R.

If Th is a theory (i.e., a collection of formulas) then $\mathsf{Th} \vdash A$ means that A can be derived from Th in classical logic with equality. Given two theories Th_1 and Th_2, we write $\mathsf{Th}_1 \subseteq \mathsf{Th}_2$ iff the theorems of Th_1 are also provable in Th_2. The theories Th_1 and Th_2 are called proof-theoretically equivalent—in symbols $\mathsf{Th}_1 \equiv \mathsf{Th}_2$—iff they prove the same arithmetical sentences. The proof-theoretic ordinal of the theory Th is denoted by $|\mathsf{Th}|$ and clearly $\mathsf{Th}_1 \equiv \mathsf{Th}_2$ implies $|\mathsf{Th}_1| = |\mathsf{Th}_2|$.

Our basic system is the theory ACA_0 that comprises the defining axioms for all primitive recursive functions and relations, the axiom schema of *arithmetical comprehension*

$$\exists Y \forall x (x \in Y \leftrightarrow A[x])$$

for all arithmetical formulas $A[x]$, and the *induction axiom*

$$0 \in X \land \forall x (x \in X \to x+1 \in X) \to \forall x (x \in X).$$

In ACA_0 we thus can prove the schema of complete induction

$$A[0] \land \forall x (A[x] \to A[x+1]) \to \forall x A[x] \tag{$*$}$$

for all arithmetical formulas $A[x]$. ACA is the extension of ACA_0 obtained by adding $(*)$ for arbitrary formulas of $\mathcal{L}_2$.

Now we introduce a series of axiom schemas. Given such a schema (Sch) we shall write Sch_0 for the theory $\mathsf{ACA}_0 + (\mathsf{Sch})$ and Sch for $\mathsf{ACA} + (\mathsf{Sch})$. All formulas in the following schemas may contain other free variables than those displayed.

Δ^1_1 comprehension. For all Σ^1_1 formulas $A[x]$ and Π^1_1 formulas $B[x]$:

$$\forall x (A[x] \leftrightarrow B[x]) \to \exists Y \forall x (x \in Y \leftrightarrow A[x]). \tag{Δ^1_1-CA}$$

Σ^1_1 axiom of choice. For all Σ^1_1 formulas $A[x,Y]$:

$$\forall x \exists Y A[x,Y] \to \exists Y \forall x A[x,(Y)_x]. \tag{Σ^1_1-AC}$$

Fixed points of positive arithmetical clauses. For all X-positive arithmetical formulas $A[X,y]$:

$$\exists X \forall y (y \in X \leftrightarrow A[X,y]). \tag{FP}$$

Arithmetical transfinite recursion. For all arithmetical formulas $A[X,y]$:

$$WO[R] \to \exists X \mathcal{H}_A[R,X]. \tag{ATR}$$

Π^1_1 and Σ^1_1 reduction. For all Σ^1_1 formulas $A[x]$ and Π^1_1 formulas $B[x]$:

$$\forall x(A[x] \to B[x]) \to \exists Y(\forall x(A[x] \to x \in Y) \land \forall x(x \in Y \to B[x])). \qquad (\Pi^1_1\text{-Red})$$

$$\forall x(B[x] \to A[x]) \to \exists Y(\forall x(B[x] \to x \in Y) \land \forall x(x \in Y \to A[x])). \qquad (\Sigma^1_1\text{-Red})$$

These forms (Π^1_1-Red) and (Σ^1_1-Red) of Π^1_1 reduction and Σ^1_1 reduction are simply reformulations of the respective schemas of Σ^1_1 separation and Π^1_1 separation in Simpson [18]. We prefer our terminology in order to avoid any confusion with the separation principles used in the context of set theory in the next section.

The following theorem recalls some well-known results, and we refer to Simpson [18] and Avigad [1] for all details.

Theorem 9.1.

1. (ATR), (FP), *and* (Π^1_1-Red) *are equivalent over* ACA_0.
2. ATR_0 *proves all instances of* (Δ^1_1-CA) *and* (Σ^1_1-AC).
3. Σ^1_1-AC_0 *proves all instances of* (Σ^1_1-Red).
4. *For any* Σ^1_1 *formula* $A[X]$, ACA_0 *proves*

$$\neg\forall X(A[X] \leftrightarrow WO[X]).$$

It will have some technical advantages later to work with the following extensions of (Π^1_1-Red) and (Σ^1_1-Red) to Π^1 and Σ^1 formulas.

Π^1 and Σ^1 reduction. For all Σ^1 formulas $A[x]$ and Π^1 formulas $B[x]$:

$$\forall x(A[x] \to B[x]) \to \exists Y(\forall x(A[x] \to x \in Y) \land \forall x(x \in Y \to B[x])). \qquad (\Pi^1\text{-Red})$$

$$\forall x(B[x] \to A[x]) \to \exists Y(\forall x(B[x] \to x \in Y) \land \forall x(x \in Y \to A[x])). \qquad (\Sigma^1\text{-Red})$$

However, as stated in the following corollary, Π^1 and Σ^1 reduction are equivalent to Π^1_1 and Σ^1_1 reduction in all interesting cases.

Corollary 9.2. *In* Σ^1_1-AC_0 *and therefore also in* ATR_0, *every* Σ^1 *formula is provably equivalent to a* Σ^1_1 *formula and every* Π^1 *formula to a* Π^1_1 *formula. In particular, we have the following equivalences:*

1. (Π^1-Red) *is equivalent to* (Π^1_1-Red) *over* ACA_0.
2. (Σ^1-Red) *is equivalent to* (Σ^1_1-Red) *over* Σ^1_1-AC_0.

We conclude this subsection with a further remark summarizing some obvious facts that will play a role later.

Remark 9.3. Let $A[X,y]$ be an arithmetical formula with the displayed free variables and possibly further parameters and set

$$\mathcal{H}^{\circ}_{A}[a,S] \ := \ (\forall x < a)\forall y(y \in (S)_x \ \leftrightarrow \ A[\bigcup_{z<x}(S)_z, y]).$$

Then we have:

1. $\mathsf{ACA}_0 \ \vdash \ \exists X \mathcal{H}^{\circ}_{A}[n,X]$ for all natural numbers n.
2. $\mathsf{ACA} \ \vdash \ \forall x \exists X \mathcal{H}^{\circ}_{A}[x,X]$.
3. $\mathsf{ATR}_0 \ \vdash \ \forall x \exists X \mathcal{H}^{\circ}_{A}[x,X]$.

The first assertion is easily obtained by complete induction from outside, whereas the second follows by complete induction inside ACA for the Σ^1_1 formula $\exists X \mathcal{H}^{\circ}_{A}[x,X]$. To see why the third assertion is true, we work in ATR_0 and introduce the set $R := \{(x,y) : x \le y\}$ by arithmetical comprehension. Then the induction axiom implies $WO[R]$, and with the schema (ATR) we obtain what we want.

9.2.2 Δ^1_1 *Transfinite Recursion and Its Relatives*

The schemas (ATR) and (FP) are formulated for arithmetical formulas. Now we generalize these two schemas to Δ^1_1 formulas and, in the case of the fixed point axioms, only ask for monotonicity instead of positivity.

Δ^1_1 **transfinite recursion.** For all Σ^1_1 formulas $A[X,y]$ and Π^1_1 formulas $B[X,y]$:

$$\forall X \forall y(A[X,y] \leftrightarrow B[X,y]) \wedge WO[R] \ \rightarrow \ \exists X \mathcal{H}_A[R,X]. \qquad (\Delta^1_1\text{-TR})$$

Fixed points of monotone Δ^1_1 clauses. For all Σ^1_1 formulas $A[X,y]$ and Π^1_1 formulas $B[X,y]$:

$$C_{A,B} \ \rightarrow \ \exists X \forall y(y \in X \ \leftrightarrow \ A[X,y]), \qquad (\mathsf{M}\Delta^1_1\text{-FP})$$

where $C_{A,B}$ stands for the formula

$$\forall X \forall y(A[X,y] \leftrightarrow B[X,y]) \ \wedge \ \forall X,Y(X \subseteq Y \ \rightarrow \ \forall z(A[X,z] \ \rightarrow \ A[Y,z]))$$

that states that $A[X,y]$ is Δ^1_1 and monotone in X.

The following remark tells us that in $\mathsf{M}\Delta^1_1\text{-FP}_0$ all monotone Σ^1_1 definable functions have fixed points. We leave it to the reader to check that also the converse is true: From the existence of fixed points of monotone Σ^1_1 definable functions, the principle ($\mathsf{M}\Delta^1_1$-FP) can be derived.

Remark 9.4. Let $A[X,Y]$ be a Σ^1_1 formula. Then $\mathsf{M}\Delta^1_1\text{-FP}_0$ proves that from

(1) $\forall X \exists! Y A[X,Y]$,
(2) $\forall X_0, X_1, Y_0, Y_1(A[X_0,Y_0] \ \wedge \ A[X_1,Y_1] \ \wedge \ X_0 \subseteq X_1 \ \rightarrow \ Y_0 \subseteq Y_1)$

we obtain that $\exists X A[X,X]$.

Proof. Indeed, assume (1) and (2) and define

$$B_0[Z,z] \ := \ \exists X(A[Z,X] \ \land \ z \in X),$$
$$B_1[Z,z] \ := \ \forall X(A[Z,X] \ \rightarrow \ z \in X).$$

Then $B_0[Z,z]$ is equivalent to a Σ^1_1 formula, $B_1[Z,z]$ to a Π^1_1 formula. We also have $\forall X \forall y(B_0[X,y] \leftrightarrow B_1[X,y])$ and

$$\forall X,Y(X \subseteq Y \ \rightarrow \ \forall z(B_0[X,z] \ \rightarrow \ B_0[Y,z])).$$

Hence ($\mathsf{M}\Delta^1_1$-FP) provides a set S such that $\forall x(x \in S \ \leftrightarrow \ B_0[S,x])$. But this implies $A[S,S]$. □

Now we make use of a straightforward extension of the proof in Simpson [18] that Π^1_1-Red_0—Σ^1_1-Sep_0 in Simpson's terminology—proves all instances of (ATR) in order to show that Π^1_1-Red_0 even proves all instances of (Δ^1_1-TR). For this end we define for any formula $A[X,y]$ with the distinguished free variables X and y (possibly containing further parameters) the following auxiliary formulas:

$$\mathcal{H}_A[a,R,S] \ := \ LO[R] \ \land \ (\forall x <_R a)\forall y(y \in (S)_x \ \leftrightarrow \ A[(S)_{Rx},y]),$$
$$\mathcal{H}^+_A[a,R,S] \ := \ LO[R] \ \land \ (\forall x \leq_R a)\forall y(y \in (S)_x \ \leftrightarrow \ A[(S)_{Rx},y]).$$

Then it is obvious that $\mathcal{H}_A[R,S]$ is equivalent to $\forall a \mathcal{H}^+_A[a,R,S]$. Moreover, these hierarchies satisfy the following uniqueness property.

Lemma 9.5. *For any formula $A[X,y]$ we can prove in* ACA_0 *that*

$$WO[R] \ \land \ \mathcal{H}_A[a,R,S] \ \land \ \mathcal{H}_A[a,R,T] \ \rightarrow \ (\forall b <_R a)((S)_b = (T)_b).$$

Indeed, it is straightforward that from the given assumptions the arithmetical assertion

$$C[z] \ := \ z <_R a \ \rightarrow \ (S)_z = (T)_z$$

can be proved by induction on $<$.

Theorem 9.6. Δ^1_1-$\mathsf{TR}_0 \ \subseteq \ \Pi^1_1$-$\mathsf{Red}_0$.

Proof. Let $A[X,y]$ be a Σ^1_1 formula and $B[X,y]$ a Π^1_1 formula. Working in Π^1_1-Red_0 we assume

$$\forall X \forall y(A[X,y] \leftrightarrow B[X,y]) \ \land \ WO[R]. \tag{1}$$

Depending on $A[Z,z]$ and R we now set

$$C_0[z] \ := \ \exists b,x \exists Y(Fd[R,b] \ \land \ z = (b,x) \ \land \ \mathcal{H}_A[b,R,(Y)_{Rb}] \ \land \ A[(Y)_{Rb},x]),$$
$$C_1[z] \ := \ \forall b,x \forall Y(Fd[R,b] \ \land \ z = (b,x) \ \land \ \mathcal{H}_A[b,R,(Y)_{Rb}] \ \rightarrow \ A[(Y)_{Rb},x]).$$

In view of (1) it is easy to see that $C_0[z]$ is equivalent to a Σ^1 formula and $C_1[z]$ to a Π^1 formula, provable in Π^1_1-Red_0. Furthermore, from Lemma 9.5 we deduce that

$$\forall z(C_0[z] \rightarrow C_1[z]).$$

By Π^1_1 reduction, cf. Theorem 9.1 and Corollary 9.2, we thus obtain a set S such that

$$\forall z(C_0[z] \rightarrow z \in S), \tag{2}$$

$$\forall z(z \in S \rightarrow C_1[z]). \tag{3}$$

Now we want to show that

$$\forall a \mathcal{H}^+_A[a, R, S]. \tag{4}$$

Since $\mathcal{H}^+_A[a, R, S]$ is equivalent to a Σ^1_1 and a Π^1_1 formula, R is a well-ordering, and Π^1_1-Red_0 comprises Δ^1_1-CA, this can be done by transfinite induction on R. So let us assume

$$(\forall b <_R a)\mathcal{H}^+_A[b, R, S]. \tag{5}$$

Then we clearly have

$$\mathcal{H}_A[a, R, S], \tag{6}$$

$$\mathcal{H}_A[a, R, (S)_{Ra}]. \tag{7}$$

Thus we have, for any x,

$$A[(S)_{Ra}, x] \rightarrow C_0[(a, x)].$$

Hence, because of (2), also

$$A[(S)_{Ra}, x] \rightarrow x \in (S)_a. \tag{8}$$

On the other hand, from (3) we obtain

$$x \in (S)_a \rightarrow C_1[(a, x)].$$

Together with $\mathcal{H}_A[a, R, (S)_{Ra}]$ and the uniqueness assertion of Lemma 9.5 we thus have

$$x \in (S)_a \rightarrow A[(S)_{Ra}, x]. \tag{9}$$

The assertions (6), (8), and (9) give us $\mathcal{H}^+_A[a, R, S]$, finishing the proof of (4). As remarked earlier, (4) immediately implies $\mathcal{H}_A[R, S]$, and so S is a suitable witness for our Δ^1_1 transfinite recursion. □

We will show that both schemas, (Δ^1_1-TR) and (MΔ^1_1-FP), are equivalent to (ATR) over ACA_0. For doing that, we introduce a further principle, which is interesting on its own.

Weak Σ^1_1 transfinite dependent choice. For all Σ^1_1 formulas $A[X,Y]$:

$$\begin{aligned} &\forall X \exists! Y A[X,Y] \land WO[R] \rightarrow \\ &\qquad \exists Z \forall x (Fd[R,x] \rightarrow A[(Z)_{Rx},(Z)_x]). \end{aligned} \tag{weak-Σ^1_1-TDC}$$

Our next step in establishing the desired equivalences is to show that all instances of weak Σ^1_1 transfinite dependent choice are provable in the theory Δ^1_1-TR_0.

Theorem 9.7. weak-Σ^1_1-TDC_0 $\subseteq$ Δ^1_1-TR_0.

Proof. We work in Δ^1_1-TR_0 and assume that

$$\forall X \exists! Y A[X,Y] \tag{$*$}$$

for some Σ^1_1 formula $A[X,Y]$ and that $WO[R]$. Now we define

$$\begin{aligned} B[Z,z] &:= \exists Y (A[Z,Y] \land z \in Y), \\ C[Z,z] &:= \forall Y (A[Z,Y] \rightarrow z \in Y) \end{aligned}$$

and observe that $(*)$ yields

$$\forall X \forall y (B[X,y] \leftrightarrow C[X,y]).$$

Hence we can apply (Δ^1_1-TR) and obtain a set S such that $\mathcal{H}_B[R,S]$, i.e.,

$$\forall x (Fd[R,x] \rightarrow \forall y (y \in (S)_x \leftrightarrow B[(S)_{Rx},y])).$$

In view of $(*)$ and the definition of $B[Z,z]$ we thus have $A[(S)_{Rx},(S)_x]$ for all x from the field of R, as was required for validating (weak-Σ^1_1-TDC). □

A further and easy observation tells us that weak Σ^1_1 transfinite dependent choice proves all instances of (ATR). From this proof it is clear that instances of (weak-Σ^1_1-TDC) with arithmetical matrices $A[X,Y]$ are sufficient.

Lemma 9.8. ATR_0 $\subseteq$ weak-Σ^1_1-TDC_0.

Proof. Given an arithmetical formula $A[X,y]$ we set

$$B[X,Y] := \forall z (z \in Y \leftrightarrow A[X,z])$$

and observe that $\forall X \exists! Y B[X,Y]$. Given any R with $WO[R]$, weak Σ^1_1 transfinite dependent choice provides the existence of a set S such that

$$\forall x (Fd[R,x] \rightarrow B[(S)_{Rx},(S)_x]).$$

However, this implies

$$\forall x(Fd[R,x] \rightarrow \forall y(y \in (S)_x \leftrightarrow A[(S)_{Rx}, y])),$$

yielding $\mathcal{H}_A[R,S]$ and finishing our proof. □

Remark 9.9. It is an immediate consequence of Theorem 9.1 and the previous theorem that in weak-Σ^1_1-TDC_0 every Σ^1 formula is equivalent to a Σ^1_1 formula and every Π^1 formula is equivalent to a Π^1_1 formula.

In order to show that every instance of ($\mathsf{M}\Delta^1_1$-FP) can be proved in weak-Σ^1_1-TDC_0 we only have to adapt Avigad's proof that FP_0 is contained in ATR_0, see Avigad [1].

Lemma 9.10. *Assume that $A[X,y]$ is Σ^1_1 and that $B[X,y]$ is Π^1_1. Working in* weak-Σ^1_1-TDC_0 *we assume that*

(1) $\forall X \forall y(A[X,y] \leftrightarrow B[X,y])$,
(2) $\forall X, Y(X \subseteq Y \rightarrow \forall z(A[X,z] \rightarrow A[Y,z]))$.

Also, we let $C[R,Z]$ be the conjunction of the following (C1) – (C4)*:*

(C1) $LO[R]$,
(C2) $\forall a(Fd[R,a] \rightarrow (Z)_a = \{x : A[\bigcup\{(Z)_b : b <_R a\}, x]\})$,
(C3) $\forall a, b(a <_R b \rightarrow (Z)_a \subseteq (Z)_b)$,
(C4) $\forall a, x(Fd[R,a] \wedge x \in (Z)_a \rightarrow$
$\exists b(Fd[R,b] \wedge x \in (Z)_b \wedge x \notin \bigcup\{(Z)_c : c <_R b\}))$.

Then we have for any set R that

$$WO[R] \rightarrow \exists Z C[R,Z].$$

Proof. To see why this is the case, set

$$D[Z_0, Z_1] \ := \ \exists X(X = \{x : \exists a((a,x) \in Z_0)\} \wedge Z_1 = \{x : A[Z_0, x]\})$$

and recall from Remark 9.9 that $D[Z_0, Z_1]$ is equivalent to a Σ^1_1 formula. Furthermore, since Δ^1_1 comprehension is available in weak-Σ^1_1-TDC_0 according to Theorem 9.1 and Lemma 9.8, we see that $\forall X \exists! Y D[X,Y]$. Given any R with $WO[R]$, we can, therefore, apply (weak-Σ^1_1-TDC) and obtain a set S such that

$$\forall x(Fd[R,x] \rightarrow D[(S)_{Rx}, (S)_x]). \qquad (*)$$

It remains to check that $C[R,S]$: Condition (C1) follows from $WO[R]$. The definition of D plus $(*)$ immediately give us (C2) and—because of the monotonicity of A—also (C3). Finally, (C4) is a direct consequence of $WO[R]$. □

Theorem 9.11. $\mathsf{M}\Delta^1_1$-$\mathsf{FP}_0 \subseteq$ weak-Σ^1_1-TDC_0.

Proof. As in the previous lemma we assume that $A[X,y]$ is Σ^1_1, that $B[X,y]$ is Π^1_1, and that

(1) $\forall X \forall y(A[X,y] \leftrightarrow B[X,y])$,
(2) $\forall X, Y(X \subseteq Y \rightarrow \forall z(A[X,z] \rightarrow A[Y,z]))$

are provable in weak-Σ^1_1-TDC_0. According to the previous lemma we then have, always working in weak-Σ^1_1-TDC_0, that

$$\forall X(WO[X] \rightarrow \exists Z C[X,Z]),$$

where $C[X,Z]$ is defined as in the previous lemma. In view of (1) and Remark 9.9 we know that $\exists Z C[X,Z]$ is equivalent to a Σ^1_1 formula. Hence Theorem 9.1 implies that there exist a linear ordering R and a set S such that

$$C[R,S] \wedge \neg WO[R].$$

Since R is not a well-ordering, there exists a non-empty subset W of its field with no R-minimal element; without loss of generality we may also assume that W is upwards closed. Furthermore, define (by arithmetical comprehension)

$$V := \{x : (\forall y \in W)(x <_R y)\}.$$

We make use of arithmetical comprehension once more and introduce the sets

$$T_0 := \bigcap_{a \in W} (S)_a \qquad \text{and} \qquad T_1 := \bigcup_{b \in V} (S)_b.$$

We claim that $T_0 = T_1$. Indeed, $T_1 \subseteq T_0$ is immediate from (C3). For the converse inclusion, assume that $x \in (S)_a$ for all $a \in W \neq \emptyset$. Because of (C4) there exists a b such that $x \in (S)_b$ and $x \notin \bigcup\{(S)_c : c <_R b\}$. Since W does not have an R-minimal element, this implies $b \in V$, hence $x \in T_1$, and our claim is proved. Next we will show:

(3) $\forall x(A[T_0,x] \rightarrow x \in T_0)$,
(4) $\forall x(x \in T_1 \rightarrow A[T_1,x])$.

In order to prove (3) assume $A[T_0,x]$ and let c be an element of W. Then we have $T_0 \subseteq \bigcup\{(S)_b : b <_R c\}$, and the monotonicity of A and (C2) yield $x \in (S)_c$. Since this is true for all $c \in W$ and all x we have (3). It remains to show (4). Pick an arbitrary $c \in V$. Then $\bigcup\{(S)_b : b <_R c\} \subseteq T_1$. Hence the monotonicity of A and (C2) give us $(S)_c \subseteq \{x : A[T_1,x]\}$. Therefore, also (4) is established. Together with $T_0 = T_1$ we obtain from (3) and (4) that

$$\forall x(x \in T_0 \leftrightarrow A[T_0,x]).$$

Hence T_0 is the required fixed point, and our theorem is proved. □

Summarizing Theorem 9.1, Theorem 9.6, Theorem 9.7, and Theorem 9.11 we obtain the following intermediate result.

Corollary 9.12. *The theories* ATR_0, FP_0, Π^1_1-Red_0, Δ^1_1-TR_0, $\mathsf{M}\Delta^1_1$-FP_0, *and* weak-Σ^1_1-TDC_0 *prove the same* $\mathcal{L}_2$ *formulas.*

9.3 Subsystems of Set Theory

In this section we consider several subsystems of set theory above the natural numbers as urelements, related to the theories of the previous section. Since we are interested in the relationship between subsystems of second order arithmetic and set theory, this is a very natural choice. It has the further advantage that we can distinguish in a natural way between two forms of induction: complete induction on the natural numbers and $\in$-induction. This offers a further possibility to calibrate subsystems by restricting these forms of induction separately.

9.3.1 Basic Set Theory BS

$\mathcal{L}^* = \mathcal{L}_1(\in, N, S, Ad)$ is the first order language that is obtained from $\mathcal{L}_1$ by adding the membership symbol $\in$, the constant N for the set of natural numbers as well as the unary relation symbols S and Ad to express that an object is a set and an admissible set, respectively. For notational simplicity, we drop all symbols for primitive recursive functions.

The definitions of terms and formulas of $\mathcal{L}^*$ are standard. $\mathcal{L}^*$ and $\mathcal{L}_2$ diverge in the intended interpretations of the first order variables: in $\mathcal{L}^*$ they are supposed to range over natural numbers and sets, not just over natural numbers as in $\mathcal{L}_2$.

We extend the use of metavariables and allow r, s, t and A, B, C also to stand for terms and formulas of $\mathcal{L}^*$. It will always be clear from the context whether we speak about $\mathcal{L}_2$ or $\mathcal{L}^*$ terms and formulas.

The Δ_0, Σ_n, Π_n, Σ, and Π formulas of $\mathcal{L}^*$ are defined as usual. Also, we write A^r for the result of replacing all unbounded quantifiers $\exists x(\ldots)$ and $\forall x(\ldots)$ in A by $(\exists x \in r)(\ldots)$ and $(\forall x \in r)(\ldots)$, respectively. Equality of objects is defined as

$$r = s \;:=\; \begin{cases} (r \in N \wedge s \in N \wedge r =_N s) \vee \\ (S(r) \wedge S(s) \wedge r \subseteq s \wedge s \subseteq r), \end{cases}$$

where $=_N$ is the relation symbol for the primitive recursive equality relation and $r \subseteq s$ is short for $(\forall x \in r)(x \in s)$. Hence two objects are equal iff they are primitive recursively equal natural numbers or sets which contain the same elements.

We begin with fixing the theory BS of basic set theory. It is formulated in classical first order logic (with equality axioms), and its non-logical axioms are grouped as follows:

Number-theoretic axioms

(PA) A^N for all closed axioms of Peano arithmetic PA.

Complete induction ($\mathcal{L}^*$-I_N) on N. For any $\mathcal{L}^*$ formula A and with $R_{\mathcal{S}}$ being the relation symbol for the primitive recursive successor relation $\mathcal{S}$,

$$A[0] \wedge (\forall x, y \in N)(A[x] \wedge R_{\mathcal{S}}(x, y) \rightarrow A[y]) \rightarrow (\forall x \in N)A[x].$$

Ontological axioms

(O.1) $a = b \rightarrow (A[a] \rightarrow A[b])$ for all atomic formulas A.
(O.2) $S(a) \leftrightarrow a \notin N$.
(O.3) $a \in N \rightarrow b \notin a$.
(O.4) $n \in N$ for all natural numbers n.
(O.5) $R_{\mathcal{Z}}(a_1, \ldots, a_n) \rightarrow a_1, \ldots, a_n \in N$, if $R_{\mathcal{Z}}$ is the relation symbol for the n-ary primitive recursive relation $\mathcal{Z}$.

Set-theoretic axioms

Pairing. $\exists x(a \in x \wedge b \in x)$.

Union. $\exists x(\forall y \in a)(\forall z \in y)(z \in x)$.

Δ_0 Separation (Δ_0-Sep). For any Δ_0 formula A,

$$\exists x(S(x) \wedge (\forall y \in x \rightarrow y \in a \wedge A[y]) \wedge (\forall y \in a)(A[y] \rightarrow y \in x)).$$

Full $\in$-induction ($\mathcal{L}^*$-$\mathsf{I}_\in$). For any $\mathcal{L}^*$ formula A,

$$\forall x((\forall y \in x)A[y] \rightarrow A[x]) \rightarrow \forall x A[x].$$

In the following, subsystems of BS without $\in$-induction will play a central role: BS^0 is obtained from BS by restricting complete induction on N to Δ_0 formulas—denoted (Δ_0-I_N)—and deleting $\in$-induction, whereas BS^1 simply is BS without $\in$-induction. Of course, more variations of restricting inductions would be possible—for example restricting both forms of induction to Δ_0 formulas—but such systems are not relevant for us later.

Remark 9.13. Please observe that all non-logical axioms of BS^0 and BS^1 are Σ formulas.

Clearly, there is a canonical embedding of the language $\mathcal{L}_2$ into the language $\mathcal{L}^*$. Let $u_0, u_1, \ldots$ and $U_0, U_1, \ldots$ be enumerations of the first and second order variables of $\mathcal{L}_2$, and let $v_0, v_1, \ldots$ be an enumeration of the variables of $\mathcal{L}^*$. Now we set $\hat{u}_i := v_{2i}$ and $\hat{U}_i := v_{2i+1}$. The translation of the $\mathcal{L}_2$ formulas A into the $\mathcal{L}^*$ formulas $\hat{A}$ is then as follows: (i) Replace all first order variables x and X of A by $\hat{x}$ and $\hat{X}$, respectively; (ii) replace all number quantifiers $\exists x(\ldots)$ and $\forall x(\ldots)$ by $(\exists \hat{x} \in N)(\ldots)$ and $(\forall \hat{x} \in N)(\ldots)$, respectively; (iii) replace all set quantifiers $\exists X(\ldots)$ and $\forall X(\ldots)$ by $(\exists \hat{X} \subseteq N)(\ldots)$ and $(\forall \hat{X} \subseteq N)(\ldots)$, respectively.

It is obvious that the translation of every arithmetical $\mathcal{L}_2$ formula is a Δ_0 formula of $\mathcal{L}^*$; furthermore, if A is Σ^1_1, then $\hat{A}$ is Σ_1, and if A is Σ^1, then $\hat{A}$ is Σ; analogously for Π^1_1 and Π^1 formulas.

From now on we often identify $\mathcal{L}_2$ formulas A with their translations $\hat{A}$ and consider $\mathcal{L}_2$ as a sublanguage of $\mathcal{L}^*$. With this convention in mind, the following lemma is easily verified.

Lemma 9.14. $\mathsf{ACA}_0 \subseteq \mathsf{BS}^0$ *and* $\mathsf{ACA} \subseteq \mathsf{BS}^1$.

9.3.2 Theories of Admissible Sets

We enter the field of admissible set theory if we strengthen basic set theory by adding the schema of Δ_0 collection, i.e., for any Δ_0 formula A,

$$(\forall x \in a)\exists y A[x, y] \to \exists z(\forall x \in a)(\exists y \in z)A[x, y]. \qquad (\Delta_0\text{-Col})$$

The textbook Barwise [2] provides an excellent introduction to Kripke-Platek set theory with and without urelements. The theories for admissible sets above the natural numbers as urelements, which play a role now, are presented in detail in, for example, Jäger [7, 8, 9, 11, 12, 14].

Kripke-Platek set theory KPu above the natural numbers as urelements is BS + (Δ_0-Col). Its canonical models are the structures $L(\alpha)_{\mathbb{N}}$ where α is an admissible ordinal greater than ω; see Barwise [2] for further reading.

As in the case of BS we obtain natural subsystems of KPu by restricting induction on N and eliminating $\in$-induction:

$$\mathsf{KPu}^0 := \mathsf{BS}^0 + (\Delta_0\text{-Col}) \quad \text{and} \quad \mathsf{KPu}^1 := \mathsf{BS}^1 + (\Delta_0\text{-Col}).$$

Such restrictions have dramatic consequences. For example, in KPu^0 we cannot prove the existence of ω. The following characterizations of the proof-theoretic strengths of KPu and its two subsystems follow from Jäger [7, 9, 12].

Theorem 9.15. *We have the following proof-theoretic equivalences:*

$$\mathsf{KPu}^0 \equiv \Delta^1_1\text{-}\mathsf{CA}_0, \quad \mathsf{KPu}^1 \equiv \Delta^1_1\text{-}\mathsf{CA}, \quad \mathsf{KPu} \equiv \mathsf{ID}_1.$$

Thus far the predicate Ad did not play a role. It will be used now to speak about admissible sets and its meaning is given by the following Ad-axioms.

***Ad*-axioms**

(Ad.1) $Ad(d) \to N \in d \land d$ is transitive.
(Ad.2) $Ad(d) \to A^d$ for any closed instance A of an axiom of KPu^0.
(Ad.3) $Ad(d_1) \land Ad(d_2) \to d_1 \in d_2 \lor d_1 = d_2 \lor d_2 \in d_1$.

These Ad-axioms determine closure properties of admissible sets; however, they do not provide for the existence of admissible sets. One possibility to overcome this "problem" is to add the limit axiom

$$\forall x \exists y(x \in y \land Ad(y)). \qquad (\mathsf{Lim})$$

This axiom pins down a universe that is a limit of admissible sets. We add it to basic set theory, to Kripke-Platek set theory, and to their subsystems without ($\in$-induction) and obtain the following theories:

$$\begin{aligned}
&\mathsf{KPl}^0 := \mathsf{BS}^0 + (\mathsf{Lim}), && \mathsf{KPi}^0 := \mathsf{KPu}^0 + (\mathsf{Lim}),\\
&\mathsf{KPl}^1 := \mathsf{BS}^1 + (\mathsf{Lim}), && \mathsf{KPi}^1 := \mathsf{KPu}^1 + (\mathsf{Lim}),\\
&\mathsf{KPl} := \mathsf{BS} + (\mathsf{Lim}), && \mathsf{KPi} := \mathsf{KPu} + (\mathsf{Lim}).
\end{aligned}$$

The full systems KPl and KPi belong to a realm of proof theory that is not relevant for what we study in this article. To see why, we simply mention a result that characterizes their proof-theoretic strengths; for proofs see Jäger [10, 11, 12] and Jäger and Pohlers [14].

Theorem 9.16. *We have the following proof-theoretic equivalences:*

$$\mathsf{KPl} \equiv \Pi^1_1\text{-}\mathsf{CA} + (\mathsf{BI}) \quad \textit{and} \quad \mathsf{KPi} \equiv \Delta^1_2\text{-}\mathsf{CA} + (\mathsf{BI}).$$

However, as soon as $\in$-induction is dropped, we are in the range of predicativity/metapredicativity. For a discussion of the general context and the relationship between our theories and transfinitely iterated fixed point theories we refer, for example, to Jäger [11], Jäger, Kahle, Setzer, and Strahm [13], and Strahm [19].

Theorem 9.17. *We have the following proof-theoretic equivalences:*

1. $\mathsf{KPl}^0 \equiv \mathsf{KPi}^0 \equiv \mathsf{ATR}_0, \quad |\mathsf{KPl}^0| = |\mathsf{KPi}^0| = \Gamma_0.$
2. $\mathsf{KPl}^1 \equiv \mathsf{ATR} \equiv \widehat{\mathsf{ID}}_\omega, \quad |\mathsf{KPl}^1| = \Gamma_{\varepsilon_0}.$
3. $\mathsf{KPi}^1 \equiv \widehat{\mathsf{ID}}_{<\varepsilon_0}, \quad |\mathsf{KPi}^1| = \varphi 1 \varepsilon_0 0.$

In the articles mentioned above you also find all the necessary information about the ordinal notations based on the binary and ternary Veblen functions occurring in this theorem.

9.3.3 Σ and Π Reduction

The set-theoretic siblings of Π^1 and Σ^1 reduction are the schemas (Π-Red) and (Σ-Red), and it is clear that (modulo our embedding of $\mathcal{L}_2$ into $\mathcal{L}^*$) the instances of (Π^1-Red) and (Σ^1-Red) are special cases of (Π-Red) and (Σ-Red).

Π and Σ reduction. For all Σ formulas $A[x]$ and Π formulas $B[x]$:

$$(\forall x \in a)(A[x] \to B[x]) \to \exists y((\forall x \in a)(A[x] \to x \in y) \land (\forall x \in y)(x \in a \land B[x])). \tag{Π-Red}$$

$$(\forall x \in a)(B[x] \to A[x]) \to \exists y((\forall x \in a)(B[x] \to x \in y) \land (\forall x \in y)(x \in a \land A[x])). \tag{Σ-Red}$$

It is also straightforward that (Π-Red) and (Σ-Red) yield Δ separation, i.e., for all Σ formulas $A[x]$ and Π formulas $B[x]$,

$$(\forall x \in a)(A[x] \leftrightarrow B[x]) \to \exists y \forall x(x \in y \leftrightarrow x \in a \land A[x]).$$

From that lower bounds for (Σ-Red) over BS^0 and BS^1 can be read off immediately.

Lemma 9.18. *We have the following inclusions:*

$$\Delta^1_1\text{-CA}_0 \subseteq \text{BS}^0 + (\Sigma\text{-Red}) \quad \textit{and} \quad \Delta^1_1\text{-CA} \subseteq \text{BS}^1 + (\Sigma\text{-Red}).$$

In the next subsection, we will show that these bounds are sharp as far as proof-theoretic strength is concerned. If (Π-Red) is available, we can do more.

Lemma 9.19. *We have the following inclusions:*

$$\text{ATR}_0 \subseteq \text{BS}^0 + (\Pi\text{-Red}) \quad \textit{and} \quad \text{ATR} \subseteq \text{BS}^1 + (\Pi\text{-Red}).$$

This is an immediate consequence of Theorem 9.1 since (Π-Red) yields (Π^1_1-Red). In the next subsection we will also show that the respective theories are equivalent in a strong sense.

Turning to Kripke-Platek set theory, the first observation is that (Σ-Red) is not interesting in this context.

Theorem 9.20. KPu^0 *proves all instances of* (Σ-Red).

Proof. We work in KPu^0 and suppose that $(\forall x \in a)(A[x] \to B[x])$ for some Π formula $A[x]$ and Σ formula $B[x]$. By applying Σ reflection we obtain a set b such that

$$(\forall x \in a)(A^b[x] \to B^b[x]). \tag{$*$}$$

Now we introduce the set $c := \{x \in a : B^b[x]\}$ by Δ_0 separation. We claim that c is a suitable witness for (Σ-Red).

Indeed, from $x \in a$ and $A[x]$ we obtain $A^b[x]$ from the downward persistence of Π formulas. Therefore, we have $B^b[x]$ because of $(*)$, and thus $x \in c$. On the other hand, if $x \in c$, then $x \in a$ and $B^b[x]$. So we also have $B[x]$ because of the upward persistence of Σ formulas. □

The situation is different for (Π-Red). Clearly, (Π-Red) is provable in BS^0 + (Σ-Sep) and thus also in KPu^0+(Σ_1-Sep). However, the exact proof-theoretic analysis of (Π-Red) will not be given here and is postponed to a later publication.

9.3.4 Reducing BS + (Σ-Red), BS^0 + (Π-Red), *and* BS + (Π-Red)

In this subsection we provide the proof-theoretic upper bounds for the theories BS + (Σ-Red), BS^0 + (Π-Red), and BS + (Π-Red). This will be achieved by reducing these set theories to the theories Σ^1_1-AC, ATR_0, and ATR. For these reductions we follow the pattern of Simpson's reduction of his theory $\text{ATR}^{\text{set}}_0$ to ATR_0 in Simpson [18]. Some additional considerations are necessary for dealing with the natural numbers as urelements and for obtaining reductions to Σ^1_1-AC_0 and Σ^1_1-AC.

We adopt the terminology of [18], II.2, and take over Simpson's definition of *finite sequence of natural numbers*. We write *Seq*—called the set of sequence numbers—for the set of all codes of finite sequences of natural numbers. From now on we let σ, τ (possibly with indices) range over elements of *Seq*.

Also, we write $lh(\sigma)$ for the length of σ and use notations such as

$$\sigma = \langle s_0, \ldots, s_{lh(\sigma)-1} \rangle$$

and then $(\sigma)_x$ for s_x if $x < lh(\sigma)$. Accordingly, $\langle\rangle$ stands for the code of the empty sequence.

We shall identify a finite sequence of natural numbers with its code whenever convenient. If σ is $\langle s_0, \ldots s_{x-1}, \rangle$ and τ is $\langle t_0, \ldots, t_{y-1} \rangle$ we denote the concatenation of the sequence numbers σ and τ by $\sigma * \tau$,

$$\sigma * \tau \;=\; \langle s_0, \ldots s_{x-1}, t_0, \ldots, t_{y-1} \rangle.$$

Hence, $lh(\sigma * \tau) = lh(\sigma) + lh(\tau)$. In particular,

$$\sigma * \langle s \rangle = \langle (\sigma)_0, \ldots, (\sigma)_{lh(\sigma)-1}, s \rangle$$

and $lh(\sigma * \langle s \rangle) = lh(\sigma) + 1$. We write $\sigma \subseteq \tau$ to mean that σ is an initial segment of τ, i.e., $lh(\sigma) \leq lh(\tau)$ and $(\forall x < lh(\sigma))((\sigma)_x = (\tau)_x)$.

A *tree* is a non-empty subset T of Seq such that $\sigma \subseteq \tau$ and $\tau \in T$ implies $\sigma \in T$. Our next definition introduces the trees that we will use for the interpretation of the objects, i.e., urelements and sets, of our set theories.

Definition 9.21. In ACA_0 we introduce the notions of *u-tree* and *representation tree*.

1. T is a u-tree, in symbols $Tree[u, T]$, iff
   (i) T is a tree,
   (ii) $(\forall \sigma \in T)(lh(\sigma) \leq u)$,
   (iii) $(\forall \sigma, x, y)(\sigma * \langle 2x+1 \rangle \in T \wedge y \neq 2x+1 \;\rightarrow\; \sigma * \langle y \rangle \notin T)$,
   (iv) $(\forall \sigma, x)(\sigma * \langle x \rangle \in T \;\rightarrow\; (\forall y < lh(\sigma))((\sigma)_y \text{ is even}))$.
2. T is a representation tree if it is a u-tree for some natural number u,

$$Rep[T] \;:=\; \exists u Tree[u, T].$$

3. If T is a representation tree and $\sigma \in T$, we put $T^\sigma := \{\tau : \sigma * \tau \in T\}$.

Obviously, every u-tree is well-founded. The u-trees will represent the objects of our set theories. The basic idea is that the natural number x is represented by the tree $\{\langle\rangle, \langle 2x+1 \rangle\}$ and that the elements of a set represented by the tree T are the sets represented by the immediate subtrees of T of the form $T^{\langle 2y \rangle}$ with $\langle 2y \rangle \in T$. However, in order to validate extensionality, we have to close this treatment of elementhood under isomorphisms; see Definition 9.23 below.

Definition 9.22. If T is a u-tree for some u, we write $Iso[X, T]$ to state that $X \subseteq T \times T$ and, for all $\sigma, \tau \in T$, $(\sigma, \tau) \in X$ iff each of the following four properties is satisfied:

(i) $\forall y(\sigma * \langle 2y \rangle \in T \;\rightarrow\; \exists z((\sigma * \langle 2y \rangle, \tau * \langle 2z \rangle) \in X))$,
(ii) $\forall z(\tau * \langle 2z \rangle \in T \;\rightarrow\; \exists y((\sigma * \langle 2y \rangle, \tau * \langle 2z \rangle) \in X))$,

(iii) $\forall y(\sigma * \langle 2y+1\rangle \in T \;\rightarrow\; (\sigma * \langle 2y+1\rangle, \tau * \langle 2y+1\rangle) \in X)$,
(iv) $\forall y(\tau * \langle 2y+1\rangle \in T \;\rightarrow\; (\sigma * \langle 2y+1\rangle, \tau * \langle 2y+1\rangle) \in X)$.

Following Simpson [18] this notion of isomorphism is now directly incorporated into the definitions of the equality and epsilon relations on representation trees.

Definition 9.23. In ACA_0 we define for all representation trees S and T:

1. $S \oplus T \;:=\; \{\langle\rangle\} \cup \{\langle 0\rangle * \sigma : \sigma \in S\} \cup \{\langle 2\rangle * \tau : \tau \in T\}$,
2. $S =^\star T \;:=\; \exists X(\mathit{Iso}[X, S \oplus T] \wedge (\langle 0\rangle, \langle 2\rangle) \in X)$,
3. $S \in^\star T \;:=\; \exists X(\mathit{Iso}[X, S \oplus T] \wedge \exists x((\langle 0\rangle, \langle 2, x\rangle) \in X))$.

These definitions of $=^\star$ and $\in^\star$ are so that the following properties of suitable trees can be proved in ACA_0.

Lemma 9.24. *The following is provable in* ACA_0. *Let R, S, T be representation trees and assume that $\mathit{Iso}[X, T]$, $\mathit{Iso}[Y, S \oplus T]$, and $\sigma, \tau \in T$. Then we have:*

1. $T^\sigma =^\star T^\tau \;\leftrightarrow\; (\sigma, \tau) \in X$.
2. $T^\sigma \in^\star T^\tau \;\leftrightarrow\; \exists x((\sigma, \tau * \langle 2x\rangle) \in X)$.
3. $S =^\star T \;\leftrightarrow\; \forall Z(Z \in^\star S \leftrightarrow Z \in^\star T)$.
4. $S =^\star T \wedge S \in^\star R \;\rightarrow\; T \in^\star R$.
5. $\mathit{Tree}[u, S] \wedge \mathit{Tree}[v, T] \wedge S \in^\star T \;\rightarrow\; u < v$.

For the following interpretability results we have to translate the $\mathcal{L}^*$ formulas into $\mathcal{L}_2$ formulas. To this end we fix enumerations $v_0, v_1, \ldots$ of the variables of $\mathcal{L}^*$ and $V_0, V_1, \ldots$ of the set variables of $\mathcal{L}_2$.

Definition 9.25. We begin with translating the terms t of $\mathcal{L}^*$ into $\mathcal{L}_2$:

$$\begin{aligned} v_i^\star &:= V_i \text{ for all indices } i, \\ n^\star &:= \{\langle\rangle, \langle 2n+1\rangle\} \text{ for all natural numbers } n, \\ N^\star &:= \{\langle\rangle\} \cup \{\sigma : \exists x(\sigma = \langle 2x\rangle \vee \sigma = \langle 2x, 2x+1\rangle)\}. \end{aligned}$$

Then the atomic formulas of $\mathcal{L}^*$ are dealt with as follows:

$$(s \in t)^\star := s^\star \in^\star t^\star, \qquad S(t)^\star := t^\star \notin^\star N^\star,$$
$$R_{\mathcal{Z}}(t)^\star := \exists x(t^\star = \{\langle\rangle, \langle 2x+1\rangle\} \wedge R_{\mathcal{Z}}(x))$$

if $R_{\mathcal{Z}}$ is the relation symbol for the unary primitive recursive relation $\mathcal{Z}$; n-ary primitive recursive relations are treated accordingly. The propositional connectives commute with $\star$, and for the quantifiers we set

$$(\exists v_i A)^\star := \exists V_i(\mathit{Rep}[V_i] \wedge A^\star) \quad\text{and}\quad (\forall v_i A)^\star := \forall V_i(\mathit{Rep}[V_i] \rightarrow A^\star).$$

Bounded quantifiers are treated accordingly.

The following lemma shows that we have Σ^1-Red on the collection of representation trees.

Lemma 9.26. *Assume that $A[X]$ is a Σ^1 formula of $\mathcal{L}_2$ and $B[X]$ a Π^1 formula of $\mathcal{L}_2$. Working in Σ^1_1-AC_0, we assume that S is a representation tree such that*

$$\forall x(S^{\langle 2x\rangle} \in^\star S \land B[S^{\langle 2x\rangle}] \rightarrow A[S^{\langle 2x\rangle}]).$$

Then there exists a representation tree T such that:

(i) $\forall x(\langle 2x\rangle \in S \land B[S^{\langle 2x\rangle}] \rightarrow \langle 2x\rangle \in T \land S^{\langle 2x\rangle} = T^{\langle 2x\rangle})$,
(ii) $\forall x(\langle 2x\rangle \in T \rightarrow \langle 2x\rangle \in S \land S^{\langle 2x\rangle} = T^{\langle 2x\rangle} \land A[T^{\langle 2x\rangle}])$.

Proof. Since we work in Σ^1_1-AC_0, all instances of (Σ^1-Red) are, according to Theorem 9.1 and Corollary 9.2, at our disposal. So we only have to apply Σ^1 reduction to our assumption and are rewarded with a set Y which satisfies:

- $\forall x(S^{\langle 2x\rangle} \in^\star S \land B[S^{\langle 2x\rangle}] \rightarrow x \in Y)$,
- $\forall x(x \in Y \rightarrow A[S^{\langle 2x\rangle}])$.

Now it is easy to check that

$$T := \{\langle\rangle\} \cup \{\sigma \in S : 1 \leq lh(\sigma) \land (\sigma)_0 \in Y\}$$

is a representation tree with the required properties. □

Our next task is to show that the translation $\in^\star$ of the $\in$-relation does not increase the complexity too much. To achieve this aim in the context of all representation trees, we have to work in ACA or ATR_0, and we show that every representation tree has a unique isomorphism then. In the next subjection we will see how to proceed in the context of Σ^1_1-AC_0.

Lemma 9.27 (Existence and uniqueness of isomorphisms). ACA *and* ATR_0 *prove that*

$$Rep[T] \rightarrow \exists! X Iso[X,T].$$

Proof. The existence of the isomorphism X on the representation tree T is proved by means of the hierarchy building principles mentioned in point two and three of Remark 9.3. The uniqueness of X follows by complete induction starting at the leaves of X. □

The proof of this lemma requires complete induction for Σ^1_1 formulas or the principle (ATR); we see no possibility to carry it through in Σ^1_1-AC_0. Lemma 9.32 below presents what we can do in Σ^1_1-AC_0.

The previous lemma is the decisive step in showing that $\in^\star$ is Δ^1_1 on all representation trees. It immediately implies the following equivalence.

Lemma 9.28. ACA *and* ATR_0 *prove*

$$\begin{aligned}&Rep[S] \land Rep[T] \rightarrow \\ &\qquad (S \in^\star T \leftrightarrow \forall X(Iso[X, S \oplus T] \rightarrow \forall x((\langle 0\rangle, \langle 2, x\rangle) \in X))).\end{aligned}$$

From now on we will freely deal with the correspondence between the variables of $\mathcal{L}^*$ and the set variables of $\mathcal{L}_2$. It should always be clear from the context what we have in mind. Besides that, if $\vec{U}$ is the list of variables $U_0, \ldots, U_n$ we use $\vec{U} \in Rep$ as shorthand notation for

$$Rep[U_0] \wedge \cdots \wedge Rep[U_n].$$

If A is a formula of $\mathcal{L}^*$ with at most the variables $\vec{u}$ free, we first translate it into the formula $A^\star$ with $\vec{U}$ listing the set variables linked to $\vec{u}$ and then set

$$|A| \ := \ (\vec{U} \in Rep \ \rightarrow \ A^\star).$$

Remark 9.29. From the definition of $\in^\star$ and Lemma 9.28 we conclude that in Σ^1_1-AC and in ATR_0 the translation $|A|$ of a Δ_0 formula A is provably equivalent to a Σ^1 and a Π^1 formula. In view of Corollary 9.2, A is even provably equivalent in these systems to a Σ^1_1 and a Π^1_1 formula.

Keep in mind that this remark does not hold for the weaker system Σ^1_1-AC_0. Now we are ready for our first reduction theorem.

Theorem 9.30. *We have for all formulas A of* $\mathcal{L}^*$*:*

1. $\mathsf{BS} + (\Sigma\text{-}\mathsf{Red}) \vdash A \quad \Longrightarrow \quad \Sigma^1_1\text{-}\mathsf{AC} \vdash |A|$.
2. $\mathsf{BS}^0 + (\Pi\text{-}\mathsf{Red}) \vdash A \quad \Longrightarrow \quad \mathsf{ATR}_0 \vdash |A|$.
3. $\mathsf{BS} + (\Pi\text{-}\mathsf{Red}) \vdash A \quad \Longrightarrow \quad \mathsf{ATR} \vdash |A|$.

Proof. We first show that Σ^1_1-AC proves $|A|$ for all axioms A of BS. This is clear for all logical axioms, the number-theoretic axioms and the ontological axioms (O.2) – (O.5). The proofs of the translations of the (O.1)-axioms follow from Lemma 9.24. To handle pairing, let S and T be the representation trees that translate the sets a and b. Then $R := S \oplus T$ is a representation tree with $S \in^\star R$ and $T \in^\star R$. Similarly for (Union).

Now let $A[x]$ be a Δ_0 formula. Then (Δ_0-Sep) claims in BS the existence of the set $b := \{x \in a : A[x]\}$; for simplicity of notation we suppress all further parameters. We represent set a by the representation tree S and use (Δ^1_1-CA) to introduce the set

$$Z \ := \ \{x : S^{\langle 2x \rangle} \in^\star S \wedge A^\star[S^{\langle 2x \rangle}]\}.$$

This is possible since $(S^{\langle 2x \rangle} \in^\star S \wedge A^\star[S^{\langle 2x \rangle}])$ is (provably in Σ^1_1-AC) equivalent to a Σ^1_1 and a Π^1_1 formula. Now it only remains to define the representation tree

$$T \ := \ \{\langle\rangle\} \cup \{\sigma \in S : 0 < lh(\sigma) \wedge (\sigma)_0 \in Z\}$$

and to verify that it is a suitable representation of b.

Every instance of ($\mathcal{L}^*$-I_N) in BS clearly translates into an instance of the schema of complete induction of Σ^1_1-AC. Moreover, to see why the translations of the instances of ($\mathcal{L}^*$-$\mathsf{I}_\in$) hold in Σ^1_1-AC, recall from Lemma 9.24 that, for any u-tree S and v-tree T,

$$S \in^\star T \ \rightarrow \ u < v.$$

Hence the translation of any instance of $(\mathcal{L}^*\text{-}\mathsf{I}_\in)$ is provable in Σ^1_1-AC since complete induction is available there for arbitrary $\mathcal{L}_2$ formulas.

For completing the proof of the first assertion of our theorem, it remains to handle (Σ-Red). But this is exactly what Lemma 9.26 does.

The proof of the third assertion of our theorem is practically the same, with the only difference that the translations of the instances of (Π-Red) are taken care of by Π^1-Red, which is available in ATR in view of Theorem 9.1 and Corollary 9.2.

The second assertion is different since there we have to embed into a second order system that does not provide complete induction for arbitrary formulas. But in BS^0 complete induction on N is restricted to Δ_0 formulas. Therefore, according to Remark 9.29, each such instance of complete induction translates into a formula that is equivalent to a Σ^1_1 and a Π^1_1 formula. Moreover, ATR_0 proves all instances of (Δ^1_1-CA), as stated in Theorem 9.1, and thus provides for complete induction with respect to all provably Δ^1_1 predicates. □

Corollary 9.31. *We have the following proof-theoretic equivalences:*

1. Σ^1_1-AC $\equiv$ $\mathsf{BS}^1 + (\Sigma\text{-Red})$ $\equiv$ $\mathsf{BS} + (\Sigma\text{-Red})$.
2. ATR_0 $\equiv$ $\mathsf{BS}^0 + (\Pi\text{-Red})$.
3. ATR $\equiv$ $\mathsf{BS}^1 + (\Pi\text{-Red})$ $\equiv$ $\mathsf{BS} + (\Pi\text{-Red})$.

This corollary is an immediate consequence of Lemma 9.18, Lemma 9.19, and our previous reduction theorem.

9.3.5 *The Case* $\mathsf{BS}^0 + (\Sigma\text{-Red})$

It remains to analyze the theory $\mathsf{BS}^0 + (\Sigma\text{-Red})$. As mentioned earlier, the important point is that in Σ^1_1-AC_0 we do not have the existence of isomorphisms for arbitrary representation trees. However, by simply replacing assertions two and three of Remark 9.3 by assertion one, we obtain the following weaker existence and uniqueness property.

Lemma 9.32 (Existence and uniqueness of isomorphisms on k-trees). ACA_0 *proves for every standard natural number k that*

$$Tree[k,T] \;\rightarrow\; \exists! X Iso[X,T].$$

Consequently, $\in^\star$ restricted to k-trees for any fixed standard natural number k is Δ^1_1.

Lemma 9.33. ACA_0 *proves for all standard natural numbers k that*

$$\begin{aligned}&Tree[k,S] \wedge Tree[k,T] \;\rightarrow\\ &\qquad (S \in^\star T \;\leftrightarrow\; \forall X(Iso[X,S\oplus T] \;\rightarrow\; \forall x((\langle 0\rangle,\langle 2,x\rangle) \in X))).\end{aligned}$$

What we will show now is that suitable fragments of BS^0 + (Σ-Red) can be reduced to the substructures that are provided in Σ^1_1-AC_0 by the k-trees for fixed natural numbers k. This is achieved by a combination of partial cut elimination and a technique that is often called the *method of asymmetric interpretations*; cf. Cantini [4].

For partial cut elimination and the asymetric interpretations we work with a reformulation of BS^0 + (Σ-Red) as a Tait-style one sided sequent calculus. This auxiliary system, we call it H, derives finite sets $(\Gamma, \Delta, \dots)$ of $\mathcal{L}^*$ formulas in negation normal form. If A is such an $\mathcal{L}^*$ formula then Γ, Δ, A is short for $\Gamma \cup \Delta \cup \{A\}$, and similar for expressions of the form Γ, A, B.

The axioms of H are all the sets Γ, A where A is the negation normal form of an axiom of BS^0. To simplify the notation we assume from now on that all $\mathcal{L}^*$ formulas are in negation normal form and write $\overline{A}$ for the negation normal form of $\neg A$.

The rules of H include the standard rules for the propositional connectives and quantifiers plus the cut rules

$$\frac{\Gamma, A \qquad \Gamma, \overline{A}}{\Gamma} \qquad \text{(cut)}$$

for all Γ and A; the formulas A and $\overline{A}$ are the cut formulas of this cut. In addition, we have rules, for all Σ formulas A and Π formulas B, of the form

$$\frac{\Gamma, (\forall x \in a)(\overline{B}[x] \vee A[x])}{\Gamma, \exists y((\forall x \in a)(\overline{B}[x] \vee x \in y) \wedge (\forall x \in y)(x \in a \wedge A[x]))} \qquad (\Sigma\text{-Red})^r$$

It should be clear that $(\Sigma\text{-Red})^r$ is the reformulation of the schema (Σ-Red) as a rule.

Definition 9.34. We recursively assign a *rank* $rk(A)$ to all formulas (in negation normal form) as follows:

1. If A is a Σ or a Π formula, then $rk(A) := 0$,
2. If A is neither a Σ nor a Π formula, then

   a. $rk(A) := \max(rk(B), rk(C)) + 1$ if A is of the form $B \vee C$ or $B \wedge C$.
   b. $rk(A) := rk(B) + 1$ if A is of the form $\exists x B$, $\forall x B$, $(\exists x \in s)B$, or $(\forall x \in s)B$.

By $\mathsf{H} \vdash^m_n \Gamma$ we mean that there exists a derivation of Γ in H whose depth is bounded by m such that all cut formulas occurring in this derivation have rank less than n. Hence $\mathsf{H} \vdash^m_1 \Gamma$ says that there is a derivation of Γ in H of depth bounded by m whose cut formulas are Σ or Π formulas.

It is easy to check that the theory BS^0+(Σ-Red) can be embedded into its Tait-style version H.

Lemma 9.35 (Embedding). *If* BS^0 + (Σ-Red) *proves A, then there exist natural numbers m and n such that* $\mathsf{H} \vdash^m_n A$.

The idea now is to perform a partial cut elimination for H that only leaves us with cuts on Σ and Π formulas. Please observe that the main formulas of all axioms of H and the main formula of the conclusion of any instance of (Σ-Red)^r are Σ formulas. By standard proof-theoretic techniques we can therefore prove a partial cut elimination theorem, as desired.

Lemma 9.36 (Partial cut elimination). *Assume that* $\mathsf{H} \vdash_n^m \Gamma$ *for some natural numbers m and n. Then there exists a natural number k such that* $\mathsf{H} \vdash_1^k \Gamma$

The next step is to carry through an asymmetric interpretation of the $(\Sigma \cup \Pi)$-fragment of H into Σ^1_1-AC_0. To this end we formulate, for all standard natural numbers k, translations $A^{[k]}$ of the formulas of $\mathcal{L}^*$.

Definition 9.37. We translate the terms t of $\mathcal{L}^*$ into terms $t^\star$ of $\mathcal{L}_2$ as in Definition 9.25. For any standard natural number k and $\mathcal{L}^*$ formula A we recursively define its translation $A^{[k]}$ by:

(k.1) The atomic formulas of $\mathcal{L}^*$ are dealt with as follows:

$$(s \in t)^{[k]} := s^\star \in^\star t^\star, \qquad S(t)^{[k]} := t^\star \notin^\star N^\star,$$
$$R_{\mathcal{Z}}(t)^{[k]} := \exists x(t^\star = \{\langle\rangle, \langle 2x+1\rangle\} \wedge R_{\mathcal{Z}}(x))$$

if $R_{\mathcal{Z}}$ is the relation symbol for the unary primitive recursive relation $\mathcal{Z}$; n-ary primitive recursive relations are treated accordingly.

(k.2) The propositional connectives commute with $[k]$, and for the quantifiers we set

$$(\exists v_i A)^{[k]} := \exists V_i(\mathit{Tree}[k, V_i] \wedge A^{[k]}),$$
$$(\forall v_i A)^{[k]} := \forall V_i(\mathit{Tree}[k, V_i] \rightarrow A^{[k]}).$$

Bounded quantifiers are treated accordingly.

These translations are as the translation introduced in Definition 9.25 with the only difference that the sets are no longer interpreted as arbitrary representation trees but as k-trees. Also, following the pattern of the proof of Lemma 9.26, we see that Σ^1-Red holds in Σ^1_1-AC_0 on the collections of k-trees.

Lemma 9.38. *Assume that $A[X]$ is a Σ^1 formula of $\mathcal{L}_2$ and $B[X]$ a Π^1 formula of $\mathcal{L}_2$. Let k be a natural number. Working in Σ^1_1-AC_0, we assume that S is a k-tree such that*

$$\forall x(S^{\langle 2x\rangle} \in^\star S \wedge B[S^{\langle 2x\rangle}] \rightarrow A[S^{\langle 2x\rangle}]).$$

Then there exists a k-tree T such that:

(i) $\forall x(\langle 2x\rangle \in S \wedge B[S^{\langle 2x\rangle}] \rightarrow \langle 2x\rangle \in T \wedge S^{\langle 2x\rangle} = T^{\langle 2x\rangle})$,
(ii) $\forall x(\langle 2x\rangle \in T \rightarrow \langle 2x\rangle \in S \wedge S^{\langle 2x\rangle} = T^{\langle 2x\rangle} \wedge A[T^{\langle 2x\rangle}])$.

It is clear from these definitions and Lemma 9.33 that the k-translation of every Σ and Π formula is provably equivalent in Σ^1_1-AC_0 to a Σ^1_1 and Π^1_1 formula, respectively. In particular, the translation $A^{[k]}$ of a Δ_0 formula A is Δ^1_1 in Σ^1_1-AC_0.

The following notations are useful for the formulation and the proof of Theorem 9.41.

Definition 9.39. Let Γ be the finite set $\{A_1, \dots, A_n\}$ of $\mathcal{L}^*$ formulas in negation normal form and let k be a natural number.

1. The *($\geq k$)-instances* of Γ are all sets of $\mathcal{L}_2$ formulas of the form
$$\{A_1^{[k_1]}, \dots, A_n^{[k_n]}\}$$
with $k \leq k_1, \dots, k_n$.
2. The *($\leq k$)-instances* of Γ are all sets of $\mathcal{L}_2$ formulas of the form
$$\{A_1^{[k_1]}, \dots, A_n^{[k_n]}\}$$
with $k_1, \dots, k_n \leq k$.

Remark 9.40. For all finite sets Γ of $\mathcal{L}^*$ formulas and all natural numbers k_1, k_2 with $k_1 \leq k_2$ we have:

1. If Φ is a ($\geq k_2$)-instance of Γ, then it is also a ($\geq k_1$)-instance of Γ.
2. If Φ is a ($\leq k_1$)-instance of Γ, then it is also a ($\leq k_2$)-instance of Γ.

Theorem 9.41 (Asymmetric interpretation). *We assume:*

(i) Φ_1 *is a finite set of* Π *formulas of* $\mathcal{L}^*$ *in negation normal form.*
(ii) Φ_2 *is a finite set of* Σ *formulas of* $\mathcal{L}^*$ *in negation normal form.*
(iii) *At most the variables* $u_1, \dots, u_n$ *are free in* Φ_1, Φ_2.
(iv) $\mathsf{H} \vdash_1^m \Phi_1, \Phi_2$ *for some natural number* m.

Then we have for all natural numbers k*, all* ($\leq k$)*-instances* Ψ_1 *of* Φ_1*, and all* ($\geq k + 2^m$)*-instances* Ψ_2 *of* Φ_2 *that*

$$\Sigma_1^1\text{-}\mathsf{AC}_0 \vdash \bigvee(\neg Tree[k, U_1], \dots, \neg Tree[k, U_n], \Psi_1, \Psi_2).$$

Proof. We show this assertion by induction on m and distinguish the following cases:

(C1) Φ_1, Φ_2 is an axiom of H. Then the assertion is easily verified.

(C2) Φ_1, Φ_2 is the conclusion of an inference rule for a propositional connective or a quantifier. Then the assertion follows more or less directly from the induction hypothesis.

(C3) Φ_1, Φ_2 is the conclusion of a cut with cut formulas A and $\overline{A}$. Without loss of generality we can thus assume that A is a Σ formula and $\overline{A}$ a Π formula. Also, there exist natural numbers $m_0, m_1 < m$ such that

$$\mathsf{H} \vdash_1^{m_0} \Phi_1, \Phi_2, A, \tag{1}$$
$$\mathsf{H} \vdash_1^{m_1} \Phi_1, \Phi_2, \overline{A}. \tag{2}$$

Since Ψ_2 is a ($\geq k + 2^m$)-instance of Φ_2, it follows that $\Psi_2, A^{[k+2^{m_0}]}$ is a ($\geq k + 2^{m_0}$)-instance of Φ_2, A. Therefore, the induction hypothesis implies that

$$\Sigma^1_1\text{-}\mathsf{AC}_0 \vdash \bigvee(\neg \mathit{Tree}[k,U_1], \ldots, \neg \mathit{Tree}[k,U_n], \Psi_1, \Psi_2, A^{[k+2^{m_0}]}). \tag{3}$$

Furthermore, simple calculations show that $\Psi_1, \overline{A}^{[k+2^{m_0}]}$ is a $(\leq k + 2^{m_0})$-instance of $\Phi, \overline{A}$ and Ψ_2 a $(\geq k + 2^{m_0} + 2^{m_1})$-instance of Φ_2. Hence the induction hypothesis applied to (2)—with k replaced by $k + 2^{m_0}$—yields

$$\Sigma^1_1\text{-}\mathsf{AC}_0 \vdash \bigvee(\neg \mathit{Tree}[k,U_1], \ldots, \neg \mathit{Tree}[k,U_n], \Psi_1, \Psi_2, \overline{A}^{[k+2^{m_0}]}). \tag{4}$$

Now the assertion follows from (3) and (4) by simple logical reasoning within Σ^1_1-AC_0.

(C4) Φ_1, Φ_2 is the conclusion of the rule $(\Sigma\text{-}\mathsf{Red})^r$. In this case we obtain our assertion directly with the help of Lemma 9.38. □

To finish this subsection we only have to collect what we have proved so far. The embedding lemma, partial cut elimination and the asymmetric interpretation of H yield the following theorem.

Theorem 9.42. *If A is a Σ sentence that is provable in* $\mathsf{BS}^0 + (\Sigma\text{-}\mathsf{Red})$, *then there exists a natural number m such that Σ^1_1-AC_0 proves $A^{[m]}$.*

Since Δ^1_1-CA_0 and Σ^1_1-AC_0 are proof-theoretically equivalent to Peano arithmetic according to, for example, Barwise and Schlipf [3] and Feferman and Sieg [6], and since $\mathsf{BS}^0 + (\Sigma\text{-}\mathsf{Red})$ contains Δ^1_1-CA_0 according to Lemma 9.18, this theorem yields the following proof-theoretic equivalences.

Corollary 9.43. *We have the following proof-theoretic equivalences:*

$$\mathsf{PA} \equiv \Delta^1_1\text{-}\mathsf{CA}_0 \equiv \Sigma^1_1\text{-}\mathsf{AC}_0 \equiv \mathsf{BS}^0 + (\Sigma\text{-}\mathsf{Red}).$$

Bärtschi is also interested in uniform versions of the second order systems studied in this paper. For example, the uniform version UFP_0 is obtained by adding for every X-positive arithmetical formula $A[X,Y,x,y]$ with the indicated free variables a fresh functional $\mathcal{F}_A$. Then the uniform version of the fixed point axiom for the operator form $A[X,Y,x,y]$ is

$$\forall Y \forall x, y(x \in \mathcal{F}_A(Y,y) \leftrightarrow A[\mathcal{F}(Y,y), Y, x, y]).$$

Similarly for ATR_0. This leads to some interesting new relationships; it is work in preparation and will be presented in full details elsewhere.

Acknowledgements This publication was made possible through the support of a grant from the John Templeton Foundation. The opinions expressed in this publication are those of the authors and do not necessarily reflect the views of the John Templeton Foundation.

References

1. J. Avigad. On the relationship between ATR_0 and $\widehat{\mathsf{ID}}_{<\omega}$. *The Journal of Symbolic Logic*, 61(3):768–779, 1996.
2. K. J. Barwise. *Admissible Sets and Structures*, volume 7 of *Perspectives in Mathematical Logic*. Springer, 1975.
3. K. J. Barwise and J. Schlipf. On recursively saturated models of arithmetic. In D.H. Saracino and V.B. Weispfenning, editors, *Model Theory and Algebra : a Memorial Tribute to Abraham Robinson*, volume 498 of *Lecture Notes in Mathematics*, pages 42–55. Springer, 1975.
4. A. Cantini. On the relation between choice and comprehension principles in second order arithmetic. *The Journal of Symbolic Logic*, 51(2):360—373, 1986.
5. S. Feferman. Systems of predicative analysis. *The Journal of Symbolic Logic*, 29(1):1–30, 1964.
6. S. Feferman and W. Sieg. Proof-theoretic equivalences between classical and constructive theories for analysis. In W. Buchholz, S. Feferman, W. Pohlers, and W. Sieg, editors, *Iterated Inductive Definitions and Subsystems of Analysis: Recent Proof-theoretical Studies*, volume 897 of *Lecture Notes in Mathematics*. Springer, 1981.
7. G. Jäger. Beweistheorie von KPN. *Archive for Mathematical Logic*, 20(1–2):53–63, 1980.
8. G. Jäger. Iterating admissibility in proof theory. In J. Stern, editor, *Proceedings of the Herbrand Symposium, Logic Colloquium '81*, volume 107 of *Studies in Logic and the Foundations of Mathematics*, pages 137–146. Elsevier, 1982.
9. G. Jäger. Zur Beweistheorie der Kripke-Platek-Mengenlehre über den natürlichen Zahlen. *Archiv für Mathematische Logik und Grundlagenforschung*, 22(3–4):121–139, 1982.
10. G. Jäger. A well-ordering proof for Feferman's theory T_0. *Archiv für Mathematische Logik und Grundlagenforschung*, 23(1):65–77, 1983.
11. G. Jäger. The strength of admissibility without foundation. *The Journal of Symbolic Logic*, 49(3):867–879, 1984.
12. G. Jäger. *Theories for Admissible Sets: A Unifying Approach to Proof Theory*, volume 2 of *Studies in Proof Theory, Lecture Notes*. Bibliopolis, 1986.
13. G. Jäger, R. Kahle, A. Setzer, and T. Strahm. The proof-theoretic analysis of transfinitely iterated fixed point theories. *The Journal of Symbolic Logic*, 64(1):53–67, 1999.
14. G. Jäger and W. Pohlers. Eine beweistheoretische Untersuchung von $(\Delta^1_2\text{-}\mathsf{CA}) + (\mathsf{BI})$ und verwandter Systeme. *Sitzungsberichte der Bayerischen Akademie der Wissenschaften, Mathematisch-Naturwissenschaftliche Klasse*, 1:1–28, 1982.
15. K. Schütte. Eine Grenze für die Beweisbarkeit der transfiniten Induktion in der verzweigten Typenlogik. *Archiv für Mathematische Logik und Grundlagenforschung*, 7(1–2):45–60, 1964.
16. K. Schütte. Predicative well-orderings. In J.N Crossley and M.A.E. Dummett, editors, *Formal Systems and Recursive Functions*, Studies in Logic and the Foundations of Mathematics, pages 280–303. North-Holland, 1965.
17. K. Schütte. *Proof Theory*, volume 225 of *Grundlehren der mathematischen Wissenschaften*. Springer, 1977.
18. S.G. Simpson. *Subsystems of Second Order Arithmetic (2nd Edition)*. Perspectives in Logic. Cambridge University Press, 2009.
19. T. Strahm. Autonomous fixed point progressions and fixed point transfinite recursion. In S. Buss, P. Hajek, and P. Pudlak, editors, *Logic Colloquium '98*, volume 13 of *ASL Lecture Notes in Logic*, pages 449–464. A K Peters, 2000.

Chapter 10
The Limits of Predicativity Revisited

Wolfram Pohlers

Abstract Inspired by a remark of Kurt Gödel in his contribution to "The Philosophy of Bertrand Russell" we revisit Schütte's famous result on the limits of predicativity from a set-theoretic point of view.

10.1 Introduction

One of the famous proof theoretic results of the sixties of the previous century is the exact determination of the limits of predicativity in terms of the ordinal Γ_0 by Kurt Schütte [20], [19] and Solomon Feferman [4]. Though similar in their core ideas both approaches differ essentially in presentation. Feferman emphasizes the mathematical aspects of predicativity. He refers to—and also partly uses—the then known results in abstract recursion theory by Kleene and Spector while Schütte, in his proven technical manner, gives a complete and self contained straightforward proof chain for the exact limit not leaving out a single detail.

My motivation to revisit this issue was raised by a recent rereading of Kurt Gödel's article on *Russell's mathematical logic* [9]. The key passage which attracted my attention is on pp. 146/147 of this paper:

> The theory of orders proves more fruitful if considered from a purely mathematical standpoint, independent of the philosophical question whether impredicative definitions are admissible. Viewed in this manner, i.e., as a theory built up within the framework of ordinary mathematics, where impredicative definitions are admitted, there is no objection to extending it to arbitrarily high transfinite orders. Even if one rejects impredicative definitions, there would, I think, be no objection to extend it to such transfinite ordinals as can be constructed within the framework of finite orders. The theory itself seems to demand such an extension since it leads automatically to the consideration of functions in whose definition one refers to all functions of finite orders, and these would be functions of order ω. Admitting transfinite

W. Pohlers
Institut für Mathematische Logik und Grundlagenforschung, Einsteinstraße 62, 48149 Münster, Germany, e-mail: pohlers@uni-muenster.de

R. Kahle and M. Rathjen (eds.), *The Legacy of Kurt Schütte*,
https://doi.org/10.1007/978-3-030-49424-7_10

> orders, an axiom of reducibility can be proved.[1] This, however, offers no help to the original purpose of the theory, because the ordinal α—such that every propositional function is extensionally equivalent to a function of order α—is so great, that it presupposes impredicative totalities. Nevertheless, so much can be accomplished in this way, that all impredicativities are reduced to one special kind, namely the existence of certain large ordinals (or, well ordered sets) and the validity of recursive reasoning for them.

I am not a predicativist and the main concern of all my research has been "impredicative proof theory". So I agree with Gödel's opinion that the theory of ordinals is *"more fruitful if considered from a purely mathematical standpoint, independent of philosophical questions"*. Therefore I want to review the Feferman-Schütte result for the limits of predicativity with an emphasis on ideas inspired by Gödel's remark.[2]

The central notion of the present paper are thus the ordinals which are needed in the construction of the constructible universe introduced by Gödel in [8]. Already Kreisel in [11] remarks that the definition of the constructible hierarchy is locally predicative.[3] Iterating it through all the ordinals leads to the constructible universe which is a model of full Zermelo-Fraenkel set theory, a theory that is certainly impredicative. So it becomes obvious that local predicativity is not sufficient to guarantee global predicativity. The crucial point is the "number" of iteration steps. This observation has already been made by Kreisel in [11] where he remarks in a footnote that the notion of well-foundedness is the impredicative element in the definition of his hierarchies ($\Sigma_{\mathrm{i}}^{\beta}$).[4]

Our aim is to (re)characterize the segment of (set theoretical) ordinals which are predicatively accessible. In order not to conflict with the—nowadays familiar—notions of accessible and inaccessible ordinals (or cardinals) we prefer to call them *predicatively attainable* (or for short just *attainable*) ordinals. Again this notion is not completely new. It is very similar to the notion of an autonomous well-ordering as given by Feferman in [4].

---

[1] A remark, which I did not understand in first readings. It became only more clear after reading [10] and R. Solovay's introduction [22] to [10]. I owe the hint to look at these passages to Godehard Link.

[2] I am uncertain about Schütte's true motivation to tackle the limits of predicativity. I do not know if he knew Gödel's paper. At least he does not mention it in the bibliography. I am also not sure if he knew about Kreisel's paper [11]. This, too, does not show up in the bibliography. Apparently his motivation differed from Feferman's motivation. Schütte hardly touches the mathematical impact of predicativity and impredicativity. All he mentions in the introduction to [19] is that *"ramified type theory has been introduced to avoid impredicative conceptions"*. He then directly turns to the definition of his semi-formal system for ramified analysis.

[3] *"[. . .] la théorie des ensembles constructibles de Gödel, est basé sur une notion semi-prédicative"*. In [11] page 374.

[4] *"La notion de bon-ordre est l'élément imprédicative dans cette caractérisation des hiérarchies."* In [11] page 376.

10.2 The Vicious Circle Principle

The notions of predicativity and non-predicativity (now impredicativity) go back to Russell [18] and have been worked out further by H. Poincaré (see e.g. [17]) and H. Weyl in [25], [26], and [27]. We do not want to recapitulate the history of the emergence of predicativity and impredicativity in mathematics and the philosophy of mathematics but refer for the history to the article by Feferman [6].

Nevertheless we have to clarify what we want to understand by impredicativity. Here we follow Hermann Weyl as he has put it in [26].[5] Translating this into a mathematical language this roughly says:

Given a collection A of sets build the set $\mathfrak{E}_A \in A$ *by defining*
$x \in \mathfrak{E}_A \quad \Leftrightarrow \quad P(A, x)$,

where P symbolizes an allowed "construction principle".

For Weyl it means an apparent vicious circle to include such a principle in the allowed construction principles and he complains that exactly this vicious circle is used in analysis. We will thus take the following reformulation of the vicious circle principle as "the definition" of impredicativity.

(VCP) *A notion N is impredicative if its definition refers to a totality to which N itself belongs.*

A famous example for (VCP) is the Russell set

$$R := \{x \,|\, x \notin x\}.$$

Another example is the definition of the least fixed point of a monotone operator $\Gamma\colon \mathrm{Pow}(A) \longrightarrow \mathrm{Pow}(A)$. Define the property

$$\mathfrak{C}(X) \quad :\Leftrightarrow \quad X \subseteq A \wedge \Gamma(X) \subseteq X$$

of being closed under the operator Γ. Then we obtain the least fixed-point of Γ by

$$I_\Gamma := \bigcap \{X \,|\, \mathfrak{C}(X)\} \text{ i.e. } x \in I_\Gamma \quad :\Leftrightarrow \quad (\forall X)[\mathfrak{C}(X) \rightarrow x \in X].$$

We easily verify $\mathfrak{C}(I_\Gamma)$ which entails that I_Γ is defined via (VCP).

If the comprehension principle

(CP) *Given any property* $E(x)$ *for a set x then* $\{x \,|\, E(x)\}$ *is a set and we have* $a \in \{x \,|\, E(x)\}$ *iff* $E(a)$

is among the "allowed construction principles" the first example leads immediately to the well-known Russell paradox

$$R \in R \quad \Leftrightarrow \quad R \notin R.$$

---

[5] *"Ist A eine Eigenschaft von Eigenschaften, so bilde man diejenige Eigenschaft* $\mathfrak{E}_A$, *welche einem Gegenstande x dann und nur dann zukommt, wenn es eine mittels dieser Prinzipien zu konstruierende Eigenschaft gibt, welche dem x zukommt und selber die Eigenschaft A besitzt".* In [26] page 88.

Because of its immediate inconsistency the first example calls for an immediate solution. One possible answer is the axiomatization by the axioms of Zermelo-Fraenkel set theory which avoid (CP). Another solution is the introduction of classes, especially the class $\mathcal{U}$ of all sets, and the modification of the comprehension principle to:

(CPU) *Given any property* $E(a)$ *for a set a then* $\{x \mid E(x)\}$ *is a class and we have* $a \in \{x \mid E(x)\}$ *iff* $a \in \mathcal{U} \wedge E(a)$.

Then the Russell paradox shows that the assumption $R \in \mathcal{U}$ leads to the contradiction $R \in R \quad \Leftrightarrow \quad R \notin R$ which in turn entails that there are classes which are not sets and R is an example for such a class. Iterating this procedure we can form classes of classes, classes of classes of classes etc. This leads to a theory of types which, when built upon the sets of natural numbers, is known as simple type theory.

However, although (CPU) avoids the Russell paradox it does not avoid (VCP). It rather is an application of (VCP) since it describes a class with reference to the class of all sets. For Gödel this form of application of (VCP) means no problem. On page 136 of [9] he writes *"If, however, it* [i.e. the totality of things] *is a question of objects that exist independently of our constructions, there is nothing in the least absurd in the existence of totalities containing members, which can be described (i.e. uniquely characterized) only by reference to this totality."*

A modification of (CP) which safely avoids (VCP) is obtained by a ramification. In the ramification approach we assume that comprehension leads to sets of higher level. Given the collection L_α of sets of level α we can form the set $\{x \in L_\alpha \mid E(x)\}$ of level $\alpha + 1$ provided that $E(x)$ is a property which talks only about sets of levels less than or equal to α. Making this approach precise apparently needs a fixed language. Restricting the language to the formal language of set theory (with parameters) leads to the definition of constructible sets. When building on sets of natural numbers (which can be identified with reals) such systems are known as "ramified analysis".

Ramifications already occur in the Principia Mathematica [28]. Since Whitehead and Russell tried to follow the logicist program, which e.g. requires a definition of the natural numbers by purely logical terms, the ramification made proofs of even elementary properties of natural numbers by complete induction impossible.[6] As a remedy the authors introduced the notorious "Axiom of Reducibility" which—in consequence—vitiates ramification.

In contrast to the first example the definition of the least fixed-point in the second example provokes no immediate contradiction. The definition of the fixed-point I_Γ can even be resolved into stages. Given the stage I_Γ^α for all $\alpha < \beta$ we can define $I_\Gamma^\beta := \Gamma(\bigcup_{\alpha<\beta} I_\Gamma^\alpha)$. A simple cardinality argument shows that for a given operator $\Gamma\colon \mathrm{Pow}(A) \longrightarrow \mathrm{Pow}(A)$ the hierarchy of stages becomes stationary at some ordinal $\gamma < \overline{\overline{A}}^{+}$ and I_Γ^γ is the least fixed-point I_Γ.

If we assume that $\Gamma(X)$ is predicatively definable from X we obtain the fixed-point as an iteration of locally predicative definition steps. Since the result I_Γ of

[6] Cf. the corresponding passage in [6].

the iteration is an impredicative notion it corroborates the observation that local predicativity is not sufficient for guaranteeing also global predicativity.

The number of iteration steps which are needed to reach the fixed-point of an operator Γ depends on the complexity of Γ. It is a result of abstract recursion theory that for predicatively definable operators on a countable set A (paradigmaticly $\mathbb{N}$, the set of natural numbers) less than ω_1^{CK}-many steps suffice, where ω_1^{CK} denotes the first ordinal which cannot be represented by a recursive well-ordering on $\mathbb{N}$. Kreisel in [11] defines such a hierarchy of stages Σ_i^α for $i = 1, 2, 3$ using a ramified second order language in the definition of the operator Γ. He notices that exactly the ordinals are the impredicative feature in this definition (cf. footnote on page 130). His remedy is to restrict the ordinals to a set A of ordinals satisfying the condition that $\alpha \in A$ needs a well-ordering of order-type α which is definable by a formula in Σ_1^β for $\beta < \alpha$. He then expresses his conviction that with these restrictions all the requirements for predicativity are fulfilled.[7] Today we know that predicative *definability* of the ordinals does not suffice. Besides predicative definability we also need a predicative *justification* for the ordinals used in the iteration. However, Kreisel seems to have been aware of this problem. In section (b) of the last part on open problems in [11] he poses the question if the recursive ordinals coincide with the predicative ordinals. He mentions a tentative—not yet justified—notion of an autonomous hierarchy of formal systems for predicativity in [12] by which he could obtain an ordinal α_0 less than ω_1^{CK} as an upper bound for the predicative ordinals—a problem which later has been solved by Schütte and Feferman.

In the present paper we want to review the problem and its solution from a set theoretical angle with emphasis on the locally predicative definition of the hierarchy of constructible sets.

10.3 Constructible Sets

We now turn to the more technical parts of the paper. We work in a set theoretical framework. The language of set theory is the language of first order predicate logic with equality and the binary $\in$-symbol as the only non-logical symbol. We are not going to repeat the basics of set theory but will to a large extend use the more or less familiar definitions and abbreviations as e.g. presented in [14] which mostly coincide with those given in [1]. Of special interest is the notion of a Δ_0-formula in which all quantifiers are bounded. Δ_0-formulae play a prominent role because of their absoluteness. They do not change their meaning even if the universe is transitively expanded or shrunk.

For a formula F in the language of set theory we use the notion F^a to denote the formula obtained from F by restricting all unbounded quantifiers in F to a.

[7] "On remarque que ces hiérachies de définitions expriment directement *l'idée de prédicativité*: dans une définition prédicative on n'utilise que des quantificateurs portant sur des ensembles déjà construits". (Italics in the original.)

10.3.1 Ordinals

The original understanding of ordinals are order-types of well-orderings, where order-types are equivalence classes of isomorphic well-orderings. Since equivalence classes of well-orderings are in general not sets this understanding is not immediately expressible in formal set theory. Therefore it has become good practice to represent ordinals in set theory by sets that are well-ordered by the membership relation $\in$. Assuming the well-foundedness of the set theoretic universe all hereditarily transitive sets are thus ordinals. The modern (set theoretical) understanding of an ordinal is rather that of a well-founded hereditarily transitive set than that of the order-type of a well-ordering. In this text we will follow this understanding.

Transitivity and hereditary transitivity can be expressed by a Δ_0-formula. Therefore the notion of an hereditarily transitive set is Δ_0, hence absolute. We have

$$Tran(a) \;:\Leftrightarrow\; (\forall x \in a)[x \subseteq a]$$

and

$$Ord(a) \;:\Leftrightarrow\; Tran(a) \wedge (\forall x \in a)[Tran(x)].$$

The first order language of set theory, however, is too weak to express the well-foundedness of a hereditarily transitive set. We shall treat that in Section 10.4.

10.3.2 The Constructible Hierarchy

The hierarchy of constructible sets is based on the following construction principle.

Definition 10.1. A set a is definable from a set M with parameters if there is a formula $F(x, x_1, \ldots, x_n)$ in the language of set theory and sets $a_1, \ldots, a_n \in M$ such that

$$a = \{x \in M \,|\, M \models F(x, a_1, \ldots, a_n)\}.$$

Let $Def(M)$ be the collection of sets that are definable from M with parameters.

Observe that the definition "definable from M" avoids (VCP) since it only refers to sets in M. So $a \in Def(M) \;:\Leftrightarrow\;$ *a is definable from* M avoids (VCP) and is therefore predicative. Moreover it is well-known that $Def(M)$ can be (set-)primitive—recursively computed from M.[8] Thus $Def(M)$ is predicatively definable from M. The constructible hierarchy is now built by iterating the operator Def along the ordinal axis.

Definition 10.2. By recursion on the ordinal α we define the stages L_α of the *constructible hierarchy*

[8] Cf. e.g. [3].

$$
\begin{aligned}
L_0 &:= \emptyset \\
L_{\alpha+1} &:= Def(L_\alpha) \\
L_\lambda &:= \bigcup_{\xi<\lambda} L_\xi \quad \text{for limit ordinals } \lambda.
\end{aligned}
$$

Let $L := \bigcup_{\xi \in On} L_\xi$.

The definition of the stages L_α is predicative in every successor step. Therefore the definition of L_α is locally predicative. As stated by Kreisel, the impredicativity of a stage L_α is due to the ordinal α. Kreisel's suggestion in [11] to restrict the ordinals to predicatively definable ordinals would lead to the segment $L_{\omega_1^{CK}}$ as the predicatively justifiable segment of the constructible hierarchy. Since $L_{\omega_1^{CK}}$ is the first admissible set above L_ω it is in fact a distinguished set but not a predicatively justifiable segment of the constructible sets.[9]

It is well-known that the constructible hierarchy L is a model of full ZFC. But already initial segments of L model many of the familiar axioms of set theory.[10] E.g, we have at every limit ordinal above ω a model for Basic Set Theory which is the fragment of Kripke Platek set theory which contains the axioms for extensionality, pair, union, absolute separation—i.e. separation for Δ_0-formulae, the axiom of infinity and the scheme of foundation.

Sometimes it is convenient to have names for the sets in L. We therefore introduce the notion of an L-term. We define inductively that for every ordinal σ the symbol L_σ is an L-term of stage σ and if $F(x_1, \ldots, x_n)$ is a formula in the language of set theory whose free variables are all in the list $x_1, \ldots, x_n$ and $s_1, \ldots, s_n$ are L-terms of stages less than σ then $\{x \in L_\sigma \mid F(s_1, \ldots, s_n)^{L_\sigma}\}$ is an L-term of stage σ.

The interpretation of L-terms as constructible set is the obvious one. Conversely there is for every constructible set s an L-term $\underline{s}$ such that s is the interpretation of $\underline{s}$. Therefore we will in general not strictly distinguish between constructible sets and their notation as L-terms but talk sloppily about "parameters from L". Parameters from L_σ are L-terms of stages less than σ. We say that two L-terms coincide iff they coincide syntactically.

10.4 Ramified Morse-Kelly Set Theory

In exploring the limits of predicativity we cannot anticipate the global well-foundedness of the universe of sets. The definition of an ordinal as a hereditarily transitive set will therefore not suffice to characterize the ordinals. However, well-foundedness is in general not expressible in the first order language of set theory. All we can express is well-foundedness in the form that every *set* possesses an $\in$-minimal element. To obtain "real" well-foundedness we need to express that every

[9] This follows from Schütte's and Feferman's results, since it contains well-orderings of oder-types $\geq \Gamma_0$.

[10] Cf. [1] for details.

class possesses an $\in$-minimal element, where the notion class is to be understood in the widest sense. We thus need a theory of classes, i.e., a second order theory above the theory of sets. Using capital letters as syntactical variables for classes we have to modify the definition of the class of ordinals by

$$On = \{x \mid Tran(x) \land (\forall y \in x)[Tran(y)] \land \\ \land (\forall X)\big[X \cap x \neq \emptyset \rightarrow (\exists y)[y \in X \cap x \land (\forall z \in y)[z \notin X \cap x]]\big]\}.$$

This immediately brings back the (VCP) problem. The class of ordinals is defined by reference to all classes. To avoid (VCP) we therefore introduce ramified classes and develop a system of ramified Morse-Kelly class theory above a segment of the constructible hierarchy.

10.4.1 A Semi-formal System for Ramified Morse-Kelly Class Theory

We will formalize Morse-Kelly class theory not by a formal but by a semi-formal system which allows for infinite proof figures.

Definition 10.3. We define the language $\mathbb{L}$ of a ramified Morse-Kelly class theory above the constructible universe L.

- Every L-sentence is an $\mathbb{L}$-formula of level 0.
- The variable X is a class term of level 0.
- If s is a parameter from L and S a class term of level α then then $s \in S$ and $s \notin S$ are $\mathbb{L}$-formulae of level α.
- If A and B are $\mathbb{L}$-formulae then $A \land B$ and $A \lor B$ are $\mathbb{L}$-formulae whose level is the maximum of the levels of a and B.
- If s is a parameter from L and $F(s)$ is an $\mathbb{L}$-formula of level ς then $(\forall x)F(x)$ and $(\exists x)F(x)$ are $\mathbb{L}$-formulae of level ς.
- If $F(X)$ is an $\mathbb{L}$-formula and $\alpha > 0$ an ordinal, then $(\forall Y^\alpha)F(Y)$ and $(\exists Z^\alpha)F(Z)$ are $\mathbb{L}$-formulae whose level is the maximum of α and the level of $F(X)$. Here we do not require that all X are replaced by Y or Z, respectively.
- If $F(s)$ is an $\mathbb{L}$-formula of level α and s a parameter from L then $\{x \mid F(x)\}$ is a class term of level α.

Atomic formulae are formulae of the form $s \in t$, $s \notin t$, $s = t$, $\neq t$, $s \in X$ and $s \notin X$ for L-parameters s and t.

By $lev(S)$ and $lev(F)$ we denote the level of a class term S or an $\mathbb{L}$-sentence, respectively.

Remark 10.4. The second order variable X serves as an "oracle" for arbitrary classes. The quantified variables $(\forall Y^\alpha)$ and $(\exists Z^\alpha)$ are just "replacement characters" for class terms which are to replaced by class terms in the definition of the characteristic sets.

Observe that the negation symbol is not among the basic symbols of $\mathbb{L}$. Instead we have equality $=$, inequality $\neq$, the membership symbol $\in$ and its negation $\notin$ among the basic symbols. So negation becomes definable via deMorgan's laws.

Following the idea that we can accept the segment L_σ of L as soon as we have accepted σ as an ordinal we will now establish a semi-formal system $\mathbb{L}^\sigma$ of Morse-Kelly class theory above the constructible segment L_σ. In $\mathbb{L}^\sigma$ all the true sentences of L_σ are treated as axioms. This is obtained by the definition of the characteristic set $\mathrm{CS}_\sigma(F)$ for $\mathbb{L}$-formulae F, the $\bigwedge$-type$_\sigma$ and the $(\bigwedge)$-rule below. The language of $\mathbb{L}^\sigma$ is that of $\mathbb{L}$.

In designing the semi-formal system we follow the pattern developed in [14] and [15]. We decorate each formula F with a characteristic set $\mathrm{CS}_\sigma(F)$, which should be viewed as the set of formulae that are needed to secure F logically. A formula F is in $\bigwedge$-type$_\sigma$ (*conjunction type*) if all the formulae in $\mathrm{CS}_\sigma(F)$ are needed to secure F and in $\bigvee$-type$_\sigma$ (*disjunction type*) if already some of the formulae in $\mathrm{CS}_\sigma(F)$ suffice. Formulae that cannot be logically secured obtain no type. This allows for the simple formulation of the inference rules in Definition 10.12 below. We are going to formulate the semi-formal system $\mathbb{L}^\sigma$ in "Tait style" as a one-sided sequent calculus. I.e., we do not derive formulae but finite sets of formulae which are to be interpreted as finite disjunctions.

Lower case Latin letters will always stand for parameters in L. Class terms are commonly denoted by upper case Latin letters.

By an L_σ-sentence we understand a Δ_0-sentence in the language of set theory which contains at most parameters from $L_\sigma \cup \{L_\sigma\}$. Call an $\mathbb{L}$-formula *genuine* if it is not an L_σ-sentence. The first step in establishing the semi-formal system $\mathbb{L}^\sigma$ is to fix $\bigvee$-type$_\sigma$ and $\bigwedge$-type$_\sigma$ for the language of $\mathbb{L}^\sigma$.

Definition 10.5.

(Atom) Genuine atomic formulae do not belong to a type.

$(\bigvee)$ The $\bigvee$-type$_\sigma$ comprises all false L_σ-sentences and all genuine $\mathbb{L}$-formulae of the form $s \in S$, if S is not X, $A \vee B$, $(\exists x)F(x)$ and $(\exists X^\alpha)F(X)$.

$(\bigwedge)$ The $\bigwedge$-type$_\sigma$ comprises all true L_σ-sentences and all genuine $\mathbb{L}$-formulae of the form $s \notin S$ for S not X, $A \wedge B$, $(\forall x)F(x)$ and $(\forall X^\alpha)F(X)$.

Definition 10.6. To obtain a decoration for the $\mathbb{L}$-formulae we define the characteristic subformula sets by the following clauses:

- For an L_σ-sentence F let $\mathrm{CS}_\sigma(F) = \emptyset$.

For the coming clauses assume that all formulae are not L_σ-sentences.

- If F is atomic then $\mathrm{CS}_\sigma(F) = \emptyset$.
- $\mathrm{CS}_\sigma(s \in \{x \mid F(x)\}) = \{F(s)\}$
- $\mathrm{CS}_\sigma(s \notin \{x \mid F(x)\}) = \{\neg F(s)\}$
- $\mathrm{CS}_\sigma((\forall x)F(x)) = \mathrm{CS}_\sigma((\exists x)F(x)) = \{F(s) \mid s \in L\}$
- $\mathrm{CS}_\sigma(F \wedge G) = \mathrm{CS}_\sigma(F \vee G) = \{F, G\}$.
- $\mathrm{CS}_\sigma((\forall X^\alpha)F(X)) = \mathrm{CS}_\sigma((\exists X^\alpha)F(X)) = \{F(S) \mid lev(S) < \alpha\}$.

We use the characteristic sets in the definition of the semantics for the language $\mathbb{L}$.

Definition 10.7. For an $\mathbb{L}$-formula F and an arbitrary class S we define the relation $\mathbb{RL}[S] \models F$ by the following clauses.

(L) If F is an L-sentence then $\mathbb{RL}[S] \models F$ iff $L \models F$.

(Atom) If $F \equiv s \in X$ the $\mathbb{RL}[S] \models F$ iff $s \in S$. If $F \equiv s \notin X$ the $\mathbb{RL}[S] \models F$ iff $s \notin S$.

($\bigvee$) If $F \in \bigvee$-type$_\sigma$ and $\mathbb{RL}[S] \models G$ holds true for some $G \in \mathrm{CS}_\sigma(F)$ then $\mathbb{RL}[S] \models F$.

($\bigwedge$) If $F \in \bigwedge$-type$_\sigma$ and $\mathbb{RL}[S] \models G$ holds true for every $G \in \mathrm{CS}_\sigma(F)$ then $\mathbb{RL}[S] \models F$.

Definition 10.8. Let F be an $\mathbb{L}$-formula and $\alpha_0 > \alpha_1 > \cdots > \alpha_n > 0$ all the levels of quantifiers occurring in F. Denote by $|\alpha_i|$ the number of quantifiers of level α_i occurring in F. Then we define its complexity as

$$comp(F) := \omega^{\alpha_0} \cdot |\alpha_0| + \omega^{\alpha_1} \cdot |\alpha_1| + \cdots + \omega \cdot |\alpha_n| + lh(F)$$

where $lh(F)$ is the number of logical symbols occurring in F.

Observe that we have $comp(F(S)) < comp((\mathsf{Q}X^\alpha)F(X))$ for any class term S of level less than α.

Definition 10.9. For an $\mathbb{L}$-formula let

$$deg_\sigma(F) := \sup\{deg_\sigma(G) + 1 \mid G \in \mathrm{CS}_\sigma(F)\}.$$

This is well-defined since $comp(G) < comp(F)$ holds true for all $G \in \mathrm{CS}_\sigma(F)$.
For a class term $S = \{x \mid F(x)\}$ let $deg_\sigma(S) = deg_\sigma(F)$.

Lemma 10.10. *Let F be an $\mathbb{L}$-formula such that $CS_\sigma(F) \neq \emptyset$. Then $deg_\sigma(G) < deg_\sigma(F)$ holds true for all $G \in CS_\sigma(F)$.*

Proof. This is obvious from Definition 10.6 and Definition 10.9. □

Lemma 10.11. *For any $\mathbb{L}$-formula F we have $deg_\sigma(F) < \omega^{lev(F)+1}$.*

Proof. This follows from $deg_\sigma(F) \leq comp(F) < \omega^{lev(F)+1}$. □

Definition 10.12. The semi-formal calculus for $\mathbb{L}^\sigma$ is given by the rules

(X) If $\{s \notin X, s \in X\} \subseteq \Delta$ for some L-term s then

$$\mathbb{L}^\sigma \mathrel{\substack{\beta \\ \vdash \\ \rho}} \Delta$$

holds true for all ordinals β and ρ.

(X_σ) If s and t are L_σ-terms such that $L_\sigma \models s = t$ and $\{t \notin X, s \in X\} \subseteq \Delta$ then

$$\mathbb{L}^\sigma \mathrel{\substack{\beta \\ \vdash \\ \rho}} \Delta$$

holds true for all ordinals α and ρ.

($\bigvee$) If $F \in \Delta \cap \bigvee$-type$_\sigma$ and $\mathbb{L}^\sigma \mathrel{\substack{\beta_0 \\ \vdash \\ \rho}} \Delta, G$ holds true for some $G \in \mathrm{CS}_\sigma(F)$ then we get $\mathbb{L}^\sigma \mathrel{\substack{\beta \\ \vdash \\ \rho}} \Delta$ for all $\beta > \beta_0$.

($\bigwedge$) If $F \in \Delta \cap \bigwedge$-type$_\sigma$ such that $\mathbb{L}^\sigma \mathrel{\substack{\beta_G \\ \vdash \\ \rho}} \Delta, G$ and $\beta_G < \beta$ for all $G \in \mathrm{CS}_\sigma(F)$ then we conclude $\mathbb{L}^\sigma \mathrel{\substack{\beta \\ \vdash \\ \rho}} \Delta$.

(Cut) $\mathbb{L}^\sigma \mathrel{\frac{\beta_0}{\rho}} \Delta, F$, $\mathbb{L}^\sigma \mathrel{\frac{\beta_0}{\rho}} \Delta, \neg F$ and $deg_\sigma(F) < \rho$ imply $\mathbb{L}^\sigma \mathrel{\frac{\beta}{\rho}} \Delta$ for all $\beta > \beta_0$.

The main-formula in inferences ($\bigvee$) and ($\bigwedge$) is the formula F, the main formulae in inferences (X) and (X_σ) are the formulae $s \in X$, $s \notin X$ and $t \notin X$, respectively.

Theorem 10.13. *Let Δ be a finite set of $\mathbb{L}$-formulae such that $\mathbb{L}^\sigma \mathrel{\frac{\alpha}{\rho}} \Delta$. Then $\mathbb{RL}[S] \models \bigvee\Delta$ for any class S.*

Proof. Induction on α. □

10.4.1.1 Some Basic Properties of $\mathbb{L}^\sigma$

The semi-formal system $\mathbb{L}^\sigma$ possesses all the familiar properties of semi-formal systems which we collect in this section.

Lemma 10.14.

(Weakening) *From $\mathbb{L}^\sigma \mathrel{\frac{\alpha}{\rho}} \Delta$ we obtain $\mathbb{L}^\sigma \mathrel{\frac{\beta}{\upsilon}} \Gamma$ for all $\beta \geq \alpha$, $\upsilon \geq \rho$ and $\Gamma \supseteq \Delta$.*

($\bigwedge$-inversion) *From $\mathbb{L}^\sigma \mathrel{\frac{\alpha}{\rho}} \Delta, F$, $F \in \bigwedge$-type and $CS_\sigma(F) \neq \emptyset$ we obtain $\mathbb{L}^\sigma \mathrel{\frac{\alpha}{\rho}} \Delta, G$ for all $G \in CS_\sigma(F)$.*

($\in$-inversion) *From $\mathbb{L}^\sigma \mathrel{\frac{\alpha}{\rho}} \Delta, s \in \{x \mid F(x)\}$ we obtain $\mathbb{L}^\sigma \mathrel{\frac{\alpha}{\rho}} \Delta, F(s)$.*

($\notin$-inversion) *From $\mathbb{L}^\sigma \mathrel{\frac{\alpha}{\rho}} \Delta, s \notin \{x \mid F(x)\}$ we obtain $\mathbb{L}^\sigma \mathrel{\frac{\alpha}{\rho}} \Delta, \neg F(s)$.*

($\vee$-exportation) *From $\mathbb{L}^\sigma \mathrel{\frac{\alpha}{\rho}} \Delta, A \vee B$ we obtain $\mathbb{L}^\sigma \mathrel{\frac{\alpha}{\rho}} \Delta, A, B$.*

Lemma 10.15. *(Detachment Lemma) Assume $\mathbb{L}^\sigma \mathrel{\frac{\alpha}{\rho}} \Delta, \Gamma$ for a finite set Γ of false L_σ-sentences. Then we can detach Γ and obtain $\mathbb{L}^\sigma \mathrel{\frac{\alpha}{\rho}} \Delta$.*

Proof. Straightforward by induction on α. □

Lemma 10.16. $\mathbb{L}^\sigma \mathrel{\frac{\alpha}{\rho}} \Delta$ *and* $\sigma \leq \tau$ *imply* $\mathbb{L}^\tau \mathrel{\frac{\alpha}{\rho}} \Delta$.

Proof. Induction on α. The only subtle point is that $\mathrm{CS}_\sigma(F)$ and $\mathrm{CS}_\tau(F)$ differ. We demonstrate this using the example of an $\bigwedge$-rule. Let $F \in \bigwedge\text{-type}_\sigma \cap \Delta$ be the main formula of the last inference. If $\mathrm{CS}_\sigma(F) = \emptyset$ then F is a true L_σ-sentence, hence by upwards persistency also a true L_τ sentence, and we obtain $\mathbb{L}^\tau \mathrel{\frac{\alpha}{\rho}} \Delta$ by an inference ($\bigwedge$). If F is a sentence $(\forall x)G(x)$ which is not an L_σ-sentence we obtain by induction hypothesis $\mathbb{L}^\tau \mathrel{\frac{\alpha_a}{\rho}} \Delta, G(a)$, for all $a \in L$. If F is not an L_τ sentence then $\mathrm{CS}_\tau(F) = \{G(a) \mid a \in L\}$ and we obtain the claim by an ($\bigwedge$)-clause. If, however, F is an L_τ-sentence then $\mathrm{CS}_\tau(F) = \emptyset$. If $L_\tau \not\models G(a)$ for some $a \in L_\tau$ we obtain $\mathbb{L}^\tau \mathrel{\frac{\alpha_a}{\rho}} \Delta$ by Lemma 10.15, hence $\mathbb{L}^\tau \mathrel{\frac{\alpha}{\rho}} \Delta$ by weakening. If $L_\tau \models G(a)$ for all $a \in L_\eta$ we have $L_\tau \models (\forall x)G(a)$ and obtain $\mathbb{L}^\tau \mathrel{\frac{\alpha}{\rho}} \Delta$ by an inference ($\bigwedge$) with empty premise. In all other cases we get the claim from the induction hypothesis by the same inference. □

Definition 10.17. Call two $\mathbb{L}$-formulae L_σ-equivalent if they differ at most in L_σ-terms $s_1, \ldots, s_n$ and $t_1, \ldots, t_n$ for which we have $L_\sigma \models s_i = t_i$ for $i = 1, \ldots, n$.

Lemma 10.18. *(Tautology) Let F and G be L_σ-equivalent $\mathbb{L}$-formulae. Then we get* $\mathbb{L}^\sigma \mathrel{\frac{|2\cdot deg_\sigma(F)}{0}} \Delta, \neg G, F$.

Proof. Induction on $deg_\sigma(F)$. □

The following observations are simple corollaries of the Tautology Lemma and Lemma 10.11.

Lemma 10.19. *Let S be a class-term of level less than η and $\varsigma := \max\{\eta, lev(F(X)) + 1\}$. Then* $\mathbb{L}^\sigma \mathrel{\frac{|{<}\omega^\varsigma}{0}} \Delta, \neg(\forall X^\eta)F(X), F(S)$.

Lemma 10.20. *For $\varsigma = \max\{\xi, lev(F(X)) + 1\}$ and $\xi \leq \eta$ we have*

$$\mathbb{L}^\sigma \mathrel{\frac{|\omega^\varsigma}{0}} \Delta, \neg(\forall X^\eta)F(X), (\forall X^\xi)F(X).$$

Lemma 10.21. *(Persistency)*

(Upwards) *From* $\mathbb{L}^\sigma \mathrel{\frac{|\alpha}{\rho}} \Delta, (\exists X^\varsigma)F(X)$ *and* $\tau \geq \varsigma$ *we get* $\mathbb{L}^\sigma \mathrel{\frac{|\alpha}{\rho}} \Delta, (\exists X^\tau)F(X)$.
From $\mathbb{L}^\sigma \mathrel{\frac{|\alpha}{\rho}} \Delta, (\exists x \in L_\mu)F(x)$ *and* $\nu \geq \mu$ *we get* $\mathbb{L}^\sigma \mathrel{\frac{|\alpha}{\rho}} \Delta, (\exists x \in L_\nu)F(x)$.
(Downwards) *From* $\mathbb{L}^\sigma \mathrel{\frac{|\alpha}{\rho}} \Delta, (\forall X^\tau)F(X)$ *and* $\tau \geq \varsigma$ *we get* $\mathbb{L}^\sigma \mathrel{\frac{|\alpha}{\rho}} \Delta, (\forall X^\varsigma)F(X)$.
From $\mathbb{L}^\sigma \mathrel{\frac{|\alpha}{\rho}} \Delta, (\forall x \in L_\mu)F(x)$ *and* $\nu \leq \mu$ *we get* $\mathbb{L}^\sigma \mathrel{\frac{|\alpha}{\rho}} \Delta, (\forall x \in L_\nu)F(x)$.

Proof. Induction on α. If the main formula of the last inference is different from $(\exists X^\varsigma)F(X)$ we obtain the claim from the induction hypothesis. Otherwise we have the premise $\mathbb{L}^\sigma \mathrel{\frac{|\alpha_0}{\rho}} \Delta, (\exists X^\varsigma)F(X), F(S)$ for a class term S of level less than $\varsigma \leq \tau$. Hence $\mathbb{L}^\sigma \mathrel{\frac{|\alpha_0}{\rho}} \Delta, (\exists X^\tau)F(X), F(S)$ by induction hypothesis and the claim follows by an inference $\bigvee$.

The proofs of the remaining claims are analogous. □

Lemma 10.22. *(Lifting Lemma) Let τ be a limit ordinal and $\langle \varsigma_\iota \mid \iota \in I \rangle$ a sequence that is cofinal in τ. If* $\mathbb{L}^\sigma \mathrel{\frac{|\alpha_\iota}{\rho}} \Delta, (\forall X^{\varsigma_\iota})F(X)$ *holds true for all $\iota \in I$ we obtain* $\mathbb{L}^\sigma \mathrel{\frac{|\alpha}{\rho}} \Delta, (\forall X^\tau)F(X)$ *for every* $\alpha \geq \sup\{\alpha_\iota + 1 \mid \iota \in I\}$.

Proof. Let S be a class term of level $\eta < \tau$. Then $\eta < \varsigma_\iota < \tau$ for some $\iota \in I$ and we obtain by inversion $\mathbb{L}^\sigma \mathrel{\frac{|\alpha_\iota}{\rho}} \Delta, F(S)$ with $\alpha_\iota < \alpha$. By an inference $(\bigwedge)$ we obtain the claim. □

10.4.1.2 Cut Elimination for $\mathbb{L}^\sigma$

Lemma 10.23. *Assume* $\mathbb{L}^\sigma \mathrel{\frac{|\alpha}{0}} \Delta, s \in X$ *and* $\mathbb{L}^\sigma \mathrel{\frac{|\beta}{0}} \Gamma, t \notin X$ *where s and t are L-terms such that either s and t coincide or $L_\sigma \models s = t$. Then we get* $\mathbb{L}^\sigma \mathrel{\frac{|\alpha+\beta}{0}} \Delta, \Gamma$.

Proof. Induction on $\alpha + \beta$. □

Lemma 10.24. *(Reduction Lemma) Assume that F is an $\mathbb{L}$-formula in $\bigwedge$-type$_\sigma$ with $deg_\sigma(F) = \rho$. Then* $\mathbb{L}^\sigma \mathrel{\frac{|\alpha}{\rho}} \Delta, F$ *and* $\mathbb{L}^\sigma \mathrel{\frac{|\beta}{\rho}} \Delta, \neg F$ *entail* $\mathbb{L}^\sigma \mathrel{\frac{|\alpha+\beta}{\rho}} \Delta, \Gamma$.

Proof. Induction on β. □

Remark 10.25. The sequel needs some familiarity with the Veblen functions φ_ξ. We give a brief reminder.

An ordinal α is additively indecomposable iff $\xi, \eta < \alpha$ entail $\xi + \eta < \alpha$. The additively indecomposable ordinals are enumerated by the powers of ω. So $1 = \omega^0$ is the first, $\omega = \omega^1$ the next additively indecomposable ordinal etc.

The function φ_0 enumerates the additively indecomposable ordinals, i.e., $\varphi_0(\alpha) = \omega^\alpha$, and for $\xi > 0$ the function φ_ξ enumerates the common fixed-points of all functions φ_ζ for $\zeta < \xi$.

Hence $\alpha < \beta$ implies $\varphi_\xi(\alpha) < \varphi_\xi(\beta)$ for any ξ and $\zeta < \xi$ implies $\varphi_\zeta(\varphi_\xi(\alpha)) = \varphi_\xi(\alpha)$.

Ordinals γ which are closed under the Veblen functions, i.e., for which $\xi, \eta < \gamma$ entail $\varphi_\xi(\eta) < \gamma$, are *strongly critical*. An ordinal γ is strongly critical iff $\varphi_\gamma(0) = \gamma$. The enumerating function of the strongly critical ordinals is commonly denoted by Γ. So Γ_0 is the first strongly critical ordinal. More profound studies can be found e.g. in [21] or [14].

The well-known consequence of Lemmata 10.24 and 10.23 is the famous Predicative Cut Elimination Theorem by W. Tait (cf. [23]).

Theorem 10.26. $\mathbb{L}^\sigma \vdash^{\alpha}_{\beta+\omega^\rho} \Delta$ *implies* $\mathbb{L}^\sigma \vdash^{\varphi_\rho(\alpha)}_{\beta} \Delta$.

Proof (sketch). We use main induction on ρ with side induction on α. The critical case is that the last inference is a cut of degree ν with $\beta \leq \nu < \beta + \omega^\rho$. Then we have the premises $\mathbb{L}^\sigma \vdash^{\alpha_0}_{\beta+\omega^\rho} \Delta, F$ and $\mathbb{L}^\sigma \vdash^{\alpha_0}_{\beta+\omega^\rho} \Delta, \neg F$ for some $\alpha_0 < \alpha$ and formula F with $deg_\sigma(F) = \nu$. By the side induction hypothesis we get $\mathbb{L}^\sigma \vdash^{\varphi_\rho(\alpha_0)}_{\beta} \Delta, F$ and $\mathbb{L}^\sigma \vdash^{\varphi_\rho(\alpha_0)}_{\beta} \Delta, \neg F$. If $\rho = 0$ we use Lemma 10.23 or Lemma 10.24, respectively, to obtain $\mathbb{L}^\sigma \vdash^{\varphi_0(\alpha_0)+\varphi_0(\alpha_0)}_{\beta} \Delta$. Since $\varphi_0(\alpha_0) + \varphi_0(\alpha_0) < \varphi_0(\alpha)$ the claim follows by weakening. If $\rho > 0$ and $\nu > \beta$ we get $\nu = \beta + \nu_0$ and decompose $\nu_0 = \omega^{\nu_1} + \cdots + \omega^{\nu_n}$ in Cantor normal form. Then $\beta < \nu < \beta + \omega^{\nu_1} \cdot n + \omega^{\nu_1}$ for $\nu_1 < \rho$ and we obtain $\mathbb{L}^\sigma \vdash^{\varphi_\rho(\alpha_0)+1}_{\beta+\omega^{\nu_1}\cdot n+\omega^{\nu_1}} \Delta$ by cut. By the main induction hypothesis we get $\mathbb{L}^\sigma \vdash^{\varphi_{\nu_1}(\varphi_\rho(\alpha_0)+1)}_{\beta+\omega^{\nu_1}\cdot n} \Delta$. Because of $\varphi_{\nu_1}(\varphi_\rho(\alpha_0) + 1) < \varphi_\rho(\alpha)$ and $\varphi_{\nu_1}(\varphi_\rho(\alpha)) = \varphi_\rho(\alpha)$ the claim follows by weakening and iterated application of the main induction hypothesis. □

10.5 The Boundedness Theorem

10.5.1 Some Preliminaries

For the following studies it is convenient to replace the notion of well-foundedness in the definition of *On* by the principle of $\in$-induction. To simplify notations we use lower case Greek letters as syntactical variables for hereditarily transitive sets. So $(\forall\xi)[\ \cdots\ ]$ stands for $(\forall\xi)[Ord(\xi) \to \cdots\ ]$ and $(\exists\xi)[\ \cdots\ ]$ stands for $(\exists\xi)[Ord(\xi) \land \cdots\ ]$.

We commonly also write $\alpha < \beta$ instead of $\alpha \in \beta$ for ordinals α and β.

Definition 10.27. Let

$$Prog(X) \quad :\Leftrightarrow \quad (\forall\xi)[\xi \subseteq X \to \xi \in X]$$

and define

$$Acc^{\varsigma} := \{\xi \mid Ord(\xi) \land (\forall X^{\varsigma})[Prog(X) \to \xi \subseteq X]\}.$$

Then Acc^{ς} is a class of level ς. The classes Acc^1 and *On* are extensionally equal. We will often write $Acc^{\varsigma}(\alpha)$ instead of $\alpha \in Acc^{\varsigma}$.

Although we have $L_\sigma \models (\forall\xi)[Acc^1(\xi)]$ this is not an axiom because it is not a first order sentence. In a first step we thus investigate whether the system $\mathbb{L}^\sigma$ is strong enough to prove the well-foundedness of the hereditarily transitive sets in L_σ. As a preparation we need a lemma.

Lemma 10.28. *For any ordinal* η *and class-term* S *we have*

$$\mathbb{L}^\sigma \overset{\alpha+8(\eta+1)}{\underset{0}{\vdash}} \neg Prog(S), \eta \notin L_\sigma, \neg Ord(\eta), \eta \subseteq S$$

for $\alpha = 2 \cdot deg_\sigma(S)$.

Proof. Induction on η. □

The next theorem is an immediate consequence of Lemma 10.28.

Theorem 10.29.

$$\mathbb{L}^\sigma \overset{\omega^{\varsigma}+8\cdot(\eta+2)}{\underset{0}{\vdash}} \eta \notin L_\sigma, Acc^{\varsigma}(\eta)\,.$$

Corollary 10.30. *For any limit ordinal* σ *we have*

$$\mathbb{L}^\sigma \overset{\omega+\sigma}{\underset{0}{\vdash}} (\forall\xi \in \mathsf{L}_\sigma)[Acc^1(\xi)].$$

By Corollary 10.30 $\mathbb{L}^\sigma$ confirms that all ordinals in L_σ are well-founded. But observe that this confirmation needs a derivation of height at least σ. In the next section we show that this is necessarily so.

10.5.2 Boundedness

Definition 10.31. For a set B of ordinals we denote by $coen_B$ the function that enumerates the complement of B. Let

$$B^\alpha = B \cup \{\xi \,|\, \xi \leq coen_B(\alpha)\} \text{ and } B^{<\alpha} := \begin{cases} B & \text{if } \alpha = 0, \\ \bigcup \{B^\xi \,|\, \xi < \alpha\} & \text{for } \alpha > 0. \end{cases}$$

Lemma 10.32. *Let B be a finite set of ordinals. Then*

$$coen_{B\cup\{\beta\}}(\alpha) \leq coen_B(\alpha + 1). \tag{10.1}$$

$$(B \cup \{\beta\})^\alpha \subseteq B^{\alpha+1} \cup \{\beta\}. \tag{10.2}$$

$$\beta \subseteq B^{<\alpha} \Rightarrow \beta \in B^\alpha. \tag{10.3}$$

Proof. Eq. (10.1) is obvious if $\beta \in B$. Otherwise $\beta = coen_B(\alpha_0)$ and for $\alpha < \alpha_0$ we have $coen_{B\cup\{\beta\}}(\alpha) = coen_B(\alpha)$ while for $\alpha = \alpha_0 + \xi$ it is $coen_{B\cup\{\beta\}}(\alpha_0 + \xi) = coen_B(\alpha_0 + 1 + \xi)$ and $\alpha_0 + 1 + \xi \leq \alpha + 1$.

Eq. (10.2) follows immediately from (10.1).

Eq. (10.3) holds trivially for $\beta \in B$. If $\beta \notin B$ then $\beta = coen_B(\xi)$ for some ordinal ξ. If $\xi \leq \alpha$ we have $\beta \in B^\alpha$. So assume $\alpha < \xi$. Then $coen_B(\alpha) < coen_B(\xi) = \beta$ and therefore $coen_B(\alpha) \in B^{<\alpha}$. Since $coen_B(\alpha) \notin B$ this leads to the contradiction $coen_B(\alpha) < coen_B(\alpha)$. □

Lemma 10.33. *(Boundedness Lemma) Let α and $\beta_1,\ldots,\beta_n$ be ordinals and $\Delta(X)$ a finite set of $\mathbb{L}$-formulae of level 0 in which X only occurs positively. Put $B := \{\beta_1,\ldots,\beta_n\}$. Then*

$$\mathbb{L}^\sigma \,\frac{\alpha}{0}\, \neg(\forall\xi)[\xi \subseteq X \rightarrow \xi \in X], \beta_1 \notin X, \ldots, \beta_n \notin X, \Delta(X)$$

implies $\mathbb{RL}[B^{<\alpha}] \models \bigvee\Delta$.

Proof. We induct on α. The interesting case is that the main formula of the last inference is $\neg(\forall\xi)[\xi \subseteq X \rightarrow \xi \in X]$, i.e., $\neg Prog(X)$.

Then we have the premise

$$\mathbb{L}^\sigma \,\frac{\alpha_0}{0}\, \neg Prog(X), Ord(\xi) \wedge \xi \subseteq X \wedge \xi \notin X, \beta_1 \notin X, \ldots, \beta_n \notin X, \Delta(X)\,.$$

By $\bigwedge$-inversion this entails

(i) $\mathbb{L}^\sigma \,\frac{\alpha_0}{0}\, \neg Prog(X), \xi \notin X, \beta_1 \notin X, \ldots, \beta_n \notin X, \Delta(X)$

and

(ii) $\mathbb{L}^\sigma \,\frac{\alpha_0}{0}\, \neg Prog(X), \beta_1 \notin X, \ldots, \beta_n \notin X, \xi \subseteq X, \Delta(X)\,.$

Assuming $\mathbb{RL}[B^{<\alpha}] \not\models \bigvee\Delta$ entails also $\mathbb{RL}[B^{<\alpha_0}] \not\models \bigvee\Delta$ by X-positivity and we get

(iii) $\xi \subseteq B^{<\alpha_0}$, hence $\xi \in B^{\alpha_0}$

by (ii) and (10.3). From (i) we obtain

(iv) $\mathbb{RL}[(B \cup \{\xi\})^{<\alpha_0}] \models \bigvee\Delta$.

By (10.2) we get $(B \cup \{\xi\})^{<\alpha_0} \subseteq B^{<\alpha} \cup \{\xi\} \subseteq B^{<\alpha}$. So (iv) contradicts our assumption $\mathbb{RL}[B^{<\alpha}] \not\models \bigvee\Delta$. □

Since $coen_\emptyset(\xi) = \xi$ it is $\emptyset^{<\alpha} \leq \alpha$ and we obtain as a corollary to the Boundedness Lemma

Theorem 10.34. *(Boundedness Theorem)* $\mathbb{L}^\sigma \mid\!\frac{\alpha}{0}\, Acc^1(\eta)$ *entails* $\eta \leq \alpha$.

10.6 Attainability

Our next aim is to secure a segment of the ordinals from below. It follows from the Boundedness Theorem that this can only be achieved by using cuts. To come to a segment of the ordinals that is predicatively secured we have to start with a segment that is "obviously" predicative and try to expand it in a predicative way.

The problem is of course the starting segment. The natural attempt is to start from scratch, i.e., from L_0. However, such an approach will apparently not go beyond the finite ordinals and our approach is probably not sufficiently fine-grained to study the predicatively meaningful natural numbers. This is also not our aim.

What we want to study are *infinite* predicative processes. Therefore we have to accept an infinity axiom, which we will express by the fact that we accept all ordinals $\leq \omega$. We will now study which ordinals we can access from this starting segment. In order not to conflict with the well-established notion of accessible or inaccessible ordinals, we prefer to talk about attainable ordinals.

Convention. Working with the finite formulae-sets of the Tait calculus needs some customization. The calculus looks less bizarre if we interpret finite sets of formulae as finite disjunctions and use the familiar "abbreviations" $A \to B$ for $\neg A \vee B$. So the formula $A \to B$ corresponds to the set $\{\neg A, B\}$. Iterating this procedure we obtain to every $\mathbb{L}$-formula F its "Tait-set" $T(F)$. We define $T(F) := \{F\}$ if the outermost logical symbol of F is different from $\vee$ and $T(F) = T(A) \cup T(B)$ if $F \equiv A \vee B$. Conversely we can translate every finite set of $\mathbb{L}$-formulae into a single $\mathbb{L}$-formula. Using a finite number of $\bigvee$-inferences or $\bigvee$-exportations, respectively, we obtain for limit ordinals α

$$\mathbb{L}^\sigma \mid\!\frac{<\alpha}{\rho}\, F \quad \Leftrightarrow \quad \mathbb{L}^\sigma \mid\!\frac{<\alpha}{\rho}\, T(F)$$

for every $\mathbb{L}$-formula F. Therefore we will not distinguish between a formula F and its Tait set $T(F)$ but always assume tacitly that we only derive the Tait set $T(F)$ even if we write $\mathbb{L}^\sigma \mid\!\frac{\alpha}{\rho}\, F$. This convention allows for inferences of the form

$$\mathbb{L}^\sigma \mid\!\frac{\alpha_S}{\rho}\, A \to F(S) \textit{ and } \alpha_S < \alpha \textit{ for all } S \textit{ of level less than } \varsigma$$
$$\Rightarrow \quad \mathbb{L}^\sigma \mid\!\frac{\alpha}{\rho}\, A \to (\forall X^\varsigma)F(X)$$

although a few more inferences would be needed to convert the set $\{\neg A, (\forall X^{\varsigma})F(X)\}$ into the single formula $A \to (\forall X^{\varsigma})F(X)$.

Definition 10.35. We inductively define the notion of an attainable ordinal by the following clauses

(AT 1) The ordinal $\omega + \omega$ is fully attainable.

(AT 2) An ordinal $\alpha > \omega + \omega$ is attainable if there is an ordinal $\gamma < \alpha$ which is attainable such that $\mathbb{L}^{\alpha} \,\big|\!\frac{\gamma}{\gamma+1}\, (\forall \xi \in \mathsf{L}_{\alpha})[Acc^{1}(\xi)]$.

An ordinal is *preattainable* if all ordinals less than α are attainable.

An ordinal α is *fully attainable* if α is preattainable and attainable.

As a consequence of cut elimination and boundedness we obtain a first unattainability result.

Theorem 10.36. *(K. Schütte) The Schütte-Feferman ordinal* Γ_0 *is unattainable.*

Proof. Assume that $\mathbb{L}^{\Gamma_0} \,\big|\!\frac{\alpha}{\alpha+1}\, (\forall \eta \in L_{\Gamma_0})[Acc^{1}(\eta)]$ holds true for an attainable ordinal $\alpha < \Gamma_0$. By $\bigwedge$-inversion, $\bigvee$-exportation and cut elimination this entails

$$\mathbb{L}^{\Gamma_0} \,\big|\!\frac{\varphi_{\alpha}(\varphi_0(\alpha))}{0}\, \neg(\forall \xi)[\xi \subseteq X \to \xi \in X], \eta \subseteq X$$

for every $\eta \in L_{\Gamma_0} \cap Ord = \Gamma_0$. The Boundedness Theorem, however, entails $\eta \leq \varphi_{\alpha}(\varphi_0(\alpha)) < \Gamma_0$ which shows that our assumption is absurd. □

10.7 The Attainability Proof

It follows from Theorem 10.36 that the only possible candidates for attainable ordinals are ordinals less than Γ_0. The aim of this section is thus to prove the main result

Theorem 10.37. *(K. Schütte, S. Feferman) The ordinal* Γ_0 *is preattainable.*

Our starting hypothesis is to regard $\omega + \omega$ as fully attainable. The problem is to extend the attainable ordinal beyond ω. As a first result we show that we can at least extend the realm of attainable ordinals a little beyond a limit ordinal $\alpha > \omega$ once α is secured to be fully attainable.

Lemma 10.38. *If* $\alpha > \omega + \omega$ *is a fully attainable limit ordinal then* $\alpha + \omega$ *is preattainable.*

Proof. Let

$$A^{1}(L_{\alpha}) \;:\Leftrightarrow\; (\forall \xi \in L_{\alpha})[Acc^{1}(\xi)]$$

Then we have

(i) $\mathbb{L}^\alpha \mathrel{|\!\!\frac{<\omega}{0}} A^1(L_\alpha) \wedge Prog(S) \wedge Ord(\alpha) \wedge \alpha \in L_\alpha \wedge \zeta < \alpha \to Ord(\zeta) \wedge \zeta \subseteq S$

for any class term S of level 0. Moreover we have

$$\mathbb{L}^\alpha \mathrel{|\!\!\frac{<\omega}{0}} Prog(S) \wedge Ord(\zeta) \wedge \zeta \subseteq S \to \zeta \in S$$

and the height of this proof does not depend on ζ. Together with (i) this implies

$$\mathbb{L}^\alpha \mathrel{|\!\!\frac{<\omega}{<\omega}} A^1(L_\alpha) \wedge Prog(S) \wedge Ord(\alpha) \wedge \zeta < \alpha \to \zeta \in S,$$

i.e.,

(ii) $\mathbb{L}^\alpha \mathrel{|\!\!\frac{<\omega}{<\omega}} A^1(L_\alpha) \wedge Ord(\alpha) \wedge Prog(S) \to \alpha \subseteq S$.

Since $L_{\alpha+1} \models Ord(\zeta) \to \zeta < \alpha \vee \zeta = \alpha$ we get by (i) and (ii)

$$\mathbb{L}^{\alpha+1} \mathrel{|\!\!\frac{<\omega}{<\omega}} A^1(L_\alpha) \wedge \zeta \in L_{\alpha+1} \wedge Ord(\zeta) \wedge Prog(S) \to \zeta \subseteq S,$$

i.e.

$$\mathbb{L}^{\alpha+1} \mathrel{|\!\!\frac{<\omega}{<\omega}} A^1(L_\alpha) \to A^1(L_{\alpha+1}).$$

More generally we have

$$L_{\alpha+n+1} \models Ord(\zeta) \to \zeta < \alpha + n \vee \zeta = \alpha + n$$

and can thus iterate this procedure to obtain

$$\mathbb{L}^{\alpha+n+1} \mathrel{|\!\!\frac{<\omega}{<\omega}} A^1(L_{\alpha+n}) \to A^1(L_{\alpha+n+1}).$$

Observe that $deg_\sigma(A^1(L_{\alpha+n})) < \omega + \omega \leq \alpha$. Therefore we obtain by cuts

$$\mathbb{L}^{\alpha+n} \mathrel{|\!\!\frac{<\omega}{\alpha}} (\forall \xi \in L_\alpha)[Acc^1(\xi)] \to (\forall \xi \in L_{\alpha+n})[Acc^1(\xi)] \text{ for any finite } n.$$

Since $\omega < \alpha$ is fully attainable we have $\mathbb{L}^\alpha \mathrel{|\!\!\frac{\gamma}{\gamma+1}} (\forall \xi \in L_\alpha)[Acc^1(\xi)]$ for some ordinal $\gamma < \alpha$ and we obtain by (cut) and an inference $\bigwedge$

$$\mathbb{L}^{\alpha+n+1} \mathrel{|\!\!\frac{<\alpha}{\alpha}} (\forall \zeta \in L_{\alpha+n})[Acc^1(\zeta)],$$

which shows that $\alpha + n$ is attainable for all finite ordinals n. □

Remark 10.39. Though important, Lemma 10.38 cannot help us to extend the attainable ordinals beyond ω. We must not postulate axiomatically that there is a finite ordinal γ such that $\mathbb{L}^\omega \mathrel{|\!\!\frac{\gamma}{\gamma+1}} (\forall \xi \in L_\omega)[Acc^1(\xi)]$ since this is obviously false. But this would be needed to apply Lemma 10.38 to the case $\alpha = \omega$. To cross the border ω and to prove Theorem 10.37 we see no other way as to assume the preattainability of $\omega + \omega$ and to use notations for ordinals beyond ω which are definable in L_ω.

The development of ordinal notations for ordinals below Γ_0 and thus the proof of Theorem 10.37 bases on a feature of the ordinal Γ_0 which has already been observed by O. Veblen ([24]). We use the notation $\eta =_{NF} \xi_1 + \cdots + \xi_n$ to denote that η is decomposed in Cantor normal form, i.e., that $\xi_1, \ldots, \xi_n$ are additively indecomposable ordinals such that $\eta = \xi_1 + \cdots + \xi_n$ and $\xi_1 \geq \xi_2 \geq \cdots \geq \xi_n$.

Theorem 10.40. *For any ordinal $\eta < \Gamma_0$ there are ordinals $\xi_1, \ldots, \xi_n$ and $\eta_1, \ldots, \eta_n$ such that*

$$\eta =_{NF} \varphi_{\xi_1}(\eta_1) + \cdots + \varphi_{\xi_n}(\eta_n)$$

and $\xi_i, \eta_i < \varphi_{\xi_i}(\eta_i)$ for $i = 1, \ldots, n$.

Theorem 10.40 opens the possibility to develop already in L_ω codes for the ordinals below Γ_0. Before we return to this possibility in Section 10.7.3 we provide some preparations.

10.7.1 Some Preparations and Notations

Convention. *Deriving formulae in $\mathbb{L}^\sigma$ is tedious since we have to control derivation height and cut degrees. In general, however, we do not need the exact proof height and cut degrees but can do with comparatively rough upper bounds. E.g., having a limit ordinal as upper bound for a derivation height leaves ample space for additional inferences (with finitely many premises). Therefore we can be more liberal and introduce the notion*

$$\mathbb{L}^\sigma_\alpha \vdash \Delta$$

to denote that there are ordinals $\beta < \alpha$ and $\rho \le \alpha$ such that $\mathbb{L}^\sigma \mathrel{\vert\!\frac{\beta}{\rho}} \Delta$.

If we have $\mathbb{L}^\sigma \mathrel{\vert\!\frac{\alpha_\iota}{\rho_\iota}} \Delta_\iota$ for a family $\iota \in I$ and there is a $\beta < \alpha$ such that $\alpha_\iota < \beta$ and $\rho_\iota \le \beta$ holds true for all $\iota \in I$ we denote that by $\mathbb{L}^\sigma_{<\alpha} \vdash \Delta_\iota$.

Definition 10.41. Regarding atomic formulae and formulae starting with a quantifier as sentential atoms we call an $\mathbb{L}$-formula *sententially valid*, if it becomes true under any assignment of truth values to its sentential atoms.

An inference $A \vdash B$ is a *sentential inference* if the formula $A \to B$ is sententially valid.

Lemma 10.42. *Let F be a formula of level ς which is sententially valid. Then*

$$\mathbb{L}^\sigma_\alpha \vdash F$$

with $\alpha = \omega^{\varsigma+1}$.

Proof. Since every sententially valid formula is derivable by finitely many finite inferences from a tautology, this follows from the Tautology Lemma 10.18 and Lemma 10.11. □

Lemma 10.43. *If A and B are $\mathbb{L}$-formulae of level ς, $A \vdash B$ is a sentential inference and $\mathbb{L}^\sigma_\alpha \vdash A$ for an ordinal $\alpha < \omega^{\varsigma+1}$ then $\mathbb{L}^\sigma_{\omega^{\varsigma+1}} \vdash B$.*

Proof. This is an immediate consequence of Lemma 10.42. □

10.7.2 Definable Well-orderings

We have to develop codes for ordinals below Γ_0 in L_ω. Ordinals can be coded by well-orderings. We therefore briefly recapitulate the notion of a well-ordering in the framework of L.

Recall the notions $\langle a, b\rangle$ of an ordered pair, $Rel(a)$ denoting that a is a relation, i.e., a set of ordered pairs, and $field(\prec)$ to denote the field of the relation $\prec$. For a relation $\prec$ we write $x \prec y$ for $\langle x, y\rangle \in \prec$. Then $x \prec y$ entails $\{x, y\} \subseteq field(\prec)$. Let

$$TRel(\prec) \;:\Leftrightarrow\; Rel(\prec) \wedge (\forall x)(\forall y)(\forall z)[x \prec y \wedge y \prec z \to x \prec z]$$

denote that $\prec$ is a transitive relation and

$$Lin(\prec) \;:\Leftrightarrow\; (\forall x)(\forall y)[x \in field(\prec) \wedge y \in field(\prec) \to \neg(x \prec x) \wedge (x \prec y \vee y \prec x \vee x = y)]$$

denote that $\prec$ is linearly ordered.

By

$$LO(\prec) \;:\Leftrightarrow\; TRel(\prec) \wedge Lin(\prec)$$

we thus express that $\prec$ is a linear ordering, an *order relation* for short.

Let

$$\prec_a := \{x \mid x \prec a\}$$

and

$$Prog_\prec(S) \;:\Leftrightarrow\; (\forall x)[x \in field(\prec) \wedge \prec_x \subseteq S \to x \in S].$$

Observe that all these formulae are of the same level as the relation $\prec$.

Definition 10.44. Let

$$Acc^\varsigma_\prec(a) \;:\Leftrightarrow\; (\forall X^\varsigma)[a \in field(\prec) \wedge LO(\prec) \wedge Prog_\prec(X) \to \prec_a \subseteq X].$$

Remark 10.45. The definition of $Acc^\varsigma_\prec(a)$ looks weird. More natural is $Acc^\varsigma_\prec(a) \;:\Leftrightarrow$ $a \in field(\prec) \wedge LO(\prec) \wedge (\forall X^\varsigma)[Prog_\prec(X) \to \prec_a \subseteq X]$. But in some (rare) situations (e.g. Lemma 10.69) we need $deg_\sigma(Acc^\xi_\prec(a)) = \omega^\varsigma$ which would be false in the more natural definition.

As an immediate consequence of its definition we obtain

Lemma 10.46. *Let $\prec$ be an ordering that is definable in L_σ. Then*

$$\mathbb{L}^\sigma_{\omega^\varsigma+1} \vdash b \preceq a \wedge Acc^\varsigma_\prec(a) \to Acc^\varsigma_\prec(b).$$

Given a well-ordering $\prec$ which is definable in L_σ. We can define its collapsing function $otp_\prec\colon field(\prec) \longrightarrow On$ by $otp_\prec(x) := \{otp_\prec(y) \mid y \prec x\}$. For $x \in field(\prec)$ we then obtain $Ord(otp_\prec(x))$ and put $\mathrm{otyp}(\prec) := \{otp_\prec(x) \mid x \in field(\prec)\}$. We then have $Ord(\mathrm{otyp}(\prec))$ and the following theorem.

Theorem 10.47. *Let $<$ be a well-ordering that is definable in L_σ of order-type $\alpha \geq \sigma$. Then*

$$L_\alpha \models (\forall x)\big[(x \in \mathit{field}(<) \to \mathit{Ord}(\mathit{otp}_<(x))) \\ \wedge (\forall y)[y \in \mathit{otp}_<(x) \leftrightarrow (\exists z)[z < x \wedge y = \mathit{otp}_<(z)]]\big].$$

For a well-founded order relation $<$ and a class S let

$$S_< := \{z \in \mathit{field}(<) \,|\, \mathit{otp}_<(z) \in S\}. \tag{10.4}$$

Lemma 10.48. *Let $<$ be a well-ordering that is definable in L_σ and $\eta = \mathrm{otyp}(<) \geq \sigma$. For any class term S of level less than ς we obtain*

$$\mathbb{L}^\eta_{\omega^\varsigma} \vdash \mathit{Prog}(S) \to \mathit{Prog}_<(S_<)^{L_\sigma}.$$

Proof. Let $\alpha := \omega^\varsigma$. We get from Tautology

(i) $\mathbb{L}^\eta_\alpha \vdash \mathit{Prog}(S) \wedge x \in \mathit{field}(<) \wedge \mathit{otp}_<(x) \subseteq S \to \mathit{otp}_<(x) \in S.$

Again from Tautology we have

$$\mathbb{L}^\eta_\alpha \vdash x \in \mathit{field}(<) \wedge (\forall z)[z < x \to \mathit{otp}_<(z) \in S] \wedge \\ y \in L_\sigma \wedge \xi = \mathit{otp}_<(y) \wedge y < x \to \xi \in S.$$

This implies

$$\mathbb{L}^\eta_\alpha \vdash x \in \mathit{field}(<) \wedge (\forall z)[z < x \to \mathit{otp}_<(z) \in S] \wedge \\ (\exists y \in L_\sigma)[\xi = \mathit{otp}_<(y) \wedge y < x] \to \xi \in S.$$

By Theorem 10.47 we have

$$L_\eta \models x \in \mathit{field}(<) \wedge \xi < \mathit{otp}_<(x) \to (\exists y \in L_\sigma)[\xi = \mathit{otp}_<(y) \wedge y < x].$$

By a cut—whose complexity is below α—and a an $(\bigwedge)$-rule we thus get

$$\mathbb{L}^\eta_\alpha \vdash x \in \mathit{field}(<) \wedge (\forall z)[z < x \to \mathit{otp}_<(z) \in S] \to \mathit{otp}_<(x) \subseteq S.$$

Cutting with (i) this entails

$$\mathbb{L}^\eta_\alpha \vdash \mathit{Prog}(S) \wedge x \in \mathit{field}(<) \to \\ (\forall z \in L_\sigma)[z < x \to \mathit{otp}_<(z) \in S] \to \mathit{otp}_<(x) \in S.$$

which entails

$$\mathbb{L}^\eta_\alpha \vdash \mathit{Prog}(S) \\ \to (\forall x \in L_\sigma)[x \in \mathit{field}(<) \wedge (\forall z \in L_\sigma)[z < x \to z \in S_<] \to x \in S_<]$$

i.e.

$$\mathbb{L}^\eta_\alpha \vdash \mathit{Prog}(S) \to \mathit{Prog}_<(S_<)^{L_\sigma}.$$ □

The next theorem is a direct consequence of Lemma 10.48.

Theorem 10.49. *Let $<$ be a well-ordering that is definable in L_σ of order-type $\eta \geq \sigma$. Then $\mathbb{L}^\eta_{\omega^\varsigma+1} \vdash \mathit{Acc}^\varsigma_<(b)^{L_\sigma} \to \mathit{Acc}^\varsigma(\mathit{otp}_<(b))$.*

Proof. By Lemma 10.19 we have for every class term S of level less than ς

$$\mathbb{L}^{\eta}_{\beta} \vdash Acc^{S}_{<}(b)^{L_{\sigma}} \rightarrow Prog_{<}(S_{<})^{L_{\sigma}} \rightarrow <_{b} \subseteq S_{<}$$

with $\beta = \omega^{\varsigma}$. Since $L_{\eta} \models (<_{b} \subseteq S_{<} \leftrightarrow otp_{<}(b) \subseteq S)$ cutting with Lemma 10.48 yields

$$\mathbb{L}^{\eta}_{\beta} \vdash Acc^{S}_{<}(b)^{L_{\sigma}} \rightarrow Prog(S) \rightarrow otp_{<}(b) \subseteq S.$$

Hence

$$\mathbb{L}^{\eta}_{\beta+1} \vdash Acc^{S}_{<}(b)^{L_{\sigma}} \rightarrow Acc^{S}(otp_{<}(b))$$

by a $(\bigwedge)$-rule. □

Theorem 10.49 has a corollary which is essential for the the attainability proof.

Corollary 10.50. *Let $\alpha \geq \omega^{\varsigma+1}$ be a limit ordinal that is preattainable, $<$ an ordering that is definable in L_{σ} for $\sigma \leq \alpha$ and $a \in field(<)$. If $\mathbb{L}^{\sigma}_{\alpha} \vdash Acc^{S}_{<}(a)$ then $otp_{<}(a)$ is fully attainable.*

Proof. >From $\mathbb{L}^{\sigma}_{\alpha} \vdash Acc^{S}_{<}(a)$ we get from Lemmata 10.20 and 10.46

$$\mathbb{L}^{\sigma}_{\alpha} \vdash b < a \rightarrow Acc^{1}_{<}(b) \tag{i}$$

for every b in L_{σ} by cutting with $Acc^{S}_{<}(a)$. For $\delta := otp_{<}(a)$ we have by Theorem 10.49

$$\mathbb{L}^{\delta}_{\omega+\omega} \vdash b < a \wedge Acc^{1}_{<}(b)^{L_{\sigma}} \rightarrow Acc^{1}(otp_{<}(b))$$

hence

$$\mathbb{L}^{\delta}_{\alpha} \vdash b < a \wedge otp_{<}(b) = \xi \wedge Acc^{1}_{<}(b)^{L_{\sigma}} \rightarrow Acc^{1}(\xi). \tag{ii}$$

Cutting (i) and (ii) yields

$$\mathbb{L}^{\delta}_{\alpha} \vdash b < a \wedge \xi = otp_{<}(b) \rightarrow Acc^{1}(\xi)$$

and thus

$$\mathbb{L}^{\delta}_{\alpha} \vdash (\exists z \in L_{\sigma})[z < a \wedge \xi = otp_{<}(z)] \rightarrow Acc^{1}(\xi) \tag{iii}$$

For any ξ we have

$$L \models \xi \in L_{\delta} \wedge Ord(\xi) \rightarrow (\exists z \in L_{\sigma})[z < a \wedge \xi = otp_{<}(z)]. \tag{iv}$$

By cutting (iv) and (iii) and an inference $\bigwedge$ it follows

$$\mathbb{L}^{\delta}_{\alpha} \vdash (\forall \xi \in L_{\delta})[\xi \in Acc^{1}].$$

Since α is preattainable this entails that δ is attainable.

This proof holds uniformly for any $a \in field(<)$. Therefore we see that all ordinals less than δ are attainable. So δ is fully attainable. □

10.7.3 Ordinal Notations and Ordinals

We now return to Theorem 10.40 and introduce codes for the ordinals below Γ_0. We will, however, not ruminate all the facts about the Veblen function but refer to [14] for details.

Definition 10.51. By $\in$-recursion we simultaneously define a class $\mathsf{OT} \subseteq L_\omega$ and a relation $\prec$ of level 0 on the elements of OT.

$\langle 0 \rangle \in \mathsf{OT}$,
$a_1, \dots, a_n \in \mathsf{HT}, 1 < n$ and $a_1 \succeq \dots \succeq a_n$ imply $\langle 1, a_1, \dots, a_n \rangle \in \mathsf{OT}$,
$a_1 \in \mathsf{OT}$ and $a_2 \in \mathsf{OT}$ imply $\langle 2, a_1, a_2 \rangle \in \mathsf{HT}$,
$\mathsf{HT} \subseteq \mathsf{OT}$.

The relation $\prec$ is defined by

$\langle 0 \rangle \prec a$ iff $a \in \mathsf{OT}$ and $a \neq \langle 0 \rangle$,
$\langle 1, a_1, \dots, a_n \rangle \prec \langle 1, b_1, \dots, b_m \rangle$ iff $a_i = b_i$ for $i = 1, \dots, k \leq n$ and $k = n < m$ or $a_{k+1} \prec b_{k+1}$,
$\langle 1, a_1, \dots, a_n \rangle \prec \langle 2, b, c \rangle$ iff $a_1 \prec \langle 2, b, c \rangle$,
$\langle 2, a, b \rangle \prec \langle 1, a_1, \dots, a_n \rangle$ iff $\langle 2, a, b \rangle \preceq a_1$,

$$\langle 2, a_1, a_2 \rangle \prec \langle 2, b_1, b_2 \rangle \text{ iff } \begin{cases} a_1 = b_1 \wedge a_2 \prec b_2, \\ a_1 \prec b_1 \wedge a_2 \prec \langle 2, b_1, b_2 \rangle, \\ b_1 \prec a_1 \wedge \langle 2, a_1, b_1 \rangle \preceq b_2. \end{cases}$$

The relation $a \preceq b$ stands for $a \prec b \vee a = b$.

Remark 10.52. We know from the theory of ordinals (cf. e.g. [21] or [14]) that the set OT and the relation $\preceq$ on OT code a copy of the ordinals below Γ_0 together with the *"less than or equal to"* relation on these ordinals. Anticipating the well-foundedness of the relation $\prec$ we obtain for its collapsing function $otp\colon \mathsf{OT} \longrightarrow On$:

$$otp(\langle 0 \rangle) = 0,$$
$$otp\langle 1, a_1, \dots, a_n \rangle =_{NF} otp(a_1) + \dots + otp(a_n),$$
$$otp\langle 2, a, b \rangle = \overline{\varphi}_{otp(a)}(otp(b))$$

and $a \prec b$ iff $otp(a) < otp(b)$ for the Mostowski collapse of the relation $\prec$. Here $\overline{\varphi}_\alpha$ stands for the fixed-point free version of the Veblen-function which is defined by

$$\overline{\varphi}_\alpha(\beta) = \begin{cases} \varphi_\alpha(\beta + 1) & \text{if } \beta = \gamma + n \text{ and } \varphi_\alpha(\gamma) = \gamma \text{ or } \varphi_\alpha(\gamma) = \alpha, \\ \varphi_\alpha(\beta) & \text{otherwise.} \end{cases}$$

Having this in mind facilitates the reading although we must not yet anticipate the well-foundedness of the relation $\prec$.

To mimic ordinal addition we define

$$a \oplus b = \begin{cases} a & \text{if } b = \langle 0 \rangle, \\ b & \text{if } b \in \mathsf{HT} \text{ and } a \prec b, \\ \langle 1, a, b \rangle & \text{if } a, b \in \mathsf{HT} \wedge b \preceq a, \\ \langle 1, a_1, \dots, a_k, b \rangle & \text{if } a = \langle 1, a_1, \dots, a_n \rangle,\ b \in \mathsf{HT} \\ & \quad \text{and } a_{k+1} \prec b \preceq a_k, \\ \langle 1, a_1, \dots, a_k, b_1, \dots, b_m \rangle & \text{if } a = \langle 1, a_1, \dots, a_n \rangle \succeq b = \langle 1, b_1, \dots, b_m \rangle \\ & \quad \text{and } a_{k+1} \prec b_1 \preceq a_k \\ \langle 1, a_1, \dots, a_k, b_1, \dots, b_m \rangle & \text{if } a = \langle 1, a_1, \dots, a_n \rangle \prec b = \langle 1, b_1, \dots, b_m \rangle \\ & \quad \text{and } a_i = b_i \text{ for } i = 1, \dots k \geq 0 \\ & \quad \text{and } a_{k+1} \prec b_1. \end{cases}$$

Anticipating again the well-foundedness of $\prec$ we obtain $otp(a \oplus b) = otp(a) + otp(b)$.

10.7.4 The Attainability Proof

10.7.4.1 The Basic Facts

We subsume some basic facts of the ordinal notations in the structure L_ω.

Lemma 10.53.

$$L_\omega \models LO(\prec) \wedge (\forall x \in \mathsf{OT})(\forall y)[y \prec x \rightarrow y \in \mathsf{OT}].$$

Lemma 10.54.

$$L_\omega \models a \in \mathsf{OT} \wedge b \in \mathsf{OT} \wedge a \preceq b \prec a \oplus c \rightarrow (\exists x)[b = a \oplus x \wedge x \prec c]$$

Lemma 10.55. *For $a \in \mathsf{OT}$ and $b \in \mathsf{OT}$ we get*

$$\mathbb{L}^{\omega}_{\omega\varsigma} \vdash a \in \mathsf{OT} \wedge b \preceq a \wedge \prec_a \subseteq S \rightarrow \prec_b \subseteq S$$

for every class term S of level less than ς.

For a class term S we define its jump

$$\mathcal{J}(S) := \{a \,|\, a \in \mathsf{OT} \wedge (\forall x)[x \in \mathsf{OT} \wedge \prec_x \subseteq S \rightarrow \prec_{x \oplus \langle 2, \langle 0 \rangle, a \rangle} \subseteq S]\}.$$

Observe that $\mathcal{J}(S)$ has the same level as S.

Lemma 10.56. *Let S be a class term of level less than ς. Then*

$$\mathbb{L}^{\omega}_{\omega\varsigma} \vdash Prog_{\prec}(S) \rightarrow Prog_{\prec}(\mathcal{J}(S)).$$

Proof. This is essentially the Gentzen proof (cf. e.g. [14] Sec. 7.4). □

Lemma 10.56 has three obvious corollaries.

Corollary 10.57.

$$\mathbb{L}^{\omega}_{\omega^{\varsigma}+1} \vdash Acc^{\varsigma}_{\prec}(a) \to Acc^{\varsigma}_{\prec}(\langle 2, \langle 0\rangle, a\rangle)$$

Corollary 10.58.

$$\mathbb{L}^{\omega}_{\omega^{\varsigma}+1} \vdash Acc^{\varsigma}_{\prec}(a) \wedge Acc^{\varsigma}_{\prec}(b) \to Acc^{\varsigma}_{\prec}(a \oplus b).$$

For an ordinal term $a = \langle 1, a_1, \ldots, a_n\rangle$ let $h(a) := a_1$. Then we have $a_i \preceq h(a)$ for $i \in \{1, \ldots, n\}$. For completeness put $h(a) := a$ for all other ordinal terms. As a corollary of Lemma 10.46 and Corollary 10.58 we then obtain

Corollary 10.59.

$$\mathbb{L}^{\omega}_{\omega^{\varsigma}+\omega} \vdash a \in \mathsf{OT} \wedge Acc^{\varsigma}_{\prec}(h(a)) \to Acc^{\varsigma}_{\prec}(a).$$

For simpler reading of the next lemma observe that—the well-foundedness of $\prec$ anticipated—we have $otp(\langle 2, \langle 2, \langle 0\rangle, \langle 0\rangle\rangle, a\rangle) = \overline{\varphi}_1(otp(a)) = \overline{\varepsilon}_{otp(a)}$.[11]

Lemma 10.60. *For every ordinal ς we have*

$$\mathbb{L}^{\omega}_{\omega^{\varsigma+2}} \vdash Acc^{\varsigma}_{\prec}(a) \to Acc^{\varsigma}_{\prec}(\langle 2, \langle 2, \langle 0\rangle, \langle 0\rangle\rangle, a\rangle).$$

Proof. Before we start we introduce two notions

$$Lim(a) \;:\Leftrightarrow\; Ord(a) \wedge (\forall x \in a)(\exists y \in a)[x \in y] \text{ and}$$
$$a \in \omega \;:\Leftrightarrow\; Ord(a) \wedge \neg Lim(a) \wedge (\forall y \in a)[\neg Lim(y)].$$

Then we obtain

(o) $L_\omega \models a \in \omega \to a = 0 \vee (\exists z)[z \in \omega \wedge a = z \cup \{z\}]$.

Using the familiar abbreviation $z + 1 := z \cup \{z\}$ we define by $\in$-recursion a function $e\colon \omega \longrightarrow L_\omega$ by

$$e(0) := a \text{ and } e(n+1) := \langle 2, \langle 0\rangle, e(n)\rangle.$$

Then we obtain

$$L_\omega \models a \in \mathsf{OT} \to (\forall x)[x \in \omega \to e(x) \in \mathsf{OT}]$$

by $\in$-induction on x. Let $E_\varsigma := \{x \mid x \in \omega \to Acc^{\varsigma}_{\prec}(e(x))\}$. Then E_ς is a class-term of level ς and we have by Corollary 10.57

(i) $\mathbb{L}^{\omega}_{\omega^{\varsigma}+\omega} \vdash Acc^{\varsigma}_{\prec}(a) \to (\forall x)[x \in \omega \wedge x \in E_\varsigma \to x + 1 \in E_\varsigma]$.

By (o) we obtain

$$\mathbb{L}^{\omega}_{\omega^{\varsigma}} \vdash b \subseteq E_\varsigma \wedge b \in \omega \to b = 0 \vee (\exists x \in b)[b = x + 1].$$

[11] By $\overline{\varepsilon}$ we understand the fix-point free version of the enumerating function ε of the ordinals closed under ω-powers.

Together with (i) we therefore have

$$\mathbb{L}^{\omega}_{\omega^{\varsigma}+\omega} \vdash Acc^{\varsigma}_{\prec}(a) \to (b \in \omega \wedge b \subseteq E_{\varsigma} \to b \in E_{\varsigma})$$

and, since $L_{\omega} \models Ord(b) \to b \in \omega$,

$$\mathbb{L}^{\omega}_{\omega^{\varsigma}+\omega} \vdash Acc^{\varsigma}_{\prec}(a) \to Prog(E_{\varsigma}).$$

Therefore we obtain by Lemma 10.19

$$\mathbb{L}^{\omega}_{\omega^{\varsigma+1}} \vdash Acc^{\varsigma+1}(\omega) \wedge Acc^{\varsigma}_{\prec}(a) \to \omega \subseteq E_{\varsigma}$$

i.e.

$$\mathbb{L}^{\omega}_{\omega^{\varsigma+1}} \vdash Acc^{\varsigma+1}(\omega) \wedge Acc^{\varsigma}_{\prec}(a) \to (\forall x \in \omega)[Acc^{\varsigma}_{\prec}(e(x))].$$

By Lemma 10.46 we have

$$\mathbb{L}^{\omega}_{\omega^{\varsigma}+\omega} \vdash Acc^{\varsigma}_{\prec}(e(x)) \wedge c \prec e(x) \to Acc^{\varsigma}_{\prec}(c)$$

hence

$$\mathbb{L}^{\omega}_{\omega^{\varsigma}+\omega} \vdash (\forall x \in \omega)[Acc^{\varsigma}_{\prec}(e(x))] \to (\exists x \in \omega)[c \prec e(x)] \to Acc^{\varsigma}_{\prec}(c).$$

So we have

$$\mathbb{L}^{\omega}_{\omega^{\varsigma+1}} \vdash Acc^{\varsigma+1}(\omega) \wedge Acc^{\varsigma}_{\prec}(a) \to (\exists x \in \omega)[c \prec e(x)] \to Acc^{\varsigma}_{\prec}(c).$$

By $\in$-induction on $TC(c)$, the transitive hull of c, we easily obtain

$$L_{\omega} \models c \prec \langle 2, \langle 2, \langle 0 \rangle, \langle 0 \rangle \rangle, a \rangle \to (\exists x)[c \prec e(x)].$$

By cut it follows

$$\mathbb{L}^{\omega}_{<\omega^{\varsigma+1}} \vdash Acc^{\varsigma+1}(\omega) \wedge Acc^{\varsigma}_{\prec}(a) \wedge c \prec \langle 2, \langle 2, \langle 0 \rangle, \langle 0 \rangle \rangle, a \rangle \to Acc^{\varsigma}_{\prec}(c)$$

for all $c \in L_{\omega}$ which by an inference $(\bigwedge)$ entails

$$\mathbb{L}^{\omega}_{\omega^{\varsigma+1}} \vdash Acc^{\varsigma+1}(\omega) \wedge Acc^{\varsigma}_{\prec}(a) \to Acc^{\varsigma}_{\prec}(\langle 2, \langle 2, \langle 0 \rangle, \langle 0 \rangle \rangle, a \rangle).$$

By Theorem 10.29 we have

$$\mathbb{L}^{\omega}_{\omega^{\varsigma+1}+\omega\cdot 2} \vdash Acc^{\varsigma+1}(\omega).$$

Hence by cut

$$\mathbb{L}^{\omega}_{\omega^{\varsigma+2}} \vdash Acc^{\varsigma}_{\prec}(a) \to Acc^{\varsigma}_{\prec}(\langle 2, \langle 2, \langle 0 \rangle, \langle 0 \rangle \rangle, a \rangle). \qquad \square$$

Let

$$\ulcorner \varepsilon_0 \urcorner := \langle 2, \langle 2, \langle 0 \rangle, \langle 0 \rangle \rangle, \langle 0 \rangle \rangle.$$

Corollary 10.61. *For every ordinal ς we have*

$$\mathbb{L}^{\omega}_{\omega^{\varsigma+2}} \vdash Acc^{\varsigma}_{\prec}(\ulcorner \varepsilon_0 \urcorner).$$

Proof. We have $L_{\omega} \models Lin(\prec) \wedge \langle 0 \rangle \in \mathsf{OT} \wedge \neg c \prec \langle 0 \rangle$ for any L-term c. Hence

$$\mathbb{L}^{\omega}_{\omega} \vdash Lin(\prec) \wedge \langle 0 \rangle \in \mathsf{OT} \wedge (c \prec \langle 0 \rangle \to c \in S)$$

for any class term S of level less than ς. This entails

$$\mathbb{L}^{\omega}_{\omega} \vdash Acc^{S}_{<}(\langle 0\rangle).$$

Together with Corollary 10.57 and Lemma 10.60 we thus get

$$\mathbb{L}^{\omega}_{\omega^{\varsigma+2}} \vdash Acc^{S}_{<}(\ulcorner\varepsilon_0\urcorner).$$ □

For $a \in \mathsf{OT}$ let

$$\mathcal{V}_{\varsigma}(a) := \{b \mid b \in \mathsf{OT} \wedge (Acc^{S}_{<}(b) \to Acc^{S}_{<}(\langle 2, a, b\rangle))\}. \tag{10.5}$$

Then $\mathcal{V}_{\varsigma}(a)$ is a class term of level ς and we have

Lemma 10.62.

$$\mathbb{L}^{\omega}_{\omega^{\varsigma+1}} \vdash a \in \mathsf{OT} \wedge b \in \mathsf{OT} \wedge <_b \subseteq \mathcal{V}_{\varsigma}(a) \wedge Acc^{S}_{<}(b) \to (\forall x)[x < b \to Acc^{S}_{<}(\langle 2, a, x\rangle)].$$

Proof. This follows from Lemma 10.18 together with some $\bigwedge$-inversions sententially. □

Let

$$A_{\varsigma}(a) \;:\Leftrightarrow\; a \in \mathsf{OT} \wedge (\forall x)(\forall y)[y \in \mathsf{OT} \wedge x < a \to y \in \mathcal{V}_{\varsigma}(x)],$$
$$B_{\varsigma}(a,b) \;:\Leftrightarrow\; a \in \mathsf{OT} \wedge b \in \mathsf{OT} \wedge (\forall y)[y < b \to Acc^{S}_{<}(\langle 2, a, y\rangle)]$$

abbreviate two hypotheses for the next lemma. By $TC(c)$ we denote again the transitive hull of a set c.

Lemma 10.63.

$$\mathbb{L}^{\omega}_{\omega^{\varsigma+1}} \vdash A_{\varsigma}(a) \wedge B_{\varsigma}(a,b) \wedge Acc^{S}_{<}(b) \wedge c < \langle 2, a, b\rangle \to Acc^{S}_{<}(c).$$

Proof. We proceed by $\in$-induction on $TC(c)$ and have the induction hypothesis

$$\mathbb{L}^{\omega}_{\omega^{\varsigma+1}} \vdash A_{\varsigma}(a) \wedge B_{\varsigma}(a,b) \wedge Acc^{S}_{<}(b) \wedge d < \langle 2, a, b\rangle \wedge d \in TC(c) \to Acc^{S}_{<}(d).$$

We distinguish cases on the shape of $c \in \mathsf{OT}$.

First we easily obtain

(I) $\mathbb{L}^{\omega}_{\omega} \vdash Acc^{S}_{<}(\langle 0\rangle)$.

Assuming $c = \langle 1, c_1, \ldots, c_n\rangle$ we have

$$L_{\omega} \models c < \langle 2, a, b\rangle \wedge c = \langle 1, c_1, \ldots, c_n\rangle \to h(c) \in TC(c) \wedge h(c) < \langle 2, a, b\rangle.$$

Together with the induction hypothesis this entails

$$\mathbb{L}^{\omega}_{\omega^{\varsigma+1}} \vdash A_{\varsigma}(a) \wedge B_{\varsigma}(a,b) \wedge Acc^{S}_{<}(b) \wedge c = \langle 1, c_1, \ldots, c_n\rangle \wedge c < \langle 2, a, b\rangle \to Acc^{S}_{<}(h(c)).$$

By Lemma 10.58 we have

$$\mathbb{L}^{\omega}_{\omega^{\varsigma}+\omega} \vdash Acc^{S}_{<}(h(c)) \to Acc^{S}_{<}(c).$$

By cut it follows

$$\mathbb{L}^{\omega}_{\omega^{\varsigma+1}} \vdash A_\varsigma(a) \wedge B_\varsigma(a,b) \wedge Acc^{\varsigma}_{\prec}(b) \wedge c = \langle 1, c_1, \ldots, c_n \rangle \\ \wedge\, c \prec \langle 2, a, b \rangle \rightarrow Acc^{\varsigma}_{\prec}(c).$$

Hence

$$\text{(II)}\ \mathbb{L}^{\omega}_{\omega^{\varsigma+1}} \vdash A_\varsigma(a) \wedge B_\varsigma(a,b) \wedge Acc^{\varsigma}_{\prec}(b) \wedge c \prec \langle 2, a, b \rangle \\ \wedge\, (\exists x_1) \ldots (\exists x_n)[c = \langle 1, x_1, \ldots, x_n \rangle] \wedge \;\rightarrow Acc^{\varsigma}_{\prec}(c).$$

In the next case assume $c = \langle 2, c_1, c_2 \rangle$. Then we have the following subcases:

$$c_1 = a \wedge c_2 \prec b.$$

In this subcase we have

$$\mathbb{L}^{\omega}_{\omega^{\varsigma+1}} \vdash B_\varsigma(a,b) \wedge c_2 \prec b \rightarrow Acc^{\varsigma}_{\prec}(\langle 2, a, c_2 \rangle)$$

hence

$$\text{(A)}\ \mathbb{L}^{\omega}_{\omega^{\varsigma+1}} \vdash A_\varsigma(a) \wedge B_\varsigma(a,b) \wedge Acc^{\varsigma}_{\prec}(b) \wedge c = \langle 2, c_1, c_2 \rangle \\ \wedge\, c_1 = a \wedge c_2 \prec b \rightarrow Acc^{\varsigma}_{\prec}(c).$$

The next subcase is $c_1 \prec a \wedge c_2 \prec \langle 2, a, b \rangle$. Then we have

$$L_\omega \models c \in \mathsf{OT} \wedge c = \langle 2, c_1, c_2 \rangle \rightarrow c_2 \in TC(c) \wedge c_2 \in \mathsf{OT}$$

and

$$\mathbb{L}^{\omega}_{\omega^{\varsigma+1}} \vdash A_\varsigma(a) \wedge c_1 \prec a \wedge c_2 \in \mathsf{OT} \rightarrow c_2 \in \mathcal{V}_\varsigma(c_1).$$

By definition of $\mathcal{V}_\varsigma(c_1)$ we have

$$\mathbb{L}^{\omega}_{\omega^{\varsigma+1}} \vdash c_2 \in \mathcal{V}_\varsigma(c_1) \wedge Acc^{\varsigma}_{\prec}(c_2) \rightarrow Acc^{\varsigma}_{\prec}(\langle 2, c_1, c_2 \rangle).$$

By induction hypothesis it holds

$$\mathbb{L}^{\omega}_{\omega^{\varsigma+1}} \vdash A_\varsigma(a) \wedge B_\varsigma(a,b) \wedge Acc^{\varsigma}_{\prec}(b) \wedge c_2 \in TC(c) \\ \wedge\, c_2 \prec \langle 2, a, b \rangle \rightarrow Acc^{\varsigma}_{\prec}(c_2).$$

By cuts these imply

$$\text{(B)}\ \mathbb{L}^{\omega}_{\omega^{\varsigma+1}} \vdash A_\varsigma(a) \wedge B_\varsigma(a,b) \wedge Acc^{\varsigma}_{\prec}(b) \wedge c = \langle 2, c_1, c_2 \rangle \\ \wedge\, c_1 \prec a \wedge c_2 \prec \langle 2, a, b \rangle \rightarrow Acc^{\varsigma}_{\prec}(c).$$

The last case is $a \prec c_1 \wedge c \preceq b$. By Lemma 10.53 we have

$$\mathbb{L}^{\omega}_{\omega^{\varsigma+1}} \vdash Acc^{\varsigma}_{\prec}(b) \wedge c \preceq b \rightarrow Acc^{\varsigma}_{\prec}(c).$$

Hence

$$\text{(C)}\ \mathbb{L}^{\omega}_{\omega^{\varsigma+1}} \vdash A_\varsigma(a) \wedge B_\varsigma(a,b) \wedge Acc^{\varsigma}_{\prec}(b) \wedge c = \langle 2, c_1, c_2 \rangle \\ \wedge\, a \prec c_1 \wedge c \preceq b \rightarrow Acc^{\varsigma}_{\prec}(c).$$

Since

$$L_\omega \models a \in \mathsf{OT} \wedge b \in \mathsf{OT} \wedge c \prec \langle 2,a,b\rangle \wedge c = \langle 2,c_1,c_2\rangle \\ \rightarrow c_1 \in \mathsf{OT} \wedge c_2 \in \mathsf{OT} \wedge [(c_1 = a \wedge c_2 \prec b) \\ \vee (c_1 \prec a \wedge c_2 \prec \langle 2,a,b\rangle) \vee (a \prec c_1 \wedge c \preceq b)]$$

we get from (A), (B) and (C)

(III) $$\mathbb{L}^{\omega}_{\omega^{\varsigma+1}} \vdash A_\varsigma(a) \wedge B_\varsigma(a,b) \wedge Acc^{S}_{\prec}(b) \wedge c \prec \langle 2,a,b\rangle \\ \wedge (\exists x_1)(\exists x_2)[c = \langle 2,x_1,x_2\rangle] \rightarrow Acc^{S}_{\prec}(c).$$

>From

$$L_\omega \models a \in \mathsf{OT} \wedge b \in \mathsf{OT} \wedge c \prec \langle 2,a,b\rangle \rightarrow c = \langle 0\rangle \\ \vee (\exists x_1)\ldots(\exists x_n)[c = \langle 1,x_1,\ldots,x_n\rangle] \vee (\exists x_1)(\exists x_2)[c = \langle 2,x_1,x_2\rangle]$$

we finally get from (I), (II) and (III)

$$\mathbb{L}^{\omega}_{\omega^{\varsigma+1}} \vdash A_\varsigma(a) \wedge B_\varsigma(a,b) \wedge Acc^{S}_{\prec}(b) \wedge c \prec \langle 2,a,b\rangle \rightarrow Acc^{S}_{\prec}(c).$$ □

Lemma 10.64.

$$\mathbb{L}^{\omega}_{\omega^{\varsigma}+1} \vdash a \in \mathsf{OT} \wedge (\forall x)[x \prec a \rightarrow Acc^{S}_{\prec}(x)] \rightarrow Acc^{S}_{\prec}(a).$$

Proof. Let S be a class-term of level less than ς. For any $b \in L_\omega$ we have

$$\mathbb{L}^{\omega}_{\omega^{\varsigma}} \vdash Acc^{S}_{\prec}(b) \wedge Prog_{\prec}(S) \rightarrow \prec_b \subseteq S,$$

hence

$$\mathbb{L}^{\omega}_{\omega^{\varsigma}} \vdash Acc^{S}_{\prec}(b) \wedge Prog_{\prec}(S) \rightarrow b \in S$$

and thus

$$\mathbb{L}^{\omega}_{\omega^{\varsigma}} \vdash (b \prec a \rightarrow Acc^{S}_{\prec}(b)) \wedge Prog_{\prec}(S) \rightarrow (b \prec a \rightarrow b \in S).$$

Hence

$$\mathbb{L}^{\omega}_{\omega^{\varsigma}} \vdash (\forall x)[x \prec a \rightarrow Acc^{S}_{\prec}(x)] \rightarrow Prog_{\prec}(S) \rightarrow \prec_a \subseteq S.$$

Since this holds true for any class-term S of level less than ς we finally obtain

$$\mathbb{L}^{\omega}_{\omega^{\varsigma}+1} \vdash a \in \mathsf{OT} \wedge (\forall x)[x \prec a \rightarrow Acc^{S}_{\prec}(x)] \rightarrow Acc^{S}_{\prec}(a).$$ □

Lemma 10.65.

$$\mathbb{L}^{\omega}_{\omega^{\varsigma+1}} \vdash A_\varsigma(a) \rightarrow Prog_{\prec}(\mathcal{V}_\varsigma(a)).$$

Proof. By Lemma 10.62 we have

$$\mathbb{L}^{\omega}_{\omega^{\varsigma+1}} \vdash a \in \mathsf{OT} \wedge b \in \mathsf{OT} \wedge \prec_b \subseteq \mathcal{V}_\varsigma(a) \wedge Acc^{S}_{\prec}(b) \\ \rightarrow (\forall x)[x \prec b \rightarrow Acc^{S}_{\prec}(\langle 2,a,x\rangle)],$$

i.e.

$$\mathbb{L}^{\omega}_{\omega^{\varsigma+1}} \vdash a \in \mathsf{OT} \wedge b \in \mathsf{OT} \wedge \prec_b \subseteq \mathcal{V}_\varsigma(a) \wedge Acc^{S}_{\prec}(b) \rightarrow B_\varsigma(a,b).$$

By Lemmata 10.63 and 10.64 we have

$$\mathbb{L}^{\omega}_{\omega^{\varsigma+1}} \vdash A_\varsigma(a) \wedge B_\varsigma(a,b) \wedge Acc^{\varsigma}_{\prec}(b) \to Acc^{\varsigma}_{\prec}(\langle 2,a,b\rangle).$$

By a cut we therefore obtain

$$\mathbb{L}^{\omega}_{\omega^{\varsigma+1}} \vdash A_\varsigma(a) \wedge \prec_b \subseteq \mathcal{V}_\varsigma(a) \wedge Acc^{\varsigma}_{\prec}(b) \to Acc^{\varsigma}_{\prec}(\langle 2,a,b\rangle)$$

hence

$$\mathbb{L}^{\omega}_{\omega^{\varsigma+1}} \vdash A_\varsigma(a) \wedge \prec_b \subseteq \mathcal{V}_\varsigma(a) \to b \in \mathcal{V}_\varsigma(a)$$

which entails the claim. □

Lemma 10.66.

$$\mathbb{L}^{\omega}_{\omega^{\varsigma+1}} \vdash A_\varsigma(a) \wedge Acc^{\varsigma+1}_{\prec}(b) \to Acc^{\varsigma}_{\prec}(\langle 2,a,b\rangle).$$

Proof. >From Lemma 10.65 and Tautology (Lemma 10.18) we obtain sententially

$$\mathbb{L}^{\omega}_{\omega^{\varsigma+1}} \vdash A_\varsigma(a) \to (Prog_{\prec}(\mathcal{V}_\varsigma(a)) \to \prec_b \subseteq \mathcal{V}_\varsigma(a)) \to \prec_b \subseteq \mathcal{V}_\varsigma(a).$$

By an inference $\bigvee$ this entails

$$\mathbb{L}^{\omega}_{\omega^{\varsigma+1}} \vdash A_\varsigma(a) \wedge Acc^{\varsigma+1}_{\prec}(b) \to \prec_b \subseteq \mathcal{V}_\varsigma(a).$$

Hence by Lemma 10.65

$$\mathbb{L}^{\omega}_{\omega^{\varsigma+1}} \vdash A_\varsigma(a) \wedge Acc^{\varsigma+1}_{\prec}(b) \to b \in \mathcal{V}_\varsigma(a)$$

and thus

$$\mathbb{L}^{\omega}_{\omega^{\varsigma+1}} \vdash A_\varsigma(a) \wedge Acc^{\varsigma+1}_{\prec}(b) \wedge Acc^{\varsigma}_{\prec}(b) \to Acc^{\varsigma}_{\prec}(\langle 2,a,b\rangle).$$

By Lemma 10.20 we have

$$\mathbb{L}^{\omega}_{\omega^{\varsigma}} \vdash Acc^{\varsigma+1}_{\prec}(b) \to Acc^{\varsigma}_{\prec}(b).$$

Hence

$$\mathbb{L}^{\omega}_{\omega^{\varsigma+1}} \vdash A_\varsigma(a) \wedge Acc^{\varsigma+1}_{\prec}(b) \to Acc^{\varsigma}_{\prec}(\langle 2,a,b\rangle).$$ □

Observe that it is Lemma 10.66 that forces us to raise the levels of classes in the attainability proof.

Lemma 10.67. *Let $a \in \mathsf{OT}$ and $otp(a) = \nu$. Then we get*

$$\mathbb{L}^{\omega}_{\varphi_1(\omega^{\nu+1}\cdot\eta)+1} \vdash Acc^{1}_{\prec}(a) \wedge Acc^{\omega^{\nu+1}\cdot\eta}_{\prec}(b) \to Acc^{\omega^{\nu+1}\cdot\eta}_{\prec}(\langle 2,a,b\rangle)$$

for any ordinal $\eta \neq 0$.

Proof. We proceed by induction on ν. Since $\omega^{\omega\cdot\eta}+1 \leq \varphi_1(\omega\cdot\eta)+1$ the case $\nu = 0$ is covered by Corollary 10.57. For $c \prec a$ we have $\mu := otp(c) < otp(a) = \nu$. Since $\nu = \mu + 1 + \delta$ we have $\omega^\nu \cdot \xi = \omega^{\mu+1+\delta} \cdot \xi = \omega^{\mu+1} \cdot \omega^\delta \cdot \xi$ for any ordinal $\xi \neq 0$ and obtain by induction hypothesis

$$\mathbb{L}^{\omega}_{\varphi_1(\omega^{\nu}\cdot\xi)+1} \vdash Acc^{1}_{\prec}(c) \wedge c \prec a \wedge Acc^{\omega^{\nu}\cdot\xi}_{\prec}(b) \to Acc^{\omega^{\nu}\cdot\xi}_{\prec}(\langle 2,c,b\rangle).$$

By Lemma 10.46 we have

$$\mathbb{L}^{\omega}_{\omega+1} \vdash Acc^{1}_{<}(a) \wedge c < a \rightarrow Acc^{1}_{<}(c)$$

and obtain by cut

$$\mathbb{L}^{\omega}_{<\varphi_1(\omega^{\nu}\cdot\xi)+\omega} \vdash Acc^{1}_{<}(a) \wedge c < a \wedge Acc^{\omega^{\nu}\cdot\xi}_{<}(b) \rightarrow Acc^{\omega^{\nu}\cdot\xi}_{<}(\langle 2, c, b\rangle)$$

which is

$$\mathbb{L}^{\omega}_{<\varphi_1(\omega^{\nu}\cdot\xi)+\omega} \vdash Acc^{1}_{<}(a) \wedge b \in \mathsf{OT} \wedge c < a \rightarrow b \in \mathcal{V}_{\omega^{\nu}\cdot\xi}(c).$$

for all parameters b in L_{ω}. By an inference $\bigwedge$ this entails

$$\mathbb{L}^{\omega}_{\varphi_1(\omega^{\nu}\cdot\xi)+\omega} \vdash Acc^{1}_{<}(a) \rightarrow (\forall y)[y \in \mathsf{OT} \wedge c < a \rightarrow y \in \mathcal{V}_{\omega^{\nu}\cdot\xi}(c)].$$

Since this holds true for all $c \in L_{\omega}$ we obtain by an inference $\bigwedge$

$$\mathbb{L}^{\omega}_{\varphi_1(\omega^{\nu}\cdot\xi+1)} \vdash Acc^{1}_{<}(a) \rightarrow (\forall x)(\forall y)[y \in \mathsf{OT} \wedge x < a \rightarrow y \in \mathcal{V}_{\omega^{\nu}\cdot\xi}(x)],$$

i.e.,

$$\mathbb{L}^{\omega}_{\varphi_1(\omega^{\nu}\cdot\xi+1)} \vdash Acc^{1}_{<}(a) \rightarrow A_{\omega^{\nu}\cdot\xi}(a).$$

By Lemma 10.66 we have

$$\mathbb{L}^{\omega}_{\varphi_0(\omega^{\nu}\cdot\xi+1)} \vdash A_{\omega^{\nu}\cdot\xi}(a) \wedge Acc^{\omega^{\nu}\cdot\xi+1}_{<}(b) \rightarrow Acc^{\omega^{\nu}\cdot\xi}_{<}(\langle 2, a, b\rangle).$$

By cut we get

$$\mathbb{L}^{\omega}_{\varphi_1(\omega^{\nu}\cdot\xi+1)} \vdash Acc^{1}_{<}(a) \wedge Acc^{\omega^{\nu}\cdot\xi+1}_{<}(b) \rightarrow Acc^{\omega^{\nu}\cdot\xi}_{<}(\langle 2, a, b\rangle).$$

For $\xi < \omega \cdot \eta$ we obtain $\omega^{\nu} \cdot \xi + 1 < \omega^{\nu+1} \cdot \eta$ and thus by upwards persistency

$$\mathbb{L}^{\omega}_{\varphi_1(\omega^{\nu}\cdot\xi+1)} \vdash Acc^{1}_{<}(a) \wedge Acc^{\omega^{\nu+1}\cdot\eta}_{<}(b) \rightarrow Acc^{\omega^{\nu}\cdot\xi}_{<}(\langle 2, a, b\rangle)$$

for all $\xi < \omega \cdot \eta$. Assume $\sigma < \omega^{\nu+1} \cdot \eta$. Dividing σ by ω^{ν} yields $\sigma = \omega^{\nu} \cdot \eta_0 + \delta$ for $\delta < \omega^{\nu}$. Hence $\sigma < \omega^{\nu} \cdot (\eta_0 + 1)$. From $\omega^{\nu} \cdot \eta_0 \le \sigma < \omega^{\nu} \cdot (\omega \cdot \eta)$ we get $\eta_0 + 1 < \omega \cdot \eta$. Hence $\omega^{\nu} \cdot \eta_0 \le \sigma < \omega^{\nu} \cdot (\eta_0 + 1) < \omega^{\nu+1} \cdot \eta$ which shows that the sequence $\langle \omega^{\nu} \cdot \xi \,|\, 0 \neq \xi < \omega \cdot \eta \rangle$ is cofinal in $\omega^{\nu+1} \cdot \eta$. Using Lemma 10.22 we finally obtain

$$\mathbb{L}^{\omega}_{\varphi_1(\omega^{\nu+1}\cdot\eta)+1} \vdash Acc^{1}_{<}(a) \wedge Acc^{\omega^{\nu+1}\cdot\eta}_{<}(b) \rightarrow Acc^{\omega^{\nu+1}\cdot\eta}_{<}(\langle 2, a, b\rangle). \qquad \square$$

Remark 10.68. Observe that Lemmata 10.66 and 10.67 are the crucial items of the attainability proof. Lemma 10.66 forces us to raise the levels of the class terms used in the proof while Lemma 10.67 uses the full infiniteness strength of the semi-formal system as it is manifested in the Lifting Lemma (Lemma 10.22).

10.7.4.2 The Bootstrap Procedure

If z is a term in OT such that $otp(z) = \zeta$, let us denote z by $\ulcorner\zeta\urcorner$. This not only improves reading but also shortens the notations. By the attainability theorem (Corollary 10.50) we then obtain that $\mathbb{L}^{\omega}_{\alpha} \vdash Acc^{\varsigma}_{\prec}(\ulcorner\gamma\urcorner)$ for $\alpha \geq \omega^{\varsigma+1}$ ensures the full attainability of γ.

Let

$$z_0 := \langle 2, \langle 2, \langle 0\rangle, \langle 0\rangle\rangle, \langle 0\rangle\rangle,$$
$$z_{n+1} := \langle 2, z_n, \langle 0\rangle\rangle.$$

Then we obtain $L_\omega \models z_n \in \mathsf{OT}$ for all $n < \omega$.

Putting

$$\zeta_0 := \overline{\varphi}_1(0) = \varphi_1(0) = \varepsilon_0$$

and

$$\zeta_{n+1} = \varphi_{\zeta_n}(0) = \overline{\varphi}_{\zeta_n}(0)$$

we obtain $\ulcorner\zeta_n\urcorner = z_n$ and $\sup_{n\in\omega} \zeta_n = \Gamma_0$.

Lemma 10.69. *The ordinal ω^ω is fully attainable.*

Proof. We trivially have $\mathbb{L}^{\omega}_{\omega} \vdash Acc^{\varsigma}_{\prec}(\ulcorner 0\urcorner)$ for any ordinal ς. It is $\ulcorner 1\urcorner = \langle 2, \langle 0\rangle, \langle 0\rangle\rangle$. Since $deg_\omega(Acc^1_{\prec}(\langle 0\rangle)) = \omega$ we obtain from Corollary 10.57 $\mathbb{L}^{\omega}_{\omega+1} \vdash Acc^1_{\prec}(\ulcorner 1\urcorner)$ and by twofold iteration also $\mathbb{L}^{\omega}_{\omega+1} \vdash Acc^1_{\prec}(\ulcorner\omega^\omega\urcorner)$. Since we postulated ω to be fully attainable we get by Corollary 10.50 that ω^ω is fully attainable. □

Theorem 10.70. *The ordinal ε_0 is fully attainable.*

Proof. >From Lemma 10.61 we obtain

$$\mathbb{L}^{\omega}_{\omega^3} \vdash Acc^1_{\prec}(\ulcorner\varepsilon_0\urcorner). \tag{10.6}$$

Since ω^ω is fully attainable by Lemma 10.69, Corollary 10.50 shows that ε_0 is fully attainable. □

Lemma 10.71. *For all n we have*

$$\mathbb{L}^{\omega}_{\varphi_1(\zeta_n\cdot\omega)+1} \vdash Acc^1_{\prec}(z_n) \to Acc^{\zeta_n\cdot\omega}_{\prec}(z_{n+1}).$$

Proof. Since $L_\omega \models z_n \in \mathsf{OT}$ we have by Lemma 10.67

$$\mathbb{L}^{\omega}_{\varphi_1(\zeta_n\cdot\omega)+1} \vdash Acc^1_{\prec}(z_n) \wedge Acc^{\zeta_n\cdot\omega}_{\prec}(\langle 0\rangle) \to Acc^{\zeta_n\cdot\omega}_{\prec}(\langle 2, z_n, \langle 0\rangle\rangle).$$

From

$$\mathbb{L}^{\omega}_{\omega} \vdash Acc^{\zeta_n\cdot\omega}_{\prec}(\langle 0\rangle)$$

we obtain by cut

$$\mathbb{L}^{\omega}_{\varphi_1(\zeta_n\cdot\omega)+1} \vdash Acc^1_{\prec}(z_n) \to Acc^{\zeta_n\cdot\omega}_{\prec}(z_{n+1}).$$

□

Corollary 10.72. *For all n we have*

$$\mathbb{L}^{\omega}_{\varphi_1(\zeta_n\cdot\omega)+1} \vdash Acc^{\zeta_n\cdot\omega}_{<}(z_{n+1}).$$

Proof. This follows from Theorem 10.70, Lemma 10.20 and Lemma 10.71 by induction on n. □

Lemma 10.73. *All ordinals $\varphi_1(\zeta_n\cdot\omega)$ are fully attainable.*

Proof. We induct on n. By Theorem 10.70 ε_0, i.e. ζ_0, is fully attainable. Since all the ordinals ζ_n are ε-numbers we have

$$\zeta_n\cdot\omega = \omega^{\zeta_n}\cdot\omega = \omega^{\zeta_n+1} = \varphi_0(\zeta_n+1) = \overline{\varphi}_0(\zeta_n)$$

and thus

$$\varphi_1(\zeta_n\cdot\omega) = \overline{\varphi}_1(\overline{\varphi}_0(\zeta_n)).$$

Hence

$$\ulcorner\varphi_1(\zeta_n\cdot\omega)\urcorner = \langle 2,\langle 2,\langle 0\rangle,\langle 0\rangle\rangle,\langle 2,\langle 0\rangle,\ulcorner\zeta_n\urcorner\rangle\rangle.$$

By Corollary 10.61 we have

$$\mathbb{L}^{\omega}_{\omega^3} \vdash Acc^{1}_{<}(\ulcorner\zeta_0\urcorner).$$

By Lemmata 10.57 and 10.60 we thus get

$$\mathbb{L}^{\omega}_{\omega^3} \vdash Acc^{1}_{<}(\ulcorner\varphi_1(\zeta_0\cdot\omega)\urcorner)$$

which entails by Corollary 10.50 that $\varphi_1(\zeta_0\cdot\omega)$ is fully attainable.

For the induction step we have by Lemma 10.71

$$\mathbb{L}^{\omega}_{\varphi_1(\zeta_n\cdot\omega)+1} \vdash Acc^{\zeta_n\cdot\omega}_{<}(\ulcorner\zeta_{n+1}\urcorner).$$

By Lemmata 10.57 and 10.60 we thus obtain

$$\mathbb{L}^{\omega}_{\varphi_1(\zeta_n\cdot\omega)+\omega} \vdash Acc^{\zeta_n\cdot\omega}_{<}(\ulcorner\overline{\varphi}_1(\overline{\varphi}_0(\zeta_{n+1}))\urcorner).$$

Since $\varphi_1(\zeta_n\cdot\omega)$ is fully attainable by induction hypothesis, hence $\varphi_1(\zeta_n\cdot\omega)+\omega$ preattainable by Lemma 10.38, we obtain by Corollary 10.50 the full attainability of $\varphi_1(\zeta_{n+1}\cdot\omega)$. □

Because $\langle\varphi_1(\zeta_n\cdot\omega)\,|\,n\in\omega\rangle$ is cofinal in Γ_0 the main result of this section, Theorem 10.37, is an immediate corollary of Lemma 10.73.

10.8 Conclusion

We have revisited the Schütte-Feferman result on the limits of predicativity from a set theoretical angle and have shown that all the initial segments L_η for $\eta < \Gamma_0$ of the constructible sets are predicatively justifiable. The methods of our study differ essentially from the methods used by Feferman and also slightly from the methods used by Schütte. We tried to follow Gödel's suggestion and to argue from a purely mathematical standpoint. So—in contrast to Feferman's approach—we completely neglected the philosophical aspects of predicativity and its impact on the foundations of mathematics. Schütte's approach seems to be be in a similar spirit. He, too, hardly responds to the philosophical aspects of predicativity and its influence on the foundations of mathematics. In the proof of the preattainability of Γ_0 he uses in [20] a formal system for ramified analysis.[12] So our approach differs at least technically from Schütte's approach. We belief, however, that from a purely mathematical standpoint the use of semi-formal system is the most canonical approach. The system $\mathbb{L}^\sigma$ uses only pure logic above L_σ and needs no artificial extra axioms. It takes its power solely from its infinite inference rules which are permissible as soon as their heights and cut ranks are predicatively justified.

The studies in the present paper lead to the suspicion that the role of Γ_0 as the limit of predicativity might not depend on deep philosophical but rather on technical reasons.[13]

The reason for Γ_0 to be unattainable is the fact that cut elimination for the systems $\mathbb{L}^\sigma$ (and all similar "predicative" systems) is ruled by the Veblen functions.

The critical lemma in the proof of the preattainability of Γ_0 is Lemma 10.67. Initially, it is based on properties of the Veblen functions φ_α which already have been studied by O. Veblen in [24]. The main observation in this regard is Theorem 10.40 which opens the possibility to define a notation system for the ordinals below Γ_0 already in L_ω.[14] Except in Theorem 10.49 we never used the strength of the segments L_σ for $\sigma > \omega$. On the other hand it follows from the Boundedness Theorem (Theorem 10.34) that Γ_0 would be the upper bound even for well-orderings definable in L_σ for $\omega < \sigma < \Gamma_0$. So we cannot gain strength by looking for well-orderings whose definitions need segments of the constructible sets that are larger than L_ω but still below L_{Γ_0}.

---

[12] Schütte has a different terminology. He clearly does not talk about attainability and his notion for Γ_0 is $\varkappa_0$. However, the identification of Γ_0 with the limit of predicativity is nowadays so common, that we did not want to change it.

[13] However, Solomon Feferman in a few papers (cf. e.g. [5] and [2] and the references there) introduced formal systems for predicativity based on more philosophical considerations. These systems have proof-theoretic ordinal Γ_0 or are reducible to systems with proof-theoretic ordinal Γ_0 (cf. the references in [2]). Whether this sheds another light on the role of Γ_0 as the limit of predicativity was not the concern of this paper.

[14] However, it follows from Veblen's results that it is possible to define much farer reaching notation systems in a recursive way.

The further basis of Lemma 10.67 is the interaction of the order-type of an element in the notation system with the ramification level and the height of the infinitary derivation. This, too, seems to be of technical nature.

Starting from the assumption that ω is fully attainable we could show that Γ_0 is preattainable but not attainable. From a purely mathematical standpoint it is now interesting to ask how far we can go by starting from the assumption that Γ_0 is fully attainable and then working in $\mathbb{L}^{\Gamma_0}$. This will certainly secure the "preattainability" of the next strongly critical ordinal Γ_1. Iterating this procedure would probably secure the preattainability of all the strongly critical ordinals Γ_n for $n \in \omega$. An approach in this vein is the topic of a forthcoming paper [16].

Acknowledgements I am deeply indebted to the anonymous referee for not only detecting many typos as well as mathematical and linguistic errors but also making suggestions for improvements.

10.9 Erratum to "Semi-Formal Calculi and Their Applications" in [15]

Theorem 5.19 in [15] is false in the there claimed generality. There are counterexamples which are countable elementary Skolem substructures of uncountable structures in which well–foundedness is elementarily definable.[15]

Although these structures are acceptable the search tree $\mathsf{S}^{\mathsf{L}(\mathfrak{M})}_{F(X)}$ for an $\mathsf{L}(\mathfrak{M})$-formula $F(X)$ is in general not elementarily definable.

As a remedy we call an acceptable structure $\mathfrak{M}$ *strictly acceptable* if there is an elementarily definable function $f\colon \mathcal{N} \xrightarrow{onto} |\mathfrak{M}|$, where $\mathcal{N}$ stands for the the copy of the natural numbers in $\mathfrak{M}$. Since $\mathfrak{M}$ is acceptable we can assign a Gödel number $\ulcorner E \urcorner$ to every expression E in the elementary language $\mathsf{L}(\mathfrak{M})$ of $\mathfrak{M}$ and extend the semi-decorated language $\mathsf{L}(\mathfrak{M})$ to a fully decorated language $\mathsf{L}(\mathfrak{M})_{\mathfrak{M}}$ in which every characteristic sequence is elementarily definable. E.g. we put $\mathsf{CS}((\mathsf{Q}x)F(x)) := \langle \ulcorner F(z) \urcorner \mid n \in \mathcal{N} \wedge f(n) = z \rangle$ for first order quantifiers Q. Using the elementary definability of the characteristic sequences we easily obtain the next lemma by recursion on $\mathcal{N}$.

Lemma 10.74. *Let $\mathfrak{M}$ be a strictly acceptable structure. Then for any finite set Δ of $\mathsf{L}(\mathfrak{M})_{\mathfrak{M}}$-formulas the search tree $\mathsf{S}^{\mathsf{L}(\mathfrak{M})}_{\Delta}$ is elementarily definable.*

As a consequence of Lemma 10.74 we obtain a modified form of Theorem 5.19 in [15].

Theorem 10.75. *Let $\mathfrak{M}$ be a strictly acceptable structure. Then $\delta^1_0(\mathfrak{M}) = \pi^{\mathfrak{M}}$.*

According to Exercise VI 5.17 in [1] no nonstandard model of arithmetic is acceptable. Therefore the structure $\mathfrak{M}$ constructed in in [13] Sec.8D has to be an

[15] Cf. [13] Sec. 8D.

acceptable standard structure. According to Theorem 10.75 it cannot be a strictly acceptable structure. The obvious question to ask is if there are acceptable structures $\mathfrak{M}$ satisfying $\delta_0^1(\mathfrak{M}) = \pi^{\mathfrak{M}}$ which are not strictly acceptable?

The alteration of Theorem 5.19 also effects Corollary 5.21. The equation $\delta_0^1(\mathfrak{M}) = \pi^{\mathfrak{M}} = O(\mathfrak{M})$ only holds true for strictly acceptable structures. The equation $\delta_1^1(\mathfrak{M}) = \sigma_1^1(\mathfrak{M}) = \pi^{\mathfrak{M}} = O(\mathfrak{M})$ remains true for countable acceptable structures.

References

1. Jon Barwise. *Admissible Sets and Structures*. Perspectives in Mathematical Logic. Springer-Verlag, Berlin/Heidelberg/New York, 1975.
2. Laura Crosilla. Predicativity and Feferman. In Gerhard Jäger and Wilfried Sieg, editors, *Feferman on Foundations*, Outstanding Contributions to Logic, pages 423–447. Springer-Verlag, Heidelberg, Berlin, 2017.
3. Keith J. Devlin. *Constructibility*. Perspectives in Mathematical Logic. Springer-Verlag, Heidelberg/New York, 1984.
4. Solomon Feferman. Systems of predicative analysis. *Journal of Symbolic Logic*, 29:1–30, 1964.
5. Solomon Feferman. A more perspicuous system for predicativity. volume 1: Spezielle Wissenschaftstheorie, pages 58–93. De Gruyter, Berlin, 1979.
6. Solomon Feferman. Predicativity. In Steward Shapiro, editor, *The Oxford Handbook of Philosophy of Mathematics and Logic*, Oxford Handbooks in Philosophy, pages 590–624. Oxford University Press, Oxford, 2007.
7. Solomon Feferman, John Dawson, Warren Goldfarb, Charles Parsons, and Robert Solovay, editors. *Kurt Gödel Collected Works*, volume III, Oxford, New York, 1995. Oxford University Press.
8. Kurt Gödel. The consistency of the axiom of choice and of the generalized continuum-hypothesis. *Proceedingss of the National Academy of Sciences of the U.S.A.*, 24:556–557, 1938.
9. Kurt Gödel. The philosophy of Bertrand Russell. In Paul Arthur Schilpp, editor, *The Philosophy of Bertrand Russell*, pages 125–153. Northwestern University, Evanstown, 1944.
10. Kurt Gödel. Vortrag Göttingen. In Feferman et al. [7], pages 126–134.
11. Georg Kreisel. La prédicativité. *Bulletin de la Societe Mathematique de France*, 88:371–391, 1960.
12. Georg Kreisel. Set theoretic problems suggested by the notion of potential totality. In *Proceedings of the Symposium on Foundations of Mathematics*, pages 103–140. Pergamon Press, Oxford, 1961.
13. Yiannis N. Moschovakis. *Elementary Induction on Abstract Structures*, volume 77 of *Studies in Logic and the Foundations of Mathematics*. North-Holland Publishing Company, Amsterdam, 1974.
14. Wolfram Pohlers. *Proof Theory. The first step into impredicativity*. Universitext. Springer-Verlag, Berlin/Heidelberg/New York, 2009.
15. Wolfram Pohlers. Semi–formal calculi and their applications. In Reinhard Kahle and Michael Rathjen, editors, *Gentzen's Centenary: The quest for consistency*, pages 317–354. Springer-Verlag, 2015.
16. Wolfram Pohlers. A note on the limits of predicativity. *submitted*, page 31, 2018.
17. Henri Poincaré. Über transfinite Zahlen. In Henri Poincaré, editor, *Sechs Vorträge über ausgewählte Gegenstände aus der reinen Mathematik und mathematischen Physik (5)*, pages 45–48. Teubner, 1910.

18. Bertrand Russell. On some difficulties in the theory of tranfinite numbers and order types. *Proceedings of the London Mathematical Society*, 4:135–164, 1906.
19. Kurt Schütte. Eine Grenze für die Beweisbarkeit der transfiniten Induktion in der verzweigten Typenlogik. *Archiv für Mathematische Logik und Grundlagenforschung*, 7:45–60, 1965.
20. Kurt Schütte. Predicative well-orderings. In John N. Crossley and Michael A. E. Dummett, editors, *Formal Systems and Recursive Functions*, Studies in Logic and the Foundations of Mathematics, pages 280–303. North-Holland Publishing Company, Amsterdam, July 1965.
21. Kurt Schütte. *Proof Theory*. Number 225 in Grundlehren der mathematischen Wissenschaften. Springer-Verlag, Heidelberg/New York, 1977.
22. Robert Solovay. Introductory note to Gödel 1939b and 1939a. In Feferman et al. [7], pages 114–126.
23. William W. Tait. Normal derivability in classical logic. In Jon Barwise, editor, *The Syntax and Semantics of Infinitary Languages*, number 72 in Lecture Notes in Mathematics, pages 204–236. Springer-Verlag, Heidelberg/New York, December 1968.
24. Oswald Veblen. Continuous increasing functions of finite and transfinite ordinals. *Transactions of the American Mathematical Society*, 9:280–292, 1908.
25. Hermann Weyl. *Das Kontinuum*. Veit & Co., Leipzig, 1918.
26. Hermann Weyl. Der Circulus vitiosus in der heutigen Begründung der Mathematik. *Jahresbericht der Deutschen Mathematiker-Vereinigung*, 28:85–92, 1919.
27. Hermann Weyl. Über die neue Grundlagenkrise der Mathematik. *Mathematische Zeitschrift*, 10:39–79, 1921.
28. Alfred North Whitehead and Bertrand Russell. *Principia Mathematica, Vol.1 2nd edition*. Cambridge University Press, Cambridge, 1925.

Chapter 11
A Note on (Meta)predicative Wellordering Proofs

Wilfried Buchholz

Abstract We prove a streamlined and slightly more general version of a lemma due to Schütte [3] and Feferman [1] which is crucial for (meta)predicative wellordering proofs.

11.1 Introduction

In [1] a certain lemma 5.3.1 is proved which is crucial for wellordering proofs below Γ_0. The lemma deals with predicates $I^c(\gamma) := \forall X^c \mathrm{TI}(\gamma, X^c)$ of relativized transfinite induction. It is mentioned in [1, p. 37] that for the proof of 5.3.1 one does not need this explicit definition of I^c, but it rather suffices to assume that the I^c's satisfy certain 1st order axioms which are labeled (A1)–(A4) in the present note.

With some notice one can see that Lemma 5.3.1 corresponds to Schütte's Lemma 4 in [3]. Moreover, from Schütte's proof of that lemma, which among other things is distinguished by not using fundamental sequences, one can extract a particularly neat and perspicuous proof of 5.3.1. This is the main content of the present note. Actually we prove a slightly more general version of the lemma, where the standard binary Veblen function φ is replaced by an arbitrary binary Veblen function ϕ. A consequence of this generalization is that the lemma becomes applicable also in meta-predicative contexts like e.g. [2, p. 59] (cf. Lemma 11.11 below).

W. Buchholz
Mathematisches Institut, LMU München, Germany, e-mail: buchholz@mathematik.uni-muenchen.de

R. Kahle and M. Rathjen (eds.), *The Legacy of Kurt Schütte*,
https://doi.org/10.1007/978-3-030-49424-7_11

11.2 Ordinal Theoretic Preliminaries

Definition 11.1. A function $u : On \times On \to On$ is called a *Veblen function*, if $\lambda\beta.u(0,\beta)$ is a normal function with $u(0,0) = 1$, and for each $\alpha > 0$, the function $\lambda\beta.u(\alpha,\beta)$ enumerates the class $\{\beta : \forall\xi < \alpha(u(\xi,\beta) = \beta)\}$.

By φ we denote the standard Veblen function which is determined by $\varphi 0\beta = \omega^\beta$. Further, Φ denotes the Veblen function determined by $\Phi 0\beta = \varphi\beta 0$.

Lemma 11.2. *If u is a Veblen function and $0 < \gamma < u(\alpha,\beta)$ then one of the following cases holds:*

(i) $u(0,\gamma_0) < \gamma < u(0,\gamma_0{+}1)$ *for some* γ_0*;*
(ii) $\gamma = u(\gamma_0,\gamma_1)$ *with* $\gamma_0 < \alpha$ & $\gamma_1 < \gamma$*;*
(iii) $\gamma = u(\alpha,\gamma_1)$ *with* $\gamma_1 < \beta$.

Assumptions. 1. $\prec$ is a primitive recursive well-ordering of $\mathbb{N}$, with first elements 0,1.
For each $a \in \mathbb{N}$ let $|a| \in On$ denote the order-type of $(\{m \in \mathbb{N} : m \prec a\}, \prec)$.
2. $\oplus, f_0, f_1$ are primitive recursive functions such that for all $a, b \in \mathbb{N}$, $|a \oplus b| = |a| + |b|,\ |f_0(a,b)| = \varphi(|a|,|b|),\ |f_1(a,b)| = \Phi(|a|,|b|)$.
3. ϕ is a primitive recursive function corresponding to a Veblen function u, so that $|\phi ab| = u(|a|,|b|)$ for all $a, b \in \mathbb{N}$.
4. Provably in arithmetic the following holds for $(\prec, \phi)$:

(ϕ1) $b \preceq \phi ab$
(ϕ2) If $0 \prec c \prec \phi ab$ then one of the following cases holds:
(i) $\exists c_0(\phi 0 c_0 \prec c \prec \phi 0(c_0 \oplus 1)\ \&\ \phi 0 c_0 <_{\mathbb{N}} c)$.
(ii) $\exists c_0, c_1(c = \phi c_0 c_1$ with $c_0 \prec a\ \&\ c_1 \prec c\ \&\ c_1 <_{\mathbb{N}} c)$.
(iii) $\exists c_1(c = \phi a c_1\ \&\ c_1 \prec b)$.

From now on we write φab (Φab, resp.) for $f_0(a,b)$ ($f_1(a,b)$, resp.)

11.3 The Theories T^ν

Let Z be the system of first order arithmetic formulated in a language $\mathcal{L}$ with symbols for all primitive recursive relations and functions, and with free predicate variables $X, Y, \ldots$.

Abbreviations. $\mathrm{FP}(X) := \forall x(\forall y \prec x X(y) \leftrightarrow X(x))$,
$\mathrm{Prog}(X) := \forall x(\forall y \prec x X(y) \to X(x))$,
$\mathrm{TI}(b,X) := \mathrm{Prog}(X) \to \forall y \preceq b X(y)$,
$\mathrm{Prog}^\phi_a(X) := \mathrm{Prog}(\lambda y.X(\phi a y))$,
$C^\phi_a(X) := \forall y(X(y) \to X(\phi a y))$

The following lemma comprises all we need to know about ϕ.

Lemma 11.3. $\mathsf{Z} \vdash \mathrm{FP}(X) \wedge C_0^\phi(X) \wedge \forall x \prec a\, C_x^\phi(X) \;\rightarrow\; \mathrm{Prog}_a^\phi(X)$.

Proof. Assuming (1) $\mathrm{FP}(X)$, (2) $C_0^\phi(X)$, (3) $\forall x \prec a\, C_x^\phi(X)$, (4) $\forall y \prec b X(\phi a y)$ we have to show $X(\phi ab)$.
For this we prove $\forall z \prec \phi ab\, X(z)$ by $<_{\mathbb{N}}$-induction on z. Then (1) yields $X(\phi ab)$.
Assume $z \prec \phi ab$. Then one of the following cases holds

0. $z = 0$: $\forall y \prec 0 X(y) \overset{(1)}{\Rightarrow} X(0)$.
1. $\phi 0 z_0 \prec z \prec \phi 0(z_0 \oplus 1)$ and $\phi 0 z_0 <_{\mathbb{N}} z$:

IH $\Rightarrow$ $X(\phi 0 z_0) \overset{(1)}{\Rightarrow} X(z_0 \oplus 1) \overset{(2)}{\Rightarrow} X(\phi 0(z_0 \oplus 1)) \overset{(1)}{\Rightarrow} X(z)$.

2. $z = \phi z_0 z_1$ with $z_0 \prec a$ & $z_1 \prec z$ & $z_1 <_{\mathbb{N}} z$: IH $\Rightarrow$ $X(z_1) \overset{(3)}{\Rightarrow} X(\phi z_0 z_1)$.
3. $z = \phi a z_1$ & $z_1 \prec b$: (4) $\Rightarrow X(\phi a z_1)$. □

Definition 11.4. Let I be a new binary predicate constant. The *theory* T^ν is formulated in the language $\mathcal{L}[I] := \mathcal{L} \cup \{I\}$. It is obtained from first order arithmetic (with induction for all $\mathcal{L}[I]$-formulas) by adding the following axioms, where $I^s(t)$ stands for $I(s,t)$.

(A1) $\forall c \preceq \nu \mathrm{FP}(I^c)$;
(A2) $\forall d \prec \nu \forall b (I^{d \oplus 1}(b) \rightarrow \mathrm{TI}(b, \lambda y. B(I^d, y)))$, for each $\mathcal{L}$-formula $B(X,y)$;
(A3) $\forall c \preceq \nu \forall d \prec c \forall x (I^c(x) \rightarrow I^d(x))$;
(A4) $\forall c \preceq \nu \forall x (Lim(c) \wedge \forall d \prec c I^d(x) \rightarrow I^c(x))$.

Abbreviations. $d \in [\omega^a] := \exists x (0 \prec d = \omega^a \cdot x)$, $\quad d \in [\omega^a]^{\prec c} := d \in [\omega^a] \wedge d \prec c$.
$\mathsf{T}^\nu[\phi] := \mathsf{T}^\nu + \mathcal{A}^\phi$, where $\mathcal{A}^\phi := \forall c \in [\omega^1]^{\preceq \nu} C_0^\phi(I^c)$.

Remark. $Lim(d) \Leftrightarrow d \in [\omega^1]$.

Lemma 11.5. (a) $\mathsf{T}^\nu[\phi] \vdash d \in [\omega^1]^{\preceq \nu} \wedge \forall x \prec a\, C_x^\phi(I^d) \rightarrow \mathrm{Prog}_a^\phi(I^d)$.
(b) $\mathsf{T}^\nu \vdash d \prec \nu \wedge \mathrm{Prog}_a^\phi(I^d) \rightarrow \forall y (I^{d \oplus 1}(y) \rightarrow I^d(\phi a y))$.
(c) $\mathsf{T}^\nu \vdash \forall d \in [\omega^a]^{\prec \nu} \forall y (I^{d \oplus 1}(y) \rightarrow I^d(\phi a y)) \rightarrow \forall c \in [\omega^{a \oplus 1}]^{\preceq \nu} C_a^\phi(I^c)$.

Proof. (a) follows immediately from lemma 11.3 and (A1).
(b) is an instance of (A2).
(c) Assume (1) $\forall d \in [\omega^a]^{\prec \nu} \forall y (I^{d \oplus 1}(y) \rightarrow I^d(\phi a y))$ and (2) $c \in [\omega^{a \oplus 1}]^{\preceq \nu}$.
From (2) we obtain (3) $\forall e \prec c \exists d \in [\omega^a]^{\prec c}(e \prec d)$. Then we can proceed as follows:
$I^c(y) \overset{(1),(A3)}{\Rightarrow} \forall d \in [\omega^a]^{\prec c} I^d(\phi a y) \overset{(3),(A3)}{\Rightarrow} \forall e \prec c I^e(\phi a y) \overset{(A4)}{\Rightarrow} I^c(\phi a y)$. □

Remark 11.6. Lemma 11.5a,b corresponds to [3, Lemma 2].
Lemma 11.5c corresponds to [3, Lemma 3].

Abbreviation. $Q_\nu^\phi(x) := \forall c \in [\omega^{x \oplus 1}]^{\preceq \nu}\, C_x^\phi(I^c)$.

Lemma 11.7 ([1, Lemma 5.3.1])**.** $\mathsf{T}^\nu[\phi] \vdash \mathrm{Prog}(Q_\nu^\phi)$.

Proof. Assume $\forall x \prec a\, Q_\nu^\phi(x)$. To prove: $Q_\nu^\phi(a)$.
I. $a = 0$: $\mathcal{A}^\phi \equiv \forall c \in [\omega^1]^{\preceq \nu} C_0^\phi(I^c) \equiv Q_\nu^\phi(0)$.
II. $0 \prec a$: $\forall x \prec a Q_\nu^\phi(x) \;\Rightarrow\; \forall d \in [\omega^a]^{\preceq \nu} \forall x \prec a\, C_x^\phi(I^d) \overset{\text{L.11.5a}}{\Rightarrow}$
$\forall d \in [\omega^a]^{\preceq \nu} \mathrm{Prog}_a^\phi(I^d) \overset{\text{L.11.5b,c}}{\Rightarrow} \forall c \in [\omega^{a \oplus 1}]^{\preceq \nu} C_a^\phi(I^c)$, i.e. $Q_\nu^\phi(a)$. □

Lemma 11.8. $\mathsf{T}^\nu \vdash \mathcal{A}^\varphi$ *and thus* $\mathsf{T}^\nu = \mathsf{T}^\nu[\varphi]$.

Proof. Abb.: $\overline{X}(y) := \forall x(X(x) \to X(x \oplus \omega^y))$.

The following is a standard result of proof theory: (G) $\mathsf{Z} \vdash \mathrm{FP}(X) \to \mathrm{Prog}(\overline{X})$. Making use of (G) and (A2) we obtain $\mathcal{A}^\varphi$ as follows:

Assume $c \in [\omega^1]$. $I^c(b) \wedge d \prec c \overset{(A3)}{\Rightarrow} I^{d\oplus 1}(b) \overset{(A2)}{\Rightarrow} \mathrm{TI}(b, \overline{I^d}) \overset{(G),(A1)}{\Rightarrow} \overline{I^d}(b) \Rightarrow I^d(\omega^b)$.

Hence: $I^c(b) \Rightarrow \forall d \prec c I^d(\omega^b) \overset{(A4)}{\Rightarrow} I^c(\omega^b)$. □

Corollary 11.9. $\mathsf{T}^\nu \vdash \mathrm{Prog}(Q_\nu^\varphi)$.

We are now going to show that Lemma 7 of [2] which (as the authors say) "has some similarities with Lemma 5.3.1 of Feferman [6]", actually can be obtained from Lemma 11.7 above by instantiating Φ for ϕ.

Definition 11.10 (cf. [2, p. 59]). $\mathrm{Main}_\nu(x) := \forall y, c[c \preceq \nu \wedge \omega^{1\oplus x}\!\uparrow\! c \wedge I^c(y) \to I^c(\varphi 1xy)]$,

where $I^c(a) := \forall b \prec c \forall f \in S_b \mathrm{TI}(f, a)$, $a\!\uparrow\! c := \exists d, \ell(Lim(\ell) \wedge c = d \oplus a{\cdot}\ell)$, and φ denotes the ternary Veblen function introduced on p. 54 of [2].

Now Lemma 7 of [2] states $\mathsf{SRT}_\nu \vdash \mathrm{Prog}(\mathrm{Main}_\nu)$ where SRT_ν denotes the theory of ν-times iterated self-reflecting truth introduced on p. 57 of [2].

Lemma 11.11. (a) $\mathsf{T}^\nu[\Phi] \vdash \mathrm{Prog}(\mathrm{Main}_\nu)$.
(b) $\mathsf{T}^\nu[\Phi] \subseteq \mathsf{SRT}_\nu$.

Proof. (a) One easily sees that (1) $\omega^a\!\uparrow\! c \leftrightarrow c \in [\omega^{a\oplus 1}]$.
Further we have (2) $\varphi 1xy = \Phi(1 \oplus x)y$.

From (1) and (2) we obtain (3) $\mathrm{Main}_\nu(x) \leftrightarrow Q_\nu^\Phi(1 \oplus x)$.

By Lemma 11.7 we have $\mathsf{T}^\nu[\Phi] \vdash \mathrm{Prog}(Q_\nu^\Phi)$ which yields $\mathsf{T}^\nu[\Phi] \vdash \mathrm{Prog}(\lambda x. Q_\nu^\Phi(1 \oplus x))$.

(b) This inclusion is derived from the above definition of I^c. The fact that SRT_ν proves (A1), (A3), (A4) is easily seen. For (A2) one may lean on (the proof of) Lemma 4 of [2]. Finally, since $\varphi a0 = \Phi 0a$, Lemma 5 of [2] states exactly $\mathsf{SRT}_\nu \vdash \mathcal{A}^\Phi$. □

References

1. Solomon Feferman. Reflecting on incompleteness. *Journal of Symbolic Logic*, 56:1–49, 1991.
2. Gerhard Jäger, Reinhard Kahle, Anton Setzer, and Thomas Strahm. The proof-theoretic analysis of tranfinitely iterated fixed point theories. *Journal of Symbolic Logic*, 64:53–67, 1999.
3. Kurt Schütte. Predicative well-orderings. In *Formal systems and recursive functions*, pages 279–302, Amsterdam, 1965. North-Holland.

Chapter 12
Well-ordering Principles, ω-models and Π^1_1-comprehension

Michael Rathjen and Ian Alexander Thomson

Abstract In this paper we show that over the weak theory $\mathbf{RCA}_0$ the existence of ω-models of Π^1_1-comprehension is equivalent to saying that a certain elementary endomorphism of the category of linear orders preserves well-orderings. This result continues a long line of research. Technically, among other methods, we use Schütte's search trees and a relativized version of the ordinal analysis of Π^1_1-comprehension via the Ω_n-rules akin to the presentation in Buchholz and Schütte's 1988 book [3]. A crucial challenge one faces is to show that the latter approach can be made to work in a weak background theory.

Key words: Reverse mathematics, well ordering principles, Schütte deduction chains, countable coded ω-model, Π^1_1-comprehension, Ω_n-rule, well-ordering proof

Mathematics subject classification: 03B30, 03F05, 03F15, 03F35, 03F35

12.1 Introduction

This paper will be concerned with a particular Π^1_2 statement of the form

(12.1) $$\mathbf{WOP}(f): \quad \forall X\,[\mathrm{WO}(\mathfrak{X}) \rightarrow \mathrm{WO}(f(\mathfrak{X}))]$$

where f is a standard proof-theoretic function from ordinals to ordinals and $\mathrm{WO}(\mathfrak{X})$ stands for '$\mathfrak{X}$ is a well-ordering'. There are by now several examples of functions f familiar from proof theory where the statement $\mathbf{WOP}(f)$ has turned out to be equivalent to one of the theories of reverse mathematics over a weak base theory (usually $\mathbf{RCA}_0$). The first explicit example appears to be due to Girard [7, 5.4.1 theorem] (see

M. Rathjen · I. A. Thomson
School of Mathematics, University of Leeds, Leeds, LS2 9JT, UK, e-mail: M.Rathjen@leeds.ac.uk, i.alecthomson@gmail.com

R. Kahle and M. Rathjen (eds.), *The Legacy of Kurt Schütte*,
https://doi.org/10.1007/978-3-030-49424-7_12

also [8]). However, it is also implicit in Schütte's proof of cut elimination for ω-logic [16] and ultimately has its roots in Gentzen's work, namely in his first unpublished consistency proof,[1] where he introduced the notion of a "Reduziervorschrift" [6, p. 102] for a sequent. The latter is a well-founded tree built bottom-up via "Reduktionsschritte", starting with the given sequent and passing up from conclusions to premises until an axiom is reached.

Theorem 12.1. *Over* $\mathbf{RCA}_0$ *the following are equivalent:*

(i) *Arithmetical comprehension.*
(ii) $\forall \mathfrak{X}\ [\mathrm{WO}(\mathfrak{X}) \to \mathrm{WO}(2^{\mathfrak{X}})]$.

Another characterization from [7], Theorem 6.4.1, shows that arithmetical comprehension is equivalent to Gentzen's Hauptsatz (cut elimination) for ω-logic. Connecting statements of form (12.1) to cut elimination theorems for infinitary logics will also be a major tool in this paper.

There are several more recent examples of such equivalences that have been proved by recursion-theoretic as well as proof-theoretic methods. These results give characterizations of the form (12.1) for the theories $\mathbf{ACA}_0^+$ and $\mathbf{ATR}_0$, respectively, in terms of familiar proof-theoretic functions. $\mathbf{ACA}_0^+$ denotes the theory $\mathbf{ACA}_0$ augmented by an axiom asserting that for any set X the ω-th jump in X exists while $\mathbf{ATR}_0$ asserts the existence of sets constructed by transfinite iterations of arithmetical comprehension. $\alpha \mapsto \varepsilon_\alpha$ denotes the usual ε function while φ stands for the two-place Veblen function familiar from predicative proof theory (cf. [17]). Definitions of the familiar subsystems of reverse mathematics can be found in [18].

Theorem 12.2. (Afshari, Rathjen [1]; Marcone, Montalbán [9]) *Over* $\mathbf{RCA}_0$ *the following are equivalent:*

(i) $\mathbf{ACA}_0^+$.
(ii) $\forall \mathfrak{X}\ [\mathrm{WO}(\mathfrak{X}) \to \mathrm{WO}(\varepsilon_{\mathfrak{X}})]$.

Theorem 12.3. (Friedman [4]; Rathjen, Weiermann [15]; Marcone, Montalbán [9]) *Over* $\mathbf{RCA}_0$ *the following are equivalent:*

(i) $\mathbf{ATR}_0$.
(ii) $\forall \mathfrak{X}\ [\mathrm{WO}(\mathfrak{X}) \to \mathrm{WO}(\varphi \mathfrak{X} 0)]$.

There is often another way of characterizing statements of the form (12.1) by means of the notion of countable coded ω-model.

Definition 12.4. Let T be a theory in the language of second order arithmetic, $\mathcal{L}_2$. A *countable coded ω-model of T* is a set $W \subseteq \mathbb{N}$, viewed as encoding the $\mathcal{L}_2$-model

$$\mathbb{M} = (\mathbb{N}, \mathcal{S}, \in, +, \cdot, 0, 1, <)$$

with $\mathcal{S} = \{(W)_n \mid n \in \mathbb{N}\}$ such that $\mathbb{M} \models T$ when the second order quantifiers are interpreted as ranging over $\mathcal{S}$ and the first order part is interpreted in the standard

[1] The original German version was finally published in 1974 [6]. An earlier English translation appeared in 1969 [5].

way (where $(W)_n = \{m \mid \langle n,m\rangle \in W\}$ with $\langle,\rangle$ being some primitive recursive coding function).

If T has only finitely many axioms, it is obvious how to express $\mathbb{M} \models T$ by just translating the second order quantifiers $QX \ldots X \ldots$ in the axioms by $Qx \ldots (W)_x \ldots$. If T has infinitely many axioms one needs to formalize Tarski's truth definition for $\mathbb{M}$. This definition can be made in $\mathbf{RCA}_0$ as is shown in [18], Definition II.8.3 and Definition VII.2. Some more details will be provided in Remark 12.11.

We write $X \in W$ if $\exists n\; X = (W)_n$.

The alternative characterizations alluded to above are as follows:

Theorem 12.5. *Over* $\mathbf{RCA}_0$ *the following are equivalent:*

(i) $\forall \mathfrak{X}\; [\mathrm{WO}(\mathfrak{X}) \rightarrow \mathrm{WO}(\varepsilon_{\mathfrak{X}})]$ *is equivalent to the statement that every set is contained in a countable coded* ω*-model of* **ACA**.
(ii) $\forall \mathfrak{X}\; [\mathrm{WO}(\mathfrak{X}) \rightarrow \mathrm{WO}(\varphi\mathfrak{X}0)]$ *is equivalent to the statement that every set is contained in a countable coded* ω*-model of* Δ^1_1-**CA** *(or* Σ^1_1-**DC***).*

Proof. See [12, Corollary 1.8]. □

Whereas Theorem 12.5 has been established independently by recursion-theoretic and proof-theoretic methods, there is also a result that has a very involved proof and so far has only been shown by proof theory. It connects the well-known Γ-function (cf. [17]) with the existence of countable coded ω-models of $\mathbf{ATR}_0$.

Theorem 12.6. (Rathjen [12, Theorem 1.4]) *Over* $\mathbf{RCA}_0$ *the following are equivalent:*

(i) $\forall \mathfrak{X}\; [\mathrm{WO}(\mathfrak{X}) \rightarrow \mathrm{WO}(\Gamma_{\mathfrak{X}})]$.
(ii) *Every set is contained in a countable coded* ω*-model of* $\mathbf{ATR}_0$.

The tools from proof theory employed in the above theorems involve search trees and Gentzen's cut elimination technique for infinitary logic with ordinal bounds. One could perhaps generalize and say that every cut elimination theorem in ordinal-theoretic proof theory encapsulates a theorem of this type.

The proof-theoretic ordinal functions that figure in the foregoing theorems are all familiar from so-called predicative or meta-predicative proof theory. The first ordinal representation system that is usually associated with a genuinely impredicative theory is of the Bachmann-Howard type. It was conjectured in [15] (Conjecture 7.2) that the pertaining principle (12.1) would be equivalent to the existence of countable coded ω-models of bar induction, **BI**. The relativization of the Bachmann-Howard construction, however, allows for two different approaches, yielding principles of different strength. As it turned out, only the strongest one is equivalent to the existence of ω-models of of the theory of bar induction, **BI** (cf. Definition 12.10).

Theorem 12.7. (Rathjen, Valencia Vizcaíno [13, 21]) *Over* $\mathbf{RCA}_0$ *the following are equivalent:*

(i) *Every set X is contained in a countable coded ω-model of* **BI**.
(ii) $\forall \mathfrak{X}\ [\mathrm{WO}(\mathfrak{X}) \to \mathrm{WO}(\vartheta_{\mathfrak{X}})]$.

In this paper we venture much further into the impredicative realm in that we investigate a relativized version of the representation system used for the ordinal analysis of the subsystem of second order arithmetic based on Π^1_1-comprehension, $\Omega_\omega \cdot \mathfrak{X}$. Note that Π^1_1-comprehension happens to be the strongest comprehension principle considered in the "big five" classification system of reverse mathematics (see [18]). The **main theorem** of this paper is the following.

Theorem 12.8. *Over* $\mathbf{RCA}_0$ *the following are equivalent:*

(i) *Every set X is contained in a countable coded ω-model of* Π^1_1-$\mathbf{CA}_0$.
(ii) $\forall \mathfrak{X}\ [\mathrm{WO}(\mathfrak{X}) \to \mathrm{WO}(\Omega_\omega \cdot \mathfrak{X})]$.

12.1.1 Subsystems of Second Order Arithmetic

In this subsection we introduce some theories based on the language of second order arithmetic. To set the context, we fix some notations.

Definition 12.9. The language of second order arithmetic, $\mathcal{L}_2$, consists of free numerical variables $a, b, c, d, \ldots$, bound numerical variables $x, y, z, \ldots$, free set variables $U, V, W, \ldots$, bound set variables $X, Y, Z, \ldots$, the constant 0, a symbol for each primitive recursive function, and the symbols $=$ and $\in$ for equality in the first sort and the elementhood relation, respectively. The *numerical terms* of $\mathcal{L}_2$ are built up in the usual way; $r, s, t, \ldots$ are syntactic variables for them. *Formulas* are obtained from atomic formulas $s = t$, $s \in U$ and negated atomic formulas $\neg\, s = t, \neg\, s \in U$ by closing under $\wedge, \vee$ and quantification $\forall x, \exists x, \forall X, \exists X$ over both sorts; so we stipulate that formulas are in negation normal form.

The classes of Π^1_2– and Σ^1_n*–formulae* are defined as usual (with $\Pi^1_0 = \Sigma^1_0 = \cup\{\Pi^0_n : n \in \mathbb{N}\}$). $\neg A$ is defined by de Morgan's laws; $A \to B$ stands for $\neg A \vee B$. All theories in $\mathcal{L}_2$ will be assumed to contain the axioms and rules of classical two sorted predicate calculus, with equality in the first sort. In addition, it will be assumed that they comprise the system $\mathbf{RCA}_0$. $\mathbf{RCA}_0$ contains all axioms of elementary number theory, i.e. the usual axioms for 0, $'$ (successor), the defining equations for the primitive recursive functions, the axiom schema of induction for Σ^0_1-formulae

$$F(0) \wedge \forall x\, [F(x) \to F(x+1)] \to \forall x\, F(x)$$

for Σ^0_1-formulae $F(a)$ (where $x+1$ stands for x'), and the Δ^0_1-comprehension scheme, i.e.,

$$\forall x\, [G(x) \leftrightarrow H(x)] \to \exists Y\, \forall x [x \in Y \leftrightarrow G(x)]$$

where $G(a)$ is a Σ^0_1-formula and $H(a)$ is a Π^0_1-formula.

Another important basic system is $\mathbf{ACA}_0$. In addition to the axioms of $\mathbf{RCA}_0$ it has all instances of *arithmetical comprehension* as axioms, i.e.,

$$\exists Z \, \forall x[x \in Z \leftrightarrow F(x)],$$

where $F(a)$ is an *arithmetic formula*, i.e., a formula without set quantifiers.

The theory Π^1_1-$\mathbf{CA}_0$ further extends $\mathbf{ACA}_0$ by allowing $F(a)$ to be of complexity Π^1_1 in the foregoing comprehension scheme.

Definition 12.10. For a 2-place relation $\prec$ and an arbitrary formula $F(a)$ of $\mathcal{L}_2$ we define

$\mathrm{Prog}(\prec, F) := (\forall x)[\forall y(y \prec x \rightarrow F(y)) \rightarrow F(x)]$ *(progressiveness)*
$\mathbf{TI}(\prec, F) := \mathrm{Prog}(\prec, F) \rightarrow \forall x F(x)$ *(transfinite induction)*
$\mathrm{WF}(\prec) := \forall X \mathbf{TI}(\prec, X) :=$
$\forall X(\forall x[\forall y(y \prec x \rightarrow y \in X)) \rightarrow x \in X] \rightarrow \forall x[x \in X])$ *(well-foundedness)*.

Let $\mathcal{F}$ be any collection of formulae of $\mathcal{L}_2$. For a 2-place relation $\prec$ we will write $\prec \in \mathcal{F}$, if $\prec$ is defined by a formula $Q(x, y)$ of $\mathcal{F}$ via $x \prec y := Q(x, y)$.

BI denotes the bar induction scheme, i.e., all formulae of the form

$$\mathrm{WF}(\prec) \rightarrow \mathbf{TI}(\prec, F),$$

where $\prec$ is an arithmetical relation (set parameters allowed) and F is an arbitrary formula of $\mathcal{L}_2$.

By **BI** we shall refer to the theory $\mathbf{ACA}_0 + \mathrm{BI}$.

*Remark 12.11. The statement of the main theorem 12.8 uses the notion of a countable coded ω-model of Π^1_1-$\mathbf{CA}_0$. As the stated equivalence is claimed to be provable in $\mathbf{RCA}_0$, a few comments on how this is formalized in this weak base theory are in order. The notion of a countable coded ω-model can be formalized in $\mathbf{RCA}_0$ according to [18, Definition VII.2.1]. Let $\mathbb{M}$ be a countable coded ω-model. Since **BI** is not finitely axiomatizable we have to quantify over all axioms of Π^1_1-$\mathbf{CA}_0$ to express that $\mathbb{M} \models \Pi^1_1$-$\mathbf{CA}_0$. The axioms of Π^1_1-$\mathbf{CA}_0$ (or rather their Gödel numbers) clearly form a primitive recursive set, $Ax(\Pi^1_1$-$\mathbf{CA}_0)$. To express $\mathbb{M} \models \varphi$ for $\varphi \in Ax(\Pi^1_1$-$\mathbf{CA}_0)$ we use the notion of a* valuation *for φ from [18, Definition VII.2.1]. A valuation f for φ is a function from the set of subformulae of φ into the set $\{0, 1\}$ obeying the usual Tarski truth conditions. Thus we write $\mathbb{M} \models \varphi$, if there exists a valuation f for φ such that $f(\varphi) = 1$. Whence $\mathbb{M} \models \Pi^1_1$-$\mathbf{CA}_0$ is defined by $\forall \varphi \in Ax(\Pi^1_1$-$\mathbf{CA}_0)\ \mathbb{M} \models \varphi$.*

Remark 12.12. Note that an ω-model of $\mathbf{RCA}_0$ (and hence of Π^1_1-$\mathbf{CA}_0$) is automatically a model of the full induction schema for all $\mathcal{L}_2$.

12.2 Relativizing the Ordinal for Π^1_1-comprehension

In this section we show how to relativize the construction that leads to an ordinal representation system for analyzing Π^1_1-comprehension to an arbitrary countable well-ordering. To begin with, mainly to foster intuitions, we provide a set-theoretic definition working in **ZFC**. This will then be followed by a purely formal definition that can be made in $\mathbf{RCA}_0$.

Throughout this section, we fix a non-empty countable well-ordering $\mathfrak{X} = (X, <_X)$ without a maximum element, i.e., an ordered pair $\mathfrak{X} = (X, <_X)$, where $X \neq \emptyset$ is a set of natural numbers, $<_X$ is a well-ordering relation on X, and $\forall v \in X\, \exists u \in X\; v <_X u$. We write $|\mathfrak{X}|$ for X.

Firstly, we need some ordinal-theoretic background. Let ON be the class of ordinals. Let $\mathsf{AP} := \{\xi \in \mathsf{ON} : \exists \eta \in \mathsf{ON}[\xi = \omega^\eta]\}$ be the class of additive principal numbers and let $E := \{\xi \in \mathsf{ON} : \xi = \omega^\xi\}$ be the class of ε–numbers which is enumerated by the function $\lambda\xi.\varepsilon_\xi$.

We write $\alpha =_{NF} \omega^{\alpha_1} + \cdots + \omega^{\alpha_n}$ if $\alpha = \omega^{\alpha_1} + \cdots + \omega^{\alpha_n}$ and $\alpha > \alpha_1 \geq \cdots \geq \alpha_n$. Note that by Cantor's normal form theorem, for every $\alpha \notin E \cup \{0\}$, there are uniquely determined ordinals $\alpha_1, \ldots, \alpha_n$ such that $\alpha =_{NF} \omega^{\alpha_1} + \cdots + \omega^{\alpha_n}$.

Let $\Omega_0 := 0$. For $\alpha > 0$, let $\Omega_\alpha := \aleph_\alpha$. For $u \in |\mathfrak{X}|$, let

$$\bar{u} := \max(\Omega_\omega, \sup\{\bar{v} + \Omega_\omega \mid v <_X u\}).$$

Thus, if u_0 is the smallest element of $|\mathfrak{X}|$, then $\overline{u_0} = \Omega_\omega$. Note that if $v <_X u$ and $\alpha < \Omega_\omega$ then $\bar{v} + \alpha < \bar{u}$.

In what follows we shall only be interested in ordinals $\leq \sup_{u \in X} \bar{u}$. Henceforth, unless indicated otherwise, any ordinal will be assumed to be smaller than that ordinal.

Definition 12.13. *We define sets of ordinals $C^{\mathfrak{X}}_n(\alpha)$ and ordinals $\psi^{\mathfrak{X}}_n\alpha$ for $n < \omega$ by recursion on $\alpha < \sup_{u \in X} \bar{u}$.*

$C^{\mathfrak{X}}_n(\alpha)$ is the smallest set of ordinals such that the following conditions are met:

(C0) $\bar{u} \in C^{\mathfrak{X}}_n(\alpha)$ *for all* $u \in |\mathfrak{X}|$.
(C1) $\{\Omega_m \mid m \in \mathbb{N}\} \cup \Omega_n \subseteq C^{\mathfrak{X}}_n(\alpha)$.
(C2) $\gamma_1, \ldots, \gamma_n \in C^{\mathfrak{X}}_n(\alpha) \wedge \xi =_{NF} \omega^{\gamma_1} + \cdots + \omega^{\gamma_n} < \Omega_\omega \implies \xi \in C^{\mathfrak{X}}_n(\alpha)$.
(C3) $\beta \in C^{\mathfrak{X}}_n(\alpha) \wedge \beta < \Omega_\omega \wedge u \in |\mathfrak{X}| \implies \bar{u} + \beta \in C^{\mathfrak{X}}_n(\alpha)$.
(C4) $\delta \in C^{\mathfrak{X}}_n(\alpha) \cap \alpha \wedge l < \omega \implies \psi^{\mathfrak{X}}_l\delta \in C^{\mathfrak{X}}_n(\alpha)$.

$\psi^{\mathfrak{X}}_n\alpha := \min\{\xi \mid \xi \notin C^{\mathfrak{X}}_n(\alpha)\}$.

Lemma 12.14. $\Omega_n < \psi^{\mathfrak{X}}_n\alpha < \Omega_{n+1}$.

Proof. As $\Omega_n \cup \{\Omega_n\} \subseteq C^{\mathfrak{X}}_n(\alpha)$ the definition of $\psi^{\mathfrak{X}}_n\alpha$ yields $\Omega_n < \psi^{\mathfrak{X}}_n\alpha$.

Note that the cardinality of $C^{\mathfrak{X}}_n(\alpha)$ is $\max(\Omega_1, \Omega_n)$ as $C^{\mathfrak{X}}_n(\alpha)$ is the closure of the set $\{\bar{u} \mid u \in |\mathfrak{X}|\} \cup \{\Omega_m \mid m \in \mathbb{N}\} \cup \Omega_n$ under finitary operations. Thus, as Ω_{n+1} is a regular cardinal, there exists an ordinal $\xi < \Omega_{n+1}$ such that $\xi \notin C^{\mathfrak{X}}_n(\alpha)$, whence $\psi^{\mathfrak{X}}_n\alpha < \Omega_{n+1}$. □

Lemma 12.15. 1. *If* $n \leq m$ *and* $\alpha \leq \beta$ *then* $C^{\mathfrak{X}}_n(\alpha) \subseteq C^{\mathfrak{X}}_m(\beta)$ *and* $\psi^{\mathfrak{X}}_n\alpha \leq \psi^{\mathfrak{X}}_m\beta$.
2. $\psi^{\mathfrak{X}}_m\alpha \in E$ *and* $\psi^{\mathfrak{X}}_n\alpha \notin C^{\mathfrak{X}}_n(\alpha)$.
3. *If* $\xi =_{NF} \omega^{\gamma_1} + \cdots + \omega^{\gamma_n} < \Omega_\omega$ *and* $\xi \in C^{\mathfrak{X}}_n(\alpha)$ *then* $\gamma_1, \ldots, \gamma_n \in C^{\mathfrak{X}}_n(\alpha)$.
4. *If* $u \in |\mathfrak{X}|$ *and* $\bar{u} + \beta \in C^{\mathfrak{X}}_n(\alpha)$ *then* $\beta \in C^{\mathfrak{X}}_n(\alpha)$.
5. *If* $\alpha \in C^{\mathfrak{X}}_n(\beta)$ *and* $\alpha < \beta$ *then* $\psi^{\mathfrak{X}}_n\alpha < \psi^{\mathfrak{X}}_n\beta$.
6. $\psi^{\mathfrak{X}}_n\alpha = C^{\mathfrak{X}}_n(\alpha) \cap \Omega_{n+1}$.
7. *If* $\alpha \in C^{\mathfrak{X}}_n(\alpha)$ *and* $\beta \in C^{\mathfrak{X}}_n(\beta)$ *then*

$$\alpha < \beta \Leftrightarrow \psi^{\mathfrak{X}}_n\alpha < \psi^{\mathfrak{X}}_n\beta\,.$$

8. *If* $\alpha \in C^{\mathfrak{X}}_n(\alpha)$ *and* $\beta \in C^{\mathfrak{X}}_n(\beta)$ *then*

$$\alpha = \beta \Leftrightarrow \psi^{\mathfrak{X}}_n\alpha = \psi^{\mathfrak{X}}_n\beta\,.$$

Proof. (1) is obvious by definition of these sets.

(2) follows from closure of $C^{\mathfrak{X}}_m(\alpha) \cap \Omega_\omega$ under ordinal addition, the fact that $\psi^{\mathfrak{X}}_m\alpha < \Omega_{m+1}$ and the definition of $\psi^{\mathfrak{X}}_m\alpha$.

For (3) one proceeds by the inductive definition of $C^{\mathfrak{X}}_n(\alpha)$. If $\xi < \Omega_n$ then this is obvious. As $\xi \notin E$, the only other clause to consider is (C2), and the claim is immediate.

For (4) notice that if $\beta \neq 0$ then the only clause by which $\bar{u}+\beta$ could have entered $C^{\mathfrak{X}}_n(\alpha)$ is (C3), whence $\beta \in C^{\mathfrak{X}}_n(\alpha)$.

(5) is obvious by (1) and the fact that $\psi^{\mathfrak{X}}_\alpha \in C^{\mathfrak{X}}_n(\beta)$ whereas $\psi^{\mathfrak{X}}_n\beta \notin C^{\mathfrak{X}}_n(\beta)$.

For (6) we proceed by the inductive definition of $C^{\mathfrak{X}}_n(\alpha)$ to show that for $\gamma \in C^{\mathfrak{X}}_n(\alpha) \cap \Omega_{n+1}$ we have $\gamma < \psi^{\mathfrak{X}}_n\alpha$. If $\gamma \leq \Omega_n$ then this is obvious. If $\gamma =_{NF} \omega^{\gamma_1} + \cdots + \omega^{\gamma_r}$, $\gamma_1, \ldots, \gamma_r \in C^{\mathfrak{X}}_n(\alpha)$ and inductively we have $\gamma_1, \ldots, \gamma_r < \psi^{\mathfrak{X}}_n\alpha$, then $\gamma < \psi^{\mathfrak{X}}_n\alpha$ by (2). If $\gamma = \psi^{\mathfrak{X}}_m\delta$ for some $\delta \in C^{\mathfrak{X}}_n(\alpha)$ with $\delta < \alpha$, then $\gamma < \psi^{\mathfrak{X}}_n\alpha$ follows from (5).

(7) follows from (5) and (1). (8) is an immediate consequence of (7). □

Define $[\eta, \delta) := \{\zeta \mid \eta \leq \zeta < \delta\}$.

Lemma 12.16. *If* $\alpha < \beta$ *and* $C^{\mathfrak{X}}_n(\alpha) \cap [\alpha, \beta) = \emptyset$ *then* $C^{\mathfrak{X}}_n(\alpha) = C^{\mathfrak{X}}_n(\beta)$.

Proof. By induction on the buildup of $C^{\mathfrak{X}}_n(\beta)$ we show that $\gamma \in C^{\mathfrak{X}}_n(\beta)$ implies $\gamma \in C^{\mathfrak{X}}_n(\alpha)$. If $\gamma \in C^{\mathfrak{X}}_n(\beta)$ holds by (C0) or (C1) this is obvious, and in the cases of (C2) and (C3) it follows immediately from the induction hypothesis. So assume $\gamma \in C^{\mathfrak{X}}_n(\beta)$ holds by (C4), i.e., there exists $\delta \in C^{\mathfrak{X}}_n(\beta) \cap \beta$ such that $\gamma = \psi^{\mathfrak{X}}_l\delta$ and inductively $\delta \in C^{\mathfrak{X}}_n(\alpha)$. Since $C^{\mathfrak{X}}_n(\alpha) \cap [\alpha, \beta) = \emptyset$ we infer that $\delta < \alpha$, whence $\gamma = \psi^{\mathfrak{X}}_l\delta \in C^{\mathfrak{X}}_n(\alpha)$. □

Lemma 12.17. *Let* $n \leq m$ *and* $\beta \leq \alpha$. *If* $\gamma \in C^{\mathfrak{X}}_n(\alpha)$ *and*

$$\delta = \min\{\xi \mid \gamma \leq \xi \in C^{\mathfrak{X}}_m(\beta)\}$$

then $\delta \in C^{\mathfrak{X}}_n(\alpha)$.

Proof. We proceed by induction on the definition of $\gamma \in C_n^{\mathfrak{X}}(\alpha)$.

If $\gamma \in C_n^{\mathfrak{X}}(\alpha)$ holds by (C0) or (C1), or $\gamma \leq \Omega_m$, then $\delta = \gamma$, thus $\delta \in C_n^{\mathfrak{X}}(\alpha)$.

Suppose $\gamma \in C_n^{\mathfrak{X}}(\alpha)$ holds by (C2). Then $\gamma =_{NF} \omega^{\gamma_1} + \cdots + \omega^{\gamma_r}$, $\gamma_1, \ldots, \gamma_r \in C_n^{\mathfrak{X}}(\alpha)$ and inductively we have $\delta_i := \min\{\xi \mid \gamma_i \leq \xi \in C_m^{\mathfrak{X}}(\beta)\} \in C_n^{\mathfrak{X}}(\alpha)$ for $i = 1, \ldots, r$. We shall employ a subsidiary induction on r. First assume $r = 1$. If $\gamma_1 \in C_m^{\mathfrak{X}}(\beta)$ then $\gamma = \delta$ and thus $\delta \in C_n^{\mathfrak{X}}(\alpha)$. If $\gamma_1 \notin C_m^{\mathfrak{X}}(\beta)$, then $\delta_1 > \gamma_1$which entails that $\delta = \omega^{\delta_1}$, whence $\delta \in C_n^{\mathfrak{X}}(\alpha)$. Now assume $r > 1$ and inductively that $\tilde{\delta} = \min\{\xi \mid \omega^{\gamma_2} + \cdots + \omega^{\gamma_r} \leq \xi \in C_m^{\mathfrak{X}}(\beta)\}$. If $\gamma_1 \in C_m^{\mathfrak{X}}(\beta)$ then $\tilde{\delta} < \omega^{\gamma_1+1}$ and thus $\delta = \omega^{\gamma_1} + \tilde{\delta} \in C_n^{\mathfrak{X}}(\alpha)$. If $\gamma_1 \notin C_m^{\mathfrak{X}}(\beta)$, then $\delta = \omega^{\delta_1}$ and hence $\delta \in C_n^{\mathfrak{X}}(\alpha)$.

Finally suppose that $\gamma = \psi_k^{\mathfrak{X}}\eta$ where $k \geq m$, $\eta \in C_n^{\mathfrak{X}}(\alpha) \cap \alpha$ and

$$\eta^* := \min\{\xi \mid \eta \leq \xi \in C_m^{\mathfrak{X}}(\beta)\} \in C_n^{\mathfrak{X}}(\alpha).$$

Clearly (in view of Lemma 12.14(3)) δ must be in E. If δ is of the form Ω_i then $\delta \in C_n^{\mathfrak{X}}(\alpha)$. Otherwise $\delta = \psi_k^{\mathfrak{X}}\eta'$ for some $\eta' \in C_m^{\mathfrak{X}}(\beta) \cap \beta$. First suppose $\eta' \leq \eta$. Then $\gamma = \delta$ and hence $\delta \in C_n^{\mathfrak{X}}(\alpha)$. So let us assume $\eta < \eta'$. Then $\eta \leq \eta^* \leq \eta'$. Thus $\eta^* < \beta$ and therefore $\psi_k^{\mathfrak{X}}\eta^* \in C_m^{\mathfrak{X}}(\beta)$. Together with $\psi_k^{\mathfrak{X}}\eta \leq \psi_k^{\mathfrak{X}}\eta^* \leq \psi_k^{\mathfrak{X}}\eta'$ and the definition of $\psi_k^{\mathfrak{X}}\eta'$ this yields $\psi_k^{\mathfrak{X}}\eta^* = \psi_k^{\mathfrak{X}}\eta'$. Thus $\delta = \psi_k^{\mathfrak{X}}\eta' \in C_n^{\mathfrak{X}}(\alpha)$ as $\eta^* \in C_n^{\mathfrak{X}}(\alpha)$ and $\eta^* < \beta \leq \alpha$. □

Lemma 12.18. *Suppose $\psi_m^{\mathfrak{X}}\zeta \in C_n^{\mathfrak{X}}(\alpha)$ and $\zeta \in C_n^{\mathfrak{X}}(\alpha) \cap \alpha$ where $m \geq n$. Then there is a uniquely determined $\zeta_0 < \alpha$ such that $\psi_m^{\mathfrak{X}}\zeta = \psi_m^{\mathfrak{X}}\zeta_0$, $\zeta_0 \in C_m^{\mathfrak{X}}(\zeta_0)$ and $\zeta_0 \in C_n^{\mathfrak{X}}(\alpha)$.*

Proof. By Lemma 12.17, we have $\zeta_0 := \min\{\xi \mid \zeta \leq \xi \in C_m^{\mathfrak{X}}(\zeta)\} \in C_n^{\mathfrak{X}}(\alpha)$. Thus, by Lemma 12.16, $C_m^{\mathfrak{X}}(\zeta_0) = C_m^{\mathfrak{X}}(\zeta)$ and $\psi_m^{\mathfrak{X}}\zeta = \psi_m^{\mathfrak{X}}\zeta_0$. $\zeta_0 \in C_m^{\mathfrak{X}}(\zeta_0)$ holds by definition of ζ_0.

If $\zeta = \zeta_0$ then $\zeta_0 < \alpha$. Now suppose $\zeta < \zeta_0$. Then $\zeta \notin C_m^{\mathfrak{X}}(\zeta) = C_m^{\mathfrak{X}}(\zeta_0)$. As $n \leq m$ this implies $\zeta \notin C_n^{\mathfrak{X}}(\zeta_0)$. But as $\zeta \in C_n^{\mathfrak{X}}(\alpha)$ this entails that $\zeta_0 < \alpha$.

The uniqueness follows from Lemma 12.15 (8). □

Definition 12.19. *Inductive definition of a set* $\mathrm{OT}(\Omega_\omega{\cdot}\mathfrak{X})$ *of ordinals and a natural number $G_{\mathfrak{X}}\alpha$ for $\alpha \in \mathrm{OT}(\Omega_\omega{\cdot}\mathfrak{X})$.*

1. $\Omega_m \in \mathrm{OT}(\Omega_\omega{\cdot}\mathfrak{X})$ *for every* $m \in \mathbb{N}$ *and* $\bar{u} \in \mathrm{OT}(\Omega_\omega{\cdot}\mathfrak{X})$ *for every* $u \in |\mathfrak{X}|$. $G_{\mathfrak{X}}\Omega_m := G_{\mathfrak{X}}\bar{u} := 0$.
2. *If* $\alpha =_{NF} \omega^{\alpha_1} + \cdots + \omega^{\alpha_n}$, $\alpha_1, \ldots, \alpha_n \in \mathrm{OT}(\Omega_\omega{\cdot}\mathfrak{X})$ *and* $\alpha_1, \ldots, \alpha_n < \Omega_\omega$ *then* $\alpha \in \mathrm{OT}(\Omega_\omega{\cdot}\mathfrak{X})$ *and* $G_{\mathfrak{X}}\alpha := \max\{G_{\mathfrak{X}}\alpha_1, \ldots, G_{\mathfrak{X}}\alpha_n\} + 1$.
3. *If* $0 < \beta < \Omega_\omega$ *and* $\beta \in \mathrm{OT}(\Omega_\omega{\cdot}\mathfrak{X})$ *then, for every* $u \in |\mathfrak{X}|$, $\bar{u} + \beta \in \mathrm{OT}(\Omega_\omega{\cdot}\mathfrak{X})$ *and* $G_{\mathfrak{X}}(\bar{u} + \beta) := G_{\mathfrak{X}}\beta + 1$.
4. *If* $\alpha = \psi_m^{\mathfrak{X}}\alpha_1$, $\alpha_1 \in \mathrm{OT}(\Omega_\omega{\cdot}\mathfrak{X})$ *and* $\alpha_1 \in C_m^{\mathfrak{X}}(\alpha_1)$, *then* $\alpha \in \mathrm{OT}(\Omega_\omega{\cdot}\mathfrak{X})$ *and* $G_{\mathfrak{X}}\alpha := G_{\mathfrak{X}}\alpha_1 + 1$.

Observe that according to Lemma 12.15 the function $G_{\mathfrak{X}}$ is well-defined. Each ordinal $\alpha \in \mathrm{OT}(\Omega_\omega{\cdot}\mathfrak{X})$ has a unique normal form using the symbols

$$\Omega_m, +, \omega, \psi^{\mathfrak{X}}_m, \bar{u}$$

for $m \in \mathbb{N}$ and $u \in |\mathfrak{X}|$.

In order to arrive at an ordinal representation system with an ordering relation that is primitive recursive in $\mathfrak{X}$ we have to show that the "normal form" condition $\alpha \in C^{\mathfrak{X}}_n(\beta)$ can be determined from the buildup of α alone. This will be achieved by the following definition.

Definition 12.20. *Inductive definition of* $\mathsf{K}_n\beta$ *for* $\beta \in \mathrm{OT}(\Omega_\omega{\cdot}\mathfrak{X})$.

1. $\mathsf{K}_n\Omega_m := \mathsf{K}_n\bar{u} := \emptyset$.
2. *If* $\alpha =_{NF} \omega^{\alpha_1} + \cdots + \omega^{\alpha_r}$, $\alpha_1, \ldots, \alpha_r \in \mathrm{OT}(\Omega_\omega{\cdot}\mathfrak{X})$ *and* $\alpha_1, \ldots, \alpha_r < \Omega_\omega$ *then* $\mathsf{K}_n\alpha := \bigcup_{1 \le i \le r} \mathsf{K}_n\alpha_i$.
3. *If* $\alpha = \bar{u} + \beta$, *where* $0 < \beta < \Omega_\omega$, $\beta \in \mathrm{OT}(\Omega_\omega{\cdot}\mathfrak{X})$ *and* $u \in |\mathfrak{X}|$, *then* $\mathsf{K}_n\alpha := \mathsf{K}_n\beta$.
4. *If* $\alpha = \psi^{\mathfrak{X}}_m\alpha_1$, $\alpha_1 \in \mathrm{OT}(\Omega_\omega{\cdot}\mathfrak{X})$ *and* $\alpha_1 \in C^{\mathfrak{X}}_m(\alpha_1)$, *then* $\mathsf{K}_n\alpha := \emptyset$ *if* $m < n$ *and* $\mathsf{K}_n\alpha := \{\alpha_1\} \cup \mathsf{K}_n\alpha_1$ *if* $n \le m$.

Note that $\mathsf{K}_n\alpha$ *is a finite set of ordinals of* $\mathrm{OT}(\Omega_\omega{\cdot}\mathfrak{X})$. *Below we shall write* $\mathsf{K}_n\alpha < \gamma$ *to convey that* $\forall\xi \in \mathsf{K}_n\alpha\ \xi < \gamma$.

Lemma 12.21. *Let* $\alpha \in \mathrm{OT}(\Omega_\omega{\cdot}\mathfrak{X})$. *Then we have*

$$\alpha \in C^{\mathfrak{X}}_n(\gamma) \Leftrightarrow \mathsf{K}_n\alpha < \gamma. \tag{12.2}$$

Proof. The proof proceeds by induction on $G_{\mathfrak{X}}\alpha$. In view of Lemma 12.15, (3) and (4), the crucial case occurs when $\alpha = \psi^{\mathfrak{X}}_m\alpha_1$, where $\alpha_1 \in C^{\mathfrak{X}}_m(\alpha_1)$. Now if $m < n$ then the equivalence (12.2) clearly holds since $\mathsf{K}_n\alpha = \emptyset$ and $\alpha \in C^{\mathfrak{X}}_n(\gamma)$.

So assume $n \le m$. As $G_{\mathfrak{X}}\alpha_1 < G_{\mathfrak{X}}\alpha$ the induction hypothesis yields $\mathsf{K}_n\alpha_1 < \gamma \Leftrightarrow \alpha_1 \in C^{\mathfrak{X}}_n(\gamma)$. Thus $\mathsf{K}_n\alpha < \gamma \Leftrightarrow \alpha_1 < \gamma \wedge \alpha_1 \in C^{\mathfrak{X}}_n(\gamma)$, and hence

$$\mathsf{K}_n\alpha < \gamma \Rightarrow \alpha \in C^{\mathfrak{X}}_n(\gamma).$$

Now assume that $\alpha \in C^{\mathfrak{X}}_n(\gamma)$. By Lemma 12.18 there exists a uniquely determined α_0 such that $\alpha = \psi^{\mathfrak{X}}_m\alpha_0$, $\alpha_0 \in C^{\mathfrak{X}}_M(\alpha_0)$ and $\alpha_0 \in C^{\mathfrak{X}}_n(\gamma) \cap \gamma$. But then $\alpha_0 = \alpha_1$ must hold by Lemma 12.15 (8). Hence $\mathsf{K}_n\alpha < \gamma$. In sum, $\mathsf{K}_n\alpha < \gamma \Leftrightarrow \alpha \in C^{\mathfrak{X}}_n(\gamma)$. □

Lemma 12.22. $\mathrm{OT}(\Omega_\omega{\cdot}\mathfrak{X}) = \bigcup\{C^{\mathfrak{X}}_0(\alpha) \colon \alpha < \sup_{u\in X}\bar{u}\} = C^{\mathfrak{X}}_0(\sup_{u\in X}\bar{u})$.

Proof. Obviously $\beta < \sup_{u\in X}\bar{u}$ holds for all $\beta \in \mathrm{OT}(\Omega_\omega{\cdot}\mathfrak{X})$.

$$\beta \in \mathrm{OT}(\Omega_\omega{\cdot}\mathfrak{X}) \Rightarrow \beta \in C^{\mathfrak{X}}_n(\sup_{u\in X}\bar{u})$$

is then shown by induction on $G_{\mathfrak{X}}\beta$.

The inclusion $C^{\mathfrak{X}}_0(\sup_{u\in X}\bar{u}) \subseteq \mathrm{OT}(\Omega_\omega{\cdot}\mathfrak{X})$ follows from the fact that $\mathrm{OT}(\Omega_\omega{\cdot}\mathfrak{X})$ is closed under the clauses (Ci) for $i = 0, 1, 2, 3$. Since $\mathfrak{X}$ is an ordering without a maximal element it is also clear that $\bigcup\{C^{\mathfrak{X}}_0(\alpha) \colon \alpha < \sup_{u\in X}\bar{u}\} = C^{\mathfrak{X}}_0(\sup_{u\in X}\bar{u})$. □

If for $\alpha, \beta \in \mathrm{OT}(\Omega_\omega \cdot \mathfrak{X})$ represented in their normal forms, we wanted to determine whether $\alpha < \beta$, we could do this by deciding $\alpha_0 < \beta_0$ for ordinals α_0 and β_0 that appear in these representations and, in addition, satisfy $G_{\mathfrak{X}}\alpha_0 + G_{\mathfrak{X}}\beta_0 < G_{\mathfrak{X}}\alpha + G_{\mathfrak{X}}\beta$. This follows from Lemma 12.15 (5) and the recursive procedure for comparing ordinals in Cantor normal form. So we come to see that after a straightforward coding in the natural numbers, we may represent $\langle \mathrm{OT}(\Omega_\omega \cdot \mathfrak{X}), <\restriction \mathrm{OT}(\Omega_\omega \cdot \mathfrak{X})\rangle$ via an ordinal representation system primitive recursive in $\mathfrak{X}$. How this ordinal representation system can be directly defined in $\mathbf{RCA}_0$ will be spelled out in the next subsection.

12.2.1 Defining $\mathbf{OT}(\Omega_\omega \cdot \mathfrak{X})$ *in* $\mathbf{RCA}_0$

We shall provide a definition of $\mathrm{OT}(\Omega_\omega \cdot \mathfrak{X})$ as a term structure in $\mathbf{RCA}_0$ which is primitive recursive in $\mathfrak{X}$. Of course, formally terms or strings of symbols have to be treated as coded by natural numbers since $\mathbf{RCA}_0$ only talks about numbers and sets of numbers. Though, as it is well-known how to do this, we can't be bothered with these niceties.

Definition 12.23. Given a total ordering $\mathfrak{X} = (X, <_X)$, i.e., an ordered pair $\mathfrak{X}$ in which X is a set of natural numbers and $<_X$ is a total ordering relation on X, we define inductively a binary relational structure $(\mathrm{OT}(\Omega_\omega \cdot \mathfrak{X}), <)$ together with auxiliary functions $\mathsf{K}_n : \mathrm{OT}(\Omega_\omega \cdot \mathfrak{X}) \to$ finite subsets of $\mathrm{OT}(\Omega_\omega \cdot \mathfrak{X})$ for $n \in \mathbb{N}$:

1. $\Omega_m, \bar{u} \in \mathrm{OT}(\Omega_\omega \cdot \mathfrak{X})$ and $\mathsf{K}_n\Omega_m := \mathsf{K}_n\bar{u} := \emptyset$ for all $m \in \mathbb{N}$ and $u \in X$.
   Below, though, we prefer to write 0 for Ω_0 and Ω_ω for $\overline{u_{min}}$, where u_{min} is the smallest element of X with regard to $<_X$.
2. If $\alpha \in \mathrm{OT}(\Omega_\omega \cdot \mathfrak{X})$ and $0 \neq \alpha$ then $0 < \alpha$.
3. For every $u \in X$ and $\alpha \in \mathrm{OT}(\Omega_\omega \cdot \mathfrak{X})$, if α is not of the form $\bar{v}$ nor of the form $\bar{v} + \beta$ for some $v \in X$ then $\alpha < \bar{u}$.
4. If $u \in X$, $\beta \in \mathrm{OT}(\Omega_\omega \cdot \mathfrak{X})$ and $\beta < \Omega_\omega$, then $\bar{u} + \beta \in \mathrm{OT}(\Omega_\omega \cdot \mathfrak{X})$ and $\mathsf{K}_n(\bar{u} + \beta) = \mathsf{K}_n\beta$.
5. If $u, v \in X$, $u <_X v$, $\alpha, \beta < \Omega_\omega$ and $\alpha, \beta \in \mathrm{OT}(\Omega_\omega \cdot \mathfrak{X})$ then $\bar{u} < \bar{v}$, $\bar{u} + \alpha < \bar{v}$, $\bar{u} < \bar{v} + \beta$, and $\bar{u} + \alpha < \bar{v} + \beta$.
6. $\Omega_l < \Omega_k$ whenever $l, k \in \mathbb{N}$ and $l < k$.
7. If $\alpha \in \mathrm{OT}(\Omega_\omega \cdot \mathfrak{X})$, $\alpha < \Omega_\omega$ and α is neither of the form Ω_l nor of the form $\psi^{\mathfrak{X}}_m\beta$, then $\omega^\alpha \in \mathrm{OT}(\Omega_\omega \cdot \mathfrak{X})$ and $\mathsf{K}_n\omega^\alpha = \mathsf{K}_n\alpha$.
8. Let $\alpha_1, \ldots, \alpha_r \in \mathrm{OT}(\Omega_\omega \cdot \mathfrak{X})$, $\alpha_1 < \Omega_\omega$ and $\alpha_1 \geq \ldots \geq \alpha_r$ with $r \geq 1$. Let E be the elements of $\mathrm{OT}(\Omega_\omega \cdot \mathfrak{X})$ of either form Ω_{i+1} or $\psi^{\mathfrak{X}}_k\alpha$. If $r > 1$ or $r = 1 \wedge \alpha_1 \notin \mathsf{E}$, then
   $$\omega^{\alpha_1} + \omega^{\alpha_2} + \cdots + \omega^{\alpha_r} \in \mathrm{OT}(\Omega_\omega \cdot \mathfrak{X})$$
   and $\mathsf{K}_n(\omega^{\alpha_1} + \omega^{\alpha_2} + \cdots + \omega^{\alpha_r}) := \bigcup_{1 \leq i \leq r} \mathsf{K}_n\alpha_i$.
9. Let $\alpha = \omega^{\alpha_1} + \cdots + \omega^{\alpha_r} \in \mathrm{OT}(\Omega_\omega \cdot \mathfrak{X})$ and $\beta \in \mathrm{OT}(\Omega_\omega \cdot \mathfrak{X})$, where β is of the form $\psi^{\mathfrak{X}}_m\gamma$ or Ω_m for some $m \in \mathbb{N}$ with $m > 0$.
   (i) If $\alpha_1 < \beta$, then $\omega^{\alpha_1} + \cdots + \omega^{\alpha_r} < \beta$.
   (ii) If $\beta \leq \alpha_1$, then $\beta < \omega^{\alpha_1} + \cdots + \omega^{\alpha_r}$.

10. If $\omega^{\alpha_1} + \cdots + \omega^{\alpha_r}, \omega^{\beta_1} + \cdots + \omega^{\beta_m} \in \mathrm{OT}(\Omega_\omega \cdot \mathfrak{X})$ then
    $\omega^{\alpha_1} + \cdots + \omega^{\alpha_r} < \omega^{\beta_1} + \cdots + \omega^{\beta_m}$ iff
    $r < m \wedge \forall i \leq r\ \alpha_i = \beta_i$ or
    $\exists i \leq \min(r, m)\ [\forall j < i\ \alpha_j = \beta_j \wedge \alpha_i < \beta_i]$.
11. If $\alpha \in \mathrm{OT}(\Omega_\omega \cdot \mathfrak{X})$ and $\mathsf{K}_m\alpha < \alpha$ then $\psi^{\mathfrak{x}}_m\alpha \in \mathrm{OT}(\Omega_\omega \cdot \mathfrak{X})$. $\mathsf{K}_n\psi^{\mathfrak{x}}_m\alpha := \emptyset$ if $m < n$ and $\mathsf{K}_n\psi^{\mathfrak{x}}_m\alpha := \{\alpha\} \cup \mathsf{K}_n\alpha$ if $m \geq n$.
12. If $\psi^{\mathfrak{x}}_m\alpha, \psi^{\mathfrak{x}}_l\beta \in \mathrm{OT}(\Omega_\omega \cdot \mathfrak{X})$ and $m < l$ then $\psi^{\mathfrak{x}}_m\alpha < \psi^{\mathfrak{x}}_l\beta$.
13. If $\psi^{\mathfrak{x}}_m\alpha \in \mathrm{OT}(\Omega_\omega \cdot \mathfrak{X})$ then $\psi^{\mathfrak{x}}_m\alpha < \Omega_l$ if $m < l$ and $\Omega_l < \psi^{\mathfrak{x}}_m\alpha$ if $l \leq m$.
14. If $\psi^{\mathfrak{x}}_m\alpha, \psi^{\mathfrak{x}}_m\beta \in \mathrm{OT}(\Omega_\omega \cdot \mathfrak{X})$ then $\psi^{\mathfrak{x}}_m\alpha < \psi^{\mathfrak{x}}_m\beta$ iff $\alpha < \beta$.

Lemma 12.24. (i) *The set* $\mathrm{OT}(\Omega_\omega \cdot \mathfrak{X})$ *equipped with the relation* $<$ *and the functions* K_n *is a structure primitive recursive in* $\mathfrak{X} = (X, <_X)$.
(ii) $<$ *is a total and linear ordering on* $\mathrm{OT}(\Omega_\omega \cdot \mathfrak{X})$.

Proof. Straightforward but (ii) requires quite a number of cases to look at. □

It is convenient to extend the operations $+$ and $x \mapsto \omega^x$ to all elements of $\mathrm{OT}(\Omega_\omega \cdot \mathfrak{X})$ smaller than Ω_ω.

Definition 12.25. Let $\alpha, \beta \in \mathrm{OT}(\Omega_\omega \cdot \mathfrak{X})$. Define $0 + \alpha := \alpha + 0 := \alpha$.
For $\alpha < \Omega_\omega$ with $\alpha \in \mathsf{E}$, define $\omega^\alpha := \alpha$.

Recall that E is the set of elements of $\mathrm{OT}(\Omega_\omega \cdot \mathfrak{X})$ of either form Ω_i or $\psi^{\mathfrak{x}}_k\alpha$.
For $\beta < \Omega_\omega$ we define $\alpha + \beta$ as follows:

1. If $\beta \in \mathsf{E}$ and $\alpha < \beta$ then $\alpha + \beta := \beta$.
2. If $\alpha, \beta \in \mathsf{E}$ and $\beta \leq \alpha$, then $\alpha + \beta := \omega^\alpha + \omega^\beta$.
3. If $\alpha \in \mathsf{E}$, $\beta = \omega^{\beta_1} + \cdots + \omega^{\beta_m}$ and $\alpha < \beta_1$, then $\alpha + \beta := \beta$.
4. If $\alpha \in \mathsf{E}$, $\beta = \omega^{\beta_1} + \cdots + \omega^{\beta_m}$ and $\alpha \geq \beta_1$, then $\alpha + \beta := \omega^\alpha + \omega^{\beta_1} + \cdots + \omega^{\beta_m}$.
5. If $\alpha = \omega^{\alpha_1} + \cdots + \omega^{\alpha_r}$, $\beta \in \mathsf{E}$, and there exists an index i such that $1 \leq i \leq r$, $\alpha_i \geq \beta$ and $\alpha_j < \beta$ holds for all $i < j \leq r$, then $\alpha + \beta := \omega^{\alpha_1} + \cdots + \omega^{\alpha_i} + \omega^\beta$.
6. If $\alpha = \omega^{\alpha_1} + \cdots + \omega^{\alpha_r}$, $\beta = \omega^{\beta_1} + \cdots + \omega^{\beta_m}$, and $\alpha_1 < \beta_1$, then $\alpha + \beta := \beta$.
7. If $\alpha = \omega^{\alpha_1} + \cdots + \omega^{\alpha_r}$, $\beta = \omega^{\beta_1} + \cdots + \omega^{\beta_m}$, and there exists an index i such that $1 \leq i \leq r$, $\alpha_i \geq \beta_1$ and $\alpha_j < \beta_1$ holds for all $i < j \leq r$, then $\alpha + \beta := \omega^{\alpha_1} + \cdots + \omega^{\alpha_i} + \omega^{\beta_1} + \cdots + \omega^{\beta_m}$.

12.2.2 Majorization and Fundamental Functions

In order to properly carry out the ordinal analyses of theories with Π^1_1-comprehension, we will use the treatment of Buchholz and Schütte's book [3] via the infinitary Ω_n-rules. To express and control the latter rules in the infinitary calculi $T^\diamond_Q$ we require a majorization relation and fundamental functions as in [3]. We also have to ensure that we can develop the latter in a fairly weak background theory. In [3] it is sometimes not clear whether their properties depend in any way on the set-theoretic modeling in **ZFC** (or fragment thereof) via the sets $C^{\mathfrak{x}}_n(\alpha)$ or whether they follow intrinsically from the formal definition of the representation system $T(\Omega)$ of [3], chapter I, section 3. In consequence, we shall revisit some of the results of [3] from a purely syntactic point of view.

In what follows we will work entirely with the formal definition of $\mathrm{OT}(\Omega_\omega{\cdot}\mathfrak{X})$ as given in Definition 12.23. Ordinal variables are assumed to range over the latter set. In this context the sets $C_n^{\mathfrak{X}}(\alpha)$ are defined by (in keeping with Lemma 12.21)

$$C_n^{\mathfrak{X}}(\alpha) := \{\gamma \in \mathrm{OT}(\Omega_\omega{\cdot}\mathfrak{X}) \mid \mathsf{K}_n\gamma < \alpha\}.$$

With this definition the properties (C0)–(C4) of Definition 12.13 and the facts stated in Lemma 12.14 and Lemma 12.15 become easily deducible in $\mathbf{RCA}_0$. Below we shall will use them often and mostly tacitly.

The first results to be revisited are Lemmata 12.16 and 12.17, i.e., Lemmata 2.7 and 2.8 from chapter I of [3]. Henceforth, $C_n^{\mathfrak{X}}(\xi) \cap [\delta,\gamma)$ stands for the set $\{\eta \in C_n^{\mathfrak{X}}(\xi) \mid \delta \leq \eta < \gamma\}$.

Lemma 12.26. *If $\alpha < \beta$ and $C_n^{\mathfrak{X}}(\alpha) \cap [\alpha,\beta) = \emptyset$ then $C_n^{\mathfrak{X}}(\alpha) = C_n^{\mathfrak{X}}(\beta)$.*

Proof. $C_n^{\mathfrak{X}}(\alpha) \subseteq C_n^{\mathfrak{X}}(\beta)$ is obvious. For the reverse inclusion we show

$$\gamma \in C_n^{\mathfrak{X}}(\beta) \Rightarrow \gamma \in C_n^{\mathfrak{X}}(\alpha)$$

by induction on the buildup of γ. The only interesting case arises when γ is of the form $\psi_m^{\mathfrak{X}}\rho$. $\gamma \in C_n^{\mathfrak{X}}(\beta)$ means that $\mathsf{K}_n\gamma < \beta$. If $m < n$ then $\gamma \in C_n^{\mathfrak{X}}(\alpha)$ follows immediately. So suppose $m \geq n$. Thus $\mathsf{K}_n\gamma = \{\rho\} \cup \mathsf{K}_n\rho < \beta$. From $\mathsf{K}_n\rho < \beta$ we inductively conclude $\rho \in C_n^{\mathfrak{X}}(\alpha)$. Moreover, $\rho < \beta$, so that by the assumption $C_n^{\mathfrak{X}}(\alpha) \cap [\alpha,\beta) = \emptyset$ we also have $\rho < \alpha$, whence $\gamma \in C_n^{\mathfrak{X}}(\alpha)$. □

Lemma 12.27. *If $\beta = \min\{\xi \in C_n^{\mathfrak{X}}(\alpha) \mid \alpha \leq \xi\}$ then $C_n^{\mathfrak{X}}(\alpha) = C_n^{\mathfrak{X}}(\beta)$ and $\beta \in C_n^{\mathfrak{X}}(\beta)$, whence $\psi_m^{\mathfrak{X}}\beta \in \mathrm{OT}(\Omega_\omega{\cdot}\mathfrak{X})$.*

Proof. Immediate by Lemma 12.26. □

Note that to claim the existence of β in the previous Lemma one would have to go beyond $\mathbf{RCA}_0$ in that it requires $\mathrm{OT}(\Omega_\omega{\cdot}\mathfrak{X})$ to be well-founded.

12.2.2.1 Majorization

We introduce the important concept of majorization from [3]. The background theory adopted from now on will be $\mathbf{ACA}_0$ plus the statement that $\mathrm{OT}(\Omega_\omega{\cdot}\mathfrak{X})$ is a well-ordering.

Definition 12.28. 1. $\alpha \lhd_\tau \beta$ if $\alpha < \beta$ and for all δ, k, η we have:

$$\alpha \leq \delta \leq \min\{\beta,\eta\} \wedge \delta,\tau \in C_k^{\mathfrak{X}}(\eta) \implies \alpha \in C_k^{\mathfrak{X}}(\eta). \tag{12.3}$$

2. $\alpha \lhd \beta$ (α is *majorized* by β) if $\alpha \lhd_0 \beta$.
3. $\alpha \unlhd \beta$ if either $\alpha \lhd \beta$ or $\alpha = \beta$.

With one exception the following basic properties are immediate consequences of the definition.

Lemma 12.29 (Basic properties)**.** (See [3] Lemma 4.1.)

1. *If* $\alpha \lhd \beta$ *then* $\alpha \lhd_\tau \beta$.
2. *If* $\alpha < \beta$ *then* $\alpha \lhd_\alpha \beta$.
3. *If* $\alpha < \beta < \gamma$ *and* $\alpha \lhd_\tau \gamma$ *then* $\alpha \lhd_\tau \beta$.
4. *If* $\alpha < \beta < \Omega_1$ *then* $\alpha \lhd \beta$.
5. *If* $\alpha \lhd \beta$ *then* $\alpha + 1 \unlhd \beta$.
6. *If* $\Omega_i < \Omega_j$ *then* $\Omega_i \lhd \Omega_j$.
7. *If* $\mathfrak{E}_u < \mathfrak{E}_v$ *then* $\mathfrak{E}_u \lhd \mathfrak{E}_v$
8. *For all* $i \leq \omega$ *and all* $u \in \mathfrak{X}, \Omega_i \lhd \mathfrak{E}_u$.

Proof. (1) This is trivial.

(2) Obviously, when $\tau = \alpha$ we get that $\alpha \in C_k^{\mathfrak{X}}(\eta)$ implies $\alpha \in C_k^{\mathfrak{X}}(\eta)$, so if $\alpha < \beta$ then $\alpha \lhd_\alpha \beta$.

(3) Suppose $\alpha < \gamma < \beta$ and $\alpha \lhd_\tau \gamma$. Then, we have for all δ, k, τ,

$$\alpha \leq \delta \leq min\{\gamma, \eta\} \ \wedge \ \delta, \tau \in C_k^{\mathfrak{X}}(\eta) \ \Rightarrow \ \alpha \in C_k^{\mathfrak{X}}(\eta).$$

Since $\gamma < \beta$, if we assume that $\alpha \leq \delta \leq \min\{\beta, \eta\} \ \wedge \ \delta \in C_k^{\mathfrak{X}}(\eta)$, then this is merely a stronger version of the same condition, and hence $\alpha \in C_k^{\mathfrak{X}}(\eta)$ as desired.

(4) Let $\alpha < \beta < \Omega_1$. Suppose $\alpha \leq \delta \leq \min\{\beta, \eta\} \ \wedge \ \delta \in C_k^{\mathfrak{X}}(\eta))$. To show $\alpha \lhd \beta$ we prove $\alpha \in C_k^{\mathfrak{X}}(\eta)$ by induction on the construction of α. If $\alpha = 0$, then $\alpha \in C_k^{\mathfrak{X}}(\eta)$ holds trivially. If $\alpha =_{NF} \omega^{\alpha_1} + \cdots + \omega^{\alpha_n}$, then $\alpha_1, \ldots, \alpha_n \leq \delta \leq \min\{\beta, \eta\}$ so that by induction hypothesis we get $\alpha_0, \ldots, \alpha_n \in C_k^{\mathfrak{X}}(\eta)$ and hence $\alpha \in C_k^{\mathfrak{X}}(\eta)$. The remaining and critical case is when $\alpha = \psi_0^{\mathfrak{X}}(\alpha_0)$, with $\alpha_0 \in C_0^{\mathfrak{X}}(\alpha_0)$. To deal with this case we employ a subsidiary induction on the construction of δ. If $\delta =_{NF} \omega^{\delta_1} + \cdots + \omega^{\delta_n}$ then $\alpha \leq \delta_1$ and inductively we can conclude that $\alpha \in C_k^{\mathfrak{X}}(\eta)$. If $\delta = \psi_0^{\mathfrak{X}}\delta_0$ then $\alpha_0 \leq \delta_0 < \eta$ and hence $\alpha_0 \in C_k^{\mathfrak{X}}(\eta)$, whence $\alpha \in C_k^{\mathfrak{X}}(\eta)$.

(5) Assume $\alpha \lhd \beta$. Then, $\alpha < \beta$, so either $\alpha + 1 = \beta$ in which case we are done, or $\alpha + 1 < \beta$. We then observe that if $\alpha \in C_k^{\mathfrak{X}}(\eta)$, then $\alpha + 1 \in C_k^{\mathfrak{X}}(\eta)$ by closure under addition.

(6), (7), and (8) are immediate since $\Omega_i, \mathfrak{E}_u \in C_k^{\mathfrak{X}}(\eta)$. □

Lemma 12.30. $\alpha \lhd_\tau \beta$ *and* $\beta \lhd_\tau \gamma \implies \alpha \lhd_\tau \gamma$.

Proof. Clearly, we have $\alpha < \gamma$. Suppose $\alpha \leq \delta \leq min\{\gamma, \eta\}$, and $\delta, \tau \in C_k^{\mathfrak{X}}(\eta)$. If $\delta \leq \beta$, then by $\alpha \lhd_\tau \beta$ we have $\alpha \in C_k^{\mathfrak{X}}(\eta)$. If $\beta < \delta$, then by $\beta \lhd_\tau \gamma$ we have $\beta < min\{\gamma, \eta\}$, which means $\beta \in C_k^{\mathfrak{X}}(\eta)$. By $\alpha \lhd_\tau \beta$, it again follows that $\alpha \in C_k^{\mathfrak{X}}(\eta)$. Hence, $\alpha \lhd_\tau \gamma$. □

Lemma 12.31. *If* $\alpha \lhd_\tau \beta$ *and* $\beta < \omega^{\gamma+1}$, *then* $\omega^\gamma + \alpha \lhd_\tau \omega^\gamma + \beta$.

Proof. By assumption, we have $\omega^\gamma + \alpha \leq \omega^\gamma + \beta$. Suppose $\omega^\gamma + \alpha \leq \delta \leq min\{\omega^\gamma + \beta, \eta\}$ and $\delta, \tau \in C_k^{\mathfrak{X}}(\eta)$. By normal form, we know that $\delta = \omega^\gamma + \delta_0$, where $\alpha \leq \delta_0 \leq \min\{\beta, \eta\}$. By definition of $C_k^{\mathfrak{X}}(\eta)$, it follows that $\gamma, \delta_0 \in C_k^{\mathfrak{X}}(\eta)$, and by $\alpha \lhd_\tau \beta$ we have $\alpha \in C_k^{\mathfrak{X}}(\eta)$. It thus follows that $\omega^\gamma + \alpha \in C_k^{\mathfrak{X}}(\eta)$. □

Corollary 12.32. $\omega^\alpha \cdot n \lhd \omega^\alpha \cdot (n+1)$.

Proof. This follows easily by induction on n from the previous lemma, with the base case $0 \lhd \omega^\alpha$ being obvious. □

Lemma 12.33. *If $\alpha \lhd_\tau \beta$ then $\omega^\alpha \cdot n \lhd_\tau \omega^\beta$.*

Proof. Clearly, we have $\omega^\alpha \cdot n < \omega^\beta$. When $n = 0$, this is immediate. Otherwise, assume $\omega^\alpha \cdot n \leq \delta \leq \min\{\omega^\beta, \eta\}$ with $\delta, \tau \in C_k^{\mathfrak{X}}(\eta)$. By normal form, we know $\delta = \omega^{\delta_1} + \delta_2$, where $\alpha \leq \delta_1 \leq \min\{\beta, \eta\}$, and $\delta_1 \in C_k^{\mathfrak{X}}(\eta)$. Hence, by $\alpha \lhd_\tau \beta$, we know $\alpha \in C_k^{\mathfrak{X}}(\eta)$, and thus $\omega^\alpha \cdot n \in C_k^{\mathfrak{X}}(\eta)$ □

Lemma 12.34. (See [3] Lemma 4.6.) *If $\alpha \lhd_\tau \beta$, with $\tau \in C_k^{\mathfrak{X}}(\alpha)$ and $\beta \in C_k^{\mathfrak{X}}(\beta)$ then $\alpha \in C_k^{\mathfrak{X}}(\alpha)$ and $\psi_k^{\mathfrak{X}}(\alpha) \lhd_\tau \psi_k^{\mathfrak{X}}(\beta)$.*

Proof. First, we shall show that $\alpha \in C_k^{\mathfrak{X}}(\alpha)$. By Lemma 12.16 we know that for $\gamma = \min\{\xi | \alpha \leq \xi \in C_k^{\mathfrak{X}}(\alpha)\}$, we have $\gamma \in C_k^{\mathfrak{X}}(\gamma) = C_k^{\mathfrak{X}}(\alpha)$. By assumption, we also have $\tau \in C_k^{\mathfrak{X}}(\gamma)$. Since $\beta \in C_k^{\mathfrak{X}}(\beta)$ we have $\alpha \leq \gamma = \min\{\beta, \gamma\}$. Thus, by definition of $\alpha \lhd_\tau \beta$, we have $\alpha \in C_k^{\mathfrak{X}}(\gamma) = C_k^{\mathfrak{X}}(\alpha)$.

Since $\alpha \in C_k^{\mathfrak{X}}(\alpha)$ and $\beta \in C_k^{\mathfrak{X}}(\beta)$, with $\alpha < \beta$, it follows that $\psi_k^{\mathfrak{X}}\alpha < \psi_k^{\mathfrak{X}}\beta$. Now, suppose $\psi_k^{\mathfrak{X}}\alpha \leq \delta \leq \min\{\psi_k^{\mathfrak{X}}\beta, \eta\}$ and $\delta, \tau \in C_m^{\mathfrak{X}}(\eta)$. We shall prove $\psi_k^{\mathfrak{X}}\alpha \in C_m^{\mathfrak{X}}(\eta)$ by induction on the construction of δ. Since $\psi_k^{\mathfrak{X}}\alpha \leq \delta \leq \psi_k^{\mathfrak{X}}\beta$, we know $\Omega_k \leq \delta < \Omega_{k+1}$. Thus, we must consider the following cases:

1. If $\delta < \Omega_m$ then $\psi_k^{\mathfrak{X}}\alpha < \Omega_m$, and thus $\psi_k^{\mathfrak{X}}\alpha \in C_m^{\mathfrak{X}}(\eta)$.
2. If $\Omega_m < \delta =_{NF} \omega^{\delta_0} \cdots + \omega^{\delta_n}$ for $\delta_0, \ldots, \delta_n \in C_m^{\mathfrak{X}}(\eta)$, then $\psi_k^{\mathfrak{X}}\alpha \leq \delta_0$, and thus $\psi_k^{\mathfrak{X}}\alpha \in C_m^{\mathfrak{X}}(\eta)$.
3. If $\Omega_m < \delta =_{NF} \psi_m^{\mathfrak{X}}\delta_0$, then $\delta_0 \in C_m^{\mathfrak{X}}(\eta)$ and $\delta_0 < \eta$. Now, if $k < m$ then $\psi_k^{\mathfrak{X}}\alpha < \Omega_m$ and thus $\psi_k^{\mathfrak{X}}\alpha \in C_m^{\mathfrak{X}}(\eta)$. Otherwise, $k = m$. Then $\alpha < \delta_0 < \eta$. Furthermore, since $\psi_k^{\mathfrak{X}}\delta_0 < \psi_k^{\mathfrak{X}}\beta$, we have $\delta_0 < \beta$. Since $\alpha \lhd_\tau \beta$ we have $\alpha \in C_m^{\mathfrak{X}}(\eta)$, and since $\alpha < \eta$ we have $\psi_k^{\mathfrak{X}}\alpha \in C_m^{\mathfrak{X}}(\eta)$. □

Corollary 12.35. $\alpha = \alpha_0 + 1 \in C_k^{\mathfrak{X}}(\alpha)$ *implies* $\alpha_0 \in C_k^{\mathfrak{X}}(\alpha_0)$ *and* $\psi_k^{\mathfrak{X}}(\alpha_0) \lhd \psi_k^{\mathfrak{X}}(\alpha)$.

Proof. The proof follows immediately from the preceding lemma with $\tau = 0$. We need only note that $\alpha_0 \lhd \alpha$. □

12.2.2.2 Fundamental Functions

Definition 12.36. A partial recursive function $f : \text{dom}(f) \to \text{OT}(\Omega_\omega{\cdot}\mathfrak{X})$ with the domain $\text{dom}(f) \subseteq \{\alpha \in \text{OT}(\Omega_\omega{\cdot}\mathfrak{X}) \mid \alpha < \Omega_\omega\}$ is a *fundamental function* if the following hold:

1. If $\beta \in \text{dom}(f)$ and $\alpha < \beta$, then $\alpha \in \text{dom}(f)$ and $f(\alpha) \lhd_\alpha f(\beta)$.
2. If $\beta \in \text{dom}(f)$ and $f(0) \leq \delta < f(\beta)$ then there is an $\alpha \in \text{dom}(f)$ such that $f(\alpha) \leq \delta < f(\alpha+1)$ and $f(\alpha) \lhd f(\alpha+1)$.
3. If $\alpha \in \text{dom}(f)$ and $f(\alpha) \in C_k^{\mathfrak{X}}(\eta)$ then $\alpha \in C_k^{\mathfrak{X}}(\eta)$.

Lemma 12.37. *If f is a fundamental function and $\alpha \in dom(f)$, then $\alpha \leq f(\alpha)$.*

Proof. Suppose for a contradiction that α is the least ordinal in $\text{dom}(f)$ such that $f(\alpha) < \alpha$. Then, by property (1) of fundamental functions, $f(\alpha) \in \text{dom}(f)$ and $f(f(\alpha)) \lhd_{f(\alpha)} f(\alpha)$. So in particular, $f(f(\alpha)) < f(\alpha)$, but this contradicts our initial assumption. Thus, $\alpha \leq f(\alpha)$ for all $\alpha \in \text{dom}(f)$.

We used classical reasoning here, but it's not necessary as we can prove this "positively" by induction on the ordering of $\text{OT}(\Omega_\omega{\cdot}\mathfrak{X})$. □

Definition 12.38. For $\beta < \Omega_\omega$ let Id_β be the identity function with $\text{dom}(Id_\beta) = \{\alpha \in \text{OT}(\Omega_\omega{\cdot}\mathfrak{X}) \mid \alpha \leq \beta\}$ and $Id_\beta(\alpha) = \alpha$ for all $\alpha \in \text{dom}(Id_\beta)$.

Lemma 12.39. *Id_β is a fundamental function.*

Proof. This is obvious from the definition of a fundamental function. □

Definition 12.40. Let f be a fundamental function.

1. For $\gamma < \Omega_\omega$ $\omega^\gamma + f$ is the function with $\text{dom}(\omega^\gamma + f) = \{\alpha \in \text{dom}(f) \mid f(\alpha) < \omega^{\gamma+1}\}$ and $(\omega^\gamma + f)(\alpha) = \omega^\gamma + f(\alpha)$ for all $\alpha \in \text{dom}(\omega^\gamma + f)$.
2. ω^f is the function with $\text{dom}(\omega^f) = \{\xi \in \text{dom}(f) \mid f(\xi) < \Omega_\omega\}$ and $(\omega^f)(\alpha) = \omega^{f(\alpha)}$ for all $\alpha \in \text{dom}(\omega^f)$.
3. $\Omega_{n+1} + f$ is the function with $\text{dom}(\Omega_{n+1} + f) = \{\xi \in \text{dom}(f) \mid f(\xi) \leq \Omega_{n+1}\}$ and $(\Omega_{n+1} + f)(\xi) = \Omega_{n+1} + f(\xi)$ for $\xi \in \text{dom}(\Omega_{n+1} + f)$.
4. For $u \in \mathfrak{X}$, $\bar{u}+f$ is the function with $\text{dom}(\bar{u}+f) = \text{dom}(f)$ and $(\bar{u}+f)(\xi) = \bar{u}+f(\xi)$ for $\xi \in \text{dom}(\bar{u}+f)$.
5. Let $\psi_k^{\mathfrak{X}} f$ be the function with $\text{dom}(\psi_k^{\mathfrak{X}} f) = \{\alpha \in \text{dom}(f) \mid \alpha < \Omega_{k+1}, f(\alpha) \in C_k^{\mathfrak{X}}(f(\alpha))\}$ and $(\psi_k^{\mathfrak{X}} f)(\alpha) = \psi_k^{\mathfrak{X}}(f(\alpha))$ for all $\alpha \in \text{dom}(\psi_k^{\mathfrak{X}} f)$.

Lemma 12.41. See [3], I. Lemma 5.3. *If f is a fundamental function, then so are* (a) $\omega^\gamma + f$, (b) ω^f, (c) $\Omega_{n+1} + f$, (d) $\bar{u} + f$, *and* (e) $\psi_k^{\mathfrak{X}} f$.

Proof. We shall briefly sketch the proofs for the first two functions. The cases (c) and (d) are straightforward. The proof for $\psi_k^{\mathfrak{X}}$ is considerably more involved, and will be handled in full. Below by Property (k) we mean the k's property of a fundamental function in Definition 12.36.

(a) Property (1) follows by virtue of Lemma 12.31. For property (2), we observe that if $\omega^\gamma + f(0) \leq \delta < \omega^\gamma + f(\beta)$, then $\delta = \omega^\gamma + \delta_0$, with $f(0) \leq \delta < f(\beta)$, and hence α can be found such that $f(\alpha) \leq \delta_0 < f(\alpha + 1)$. Property (3) is shown by first noting that if $\omega^\gamma + f(\alpha) \in C_k^{\mathfrak{X}}(\eta)$ then $f(\alpha) \in C_k^{\mathfrak{X}}(\eta)$. $\alpha \in C_k^{\mathfrak{X}}(\eta)$ then follows because f is fundamental.

(b) Property (1) holds due to Lemma 12.33, where $n = 1$. For property (2), we note that if $\omega^{f(0)} \leq \delta < \omega^{f(\beta)}$ then $\delta = \omega^{\delta_0} + \delta_1$, with $f(0) \leq \delta_0 < f(\beta)$. Since f is fundamental, we can find α such that $f(\alpha) \leq \delta_0 < f(\alpha + 1)$, and thus $\omega^{f(\alpha)} \leq \delta < \omega^{f(\alpha+1)}$. Property (3) is proven noting that if $\omega^{f(\alpha)} \in C_k^{\mathfrak{X}}(\eta)$ then $f(\alpha) \in C_k^{\mathfrak{X}}(\eta)$. $\alpha \in C_k^{\mathfrak{X}}(\eta)$ then follows because f is fundamental.

(e) We now consider the $\psi_k^{\mathfrak{X}}$ function.

1. Suppose $\beta \in \mathrm{dom}(\psi_k^{\mathfrak{X}} f)$ and $\alpha < \beta$. We must show $\alpha \in \mathrm{dom}(\psi_k^{\mathfrak{X}} f)$ and $\psi_k^{\mathfrak{X}} f(\alpha) \lhd_\alpha \psi_k^{\mathfrak{X}} f(\beta)$. Since f is a fundamental function and $\alpha < \beta$ we have $\alpha \in \mathrm{dom}(f)$ and $f(\alpha) \lhd_\alpha f(\beta)$. We also have $f(\beta) \in C_k^{\mathfrak{X}}(f(\beta))$ and $\beta < \Omega_{k+1}$ by the definition of $\mathrm{dom}(\psi_k^{\mathfrak{X}} f)$. By Lemma 12.27 there is $\gamma = \min\{\xi \mid f(\alpha) \leq \xi \in C_k^{\mathfrak{X}}(f(\alpha))\}$ such that $C_k^{\mathfrak{X}}(\gamma) = C_k^{\mathfrak{X}}(f(\alpha))$. Since $f(\beta) \in C_k^{\mathfrak{X}}(f(\beta))$, it follows that $f(\alpha) \leq \gamma \leq f(\beta)$.
   If $\gamma = f(\beta)$, then $f(\beta) \in C_k^{\mathfrak{X}}(f(\beta)) = C_k^{\mathfrak{X}}(f(\alpha))$. By property (3) of fundamental functions, this implies $\beta \in C_k^{\mathfrak{X}}(f(\alpha))$. Since $\alpha < \beta < \Omega_{k+1}$, and $\beta \in C_k^{\mathfrak{X}}(f(\alpha))$ it follows that $\alpha \in C_k^{\mathfrak{X}}(f(\alpha))$.
   If $\gamma < f(\beta)$, then there exists $\tau < \beta$ such that $f(\tau) \leq \gamma < f(\tau + 1)$, $\alpha \leq \tau$ and $f(\tau) \lhd f(\tau + 1)$. Since $\gamma \in C_k^{\mathfrak{X}}(\gamma)$, it follows that $f(\tau) \in C_k^{\mathfrak{X}}(\gamma)$. By property (3) of fundamental functions this means $\tau \in C_k^{\mathfrak{X}}(\gamma)$ and since $\alpha \leq \tau$ we get $\alpha \in C_k^{\mathfrak{X}}(\tau) = C_k^{\mathfrak{X}}(f(\alpha))$.
   Thus, we have $\alpha \in C_k^{\mathfrak{X}}(f(\alpha))$. Combining this with the fact that $f(\alpha) \lhd_\alpha f(\beta)$, it follows that $f(\alpha) \in C_k^{\mathfrak{X}}(f(\alpha))$. Hence $\alpha \in \mathrm{dom}(\psi_k^{\mathfrak{X}} f)$ and $\psi_k^{\mathfrak{X}} f(\alpha) \lhd \psi_k^{\mathfrak{X}} f(\beta)$ by Lemma 12.34.
2. Suppose $\beta \in \mathrm{dom}(\psi_k^{\mathfrak{X}} f)$ and $\psi_k^{\mathfrak{X}} f(0) \leq \delta < \psi_k^{\mathfrak{X}} f(\beta)$. We shall prove there exists $\alpha < \beta$ such that $f(\alpha) \leq \delta < f(\alpha + 1)$, using induction on the construction of δ.
   For the base case, if $\delta = 0, \Omega_m$, or $\mathfrak{E}_u$ then δ cannot be between $\psi_k^{\mathfrak{X}} f(0)$ and $\psi_k^{\mathfrak{X}} f(\beta)$, so the statement holds vacuously.
   If $\delta =_{NF} \omega^{\delta_1} + \cdots + \omega^{\delta_n}$, then by induction hypothesis, there is α such that $\psi_k^{\mathfrak{X}} f(\alpha) \leq \delta_0 < \psi_k^{\mathfrak{X}} f(\alpha + 1)$, and thus $\psi_k^{\mathfrak{X}} f(\alpha) \leq \delta < \psi_k^{\mathfrak{X}} f(\alpha + 1)$.
3. Suppose $\beta \in \mathrm{dom}(\psi_k^{\mathfrak{X}} f)$. If $k < m$ then obviously $\beta \in C_m^{\mathfrak{X}}(\eta)$ since $\beta < \Omega_{k+1}$. If $k = m$ and $\psi_k^{\mathfrak{X}} f(\beta) \in C_m^{\mathfrak{X}}(f(\eta))$ then $f(\beta) \in C_m^{\mathfrak{X}}(\eta)$ by definition, and since f is fundamental, it follows that $\beta \in C_m^{\mathfrak{X}}(\eta)$. □

Lemma 12.42. (See [3] Lemma 5.4.) *If f is a fundamental function with $\alpha, \Omega_{k+1} \in$ dom(f), $\alpha < \beta = \psi_k^{\mathfrak{X}}(f(\alpha))$, and $f(\alpha) \lhd f(\Omega_{k+1})$, then $f(\beta) \lhd f(\Omega_{k+1})$.*

Proof. Since $\beta = \psi_k^{\mathfrak{X}} f(\alpha) < \Omega_{k+1}$, by property (3) of fundamental functions we obtain $f(\beta) \lhd_\beta f(\Omega_{k+1})$. Now suppose that $f(\beta) \leq \delta \leq \min\{\Omega_{k+1}, \eta\}$ and $\delta \in C_t^{\mathfrak{X}}(\eta)$, for some t. We will prove that $\beta \in C_t^{\mathfrak{X}}(\eta)$.

Since $\alpha < \beta$, we get $f(\alpha) < f(\beta) < \eta$. Moreover, $f(\alpha) \leq \delta \leq \min\{f(\Omega_{k+1}), \eta\}$, so by $f(\alpha) \lhd f(\Omega_{k+1})$, we have $f(\alpha) \in C^{\mathfrak{x}}_t(\eta)$.

Since $f(\alpha) \in C^{\mathfrak{x}}_t(\eta)$ and $f(\alpha) < \eta$, it follows that $\beta = \psi^{\mathfrak{x}}_k f(\alpha) \in C^{\mathfrak{x}}_t(\eta)$, as desired. Applying $f(\beta) \lhd_\beta f(\Omega_{k+1})$, we get $f(\beta) \in C^{\mathfrak{x}}_t(\eta)$. Thus $f(\beta) \lhd f(\Omega_{k+1})$. □

Readers may note that the above proof is somewhat simpler than that presented in [3] . This is because we are using only a fragment of the full ordinal representation system presented in that book. In the full system, there is a function $\Omega_x : \tau \mapsto \Omega_\tau$, where τ is itself an ordinal term. When dealing with larger τ, one has to be careful, and ensure that $\tau \in C^{\mathfrak{x}}_t(\eta)$ before applying the corresponding $\psi^{\mathfrak{x}}_\tau$ function.

Corollary 12.43. (See [3], I. Corollary 5.4.) *If f is a fundamental function with $\Omega_{k+1} \in dom(f)$ then $f(\psi^{\mathfrak{x}}_k f(0)) \lhd f(\Omega_{k+1})$.*

Proof. This is a direct application of the preceding lemma, with $\alpha = 0$, observing that $f(0) \lhd f(\Omega_{k+1})$. □

Lemma 12.44. (See [3], I. Lemma 5.5.) *Let f be a fundamental function, where $\Omega_{k+1} \in dom(f), j \leq k, f(\Omega_{k+1}) \in C^{\mathfrak{x}}_j(f(\Omega_{k+1}))$, and let (β_n) be the sequence where $\beta_0 = 0$ and $\beta_{n+1} = \psi^{\mathfrak{x}}_k(f(\beta_n))$. Let g be the function with $dom(g) = \{\alpha \mid \alpha \leq \omega\}$, where $g(n) = \psi^{\mathfrak{x}}_j(f(\beta_n))$. Then g is a fundamental function.*

Proof. To prove the first property of fundamental functions, we begin by noting that if $\gamma \in \mathrm{dom}(g) = \{\alpha \mid \alpha \leq \omega\}$, then for all $\alpha < \gamma, \alpha \in \mathrm{dom}(g)$.

Next, we shall prove

$$(1)\ \beta_n < \beta_{n+1} \text{ and } f(\beta_n) \lhd f(\Omega_{k+1})$$

using induction on n.

When $n = 0$ we have $0 < \psi^{\mathfrak{x}}_k(f(0))$ and by part 4 of Lemma 12.29, $0 \lhd f(\Omega_{k+1})$.

For the induction step, assume $\beta_n < \beta_{n+1}$ and $f(\beta_n) \lhd f(\Omega_{k+1})$. Note that $f(\Omega_{k+1}) \in C^{\mathfrak{x}}_j(f(\Omega_{k+1})) \subseteq C^{\mathfrak{x}}_k(f(\Omega_{k+1}))$. Combining this with $f(\beta_n) \lhd f(\Omega_{k+1})$, we apply Lemma 12.34 to obtain $f(\beta_n) \in C^{\mathfrak{x}}_k(f(\beta_n))$. Since f is fundamental, and $\beta_n < \beta_{n+1}$ we get $f(\beta_n) < f(\beta_{n+1})$. Hence $\psi^{\mathfrak{x}}_k(f(\beta_n)) = \beta_{n+1} \in C^{\mathfrak{x}}_k(f(\beta_{n+1}))$, and thus

$$\beta_{n+1} = \psi^{\mathfrak{x}}_k(f(\beta_n)) < \psi^{\mathfrak{x}}_k(f(\beta_{n+1})) = \beta_{n+2}.$$

Since $\beta_n < \beta_{n+1}$ and $f(\beta_n) \lhd f(\Omega_{k+1})$, then by Lemma 12.42 we get $f(\psi^{\mathfrak{x}}_k(f(\beta_n))) = f(\beta_{n+1}) \lhd f(\Omega_{k+2})$, which completes the inductive proof of (1).

Next, using $f(\beta_n) \lhd f(\Omega_{k+1})$ and $f(\Omega_{k+1}) \in C^{\mathfrak{x}}_j(f(\Omega_{k+1}))$ we apply Lemma 12.34 to get $f(\beta_n) \in C^{\mathfrak{x}}_j(f(\beta_n))$. Combined with $f(\beta_n) < f(\beta_{n+1})$, this yields

$$(2)\ g(n) = \psi^{\mathfrak{x}}_j(f(\beta_n)) < \psi^{\mathfrak{x}}_j(f(\beta_{n+1})) = g(n+1).$$

Applying 12.34 once more yields

$$(3)\ g(n) = \psi^{\mathfrak{x}}_j(f(\beta_n)) \lhd \psi^{\mathfrak{x}}_j(f(\Omega_{k+1})) = g(\omega).$$

Hence, if $\gamma \in \mathrm{dom}(g)$ and $\alpha < \gamma$, then $g(\alpha) \lhd \psi_j^{\mathfrak{x}}(f(\Omega_{k+1}))$. By part 3 of Lemma 12.29, $g(\alpha) \lhd g(\gamma)$, and thus $g(\alpha) \lhd_\alpha g(\gamma)$. Thus, we have proven the first property of fundamental functions for g.

For the second property, we begin by proving

$$\text{(4) if } \gamma < \psi_k^{\mathfrak{x}}(f(\Omega_{k_1})), \text{ then there is } n \text{ such that } \beta_n \leq \gamma < \beta_{n+1},$$

using induction on the construction of γ.

For the base case, if $\gamma < \beta$ then $\beta_0 = 0 \leq \gamma < \beta_1$. Otherwise, assume $\beta_1 = \psi_k^{\mathfrak{x}}(f(0)) \leq \gamma < \psi_k^{\mathfrak{x}}(f(\Omega_{k+1}))$. If $\gamma =_{NF} \omega^{\gamma_1} + \cdots + \omega\gamma_n$, then the assertion follows immediately from the induction hypothesis.

This leaves the case where $\gamma = \psi_k^{\mathfrak{x}}(\gamma_0)$, with $\gamma_0 \in C_k^{\mathfrak{x}}(\gamma_0)$, with $f(0) \leq \gamma_0 < f(\Omega_{k+1})$. Since f is fundamental, we may find $\alpha < \Omega_{k+1}$ such that $f(\alpha) \leq \gamma_0 < f(\alpha+1)$ and $f(\alpha) \lhd f(\alpha + 1)$. Applying part 3 of Lemma 12.29 yields $f(\alpha) \lhd \gamma_0 <$ and since $\gamma_0 \in C_k^{\mathfrak{x}}(\gamma_0)$, we may apply Lemma 12.34 to get $f(\alpha) \in C_k^{\mathfrak{x}}(f(\alpha)) \subseteq C_k^{\mathfrak{x}}(\gamma_0)$. Since f is fundamental, $\alpha \in C_k^{\mathfrak{x}}(\gamma_0)$ and since $\alpha < \Omega_{k+1}$ we know $\alpha < \psi_k^{\mathfrak{x}}(\gamma_0)$. Applying our induction hypothesis, we get $\beta_n \leq \alpha < \alpha + 1 < \beta_{n+1}$. Hence

$$f(\beta_n) \leq f(\alpha) \leq \gamma_0 < f(\alpha + 1) < f(\beta_{n+1})$$

and thus $\beta_{n+1} = \psi_k^{\mathfrak{x}}(f(\beta_n)) \leq \psi_k^{\mathfrak{x}}(\gamma_0) = \gamma < \psi_k^{\mathfrak{x}}(f(\beta_{n+1})) = \beta_{n+1}$. This completes the proof of (4).

Next, by induction on the construction of δ we shall prove

$$\text{(5) if } g(0) \leq \delta < g(\omega), \text{ then there is } n \text{ such that } g(n) \leq \delta < g(n + 1).$$

If $\delta =_{NF} \omega^{\delta_1} + \cdots + \omega^{\delta_n}$, then this follows from the induction hypothesis, much like in the proof of (4).

Otherwise, $\delta = \psi_j^{\mathfrak{x}}(\delta)$ with $\delta_0 \in C_j^{\mathfrak{x}}(\delta_0)$ and $f(0) \leq \delta_0 < f(\Omega_{k+1})$. Since f is fundamental, we may find $\alpha < \Omega_{k+1}$ such that $f(\alpha) \leq \delta_0 < f(\alpha + 1)$, with $f(\alpha) \lhd f(\alpha + 1)$. Much like the proof of (4), this yields, $f(\alpha) \in C_j^{\mathfrak{x}}(\delta_0)$ and hence $\alpha \in C_j^{\mathfrak{x}}(\delta_0) \subseteq C_j^{\mathfrak{x}}(f(\Omega_{k+1}))$. This, in turn, gives us $\alpha < \psi_k^{\mathfrak{x}}(f(\Omega_{k+1}))$ so we may apply (4), to find $\beta_n \leq \alpha < \alpha + 1 < \beta_{n+1}$. Thus,

$$f(\beta_n) \leq f(\alpha) \leq \delta_0 < f(\alpha + 1) < f(\beta_{n+1})$$

which leads to

$$g(n) = \psi_j^{\mathfrak{x}}(f(\beta_n)) \leq \psi_j^{\mathfrak{x}}(\delta_0) < \psi_j^{\mathfrak{x}}(f(\beta_{n+1})) = g(n + 1).$$

And since $g(n) \lhd g(n + 1)$ follows from (3) by use of part 3 of Lemma 12.29, this completes the proof of property 2.

For the third property of fundamental functions note that if $\gamma \leq \omega$, then $\gamma \in C_h^{\mathfrak{x}}(\eta)$ for all h and all η. □

12.3 A Well-ordering Proof

In this section we investigate in which theories the representation system $\mathrm{OT}(\Omega_\omega\cdot\mathfrak{X})$ can be proved to be well-ordered if $\mathfrak{X}$ is a well-ordering. As a consequence, we will prove the part (i) $\Rightarrow$ (ii) of the main Theorem 12.8.

$\alpha,\beta,\gamma,\delta,\xi,\zeta,\rho$ will always denote elements of $\mathrm{OT}(\Omega_\omega\cdot\mathfrak{X})$. λ will be reserved for those of limit type.

By $\mathfrak{R}_0$ we shall denote the set $\{\Omega_n \mid n \in \mathbb{N}\}$. Variables ν,μ,τ are understood to range exclusively over elements of $\mathfrak{R}_0$. Let $\mu^+ := \Omega_{n+1}$ if $\mu = \Omega_n$. (Recall that $0 := \Omega_0$.)

Definition 12.45. *We assign a* level $\mathsf{S}\alpha \in \mathfrak{R}_0 \cup \{\Omega_\omega\}$ *to every* $\alpha \in \mathrm{OT}(\Omega_\omega\cdot\mathfrak{X})$.

1. $\mathsf{S}\Omega_n = \Omega_n$; $\mathsf{S}\alpha = \Omega_\omega$ *if* $\alpha \geq \Omega_\omega$.
2. *If* $\alpha = \omega^{\alpha_1} + \cdots + \omega^{\alpha_n}$, *then* $\mathsf{S}\alpha := \mathsf{S}\alpha_1$.
3. *If* $\alpha = \psi^{\mathfrak{X}}_m\beta$, *then* $\mathsf{S}\alpha = \Omega_m$.

Note that if $\alpha < \Omega_\omega$ *and* $\mathsf{S}\alpha = \Omega_n$, *then* $\Omega_n \leq \alpha < \Omega_{n+1}$.

For $\alpha < \Omega_\omega$ *let* $\alpha^+ = \Omega_{n+1}$ *if* $\mathsf{S}\alpha = \Omega_n$.

We use the following conventions: (α,β), $(\alpha,\beta]$, $[\alpha,\beta)$, and $[\alpha,\beta]$ denote the intervals of ordinals between α and β in the obvious sense. For a set of ordinals A we use the abbreviations $A < \alpha := (\forall\eta \in A)\,\eta < \alpha$, $A \leq \alpha := (\forall\eta \in A)\,\eta \leq \alpha$, $\alpha < A := (\exists\eta \in A)\,\alpha < \eta$, and $\alpha \leq A = (\exists\eta \in A)\,\alpha \leq \eta$.

12.3.1 Distinguished Sets

By a well-ordering proof in a given theory T we mean a proof formalizable in T which shows that a certain ordinal representation system (or a subset of it) is well-ordered. The notion of a *distinguished set* (of ordinals) (in German: ausgezeichnete Menge) will be central to carrying out well ordering proofs in the various subtheories of second order arithmetic introduced in earlier sections. The presentation in this chapter is very close to section 10 of [10] and section 11 of [11]. As a base theory in which all the results of this section can be proved one can take Π^1_1-$\mathbf{CA}_0$.

We start with a crucial definition.

Definition 12.46. *Let* E *be the set of elements of* $\mathrm{OT}(\Omega_\omega\cdot\mathfrak{X})$ *of either form* Ω_{m+1} *or* $\psi^{\mathfrak{X}}_m\alpha$.

The set of critical subterms of level μ *of* α, $\mathsf{E}_\mu(\alpha)$, *is defined inductively as follows:*

1. $\mathsf{E}_\mu(\Omega_0) = \emptyset$. $\mathsf{E}_\mu(\bar{u} + \alpha) = \mathsf{E}_\mu(\alpha)$ *for* $u \in X$.
2. $\mathsf{E}_\mu(\alpha) = \{\alpha\}$ *if* $\alpha \in \mathsf{E} \cap \mu^+$.
3. $\mathsf{E}_\mu(\alpha) = \bigcup\{\mathsf{E}_\mu(\alpha_i) \mid i = 1,\ldots,n\}$ *if if* $\alpha = \omega^{\alpha_1} + \cdots + \omega^{\alpha_n}$.
4. $\mathsf{E}_\mu(\Omega_n) = \emptyset$ *if* $\mu^+ \leq \Omega_n$.
5. $\mathsf{E}_\mu(\alpha) = \mathsf{E}_\mu(\gamma)$ *if* $\alpha = \psi^{\mathfrak{X}}_n\gamma$ *and* $\mu^+ \leq \alpha$.

Definition 12.47. *Let $U \subseteq \mathrm{OT}(\Omega_\omega \cdot \mathfrak{X})$ and $F(a)$ be an $\mathcal{L}_2$-formula.*

(i) $U \cap \alpha := \{\eta \in U \mid \eta < \alpha\}$.
(ii) $U \cap \alpha \subseteq F \;:\Leftrightarrow\; (\forall \eta \in U \cap \alpha)\, F(\eta)$.
(iii) $\mathrm{Prg}(U, F) \;:\Leftrightarrow\; \forall \eta \in U\, [U \cap \eta \subseteq F \to F(\eta)]$.
(iv) $\mathrm{W}[U] := \{\eta \in U \mid \forall Y[\mathrm{Prg}(U, Y) \to U \cap \eta \subseteq Y]\}$.
(v) $\mathrm{M}^U_\mu := \{\eta < \mu^+ \mid (\forall \nu \in U \cap \mu)\mathsf{E}_\nu(\eta) \subseteq U\}$.
(vi) $\mathrm{W}^U_\mu := \mathrm{W}[\mathrm{M}^U_\mu]$.

Remark 12.48. (i) *If $<_U$ denotes the restriction of $<$ to U and $F_U(a)$ is the formula $a \in U \to F(a)$ then* $\mathrm{Prg}(U, F) \leftrightarrow$ *Prog*$(<_U, F_U)$ *holds with Prog defined in section 12.1.1.*

(ii) M^U_μ *is always a set by arithmetical comprehension. To show that* $\mathrm{W}[U]$ *and* W^U_μ *are sets one can use Π^1_1 comprehension.* $\mathrm{W}[U]$ *and* W^U_μ *can also be shown to be sets in any theory which proves that the accessible part of an ordering R on $\mathbb{N}$ (where R is assumed to be a set) is a set.*

The next Lemma lists basic properties of $\mathrm{W}[U]$, M^U_μ and W^U_μ.

Lemma 12.49. (i) $\mathrm{Prg}(U, S) \to \mathrm{W}[U] \subseteq S$.
(ii) $\mathrm{Prg}(U, \mathrm{W}[U])$.
(iii) $U \subseteq V \;\wedge\; \mathrm{Prg}(U, S) \to \mathrm{Prg}(V, \{\eta \mid \eta \in U \to \eta \in S\})$.
(iv) $\mathrm{Prg}(\mathrm{W}[U], S) \;\to\; \mathrm{W}[U] \subseteq S$.
(v) $\mathrm{W}[\mathrm{W}[U]] = \mathrm{W}[U]$.
(vi) $\mathrm{W}[U \cap \alpha] \subseteq \mathrm{W}[U]$.
(vii) $U \cap \mu = V \cap \mu \;\to\; \mathrm{M}^U_\mu = \mathrm{M}^V_\mu \;\wedge\; \mathrm{W}^U_\mu = \mathrm{W}^V_\mu$.
(viii) $\alpha \in \mathrm{W}^U_\mu \;\leftrightarrow\; \alpha \in \mathrm{M}^U_\mu \;\wedge\; \mathrm{M}^U_\mu \cap \alpha \subseteq \mathrm{W}^U_\mu$.

Proof. (i) and (vii) are immediate by going back to the definitions.

(ii) Let $\alpha \in U$ and $U \cap \alpha \subseteq \mathrm{W}[U]$. By (i) we have $U \cap \alpha \subseteq S$ for every S satisfying $\mathrm{Prg}(U, S)$. Thence $\alpha \in \mathrm{W}[U]$.

(iii) Assume that $U \subseteq V$ and $\mathrm{Prg}(U, S)$ hold and also that $\alpha \in V$ and

$$V \cap \alpha \subseteq \{\eta \mid \eta \in U \to \eta \in S\}.$$

Then $U \cap \alpha = U \cap V \cap \alpha \subseteq S$, thus $\alpha \in U \to \alpha \in S$, i.e., $\alpha \in \{\eta \mid \eta \in U \to \eta \in S\}$.

(iv) Suppose $\mathrm{Prg}(W[U], S)$. (iii) implies $\mathrm{Prg}(U, \{\eta \mid \eta \in \mathrm{W}[U] \to \eta \in S\})$. Therefore, by (i), we also have $\mathrm{W}[U] \subseteq \{\eta \mid \eta \in \mathrm{W}[U] \to \eta \in S\}$, and hence $\mathrm{W}[U] \subseteq S$.

(v) $\mathrm{W}[\mathrm{W}[U]] \subseteq \mathrm{W}[U]$ holds by definition. Using (ii) we have $\mathrm{Prg}(\mathrm{W}[U], \mathrm{W}[\mathrm{W}[U]])$, hence, by (iv), $\mathrm{W}[U] \subseteq \mathrm{W}[\mathrm{W}[U]]$.

(vi) From

$$\eta \in U \cap \alpha \;\wedge\; \forall Y(\mathrm{Prg}(U \cap \alpha, Y) \to U \cap \alpha \cap \eta \subseteq Y)$$

we deduce that $\forall Y(\mathrm{Prg}(U, Y) \to U \cap \eta \subseteq Y)$, thence $\eta \in \mathrm{W}[U]$.

(viii) By (ii) we have $\mathrm{Prg}(\mathrm{M}^U_\mu, \mathrm{W}^U_\mu)$. W^U_μ is also a set. Thus (viii) follows. □

Definition 12.50. (i) *A set* $U \subseteq \mathrm{OT}(\Omega_\omega{\cdot}\mathfrak{X}) \cap \Omega_\omega$ *is said to be* distinguished *if* (D1) *and* (D2) *are satisfied:*

(D1) $(\forall \alpha \in U)\, \mathsf{S}\alpha \in U$.
(D2) $(\forall \mu \in U)\, U \cap \mu^+ = \mathrm{W}^U_\mu$.

(ii) *We shall use the abbreviation* $\mathrm{Ds}(U)$ *to convey that* U *is a distinguished set. Variables* P *and* Q *will always refer to distinguished sets.*
(iii) $\mathfrak{W} := \{\eta \mid \exists X\, [\mathrm{Ds}(X) \wedge \eta \in X]\}$.
Note that $\mathfrak{W}$ *cannot be shown to be a set in our background theory* Π^1_1-$\mathbf{CA}_0$ *(nor in* Π^1_1-$\mathbf{CA}$*).*

Lemma 12.51. *Recall that the letters* Q *and* P *are reserved for distinguished sets.*

(i) $Q \subseteq \mathrm{W}[Q]$ *and hence* $Q = \mathrm{W}[Q]$.
(ii) $\mathrm{Prg}(Q, V) \to Q \subseteq V$.

Proof. (i) Let $\alpha \in Q$. Then $\mathsf{S}\alpha \in Q$ by (D1) and hence $Q \cap \alpha^+ = \mathrm{W}^Q_{\mathsf{S}\alpha}$. So by Lemma 12.49(v),(vi) we arrive at $\alpha \in Q \cap \alpha^+ = \mathrm{W}^Q_{\mathsf{S}\alpha} = \mathrm{W}[\mathrm{W}^Q_{\mathsf{S}\alpha}] = \mathrm{W}[Q \cap \alpha^+] \subseteq \mathrm{W}[Q]$.
(ii) is an immediate consequence of (i) and Lemma 12.49(i). □

Owing to Lemma 12.51(ii) we have transfinite induction over $<_Q := < \cap (Q \times Q)$ for arbitrary sets. Thus if we want to show that $Q \subseteq V$ holds for a set V it suffices to prove that

$$\forall \beta(\beta \in Q \wedge Q \cap \beta \subseteq V \to \beta \in V).$$

Specifically we have $\mathrm{WO}(<_Q)$.

Lemma 12.52. (i) $\nu \leq \mu \wedge \beta \in \mathsf{E}_\mu(\alpha) \to \mathsf{E}_\nu(\beta) \subseteq \mathsf{E}_\nu(\alpha)$.
(ii) $\alpha \in Q \wedge \mu \in Q \to \mathsf{E}_\mu(\alpha) \subseteq Q$.
(iii) $\mu \in \mathrm{M}^Q_\mu$ *and* $(\forall \nu \in Q)\mathsf{E}_\nu(\mu) \subseteq Q$.
(iv) $\mu \leq Q \to \mu \in Q$.

Proof. (i) follows by induction on the construction of α.
(ii) 1. Suppose $\mu < \mathsf{S}\alpha$. Then (D1) and (D2) imply that $\alpha \in Q \cap \alpha^+ = \mathrm{W}^Q_{\mathsf{S}\alpha} \subseteq \mathrm{M}^Q_{\mathsf{S}\alpha}$. As $\mu \in Q \cap \mathsf{S}\alpha$ we see that $\mathsf{E}_\mu(\alpha) \subseteq Q$ by definition of $\mathrm{M}^Q_{\mathsf{S}\alpha}$.
2. Suppose $\mu \geq \mathsf{S}\alpha$. From (D2) it then follows that $\alpha \in \mathrm{W}^Q_\mu \subseteq \mathrm{M}^Q_\mu$. For $\nu \in Q \cap \mu$ we thus have $\mathsf{E}_\nu(\alpha) \subseteq Q$, and from (i) we conclude that $(\forall \beta \in \mathsf{E}_\mu(\alpha))\, \mathsf{E}_\nu(\beta) \subseteq Q$. Therefore $\mathsf{E}_\mu(\alpha) \subseteq \{\alpha\} \cup \mathrm{M}^Q_\mu \cap \alpha$. By Lemma 12.49(viii) we get $\mathsf{E}_\mu(\alpha) \subseteq \mathrm{W}^Q_\mu \subseteq Q$ as $\alpha \in \mathrm{W}^Q_\mu$.
(iii) Since $\mu < \mu^+$ and for $\nu < \mu$ we have $\mathsf{E}_\nu(\mu) = \emptyset$, it is obvious that $\mu \in \mathrm{M}^Q_\mu$. $(\forall \nu \in Q)\mathsf{E}_\nu(\mu) \subseteq Q$ will be proved by transfinite induction on Q (i.e., $<_Q$).
1. If $\nu \in Q \cap \mu^+$ then the desired assertion follows in the case $\nu < \mu$ from the definition of M^Q_μ and in the case $\nu = \mu$ from (ii).
2. If $\nu \in Q$ and $\mu < \nu$ then by induction hypothesis we have $(\forall \tau \in Q \cap \nu)\mathsf{E}_\tau(\mu) \subseteq Q$, and consequently $\mu \in \mathrm{M}^Q_\nu$. From $\nu \in Q \cap \nu^+ = \mathrm{W}^Q_\nu$ we obtain by Lemma

12.49(viii) that $\mathrm{M}^Q_\nu \cap \nu \subseteq \mathrm{W}^Q_\nu$, whence $\mu \in \mathrm{W}^Q_\nu$. Since $\mathsf{E}_\nu(\mu) \subseteq \{\mu\}$ we arrive at the desired assertion.

(iv) follows directly from (iii). □

Lemma 12.53. $Q \cap \mu^+ \subseteq \mathrm{W}^Q_\mu$.

Proof. Let $\alpha \in Q \cap \mu^+$. Then $\alpha \in \mathrm{W}^Q_{\mathsf{S}\alpha}$ and so by Lemma 12.49(viii), $\mathrm{M}^Q_{\mathsf{S}\alpha} \cap \alpha \subseteq \mathrm{W}^Q_{\mathsf{S}\alpha}$. In view of Lemma 12.49(vi) it suffices to show that $\alpha \in \mathrm{W}[\mathrm{M}^Q_\mu \cap \alpha^+]$. Lemma 12.52(ii) yields $\alpha \in \mathrm{M}^Q_\mu \cap \alpha^+$. Using Lemma 12.49(iii), $\mathrm{Prg}(\mathrm{M}^Q_\mu \cap \alpha^+, U)$ implies

$$\mathrm{Prg}(\mathrm{M}^Q_{\mathsf{S}\alpha}, \{\eta \mid \eta \in \mathrm{M}^Q_\mu \cap \alpha^+ \to \eta \in U\}),$$

and further, by Lemma 12.49(i),

$$\mathrm{M}^Q_\mu \cap \alpha \subseteq \mathrm{M}^Q_{\mathsf{S}\alpha} \cap \alpha \subseteq \mathrm{W}^Q_{\mathsf{S}\alpha} \subseteq \{\eta \mid \eta \in \mathrm{M}^Q_\mu \cap \alpha^+ \to \eta \in U\},$$

thence $\mathrm{M}^Q_\mu \cap \alpha^+ \cap \alpha \subseteq U$. This shows $\alpha \in \mathrm{W}[\mathrm{M}^Q_\mu \cap \alpha^+]$. □

Proposition 12.54. $\mathrm{M}^Q_\mu \cap \mu \subseteq Q \to \mu \in \mathrm{W}^Q_\mu \wedge \mathrm{Ds}(\mathrm{W}^Q_\mu)$.

Proof. By Lemma 12.53, $\mathrm{M}^Q_\mu \cap \mu \subseteq Q$ implies $\mathrm{M}^Q_\mu \cap \mu = \mathrm{W}^Q_\mu \cap \mu$. Thus, by Lemma 12.49(viii), $\mu \in \mathrm{M}^Q_\mu$ implies $\mu \in \mathrm{W}^Q_\mu$.

Next we show that W^Q_μ is a distinguished set.

Ad (D1): If $\alpha \in \mathrm{W}^Q_\mu \cap \mu$ then $\mathsf{S}\alpha \in Q \cap \mu \subseteq \mathrm{W}^Q_\mu \cap \mu$. From $\alpha \in \mathrm{W}^Q_\mu$ and $\mu \leq \alpha$ we obtain $\mathsf{S}\alpha = \mu \in \mathrm{W}^Q_\mu$.

Ad (D2): For $\tau \leq \mu$ we have $(*)$ $\mathrm{W}^Q_\mu \cap \tau = Q \cap \tau$ since $\mathrm{M}^Q_\mu \cap \mu \subseteq Q$ yields $\mathrm{W}^Q_\mu \cap \tau \subseteq Q$, and so, by Lemma 12.53, $Q \cap \tau \subseteq \mathrm{W}^Q_\mu$ holds. Now let $P := \mathrm{W}^Q_\mu$ and suppose $\nu \in P$. By $(*)$, we then have $P \cap \nu = Q \cap \nu$, and thus, by Lemma 12.49(viii), $(**)$ $\mathrm{W}^P_\nu = \mathrm{W}^Q_\nu$. For $\nu < \mu$, $(*)$ entails $\nu \in Q$ and therefore $\mathrm{W}^P_\nu = \mathrm{W}^Q_\nu = Q \cap \nu^+ \stackrel{(*)}{=} \mathrm{W}^Q_\mu \cap \nu^+ = P \cap \nu^+$. If $\nu = \mu$, then $(**)$ yields $\mathrm{W}^Q_\mu = P = P \cap \mu^+$. □

Vacuously $\emptyset$ is a distinguished set. Proposition 12.54 yields the existence of non-trivial distinguished sets. For example, $\mathrm{W}^\emptyset_0$ is a distinguished set.

Lemma 12.55. $\mathrm{Prg}(P \cup Q, U) \to P \cup Q \subseteq U$.

Proof. Suppose $\mathrm{Prg}(P \cup Q, U)$. Then we have

$$P \cap \alpha \subseteq U \to \mathrm{Prg}(Q, \{\eta \mid \eta < \alpha \to \eta \in U\}), \text{ and}$$
$$P \cap \alpha \subseteq U \ \wedge\ Q \cap \alpha \subseteq U \ \wedge\ \alpha \in P \to \alpha \in U.$$

Therefore, by Lemma 12.51(ii), we have

$$P \cap \alpha \subseteq U \ \wedge\ \alpha \in P \to \alpha \in U,$$

i.e., $\mathrm{Prg}(P, U)$ holds, and consequently $P \subseteq U$ by Lemma 12.51(ii). Similarly one shows that $Q \subseteq U$. □

Lemma 12.56. $\mu \in P \cup Q \ \wedge \ \mu \leq P \ \wedge \ \mu \leq Q \rightarrow P \cap \mu^+ = Q \cap \mu^+$.

Proof. We use induction on $P \cup Q$, i.e., Lemma 12.55. Let $\mu \in P$ and suppose $\mu \leq Q$. The induction hypothesis yields $P \cap \mu = Q \cap \mu$ and, by Lemma 12.49(vii), we conclude that $\mu \in P \cap \mu^+ = \mathrm{W}^P_\mu = \mathrm{W}^Q_\mu \subseteq \mathrm{M}^Q_\mu$, and hence $\mu \in Q$ by Lemma 12.52(iv). As a result, $P \cap \mu^+ = \mathrm{W}^P_\mu = \mathrm{W}^Q_\mu = Q \cap \mu^+$. The same arguments can be used if $\mu \in Q$ and $\mu \leq P$. □

Proposition 12.57. $\alpha \in Q \rightarrow Q \cap \alpha^+ = \mathfrak{W} \cap \alpha^+$.

Proof. Let $\alpha \in Q$. $Q \cap \alpha^+ \subseteq \mathfrak{W} \cap \alpha^+$ is obvious by definition of $\mathfrak{W}$. Let $\eta \in \mathfrak{W} \cap \alpha^+$. Then there exists a distinguished set P such that $\eta \in P \cap \alpha^+$. Thus $\mathsf{S}\eta \in P \cup Q$, $\mathsf{S}\eta \leq \eta \in P$ and $\mathsf{S}\eta \leq \alpha \in Q$. Therefore $\eta \in P \cap \eta^+ = Q \cap \eta^+ \subseteq Q \cap \alpha^+$ using Lemma 12.56. □

Next we study closure properties shared by all distinguished sets.

Proposition 12.58. (i) $\alpha, \beta \in Q \rightarrow \alpha + \beta \in Q$.
(ii) $\alpha, \beta \in \mathfrak{W} \rightarrow \alpha + \beta \in \mathfrak{W}$.

Proof. (ii) is an immediate consequence of (i) in view of Proposition 12.57. In the proof of (i) let $X := \mathrm{M}^Q_{\mathsf{S}\alpha}$, $Y := \mathrm{W}^Q_{\mathsf{S}\alpha}$ and $U := \{\xi \mid \alpha + \xi \in Y\}$. Suppose $\alpha, \beta \in Q$. If $\mathsf{S}\alpha < \mathsf{S}\beta$ then $\alpha + \beta = \beta \in Q$. Now assume $\mathsf{S}\beta \leq \mathsf{S}\alpha$. Then we have $Q \cap \alpha^+ = Y$ and $\alpha, \beta \in Y$. Moreover we have

$$\eta \in X \ \wedge \ X \cap \eta \subseteq U \rightarrow \alpha + \eta \in X \ \wedge \ X \cap (\alpha + \eta) \subseteq Y,$$

so that with Lemma 12.49(viii) we get $\eta \in X \ \wedge \ X \cap \eta \subseteq U \rightarrow \alpha + \eta \in Y$. As a result, $\mathrm{Prg}(X, U)$ holds, and thus $Y \subseteq U$ by Lemma 12.49(i), hence $\alpha + \beta \in Y \subseteq Q$. □

Lemma 12.59. *Letting* $\mathfrak{F}(\beta)$ *be the formula*

$$\beta \in Q \ \wedge \ (\forall \eta \in Q \cap \beta)(\omega^\eta \in Q),$$

the following are true:

(i) $\mathfrak{F}(\beta) \ \wedge \ \mu = \mathsf{S}\beta \ \wedge \ \gamma \in \mathrm{M}^Q_\mu \cap \omega^\beta \rightarrow \gamma \in Q$.
(ii) $\mathfrak{F}(\beta) \rightarrow \omega^\beta \in Q$.

Proof. We show (i) by induction on the construction of γ. $\mathfrak{F}(\beta)$ implies $\beta \in Q \cap \mu^+ = \mathrm{W}^Q_\mu$. We distinguish cases according to the shape of γ. The assertion is trivially true if $\gamma = 0$. Let $\gamma = \omega^{\gamma_1} + \cdots + \omega^{\gamma_n}$. Then $\gamma_1, \ldots, \gamma_n \in \mathrm{M}^Q_\mu \cap \beta$, and thus by the induction hypothesis, $\gamma_1, \ldots, \gamma_n \in Q$. $\mathfrak{F}(\beta)$ also yields $\omega^{\gamma_1}, \ldots, \omega^{\gamma_n} \in Q$, and hence $\gamma \in Q$ by Proposition 12.58.

If $\gamma \in \mathsf{E}$ then $\gamma \leq \beta$, and therefore, as $\beta \in \mathrm{W}^Q_\mu$ and $\gamma \in \mathrm{M}^Q_\mu$, it follows from Lemma 12.49(viii) that $\gamma \in \mathrm{W}^Q_\mu \subseteq Q$.

(ii) By (i) we have

$$\mathfrak{F}(\beta) \ \wedge \ \mu = \mathsf{S}\beta \to \mathrm{M}^Q_\mu \cap \omega^\beta \subseteq \mathrm{W}^Q_\mu .$$

By Lemma 12.52(ii) we also have

$$\mathfrak{F}(\beta) \ \wedge \ \mu = \mathsf{S}\beta \to \omega^\beta \in \mathrm{M}^Q_\mu .$$

Thus, by Lemma 12.49(viii),

$$\mathfrak{F}(\beta) \ \wedge \ \mu = \mathsf{S}\beta \to \omega^\beta \in \mathrm{W}^Q_\mu ,$$

and hence $\mathfrak{F}(\beta) \to \omega^\beta \in Q$. □

Proposition 12.60. (i) $\beta \in Q \to \omega^\beta \in Q$.
(ii) $\beta \in \mathfrak{W} \to \omega^\beta \in \mathfrak{W}$.

Proof. Again, by Proposition 12.57, (ii) is an immediate consequence of (i). Let $V := \{\eta \in Q \mid \omega^\eta \in Q\}$. Lemma 12.59(ii) yields $\mathrm{Prg}(Q, V)$ and hence, using Lemma 12.51(ii), $Q \subseteq V$. □

Corollary 12.61. (i) $\mathsf{S}\alpha \leq \mu \ \wedge \ \mu \in Q \ \wedge \ \mathsf{E}_\mu(\alpha) \subseteq Q \to \alpha \in Q$.
(ii) $]\ \mathsf{S}\alpha \leq \mu \ \wedge \ \mu \in \mathfrak{W} \ \wedge \ \mathsf{E}_\mu(\alpha) \subseteq Q \to \alpha \in \mathfrak{W}$.

Proof. This follows from Propositions 12.58 and 12.60. □

Lemma 12.62. (i) $\beta \in Q \ \wedge \ \alpha \in \mathrm{M}^Q_{\mathsf{S}\beta} \cap \beta \to \alpha \in Q$.
(ii) $\beta \in \mathfrak{W} \ \wedge \ \alpha \in \mathrm{M}^Q_{\mathsf{S}\beta} \cap \beta \to \alpha \in \mathfrak{W}$.

Proof. (i) $\beta \in Q$ implies $\beta \in Q \cap \beta^+ = \mathrm{W}^Q_{\mathsf{S}\beta}$. Therefore, by Lemma 12.49(viii), $\alpha \in \mathrm{W}^Q_{\mathsf{S}\beta} \subseteq Q$. (ii) is an immediate consequence of (i). □

With the help of full induction one can show that every Ω_n is in $\mathfrak{W}$.

Corollary 12.63. (Π^1_1-**CA**) *For every n*, $\Omega_n \in \mathfrak{W}$.

Proof. We use induction on n. This is immediate for Ω_0. Suppose $\Omega_n \in Q$ with Q being a distinguished set. We want to show that

$$\mathrm{M}^Q_{\Omega_{n+1}} \cap \Omega_{n+1} \subseteq Q. \tag{12.4}$$

Now, $\beta \in \mathrm{M}^Q_{\Omega_{n+1}} \cap \Omega_{n+1}$ entails $\mathsf{E}_{\Omega_n}(\beta) \subseteq Q$ and $\mathsf{S}\beta \leq \Omega_n$. Thus $\beta \in Q$ follows from Lemma 12.62(i), showing (12.4).

With the help of (12.4) and Proposition 12.54 we get $\Omega_{n+1} \in \mathrm{W}^Q_{\Omega_{n+1}}$ and, moreover, that $\mathrm{W}^Q_{\Omega_{n+1}}$ is a distinguished set. □

Definition 12.64. *Let*

$$\mathfrak{B}^Q_\mu := \{\xi \mid (\forall n \in \mathbb{N})[\Omega_n \in Q \cap \mu \ \wedge \ \xi \in C^{\mathfrak{x}}_n(\xi) \rightarrow \psi^{\mathfrak{x}}_n \xi \in Q]\}.$$

Lemma 12.65. *Assume $\alpha \in \mathrm{M}^Q_\mu$, $\mathrm{M}^Q_\mu \cap \alpha \subseteq \mathfrak{B}^Q_\mu$, $\nu = \Omega_n$, $\nu \in Q \cap \mu$, $\alpha \in C^{\mathfrak{x}}_n(\alpha)$, and $\gamma \in \mathrm{M}^Q_\nu \cap \psi^{\mathfrak{x}}_n\alpha$. Then we have $\gamma \in Q$.*

Proof. We proceed by induction on the buildup of γ.

If $\gamma \le \nu$ then $\gamma \in Q$ by Lemma 12.62(i). Now let $\Omega_n < \gamma$.

1. Suppose $\gamma = \omega^{\gamma_1} + \ldots + \omega^{\gamma_n}$. By the induction hypothesis, $\gamma_1, \ldots, \gamma_n \in Q$ and hence $\gamma \in Q$ by Propositions 12.58 and 12.60.

2. Suppose $\gamma = \psi^{\mathfrak{x}}_n\eta$. Then $\eta < \alpha$. By Lemma 12.52(i), $\gamma \in \mathrm{M}^Q_\nu$ entails that

$$(\forall \tau \in Q \cap \nu)(\forall \beta \in \mathsf{E}_\nu(\eta))\, \mathsf{E}_\tau(\beta) \subseteq \mathsf{E}_\tau(\eta) \subseteq Q.$$

Since $\mathsf{E}_\nu(\eta) < \psi^{\mathfrak{x}}_n\eta < \psi^{\mathfrak{x}}_n\alpha$, the latter entails that $\mathsf{E}_\nu(\eta) \subseteq \mathrm{M}^Q_\nu \cap \psi^{\mathfrak{x}}_n\alpha$, and therefore, by the induction hypothesis, $\mathsf{E}_\nu(\eta) \subseteq Q$. As a result we have shown that

$$(\forall \tau \le \nu)[\tau \in Q \cap \mu \rightarrow \mathsf{E}_\tau(\eta) \subseteq Q]. \tag{12.5}$$

Via a subsidiary induction on Q we shall show that

$$(\forall \tau \in Q \cap \mu)\, \mathsf{E}_\tau(\eta) \subseteq Q. \tag{12.6}$$

Let $\tau \in Q \cap \mu$. In view of (12.5), we may assume that $\nu < \tau$. So τ is of the form Ω_m for some $m > n$. The subsidiary induction hypothesis yields $(\forall \tau' \in Q \cap \tau)\, \mathsf{E}_{\tau'}(\eta) \subseteq Q$, which implies $\mathsf{E}_\tau(\eta) \subseteq \mathrm{M}^Q_\tau$. Since $\nu < \tau$, $\eta \in C^{\mathfrak{x}}_n(\eta)$ and $\alpha \in C^{\mathfrak{x}}_n(\alpha)$ hold, we conclude that $\eta \in C^{\mathfrak{x}}_m(\eta)$, $\alpha \in C^{\mathfrak{x}}_m(\alpha)$, and $\mathsf{E}_\tau(\eta) < \psi^{\mathfrak{x}}_m\eta < \psi^{\mathfrak{x}}_m\alpha$. Therefore, $\mathsf{E}_\tau(\eta) \subseteq \mathrm{M}^Q_\tau \cap \psi^{\mathfrak{x}}_m\alpha$, and consequently, by applying the main induction hypothesis, $\mathsf{E}_\tau(\eta) \subseteq Q$. This completes the proof of (12.6). Finally, from (12.6) we conclude that $\eta \in \mathrm{M}^Q_\mu \cap \alpha \subseteq \mathfrak{B}^Q_\mu$, yielding $\gamma = \psi^{\mathfrak{x}}_n\eta \in Q$. □

Lemma 12.66. $\mathrm{Prg}(\mathrm{M}^Q_\mu, \mathfrak{B}^Q_\mu)$.

Proof. Let $\alpha \in \mathrm{M}^Q_\mu$ and $\mathrm{M}^Q_\mu \cap \alpha \subseteq \mathfrak{B}^Q_\mu$. We have to show $\alpha \in \mathfrak{B}^Q_\mu$. So suppose $\nu \in Q \cap \mu$, $\nu = \Omega_n$, and $\alpha \in C^{\mathfrak{x}}_n(\alpha)$. By Lemma 12.65 we have $\mathrm{M}^Q_\nu \cap \psi^{\mathfrak{x}}_n\alpha \subseteq \mathrm{W}^Q_\nu$. For $\tau \in Q \cap \nu$ it holds $\mathsf{E}_\tau(\psi^{\mathfrak{x}}_n\alpha) = \mathsf{E}_\tau(\alpha)$ and therefore, using Lemma 12.52(ii), $\mathsf{E}_\tau(\psi^{\mathfrak{x}}_n\alpha) \subseteq Q$ since $\nu \in Q$ and $\alpha \in \mathrm{M}^Q_\mu$. Thus $\psi^{\mathfrak{x}}_n\alpha \in \mathrm{M}^Q_\nu$, so that by Lemma 12.49(viii) we have $\psi^{\mathfrak{x}}_n\alpha \in \mathrm{W}^Q_\nu \subseteq Q$. This shows $\alpha \in \mathfrak{B}^Q_\mu$. □

Lemma 12.67. (i) $\alpha, \Omega_n \in Q \ \wedge \ \alpha \in C^{\mathfrak{x}}_n(\alpha) \rightarrow \psi^{\mathfrak{x}}_n\alpha \in Q$.
(ii) $\alpha \in \mathfrak{W} \ \wedge \ \alpha \in C^{\mathfrak{x}}_n(\alpha) \rightarrow \psi^{\mathfrak{x}}_n\alpha \in \mathfrak{W}$.

Proof. (ii) is a consequence of (i) and Corollary 12.63. For (i), let $\tau := \max\{\mathsf{S}\alpha, \Omega_n\}$ and $\mu := \tau^+$. By Lemmata 12.66 and 12.49(i), we have $\mathrm{W}^Q_\mu \subseteq \mathfrak{B}^Q_\mu$. Therefore, since $\tau \in Q$, we have $Q \cap \mu \subseteq \mathfrak{B}^Q_\mu$, and hence $\psi^{\mathfrak{x}}_n\alpha \in Q$. □

Definition 12.68. *Let*

$$\mathfrak{H}(u) := u \in X \;\wedge\; (\forall n \in \mathbb{N})(\forall \eta \in \mathfrak{W})\,[\eta \in C_n^{\mathfrak{x}}(\bar{u}+\eta) \rightarrow \psi_n^{\mathfrak{x}}(\bar{u}+\eta) \in \mathfrak{W}];$$

$$\mathfrak{A}^u(\alpha) := u \in X \;\wedge\; \alpha \in \mathfrak{W} \;\wedge\; \alpha \in C_n^{\mathfrak{x}}(\bar{u}+\alpha) \;\wedge\; (\forall \eta \in \mathfrak{W} \cap \alpha)(\forall m \in \mathbb{N})[\eta \in C_m^{\mathfrak{x}}(\bar{u}+\eta) \rightarrow \psi_m^{\mathfrak{x}}(\bar{u}+\eta) \in \mathfrak{W}].$$

For the remainder of this subsection we work in Π_1^1-**CA**, i.e., Π_1^1-**CA**$_0$ plus the schema of induction on naturals for the full language.

Lemma 12.69.

$$(\forall v <_X u)\,\mathfrak{H}(v) \;\wedge\; \mathfrak{A}^u(\alpha) \rightarrow (\forall \gamma)(\forall n)[\gamma \in \mathrm{M}_{\Omega_n}^{\mathfrak{W}} \cap \psi_n^{\mathfrak{x}}(\bar{u}+\alpha) \rightarrow \gamma \in \mathfrak{W}].$$

Proof. Assume the antecedent of the implication we have to verify. Let $\gamma \in \mathrm{M}_{\Omega_n}^{\mathfrak{W}} \cap \psi_n^{\mathfrak{x}}(\bar{u}+\alpha)$. We shall carry out an induction on the construction of γ in order to show $\gamma \in \mathfrak{W}$, distinguishing between the different shapes γ might take.

1. $\gamma \le \Omega_n$: Then $\gamma \in \mathfrak{W}$ follows from Lemma 12.62(ii). Henceforth assume $\gamma > \Omega_n$.
2. Let $\gamma = \omega^{\gamma_1} + \cdots + \omega^{\gamma_r}$. Then $\gamma_j \in \mathrm{M}_{\Omega_n}^{\mathfrak{W}} \cap \psi_n^{\mathfrak{x}}(\bar{u}+\alpha)$ and therefore, by the inductive assumption, $\gamma_j \in \mathfrak{W}$, thus $\gamma \in \mathfrak{W}$ by 12.58 and 12.60.
3. In the remaining cases we have $\gamma = \psi_n^{\mathfrak{x}}\gamma'$ for some $\gamma' < \bar{u}+\alpha$. Since $\gamma \in \mathrm{M}_{\Omega_n}^{\mathfrak{W}}$, Lemma 12.52(i) entails that

$$(\forall m < n)(\forall \xi \in \mathrm{E}_{\Omega_n}(\gamma'))\;\mathrm{E}_{\Omega_m}(\xi) \subseteq \mathfrak{W}.$$

The latter implies $\mathrm{E}_{\Omega_n}(\gamma') \subseteq \mathrm{M}_{\Omega_n}^{\mathfrak{W}} \cap \psi_n^{\mathfrak{x}}\gamma' \subseteq \mathrm{M}_{\Omega_n}^{\mathfrak{W}} \cap \psi_n^{\mathfrak{x}}(\bar{u}+\alpha)$. Thus

$$\mathrm{E}_{\Omega_n}(\gamma') \subseteq \mathfrak{W} \tag{12.7}$$

by the induction hypothesis. Next we show, via a subsidiary induction on k, that

$$\mathrm{E}_{\Omega_k}(\gamma') \subseteq \mathfrak{W}. \tag{12.8}$$

For $k = n$ then this follows from (12.7). If $k < n$ then $\mathrm{E}_{\Omega_k}(\gamma') \subseteq \mathrm{E}_{\Omega_k}(\gamma) \subseteq \mathfrak{W}$ since $\gamma \in \mathrm{M}_{\Omega_n}^{\mathfrak{W}}$.

Now assume that $n < k$. Since $\mathrm{E}_{\Omega_k}(\gamma') < \psi_k^{\mathfrak{x}}\gamma' < \psi_k^{\mathfrak{x}}(\bar{u}+\alpha)$, the subsidiary induction hypothesis together with Lemma 12.52(i) yields $\mathrm{E}_{\Omega_k}(\gamma') \subseteq \mathrm{M}_{\Omega_k}^{\mathfrak{W}} \cap \psi_k^{\mathfrak{x}}(\bar{u}+\alpha)$. Thus, by the main induction hypothesis, $\mathrm{E}_{\Omega_k}(\gamma') \subseteq \mathfrak{W}$. This completes the proof of (12.8).

To conclude that $\gamma \in \mathfrak{W}$ we need to consider the different shapes of γ'.

   a. Suppose $\gamma' = \bar{u}+\alpha'$ for some $\alpha' < \alpha$. (12.8) implies $\alpha' \in \mathfrak{W}$ and thus it follows from $\mathfrak{A}^u(\alpha)$ that $\gamma \in \mathfrak{W}$.
   b. Suppose $\gamma' = \bar{v}+\alpha'$ for some $v <_X u$ and $\alpha' < \Omega_\omega$. Again, by (12.8), we have $\alpha' \in \mathfrak{W}$ and thus $\gamma \in \mathfrak{W}$ follows from $\mathfrak{H}(v)$.
   c. Suppose $\gamma' < \Omega_\omega$. Then (12.8) yields $\gamma \in \mathfrak{W}$, so that, by Lemma 12.67(ii), we conclude $\gamma \in \mathfrak{W}$. □

Lemma 12.70. $u \in X \ \wedge\ (\forall u' <_X u)\mathfrak{H}(u') \rightarrow \mathfrak{H}(u)$.

Proof. Assume $u \in X$ and $(\forall u' <_X u)\mathfrak{H}(u')$. From $\mathfrak{A}^u(\alpha)$ we can infer with the help of 12.69 that $\mathrm{M}^{\mathfrak{W}}_{\Omega_n} \cap \psi^{\mathfrak{x}}_n(\bar{u}+\alpha) \subseteq \mathfrak{W}$, and hence $\psi^{\mathfrak{x}}_n(\bar{u}+\alpha) \in \mathfrak{W}$ by Lemma 12.49(viii). This shows

$$\mathfrak{A}^u(\alpha) \rightarrow \psi^{\mathfrak{x}}_n(\bar{u}+\alpha) \in \mathfrak{W}. \tag{12.9}$$

We want to show

$$(\forall n)(\forall \alpha \in \mathfrak{W})\,[\alpha \in C^{\mathfrak{x}}_n(\bar{u}+\alpha) \rightarrow \psi^{\mathfrak{x}}_n(\bar{u}+\alpha) \in \mathfrak{W}]. \tag{12.10}$$

So let Q be a distinguished set with $\Omega_n \in Q$. Since $\mathfrak{W} \cap \Omega_{n+1} = Q \cap \Omega_{n+1}$ it suffices to show that if $\alpha \in Q \cap \Omega_{n+1}$ and $\mathsf{K}_n\alpha < \bar{u}+\alpha$ hold true then $\psi^{\mathfrak{x}}_n(\bar{u}+\alpha) \in Q$. We use induction on Q with α being the variable of induction. By induction hypothesis we then have

$$(\forall \eta \in Q \cap \alpha)(\forall \Omega_m \in Q \cap \Omega_{n+1})[\eta \in C^{\mathfrak{x}}_m(\bar{u}+\eta) \rightarrow \psi^{\mathfrak{x}}_m(\bar{u}+\eta) \in Q].$$

But then (12.9) implies $\psi^{\mathfrak{x}}_n(\bar{u}+\alpha) \in Q$. □

Theorem 12.71. *In* (Π^1_1-**CA**) *plus the schema of transfinite induction along* $\mathfrak{X}$ *we can prove the following.*

(i) $(\forall \alpha < \Omega_\omega)\, \alpha \in \mathfrak{W}$.
(ii) $\mathrm{OT}(\Omega_\omega{\cdot}\mathfrak{X})$ *is a well-ordering.*

Proof. (i). From Lemma 12.70 we conclude $\forall u \in X\, \mathfrak{H}(u)$. With the help of the latter, Lemma 12.58, Lemma 12.60, and Lemma 12.67, one shows the assertion by induction on the generation of α.

(ii). Let $U \subseteq \mathrm{OT}(\Omega_\omega{\cdot}\mathfrak{X})$ be non-empty. We have to show that U contains a least element with respect to the ordering $<$ of $\mathrm{OT}(\Omega_\omega{\cdot}\mathfrak{X})$.

Case 1: If U contains an element $\rho < \Omega_\omega$, then by (i) we can choose a distinguished set Q with $\rho \in Q$. Picking the least element of $Q \cap U$ furnishes us with the least element of U.

Case 2: Otherwise all elements of U are of the form $\bar{u}+\alpha$ with $u \in X$ and $\alpha < \Omega_\omega$. Then first pick the $<_X$-least $u \in X$ such that $\bar{u}+\alpha \in U$ for some $\alpha < \Omega_\omega$, and denote it by v. There exists $\alpha < \Omega_\omega$ such that $\bar{v}+\alpha \in U$. By (i), there is a distinguished set Q such that $\alpha \in Q$. As a consequence, we may pick the least element α_0 in Q such that $\bar{v}+\alpha_0 \in U$, which by design is the least element of U. □

Corollary 12.72. *In* $\mathbf{RCA}_0$, *the statement*

Every set X is contained in a countable coded ω-model of Π^1_1-$\mathbf{CA}_0$

yields

$$\forall \mathfrak{X}\,[\mathrm{WO}(\mathfrak{X}) \rightarrow \mathrm{WO}(\Omega_\omega \cdot \mathfrak{X})].$$

Proof. We argue in $\mathbf{RCA}_0$, assuming that every set X is contained in a countable coded ω-model of Π^1_1-$\mathbf{CA}_0$. Assume WO($\mathfrak{X}$). Then pick an ω-model $\mathfrak{B}$ of Π^1_1-$\mathbf{CA}_0$ containing $\mathfrak{X}$ and another arbitrary set U. Inside $\mathfrak{B}$ the schema of transfinite induction along $\mathfrak{X}$ holds and $\mathfrak{B}$ is a model of Π^1_1-$\mathbf{CA}_0$ with full induction on $\mathbb{N}$. As a result, Theorem 12.71 holds in $\mathfrak{B}$, and thus there is a least element α_0 of $U \cap \mathrm{OT}(\Omega_\omega{\cdot}\mathfrak{X})$ in that model if $U \cap \mathrm{OT}(\Omega_\omega{\cdot}\mathfrak{X})$ is non-empty. But that's also the least element of $U \cap \mathrm{OT}(\Omega_\omega{\cdot}\mathfrak{X})$ in the ambient world on account of absoluteness. □

12.4 Deduction Chains

From now on we will be concerned with the part (ii) $\Rightarrow$ (i) of the main Theorem 12.8. An important tool will be the method of deduction chains. Given a sequent Γ and a set $Q \subseteq \mathbb{N}$, deduction chains starting at Γ are built by systematically decomposing Γ into its subformulas, and adding additionally at the nth step the formulas $\neg A_n$ and $\neg\bar{Q}(\bar{n})$, where $(A_n \mid n \in \mathbb{N})$ is an enumeration of all the instances of Π^1_1-comprehension of the form

$$\exists Y\,\forall x\,[x \in Y \leftrightarrow F(x)] \tag{12.11}$$

where $F(a)$ is a Π^1_1-formula ($F(a)$ may contain further free set variables but no number variables), and $\bar{Q}(\bar{n})$ is the atom $\bar{n} \in U_0$ if $n \in Q$ and $\bar{n} \notin U_0$ otherwise. The set of all deduction chains that can be built from the empty sequent with respect to a given set Q forms the tree $\mathcal{D}_Q$. There are two scenarios to be considered.

(i) If there is an infinite deduction chain, i.e., $\mathcal{D}_Q$ is ill-founded, then this readily yields a model of Π^1_1-comprehension that contains Q.
(ii) If each deduction chain is finite, then this yields a derivation of the empty sequent, $\bot$, in a corresponding infinitary system with the ω-rule and Buchholz's Ω_{n+1}-rules for $n \in \mathbb{N}$. The depth of this derivation is bounded by the order-type α of the Kleene-Brouwer ordering of $\mathcal{D}_Q$. By the well-ordering principle, transfinite induction on $\Omega_\omega \cdot \alpha$ is available, which allows one to transform this proof into a cut-free proof of $\bot$.

As the second alternative is impossible, the first yields the desired model.

Definition 12.73. 1. We let $U_0, U_1, \ldots, U_m, \ldots$ be an enumeration of the free set variables of $\mathcal{L}_2$ and, given a closed term t, we write $t^{\mathbb{N}}$ for its numerical value.
2. Henceforth a *sequent* will be a finite list of $\mathcal{L}_2$-formulae *without* free number variables. For a sequent Γ and formula F we write $F \in \Gamma$ to convey that F appears in the list Γ.
3. A sequent Γ is *axiomatic* if it satisfies at least one of the following conditions:

   a. Γ contains a true *literal*, i.e., a true formula of either of the forms $R(t_1, \ldots, t_n)$ or $\neg R(t_1, \ldots, t_n)$, where R is a predicate symbol in $\mathcal{L}_2$ for a primitive recursive relation and $t_1, \ldots, t_n$ are closed terms.

b. Γ contains formulae $s \in U$ and $t \notin U$ for some set variable U and terms s, t with $s^{\mathbb{N}} = t^{\mathbb{N}}$.

4. A sequent is *reducible* if it is not axiomatic and contains a formula which is not a literal.

Definition 12.74. For $Q \subseteq \mathbb{N}$ we define

$$\bar{Q}(n) \Leftrightarrow \begin{cases} \bar{n} \in U_0 & \text{if } n \in Q, \\ \bar{n} \notin U_0 & \text{otherwise.} \end{cases}$$

Definition 12.75. In what follows, we fix an enumeration of $A_0, A_2, A_3, \ldots$ of all formulae of the form (12.11). Let $Q \subseteq \mathbb{N}$.

A *Q-deduction chain* is a finite string

$$\Gamma_0, \Gamma_1, \ldots, \Gamma_k$$

of sequents Γ_i constructed according to the following rules:

1. $\Gamma_0 = \neg\bar{Q}(0), \neg A_0$.
2. Γ_i is not axiomatic for $i < k$.
3. If $i < k$ and Γ_i is not reducible, then

$$\Gamma_{i+1} = \Gamma_i, \neg\bar{Q}(i+1), \neg A_{i+1}.$$

4. Every reducible Γ_i with $i < k$ is of the form

$$\Gamma_i', E, \Gamma_i''$$

where E is not a literal and Γ_i' contains only literals. E is said to be the *redex* of Γ_i.

Let $i < k$ and Γ_i be reducible. Γ_{i+1} is obtained from $\Gamma_i = \Gamma_i', E, \Gamma_i''$ as follows:

a. If $E \equiv E_0 \vee E_1$, then

$$\Gamma_{i+1} = \Gamma_i', E_0, E_1, \Gamma_i'', \neg\bar{Q}(i+1), \neg A_{i+1}.$$

b. If $E \equiv E_0 \wedge E_1$, then

$$\Gamma_{i+1} = \Gamma_i', E_j, \Gamma_i'', \neg\bar{Q}(i+1), \neg A_{i+1}$$

where $j = 0$ or $j = 1$.

c. If $E \equiv \exists x F(x)$, then

$$\Gamma_{i+1} = \Gamma_i', F(\bar{m}), \Gamma_i'', \neg\bar{Q}(i+1), \neg A_{i+1}, E$$

where m is the first number such that $F(\bar{m})$ does not occur in $\Gamma_0, \ldots, \Gamma_i$.

d. If $E \equiv \forall x F(x)$, then

$$\Gamma_{i+1} = \Gamma_i', F(\bar{m}), \Gamma_i'', \neg\bar{Q}(i+1), \neg A_{i+1}$$

for some m.

e. If $E \equiv \exists X F(X)$, then

$$\Gamma_{i+1} = \Gamma_i', F(U_m), \Gamma_i'', \neg\bar{Q}(i+1), \neg A_{i+1}, E$$

where m is the first number such that $F(U_m)$ does not occur in $\Gamma_0, \ldots, \Gamma_i$.

f. If $E \equiv \forall X F(X)$, then

$$\Gamma_{i+1} = \Gamma_i', F(U_m), \Gamma_i'', \neg\bar{Q}(i+1), \neg A_{i+1}$$

where m is the first number such that U_m does not occur in $\Gamma_i, \neg\bar{Q}(i+1), \neg A_{i+1}$.

The set of Q-deduction chains forms a tree $\mathcal{D}_Q$ labeled with strings of sequents. We will now consider two cases.

Case I. $\mathcal{D}_Q$ is not well-founded. Then $\mathcal{D}_Q$ contains an infinite path $\mathbb{P}$. Now define a set M via

$$(M)_i = \{k \mid \bar{k} \notin U_i \text{ occurs in } \mathbb{P}\}.$$

Set $\mathbb{M} = (\mathbb{N}; \{(M)_i \mid i \in \mathbb{N}\}, \in, +, \cdot, 0, 1, <)$.

For a formula F, let $F \in \mathbb{P}$ mean that F occurs in $\mathbb{P}$, i.e., $F \in \Gamma$ for some $\Gamma \in \mathbb{P}$.

Claim. Under the assignment $U_i \mapsto (M)_i$ we have

$$F \in \mathbb{P} \quad \Rightarrow \quad \mathbb{M} \models \neg F. \tag{12.12}$$

The Claim will imply that $\mathbb{M}$ is an ω-model of Π^1_1-comprehension. Also note that $(M)_0 = Q$, thus Q is in $\mathbb{M}$. The proof of (12.12) follows by induction on F using Lemma 12.76 below. The upshot of the foregoing is that we can prove Theorem 12.8 (ii)→(i) under the assumption that $\mathcal{D}_Q$ is ill-founded for all sets $Q \subseteq \mathbb{N}$.

Lemma 12.76. *Let Q be an arbitrary subset of $\mathbb{N}$ and $\mathcal{D}_Q$ be the corresponding deduction tree. Moreover, suppose $\mathcal{D}_Q$ is not well-founded. Then $\mathcal{D}_Q$ has an infinite path $\mathbb{P}$. $\mathbb{P}$ has the following properties:*

1. *$\mathbb{P}$ does not contain literals which are true in $\mathbb{N}$.*
2. *$\mathbb{P}$ does not contain formulas $s \in U_i$ and $t \notin U_i$ for constant terms s and t such that $s^{\mathbb{N}} = t^{\mathbb{N}}$.*
3. *If $\mathbb{P}$ contains $E_0 \vee E_1$, then $\mathbb{P}$ contains E_0 and E_1.*
4. *If $\mathbb{P}$ contains $E_0 \wedge E_1$, then $\mathbb{P}$ contains E_0 or E_1.*
5. *If $\mathbb{P}$ contains $\exists x F(x)$, then $\mathbb{P}$ contains $F(\bar{n})$ for all n.*
6. *If $\mathbb{P}$ contains $\forall x F(x)$, then $\mathbb{P}$ contains $F(\bar{n})$ for some n.*
7. *If $\mathbb{P}$ contains $\exists X F(X)$, then $\mathbb{P}$ contains $F(U_m)$ for all m.*
8. *If $\mathbb{P}$ contains $\forall X F(X)$, then $\mathbb{P}$ contains $F(U_m)$ for some m.*
9. *$\mathbb{P}$ contains $\neg A_i$ for all i.*
10. *$\mathbb{P}$ contains $\neg\bar{Q}(m)$ for all m.*

Proof. Standard. □

Corollary 12.77. *If $\mathcal{D}_Q$ is ill-founded, then there exists a countable coded ω-model of Π^1_1-comprehension which contains Q.*

For our purposes it is important that Corollary 12.77 can be proved in $T_0 := \mathbf{RCA}_0 + \forall\mathfrak{X}\,(\mathrm{WO}(\mathfrak{X}) \to \mathrm{WO}(\Omega_\omega \cdot \mathfrak{X})$. To this end we need to show that the semantics of ω-models can be handled in the latter theory, i.e., for every formula F of $\mathcal{L}_2$ there exists a valuation for F in the sense of [18, VII.2.1]. It is easily seen that the principle $\forall\mathfrak{X}\,(\mathrm{WO}(\mathfrak{X}) \to \mathrm{WO}(\Omega_\omega \cdot \mathfrak{X})$ implies

$$\forall\mathfrak{X}\,(\mathrm{WO}(\mathfrak{X}) \to \mathrm{WO}(\varepsilon_{\mathfrak{X}}))$$

(see [1, Definition 2.1]) and thus, by [1, Theorem 4.1], T_0 proves that every set is contained in an ω-model of **ACA**. Now take an ω-model containing $\mathcal{D}_Q$ and an infinite branch of $\mathcal{D}_Q$. In this ω-model we find a valuation for every formula by [18, VII.2.2]. And hence Corollary 12.77 holds in the model, but then it also holds in the world at large by absoluteness.

Case II. $\mathcal{D}_Q$ is well-founded. Below we shall show that this is impossible.

12.5 Proof of the Main Theorem: the Hard Direction

The remainder of the paper will be devoted to ruling out the possibility that for some Q, $\mathcal{D}_Q$ could be a well-founded tree. This is the place where the principle $\forall\mathfrak{X}\,(\mathrm{WO}(\mathfrak{X}) \to \mathrm{WO}(\mathrm{OT}(\Omega_\omega{\cdot}\mathfrak{X})))$ in the guise of cut elimination for an infinitary proof system enters the stage. Aiming at a contradiction, suppose that $\mathcal{D}_Q$ is a well-founded tree. Let $\mathfrak{X}$ be the Kleene-Brouwer ordering on $\mathcal{D}_Q$ (see [18, Definition V.1.2]). Then $\Omega_\omega \cdot \mathfrak{X}$ is a well-ordering. In a nutshell, the idea is that a well-founded $\mathcal{D}_Q$ gives rise to a derivation of the empty sequent (contradiction) in an infinitary proof system.

12.5.1 The Infinitary Calculus $T^\diamond_Q$

We fix a set $Q \subseteq \mathbb{N}$. Let $\mathcal{L}^Q_2$ be the language of second order arithmetic augmented by a unary predicate $\bar{Q}$. The *formulas* of $T^\diamond_Q$ arise from special $\mathcal{L}^Q_2$-formulas, called *weak formulas*, as identified in [3]. The weak formulas figure as *isolated formulas* in Takeuti's [19].

Definition 12.78. *By* $\mathrm{PV}(F)$ *we denote the set of free set variables that occur in the scope of a set quantifier in* F*. More precisely* $\mathrm{PV}(F)$ *is defined by induction on the buildup of* F *as follows:* (i) $\mathrm{PV}(F) = \emptyset$ *if* F *is atomic or a negated atom;* (ii) $\mathrm{PV}(F_1 \wedge F_2) = \mathrm{PV}(F_1 \vee F_2) = \mathrm{PV}(F_1) \cup \mathrm{PV}(F_2)$*;* (iii) $\mathrm{PV}(\exists x\, F_1(x)) = \mathrm{PV}(\forall x\, F_1(x)) = \mathrm{PV}(F_1(0))$*;* (iv) $\mathrm{PV}(\exists X\, F_1(X))$ *as well as* $\mathrm{PV}(\forall X\, F_1(X))$ *is the set of* all *free set variables occurring in the respective formula.*

The weak formulas *are defined inductively as follows:* (a) *every atomic and negated atomic formula is weak;* (b) *if* F_1 *and* F_2 *are weak, so are* $F_1 \wedge F_2$ *and* $F_1 \vee F_2$; (c) *if* $F(u)$ *is weak, so are* $\exists x F(x)$ *and* $\forall x F(x)$; (d) *if* $G(U)$ *is weak and* $U \notin \mathrm{PV}(G(U))$ *and* U *does not occur in* $\exists X\, G(X)$, *then* $\exists X\, G(X)$ *and* $\forall X\, G(X)$ *are weak formulas.*

Definition 12.79. *The language of* $T_Q^\diamond$ *has the same symbols as* $\mathcal{L}_2^Q$ *except that instead of free set variables* $U, V, W, \ldots$ *it has stratified set variables* $U^n, V^n, W^n, \ldots$ *for all* $n < \omega$.

The formulas *of* $T_Q^\diamond$ *(or* $T_Q^\diamond$*-formulas) are obtained from weak formulas of* $\mathcal{L}_2^Q$ *without free numerical variables by attaching a numerical superscript to every free set variable; i.e., if* F *is weak formula of* $\mathcal{L}_2^Q$ *without free numerical variables we obtain a formula of* $T_Q^\diamond$ *by replacing every free set variable* U *in* F *by* U^n *for some* $n < \omega$.

We are going to measure the length of $T_Q^\diamond$*-formulas by assigning a* grade $< \omega$ *to them.*

1. $gr(A) = 0$ *if* A *is atomic or a negated atom.*
2. $gr(\forall X F(X)) = gr(\exists X F(X)) = 0$.
3. $gr(A \wedge B) = gr(A \vee B) = \max(gr(A), gr(B)) + 1$.
4. $gr(\forall x H(x)) = gr(\exists x H(x)) = gr(H(0)) + 1$.

We also assign a *stage* $< \omega$ to each $T_Q^\diamond$-formula.

Definition 12.80. 1. $st(A) = 0$ *if* A *is an atomic formula or negated atomic formula without set variables.*
2. $st(t \in U^n) = st(t \notin U^n) = n$.
3. $st(\forall X F(X)) = st(F(U^0))$, $st(\exists X F(X)) = st(F(U^0)) + 1$.
4. $st(A \wedge B) = st(A \vee B) = \max(st(A), st(B))$.
5. $st(\forall x H(x)) = st(\exists x H(x)) = st(H(0))$.

According to this definition, every formula has a stage $< \omega$.

Definition 12.81. 1. *Two closed terms* s *and* t *are said to be equivalent if they yield the same value when computed.*
2. *A formula is called constant if it contains no set variables. The truth or falsity of such a formula is understood with respect to the standard structure of the integers.*
3. $\overline{0} := 0$, $\overline{m+1} := \overline{m}'$.

In the sequent calculus $T_Q^\diamond$ below we shall use the following basic rules of inference, where all exhibited formulae are assumed to be formulae of $T_Q^\diamond$:

$(\wedge)$	$\vdash \Gamma, A$ and $\vdash \Gamma, B \Longrightarrow \vdash \Gamma, A \wedge B$;
$(\vee)$	$\vdash \Gamma, A_i \Longrightarrow \vdash \Gamma, A_0 \vee A_1$ if $i \in \{0,1\}$;
$(\forall_2\text{-}n)$	$\vdash \Gamma, F(U^n) \Longrightarrow \vdash \Gamma, \forall X F(X)$, if $st(\forall X F(X)) = n$;
(ω-rule)	$\vdash \Gamma, A(\bar{n})$ for all $n \Longrightarrow \vdash \Gamma, \forall x A(x)$;
$(\exists_1)$	$\vdash \Gamma, B(t) \Longrightarrow \vdash \Gamma, \exists x B(x)$;
(Cut)	$\vdash \Gamma, C$ and $\vdash \Gamma, \neg C \Longrightarrow \vdash \Gamma$,

where in $(\forall_2\text{-}n)$ the free variable U^n is not to occur in the conclusion.

Definition 12.82. *Inductive definition of* $T^\diamond_Q \mathrel{\vert\!\frac{\alpha}{k}} \Gamma$ *for* $\alpha \in \mathrm{OT}(\Omega_\omega \cdot \mathfrak{X})$ *and* $k < \omega$.

1. *If A is a true constant prime formula or negated prime formula and* $A \in \Gamma$*, then* $T^\diamond_Q \mathrel{\vert\!\frac{\alpha}{k}} \Gamma$. *s*
2. *If* $n \in Q$ *and t is a closed term with value n and* $\bar{Q}(t)$ *is in* Γ*, then* $T^\diamond_Q \mathrel{\vert\!\frac{\alpha}{k}} \Gamma$.
3. *If* $n \notin Q$ *and t is a closed term with value n and* $\neg\bar{Q}(t)$ *is in* Γ*, then* $T^\diamond_Q \mathrel{\vert\!\frac{\alpha}{k}} \Gamma$.
4. *If* Γ *contains formulas* $A(s_1,\ldots,s_n)$ *and* $\neg A(t_1,\ldots,t_n)$ *of grade* 0*, where* s_i *and* t_i $(1 \le i \le n)$ *are equivalent terms, then* $T^\diamond_Q \mathrel{\vert\!\frac{\alpha}{k}} \Gamma$.
5. *If* $T^\diamond_Q \mathrel{\vert\!\frac{\beta}{k}} \Gamma_i$ *and* $\beta \lhd \alpha$ *hold for every premise* Γ_i *of an inference* $(\wedge)$, $(\vee)$, $(\exists_1)$, *(ω-rule)*, *($\forall_2$-n)* *or* *(Cut)* *with a cut formula having grade* $< k$*, and conclusion* Γ*, then* $T^\diamond_Q \mathrel{\vert\!\frac{\alpha}{k}} \Gamma$.
6. *(Ω_{n+1}-rule). Let f be a fundamental function satisfying*

   (a) $\Omega_{n+1} \in dom(f)$ *and* $f(\Omega_{n+1}) \unlhd \alpha$,

   (b) $T^\diamond_Q \mathrel{\vert\!\frac{f(0)}{k}} \Gamma, \forall X F(X)$*, where* $st(\forall X F(X)) = n$,

   (c) *and for every* $\beta < \Omega_{n+1}$, $T^\diamond_Q \mathrel{\vert\!\frac{\beta}{0}} \Xi, \forall X F(X)$ *implies* $T^\diamond_Q \mathrel{\vert\!\frac{f(\beta)}{k}} \Xi, \Gamma$ *for every set of formulas* Ξ *of stages* $\le n$.

   Then $T^\diamond_Q \mathrel{\vert\!\frac{\alpha}{k}} \Gamma$ *holds.*

Remark 12.83. The derivability relation $T^\diamond_Q \mathrel{\vert\!\frac{\alpha}{k}} \Gamma$ *extends one from [14] and is modeled upon the relation* $PB^* \mathrel{\vert\!\frac{\alpha}{n}} F$ *of [3], the main difference being the sequent calculus setting instead of P- and N-forms.*

Note that $T^\diamond_Q$ *has no basic rules pertaining to the existential set quantifier. To deduce* $\Gamma, \exists X F(X)$ *from* $\Gamma, F(U^n)$ *we shall have to deploy the* Ω_i*-rules.*

Remark 12.84. If one ruminates on the definition of the derivability predicate $T^\diamond_Q \mathrel{\vert\!\frac{\alpha}{k}} \Xi$ *the question arises whether it is actually a proper inductive definition. The critical point is obviously the condition* (c) *of the* Ω_{n+1}*-rule. Note that* $T^\diamond_Q \mathrel{\vert\!\frac{\beta}{0}} \Xi, \forall X F(X)$ *occurs negatively in clause* (c)*. However, since* $\beta < \Omega_{n+1}$*, the pertaining derivation does not contain any applications of* Ω_m*-rules where* $m \ge n+1$*. When defining the collection of derivations with the* Ω_{n+1}*-rule one assumes that derivability with regard to* Ω_{k+1}*-rules for* $k < n$ *has already been introduced. Thus the definition of* $T^\diamond_Q \mathrel{\vert\!\frac{\alpha}{k}} \Xi$ *proceeds via an iterated inductive definition.* Π^1_1-**CA** *will be sufficient for a formalization of this notion of derivability. It will be a non trivial issue to handle such inductive definitions in a weak background theory.*

Lemma 12.85. 1. $T^\diamond_Q \mathrel{\vert\!\frac{\alpha}{k'}} \Gamma \;\&\; \Gamma \subseteq \Delta \;\&\; \alpha \unlhd \beta \;\&\; k' \le k \implies T^\diamond_Q \mathrel{\vert\!\frac{\beta}{k}} \Delta$.

2. $T^\diamond_Q \mathrel{\vert\!\frac{\alpha}{k}} \Gamma, A \wedge B \implies T^\diamond_Q \mathrel{\vert\!\frac{\alpha}{k}} \Gamma, A \;\&\; T^\diamond_Q \mathrel{\vert\!\frac{\alpha}{k}} \Gamma, B$.
3. $T^\diamond_Q \mathrel{\vert\!\frac{\alpha}{k}} \Gamma, A \vee B \implies T^\diamond_Q \mathrel{\vert\!\frac{\alpha}{k}} \Gamma, A, B$.
4. $T^\diamond_Q \mathrel{\vert\!\frac{\alpha}{k}} \Gamma, F(t) \implies T^\diamond_Q \mathrel{\vert\!\frac{\alpha}{k}} \Gamma, F(s)$ *if t and s are equivalent.*
5. $T^\diamond_Q \mathrel{\vert\!\frac{\alpha}{k}} \Gamma, \forall x F(x) \implies T^\diamond_Q \mathrel{\vert\!\frac{\alpha}{k}} \Gamma, F(s)$ *for every closed term s.*

6. *If* $T^\diamond_Q \overset{\alpha}{\underset{k}{\vdash}} \Gamma$ *and* $\alpha \leq \Omega_{n+1}$ *then* $T^\diamond_Q \overset{\Omega_{n+1}+\alpha}{\underset{k}{\vdash}} \Gamma$.
7. *If* $T^\diamond_Q \overset{\alpha}{\underset{k}{\vdash}} \Gamma$, $\alpha \leq \Omega_\omega$ *and* $u \in \mathfrak{X}$, *then* $T^\diamond_Q \overset{\bar{u}+\alpha}{\underset{k}{\vdash}} \Gamma$.

Proof. Proceed by induction on α. These can be carried out straightforwardly. (5) requires (4). As to (6), observe that any fundamental function f in the proof can be replaced with $\Omega_{n+1} + f$ by Lemma 12.41. Similarly in (7), any fundamental function f in the proof can be replaced with $\bar{u} + f$ by Lemma 12.41. □

Lemma 12.86. $T^\diamond_Q \overset{2\cdot m}{\underset{0}{\vdash}} \Gamma, A(s_1, \ldots, s_k), \neg A(t_1, \ldots, t_k)$ *if* $m \geq gr(A(s_1, \ldots, s_k))$ *and* s_i *and* t_i *are equivalent terms.*

Proof. Proceed by induction on $gr(A(s_1, \ldots, s_k))$. Crucially note that if $gr(A(s_1, \ldots, s_k)) = 0$ then $\Gamma, A(s_1, \ldots, s_k), \neg A(t_1, \ldots, t_k)$ is an axiom according to Definition 12.82 clause (4). □

Definition 12.87. *For formulas* $F(U)$ *and* $A(a)$, $F(A)$ *denotes the result of replacing each occurrence of the form* $e \in U$ *in* $F(U)$ *by* $A(e)$. *The expression* $F(A)$ *is a formula if the bound variables in* $A(a)$ *are chosen in an appropriate way, in particular if* $F(U)$ *and* $A(a)$ *have no bound variables in common.*

Lemma 12.88. *Suppose* $\alpha < \Omega_{n+1}$ *and let* $\Delta(U^n) = \{F_1(U^n), \ldots, F_r(U^n)\}$ *be a set of weak formulas such that* U^n *doesn't occur in the scope of a second order quantifier in these formulas. Let* $\alpha < \Omega_{n+1}$. *For an arbitrary formula* $A(a)$ *we then have:*

$$T^\diamond_Q \overset{\alpha}{\underset{0}{\vdash}} \Delta(U^n) \implies T^\diamond_Q \overset{\Omega_{n+1}+\alpha}{\underset{0}{\vdash}} \Delta(A).$$

Proof. We proceed by induction on α. Suppose $\Delta(U^n)$ is an axiom. Then either $\Delta(A)$ is an axiom too, or $T^\diamond_Q \overset{k}{\underset{0}{\vdash}} \Delta(A)$ for some $k < \omega$ can be obtained through use of Lemma 12.86. Therefore, $T^\diamond_Q \overset{\Omega+\alpha}{\underset{0}{\vdash}} \Delta(A)$ by Lemma 12.85 (1) since (trivially) $k \lhd \Omega + \alpha$. If $T^\diamond_Q \overset{\alpha}{\underset{0}{\vdash}} \Delta(U^n)$ is the result of an inference, then this inference must be different from $(\exists_2)$, (Cut), and the $(\Omega_m$-rule) with $m \geq n + 1$ since $\Delta(U)$ consists of weak formulas, the derivation is cut-free and $\alpha < \Omega_{n+1}$. So let's assume that the last rule in $T^\diamond_Q \overset{\alpha}{\underset{0}{\vdash}} \Delta(U^n)$ was an Ω_{k+1}-rule with $k < n$. But then, as no formula of stage $\leq k$ contains the variable U^n, this is trivially true.

For the remaining possible inference rules the assertion follows easily from the induction hypothesis, noting that for a fundamental function f taking values $\leq \Omega_{n+1}$, $\Omega_{n+1} + f$ is a fundamental function, too. □

Lemma 12.89. *Let* Γ *be a set of formulas of stages* $\leq n$, $\forall X F(X)$ *be a formula of stage* n *and* $\alpha < \Omega_{n+1}$. *If* $T^\diamond_Q \overset{\alpha}{\underset{0}{\vdash}} \Gamma, \forall X F(X)$, *then* $T^\diamond_Q \overset{\alpha}{\underset{0}{\vdash}} \Gamma, F(U^n)$.

Proof. Use induction on α. Note that $\forall X F(X)$ cannot be a principal formula of an axiom in $\Gamma, \forall X F(X)$, since $st(\exists X \neg F(X)) = n + 1$. Also, since $\alpha < \Omega_{n+1}$, the derivation doesn't involve instances of Ω_{m+1}-rules with $m \geq n$, respectively. So let's assume that the last rule in $T^\diamond_Q \overset{\alpha}{\underset{0}{\vdash}} \Gamma, \forall X F(X)$ was an Ω_{k+1}-rule with $k < n$. But then the assertion follows immediately from the induction hypotheses. □

The role of the Ω_{n+1}-rules in our calculus $T_Q^\diamond$ is enshrined in the next lemma.

Lemma 12.90. $T_Q^\diamond \mid\frac{\Omega_{n+1}\cdot 2}{0} \exists X F(X), \neg F(A)$ *for every formula* $\forall X \neg F(X)$ *of stage* n *and arbitrary formula* $A(a)$.

Proof. Note that $f(\alpha) := \Omega_{n+1} + \alpha$ with $\mathrm{dom}(f) := \{\alpha \in \mathrm{OT}(\Omega_\omega \cdot \mathfrak{X}) \mid \alpha \leq \Omega_{n+1}\}$ is a fundamental function. Then

$$T_Q^\diamond \mid\frac{f(0)}{0} \forall X \neg F(X), \exists X F(X), \neg F(A) \tag{12.13}$$

according to Lemma 12.86. For $\alpha < \Omega_{n+1}$ and every set Θ of formulas with stages $\leq n$ we have, by Lemmata 12.88 and 12.89,

$$T_Q^\diamond \mid\frac{\alpha}{0} \Theta, \forall X \neg F(X) \implies T_Q^\diamond \mid\frac{f(\alpha)}{0} \Theta, \neg F(A).$$

Therefore, by Lemma 12.85 (1),

$$T_Q^\diamond \mid\frac{\alpha}{0} \Theta, \forall X \neg F(X) \implies T_Q^\diamond \mid\frac{f(\alpha)}{0} \Theta, \exists X F(X), \neg F(A). \tag{12.14}$$

The assertion now follows from (12.13) and (12.14) by the Ω_{n+1}-rule. □

Corollary 12.91. $T_Q^\diamond \mid\frac{\Omega_{n+1}\cdot 2+1}{k_0} \exists X \forall y (y \in X \leftrightarrow B(y))$ *for every formula* $B(a)$ *with* $st(B(0)) = n - 1$, *where* $k_0 = gr(\forall y (y \in X \leftrightarrow B(y))) + 1$.

Proof. Owing to Lemma 12.90 we have

$$T_Q^\diamond \mid\frac{\Omega_{n+1}\cdot 2}{0} \exists X \forall y (y \in X \leftrightarrow B(y)), \neg \forall y (B(y) \leftrightarrow B(y)). \tag{12.15}$$

As Lemma 12.86 yields $T_Q^\diamond \mid\frac{k}{0} \forall y (B(y) \leftrightarrow B(y))$ for some $k < \omega$, cutting with (12.15) yields $T_Q^\diamond \mid\frac{\Omega\cdot 2+1}{k_0} \exists X \forall y (y \in X \leftrightarrow B(y))$. □

12.5.2 The Reduction Procedure for $T_Q^\diamond$

Lemma 12.92. *Let* C *be a formula of grade* k. *Suppose* C *is a prime formula or of either form* $\exists X H(X)$, $\exists x G(x)$ *or* $A \vee B$. *Let* $\alpha = \omega^{\alpha_1} + \cdots + \omega^{\alpha_k}$ *with* $\delta \leq \omega^{\alpha_k} \leq \cdots \leq \omega^{\alpha_1} < \Omega_\omega$ *Then we have*

$$T_Q^\diamond \mid\frac{\alpha}{k} \Delta, \neg C \quad \textit{and} \quad T_Q^\diamond \mid\frac{\delta}{k} \Gamma, C \implies T_Q^\diamond \mid\frac{\alpha+\delta}{k} \Delta, \Gamma.$$

Proof. We proceed by induction on δ.

1. Let Γ, C be an axiom. Then there are three cases to consider.

1.1. Γ is an axiom. Then so is Δ, Γ. Hence $T_Q^\diamond \mid\frac{\alpha+\delta}{k} \Delta, \Gamma$.

1.2. C is a true constant prime formula or negated prime formula. A straightforward induction on α then yields $T_Q^\diamond \overset{\alpha}{\underset{k}{\vdash}} \Delta$, and thus $T_Q^\diamond \overset{\alpha+\delta}{\underset{k}{\vdash}} \Delta, \Gamma$ by Lemma 12.85 (1).

1.3. $C \equiv A(s_1, \ldots, s_n)$ and $gr(C) = 0$ and Γ contains a formula $\neg A(t_1, \ldots, t_n)$ where s_i and t_i are equivalent terms. From $T_Q^\diamond \overset{\alpha}{\underset{k}{\vdash}} \Delta, \neg A(s_1, \ldots, s_n)$ one can go to $T_Q^\diamond \overset{\alpha}{\underset{k}{\vdash}} \Delta, \neg A(t_1, \ldots, t_n)$ by use of Lemma 12.85 (4). Thence $T_Q^\diamond \overset{\alpha+\delta}{\underset{k}{\vdash}} \Delta, \Gamma$ follows by use of Lemma 12.85 (1), since $\neg A(t_1, \ldots, t_n) \in \Gamma$.

2. Suppose $C \equiv A \vee B$ and $T_Q^\diamond \overset{\delta_0}{\underset{k}{\vdash}} \Gamma, C, A_0$ with $A_0 \in \{A, B\}$ and $\delta_0 \lhd \delta$. Inductively we get

$$T_Q^\diamond \overset{\alpha+\delta_0}{\underset{k}{\vdash}} \Delta, \Gamma, A_0\,. \tag{12.16}$$

Next use Lemma 12.85 (2) on $T_Q^\diamond \overset{\alpha}{\underset{k}{\vdash}} \Delta, \neg A \wedge \neg B$ to obtain

$$T_Q^\diamond \overset{\alpha+\delta_0}{\underset{k}{\vdash}} \Delta, \Gamma, \neg A_0\,. \tag{12.17}$$

Whence use a cut on (12.16) and (12.17) to get the assertion.

3. Suppose $C \equiv \exists x G(x)$ and $T_Q^\diamond \overset{\delta_0}{\underset{k}{\vdash}} \Gamma, C, G(t)$ with $\delta_0 \lhd \delta$. Inductively we get

$$T_Q^\diamond \overset{\alpha+\delta_0}{\underset{k}{\vdash}} \Delta, \Gamma, G(t)\,. \tag{12.18}$$

By Lemma 12.85 (1), (5), we also get

$$T_Q^\diamond \overset{\alpha+\delta_0}{\underset{k}{\vdash}} \Delta, \Gamma, \neg G(t)\,; \tag{12.19}$$

thus (12.18) and (12.19) yield $T_Q^\diamond \overset{\alpha+\delta}{\underset{k}{\vdash}} \Delta, \Gamma$ by (Cut).

4. Let $T_Q^\diamond \overset{\delta}{\underset{k}{\vdash}} \Gamma, C$ be derived by the Ω_{n+1}-rule with fundamental function f. Then the assertion follows from the induction hypothesis by the Ω_{n+1}-rule using the fundamental function $\alpha + f$.

5. The remaining cases are such that the last inference was a basic one but C was not its principal formula. Then the assertion follows from the induction hypothesis applied to the premises, followed by reapplying the same inference.

Note that there are no rules that introduce a principal formula of second order existential form, $\exists X H(X)$. □

Lemma 12.93. *For* $\alpha < \Omega_\omega$,

$$T_Q^\diamond \overset{\alpha}{\underset{k+1}{\vdash}} \Gamma \implies T_Q^\diamond \overset{\omega^\alpha}{\underset{k}{\vdash}} \Gamma.$$

Proof. We proceed by induction on k. We only treat the crucial case when $T_Q^\diamond \overset{\alpha_0}{\underset{k+1}{\vdash}} \Gamma, D$ and $T_Q^\diamond \overset{\alpha_0}{\underset{k+1}{\vdash}} \Gamma, \neg D$, where $\alpha_0 \lhd \alpha$, and $gr(D) = k$. Inductively this becomes $T_Q^\diamond \overset{\omega^{\alpha_0}}{\underset{k}{\vdash}} \Gamma, D$ and $T_Q^\diamond \overset{\omega^{\alpha_0}}{\underset{k}{\vdash}} \Gamma, \neg D$. Since D or $\neg D$ must be of one of the forms exhibited in Lemma 12.92, we arrive at $T_Q^\diamond \overset{\omega^{\alpha_0}+\omega^{\alpha_0}}{\underset{k}{\vdash}} \Gamma$ by Lemma 12.92. As $\omega^{\alpha_0} + \omega^{\alpha_0} \lhd \omega^\alpha$, we can use Lemma 12.85 (1) to infer the desired assertion. □

Theorem 12.94 (*Collapsing Theorem*). *If Γ is a set of formulas of levels $\leq n$ and $\alpha \in C^{\mathfrak{x}}_n(\alpha)$ then*

$$T^{\diamond}_{Q} \overset{\alpha}{\underset{0}{\vdash}} \Gamma \implies T^{\diamond}_{Q} \overset{\psi^{\mathfrak{x}}_n \alpha}{\underset{0}{\vdash}} \Gamma.$$

Proof. We proceed by induction on α.

1. If Γ is an axiom, then the assertion is trivial.

2. Let $T^{\diamond}_{Q} \overset{\alpha}{\underset{0}{\vdash}} \Gamma$ be the result of an inference which is neither a (Cut) nor an Ω_{m+1}-rule. Then we have $T^{\diamond}_{Q} \overset{\alpha_0}{\underset{0}{\vdash}} \Gamma_i$ with $\alpha_0 \lhd \alpha$ and Γ_i being the i-th premise of that inference. From $\alpha_0 \lhd \alpha$ and $\alpha \in C^{\mathfrak{x}}_n(\alpha)$ we obtain $\alpha_0 \in C^{\mathfrak{x}}_n(\alpha_0)$ by Lemma 12.34. Thus we can apply th induction hypothesis to conclude $T^{\diamond}_{Q} \overset{\psi^{\mathfrak{x}}_n \alpha_0}{\underset{0}{\vdash}} \Gamma_i$. $\alpha_0 \lhd \alpha$ also implies $\psi^{\mathfrak{x}}_n \alpha_0 \lhd \psi^{\mathfrak{x}}_n \alpha$ by Lemma 12.34. Therefore $T^{\diamond}_{Q} \overset{\psi^{\mathfrak{x}}_n \alpha}{\underset{0}{\vdash}} \Gamma$ by reapplying the same inference.

3. Suppose $T^{\diamond}_{Q} \overset{\alpha}{\underset{0}{\vdash}} \Gamma$ results by the Ω_{m+1}-rule with respect to a formula $\forall X F(X)$ of stage m and a fundamental function f. Then $\Omega_{m+1} \in \mathsf{dom}(f)$ and $f(\Omega_{m+1}) \unlhd \alpha$ and

$$T^{\diamond}_{Q} \overset{f(0)}{\underset{0}{\vdash}} \Gamma, \forall X F(X), \tag{12.20}$$

and, moreover, for every set of formulas Ξ with stages $\leq m$ and $\beta < \Omega_{m+1}$,

$$T^{\diamond}_{Q} \overset{\beta}{\underset{0}{\vdash}} \Xi, \forall X F(X) \implies T^{\diamond}_{Q} \overset{f(\beta)}{\underset{0}{\vdash}} \Xi, \Gamma. \tag{12.21}$$

Case A. $m < n$. Then $\Omega_{m+1} \leq \Omega_n$ and $\Omega_{m+1} \in \mathsf{dom}(\psi^{\mathfrak{x}}_n f)$. f being a fundamental function, we have $f(\beta) \lhd_{\beta} f(\Omega_{m+1})$ for $\beta < \Omega_{m+1}$ by property (1) in Definition 12.36, whence $f(\beta) \lhd_{\beta} \alpha$ since $f(\Omega_{m+1}) \unlhd \alpha$. Since $\beta \in C^{\mathfrak{x}}_n(f(\beta))$ holds (trivially) for $\beta < \Omega_{m+1}$, we get $f(\beta) \in C^{\mathfrak{x}}_n(f(\beta))$ from Lemma 12.34. As a result, we can apply the induction hypothesis, arriving at

$$T^{\diamond}_{Q} \overset{\psi^{\mathfrak{x}}_n(f(0))}{\underset{0}{\vdash}} \Gamma, \forall X F(X), \tag{12.22}$$

and, moreover, for every set of formulas Ξ with stages $\leq m$ and $\beta < \Omega_{m+1}$,

$$T^{\diamond}_{Q} \overset{\beta}{\underset{0}{\vdash}} \Xi, \forall X F(X) \implies T^{\diamond}_{Q} \overset{\psi^{\mathfrak{x}}_n(f(\beta))}{\underset{0}{\vdash}} \Xi, \Gamma. \tag{12.23}$$

An application of the Ω_{m+1}-rule with the fundamental function $\psi^{\mathfrak{x}}_n f$ thus yields $T^{\diamond}_{Q} \overset{\psi^{\mathfrak{x}}_n \alpha}{\underset{0}{\vdash}} \Gamma$ as $\psi^{\mathfrak{x}}_n(f(\Omega_{n+1})) \unlhd \psi^{\mathfrak{x}}_n \alpha$ by Lemma 12.34.

Case B. $n \leq m$. By the same arguments as in the previous case we have $f(0) \in C^{\mathfrak{x}}_m(f(0))$, thus the induction hypothesis used on (12.20) supplies us with $T^{\diamond}_{Q} \overset{\psi^{\mathfrak{x}}_m(f(0))}{\underset{0}{\vdash}} \Gamma, \forall X F(X)$. Hence with $\Xi = \Gamma$ we get

$$T^{\diamond}_{Q} \overset{f(\psi^{\mathfrak{x}}_m(f(0)))}{\underset{0}{\vdash}} \Gamma \tag{12.24}$$

from (12.21). Now Corollary 12.43 ensures that $f(\rho) \lhd f(\Omega_{m+1})$, where $\rho = \psi^{\mathfrak{x}}_m(f(0))$. Thus, as $f(\Omega_{m+1}) \unlhd \alpha$, we have $f(\rho) \lhd \alpha$. Since $\alpha \in C^{\mathfrak{x}}_n(\alpha)$ we conclude that $f(\rho) \in C^{\mathfrak{X}}_n(f(\rho))$ by Lemma 12.34. As a result, we can apply the induction hypothesis to (12.24), obtaining

$$T^{\diamond}_{Q} \left|\frac{\psi^{\mathfrak{X}}_n(f(\rho))}{0}\right. \Gamma\,. \tag{12.25}$$

Again by Lemma 12.34, we infer $\psi^{\mathfrak{X}}_n(f(\rho)) \lhd \psi^{\mathfrak{X}}_n\alpha$, thus $T^{\diamond}_{Q} \left|\frac{\psi^{\mathfrak{X}}_n\alpha}{0}\right. \Gamma$. □

12.5.3 Embedding $\mathcal{D}_Q$ into $T^{\diamond}_{Q}$

Assuming that $\mathcal{D}_Q$ is a well-founded tree, the objective of this section is to embed $\mathcal{D}_Q$ into $T^{\diamond}_{Q}$, so as to obtain a contradiction. Let $\mathfrak{X}$ be the Kleene-Brouwer ordering of $\mathcal{D}_Q$. We write $\mathcal{D}_Q \left|\frac{u}{}\right. \Gamma$ if Γ is the sequent attached to the node u in $\mathcal{D}_Q$.

Definition 12.95. *An interpretation $(.)^*$ is a map that assigns to every free set variable U of $\mathcal{L}_2$ the same variable but with a superscript n for some $n \in \mathbb{N}$; i.e., $(U)^* = U^n$.*

For a sequent Ξ of weak $\mathcal{L}_2$-formulas without free number variables and an interpretation $(.)^*$, the sequent Ξ^* of $T^{\diamond}_{Q}$-formulas is the result of replacing every occurrence of a free set variable U in a formula of Ξ by $(U)^*$.

Recall that every sequent Ξ appearing in $\mathcal{D}_Q$ will consist of weak formulas and does not contain free number variables. So Ξ^* will only differ from Ξ with respect to the superscripts attached to free second order variables.

Theorem 12.96. *If $\mathcal{D}_Q \left|\frac{u}{}\right. \Xi$ then $T^{\diamond}_{Q} \left|\frac{\bar{u}+1}{0}\right. \Xi^*$ holds for every interpretation $(.)^*$.*

Proof. We proceed by induction on u, i.e., the Kleene-Brouwer ordering of $\mathcal{D}_Q$.

Suppose u is an end-node of $\mathcal{D}_Q$. Then Ξ must be axiomatic and therefore Ξ^* will also be an axiom of $T^{\diamond}_{Q}$, and hence $T^{\diamond}_{Q} \left|\frac{u}{0}\right. \Xi^*$.

Now assume that u is not an end-node of $\mathcal{D}_Q$. Then Ξ is not axiomatic.

If Ξ is not reducible, then there is a node u_0 immediately above u in $\mathcal{D}_Q$ such that $\mathcal{D}_Q \left|\frac{u_0}{}\right. \Xi, \neg\bar{Q}(i), \neg A_i$ for some i. Inductively we have

$$T^{\diamond}_{Q} \left|\frac{\overline{u_0}+1}{0}\right. \Xi^*, \neg\bar{Q}(i), \neg A^*_i\,,$$

and thus

$$T^{\diamond}_{Q} \left|\frac{\psi^{\mathfrak{X}}_m(\overline{u_0}+1)}{0}\right. \Xi^*, \neg\bar{Q}(i), \neg A^*_i$$

by the Collapsing Theorem 12.94 if m is not smaller than the maximum of the stages of all formulas.

We also have $T^{\diamond}_{Q} \left|\frac{0}{0}\right. \bar{Q}(i)$ and, using Corollary 12.91, $T^{\diamond}_{Q} \left|\frac{\Omega_{n+1}\cdot 2+1}{m_0}\right. A^*_i$ for sufficiently large n and m_0. We can assume $m \geq n+1$ so that $\Omega_{n+1}\cdot 2+1 \lhd \psi^{\mathfrak{X}}_m(\overline{u_0}+1)$. Thus we have $T^{\diamond}_{Q} \left|\frac{\psi^{\mathfrak{X}}_m(\overline{u_0}+1)}{m_0}\right. A^*_i$, and therefore, using two cuts, we arrive at

$$T_Q^\diamond \,\Big|\frac{\psi_m^{\mathfrak{x}}(\overline{u_0}+1)+2}{k}\; \Xi^*$$

for some $k < \omega$. Employing Lemma 12.93 k-times we conclude that

(12.26)
$$T_Q^\diamond \,\Big|\frac{\omega_k(\psi_m^{\mathfrak{x}}(\overline{u_0}+1)+2)}{0}\; \Xi^* \,.$$

(12.26) implies

(12.27)
$$T_Q^\diamond \,\Big|\frac{\overline{u_0}+\omega_k(\psi_m^{\mathfrak{x}}(\overline{u_0}+1)+2)}{0}\; \Xi^*$$

by Lemma 12.85(7). It is easy to see that $\overline{u_0} + \omega_k(\psi_m^{\mathfrak{x}}(\overline{u_0} + 1) + 2) \lhd \bar{u} + 1$ since for any $\eta \geq \overline{u_0} + \omega_k(\psi_m^{\mathfrak{x}}(\overline{u_0} + 1) + 2)$, $\overline{u_0} + \omega_k(\psi_m^{\mathfrak{x}}(\overline{u_0} + 1) + 2) \in C_l^{\mathfrak{x}}(\eta)$. Hence,

$$T_Q^\diamond \,\Big|\frac{\bar{u}+1}{0}\; \Xi^* \,.$$

Now suppose that Ξ is reducible. Ξ will be of the form

$$\Xi', E, \Xi''$$

where E is not a literal and Ξ' contains only literals.

First assume E to be of the form $\forall x\, F(x)$. Then, for each l, there is a node u_l immediately above u in $\mathcal{D}_Q$ such that

$$\mathcal{D}_Q \,\Big|\frac{u_l}{}\; \Xi', F(\bar{l}), \Xi'', \neg\bar{Q}(i), \neg A_i$$

for some i. Inductively we have

$$T_Q^\diamond \,\Big|\frac{\overline{u_l}+1}{0}\; (\Xi')^*, (F(\bar{l}))^*, (\Xi'')^*, \neg\bar{Q}(i), \neg A_i^*$$

and thus

$$T_Q^\diamond \,\Big|\frac{\psi_m^{\mathfrak{x}}\overline{u_l}+1}{0}\; (\Xi')^*, (F(\bar{l}))^*, (\Xi'')^*, \neg\bar{Q}(i), \neg A_i^*$$

by the Collapsing Theorem 12.94 if m is not smaller than the maximum of the stages of all formulas. We also have $T_Q^\diamond \,\Big|\frac{0}{0}\; \bar{Q}(i)$ and, using Corollary 12.91, $T_Q^\diamond \,\Big|\frac{\Omega_{n+1}\cdot 2+1}{m_0}\; A_i^*$ for sufficiently large n and m_0. We can assume $m \geq n+1$ so that $\Omega_{n+1}\cdot 2+1 \lhd \psi_m^{\mathfrak{x}}(\overline{u_l}+1)$ for all l. Thus we have $T_Q^\diamond \,\Big|\frac{\psi_m^{\mathfrak{x}}(\overline{u_l}+1)}{m_l}\; A_i^*$, and therefore, using two cuts, we arrive at

$$T_Q^\diamond \,\Big|\frac{\psi_m^{\mathfrak{x}}(\overline{u_l}+1)+2}{k}\; (\Xi')^*, (F(\bar{l}))^*, (\Xi'')^*$$

for some $k < \omega$. Employing Lemma 12.93 k-times we conclude that

(12.28)
$$T_Q^\diamond \,\Big|\frac{\omega_k(\psi_m^{\mathfrak{x}}(\overline{u_l}+1)+2)}{0}\; (\Xi')^*, (F(\bar{l}))^*, (\Xi'')^* \,.$$

(12.28) yields

$$T_Q^\diamond \mid\!\frac{\overline{u_0}+\omega_k(\psi_m^{\bar{x}}(\overline{u_0}+1)+2)}{0} (\Xi')^*, (F(\bar{l}))^*, (\Xi'')^* \tag{12.29}$$

by Lemma 12.85. Since $\overline{u_l} + \omega_k(\psi_m^{\bar{x}}(\overline{u_l} + 1) + 2) \lhd \bar{u}$ holds, we also get that

$$T_Q^\diamond \mid\!\frac{\bar{u}}{0} (\Xi')^*, (F(\bar{l}))^*, (\Xi'')^*$$

for all l, whence we can infer $T_Q^\diamond \mid\!\frac{\bar{u}+1}{0} (\Xi')^*, (\forall x F(x))^*, (\Xi'')^*$ via the ω-rule, i.e., $T_Q^\diamond \mid\!\frac{\bar{u}+1}{0} (\Xi)^*$.

Next assume E to be of the form $\forall X\, F(X)$. Then there is a node u_0 immediately above u in $\mathcal{D}_Q$ such that

$$\mathcal{D}_Q \mid\!\frac{u_0}{} \Xi', F(U), \Xi'', \neg\bar{Q}(i), \neg A_i$$

for a set variable U which does not occur Ξ. Inductively we have

$$T_Q^\diamond \mid\!\frac{\overline{u_0}+1}{0} (\Xi')^*, F^*(U^r), (\Xi'')^*, \neg\bar{Q}(i), \neg A_i^*$$

where $F^*(U^r)$ stands for $(F(U))^*$. Thus

$$T_Q^\diamond \mid\!\frac{\psi_m^{\bar{x}}\overline{u_0}+1}{0} (\Xi')^*, F^*(U^r), (\Xi'')^*, \neg\bar{Q}(i), \neg A_i^*$$

by the Collapsing Theorem 12.94 if m is not smaller than the maximum of the stages of all formulas. We also have $T_Q^\diamond \mid\!\frac{0}{0} \bar{Q}(i)$ and, using Corollary 12.91, $T_Q^\diamond \mid\!\frac{\Omega_{n+1}\cdot 2+1}{m_0} A_i^*$ for sufficiently large n and m_0. We can assume $m \geq n+1$ so that $\Omega_{n+1}\cdot 2+1 \lhd \psi_m^{\bar{x}}(\overline{u_0}+1)$. Thus we have $T_Q^\diamond \mid\!\frac{\psi_m^{\bar{x}}(\overline{u_0}+1)}{m_l} A_i^*$, and therefore, using two cuts, we arrive at

$$T_Q^\diamond \mid\!\frac{\psi_m^{\bar{x}}(\overline{u_0}+1)+2}{k} (\Xi')^*, F^*(U^r), (\Xi'')^*$$

for some $k < \omega$. Employing Lemma 12.93 k-times we conclude that

$$T_Q^\diamond \mid\!\frac{\omega_k(\psi_m^{\bar{x}}(\overline{u_0}+1)+2)}{0} (\Xi')^*, F^*(U^r), (\Xi'')^*\,. \tag{12.30}$$

(12.30) yields

$$T_Q^\diamond \mid\!\frac{\overline{u_0}+\omega_k(\psi_m^{\bar{x}}(\overline{u_0}+1)+2)}{0} (\Xi')^*, F^*(U^r), (\Xi'')^* \tag{12.31}$$

by Lemma 12.85. Since $\overline{u_0} + \omega_k(\psi_m^{\bar{x}}(\overline{u_0} + 1) + 2) \lhd \bar{u}$, we also get that

$$T_Q^\diamond \mid\!\frac{\bar{u}}{0} (\Xi')^*, F^*(U^r), (\Xi'')^*$$

whence we can infer $T_Q^\diamond \mid\!\frac{\bar{u}+1}{0} (\Xi')^*, (\forall X F(X))^*, (\Xi'')^*$ via an inference ($\forall_2$-r), i.e., $T_Q^\diamond \mid\!\frac{\bar{u}+1}{0} (\Xi)^*$.

Now assume E to be of the form $\exists X\, B(X)$. Then there is a node u_0 immediately above u in $\mathcal{D}_Q$ such that

$$\mathcal{D}_Q \stackrel{u_0}{\vdash} \Xi', B(U), \Xi'', \neg\bar{Q}(i), \neg A_i, E$$

for some i and set variable U. Let $\exists X\, B^*(X)$ be the formula $(\exists X\, B(X))^*$ and $U^r = (U)^*$. Inductively we then have

$$T_Q^\diamond \,\big|\!\frac{\overline{u_0}+1}{0}\, (\Xi')^*, B^*(U^r), (\Xi'')^*, \neg\bar{Q}(i), \neg A_i^*$$

and thus

$$T_Q^\diamond \,\big|\!\frac{\psi_m^{\mathfrak{x}}(\overline{u_0}+1)}{0}\, (\Xi')^*, B^*(U^r), (\Xi'')^*, \neg\bar{Q}(i), \neg A_i^*$$

by the Collapsing Theorem 12.94 if m is not smaller than the maximum of the stages of all formulas. We also have $T_Q^\diamond \,\big|\!\frac{0}{0}\, \bar{Q}(i)$ and, using Corollary 12.91, $T_Q^\diamond \,\big|\!\frac{\Omega_{n+1}\cdot 2+1}{m_0}\, A_i^*$ for sufficiently large n and m_0. We can assume $m \geq n+1$ so that $\Omega_{n+1}\cdot 2+1 \lhd \psi_m^{\mathfrak{x}}(\overline{u_0}+1)$. Thus we have $T_Q^\diamond \,\big|\!\frac{\psi_m^{\mathfrak{x}}(\overline{u_0}+1)}{m_0}\, A_i^*$, and therefore, using two cuts, we arrive at

(12.32) $$T_Q^\diamond \,\big|\!\frac{\psi_m^{\mathfrak{x}}(\overline{u_0}+1)+2}{k}\, (\Xi')^*, B^*(U^r), (\Xi'')^*$$

for some $k < \omega$. From Lemma 12.91 we also infer that

(12.33) $$T_Q^\diamond \,\big|\!\frac{\Omega_{s+1}\cdot 2}{0}\, \exists X\, B^*(X), \neg B^*(U^r)$$

where s is the stage of $\forall X\, \neg B^*(X)$. Cutting (12.32) with (12.33) yields

$$T_Q^\diamond \,\big|\!\frac{\psi_m^{\mathfrak{x}}(\overline{u_0}+1)+3}{k}\, (\Xi')^*, \exists X\, B^*(X), (\Xi'')^*$$

for some $k < \omega$. Thus we have $T_Q^\diamond \,\big|\!\frac{\psi_m^{\mathfrak{x}}(\overline{u_0}+1)+3}{k}\, \Xi^*$.

Employing Lemma 12.93 k-times we conclude that

(12.34) $$T_Q^\diamond \,\big|\!\frac{\omega_k(\psi_m^{\mathfrak{x}}(\overline{u_0}+1)+3)}{0}\, \Xi^* .$$

(12.34) yields

(12.35) $$T_Q^\diamond \,\big|\!\frac{\overline{u_0}+\omega_k(\psi_m^{\mathfrak{x}}(\overline{u_0}+1)+3)}{0}\, \Xi^*$$

by Lemma 12.85. From $\overline{u_0} + \omega_k(\psi_m^{\mathfrak{x}}(\overline{u_0} + 1) + 3) \lhd \bar{u}$ we also get that $T_Q^\diamond \,\big|\!\frac{\bar{u}+1}{0}\, \Xi^*$ as desired.

The cases where E is a redex of the form $E_0 \vee E_1$ or $E_0 \wedge E_1$ or $\exists x F(x)$ are much simpler and one proceeds in a similar manner as in the first case. □

Below $\emptyset$ stands for the empty sequent and u_b denotes the bottom node of $\mathcal{D}_Q$ which is the maximum element of the pertaining Kleene-Brouwer ordering.

Corollary 12.97. *If $\mathcal{D}_Q$ is well-founded, then $T_Q^\diamond \mid\!\frac{\rho}{0}\, \emptyset$ for some $\rho < \Omega_1$.*

Proof. We have $\mathcal{D}_Q \mid\!\frac{u_b}{}\, \neg\bar{Q}(0), \neg A_0$. By Theorem 12.96 we conclude that

$$T_Q^\diamond \mid\!\frac{\overline{u_b}+1}{0}\, \neg\bar{Q}(i), \neg A_0^*\,,$$

and thus

$$T_Q^\diamond \mid\!\frac{\psi_m^{\mathfrak{X}}(\overline{u_b}+1)}{0}\, \Xi^*, \neg\bar{Q}(i), \neg A_0^*$$

by the Collapsing Theorem 12.94 if m is not smaller than the stage of $\neg A_0^*$. We also have $T_Q^\diamond \mid\!\frac{0}{0}\, \bar{Q}(i)$ and, using Corollary 12.91, $T_Q^\diamond \mid\!\frac{\Omega_{n+1}\cdot 2+1}{m_0}\, A_0^*$ for sufficiently large n and m_0. We can assume $m \geq n+1$ so that $\Omega_{n+1}\cdot 2+1 \lhd \psi_m^{\mathfrak{X}}(\overline{u_b}+1)$. Thus we have $T_Q^\diamond \mid\!\frac{\psi_m^{\mathfrak{X}}(\overline{u_b}+1)}{m_0}\, A_0^*$, and therefore, using two cuts, we arrive at

$$T_Q^\diamond \mid\!\frac{\psi_m^{\mathfrak{X}}(\overline{u_b}+1)+2}{k}\, \emptyset$$

for some $k < \omega$. Employing Lemma 12.93 k-times we conclude that

(12.36) $$T_Q^\diamond \mid\!\frac{\omega_k(\psi_m^{\mathfrak{X}}(\overline{u_b}+1)+2)}{0}\, \emptyset\,.$$

(12.36) implies

(12.37) $$T_Q^\diamond \mid\!\frac{\overline{u_b}+\omega_k(\psi_m^{\mathfrak{X}}(\overline{u_b}+1)+2)}{0}\, \emptyset$$

by Lemma 12.85. Since $\overline{u_b}+\omega_k(\psi_m^{\mathfrak{X}}(\overline{u_b}+1)+2) \in C^{\mathfrak{X}}(\overline{u_b}+\omega_k(\psi_m^{\mathfrak{X}}(\overline{u_b}+1)+2))$ the Collapsing Theorem 12.94 applied to (12.37) yields

$$T_Q^\diamond \mid\!\frac{\psi_0^{\mathfrak{X}}(\overline{u_b}+\omega_k(\psi_m^{\mathfrak{X}}(\overline{u_b}+1)+2))}{0}\, \emptyset\,.$$

□

Corollary 12.98. *$\mathcal{D}_Q$ is not well-founded.*

Proof. If $\mathcal{D}_Q$ were well-founded, we would have

(12.38) $$T_Q^\diamond \mid\!\frac{\rho}{0}\, \emptyset$$

for some $\rho < \Omega_1$ by Corollary 12.97. But a straightforward induction on $\alpha < \Omega_1$ shows that

$$T_Q^\diamond \mid\!\frac{\alpha}{0}\, \Gamma \;\Rightarrow\; \Gamma \neq \emptyset$$

since in such a derivation there are neither cuts nor applications of Ω_{n+1}-rules, yielding that (12.38) is impossible. □

At this point we have to recall that in this (long) section we set out to prove the direction (ii) ⇒ (i) of the main Theorem 12.8. We have certainly used the assumption (ii), i.e., $\forall\mathfrak{X}\,(\mathrm{WO}(\mathfrak{X}) \rightarrow \mathrm{WO}(\mathrm{OT}(\Omega_\omega{\cdot}\mathfrak{X})))$, to prove that $\mathcal{D}_Q$ is an ill-founded tree, which by Corollary 12.77 yields that there is an ω-model of Π^1_1-$\mathbf{CA}_0$ containing Q. However, we also have to make sure that the further reasoning employed in this section can be handled in $\mathbf{RCA}_0$. This is far from clear as the the notion of derivability with the Ω_{n+1}-rules appears to require n-times iterated inductive definitions.[2]

For $\rho \leq \omega$ let $T^\diamond_Q{\restriction}_\rho$ be the restriction of $T^\diamond_Q$ where Ω_{m+1}-rules with $m+1 \geq \rho$ are disallowed. Thus $T^\diamond_Q{\restriction}_\omega$ is the same as $T^\diamond_Q$. As it turns out, it suffices to recursively define fixed points predicates D_n and a final fixed point predicate D_ω with regard to the inductive clauses that define derivability in $T^\diamond_Q{\restriction}_n$ and $T^\diamond_Q{\restriction}_\omega$, respectively. All we need is a fixed point not a proper inductive definition, i.e., to capture the notion of derivability in $T^\diamond_Q{\restriction}_\rho$ it suffices that the following holds:

(∗) $\mathcal{D}_\rho(\alpha; k; \Gamma)$ if and only if $\rho \leq \omega$, $\alpha \in \mathrm{OT}(\Omega_\omega{\cdot}\mathfrak{X})$, $k < \omega$, Γ is a sequent, and either Γ contains an axiom of $T^\diamond_Q{\restriction}_\rho$ or Γ is the conclusion of one of the inferences $(\wedge), (\vee), (\exists_1)$,($\omega$-rule),($\forall_2$-$l$), (Cut) with premises $(\Gamma_i)_{i\in I}$ such that there exists $\beta \lhd \alpha$ with $\mathcal{D}_\rho(\beta; k; \Gamma_i)$ for all $i \in I$ and if the inference is a cut it has rank $< k$, or for some $m+1 \leq \rho$, formula $\forall X\, F(X)$ and fundamental function f, Γ is the conclusion of an Ω_{m+1}-rule with f satisfying (a) $\Omega_{m+1} \in dom(f)$ and $f(\Omega_{m+1}) \unlhd \alpha$, (b) $\mathcal{D}_\rho(f(0); k; \Gamma, \forall X\, F(X))$ with $st(\forall X\, F(X)) = m$, and (c) for every $\beta < \Omega_{m+1}$, $\mathcal{D}_m(\beta; 0; \Xi, \forall X\, F(X))$ implies $\mathcal{D}_\rho(f(\beta); k; \Xi, \Gamma)$ for every finite set of formulas Ξ of stages $\leq m$.

(∗) can be viewed as a fixed-point axiom which together with transfinite induction for $\mathrm{OT}(\Omega_\omega{\cdot}\mathfrak{X})$ defines $T^\diamond_Q{\restriction}_\rho$-derivability implicitly. How can we find fixed points as described in (∗)? It follows from [12] that $\mathbf{RCA}_0 + \forall\mathfrak{X}\,[\mathrm{WO}(\mathfrak{X}) \rightarrow \mathrm{WO}(\Gamma_\mathfrak{X})]$ proves that every set is contained in a countable coded ω-model of $\mathbf{ATR}_0$. As one can show in $\mathbf{RCA}_0$ with the methods of [3, I.7] that $\mathrm{WO}(\mathrm{OT}(\Omega_\omega{\cdot}\mathfrak{X}))$ implies $\mathrm{WO}(\Gamma_\mathfrak{X})$, we can pick a countable coded ω-model $\mathfrak{B}$ of $\mathbf{ATR}_0$ which contains Q. It is known that $\mathbf{ATR}_0$ proves the Σ^1_1 axiom of choice, Σ^1_1-$\mathbf{AC}$ (see [18, Theorem V.8.3]). Furthermore, it follows from Theorem 12.5 that in $\mathfrak{B}$ every set is contained in an ω-model of Σ^1_1-$\mathbf{AC}$. As a result, arguing in $\mathfrak{B}$ on finds an ω-sequence of ω-models $\mathfrak{A}_n$ of Σ^1_1-$\mathbf{AC}$ inside $\mathfrak{B}$ such that $Q \in \mathfrak{A}_0$ and $\mathfrak{A}_n \in \mathfrak{A}_{n+1}$. Moreover, in $\mathbf{ACA}_0 + \Sigma^1_1$-$\mathbf{AC}$ one can prove for every P-positive arithmetical formula $A(u, P)$ that there is a Σ^1_1 formula $F(u)$ such that $\forall x[F(x) \leftrightarrow A(x, F)]$, where $A(x, F)$ arises from $A(x, P)$ by replacing every occurrence of the form $P(t)$ in the first formula by $F(t)$. This is known as the Second Recursion Theorem (see [2, V.2.3]). Therefore, one can model $\mathcal{D}_n$ inside $\mathfrak{A}_n$ and $\mathcal{D}_\omega$ inside $\mathfrak{B}$.

The upshot of the foregoing considerations is that $T^\diamond_Q$-derivability can be accounted for in $\mathbf{RCA}_0 + \forall\mathfrak{X}\,(\mathrm{WO}(\mathfrak{X}) \rightarrow \mathrm{WO}(\mathrm{OT}(\Omega_\omega{\cdot}\mathfrak{X})))$. This completes the proof of Theorem 12.8 $(ii) \Rightarrow (i)$ and thereby finishes the paper.

---

[2] This also emerged as a complication in connection with the proof-theoretic strength of Kruskal's theorem in [14].

Acknowledgements The first author's research was supported by a grant from the John Templeton Foundation ("A new dawn of intuitionism: mathematical and philosophical advances," ID 60842).[3]

The second author was supported by a Leeds University International Research Scholarship (LIRS).

Related results for the stronger theory Π^1_1-**CA** + **BI** were developed in the PhD thesis [20] of the second author.

References

1. B. Afshari and M. Rathjen. Reverse mathematics and well-ordering principles: A pilot study. *Annals of Pure and Applied Logic*, 160:231–237, 2009.
2. J. Barwise. *Admissible Sets and Structures*. Springer, Berlin, 1975.
3. W. Buchholz and K. Schütte. *Proof theory of impredicative subsystems of analysis*. Bibliopolis, Napoli, 1988.
4. H. Friedman, A. Montalbán, and A. Weiermann. Phi function. draft, 2007.
5. G. Gentzen. *The Collected Papers of Gerhard Gentzen*. North-Holland, Amsterdam, 1969. Translated and edited by M. E. Szabo.
6. G. Gentzen. Der erste Widerspruchsfreiheitsbeweis für die klassische Zahlentheorie. *Archiv für Mathematische Logik und Grundlagenforschung*, 16:97–118, 1974.
7. J.-Y. Girard. *Proof Theory and Logical Complexity*, volume 1. Bibliopolis, 1987.
8. Jeffry L. Hirst. Reverse mathematics and ordinal exponentiation. *Annals of Pure and Applied Logic*, 66:1–18, 1994.
9. A. Marcone and A. Montalbán. The Veblen functions for computability theorists. *Journal of Symbolic Logic 76*, pages 575–602, 2011.
10. M. Rathjen. *Untersuchungen zu Teilsystemen der Zahlentheorie zweiter Stufe und der Mengenlehre mit einer zwischen $\Delta^1_2 - CA$ und $\Delta^1_2 - CA + BI$ liegenden Beweisstärke*. PhD thesis, Publication of the Institute for Mathematical Logic and Foundational Research of the University of Münster, 1988.
11. M. Rathjen. Investigations of subsystems of second order arithmetic and set theory in strength between Π^1_1-**CA** and Δ^1_2-**CA** + **BI**: Part I. In R. Schindler, editor, *Ways of Proof Theory*, pages 363–439. Ontos Verlag, Frankfurt, 2010.
12. M. Rathjen. ω-models and well-ordering principles. In Neil Tennant, editor, *Foundational Adventures: Essays in Honor of Harvey M. Friedman*, pages 179–212. College Publications, London, 2014.
13. M. Rathjen and P. F. Valencia Vizcaíno. Well ordering principles and bar induction. In R. Kahle and M. Rathjen, editors, *Gentzen's centenary: The quest for consistency*, pages 533–561. Springer, Berlin, 2015.
14. M. Rathjen and A. Weiermann. Proof–theoretic investigations on Kruskal's theorem. *Annals of Pure and Applied Logic*, 60:49–88, 1993.
15. M. Rathjen and A. Weiermann. Reverse mathematics and well-ordering principles. In S. Cooper and A. Sorbi, editors, *Computability in Context: Computation and Logic in the Real World*, pages 351–370. Imperial College Press, 2011.
16. K. Schütte. Beweistheoretische Erfassung der unendlichen Induktion in der Zahlentheorie. *Mathematische Annalen*, 122:369–389, 1951.
17. K. Schütte. *Proof Theory*. Springer, 1977.
18. S. G. Simpson. *Subsystems of Second Order Arithmetic*. Cambridge University Press, second edition edition, 2009.
19. G. Takeuti. *Proof Theory*. North-Holland, Amsterdam, 1975.

[3] The opinions expressed in this publication are those of the authors and do not necessarily reflect the views of the John Templeton Foundation.

20. I. A. Thomson. *Well-Ordering Principles and Π^1_1-Comprehension + Bar Induction*. PhD thesis, University of Leeds, 2017. Available online at http://etheses.whiterose.ac.uk/22206/.
21. P. F. Valencia Vizcaíno. *Some Uses of Cut Elimination*. PhD thesis, University of Leeds, 2013.

Chapter 13
From Schütte's Formal Systems to Modern Automated Deduction

Wolfgang Bibel and Jens Otten

Abstract The paper traces the development of a family of high-performance proof systems out of formal first-order logical systems due to Kurt Schütte. In a first step Schütte's basic classical system is compressed by eliminating its redundancy leading to syntactic characterizations of validity exclusively in terms of formula features including connections. In a second step connection calculi for various logics are developed on the basis of this characterization. Their implementations, leanCoP for classical clausal logic, nanoCoP for classical non-clausal logic, ileanCoP for intuitionistic logic, and MleanCoP for various modal logics, turn out to be among the best performing proof systems internationally. Schütte's influences to Automated Deduction are pointed out by way of this exposition.

13.1 Introduction

In the history of mankind there has never been a period with such an unprecedented acceleration of changes for humans and human societies as the one of the last eighty years. The driving force behind these changes is the development of Information Technology (IT). It has been argued many times that one of the strongest roots of IT is Mathematical Logic (see eg. [24]).

In the 19th and the beginning 20th century the discipline of logic itself underwent a fundamental change causing a rise in its relevance for other disciplines. Among many others Boole [22] and Frege [28] have contributed substantially to make this change toward a well-founded mathematical discipline happen. In the course of these changes the discipline caused what is known as the foundational crisis of mathematics (Grundlagenkrise der Mathematik) and led to three different logical

W. Bibel
Darmstadt University of Technology, Darmstadt, Germany, e-mail: bibel@gmx.net

J. Otten
University of Oslo, Oslo, Norway, e-mail: jeotten@ifi.uio.no

R. Kahle and M. Rathjen (eds.), *The Legacy of Kurt Schütte*,
https://doi.org/10.1007/978-3-030-49424-7_13

attempts to overcome it: logicism, intuitionism, and formalism [5]. Hilbert was the main representative of formalism.

Kurt Schütte, the last doctoral student of Hilbert although advised in his PhD work mainly by Hilbert's collaborator Paul Bernays, thereby became an important representative of formalism throughout his career. He interpreted this approach as a structural theory about mathematical proofs and founded the Munich School for Proof Theory. In its initial three years the first author was one of its members and got his PhD with a thesis on cut elimination in Higher-Order Logic during that time under Schütte's supervision.

Recognizing the enormous relevance for the emerging new discipline Computer Science (CS) the first author, after completing his PhD thesis, redirected his research work towards CS, especially on the special subject of Automated Deduction (AD), also known as Automated Theorem Proving (ATP). This area aims at determining the truth of statements represented as formulas in some logic. More specifically, in one line of research within AD this is achieved by proving the validity of these formulas. Determining the validity of formulas, in turn, is realized by proof search procedures evolved from so-called formal systems. This whole endeavour has far-reaching consequences for science and its applications in general and for IT in particular. In the first author's work within this area the formal basis acquired from Schütte turned out to be an excellent framework for the development of powerful deductive systems.

The present paper traces this development, starting with Schütte's formal system GS for First-Order Logic (FOL) and thereafter presenting several internationally leading deductive systems. In more detail, we begin by introducing GS in sections 13.2.1 and 13.2.2, thereby pointing out the remarkable compression achieved in comparison with Gentzen's original system. GS's particular features can yet be found in modern AD though under different names, notably the polarities in formulas and the classification into formula types α, β, γ and δ reducing the number of rules drastically in comparison with Gentzen.

In a sequence of steps we further compress GS in order to obtain, through intermediate systems GS_1 (Sect. 13.2.3) and GS_2 (Sect. 13.2.4), a formal system GS_3 in Sect. 13.2.6 with the following remarkable feature: it has no rules at all and its axioms are any valid formulas of First-Order Logic. The validity of its axioms, ie. the axiomhood, is syntactically characterized by structural features inherent in the formulas to be proved (like multiplicities, paths, connections, substitutions). In GS_3 the popular operation of skolemization becomes superfluous through an ordering approach to unification which in a certain sense models the order of inferences in GS.

We show in Sect. 13.2.5 that from the intermediate system GS_2 there is only a trivial step toward the standard matrix representation or clausal form of formulas which forms the basis for most deductive systems in use. The discussion of such systems will be left for Section 13.3.

Further in Sect. 13.2.6 we point out the extra feature of splitting-by-need which can be integrated into GS_3 in order to further enhance its proof power. Similarly, in Sect. 13.2.7 we point out the importance of the cut for future deductive systems.

These are two of the many topics for future research in AD with the long term goal of yet more powerful proof systems.

Finally, in Sect. 13.2.8 we briefly cover non-classical formal systems and point out that only on the basis of the ordering approach to unification involved in GS_3 became it feasible to develop proof systems for modal and intuitionistic logics with a performance comparable with those for FOL.

A formal system like GS is generative in nature. It allows to generate valid formulas from its axioms by way of applying its rules while our goal rather is to analyse given formulas. For that purpose the entire Sect. 13.2 extracts from GS in a stepwise fashion a formal syntactic characterization of the validity of formulas by compressing GS into a practically redundance-free formal system such as GS_3. The resulting characterization then forms the theoretical basis for the development of analytic calculi and of proof systems based on them which, applied to some formula, test it for validity.

Such calculi along with implemented systems are the topic of Sect. 13.3. Thereby we restrict ourselves to the connection calculi which have proved to be particularly powerful. That is, in contrast to competitive methods the systems described here are based on the Connection Method (CM) [1, 11, 13] which means that they guide the proof search in a formula-oriented and goal-oriented manner determined by the given formula's connection structure.

Sect. 13.3.1 introduces the simplest such connection calculus which is based on the matrix characterization of logical validity presented in Sect. 13.2.5 and is restricted to matrices in normal form. A proof system, leanCoP, based on this calculus is described in Sect. 13.3.4. The PROLOG codes of its simplest version, consisting just of three PROLOG clauses, as well as of an advanced and more powerful version, consisting of four clauses, are presented. Comparable systems are mentioned in passing.

Typically, formulas derived from applications turn out not to be in normal form. Even the best transformations of formulas into normal or clausal form introduce undesired redundancy and thus increase the search efforts needed for finding a proof. In order to avoid these unnecessary search efforts, a non-clausal connection calculus operating on non-normal form matrices has been developed which is described in Sect. 13.3.2. Its implementation, called nanoCoP, is briefly described in Sect. 13.3.4.

Following the same methodological approach as for classical logic, connection calculi for non-classical logics, briefly addressed in Sect. 13.2.8, have been developed which are the topic of Sect. 13.3.3. These include connection calculi for intuitionistic as well as for a variety of first-order modal logics which are outlined therein. The corresponding implementations are ileanCoP for Intuitionistic Logic and MleanCoP for several first-order modal logics, briefly described in Sect. 13.3.4. Their codes are so closely related to that of leanCoP that the latter can be obtained from the former by simply deleting parts of it which demonstrates the uniform development methodology underlying our approach to building proof systems. The conclusions in Sect. 13.4 end the present chapter in this volume.

With this exposition of the development of a range of powerful proof systems out of Schütte's formal systems we demonstrate the influence of Mathematical Logic to IT in a direct way since proof systems have become an essential part in IT. For instance, the development of correct hardware and software, of the semantic web, and of knowledge systems would not be possible without these kinds of systems to mention but three out of the numerous roles they play in modern technology in one or the other way. The first author has carried this influence from Mathematical Logic into IT in person. He owes Schütte a lot in view of his entire career as this paper, the book [8] as well as his article with reminiscences in this volume show.

13.2 Schütte's Influences on the History of Automated Deduction

In the present section we outline some of the developments in the last century which eventually led to the deductive systems described in the section thereafter. This is of course not the place to give an outline of the history of the logics underlying those systems nor of the field of Automated Deduction (AD). There are excellent sources in the literature for this purpose. The book [36] gives a comprehensive historical account for the general field of Logic. The article [23] summarizes Logic's history with an emphasis on AD (as well as on its author's work in it). The chapter [16] describes the developments in the early years of AD. In order to point out Schütte's entrance into the course of events, we just mention here the following historical highlights.

Leibniz was the visionary for a computational mechanism to enhance the powers of reasoning [23, pp.2ff, p.14]. Frege's Begriffsschrift [28], with explicit reference to this vision, has laid the grounds for formal languages, logical or programming ones, as well as for logical calculi. Around the 1920's and early 1930's the works of Hilbert, Skolem, Herbrand, Gödel, Gentzen, and Jaśkowski as well as the book by Hilbert and Ackermann [33] clarified the most important logical concepts and issues such as logical calculi, consistency, completeness, decidability, Skolem functions, Skolem-Gödel-Herbrand theorem, Herbrand universe,[1] Gentzen calculi, cut elimination, and so forth. This short list leaves out many important names and contributions and is by no means to be seen as a comprehensive account of the roots of AD.

On this basis, in the mid 1950's, a proof procedure for classical First-Order Logic (FOL) by Quine [68] as well as four new and simplified completeness proofs for FOL by Beth [7], Hintikka [34], Kanger [35] and Schütte [60] were published independently. Each of these papers had an immediate and differing impact on the way early theorem provers were developed and designed. In particular, it was Prawitz et al. who succeeded in developing and implementing the first theorem prover for FOL [55]. While Beth's and Hintikka's systems used proof by contradiction, Kanger and Schütte pursued an affirmative approach. Although the difference between these two lines is completely irrelevant from a logical or deductive point of view, it is a historical coincidence that Prawitz and others decided to follow the contradictory approach which led most later researchers in AD to stay with this choice. The work

[1] For historical correctness the Herbrand universe should actually be named the "Skolem universe" [26].

reported in this chapter is one of the exceptions and has rather followed Schütte's affirmative approach. But this is only a minor point out of the many influences Schütte's work exerted on our work and on AD in general. This impact is outlined in the subsections of the current section, each covering a particular aspect relevant for AD. Most of these aspects are outlined for (classical) FOL and only the final Subsection 13.2.8 briefly addresses non-classical features.

13.2.1 The Formula Structure

Gentzen in [29] introduced the structuring of formulas in FOL in the form of "Sequenzen" (sequences). These are structures of the form $A_1, \ldots, A_m \rightarrow B_1, \ldots, B_n$ with the antecedents A_i and the succedents B_j, in which A_i and B_j are formulas. Such a sequence is meant to represent the formula $A_1 \wedge \ldots \wedge A_m \rightarrow B_1 \vee \ldots \vee B_n$ in FOL. The sequent structure led to a mathematically intuitive and symmetric set of 19 inference rules ("Schlußfiguren") in Gentzen's system LK [29, p.192f]. This structure had, however, also the disadvantageous consequence that proofs for theorems in proof theory (such as Gentzen's Hauptsatz) became rather lengthy because each of these 19 rules required a particular treatment in those.

One of Schütte's remarkable strengths consisted in his incredible ability to minimize proofs. For this purpose he first restricted the set of logical symbols in formulas to a minimal, yet functionally complete one such as $\neg, \vee, \exists$. It is well-known that any other known logical symbol in FOL can be defined with these so that this restriction is without any loss of generality. Then he introduced a different and generalized formula structure, determined by what he called positive and negative formula parts, in the following way [60, 62, 61].

Definition 13.1. Inductive definition of positive and negative parts of a formula F.

1. F is a positive part of F.
2. If $\neg A$ is a positive part of F then A is a negative part of F.
3. If $\neg A$ is a negative part of F then A is a positive part of F.
4. If $A \vee B$ is a positive part of F then A and B are positive parts of F.

If A is a positive part of F occurring at a particular position in F then this is denoted by $F[A_+]$ and similarly for negative parts using the sign "$-$" instead. Note that the A_i's in Gentzen's sequences above are negative parts and the B_j's positive parts of the corresponding formula under this definition. This shows that Schütte's concepts of positive and negative parts generalize Gentzen's concepts of antecedents and succedents. Later Smullyan in [64] used the notions of signs 0, 1 and signed formulas $\langle F, k \rangle$, $k = 0, 1$, for the same purpose and without reference to Schütte. Eventually, Wallen, eg. in [70, p.30], coined the notion *polarity* for denoting a positive or a negative occurrence of a given proposition which since then has become the standard name for pointing out this distinction of occurrences in a formula. But, except for the name, the concept positive/negative originates in Schütte's work for the first time. It has become an important conceptual feature in our field.

13.2.2 The Formal System GS

Besides the formula structure, discussed in the previous subsection, Schütte's second component for achieving short proofs consisted in his formal system for FOL, a variant of which is presented in the following definition adapted from [60, 62, 61].

Definition 13.2. Inductive definition of the derivability relation $\vdash_{GS}$, or $\vdash$ for short, in the formal system GS.

1. $\vdash F[P, \neg P]$ holds.
   Thereby F is a well-formed formula of FOL restricted to the logical operators $\neg, \vee, \exists$, in which the *prime* (or atomic) formula P (ie. without any logical symbols) occurs as a positive part of F. By generalizing the brackets notation in an obvious way for more than just one occurrence, the same is true for $\neg P$. Because the definition employs only positive and no negative parts, the index $+$ is omitted for simplicity. Prime formulas and their negated forms are also called *literals*. $\{P, \neg P\}$ is called a *complementary* pair of literals. Formulas of the particular structure depicted are called *axioms* (of GS).
2. If the premises of the following three proof rules are derivable then also the conclusion, depicted as follows.

$$\frac{F[\neg A] \quad F[\neg B]}{F[\neg(A \vee B)]}\;\wedge \qquad\qquad \frac{F[\neg A]}{F[\neg \exists a\; A]}\;\forall \qquad\qquad \frac{F[\exists x\; A \vee A\{x\backslash t\}]}{F[\exists x\; A]}\;\exists$$

   The attached rule names are chosen to reflect the underlying semantic intention. Rule ($\forall$) is subject to the *variable condition* that *eigenvariable* a must not occur in F except within the occurring part A. $\{x\backslash t\}$ denotes the standard function for (term) *substitution*, where x is substituted by t. We follow the standard notation that $A\,\sigma$ denotes the application of the substitution σ to the formula A.

Figure 13.1 shows the derivation of a simple formula in GS for illustration. The complementary pair of literals in each of the two axioms is indicated by a boldface font.

$$\cfrac{\cfrac{\cfrac{\mathbf{\neg Pa} \vee \exists z\; Pz \vee \mathbf{Pa} \vee \neg\exists b\; Qfb}{\neg Pa \vee \exists z\; Pz \vee \neg\exists b\; Qfb}\;\exists}{\neg\exists a\; Pa \vee \exists z\; Pz \vee \neg\exists b\; Qfb}\;\forall \qquad \cfrac{\cfrac{\neg\neg(\exists y\; Qy \vee \mathbf{Qfb}) \vee \exists z\; Pz \vee \mathbf{\neg Qfb}}{\neg\neg\exists y\; Qy \vee \exists z\; Pz \vee \neg Qfb}\;\exists}{\neg\neg\exists y\; Qy \vee \exists z\; Pz \vee \neg\exists b\; Qfb}\;\forall}{\neg(\exists a\; Pa \vee \neg\exists y\; Qy) \vee \exists z\; Pz \vee \neg\exists b\; Qfb}\;\wedge$$

Fig. 13.1 A derivation in GS.

GS is sound and complete, as are similar variants of Schütte's system; see eg. [62, 9, 14]. The denotation GS refers to Gentzen and Schütte.[2] In GS the number

[2] The name GS has first been introduced in [14, Def.7.1]. [67] uses the same name for the same system without giving reference to that much earlier publication.

of rules has been reduced from 19 to just three (or rather four, see Sect. 13.2.7), simplifying proofs as well as proof programs drastically. This achievement by Schütte has many more advantageous consequences as we will see in the sequel.

Although Schütte himself focused in his research more or less exclusively on proof theoretic issues, implicitly he thereby exerted a strong influence on the work in AD. We already mentioned the line of development from positive/negative parts to polarities in this context. Another important point worth mentioning in the context of GS is the close relationship between Schütte's three rules of inference and Smullyan's uniform notation for signed formulas and their classification into the types α, β, γ and δ [64]. Three of these types coincide exactly with Schütte's three rules of inference while the fourth is incorporated in Schütte's formula structure. In other words, Schütte implicitly introduced these types in the form of his formal system in [60] twelve years, and in the first edition of [62] eight years, before the first edition of Smullyan's book. It is most unlikely that Smullyan had not been influenced directly or indirectly by Schütte's work (although he does not cite it).

It is common in AD to eliminate existential quantifiers with polarity 1 through skolemization (in terms of our restriction to the three logical operators $\neg$, $\vee$, $\exists$). This is because of the derivability of a formula being implied by that of its skolemized form (see eg. [14, Lemma 4.3, p.81]). If we apply this simplification to the system GS then its rule $\forall$ becomes obsolete, leaving GS with just two remaining rules. Figure 13.2 shows the derivation for the skolemized form of our previous example formula.

$$\dfrac{\dfrac{\neg\mathbf{Pa} \vee \exists z\, Pz \vee \mathbf{Pa} \vee \neg Qfb}{\neg Pa \vee \exists z\, Pz \vee \neg Qfb}\,\exists \qquad \dfrac{\neg\neg(\exists y\, Qy \vee \mathbf{Qfb}) \vee \exists z\, Pz \vee \neg\mathbf{Qfb}}{\neg\neg\exists y\, Qy \vee \exists z\, Pz \vee \neg Qfb}\,\exists}{\neg(Pa \vee \neg\exists y\, Qy) \vee \exists z\, Pz \vee \neg Qfb}\,\wedge$$

Fig. 13.2 The derivation of the skolemized form of the formula from Figure 13.1.

These examples demonstrate that in comparison with Gentzen's original systems featuring sequences and 19 rules we have arrived at a much more compact and transparent formalism with just two remaining rules and a simple criterion for testing axioms. Having achieved this much, the natural question arises whether in view of proof search further compressions are possible which will be the topic of the following subsection.

13.2.3 Compressing the Formal System Further into GS_1

Gentzen's motivation in developing his formal systems was the modelling of human reasoning. Schütte optimized Gentzen's systems in view of short human proofs in Proof Theory. AD in contrast aims for systems supporting the search for proofs whereby different criteria apply. Especially, computers have no problems with applying rules and criteria even if these are too complicated for human use. These are the reasons why the natural next step after—but on the basis of—Schütte's contribution was done no more in the field of Mathematical Logic but rather in AD. This next step can be interpreted in the following way.

As we will see now the rule ($\wedge$) in GS can be dispensed with at the price of complicating the criterion for the axioms. We first generalize the notions of positive and negative part and thereby rename them as 0-part and 1-part for distinction.

Definition 13.3. Inductive definition of 0-parts and 1-parts of a formula F.

1. F is a 0-part of F.
2. If $\neg A$ is a 0-part of F then A is a 1-part of F.
3. If $\neg A$ is a 1-part of F then A is a 0-part of F.
4. If $A \vee B$ is a 0-part of F then A and B both are 0-parts of F.
5. If $A \vee B$ is a 1-part of F then A and B both are 1-parts of F.

It is obvious that any positive part is a 0-part and any negative part is a 1-part. For readers familiar with the notion of polarity, to be introduced in the next subsection, it is also obvious that any k-part has polarity k, $k = 0, 1$. In other words, this generalized notion has lead us a step closer to the general notion of polarity in AD.

In order to specify a more general notion of an axiom we need the following notion of a path through a formula.

Definition 13.4. A path through a formula F is a set of literals of F defined inductively as follows.

1. If F is a literal then $\{F\}$ is the only path through F.
2. If $F = \exists x A$ then the empty path $\{\}$ is the only path through F.
3. If $F = \neg\neg A$ then any path through A is also a path through F.
4. If $F = A \vee B$ then $p \cup q$ is a path through F for any path p through A and any path q through B.
5. If $F = \neg(A \vee B)$ then any path through $\neg A$ or through $\neg B$ is a path through F.

A path through F which contains literals of the form P and $\neg P$ is called *complementary*. F is called *complementary* if all paths through it are complementary.

Now let GS_1 be the formal system obtained from GS by the following three changes. First, axioms are any complementary formulas. Second, the notion $F[A]$ in the rules of GS in GS_1 refer to 0-parts rather than positive parts as in GS. Third, only the single rule $\exists$ remains.

For illustration, Fig. 13.3 shows the derivation of our running example from Fig. 13.2 in GS_1. There are two paths $\{\neg Pa, Pa, \neg Qfb\}$ and $\{Qfb, Pa, \neg Qfb\}$ through the formula in the top line, each of which contains a complementary pair of literals $\{\neg Pa, Pa\}$ and $\{Qfb, \neg Qfb\}$, respectively, indicated in boldface. Hence the formula is complementary and therefore an axiom in GS_1.

The soundness and completeness of GS_1 for FOL is a straightforward consequence of those properties holding for GS.

$$
\cfrac{\cfrac{\neg(\mathbf{Pa} \vee \neg(\exists y\ Qy \vee \mathbf{Qfb})) \vee \exists z\ Pz \vee \mathbf{Pa} \vee \neg\mathbf{Qfb}}{\neg(Pa \vee \neg(\exists y\ Qy \vee Qfb)) \vee \exists z\ Pz \vee \neg Qfb}\ \exists}{\neg(Pa \vee \neg\exists y\ Qy) \vee \exists z\ Pz \vee \neg Qfb}\ \exists
$$

Fig. 13.3 The derivation in GS_1 of the skolemized form of the formula from Fig. 13.1.

The step from GS to GS_1 may be seen under a different aspect which is elaborated by Guglielmi in [30] under the term "deep inference". Gentzen's rules were applicable only at the surface of each formula in the sequent. Schütte allowed the rules to be applied to positive parts inside the formula and this way from the viewpoint of AD took a first step in the right direction. GS_1 goes a step further in the same direction and allows inferences to be applied to 0-parts. Guglielmi allows his rules to be applied anywhere inside the formula, hence the term "deep inference".

From the point of view of proof search the derivation depicted in Fig. 13.3 seems to be extremely redundant. This is because at each step backwards from the formula to be proved just a term is associated with an existentially quantified variable while everything else is left unchanged and is therefore carried along as a completely unnecessary burden. This obviously extreme redundancy begs for an improvement by which we get rid of the superfluous burden. This is the task for the following subsection.

13.2.4 Connections and the System GS_2

The improvement just announced is based on the simple observation that all informations contained in the axioms of the derivation of Fig. 13.3 can be located already in the formula to be proved. In other words, at least in the case of skolemized formulas considered so far, we may dispense with rules altogether and just redefine the criterion for axioms to apply for general skolemized formulas in FOL right away. This is the topic of the current subsection. We begin by introducing the general notion of polarity.

Definition 13.5. Inductive definition of the polarity of the occurrence of any subformula of a formula F.

1. F itself has polarity 0.
2. If $\neg A$ is a subformula of F with polarity 0 then A has polarity 1.
3. If $\neg A$ is a subformula of F with polarity 1 then A has polarity 0.
4. If $A \vee B$ is a subformula of F with polarity 0 then A and B both have polarity 0.
5. If $A \vee B$ is a subformula of F with polarity 1 then A and B both have polarity 1.
6. If $\exists x A$ is a subformula of F with polarity 0 then A has polarity 0.
7. If $\exists a A$ is a subformula of F with polarity 1 then A has polarity 1.

Polarities are indicated in formulas by an upper index, in the case of an atom attached either to the parenthesised atom or to its predicate symbol.

As we see, polarities are a straightforward generalization of 0- and 1-parts, and hence also a generalization of Schütte's original positive and negative parts. They simply count the number of negation signs dominating the subformula occurrence in the formula and assign 0 to the subformula if this number is even, otherwise 1.

If we look at the axioms of the derivation in Fig. 13.3 we see that they still contain existential subformulas so that they could be subject to yet another, in fact to arbitrarily many more inverse applications of rule (∃), each possibly with a different term associated. In order to illustrate this important feature Fig. 13.4 shows a derivation of the formula $\neg N0 \vee \exists x \neg(\neg Nx \vee Nfx) \vee Nff0$ where a proof can only be found with two such rule applications. The formula expresses that 2 is a natural number if this holds for 0 and any successor. The prime formulas with attached polarities in the axioms are embraced by parentheses for clarity; in the examples thereafter these unnecessary parentheses will be left out.

$$\dfrac{\dfrac{\neg(\mathbf{N0})^1 \vee \exists x \neg(\neg Nx \vee Nfx) \vee \neg(\neg(\mathbf{Nf0})^0 \vee (\mathbf{Nff0})^1) \vee \neg(\neg(\mathbf{N0})^0 \vee (\mathbf{Nf0})^1) \vee (\mathbf{Nff0})^0}{\neg N0 \vee \exists x \neg(\neg Nx \vee Nfx) \vee \neg(\neg N0 \vee Nf0) \vee Nff0}\ \exists}{\neg N0 \vee \exists x \neg(\neg Nx \vee Nfx) \vee Nff0}\ \exists$$

Fig. 13.4 Derivation with double application of the rule (∃).

As we see from the example, the derived formula after the first inverse inference step is not an axiom, whatever term is chosen. A second step, ie. a second copy of the existentially quantified subformula, is needed to derive a proof for this theorem. In fact, if we would want to prove Nn for an arbitrarily large n, we would need n copies. In order to cope for this necessity we introduce the multiplicity function on formulas in the following definition.

Definition 13.6. For any formula F of FOL (built with the logical symbols $\neg, \vee, \exists$) a *multiplicity* μ is a function which assigns to each occurrence of a subformula $\exists x A$ of polarity 0 a natural number $m \geq 0$. F along with μ is also written F^μ.

With multiplicities and polarities we are now in a position to generalize also the intermediate notion of path from Def. 13.4. Since the intermediate notion will not be used any further and turns out to become a special case, we use the same name for the generalized notion.

Definition 13.7. A *path* through a skolemized formula F^μ is a set of prime (or atomic) subformulas of F along with their polarities and is defined inductively as follows.

1. If F is a prime formula with polarity k then $\{F^k\}$ is the only path through F.
2. If $F = \neg A$ then any path through A is also a path through F.
3. If $F = A \vee B$ with polarity 0 then $p \cup q$ is a path through F for any path p through A and any path q through B.
4. If $F = A \vee B$ with polarity 1 then any path through A or through B is a path through F.

5. If $F = \exists x A$ with polarity 0 and multiplicity m then $p_1 \cup \cdots \cup p_m$ is a path through F for any paths p_i through $A\{x \backslash x_i\}$, $i = 1, \ldots, m$.

An unordered pair of elements in a path through F^μ with identical predicate symbols but differing polarities is called a *connection*. A set U of connections is called *spanning* for F^μ if each path through F^μ contains at least one connection of U. If a connection consists of two identical prime formulas with differing polarities, then it is called *complementary*. A skolemized formula F is called *complementary* if there is a multiplicity μ and a substitution σ such that $p\sigma$ contains a complementary connection for any path p through F^μ.

Let GS_2 be the formal system for skolemized FOL which has complementary formulas as its axioms and no rules at all.

We first note that the concept of path in this definition generalizes that of Def. 13.4 in the following way. First, the more special concept from Def. 13.4 only applies to formulas with a multiplicity which is 0 throughout. Second, here we use polarities to identify connections rather than pairs of literals as before. From now on the notion of path will refer exclusively to the present definition unless stated otherwise.

A proof of a formula F in GS_2 is characterized by a multiplicity μ, a substitution σ, and a set U of connections, referred to as a *connection proof*. Figure 13.5 shows such a connection proof for our running example. The multiplicity for each existential subformula is chosen to be 1, in which case it is not noted explicitly in the proof by default, but only by way of the index 1 attached to the variables y and z in the substitution. As before there are two paths through F each containing exactly one of the two depicted connections, ie. these are spanning. Along with the given substitutions these connections are complementary. Hence the formula is an axiom, thus establishing the proof. The reader might wish to compare this proof with the equivalent proofs given in the figures 13.1, 13.2, and 13.3. What has remained unchanged in all four versions is of course the derived formula, but also the two connections are exactly the same in all of them. This tells us that the formula along with a multiplicity, a substitution, and a spanning and complementary set of connections, ie. a connection proof, is all that is needed for a full-fledged guarantee for validity. The connections form the propositional, the multiplicity and the substitution the first-order part of the proof.

$$\neg(P^1 a \vee \neg\exists y\, Q^0 y) \vee \exists z\, P^0 z \vee \neg Q^1 fb \text{ with } \sigma = \{y_1 \backslash fb, z_1 \backslash a\}$$

Fig. 13.5 The connection proof for the skolemized formula from Fig. 13.1.

Thus we have achieved the complete elimination of any redundancy from the proof representation without any loss of information. This way we have *compressed* the formal system in the sense defined in [15, Sect. 6]. Validity of a formula is now characterized in terms of its structural features. All theorems have become axioms.

What needs to be done in view of finding a proof in GS_2 is, first, locating the set of connections in the formula which can be done fast. From this set and for a chosen multiplicity then a spanning subset of connections and a substitution has to be determined which renders the subset complementary, a task which is trivial in the present example but may be exponentially difficult in general since proof search is co-NP-complete even in the restriction to propositional logic, as is well-known. Any calculus which solves this task with a focus on connections or connection-structures is called a *connection calculus*. The general approach to determine validity through some connection calculus is termed the *connection method* (or connection principle). Apart from the lack of redundancy the main advantage of the connection method in comparison with other proof methods is its formula-oriented approach in a search space which is clear-cut by the very structure of the given formula (rather than by many additionally generated formula parts as in competitive methods, such as resolution or instance-based methods). In consequence, the space required in larger examples amounts only to a tiny fraction of that required by other methods.

Figure 13.6 shows the connection proof for our second example. The multiplicity of the existential subformula is 2, depicted by the upper index 2 attached to the existential quantifier. We therefore get two different variables x_1 and x_2 originating from x. The formula has 4 connections of which one connects terms which are not unifiable under any substitution, hence can be ignored. The remaining three are depicted in the figure. There are four paths through the formula and each contains one of these connections, hence these are spanning. The depicted substitution renders all three connections complementary, so that the formula is in fact an axiom and thus the formula is provable in GS_2.

$$\neg N^1 0 \vee \exists^2 x \neg(\neg N^0 x \vee N^1 f x) \vee N^0 f f 0 \quad \text{with} \quad \sigma = \{x_1 \backslash 0, x_2 \backslash f0\}$$

Fig. 13.6 The connection proof for the formula from Fig. 13.4.

Recall that the formula in the figure expresses that 2 is a natural number. Had we chosen 3 instead of 2 then for a proof we would have to choose multiplicity 3 and we would get a second instance of the middle connection with the multiplicity indices of the connected literals increased by 1 in comparison with the first instance. This illustrates the fact that the original connections of a formula may lead to arbitrarily many instances of any of these in order to achieve a proof.

The following theorem is now a straightforward consequence of the soundness and completeness of GS. In fact, any proof for a skolemized formula in GS can easily be reduced to a connection proof in GS_2 as illustrated with our examples. Conversely, any connection proof in GS_2 can immediately be extended to a proof in GS. Our way of illustrating the development of GS_2 in this section has traced the close relationship between GS and GS_2.

Theorem 13.8. *GS_2 is sound and complete for skolemized FOL.*

We are not yet done completely because of the restriction to skolemized formulas, a restriction to be lifted in Sect. 13.2.6. Before turning to this alternative technique let us simplify matters somewhat in the next subsection.

13.2.5 The Matrix Representation

Quantifiers let formulas look complicated. That is why in AD for the daily practice they have been eliminated completely. We already achieved the first step in this process by eliminating (existential) quantifiers with polarity 1 through skolemization. Here we now take the second step in this process.

A skolemized formula (with the restricted variety of logical operators considered here) has only quantifiers with polarity 0 if any. Without restricting generality we can assume that for any two different quantifiers in the formula the variables bound by them are denoted differently. Otherwise this can be achieved by separating them apart through renaming of bound variables. Let us assume throughout this section that bound variables are separated apart in this way.

Now we transform any given skolemized formula into its equivalent prenex form.[3] In the case of our example formula $\neg(P^1 a \vee \neg\exists y\ Q^0 y) \vee \exists z\ P^0 z \vee \neg Q^1 fb$ this transformation results in $\exists y \exists z(\neg(P^1 a \vee \neg Q^0 y) \vee P^0 z \vee \neg Q^1 fb)$ where the ordering of the quantifiers is chosen arbitrarily. As we can see from this example the quantifiers have become redundant in this form since they are uniquely determined through the bound variables in the formula (ie. y and z in the example). Therefore the quantifiers can be deleted by default. Thus we obtain $\neg(P^1 a \vee \neg Q^0 y) \vee P^0 z \vee \neg Q^1 fb$.

Recall from the previous subsection that we are looking for a spanning set of connections in such a formula, a notion which is based on that of paths through it. It would be nice to have a representation of formulas which exhibits paths more visibly. This motivation has led to representing formulas as matrices in which disjunction is displayed horizontally and conjunction vertically (in our affirmative approach while in the contractory approach it is just the other way around).

In our formal system conjunction is represented as disjunction with polarity 1. For the purpose of more comfortable visibility aimed at in the present subsection we will reintroduce conjunction explicitly, thus yielding for our example $(\neg Pa \wedge Qy) \vee Pz \vee \neg Qfb$ (whereby polarities are dropped as obvious). As a matrix this formula looks like shown in Fig. 13.7.

All notions from Def. 13.7 carry over from formula to matrix in a straightforward way. If the represented formula is in disjunctive normal form, which happens to be the case in our example, then a path through such a matrix can be regarded as a walk through it strictly from left to right, thereby collecting exactly one literal from each column. If this is done for our example matrix, we see that it has exactly two different paths, $\{\neg Pa, Pz, \neg Qfb\}$ and $\{Qy, Pz, \neg Qfb\}$. Each of them contains a connection which is complementary under the attached substitution. Hence the two depicted

[3] In view of efficiency of proof search care has to be taken in how to carry out this transformation in detail, an issue not pursued any further here.

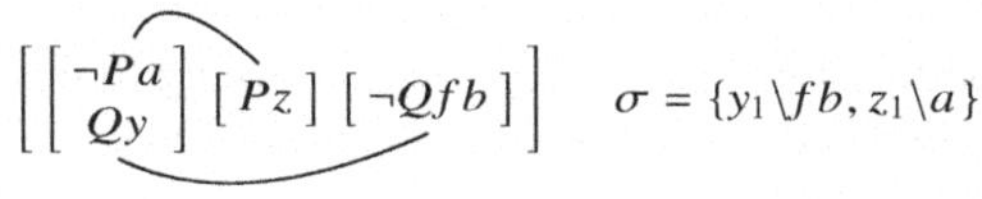

Fig. 13.7 The connection proof from Fig. 13.5 in matrix representation.

connections are spanning and complementary; thus the matrix is complementary; hence the formula from which it derives is valid.

Eliminating the quantifier and replacing the disjunction with polarity 1 by conjunction in our second example $\neg N0 \vee \exists x \neg(\neg Nx \vee Nfx) \vee Nff0$ similarly as before yields $\neg N0 \vee (Nx \wedge \neg Nfx) \vee Nff0$. It again happens to be in disjunctive normal form. Its matrix representation and a connection proof for it is shown in Fig. 13.8.

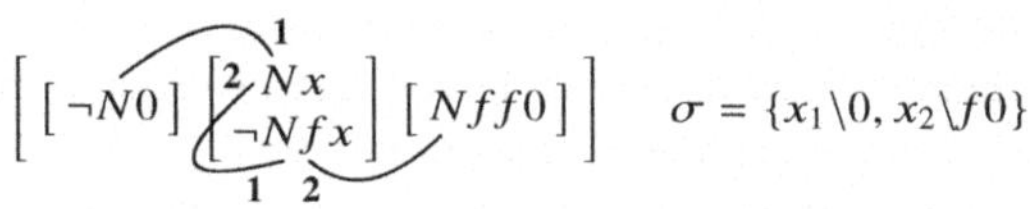

Fig. 13.8 The connection proof from Fig. 13.6 in matrix representation.

For the resulting quantifier-free formulas in disjunctive normal form each disjunct is called a *clause*. Thus in this special kind of formula a clause is a conjunction of literals in our affirmative approach (while in the popular contradictory approach it is a disjunction of literals). If formulas are restricted to this special case then we say that these as well as their matrix representations are in *clausal form*. Most popular theorem provers in AD operate on formulas in clausal form. This is without restriction of generality since any formula can equivalently be transformed into clausal form. This transformation may affect the efficiency of the prover, however.

In order to cope with this disadvantage we note that Def. 13.7 applies to arbitrary skolemized formulas, not only to those in disjunctive normal form, ie. clausal form. In general, each conjunct might therefore itself be a more involved formula rather than simply a literal. This leads to nested matrices by way of the transition to the matrix representation. A path through such a matrix is then still visualized by a literal collecting walk through the matrix from left to right, although a more complicated one observing the nesting structure of the matrix (see Sect. 13.3.2). Since the transformation to clausal form introduces redundancy into the formulas, engaging it is contrary to our main objective of excluding redundancy as much as possible, a point to be kept in mind for Sect. 13.3 where we will present proof systems both for clausal form (Sect. 13.3.1) as well as for non-clausal form (Sect. 13.3.2).

The connection method applied to formulas in matrix representation is often also called *matrix method*. The matrix method has become popular because the two-dimensional representation of formulas is very helpful for the human intuition. From a technical viewpoint, however, the difference between the representation of formulas in a one-dimensional form and that in the form of two-dimensional matrices is completely irrelevant. What counts technically is exclusively the focus on the connection structure determined by the given formula represented in whatever form. It is for this reason that we prefer "connection method" to "matrix method" as the characterizing term for this approach.

13.2.6 Alternative to Skolemization (GS_3) and Splitting by Need

The previous subsection sort of interrupted our line of development in this section in order to link it with the standard clausal form which is in common use in AD. This is also to say that the community in AD has got stuck at this form of representation although there is plenty of room for further optimization. The clause form is far simpler for developers of proof systems to work with than the formulas as they derive directly from applications in mathematics, program verification etc. However, it would be much better for the proof systems to work right with the original formulas rather than a version which just provides more comfort to the developers. This is because the clause form possibly increases the search space and thus renders the search efforts unnecessarily more complex.

Therefore, after this interruption, we resume here the line of our development up to Sect. 13.2.4 which has resulted there in the formal system GS_2. Recall that this system is already defined for arbitrary skolemized formulas. Could this result be generalized by doing even without skolemization? This is the topic for the present subsection.

Unfortunately, the technical details for this generalization of GS_2 are too involved to be presented in all details in this chapter. We refer to [12, Theorem H_3 and H_4], [14, Corollary IV.8.5 and Theorem IV.10.4] and [2, 3] for more details and for proofs. Here we try to present the gist of the solution.

Recall our example formula $\neg(\exists a\ Pa \vee \neg\exists y\ Qy) \vee \exists z\ Pz \vee \neg\exists b\ Qfb$ from Fig. 13.1 in its original form. How could a connection proof like that for its skolemized form in GS_2, shown in Figure 13.5, be obtained right away for the original formula?

In order to solve this problem we have to identify the feature in GS which takes care of the role played by skolemization. This is in fact the variable condition accompanying Rule (∀) in Def. 13.2 which applies to existential quantifiers with polarity 1. It restricts the choice of a free variable as we apply the rule in an inverse direction. This restriction in view of proof search has exactly the same effect as skolemization [9]. So the problem just stated reduces to the problem of incorporating this variable condition into the transition from GS to GS_2 in order to arrive at a formal system GS_3 with connection proofs for arbitrary formulas.

Figure 13.9 shows what could be a connection proof for our example formula. Is it a correct proof according to our envisaged GS_3? Yes, it happens to be one which gives us a first feel for where we are heading to. In order to be a correct connection proof the shown substitution σ has to fulfill a certain criterion to be discussed shortly which happens to be the case in this simple example right away.

$$\neg(\exists a\, P^1 a \vee \neg\exists y\, Q^0 y) \vee \exists z\, P^0 z \vee \neg\exists b\, Q^1 f b \quad \text{with} \quad \sigma = \{y_1 \backslash f b_1, z_1 \backslash a_1\}$$

Fig. 13.9 The connection proof for the formula from Fig. 13.1.

Due to the (eigen-)variable condition of Rule ($\forall$) in GS variables bound by existential quantifiers with polarity 1 play a completely different role from those bound by existential quantifiers with polarity 0. In order to visualize this distinction the former are denoted by $a, b, \ldots$ in GS and are called *constants* (or constant variables) while the latter are denoted by $x, y, \ldots$ and are called (free) *variables*.

If we now compare Fig. 13.1 with Fig. 13.9 we first note that the former is condensed in the latter into just two connections along with a substitution without any loss of information relevant for the proof. Where then has the relevant information, that in Fig. 13.1 the two rules ($\exists$) are above the corresponding ($\forall$) rules, gone in the connection proof? This information is indeed relevant since only then the variable condition is fulfilled. This information is hidden in the already mentioned criterion for the substitution which we now introduce with the help of the following definition.

Definition 13.9. Let $<$ denote the tree ordering among the occurrences of the subformulas of a formula which itself denotes the root of $<$. This ordering induces (via the corresponding quantified subformulas) also an ordering among the constants and variables in the formula which is denoted also by $<$.

For any substitution σ an equivalence relation $\sim$ on the set of variables in σ and a partial ordering relation $\lessdot$ on the set of constants and variables in σ is defined as follows.

1. If $x \backslash y \in \sigma$ then $x \sim y$ holds.
2. If $x \backslash t \in \sigma$ where t is not a variable, then $y \sim x$ holds for any variable y occurring in t and $c \lessdot x$ holds for any constant c occurring in t.
3. If $c \lessdot x$ and $x \sim y$ then also $c \lessdot y$.

Let $\lhd$ denote the transitive closure of the union of $<$ and $\lessdot$. We say that σ has no cycles, ie. is irreflexive, if this is the case for $\lhd$.

A formula in FOL with multiplicity μ is called *complementary* if there is a substitution σ without cycles such that $p\sigma$ contains a complementary connection for any path p through F^{μ}.

Let GS_3 be the formal system for FOL which has complementary formulas as its axioms and no rules at all.

For the substitution in Fig. 13.9 we get $b_1 \lhd y_1$ and $a_1 \lhd z_1$ according to this definition. In this ordering we find the relevant information that we mentioned just before the definition which expresses that the ($\exists$) rules must occur above the ($\forall$)

rules in the corresponding GS derivation. It is obvious that $\lhd$ in this example does not have cycles so that the formula is complementary. Hence it is also valid. This is because GS_3 is sound and complete according to the references given above, eg. [14, Coroll. 8.5]. As a matter of fact, the equivalence relation $\sim$ can be dispensed with if only idempotent substitutions are considered, a restriction that does not affect completeness.

In order to illustrate the cycle condition with a non-theorem consider the formula $\exists x \neg \exists a \neg (\neg P^1 ax \vee P^0 xa)$ with multiplicity 1. The only possible connection leads to the substitution $\sigma = \{x \backslash a\}$ (or rather to $\sigma = \{x_1 \backslash a_1\}$, but we drop the indices for simplicity here). Hence we get $a \lessdot x$, and therefore $a \lhd x$. The tree ordering of the formula gives $x < a$, and therefore $x \lhd a$. By transitivity of the ordering we get $x \lhd a \lhd x \lhd a$, ie. the ordering, and thus the substitution, is cyclic so that this connection along with the substitution does not constitute a "proof" for this obviously invalid formula.

Once again we note that this *ordering approach* to unification realized in GS_3 via the relation $\lhd$ plays exactly the same role as the interplay of skolemization with unification. The latter has dominated the work in AD since shortly after its beginnings. But only through the insight obtained by the ordering approach became it possible much later to develop effective proof systems for non-classical logics, a fact which Wallen in [70, p. 3] phrases as follows: "*Of most note is the proof-theoretical analysis of unification that emerges . . . It is this rediscovery of the rôle of unification that forms the foundation for the matrix characterisations of non-classical logics in the sequel.*".

The ordering approach was introduced by Prawitz in [54] for the first time and refined in [9] and in later papers of the first author of the present paper. The influence of Schütte's paper from 1956 on this early work by Prawitz is clearly recognizable in it and the paper is of course also cited in Prawitz's paper. The same is of course the case for the cited paper of the first author of the present paper. So there is a direct line of influences from Schütte to the ordering approach to unification.

By the way, the test for cycles required in GS_3 is no extra burden in comparison with other proof methods since any unification algorithm needed in any such a method also requires the analogue cycle test [14, Sect. IV.9], a fact which again illustrates the close relationship between these two approaches.

There is an additional possibility to minimize connection proofs and thus enhance their compactness which is known as *splitting by need*. Consider the simple formula $\exists \neg P^1 x \vee \neg(\neg P^0 a \vee_1 \neg P^0 b)$ with multiplicity 1 in which we distinguish the occurrence of one of the disjunction symbols by attaching an index 1. There are two paths through it, each containing a connection. But x_1 is not unifiable with both a and b at the same time. Thinking in terms of finding a proof in GS, however, we could first apply the inverse of Rule ($\wedge$) leading to two independent subformulas each provable immediately. In terms of GS_3 this can be simulated by including positions of disjunction symbols with polarity 1 like $\vee_1$ in the example into the relation $\lessdot$ and by considering it in the cycle test. In our example the two connections would become complementary through the additional requirement $\vee_1 \lessdot x$ generated "by need" upon noticing the conflict between a and b in the unification attempt.

In full generality this technique is quite involved. It has first been explored in [21] and in more details in [14, Sect. IV.10]. Further work on it has been done in [2, 3, 32], but in a simpler context than that provided by GS_3. If the technique is incorporated into GS_3, it allows, for instance, a straightforward 4-connection proof for the formula $\neg\exists b\neg(\neg\exists a\neg\exists z\neg(\neg\exists x\, Paxz \vee_1 \neg\exists y\, Pbyz) \vee \neg(\exists c\neg\exists u\, \neg Puuc \vee_2 \exists d\neg\exists v\, \neg Pvbd))$ which is taken from [21]. If its multiplicity is restricted to 1 then the formula has four paths and four connections resulting in a straightforward substitution. By need the requirements $\vee_2 \lessdot z$, $\vee_1 \lessdot u$ and $\vee_1 \lessdot v$ are generated with which the resulting substitution becomes free of cycles thus establishing the formula's validity. The reader might wish to compare this 4-steps proof search with that of his/her favorite proof method. Of course, the formula is artificially constructed in order just to demonstrate the effects in a small example.

13.2.7 Cuts

In Sect. 13.2.1 we referred to Gentzen's formal system LK featuring 19 rules of inference. These rules include the cut (or Schnitt) rule which has not been included in GS of Sect. 13.2.2. This omission is justified by Gentzen's Hauptsatz which states that any proof in GS (or in LK for that matter) with cut can be reduced to a proof without cut. This cut elimination leads to longer proofs, in fact exponentially longer proofs in the worst case [67, Sect. 5.2]. This increase of the proof lengths is therefore an important issue for proof search since searching for exponentially long proofs is hopeless to begin with. Can we expect any improvement of this situation from the point of view of GS_3? The present section is concerned with this question.

The cut rule in GS may be written in the following way.

$$\frac{F[\neg C] \quad G[C]}{F[\,] \vee G[\,]}\ cut$$

Here $F[\,]$ denotes the formula obtained from $F[\neg C]$ by deleting $\neg C$ and with it any symbols which in view of a well-formed formula have become superfluous thereby, and similarly for $G[\,]$.

In order to view the cut rule in terms of GS_3 let us first assume for simplicity that C is a prime formula. Now assume further that we have connection proofs in GS_3 for the two premises of the rule. Then we could construct a connection proof in GS_3 also for the conclusion in the following way. If the cut formula $\neg C$ is not contained in a connection of the connection proof for the left premise then this connection proof is also a connection proof for the conclusion (and similarly for the right premise). Otherwise, if the cut formula is contained in a connection of the connection proof for each premise, then let $c_1 = \{C_1, \neg C\}$ be such a connection in the left premise and $c_2 = \{C, \neg C_2\}$ one in the right premise. For any pair of such connections in the premises we add the connection $\{C_1, \neg C_2\}$ to the union of all connections in the premises not involving the cut formula resulting in a set of connections to be considered in the conclusion. This set along with the union of the substitutions for the premises is a connection proof of the conclusion in GS_3, as can be shown in a straightforward way by looking at the paths through F and G. This cut elimination step can be generalized to cut formulas other than literals. In this manner cuts may be eliminated from a formal system GS_3^c that includes the cut rule, thus resulting back in GS_3. (Formally, this cut elimination in GS_3^c is proved by induction on the number of cuts in the derivation and by a subinduction on the length of the cut formula; when the cut formula is a disjunction with negative polarity the cut is replaced by two cuts on the disjunctive parts; when it is an existential formula with negative polarity, it is replaced by m cuts where m denotes the multiplicity of the quantifier.)

The first observation from the point of view of this insight shows us how the exponential explosion may occur in this process of cut elimination. Namely, let m_1 be the number of connections containing the left cut formula and m_2 the one for the right one, assumed to be literals for simplicity. Then the elimination step just described yields $m_1 \cdot m_2 - (m_1 + m_2)$ additional connections in the connection proof of the conclusion as compared with the sum of connections in the two proofs of the premises. A corresponding result is obtained for non-literal cut-formulas.[4] If we have k cut rules in the derivation this amounts to an increase of the order of n^k additional connections for some n, in accordance with the exponential increase known from other formal systems discussed in the reference given above. But a precise result in this respect for the situation given by a system such as GS_3^c is currently not known.

This increase can be avoided if the set of connections containing a cut formula is handled as a connection structure in the sense of [19, Sect. 2.10]. Although there has been some work in this direction and generally on cuts in the context of proof search such as [25, 37] and the references therein, the issue has remained a major challenge for AD. We may, for instance, remind the community in this context of the conjecture posed in [15, Sect. 6.1].

13.2.8 Non-classical Formal Systems

Up to now in this section we have focused exclusively on classical FOL. In the present subsection we will briefly extend the view also to include non-classical logics. Again, Schütte was among the first logicians who presented compact completeness and soundness proofs for intuitionistic and modal first-order logics [61], shortly after Kripke had introduced his uniform semantics for these logics. This book by Schütte seems to have had no direct influence on proof systems for non-classical logics. For instance, the book [70] does not cite it.

But since this work of Schütte had its roots firmly in his own work on classical FOL, it indirectly exerted its influence also on the later work on proof systems for non-classical logics through all his contributions pointed out in the preceding subsections such as polarity, types of inferences and especially the ordering approach

[4] Cut formulas may be regarded as mathematical lemmata. What is a good lemma? In fact, $m_1 + m_2$ may be taken as a concrete measure for the goodness of the lemma (in the sense "the bigger the better").

to unification. Recall the quotation of Wallen in Sect. 13.2.6 in which he pointed to the fundamental insight through the ordering approach leading to his ingenious solutions for intuitionistic and modal proof systems.

Namely, in Gentzen-type systems for non-classical logics the role of modal operators is analogous to that of quantifiers in terms of the rules of inference and their ordering. Therefore their treatment can be integrated into the unificational part by way of the ordering approach in full analogy with the treatment of the quantifiers described in Sect. 13.2.6. Thus the non-classical features affect solely the details involved in the relation $\lhd$ to be adapted appropriately. Further details about this integration as well as the resulting proof systems for non-classical logics will be presented in Sect. 13.3.3 of the present chapter.

Thus the theoretical basis for the research program outlined in the preceding subsections can be transferred to many non-classical logics in a uniform way on the basis of Wallen's work. As far as the formal systems for those logics were concerned Wallen was directly influenced mostly by the book [27] of Fitting who of course was closely familiar with Schütte's work.

13.3 Modern Connection Calculi

The previous section has presented the theoretical basis for proof systems in FOL. It has done so in a stepwise fashion whereby at each step the underlying formal system was compressed somewhat further by eliminating redundancy left over in its predecessor. The field AD so far has not succeeded in producing a proof system for any logic based on the most compressed system GS_3. It has up to this point in time produced systems which are based on the intermediate formalism described in Sect. 13.2.5 and which are mostly restricted to clause form. This intermediate formalism is still quite redundant in comparison with GS_3 so that there remains a major challenge ahead for the field to develop systems on the basis of GS_3 including the features discussed in sections 13.2.6 and 13.2.7.

Nonetheless, the systems obtained so far on this intermediate basis are already quite impressive in their performance. One of these systems, leanCoP, will be described in the present section. It has been developed very closely in line with the research program outlined in the previous section and insofar was influenced at least indirectly by Schütte's work as well.

The first step towards a proof system on the basis of the formalism presented in Sect. 13.2.5 consists in the development of a calculus which determines the search for a spanning set of connections. Calculi serving this purpose are called *connection calculi* as already pointed out in Sect. 13.2.4. Examples of this kind of calculi are those presented in [11, 13, 14], the connection tableau calculus [39], and the model elimination calculus [40]. Proof search in a connection calculus is guided by connections, the structural elements within the formula to be proved which were introduced in the previous section. This guidance along the formula's structural features, which are crucial for establishing a proof, renders the search much more goal-oriented compared to the proof search in sequent, tableau or resolution calculi. Furthermore, compared to GS_3 presented in Sect. 13.2.5, these calculi contain, eg., representational redundancies that cannot be completely eliminated by using structure-sharing techniques. It has also been shown that a slight variation of the connection calculus in form of the *connection structure calculus* [19, Sect. 2.10] can simulate resolution in linear time. Thus future connection calculi incorporating all such additional features might have the potential to outperform any of its competitors.

This section first introduces a clausal connection calculus for classical FOL. Afterwards, a non-clausal connection calculus and connection calculi for several popular non-classical logics are presented. Finally, leanCoP and versions of leanCoP for classical and some non-classical logics that are based on the given calculi are described. leanCoP's core program consists of just four PROLOG clauses and yet features a performance competitive with some of the best provers worldwide.

13.3.1 The Clausal Calculus

The connection calculus for classical logic to be introduced now is based on the matrix characterization of logical validity presented in Sect. 13.2.5. Recall that this characterization established a *connection proof* by identifying a spanning set of unifiable connections for the given matrix as illustrated in Fig. 13.10.

$$\left[\,[\,P^0 z\,]\begin{bmatrix}P^1 a\\ Q^0 y\end{bmatrix}[\,Q^1 fb\,]\right]\quad \sigma=\{z\backslash a\},\qquad \left[\,[\,P^0 z\,]\begin{bmatrix}P^1 a\\ Q^0 y\end{bmatrix}[\,Q^1 fb\,]\right]\quad \sigma=\{z\backslash a, y\backslash fb\}$$

Fig. 13.10 The connection proof for the formula $Pz \vee (\neg Pa \wedge Qy) \vee \neg Qfb$.

Determining a spanning set of unifiable connections in general is a computationally complex task. We describe the solution to this task by way of an indeterministic logical calculus which allows the flexibility for certain strategical choices in the search. The idea behind its derivational mechanism is to identify connections in a systematic order which guarantees that eventually all paths through the matrix are covered. Once a connection is selected then only paths that do not contain this connection are investigated afterwards. If every path contains a connection, the proof search succeeds and the given formula is valid. At each step the substitution is updated to make each selected connection complementary. For example, the proof of the previous matrix consists of two inferences, which identify two connections resulting in the shown substitution.

A formal description of a connection calculus was given by Otten and Bibel [51] and is presented in the following definition.

Definition 13.10. The axiom and the rules of this *clausal connection calculus* are given in Fig. 13.11, in which $\overline{L_2}$ denotes the complement of L_2, ie. $\{L_1, L_2\}$ is a complementary connection for the substitution σ. The words of the calculus are tuples of the form "$C, M, Path$", where M is a matrix, C and $Path$ are sets of literals or ε; C is the *subgoal clause*, $Path$ is the *active path*, and σ is a *rigid* term substitution (which is applied to the whole derivation). A *clausal connection proof* of a matrix M is a clausal connection proof of $\varepsilon, M, \varepsilon$.

Axiom (A)
$$\frac{}{\{\}, M, Path}$$

Start (S)
$$\frac{C_2, M, \{\}}{\varepsilon, M, \varepsilon} \quad \text{and } C_2 \text{ is copy of } C_1 \in M$$

Reduction (R)
$$\frac{C, M, Path \cup \{L_2\}}{C \cup \{L_1\}, M, Path \cup \{L_2\}} \quad \text{and } \sigma(L_1) = \sigma(\overline{L_2})$$

Extension (E)
$$\frac{C_2 \backslash \{L_2\}, M, Path \cup \{L_1\} \qquad C, M, Path}{C \cup \{L_1\}, M, Path} \quad \text{and } C_2 \text{ is a copy of } C_1 \in M,\ L_2 \in C_2, \text{ and } \sigma(L_1) = \sigma(\overline{L_2})$$

Fig. 13.11 The clausal connection calculus for classical logic.

The proof of the matrix $M_1 = \{\{P^0z\}, \{P^1a, Q^0y\}, \{Q^1fb\}\}$ in the clausal connection calculus with $\sigma = \{z' \backslash a,\ y' \backslash fb\}$ is given in Fig. 13.12. The clausal connection calculus always uses *copies of clauses*, in which all variables are renamed. By this technique the concept of multiplicity introduced in Def. 13.6 is realized in a simplified way by expanding copies explicitly.

$$\dfrac{\dfrac{\dfrac{}{\{\}, \{\{P^0z\}, \{P^1a, Q^0y\}, \{Q^1fb\}\}, \{P^0z', Q^0y'\}}\,A \qquad \dfrac{}{\{\}, M', \{P^0z'\}}\,A}{\{\mathbf{Q^0y'}\}, \{\{P^0z\}, \{P^1a, Q^0y\}, \{\mathbf{Q^1fb}\}\}, \{P^0z'\}}\,E \qquad \dfrac{}{\{\}, M', \{\}}\,A}{\dfrac{\{\mathbf{P^0z'}\}, \{\{P^0z\}, \{\mathbf{P^1a}, Q^0y\}, \{Q^1fb\}\}, \{\}}{\varepsilon, \{\{P^0z\}, \{P^1a, Q^0y\}, \{Q^1fb\}\}, \varepsilon}\,S}\,E$$

Fig. 13.12 A proof in the (formal) connection calculus.

The presented clausal connection calculus is *correct* and *complete*, ie. a formula is valid in classical logic if and only if there is a clausal connection proof for its matrix [14]. The proof is based on the matrix characterization for classical logic.

Proof search in the clausal connection calculus is carried out by applying the rules of the calculus in an *analytic* way, ie. from bottom to top, starting with $\varepsilon, M, \varepsilon$, in which M is the matrix of the given formula. At first a start clause is selected. Afterwards, connections are successively identified in order to make sure that all paths through the matrix contain a σ-complementary connection. This process is guided by the active path, a subset of a path through M. During the proof search, backtracking might be required, ie. alternative rules or rule instances have to be considered if the chosen rule or rule instance does not lead to a proof. This might happen when choosing the clause C_1 in the start and extension rules or the literal L_2 in the reduction and extension rules. The term substitution σ is calculated step by step by one of the well-known *term unification algorithms* (see, eg. [58]) whenever a reduction or extension rule is applied.

13.3.2 The Non-clausal Calculus

Clausal connection calculi, such as the one presented in Sect. 13.3.1, require the input formula in disjunctive normal or clausal form. Formulas that are not in clausal form have to be translated into clausal form. The standard transformation translates a first-order formula F into clausal form by applying the distributivity laws. In the worst case, the size of the resulting formula grows exponentially with respect to the size of the original formula F. This increases the search space significantly. Even a definitional translation [53] that introduces definitions for subformulas introduces a significant overhead for the proof search [44] and both clausal form translations modify the structure of the original formula.

A *non-clausal* connection calculus [45] that works directly on the structure of the original formula does not have these disadvantages. Other non-clausal approaches [1, 14, 31] work only on ground formulas, ie., for formulas that do not contain variables. For first-order formulas, these approaches have to add copies of subformulas iteratively, which introduces a huge redundancy into the proof search. For a more efficient proof search, clauses have to be added dynamically during the proof search, similar to the approach used for copying clauses in clausal connection calculi. Hence, in the following the clausal connection calculus is generalized and its rules are carefully extended [45].

Definition 13.11. The *non-clausal matrix* $M(F^{pol})$ of a formula F^{pol} (where *pol* is a polarity) is a set of clauses, in which a clause is a set of literals and (sub-)matrices, and is defined inductively according to Table 13.1. In Table 13.1, x^* is a new variable, t^* is the Skolem term $f^*(x_1, \ldots, x_n)$ in which f^* is a new function symbol and $x_1, \ldots, x_n$ are the free variables in $\forall x G$ or $\exists x G$. The *non-clausal matrix* of a formula F is the matrix $M(F^0)$.

In the *graphical representation* the clauses of matrices are arranged horizontally, literals and matrices of clauses are arranged vertically. For example, the formula

$$(Pz \vee (\neg Pa \wedge Qy) \vee \neg Qfb) \wedge ((N0 \wedge \forall x(Nx \Rightarrow Nfx)) \Rightarrow Nff0)$$

Table 13.1 The definition of the non-clausal matrix.

type	F^{pol}	$M(F^{pol})$	type	F^{pol}	$M(F^{pol})$
atomic	P^0	$\{\{P^0\}\}$ (P is atomic)	β	$(A \wedge B)^0$	$\{\{M(A^0), M(B^0)\}\}$
	P^1	$\{\{P^1\}\}$ (P is atomic)		$(A \vee B)^1$	$\{\{M(A^1), M(B^1)\}\}$
α	$(\neg A)^0$	$M(A^1)$		$(A \Rightarrow B)^1$	$\{\{M(A^0), M(B^1)\}\}$
	$(\neg A)^1$	$M(A^0)$	γ	$(\forall x A)^1$	$M(A[x\backslash x^*]^1)$
	$(A \wedge B)^1$	$\{\{M(A^1)\}, \{M(B^1)\}\}$		$(\exists x A)^0$	$M(A[x\backslash x^*]^0)$
	$(A \vee B)^0$	$\{\{M(A^0)\}, \{M(B^0)\}\}$	δ	$(\forall x A)^0$	$M(A[x\backslash t^*]^0)$
	$(A \Rightarrow B)^0$	$\{\{M(A^1)\}, \{M(B^0)\}\}$		$(\exists x A)^1$	$M(A[x\backslash t^*]^1)$

has the simplified non-clausal matrix (in which redundant parentheses are omitted)

$$\{\{\ \{\{P^0 z\}, \{P^1 a, Q^0 y\}, \{Q^1 fb\}\}\ ,\ \{\{N^1 0\}, \{N^0 x, N^1 fx\}, \{N^0 ff0\}\}\ \}\}\ .$$

The graphical representation of this (non-clausal) matrix is depicted in Fig. 13.13.

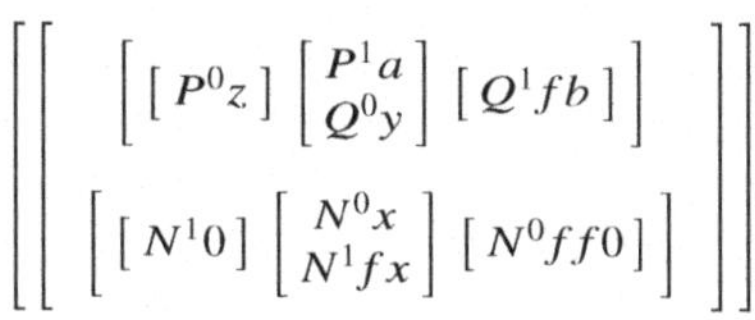

Fig. 13.13 The graphical representation of a non-clausal matrix.

The definition of paths through a non-clausal matrix is generalized in a straightforward way, see also Def. 13.4. All other concepts used for clausal matrices, eg. the definitions of connections and term substitutions, remain unchanged.

The graphical non-clausal connection proof for the previous formula represented by the matrix in Fig. 13.13 is given in Fig. 13.14.

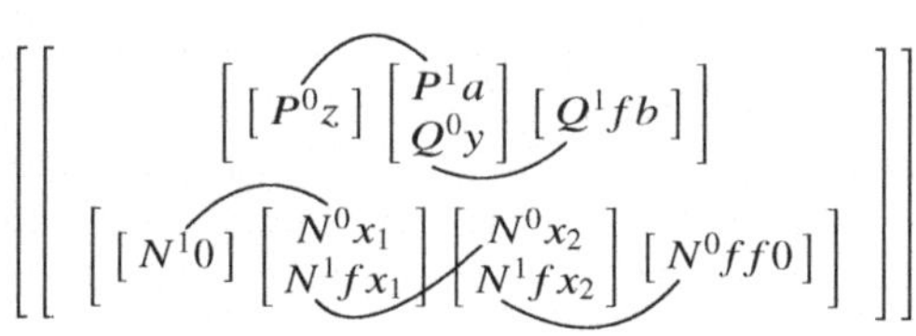

$\sigma = \{z\backslash a, y\backslash fb, x_1\backslash 0, x_2\backslash f0\}$

Fig. 13.14 A (graphical) non-clausal connection proof.

The formal non-clausal connection calculus has the same axiom, start rule, and reduction rule as the clausal connection calculus. The extension rule is slightly modified and a decomposition rule that splits subgoal clauses into their subclauses is added.

Axiom (A) $$\frac{}{\{\},\, M,\, Path}$$

Start (S) $$\frac{C_2,\, M,\, \{\}}{\varepsilon,\, M,\, \varepsilon} \quad \text{and } C_2 \text{ is copy of } C_1 \in M$$

Reduction (R) $$\frac{C,\, M,\, Path \cup \{L_2\}}{C \cup \{L_1\},\, M,\, Path \cup \{L_2\}} \quad \text{and } \sigma(L_1) = \sigma(\overline{L_2})$$

Extension (E) $$\frac{C_3,\, M[C_1 \backslash C_2],\, Path \cup \{L_1\} \qquad C,\, M,\, Path}{C \cup \{L_1\},\, M,\, Path}$$ and $C_3{:=}\beta\text{-}clause_{L_2}(C_2)$, C_2 is copy of C_1, C_1 is e-clause of M wrt. $Path \cup \{L_1\}$, C_2 contains L_2 with $\sigma(L_1) = \sigma(\overline{L_2})$

Decomposition (D) $$\frac{C \cup C_1,\, M,\, Path}{C \cup \{M_1\},\, M,\, Path} \quad \text{and } C_1 \in M_1$$

Fig. 13.15 The non-clausal connection calculus.

For example, in the above non-clausal matrix, copies of the clause $\{N^0 x, N^1 f x\}$ need to be allowed. A few definitions are required to specify which clauses can be copied when the non-clausal extension rule is applied [45].

A clause C is *α-related* to a literal L, only if L occurs in some clause C' such that C and C' are different clauses of the same matrix. I.e., in the graphical matrix representation, C occurs besides L. In the clausal connection calculus, a copy of a clause C is made by simply renaming all variables in C. In the non-clausal connection calculus, a *copy of the clause* C in the matrix M is made by renaming all *free variables* in C. $M[C_1 \backslash C_2]$ denotes the matrix M, in which the clause C_1 is replaced by the clause C_2. The *parent clause* of a clause C in a matrix M is the smallest clause in M that contains C (there are clauses that do not have a parent clause). The clause C in a matrix M is an *extension clause (e-clause) of M with respect to* a set of literals *Path*, only if either (a) C contains a literal of *Path*, or (b) C is α-related to all literals of *Path* occurring in M and if C has a parent clause, it contains a literal of *Path*. In the *β-clause* of C_2 with respect to L_2, denoted by $\beta\text{-}clause_{L_2}(C_2)$, the literal L_2 and all clauses that are α-related to L_2 are deleted from C_2, as these clauses do not need to be considered in the new subgoal clause in the premise of the extension rule.

Definition 13.12. The axiom and the rules of the *non-clausal connection calculus* are given in Fig. 13.15. The words of the calculus are tuples "$C, M, Path$", where M is a non-clausal matrix, C is a clause or ε and *Path* is a set of literals or ε, and σ is a rigid term substitution. A *non-clausal connection proof* of M is a non-clausal connection proof of $\varepsilon, M, \varepsilon$.

The non-clausal connection calculus for classical logic is *correct* and *complete*. The correctness proof is based on the non-clausal matrix characterization, completeness is proved by an embedding into the clausal connection calculus [45].

The *proof search* in the non-clausal connection calculus is carried out in the same way as in the clausal connection calculus, ie. the rules of the calculus are applied in an analytic way. Additional backtracking might be required when choosing the clause C_1 in the decomposition rule, but no backtracking is required when choosing the matrix M_1 in the decomposition rule. Again, the term substitution σ is calculated by one of the well-known algorithms for term unification. On formulas that are in clausal form, the non-clausal connection calculus behaves just like the clausal connection calculus.

If the matrices that are used in the non-clausal connection calculus are slightly modified, the start and the reduction rule are subsumed by the decomposition and the extension rule, respectively. The resulting *simplified* non-clausal calculus consists only of the axiom, the extension rule, and the decomposition rule [45]. Optimization techniques, such as positive start clauses, regularity, and restricted backtracking, can be used in the non-clausal connection calculus as well [50]. Furthermore, the non-clausal calculus can be extended to some *non-classical logics* in the same way as done for the clausal connection calculus (see Sect. 13.3.3).

13.3.3 Calculi for Non-classical Logics

By using the notion of *prefixes* the connection calculus for classical logic can be extended to intuitionistic logic and several modal logics.

To adapt the connection calculus to *intuitionistic logic*, a prefix is assigned to every subformula of a given formula. A *prefix* is a string, ie. a sequence of characters over an alphabet $\Phi \cup \Psi$, in which Φ is a set of *prefix variables* and Ψ is a set of *prefix constants*. Prefix constants and variables represent applications of the rules $\neg$*-right*, $\Rightarrow$*-right*, $\forall$*-right*, and $\neg$*-left*, $\Rightarrow$*-left*, $\forall$*-left* of the intuitionistic sequent calculus [29], respectively [69, 70]. Each subformula A is labelled with its prefix p, denoted $A : p$, which specifies the sequence of rules that have to be applied (analytically) to obtain the formula A in the sequent of a derivation. Two atomic formulas form an axiom in the intuitionistic sequent calculus if their prefixes unify. This is achieved by an *intuitionistic substitution* $\sigma_J : \Phi \rightarrow (\Phi \cup \Psi)^*$ that maps elements of Φ to strings over $\Phi \cup \Psi$. In the *intuitionistic matrix* of a formula, every literal is labelled with its prefix.

In the *matrix characterization for intuitionistic logic* it is additionally required that the prefixes of the two literals in every connection unify under σ_J [70]. For a combined term and prefix substitution (σ, σ_J), a connection $\{L_1 : p_1, L_2 : p_2\}$ is *complementary* only if $\sigma(L_1) = \sigma(\overline{L_2})$ and $\sigma_J(p_1) = \sigma_J(p_2)$. An additional *admissibility condition* ensures that σ and σ_J are mutually consistent [70].

The *skolemization* technique, originally used to eliminate *eigenvariables* in classical logic, is extended and also used for prefix constants in intuitionistic logic [42]. This allows the specification of a *clausal* matrix characterization, in which clause copies can simply be made by renaming all term and prefix variables [42].

For example, for the formula $(N0 \wedge \forall x(Nx \Rightarrow Nfx)) \Rightarrow Nf0$ the intuitionistic matrix is

$$\{\{N^1 0 : a_1 V_1\}, \{N^0 x : a_1 V_2\, a_2(x),\ N^1 fx : a_1 V_2 V_3\}, \{N^0 f0 : a_1 a_3\}\},$$

in which a_1, $a_2(x)$, a_3 are prefix constants, and V_1, V_2, V_3 are prefix variables. A graphical intuitionistic connection proof of this matrix with $\sigma = \{x\backslash 0\}$ and $\sigma_J = \{V_1\backslash a_2(0), V_2\backslash\varepsilon, V_3\backslash a_3\}$ (where ε is the empty string) is shown in Fig. 13.16.

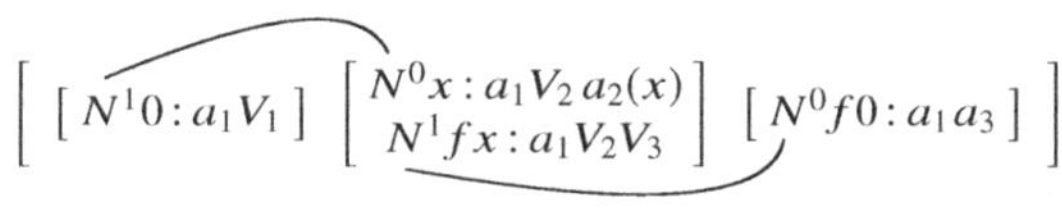

Fig. 13.16 The intuitionistic connection proof for the formula $(N0 \wedge \forall x(Nx \Rightarrow Nfx)) \Rightarrow Nf0$.

The intuitionistic matrix of the formula $N0 \vee \neg N0$ is $\{\{N^0 0 : a_1\}, \{N^1 0 : a_2 V_1\}\}$, As there is no intuitionistic substitution σ_J with $\sigma_J(a_1) = \sigma_J(a_2 V_1)$ and, hence, no connection proof of this matrix, this formula is *not* valid in intuitionistic logic.

The *clausal connection calculus for intuitionistic logic* is an extension of the clausal connection calculus for classical logic, in which a prefix has been added to each literal and an additional prefix unification is used to calculate the intuitionistic substitution σ_J. The *prefix unification* procedure takes the *prefix property* of all prefixes into account: for two prefixes $p_i = u_1 X w_1$ and $p_j = u_2 X w_2$ with $X \in \Phi \cup \Psi$, $u_1, u_2, w_1, w_2 \in (\Phi \cup \Psi)^*$, the property $u_1 = u_2$ holds. This reflects the fact that prefixes correspond to sequences of connectives and quantifiers within the same formula. Details of the intuitionistic calculus and the prefix unification algorithm can be found in [42].

For *modal logic* the classical sequent calculus is extended by rules for the modal operators $\Box$ ("necessity") and $\Diamond$ ("possibility"). Again, prefixes are used to capture the properties of these rules. In the connection calculus for modal logic, a *prefix* over an alphabet of *prefix variables* ν and *prefix constants* Π is assigned to every subformula of a given formula. Prefix variables and constants represent applications of the rules $\Box$-*left* or $\Diamond$-*right*, and $\Box$-*right* or $\Diamond$-*left*, respectively [69, 70]. Proof-theoretically, a prefix of a subformula A captures the modal context of A and specifies the sequence of modal rules that have to be applied analytically in order to obtain A in the sequent of a derivation. Semantically, a prefix denotes a specific world in a Kripke model [27, 70]. A modal substitution $\sigma_M : \nu \to (\nu \cup \Pi)^*$ is used to unify the prefixes of literals in each connection.

The core of the formal *clausal connection calculus for modal logic* [6, 46] is similar to the one for intuitionistic logic. The only difference to the intuitionistic calculus is the definition of the prefixes and the prefix unification. The prefix unification procedure depends on the specific modal logic and its domain condition. The *accessibility condition* has to be respected when calculating the modal substitution, eg. $|\sigma_M(V)| = 1$ for the modal logic D and $|\sigma_M(V)| \leq 1$ for the modal logic T for all prefix variables V; there is no restriction for the modal logics S4 and S5. The prefix unification for D is a simple pattern matching, for S4 the prefix unification for intuitionistic logic can be used, for S5 only the last character (or ε if the prefix is empty) of each prefix has to be unified [46, 47].

Both clausal calculi for intuitionistic logic and modal logic can be generalized to non-clausal form. By adding prefixes and a prefix unification, the non-clausal connection calculus presented in Sect. 13.3.2 is extended to first-order intuitionistic and first-order modal logic [49].

13.3.4 Implementations

Several automated theorem provers for classical logic that are based on clausal connection calculi have been implemented so far, such as KoMeT [18], METEOR [4], PTTP [65], and SETHEO [38].

leanCoP is a very compact PROLOG implementation of the connection calculus described in Sect. 13.3.1. leanCoP 1.0 [51] essentially implements the basic clausal connection calculus shown in Fig. 13.11. The source code of the core prover is given in Fig. 13.17 (sound PROLOG unification has to be used). PROLOG lists are used to represent sets, terms are used to represent atomic formulas, PROLOG variables represent term variables, and "-" is used to mark literals that have polarity 1. For example, the matrix $M = \{\{P^0 z\}, \{P^1 a, Q^0 y\}, \{Q^1 fb\}\}$ is represented by the PROLOG list `M=[[Pz],[-Pa,Qy],[-Qfb]]`.

```
prove(M,I) :- append(Q,[C|R],M), \+member(-_,C),
  append(Q,R,S), prove([!],[[-!|C]|S],[],I).
prove([],_,_,_).
prove([L|C],M,P,I) :- (-N=L; -L=N) -> (member(N,P);
  append(Q,[D|R],M), copy_term(D,E), append(A,[N|B],E),
  append(A,B,F), (D==E -> append(R,Q,S); length(P,K), K<I,
  append(R,[D|Q],S)), prove(F,S,[L|P],I)), prove(C,M,P,I).
```

Fig. 13.17 The source code of the leanCoP 1.0 core prover for classical first-order logic.

The prover is invoked by calling the predicate `prove(M,I)`, in which `M` is a matrix and `I` is a positive number. The predicate succeeds only if there is a connection proof for the matrix `M`, in which the size of the active path is smaller than `I`. The proof search starts by applying the start rule implemented in the first two lines. Afterwards, reduction and extension rules are repeatedly applied. These rules are implemented in the last four lines by the PROLOG predicate `prove(C,M,P,I)`, in which `C` is the subgoal clause, `M` is the matrix, `P` is the active path, and `I` is the maximum path limit. The path limit is used to perform iterative deepening on the size of the active path, which is necessary for completeness. When the extension rule is applied, the proof search continues with the left premise before the right premise is considered. The axiom is implemented in the third line. The term substitution σ is stored implicitly by PROLOG.

leanCoP 2.0 integrates additional optimization techniques into the basic connection calculus [43, 44]. The source code of the core prover is shown in Fig. 13.18.[5] The prover uses *lean PROLOG technology*, a technique that stores the clauses of the matrix in PROLOG's database. This integrates the main advantage of the "PROLOG technology" approach [65] into leanCoP by using PROLOG's fast indexing mechanism to quickly find connections. A *controlled iterative deepening* stops the overall proof search if the current path limit for the size of the active path is not reached in the current iteration. This yields a decision procedure for formulas without (free) variables and also allows for refuting some (not valid) first-order formulas. The *regularity* condition restricts the proof search such that no literal occurs more than once in the active path [39]. The *lemmata* technique reuses the subproof of a literal in order solve the same literal on other branches [39]. *Restricted backtracking* [44] cuts off alternative connections once the application of the reduction or extension rule has successfully solved a literal. A *definitional clausal form translation* is used in a preprocessing step to translate arbitrary first-order formulas into an equivalent clausal form by introducing definitions for certain subformulas [44]. Furthermore, leanCoP uses a *fixed strategy scheduling*.

The techniques described here improve the performance of leanCoP significantly, in particular for formulas containing many axioms [44]. In general, the performance of leanCoP is slightly better than the performance of Prover9 [41], the successor of the famous OTTER theorem prover.

```
prove(I,S) :- \+member(scut,S) -> prove([-(#)],[],I,[],S) ;
    lit(#,C,_) -> prove(C,[-(#)],I,[],S).
prove(I,S) :- member(comp(L),S), I=L -> prove(1,[]) ;
    (member(comp(_),S);retract(p)) -> J is I+1, prove(J,S).
prove([],_,_,_,_).
prove([L|C],P,I,Q,S) :- \+ (member(A,[L|C]), member(B,P),
    A==B), (-N=L;-L=N) -> ( member(D,Q), L==D ;
    member(E,P), unify_with_occurs_check(E,N) ; lit(N,F,H),
    (H=g -> true ; length(P,K), K<I -> true ;
    \+p -> assert(p), fail), prove(F,[L|P],I,Q,S) ),
    (member(cut,S) -> ! ; true), prove(C,P,I,[L|Q],S).
```

Fig. 13.18 The source code of the leanCoP 2.0 core prover for classical first-order logic.

[5] The full source code of the prover and additional information is available on the leanCoP website at `http://www.leancop.de`.

nanoCoP [48] is a very compact PROLOG implementation of the non-clausal connection calculus specified in Fig. 13.15.[6] In a first step the input formula F is translated into a non-clausal (indexed) matrix $M(F)$ according to Table 13.1. In the second step this matrix is written into PROLOG's database.

An additional predicate `prove_ec` is used to calculate extension clauses. An extension clause has to fulfill the conditions described in Sec. 13.3.2: it has to be (a) large enough to contain a literal of the active path or (b) small enough to be α-related to all literals of the active path in the current matrix, and again large enough that in case it has a parent clause, this contains a literal of the active path; the extension clause also has to be large enough such that the literals in the connection can be unified. Additional optimization techniques are integrated that are already used in the classical (clausal) connection prover leanCoP [44], such as regularity, lemmata, and restricted backtracking.

As nanoCoP can simulate leanCoP (in linear time), the proof time and proof size for formulas in clausal form is very similar for both provers [20]. For formulas that are not in clausal form, the proof time and size of the non-clausal nanoCoP proofs can be significantly smaller than those of the clausal leanCoP proofs [20].

ileanCoP [42, 43] is a compact PROLOG implementation of the clausal connection calculus for first-order intuitionistic logic described in Sect. 13.3.3.[7] It extends the classical connection prover leanCoP by (a) prefixes that are added to the literals in the matrix in a preprocessing step, (b) a set of prefix equations that are collected during the proof search, (c) a set of term variables together with their prefixes in order to check the admissibility condition, and (d) an additional prefix unification algorithm that unifies the prefixes of the literals in each connection. First, ileanCoP performs a classical proof search. After a classical proof is found, the prefixes of the literals in each connection are unified and the admissibility condition is checked. If the prefix unification or the admissibility condition fails, the search for alternative connections continues via backtracking. ileanCoP 1.2 proves significantly more formulas of the TPTP problem library and the ILTP problem library [57] than any other fully automated theorem prover for first-order intuitionistic logic [43].

MleanCoP [46, 47] is a compact PROLOG implementation of the clausal connection calculus for several first-order modal logics, as described in Sect. 13.3.3.[8] The source code of the core prover is similar to the source code of ileanCoP. Only the prefix unification algorithm changes according to the specific modal logic under consideration. MleanCoP supports the constant, cumulative, and varying domain variants of the first-order modal logics D, T, S4, and S5. MleanCoP 1.3 proves more formulas from the QMLTP problem library [56] than any other prover for first-order modal logic [6, 47].

Extending the non-clausal connection prover nanoCoP by prefixes and prefix unification algorithms, we obtain non-clausal provers for intuitionistic logic and several modal logics. nanoCoP-i and nanoCoP-M are such compact PROLOG implemen-

---

[6] The source code of nanoCoP is available at `http://www.leancop.de/nanocop/`.

[7] The source code of ileanCoP is available at `http://www.leancop.de/ileancop/`.

[8] The source code of MleanCoP is available at `http://www.leancop.de/mleancop/`.

tations of the *non-clausal* connection calculi for first-order intuitionistic logic and several first-order modal logics [49].[9]

13.4 Conclusions

Starting with Schütte's formal system GS for First-Order Logic we have derived characterizations of validity exclusively in terms of formula features such as its connections, paths and substitutions for a certain multiplicity. This characterization provides the basis for the development of proof calculi which determine a complementary set of connections spanning the formula thus establishing its validity. This approach guides the proof search in a formula-oriented and goal-oriented way focussed on nothing else than the given formula which is in stark contrast to approaches based on the still popular resolution or tableaux methods.

This general approach, termed Connection Method, is characterized by connection calculi which form the basis for a uniform family of implemented proof systems for a variety of different logics. Here we presented leanCoP for classical clausal logic, nanoCoP for classical non-clausal logic, ileanCoP for Intuitionistic Logic, and MleanCoP for various modal logics. Any of these turns out to be among the very best systems available for the respective logic, some of them even outperforming any competitive system by a wide margin. The implementations are in the high-level language PROLOG offering the potential for substantial further optimizations.

The line of research presented in this chapter, despite its remarkable success already in its present state of affairs, has plenty of room for further improvements. The present systems are all based on some form of GS_2 defined in Sect. 13.2.4 so that the step to GS_3 along with the possible improvements described in sections 13.2.6 and 13.2.7 are still not incorporated. In this context we refer the interested reader also to the respective literature, notably to [15, Sect. 2.7–2.8] and [14, Sect. IV.11]. Additional improvements can be achieved by special techniques for theories such as equational ones as well as for induction. The theoretical basis for work into such a direction can be found in [14, sect. V.3ff]. On top of all these potential improvements the resulting systems might be trained in their search behaviour by the deep learning approaches which resulted in such spectacular successes in other areas like the game Go. The papers [17, 20] contain, among other envisioned improvements, an elaboration of applying learning to connection provers.

As pointed out in the Introduction proof systems play an important role in modern IT. For the particular application mentioned there, namely provably correct systems, the authors have contributed a chapter concerning this application [52]. We still also see a bright future for program synthesis of the kind first outlined in [10], a topic which, however, would require a further chapter.

[9] The source code of nanoCoP-i is available at `http://www.leancop.de/nanocop-i/`, the source code of nanoCoP-M is available at `http://www.leancop.de/nanocop-m/`.

Acknowledgements We would like to thank the editors Reinhard Kahle and Michael Rathjen for inviting us to contribute to this volume. We also appreciate the comments of the anonymous referee. Thanks are also due to Hannes Bibel and Florian Hauser for bailing the first author out of a disastrous computer problem which reestablished the prerequisites for writing this chapter.

References

1. Peter B. Andrews. Theorem proving via general matings. *Journal of the ACM*, 28:193–214, 1981.
2. Roger Antonsen. *The Method of Variable Splitting*. PhD thesis, University of Oslo, 2008.
3. Roger Antonsen and Arild Waaler. Liberalized variable splitting. *J. Automated Reasoning*, 38:3–30, 2007.
4. Owen L. Astrachan and Donald W. Loveland. Meteors: High performance theorem provers using model elimination. In Robert S. Boyer, editor, *Automated Reasoning: Essays in Honor of Woody Bledsoe*, Automated Reasoning Series, pages 31–60. Kluwer Academic Publishers, 1991.
5. Paul Benacerraf and Hilary Putnam, editors. *Philosophy of Mathematics*. Cambridge University Press, 1964. ISBN 0-521-29648-X.
6. Christoph Benzmüller, Jens Otten, and Thomas Raths. Implementing and evaluating provers for first-order modal logics. In L. De Raedt et al., editor, *20th European Conference on Artificial Intelligence (ECAI 2012)*, pages 163–168, Amsterdam, 2012. IOS Press.
7. Evert Willem Beth. Semantic entailment and formal derivability. *Mededelingen van de Koninklijke Nederlandse Akademie van Wetenschappen*, 18(13):309–342, 1955.
8. L. Wolfgang Bibel. *Reflexionen vor Reflexen – Memoiren eines Forschers*. Cuvillier Verlag, Göttingen, 2017.
9. Wolfgang Bibel. An approach to a systematic theorem proving procedure in first-order logic. *Computing*, 12:43–55, 1974. First presented to the GI Annual Conference in 1971; also available as Bericht Nr. 7207, Technische Universität München, Abteilung Mathematik (1972).
10. Wolfgang Bibel. Syntax–directed, semantics–supported program synthesis. *Artificial Intelligence*, 14:243–261, 1980.
11. Wolfgang Bibel. On matrices with connections. *Journal of ACM*, 28:633–645, 1981. Also available as Bericht 79, Universität Karlsruhe, Institut für Angewandte Informatik und Formale Beschreibungsverfahren.
12. Wolfgang Bibel. Computationally improved versions of Herbrand's theorem. In J. Stern, editor, *Proceedings of the Herbrand Symposium*, pages 11–28, Amsterdam, 1982. North–Holland.
13. Wolfgang Bibel. Matings in matrices. *Comm. ACM*, 26:844–852, 1983.
14. Wolfgang Bibel. *Automated Theorem Proving*. Vieweg Verlag, Braunschweig, second edition, 1987. First edition 1982.
15. Wolfgang Bibel. Research perspectives for logic and deduction. In Oliviero Stock and Marco Schaerf, editors, *Reasoning, Action, and Interaction in AI Theories and Systems – Essays dedicated to Luigia Carlucci Aiello*, volume 4155 of *LNAI*, pages 25–43. Springer, Berlin, 2006.
16. Wolfgang Bibel. Early history and perspectives of automated deduction. In J. Hertzberg, M. Beetz, and R. Englert, editors, *Proceedings of the 30th Annual German Conference on Artificial Intelligence (KI-2007)*, volume 4667 of *LNAI*, pages 2–18, Berlin, 2007. Springer.
17. Wolfgang Bibel. A vision for automated deduction rooted in the connection method. In R. Schmidt and C. Nalon, editors, *The 26th International Conference on Automated Reasoning with Analytic Tableaux and Related Methods (TABLEAUX 2017)*, volume 10501 of *LNAI*, pages 3–21. Springer, 2017.
18. Wolfgang Bibel, Stefan Brüning, Uwe Egly, and Thomas Rath. KoMeT. In A. Bundy, editor, *Proceedings of the International Conference on Automated Deduction, CADE-94*, Lecture Notes in Artificial Intelligence, pages 783–787, Berlin, 1994. Springer.

19. Wolfgang Bibel and Elmar Eder. Methods and calculi for deduction. In D. M. Gabbay, C. J. Hogger, and J. A. Robinson, editors, *Handbook of Logic in Artificial Intelligence and Logic Programming*, volume 1, chapter 3, pages 71–193. Oxford University Press, Oxford, 1993.
20. Wolfgang Bibel and Jens Otten. Experiments with connection method provers. *Presented to the 4th Conference on Artificial Intelligence and Theorem Proving, AITP 2019, April 7-12, 2019, Obergurgl, Austria*, 2019.
21. Wolfgang Bibel and J. Schreiber. Proof search in a Gentzen–like system of first–order logic. In E. Gelenbe and D. Poitier, editors, *Proceedings of the International Computing Symposium*, pages 205–212, Amsterdam, 1975. North–Holland.
22. George Boole. *An Investigation of the Laws of Thought on Which are Founded the Mathematical Theories of Logic and Probabilities*. Macmillan, 1854.
23. Martin Davis. The Prehistory and Early History of Automated Deduction. In J. Siekmann and G. Wrightson, editors, *Automation of Reasoning 1 – Classical Papers on Computational Logic 1957–1966*, pages 1–28. Springer, Berlin, 1983.
24. Martin Davis. *Engines of Logic – Mathematicians and the Origin of the Computer*. Norton, New York, 2000.
25. Elmar Eder. The cut rule in theorem proving. In Steffen Hölldobler, editor, *Intellectics and Computational Logic*, volume 19 of *Applied Logic Series*, pages 101–123. Kluwer Academic Publishers, Dordrecht, 2000.
26. Jens Erik Fenstad and Hao Wang. Thoralf Albert Skolem. In Dov M. Gabbay and John Woods, editors, *Logic from Russell to Church*, volume 5 of *Handbook of the History of Logic*, pages 127–194. Elsevier B.V., 2009.
27. Melvin C. Fitting. *Proof methods for modal and intuitionistic logics*. D. Reidel, Dordrecht, 1983.
28. Gottlob Frege. *Begriffsschrift*. Louis Nebert, Halle, 1879.
29. Gerhard Gentzen. Untersuchungen über das logische Schließen. *Mathematische Zeitschrift*, 39:176–210 and 405–431, 1935. Engl. transl. in [66].
30. Alessio Guglielmi. A system of interaction and structure. *ACM Transactions on Computational Logic*, 8(1):1–64, January 2007.
31. Reiner Hähnle, Neil. V. Murray, and Erik Rosenthal. Linearity and regularity with negation normal form. *Theoretical Computer Science*, 328:325–354, 2004.
32. Christian Mahesh Hansen. *A Variable Splitting Theorem Prover*. PhD thesis, University of Oslo, 2012.
33. David Hilbert and W. Ackermann. *Grundzüge der Theoretischen Logik*. Springer, 1928.
34. Kaarlo J. J. Hintikka. Form and content in quantification theory. *Acta Philosophica Fennica*, 8:7–55, 1955.
35. Stig Kanger. *Provability in Logic*. PhD thesis, University of Stockholm, 1957.
36. William Kneale and Martha Kneale. *The Development of Logic*. Clarendon Press, Oxford, 1984.
37. Reinhold Letz, Klaus Mayr, and Christoph Goller. Controlled integration of the cut rule into connection tableaux calculi. *Journal of Automated Reasoning*, 13:297–337, 1994.
38. Reinhold Letz, Johannes Schumann, Stephan Bayerl, and Wolfgang Bibel. SETHEO — A high-performance theorem prover for first-order logic. *Journal of Automated Reasoning*, 8(2):183–212, 1992.
39. Reinhold Letz and Gernot Stenz. Model elimination and connection tableau procedures. In *Handbook of Automated Reasoning*, pages 2015–2112. Elsevier Science Publishers, Amsterdam, 2001.
40. Doanld W. Loveland. Mechanical theorem proving by Model Elimination. *Journal of the ACM*, pages 236–251, 1968.
41. William McCune. Release of Prover9. In *Mile high conference on quasigroups, loops and nonassociative systems*, Denver, 2005.
42. Jens Otten. Clausal connection-based theorem proving in intuitionistic first-order logic. In *TABLEAUX 2005, International Conference on Automated Reasoning with Analytic Tableaux and Related Methods*, volume 3702 of *LNCS*, pages 245–261, Berlin, 2005. Springer.

43. Jens Otten. leanCoP 2.0 and ileanCoP 1.2 – High performance lean theorem proving in classical and intuitionistic logic. In A. Armando, P. Baumgartner, and G. Dowek, editors, *IJCAR 2008*, volume 5195 of *LNCS*, pages 283–291, Heidelberg, 2008. Springer.
44. Jens Otten. Restricting backtracking in connection calculi. *AI Communications*, 23:159–182, 2010.
45. Jens Otten. A non-clausal connection calculus. In K. Brünnler and G. Metcalfe, editors, *TABLEAUX 2011*, volume 6793 of *LNAI*, pages 226–241, Heidelberg, 2011. Springer.
46. Jens Otten. Implementing connection calculi for first-order modal logics. In E. Ternovska, K. Korovin, and S. Schulz, editors, *9th International Workshop on the Implementation of Logics (IWIL 2012)*, Merida, Venezuela, 2012.
47. Jens Otten. MleanCoP: A connection prover for first-order modal logic. In Stéphane Demri, Deepak Kapur, and Christoph Weidenbach, editors, *IJCAR 2014*, volume 8562 of *LNCS*, pages 269--276, Heidelberg, 2014. Springer.
48. Jens Otten. nanoCoP: A non-clausal connection prover. In Nicola Olivetti and Ashish Tiwari, editors, *IJCAR 2016*, volume 9706 of *LNCS*, pages 302–312, Heidelberg, 2016. Springer.
49. Jens Otten. Non-clausal connection calculi for non-classical logics. In R. Schmidt and C. Nalon, editors, *26th International Conference on Automated Reasoning with Analytic Tableaux and Related Methods*, LNAI 10501, pages 209–227, Cham, Switzerland, 2017. Springer.
50. Jens Otten. Proof search optimizations for non-clausal connection calculi. In B. Konev, J. Urban, and P. Rümmer, editors, *Practical Aspects of Automated Reasoning, PAAR 2018*, volume 2162 of *CEUR Workshop Proceedings*, pages 49–57. CEUR-WS.org, 2018.
51. Jens Otten and Wolfgang Bibel. leanCoP: Lean connection-based theorem proving. *Journal of Symbolic Computation*, 36:139–161, 2003.
52. Jens Otten and Wolfgang Bibel. Advances in connection-based automated theorem proving. In Jonathan Bowen, Mike Hinchey, and Ernst-Rüdiger Olderog, editors, *Provably Correct Systems*, NASA Monographs in Systems and Software Engineering, pages 211–241. Springer, London, 2017.
53. David A. Plaisted and Steven Greenbaum. A structure-preserving clause form translation. *J. Symbolic Computation*, 2(3):293–304, 1986.
54. Dag Prawitz. An improved proof procedure. *Theoria*, 26:102–139, 1960. Also contained in [63, 159–199].
55. Dag Prawitz, H. Prawitz, and Neri Voghera. A mechanical proof procedure and its realization in an electronic computer. *J.ACM*, 7:102–128, 1960.
56. Thomas Raths and Jens Otten. The QMLTP problem library for first-order modal logics. In Bernhard Gramlich, Dale Miller, and Uli Sattler, editors, *IJCAR 2012*, volume 7364 of *LNCS*, pages 454–461. Springer, 2012.
57. Thomas Raths, Jens Otten, and Christoph Kreitz. The ILTP problem library for intuitionistic logic. *Journal of Automated Reasoning*, 38:261–271, 2007.
58. John A. Robinson. A machine-oriented logic based on the resolution principle. *Journal of ACM*, 12:23–41, 1965.
59. John Alan Robinson and Andrei Voronkov, editors. *Handbook of Automated Reasoning*. Elsevier Science Publishers, Amsterdam, 2001.
60. Kurt Schütte. Ein System des verknüpfenden Schließens. *Archiv für Mathematische Logik und Grundlagenforschung*, 2:55–67, 1956.
61. Kurt Schütte. *Vollständige Systeme Modaler und Intuitionistischer Logik*. Ergebnisse der Mathematik und ihrer Grenzgebiete, Band 42. Springer Verlag, Berlin, 1968.
62. Kurt Schütte. *Proof Theory*. Grundlehren der mathematischen Wissenschaften 225. Springer-Verlag, Berlin, 1977. English and revised version of: Beweistheorie (1960).
63. Jörg Siekmann and Graham Wrightson, editors. *Automation of Reasoning — Classical Papers on Computational Logic 1957–1966*, volume 1. Springer, Berlin, 1983.
64. Raymond M. Smullyan. *First-Order Logic*. Ergebnisse der Mathematik und ihrer Grenzgebiete. Springer-Verlag, Berlin, Heidelberg, New York, 1971.
65. Mark E. Stickel. A Prolog technology theorem prover: Implementation by an extended Prolog compiler. *Journal of Automated Reasoning*, 4:353–380, 1988.

66. M. E. Szabo, editor. *The collected papers of Gerhard Gentzen*. North-Holland, Amsterdam, 1969.
67. Anne S. Troelstra and Helmut Schwichtenberg. *Basic Proof Theory*. Cambridge Tracts in Theoretical Computer Science. Cambridge University Press, Cambridge, 2nd edition, 2000.
68. Willard van Orman Quine. A proof procedure for quantification theory. *J. Symbolic Logic*, 20:141–149, 1955.
69. Arild Waaler. Connections in nonclassical logics. In *Handbook of Automated Reasoning*, pages 1487–1578. Elsevier Science Publishers, Amsterdam, 2001. See [59].
70. Lincoln A. Wallen. *Automated Deduction in Non-Classical Logics*. MIT Press, Cambridge, Mass., 1990.

Chapter 14
Calculating Maximal Order Types for Finite Rooted Unstructured Labeled Trees

Diana Schmidt, Jeroen van der Meeren, and Andreas Weiermann

Abstract Diana Schmidt, in her Habilitationsschrift in 1979, completely classified the maximal order types of the natural tree embeddability relations for finite rooted structured labeled trees. Her results since have found interesting applications in proof theory and reverse mathematics. The question concerning the maximal order types of unstructured trees has been left open for years, and a conclusive answer will be given in this article. Moreover, we provide an answer to a question of Harvey Friedman regarding Γ_0 and binary two-labeled unstructured trees.

14.1 Introduction

Dick H. de Jongh has shown in an unpublished note for $n = 2$ and Diana Schmidt has shown in [8] for $n > 2$ that the maximum closure ordinal of an n-ary monotone function is given by the first value of the $(n + 1)$-ary Veblen function (aka Schütte-Klammerterm function). More generally, in [9] she classified the maximal order type of classes of finite rooted structured labeled trees with bounded outdegree (and in an additional chapter she furthermore dealt with labeled trees of unbounded outdegree).

In this discussion we of course tacitly assume that the labels are well quasi ordered, so that the tree embeddability relations are well quasi ordered by seminal results of Higman [6] and Kruskal [7].

In this article we calculate lower bounds for the maximal order types of unstructured rooted labeled finite trees with bounded outdegree. Our lower bounds match

D. Schmidt
Hochschule Heilbronn, Max-Planck-Str. 39, 74081 Heilbronn, Germany, e-mail: diana.schmidt@hs-heilbronn.de

J. van der Meeren · A. Weiermann
Gent Universitet, Vakgroep Wiskunde, Krijgslaan 281 Gebouw S22, 9000 Ghent, Belgium, e-mail: {jeroen.vandermeeren,andreas.weiermann}@ugent.be

R. Kahle and M. Rathjen (eds.), *The Legacy of Kurt Schütte*,
https://doi.org/10.1007/978-3-030-49424-7_14

Diana Schmidt's upper bounds on finite rooted structured labeled trees, so we get exact matches.

Our investigation is furthermore motivated by a question of Harvey M. Friedman (private communication) regarding the maximal order type for binary trees with one or two labels.

Our result will cover more generally n-branching trees for any $n \geq 2$. The case for binary unstructured unlabeled trees has been treated previously in an unpublished thesis by Floris Van Vugt (under supervision of Weiermann in 2005) and also independently by Harvey M. Friedman (private communication).

Upper bounds in the case of unstructured rooted labeled n-ary trees follow as an easy corollary from Schmidt [9], and lower bounds have been established by Van der Meeren and Weiermann (unpublished). These findings are put together in this article.

In a more general context our investigations are motivated by Feferman's natural well ordering problem, which is discussed, for example, in [1], [5], [4].

14.2 Ordinal Background

In this article we work with Veblen functions for which Schütte in [10] introduced a specific Klammerterm notation. Since only Veblen functions with finite arities are involved we avoid explicit Klammerterm notations and work entirely with $(n+1)$-ary functions. Let us fix $n > 0$. For $(n+1)$-tuples we can define the lexicographic ordering $(\alpha_n, \ldots, \alpha_0) <_{lex} (\beta_n, \ldots, \beta_0)$ to hold if there exists an $m \leq n$ such that $\alpha_m < \beta_m$ and such that $\alpha_i = \beta_i$ for $n \geq i > m$. The $(n+1)$-ary Veblen function Φ^{n+1} is defined by recursion on $<_{lex}$ as follows.

We put $\Phi^{n+1}(0, \ldots, 0, 0, \alpha) := \alpha + 1$ and $\Phi^{n+1}(0, \ldots, 0, 1, \alpha) := \omega \cdot (1 + \alpha)$.

Now assume that $0 < k \leq n$ and $\alpha_k > 0$. Then $\Phi^{n+1}(\alpha_n, \ldots, \alpha_k, 0, \ldots, 0, \alpha)$ is the α-th element of the class $\{\eta : \forall \alpha < \alpha_k \Phi^{n+1}(\alpha_n, \ldots, \alpha_{k+1}, \alpha, \eta, 0, \ldots, 0) = \eta\}$.

For technical reasons, it is convenient to work with the fixed point free version $\overline{\Phi}^{n+1}$ of Φ^{n+1}. This function is defined by the following case distinction:

If $\Phi^{n+1}(\alpha_n, \ldots, \alpha_0) = \alpha_0$ then for every $m < \omega$

$$\overline{\Phi}^{n+1}(\alpha_n, \ldots, \alpha_0 + m) := \Phi^{n+1}(\alpha_n, \ldots, \alpha_0 + m + 1).$$

If $\Phi^{n+1}(\alpha_n, \ldots, \alpha_j, 0, \ldots, 0) = \alpha_j$ and $j > 0$ then for every $m < \omega$

$$\overline{\Phi}^{n+1}(\alpha_n, \ldots, \alpha_j, 0, \ldots, 0, m) := \Phi^{n+1}(\alpha_n, \ldots, \alpha_j, 0, \ldots, 0, m + 1).$$

In all other cases $\overline{\Phi}^{n+1}(\alpha_n, \ldots, \alpha_0) := \Phi^{n+1}(\alpha_n, \ldots, \alpha_0)$.

It is easy to see that for $m > n$ we have

$$\Phi^{n+1}(\alpha_n, \ldots, \alpha_0) = \Phi^{m+1}(0, \ldots, 0, \alpha_n, \ldots, \alpha_0)$$

and
$$\overline{\Phi}^{n+1}(\alpha_n,\ldots,\alpha_0) = \overline{\Phi}^{m+1}(0,\ldots,0,\alpha_n,\ldots,\alpha_0).$$

It is well known that the closure of the singleton set $\{0\}$ under the binary function $\alpha,\beta \mapsto \Phi^2(\alpha,\beta)$ generates precisely all ordinals less than $\varepsilon_0 := \min\{\xi \in On : \xi = \omega^\xi\}$. By an unpublished result of De Jongh it is known that the maximal order type of structured binary trees without labels is equal to ε_0. The same ordinal is also the maximal order type of unstructured binary trees, as has been shown by Van Vugt and Weiermann [13]. The latter result will be reproved in this article.

Theorem 14.1. 1. *The ordinal $\Phi^{n+1}(\alpha_n,\ldots,\alpha_0)$ is always defined.*
2. $\max\{\alpha_0,\ldots,\alpha_n\} \leq \Phi^{n+1}(\alpha_n,\ldots,\alpha_0)$, *if* $\alpha_0 \neq 0$ *then* $\max\{\alpha_1,\ldots,\alpha_n\} < \Phi^{n+1}(\alpha_n,\ldots,\alpha_0)$.
3. *If* $\gamma < \alpha_q$ *and* $\gamma_0,\ldots,\gamma_{q-1} < \Phi^{n+1}(\alpha_n,\ldots,\alpha_0)$ *then*
$$\Phi^{n+1}(\alpha_n,\ldots,\alpha_{q+1},\gamma,\gamma_{q-1},\ldots,\gamma_0) < \Phi^{n+1}(\alpha_n,\ldots,\alpha_0).$$
4. *If* $\alpha_0 = sup\{\gamma_\xi : \xi < \delta\}$ *then*
$$\Phi^{n+1}(\alpha_n,\ldots,\alpha_1,\alpha_0) = \sup\{\Phi^{n+1}(\alpha_n,\ldots,\alpha_1,\gamma_\xi) : \xi < \delta\}.$$

Theorem 14.2. 1. $max\{\alpha_n,\ldots,\alpha_0\} < \overline{\Phi}^{n+1}(\alpha_n,\ldots,\alpha_0)$ *and*
2. *If* $(\alpha_n,\ldots,\alpha_0) <_{lex} (\beta_n,\ldots,\beta_0)$ *and* $\{\alpha_0,\ldots,\alpha_n\} < \overline{\Phi}^{n+1}(\beta_n,\ldots,\beta_0)$ *then* $\overline{\Phi}^{n+1}(\alpha_n,\ldots,\alpha_0) < \overline{\Phi}^{n+1}(\beta_n,\ldots,\beta_0)$.

In [10] Schütte showed that the ordinals less than $\Phi^{n+2}(1,0,\ldots,0)$ can be generated from $\{0\}$ with the function $\alpha_n,\ldots,\alpha_0 \mapsto \Phi^{n+1}(\alpha_n,\ldots,\alpha_0)$, and moreover in the same article Schütte showed the ordinals less than $\Phi^{n+2}(1,0,\ldots,0)$ can also be generated from $\{0\}$ with the function $\alpha_n,\ldots,\alpha_0 \mapsto \overline{\Phi}^{n+1}(\alpha_n,\ldots,\alpha_0)$.

Moreover, some routine calculations with ordinals yield $\Phi^3(1,0,0) = \varepsilon_0$, the proof-theoretic ordinal of first order Peano arithmetic, and $\Phi^3(2,0,0) = \Gamma_0$, the Feferman-Schütte ordinal of predicativity [3, 11]. (The corresponding order embeddings of the standard ordinal notations into the Veblen functions from this article are provided by the maps: $\omega^\alpha + \beta \mapsto \Phi^2(\alpha,\beta)$ and $\varphi(\alpha,\beta) + \gamma \mapsto \Phi^3(0,\Phi^3(1,\alpha,\beta),\gamma)$, where the left hand side is considered to be in normal form.)

14.3 Lower Bounds for the Maximal Order Types of Unstructured Trees

A quasi order $\langle X,\leq\rangle$ is a set X together with a reflexive and transitive binary relation $\leq$ on X. A quasi order $\langle X,\leq\rangle$ is called a well quasi order if for all ω-sequences $\langle x_i : i < \omega\rangle$ of elements in X there exists $i,j < \omega$ such that $i < j$ and $x_i \leq x_j$. A partial order $\langle X,\leq\rangle$ which is also a well quasi order is called a well partial order. Every well quasi order naturally induces a well partial order after an obvious factorization and we identify well quasi orders with their induced well partial orders (in our setting the difference does not cause problems).

De Jongh and Parikh showed in [2] that for every well partial order $\langle X, \leq \rangle$ there exists a linear extension $\leq^+$ of $\leq$ on the same set X such that the order type of $\langle X, \leq^+ \rangle$ is maximal possible. The order type of $\langle X, \leq^+ \rangle$ is called the maximal order type of $\langle X, \leq \rangle$ and is denoted by $o\langle X, \leq \rangle$. Moreover they calculated maximal order types for disjoint unions and cartesian products of well partial orders.

In this article we calculate lower bounds for the class of n-branching trees with labels from a well quasi ordering X. These trees are well partially ordered by results of Higman and Kruskal.

Definition 14.3. Let $X_0, \ldots, X_n$ be disjoint partial orders with $X_0 \neq \emptyset$. The set $\mathcal{T}\begin{pmatrix} X_0 & \ldots & X_n \\ 0 & \ldots & n \end{pmatrix}$ (of finite, rooted labeled trees) is defined as the least set T such that, for any $i \geq 0$, if $x_i \in X_i$ and $t_1, \ldots, t_i \in T$ then $x_i(t_1, \ldots, t_i) \in T$.

For $t, t' \in \mathcal{T}\begin{pmatrix} X_0 & \ldots & X_n \\ 0 & \ldots & n \end{pmatrix}$ we define $t \trianglelefteq t'$ recursively as follows.

1. If $t' = x_i'(t_1', \ldots, t_i')$ and if $t \trianglelefteq t_j'$ for some $j \leq i$ then $t \trianglelefteq t'$.
2. if $t = x_i(t_1, \ldots, t_i)$ and $t' = x_i'(t_1', \ldots, t_i')$ and if $x_i \leq x_i'$ (mod X_i) and if there is a permutation $\pi : \{1, \ldots, i\} \to \{1, \ldots, i\}$ such that $t_j \trianglelefteq t'_{\pi(j)}$ for $1 \leq j \leq n$ then $t \trianglelefteq t'$.

The elements of $\mathcal{T}\begin{pmatrix} X_0 & \ldots & X_n \\ 0 & \ldots & n \end{pmatrix}$ can (after an appropriate factorization) be considered as unstructured rooted labeled trees, since permuting the immediate subtrees at any node leads to equivalent trees modulo the natural equivalence relation induced by $\trianglelefteq$. Note that $\trianglelefteq$ is not yet antisymmetric, but becomes antisymmetric after the factorisation.

If all X_i are singletons one obtains the set of finite rooted trees of outdegree not bigger than n.

The elements of $\mathcal{T}\begin{pmatrix} X_0 & \ldots & X_n \\ 0 & \ldots & n \end{pmatrix}$ can also be seen as finite rooted trees t with labels in $X_0 \cup \ldots \cup X_n$ where the label of a node p in t is in X_i if p has i direct successors in t. In this situation an alternative description of the embeddability relation $t \trianglelefteq t'$ is provided by postulating the existence of an inf-preserving mapping h from the nodes of t to the nodes of t' such that $l \leq l'$ if l is the label of node n and l' is the label of node $h(n)$. The disjointness of the X_i then guarantees that the outdegrees are preserved correctly.

For the sake of completeness let us also mention the embeddability relation for structured trees.

Definition 14.4. For $t, t' \in \mathcal{T}\begin{pmatrix} X_0 & \ldots & X_n \\ 0 & \ldots & n \end{pmatrix}$ we define $t \trianglelefteq^s t'$ recursively as follows.

1. If $t' = x_i'(t_1', \ldots, t_i')$ and if $t \trianglelefteq^s t_j'$ for some j with $0 \leq j \leq i$ then $t \trianglelefteq^s t'$.
2. if $t = x_i(t_1, \ldots, t_i)$ and $t' = x_i'(t_1', \ldots, t_i')$ and if $x_i \leq x_i'$ (mod X_i) and if $t_j \trianglelefteq^s t_j'$ for $1 \leq j \leq i$ then $t \trianglelefteq^s t'$.

Let A be the tree

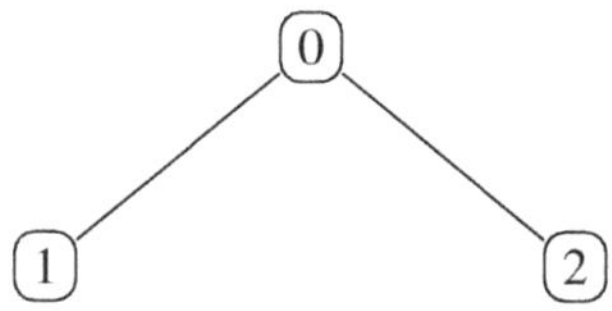

and let B be the tree

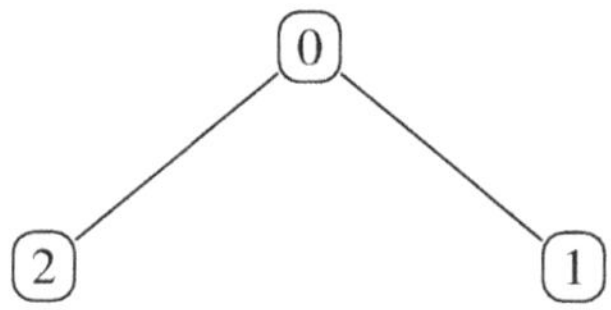

Then $A \trianglelefteq B$ but $A \trianglelefteq^s B$ does not hold.

As mentioned before by results of Higman and Kruskal the structure

$$\langle \mathcal{T}\begin{pmatrix} X_0 & \dots & X_n \\ 0 & \dots & n \end{pmatrix}, \trianglelefteq^s\rangle$$

is a well partial order if the structures $\langle X_i, \leq_i\rangle$ are well partial orders.

Since $t \trianglelefteq^s t$ implies $t \trianglelefteq t$ for $t, t' \in \mathcal{T}\begin{pmatrix} X_0 & \dots & X_n \\ 0 & \dots & n \end{pmatrix}$ the structure $\langle \mathcal{T}\begin{pmatrix} X_0 & \dots & X_n \\ 0 & \dots & n \end{pmatrix}, \trianglelefteq\rangle$ is a well-quasi order if the structures $\langle X_i, \leq_i\rangle$ are well quasi orders.

So there will exist a maximal order type for $\langle \mathcal{T}\begin{pmatrix} X_0 & \dots & X_n \\ 0 & \dots & n \end{pmatrix}, \trianglelefteq\rangle$. In fact we will show that this maximal order type can be calculated via a natural ordinal function, namely the $(n+1)$-ary Veblen (Schütte Klammerterm) function in terms of the maximal order types of the structures $\langle X_i, \leq_i\rangle$. Moreover we will even show that this maximal order type is the same as the maximal order type of $\langle \mathcal{T}\begin{pmatrix} X_0 & \dots & X_n \\ 0 & \dots & n \end{pmatrix}, \trianglelefteq^s\rangle$.

The upper bound is obtained by Diana Schmidt in her Habilitation thesis [9].

Theorem 14.5 (Schmidt). *Assume $n \geq 2$ and $\alpha_n > 0$. There exist disjoint well orders $X_0, \dots, X_n$ of order types $1 + \alpha_0, \dots, \alpha_n$ such that*

$$o\langle \mathcal{T}\begin{pmatrix} X_0 & \dots & X_n \\ 0 & \dots & n \end{pmatrix}, \trianglelefteq^s\rangle = \overline{\Phi}^{n+1}(\alpha_n, \dots, \alpha_0).$$

Schmidt's result immediately provides the following lower bound for the unstructured trees:

$$o\langle \mathcal{T}\begin{pmatrix} X_0 & \dots & X_n \\ 0 & \dots & n \end{pmatrix}, \trianglelefteq\rangle \leq \overline{\Phi}^{n+1}(\alpha_n, \dots, \alpha_0\rangle.$$

In the rest of this article we show that this upper bound is also a lower bound in the case of unstructured trees. Since we work with unstructured trees (which can be seen as non planar trees), some care is needed in dealing with permutations of immediate subtrees. The intuition is borrowed from set theory, where the ordered pair can be coded in terms of unordered pairs.

Theorem 14.6. *Assume $\alpha_n > 0$ and $n \geq 2$. There exist disjoint well orders $X_0, \ldots, X_n$ of order types $1 + \alpha_0, \alpha_1, \ldots, \alpha_n$ such that*

$$o\langle \mathcal{T}\begin{pmatrix} X_0 & \ldots & X_n \\ 0 & \ldots & n \end{pmatrix}, \trianglelefteq\rangle \geq \Phi^{n+1}(\alpha_n, \ldots, \alpha_0).$$

Proof. Put $X_0 := \{0\} \cup \{f^0_\xi : \xi < \alpha_0\}$ (assuming this union is disjoint) and $X_i := \{f^i_\xi : \xi < \alpha_i\}$ for $1 \leq i \leq n$. The ordering on these sets is induced by the ordering on the indices, where for $i = 0$ we assume that $0 < f^0_\xi$.

For each $\gamma < \Phi^{n+1}(\alpha_n, \ldots, \alpha_0) =: \Gamma$ let us define a finite tree $T(\gamma) \in \mathcal{T}\begin{pmatrix} X_0 & \ldots & X_n \\ 0 & \ldots & n \end{pmatrix}$ recursively as follows.

1. If $\gamma = 0$ then $T(\gamma) := 0$.
2. If $\gamma = \Phi^{n+1}(\alpha_n, \ldots, \alpha_{i+1}, \beta, \beta_{i-1}, \ldots, \beta_0)$ for some $\beta < \alpha_i$ and if $\beta_l < \gamma$ for $0 \leq l \leq i-1$ and if $T_l = f^n_0(T(\beta_l), \ldots, T(\beta_{i-1}), 0, \ldots, 0)$ for $0 \leq l \leq i-1$ then $T(\gamma) = f^i_\beta(T_0, \ldots, T_{i-1})$. For $i = 0$ this means $T(\gamma) = f^0_\beta$ and in the other cases we fill in the arguments from the left to the right where the subtrees become less and less complex.

We claim that $T(\gamma) \trianglelefteq T(\delta)$ implies $\gamma \leq \delta$. This shows that the ordering on Γ extends the embeddability relation so that Γ is a lower bound for the maximal order type in question.

We prove the claim by induction on the length of γ (as a term) plus the length of δ (as a term). Clearly we may assume that $\gamma \neq 0 \neq \delta$.

We may also assume that γ and δ are in normal form, which means that they are strictly bigger than all their arguments.

Assume that, for some i with $0 \leq i \leq n$, $\gamma = \Phi^{n+1}(\alpha_n, \ldots, \alpha_{i+1}, \beta, \beta_{i-1}, \ldots, \beta_0)$ for some $\beta < \alpha_i$ where $\beta_k < \gamma$ for $k < i$. Let $T_l := f^n_0(T(\beta_l), \ldots, T(\beta_{i-1}), 0, \ldots, 0)$ for $0 \leq l \leq i-1$.

Then, by definition, $T(\gamma) = f^i_\beta(T_0, \ldots, T_{i-1})$.

Further assume that, for some j with $0 \leq j \leq n$,

$$\delta = \Phi^{n+1}(\alpha_n, \ldots, \alpha_{j+1}, \beta', \beta'_{j-1}, \ldots, \beta'_0)$$

for some $\beta' < \alpha_j$ where $\beta'_l < \delta$ for all $l < j$. Let $T'_l := f^n_0(T(\beta'_l), \ldots, T(\beta'_{j-1}), 0, \ldots, 0)$ for $0 \leq l \leq j-1$.

Then, by definition, $T(\delta) = f^j_{\beta'}(T'_0, \ldots, T'_{j-1})$.

Case 1. Assume $i \neq j$.

Then $T(\gamma) \leq 0$ is not possible and from $T(\gamma) \trianglelefteq T(\delta)$ we conclude $T(\gamma) \trianglelefteq T'_l$ for some $l \leq j-1$. Hence $T(\gamma) \trianglelefteq f^n_0(T(\beta'_l), \ldots, T(\beta'_{j-1}), 0, \ldots, 0)$ for some $l \leq j-1$.

Case 1.1. Assume $i \neq n$.

Then $T(\gamma) \trianglelefteq T(\beta'_k)$ for some k. The induction hypothesis yields $\gamma \leq \beta'_k \leq \delta$.

Case 1.2. Assume $i = n \neq j$.

If $\beta > 0$ then then from $T(\gamma) \trianglelefteq T(\delta)$ we conclude $T(\gamma) \trianglelefteq T(\beta'_k)$ for some k. The induction hypothesis yields $\gamma \leq \beta'_k \leq \delta$.

Assume $\beta = 0$. Then $T(\gamma) = f^n_\beta(T_0, \dots, T_{n-1}) \trianglelefteq f^n_0(T(\beta'_l), \dots, T(\beta'_{j-1}), 0, \dots, 0)$ for some l. If $T(\gamma) \trianglelefteq T(\beta'_k)$ for some k then the induction hypothesis yields $\gamma \leq \beta'_k \leq \delta$. Assume that $T(\gamma) \trianglelefteq T(\beta'_k)$ is not the case for all k with $l \leq k \leq j-1$. We have $T_l := f^n_0(T(\beta_l), \dots, T(\beta_{i-1}), 0, \dots, 0)$ for $0 \leq l \leq i-1$. Since $j \neq n$ one of the T_k has to be mapped into 0 because of

$$T(\gamma) = f^n_\beta(T_0, \dots, T_{n-1}) \trianglelefteq f^n_0(T(\beta'_l), \dots, T(\beta'_{j-1}), 0, \dots, 0).$$

This is of course not possible since $T_l \neq 0$ for every $l < n = i$.

Case 2. $i = j$.

Case 2.1. $T(\gamma) \trianglelefteq T'_l$ for some $l \leq j-1$.

Then $T(\gamma) \trianglelefteq f^n_0(T(\beta'_l), \dots, T(\beta'_{j-1}), 0, \dots, 0)$. If $i \neq n$ then we can argue as in Case 1. Assume $i = j = n$. If $\beta > 0$ then we can argue as in Case 1.

Assume $\beta = 0$. Then $T(\gamma) = f^n_\beta(T_0, \dots, T_{n-1}) \trianglelefteq f^n_0(T(\beta'_l), \dots, T(\beta'_{n-1}), 0, \dots, 0)$.

If $T(\gamma) \trianglelefteq T(\beta'_k)$ for some k then the induction hypothesis yields $\gamma \leq \beta'_k \leq \delta$.

Assume that for all k in question $T(\gamma) \trianglelefteq T(\beta'_k)$ does not hold. If $l > 0$ then for some $r \leq n-1$ the tree T_r has to be mapped into 0 which is impossible.

So assume that $T(\gamma) = f^n_\beta(T_0, \dots, T_{n-1}) \trianglelefteq f^n_0(T(\beta'_0), \dots, T(\beta'_{n-1}))$. Then there exists an r such that $T_r = f^n_0(T(\beta_r), \dots, T(\beta_{n-1}), 0 \dots, 0) \trianglelefteq T(\beta'_{n-1})$. Then $\beta_{n-1} = \beta'_{n-1}$ is impossible since a tree is not embeddable into a direct subtree. We have $T(\beta_{n-1}) \trianglelefteq T_r \trianglelefteq T(\beta'_{n-1})$. The induction hypothesis yields $\beta_{n-1} \leq \beta'_{n-1}$ hence $\beta_{n-1} < \beta'_{n-1}$. Moreover we have $T(\beta_p) \trianglelefteq T(\gamma) \trianglelefteq T(\delta)$ for $p < n-1$. The case $\beta_p = \delta$ is impossible since $T(\beta_p)$ is a strict subtree of $T(\gamma)$. The induction hypothesis yields $\beta_p \leq \delta$ hence $\beta_p < \delta$ for $p < n-1$. With $\beta = 0$ this yields $\gamma < \delta$.

Case 2.2. Assume now that for all l in question $T(\gamma) \trianglelefteq T'_l$ does not hold. We have $T(\gamma) = f^i_\beta(T_0, \dots, T_{i-1})$ and $T(\delta) = f^i_{\beta'}(T'_0, \dots, T'_{i-1})$.

Then $\beta \leq \beta'$ and there exists a permutation $\pi : \{0, \dots, i-1\} \rightarrow \{0, \dots, i-1\}$ such that $T_l \trianglelefteq T'_{\pi(l)}$ for $0 \leq l \leq i-1$. If $i = j = 0$ then $\gamma = \Phi^{n+1}(\alpha_n, \dots, \alpha_1, \beta) \leq \Phi^{n+1}(\alpha_n, \dots, \alpha_1, \beta') = \delta$.

Assume now that $i = j > 0$. Now we carefully analyse the imposed structure induced by the specific T_k and T'_l.

We first look for the unique k such that $\pi(k) = i-1$ with $T_k \trianglelefteq T'_{i-1} = f^n_0(T(\beta'_{i-1}), 0 \dots, 0)$.

Then, after an easy case distinction, we find $T(\beta_{i-1}) \trianglelefteq T(\beta'_{i-1})$; indeed, the case $T_k \trianglelefteq 0$ is impossible. If $T_k \trianglelefteq T(\beta'_{i-1})$ then $T(\beta_{i-1}) \trianglelefteq T(\beta'_{i-1})$ since $T(\beta_{i-1})$ is a subtree of T_k.

If $T_k \trianglelefteq T(\beta'_{i-1})$ does not hold the relation $T_k \trianglelefteq T'_{i-1}$ yields $T(\beta_{i-1}) \trianglelefteq T(\beta'_{i-1})$.

The induction hypothesis yields $\beta_{i-1} \leq \beta'_{i-1}$.

Case 2.2.1. Assume first that $\beta_{i-1} < \beta'_{i-1}$. We have $T(\beta_r) \trianglelefteq T(\gamma) \trianglelefteq T(\delta)$ hence $\beta_r \leq \delta$ and hence (since $\beta_r = \delta$ is excluded since $T(\beta_r)$ is a strict subtree of $T(\gamma)$)) $\beta_r < \delta$ for all $r < i-1$.

We conclude $\gamma < \delta$.

Case 2.2.2. Assume finally that $\beta_{i-1} = \beta'_{i-1}$.

We have by construction $T_{i-1} \trianglelefteq T_k$ and by construction that $T'_{i-1} \trianglelefteq T'_{\pi(i-1)}$. Moreover, we see $T_k \trianglelefteq T'_{i-1}$. Then $T_k \trianglelefteq T'_{\pi(i-1)}$ and $T_{i-1} \trianglelefteq T'_{i-1}$.

We now exploit the last relation to see that there exists a permutation $\pi' : \{0,\dots,i-1\} \to \{0,\dots,i-1\}$ such that $\pi'(i-1) = i-1$ and $T_l \trianglelefteq T'_{\pi'(l)}$ for $0 \le l \le i-2$.

Indeed let $\pi'(i-1) = i-1$, $\pi'(l) = \pi(l)$ for $i-1 > l \neq k$, and $\pi'(k) = \pi(i-1)$.

Then there is some unique q such that $\pi'(q) = i-2$ with $T_q \trianglelefteq T'_{i-2} = f_0^n(T(\beta'_{i-2}), T(\beta'_{i-1}), 0 \dots, 0)$. Then $T_{i-2} := f_0^n(T(\beta_{i-2}), T(\beta_{i-1}), 0 \dots, 0) \trianglelefteq T_q \trianglelefteq T'_{i-2} = f_0^n(T(\beta'_{i-2}), T(\beta'_{i-1}), 0 \dots, 0)$.

A simple case distinction yields $T(\beta_{i-2}) \trianglelefteq T(\beta'_{i-2})$; indeed, the case $T_{i-2} \trianglelefteq T(\beta'_{i-1})$ is excluded since $\beta_{i-1} = \beta'_{i-1}$. Moreover, if $T_{i-2} \trianglelefteq T(\beta'_{i-2})$ then $T(\beta_{i-2}) \trianglelefteq T(\beta'_{i-2})$. In the remaining case the immediate subtrees of T_{i-2} have to be embedded into the immediate subtrees of $T'_{i-2} = f_0^n(T(\beta'_{i-2}), T(\beta'_{i-1}), 0 \dots, 0)$. If $T(\beta_{i-2}) \trianglelefteq T(\beta'_{i-2})$ we are done. Otherwise $T(\beta_{i-2}) \trianglelefteq T(\beta'_{i-1})$ and $T(\beta_{i-1}) \trianglelefteq T(\beta'_{i-2})$. But then $\beta_{i-1} = \beta'_{i-1}$ yields again $T(\beta_{i-2}) \trianglelefteq T(\beta'_{i-2})$.

If $\beta_{i-2} < \beta'_{i-2}$ we can argue as before in the case $\beta_{i-1} < \beta'_{i-1}$ to prove that $\gamma \le \delta$. Otherwise $\beta_{i-2} = \beta'_{i-2}$.

By now the pattern is clear and an iterative argument yields that we may assume that $\beta_l \le \beta'_l$ for all $l < n$. Together with $\beta \le \beta'$ we see that $\gamma \le \delta$. □

Example. Let us assume the conditions of the proof of the theorem. To visualize the construction let us assume that $n = 2$, $\beta < \alpha_2$ and that trees $T(\beta_0)$ and $T(\beta_1)$ have been already assigned. Then the tree $T(\gamma)$ for $\gamma = \Phi^3(\beta, \beta_1, \beta_0)$ has the following graphical representation.

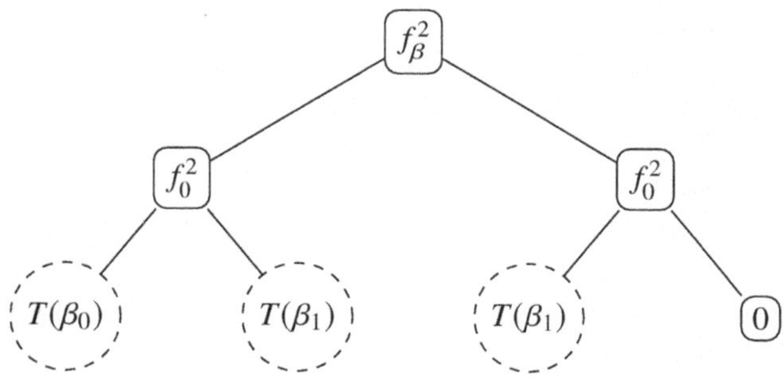

It is important to note that the tree $T(\beta_1)$ occurs as a subtree in all immediate subtrees of $T(\gamma)$ since the ordinal β_1 has higher weight in the lexicographic ordering than β_0.

Theorem 14.7. *Assume that $\alpha_n > 0$ and that $n \ge 2$. There exist disjoint well orders $X_0, \dots, X_n$ of order types $1 + \alpha_0, \dots, \alpha_n$ such that*

$$o\langle \mathcal{T}\begin{pmatrix} X_0 & \dots & X_n \\ 0 & \dots & n \end{pmatrix}, \trianglelefteq \rangle \ge \overline{\Phi}^{n+1}(\alpha_n, \dots, \alpha_0).$$

Proof. If $\Phi^{n+1}(\alpha_n,\dots,\alpha_0) = \overline{\Phi}^{n+1}(\alpha_n,\dots,\alpha_0)$ then the assertion follows by the previous theorem.

Case 1. Assume that $\Phi^{n+1}(\alpha_n,\dots,\alpha_0+1) = \overline{\Phi}^{n+1}(\alpha_n,\dots,\alpha_0)$ where for some α we have $\alpha_0 = \alpha + m$ and $\Phi^{n+1}(\alpha_n,\dots,\alpha) = \alpha$. Then $1+\alpha_0 = \alpha_0$.

To each $\gamma < \overline{\Phi}^{n+1}(\alpha_n,\dots,\alpha_0)$ let us recursively define a finite tree $T(\gamma)$ as follows.

1. If $\gamma < \alpha$ then $T(\gamma) := f^0_\gamma$. Let $0 := f^0_0$. (Since $1+\alpha_0 = \alpha_0$ we do not need a specific distinguished extra constant for 0 and can use 0 as an abbreviation of f^0_0.)
2. $T(\alpha) := f^n_0(0,\dots,0)$.
3. If $\alpha < \gamma = \Phi^{n+1}(\alpha_n,\dots,\alpha+l)$ for some $l \in \{1,\dots,m\}$ then $T(\gamma) = f^0_{\alpha+l-1}$.
4. If $\alpha < \gamma = \Phi^{n+1}(\alpha_n,\dots,\alpha_{i+1},\beta,\beta_{i-1},\dots,\beta_0)$ for some $\beta < \alpha_i$ where $\beta_j < \gamma$ for $j \in \{0,\dots,i-1\}$ then let $T_l := f^n_0(T(\beta_l),\dots,T(\beta_{i-1}),0,\dots,0)$ for $l < i$ and $T(\gamma) := f^i_\beta(T_0,\dots,T_{i-1})$. (Note that here $i > 0$ since $\alpha < \gamma$).

We claim that $T(\gamma) \trianglelefteq T(\delta)$ implies $\gamma \le \delta$.

Case 1.1. $\gamma < \alpha$. Then $T(\gamma) = f^0_\gamma$. If $T(\delta) = f^0_\delta$ and $T(\gamma) \trianglelefteq T(\delta)$ then $f^0_\gamma \trianglelefteq f^0_\delta$ and $\gamma \le \delta$. If $T(\gamma) \trianglelefteq T(\delta) = f^n_0(0,\dots,0)$ then $T(\gamma) \trianglelefteq 0$ and $\gamma = 0$, hence $\gamma \le \delta$.

If $T(\delta) = f^0_{\alpha+l-1}$ or $T(\delta) = f^i_\beta(T_0,\dots,T_{i-1})$ with $T_l = f^n_0(T(\beta_l),\dots,T(\beta_{i-1}),0,\dots,0)$ for $l < i$ then $\delta \ge \alpha > \gamma$.

Case 1.2. $\gamma = \alpha$. Then $T(\gamma) = f^n_0(0,\dots,0)$. $T(\gamma) \trianglelefteq T(\delta)$ yields that $T(\delta)$ has not the form f^0_δ for any $\delta < \alpha$. Hence $\delta \ge \alpha = \gamma$.

Case 1.3. $\gamma > \alpha$ and $T(\gamma) = f^0_{\alpha+l-1}$ where $\gamma = \Phi^{n+1}(\alpha_n,\dots,\alpha+l)$.

The case $T(\gamma) \trianglelefteq T(\delta)$ excludes the possibility that $T(\delta) = f^0_\delta$ for some $\delta < \alpha$ since $\alpha + l - 1 > \delta$. The case $T(\delta) = f^n_0(0,\dots,0)$ is impossible since $T(\gamma) \trianglelefteq T(\delta)$ would yield $T(\gamma) \trianglelefteq 0$. If $T(\delta) = f^0_{\alpha+l'-1}$ where $\delta = \Phi^{n+1}(\alpha_n,\dots,\alpha+l')$ then $T(\gamma) \trianglelefteq T(\delta)$ yields $l \le l'$ hence $\gamma = \Phi^{n+1}(\alpha_n,\dots,\alpha+l) \le \Phi(\alpha_n,\dots,\alpha+l') = \delta$.

To finish case 1.3 let us finally assume that $T(\delta) = f^i_\beta(T_0,\dots,T_{i-1})$ where $T_l = f^n_0(T(\beta_l),\dots,T(\beta_{i-1}),0,\dots,0)$. $T(\gamma) \trianglelefteq T(\delta)$ yields the relation $T(\gamma) \trianglelefteq f^n_0(T(\beta_j),\dots, T(\beta_{i-1}),0,\dots,0)$ for some j hence $T(\gamma) \trianglelefteq T(\beta_l)$ for some l. The induction hypothesis yields

$$\gamma \le \beta_l \le \Phi^{n+1}(\alpha_n,\dots,\alpha_{i+1},\beta,\beta_{i-1},\dots,\beta_0) = \delta.$$

Case 1.4. $T(\gamma) = f^i_\beta(T_0,\dots,T_{i-1})$ with $T_l = f^n_0(T(\beta_l),\dots,T(\beta_{i-1}),0,\dots,0)$.

Then the cases $T(\delta) = f^0_\delta$ and $T(\delta) = f^0_{\alpha+l-1}$ are impossible since $T(\gamma) \trianglelefteq T(\delta)$. The case $T(\delta) = f^n_0(0,\dots,0)$ is impossible since $T(\gamma) \trianglelefteq T(\delta)$ would yield $T(\gamma) \trianglelefteq 0$ or $T_0 \trianglelefteq 0$. We therefore obtain the situation $T(\delta) = f^j_{\beta'}(T'_0,\dots,T'_{j-1})$ where $T'_l = f^n_0(T(\beta'_l),\dots,T(\beta'_{j-1})),0,\dots,0)$.

We may then argue as in the proof of Theorem 14.6 to arrive at the desired conclusion.

Case 2. Let us assume that $\overline{\Phi}^{n+1}(\alpha_n,\dots,\alpha_0) = \Phi^{n+1}(\alpha_n,\dots,\alpha_0+1)$ where for some α we have $\alpha_0 = \alpha + m$ and $\Phi^{n+1}(\alpha_n,\dots,\alpha) = \alpha_j$ where $0 < j \le n$. Then $\alpha = 0, \alpha_0 = m$ and $\alpha_1,\dots,\alpha_{j-1} = 0$. This means $\Phi^{n+1}(\alpha_n,\dots,\alpha_j,0,\dots,0) = \alpha_j$.

Let $X_0 := \{0\} \cup \{f^0_l : l < m\}$ and $X_i := \{f^i_\beta : \beta < \alpha_i\}$. Note that $X_i = \emptyset$ for $i \in \{1,\dots,j-1\}$.

To each $\gamma < \overline{\Phi}^{n+1}(\alpha_n,\dots,\alpha_0)$ let us define a finite tree $T(\gamma)$ recursively as follows.

1. If $\gamma = \Phi^{n+1}(\alpha_n,\dots,\alpha_1,l)$ for some l such that $0 < l \le m$ then $T(\gamma) = f^0_{l-1}$.
2. If $\alpha_j < \gamma = \Phi^{n+1}(\alpha_n,\dots,\alpha_{i+1},\beta,\beta_{i-1},\dots,\beta_0)$ for some $\beta < \alpha_i$ and $\beta_l < \gamma$ for $l \in \{0,\dots,i-1\}$ then $T(\gamma) = f^i_\beta(T_0,\dots,T_{i-1})$ where $T_l = f^n_0(T(\beta_l),\dots,T(\beta_{i-1}), 0,\dots,0)$. In this case we necessarily have $j \le i$ since $\alpha_1,\dots,\alpha_{j-1} = 0$ and we are not in case 1.
3. If $\gamma < \alpha_j$ then $T(\gamma) = f^j_\gamma(0,\dots,0)$.
4. $T(\alpha_j) := f^n_0(f^j_0(0,\dots,0),0,\dots,0)$. Here the idea is to reserve a special tree for α_j. It has $n \ge 2$ immediate subtrees and only one of them is different from 0. The last property guarantees that the other trees in our construction which have n immediate non zero subtrees are not embeddable into $T(\alpha_j)$. The last property also guarantees that no tree in our construction which has $i \neq n$ immediate non zero subtrees is embeddable into $T(\alpha_j)$.

We claim that $T(\gamma) \trianglelefteq T(\delta)$ implies $\gamma \le \delta$.

Case 2.1. $T(\gamma) = f^0_l$ where $\gamma = \Phi^{n+1}(\alpha_n,\dots,\alpha_1,l)$

If $T(\delta) = f^0_{l'}$ where $\delta = \Phi^{n+1}(\alpha_n,\dots,\alpha_1,l')$ then $T(\gamma) \trianglelefteq T(\delta)$ yields $l \le l'$ and $\gamma \le \delta$.

If $T(\delta) = f^i_\beta(T_0,\dots,T_{i-1})$ with $T_l = f^n_0(T(\beta_l),\dots,T(\beta_{i-1}),0,\dots,0)$ where $\delta = \Phi^{n+1}(\alpha_n,\dots,\alpha_{i+1},\beta,\beta_{i-1},\dots,\beta_0)$ for some $\beta < \alpha_i$ and $\beta_l < \delta$ for $l < i$ then $T(\gamma) \trianglelefteq T(\delta)$ yields $T(\gamma) \trianglelefteq T_l$, hence $T(\gamma) \trianglelefteq T(\beta_m)$ for some m hence the induction hypothesis yields $\gamma \le \beta_m \le \delta$. If $T(\delta) = f^j_\delta(0,\dots,0)$ or if $T(\delta) = f^n_0(f^j_0(0,\dots,0),0,\dots,0)$ then $T(\gamma) \le T(\delta)$ yields $T(\gamma) \le 0$ which is impossible.

Case 2.2. $T(\gamma) = f^j_\gamma(0,\dots,0)$ with $\gamma < \alpha_j$. If $T(\delta) = f^j_\delta(0,\dots,0)$ with $\delta < \alpha_j$ then $T(\gamma) \trianglelefteq T(\delta)$ yields $\gamma \le \delta$. In all other cases $\delta \ge \alpha_j > \gamma$.

Case 2.3. $T(\gamma) = f^n_0(f^j_0(0,\dots,0),0,\dots,0)$ with $\gamma = \alpha_j$. Then $T(\delta) = f^j_\delta(0,\dots,0)$ with $\delta < \alpha_j$ is impossible since $T(\gamma) \trianglelefteq T(\delta)$ would then yield $f^n_0(0,\dots,0) \trianglelefteq 0$. In all other cases $\delta \ge \alpha_j = \gamma$.

Case 2.4.

$$T(\gamma) = f^i_\beta(T_0,\dots,T_{i-1})$$

with $T_l = f^n_0(T(\beta_l),\dots,T(\beta_{i-1}),0,\dots,0)$ where

$$\alpha_j < \gamma = \Phi^{n+1}(\alpha_n,\dots,\alpha_{i+1},\beta,\beta_{i-1},\dots,\beta_0)$$

for some $\beta < \alpha_j$ and $\beta_l < \gamma$. Then $i > j$.

Then the case $T(\delta) = f^0_l$ and also the case $T(\delta) = f^j_\delta(0,\dots,0)$ with $\delta < \alpha_j$ are impossible since $T(\gamma) \trianglelefteq T(\delta)$. If $T(\delta) = f^n_0(f^j_0(0,\dots,0),0,\dots,0)$ with $\delta = \alpha_j$ then $T(\gamma) \trianglelefteq T(\delta)$ yields also a contradiction by the specific choice of $T(\delta)$ as indicated before.

Assume finally that

$$T(\delta) = f^k_{\beta'}(T'_0,\dots,T'_{k-1})$$

with $T'_l = f^n_0(T(\beta'_l),\dots,T(\beta'_{k-1}),0,\dots,0)$ where

$$\alpha_j < \delta = \Phi^{n+1}(\alpha_n,\dots,\alpha_{k+1},\beta',\beta'_{k-1},\dots,\beta'_0)$$

for some $\beta' < \alpha_k$ and $\beta'_l < \delta$. Then $\gamma \le \delta$ follows as in the proof of Theorem 14.6. □

14.4 Applications

Our construction provides some immediate applications to the proof theory of subsystems of second order arithmetic. We use the terminology of Simpson's book [12]. Theorem 14.7 can be formalized in the standard system RCA_0 of reverse mathematics.

Theorem 14.8. *The following is provable in* RCA_0*: Assume that* $n \geq 2, \alpha_n > 0$*,* X_i *is a partial order,* α_i *is a linearization of* X_i *for* $i > 0$ *and that* $1 + \alpha_0$ *is a linearization of* X_0*. Assume that* $\langle \mathcal{T}\begin{pmatrix} X_0 & \dots & X_n \\ 0 & \dots & n \end{pmatrix}, \trianglelefteq \rangle$ *is a well partial order. Then the set of ordinal terms below* $\overline{\Phi}^{n+1}(\alpha_n, \dots, \alpha_0)$ *is well ordered.*

Theorem 14.9. *The following is provable in* RCA_0*: Assume that* $\langle \mathcal{T}\begin{pmatrix} 1 & 1 \\ 0 & 2 \end{pmatrix}, \trianglelefteq \rangle$ *is a well partial order. Then the set of ordinal terms below* $\varepsilon_0 = \overline{\Phi}^3(1,0,0)$ *is well ordered.*

Corollary 14.10. ACA_0 *does not prove that* $\langle \mathcal{T}\begin{pmatrix} 1 & 1 \\ 0 & 2 \end{pmatrix}, \trianglelefteq \rangle$ *is a well partial order.*

Theorem 14.11. *The following is provable in* RCA_0*: Assume that* $\langle \mathcal{T}\begin{pmatrix} 1 & 2 \\ 0 & 2 \end{pmatrix}, \trianglelefteq \rangle$ *is a well partial order. Then the set of ordinal terms below* $\Gamma_0 = \overline{\Phi}^3(2,0,0)$ *is well ordered.*

Corollary 14.12. ATR_0 *does not prove that* $\langle \mathcal{T}\begin{pmatrix} 1 & 2 \\ 0 & 2 \end{pmatrix}, \trianglelefteq \rangle$ *is a well partial order. In particular* ATR_0 *does not prove Kruskal's theorem.*

References

1. J. N. Crossley and J. Bridge Kister. Natural well-orderings. *Archiv für mathematische Logik und Grundlagenforschung*, 26:57–76, 1986/7.
2. D. H. J. de Jongh and R. Parikh. Well partial orderings and hierarchies. *Indigationes Mathematicae*, 39:195–206, 1977.
3. S. Feferman. Systems of predicative analysis. II. Representations of ordinals. *Journal of Symbolic Logic*, 33:193–220, 1968.
4. S. Feferman. Proof theory: A personal report. In G. Takeuti, editor, *Proof theory*, volume 81, pages 447–485. North-Holland Publishing Co., Amsterdam, second edition, 1987.
5. S. Feferman. Three conceptual problems that bug me, 1996. https://math.stanford.edu/˜feferman/papers/conceptualprobs.pdf.
6. G. Higman. Ordering by divisibility in abstract algebras. *Proceedings of the London Mathematical Society*, 3(2):326–336, 1952.
7. J. Kruskal. Well-quasi-ordering, the tree theorem, and Vázsonyi's conjecture. *Transactions of the American Mathematical Society*, 95:210–225, 1960.
8. D. Schmidt. Bound for the closure ordinals for replete monotonic increasing functions. *Journal of Symbolic Logic*, 40:305–316, 1975.

9. D. Schmidt. *Well-partial Orderings and their Maximal Order Types*. Habilitationsschrift, Heidelberg, 1979.
10. K. Schütte. Kennzeichnung von Ordnungszahlen durch rekursiv erklärte Funktionen. *Mathematische Annalen*, 127:15–32, 1954.
11. K. Schütte. *Proof Theory*. Springer, 1977.
12. S. G. Simpson. Nonprovability of certain combinatorial properties of finite trees. In *H. Friedman's Research on the Foundations of Mathematics*, volume 117 of *Stud. Logic Found. Math.*, pages 87–117. North-Holland, Amsterdam, 1985.
13. F. van Vugt. *Independence of Variations to Kruskal's Theorem in ACA_0*. Thesis written under supervision of A. Weiermann, Utrecht, 2005.

Erinnerungen von Diana Schmidt aus 1968–69 an Kurt Schütte

Nach dem Abschluss meines Bachelor-Studiums in Mathematik an der Universität Oxford bekam ich ein Promotionsstipendium, das mir (auf meinen Wunsch) ermöglichte, das erste Jahr an der Universität München unter der Anleitung von Kurt Schütte zu verbringen. Grund hierfür war, dass ich während meines Bachelor-Studiums das Wahlfach Mathematische Logik bei John Crossley belegte und mich für dieses Gebiet interessierte, auf dem John Crossley mit Kurt Schütte forschte und veröffentlichte.

Schüttes Forschungsinteressen lagen in den Gebieten Beweistheorie und transfinite Ordinalzahlen.

Ich suchte Herrn Schütte bald nach meiner Ankunft in München in seinem Büro auf: Er empfahl mir, seine Vorlesungen zu besuchen und an seinem Seminar teilzunehmen, und gab mir Abdrücke von seinen Veröffentlichungen. Ich stellte bald fest, dass mich Beweistheorie weniger interessierte, Ordinalzahlen dafür aber umso mehr, und insbesondere Schüttes Klammersymbole, mit denen er größere Ordinalzahlen darstellte, für die die damals gängigen Mittel nicht ausreichten.

Einmal hat Herr Schütte abends Studierende zu sich nach Hause eingeladen. Da ich mit deutschen Gepflogenheiten nicht vertraut war fragte ich eine befreundete Kommilitonin (nicht aus der Logik), ob man als Gast etwas mitbringen solle. Sie empfahl Blumen, also kam ich mit einem kleinen Blumenstrauß an und übergab ihn Herrn Schütte. Er war aber sichtlich bestürzt wegen meiner Unkosten, und sagte: „Ich habe doch bei der Einladung gesagt, dass die Gäste mit nichts in der Hand kommen sollen." Diesen Ausdruck hatte ich entweder überhört oder nicht verstanden, weil ich ihn noch nicht kannte; aber ich nahm Schüttes Fürsorglichkeit wahr.

Mittags ist mir (zumindest nach meiner Erinnerung) mehr als einmal Folgendes passiert: Nach dem Mittagessen in der Mensa ging ich am Mathematischen Institut vorbei, und Herr Schütte mit den Herren Diller und Czermak kamen mir entgegen, vielleicht nach einem Mittagessen in einem nahe gelegenen Lokal. Jeder trug einen Hut auf dem Kopf, und kurz bevor wir aneinander vorbeigingen nahmen sie (nach meiner Erinnerung gleichzeitig) ihren jeweiligen Hut nach vorne ab und verbeugten dabei leicht ihre Köpfe. Man stelle sich vor: Ich war gerade einundzwanzig, und drei für mich erhabene Herren verbeugten sich vor mir. Ich vermute, dass es von Herrn Schütte (als Ältestem) ausging.

Chapter 15
Cut-Elimination for SBL

Toshiyasu Arai

Abstract In this paper we give a terminating cut-elimination procedure for a logic calculus SBL. SBL corresponds to the second order arithmetic Π^1_2-Separation and Bar Induction.

15.1 Introduction

Let Π^1_2-Sep+BI denote the subsystem of second order arithmetic with Π^1_2-Separation and Bar Induction. For definition of Π^1_2-Separation, cf. Definition 15.5 below. See also p.17 of Schütte [17]. It is known that Δ^1_2-CA proves Π^1_2-Sep, i.e., Π^1_2-Sep + BI = Δ^1_2-CA + BI. This is seen from the fact that Δ^1_2-CA proves Σ^1_2-AC, cf.Theorem VII.6.9.1 and Exercise VII.6.13 of [18].

It is shown in Jäger and Pohlers [12] that Δ^1_2-CA + BI is proof-theoretically equivalent to the set theory KPi for recursively inaccessible universes. K. Schütte [17] gives an upper bound $\psi_0 I$ for the proof theoretic ordinal of Π^1_2-Sep + BI. The ordinal $\psi_0 I$ is the order type of an initial segment of the recursive notation system $T(I)$ of ordinals introduced by W. Buchholz and Schütte [8].

The analysis of the derivations in Π^1_2-Sep + BI due to Schütte is based on Buchholz' $\Omega_{\mu+1}$-rule introduced in [5], combined with the Pohlers' method of local predicativity in [14]. In the proof, the system $(T(I), <)$ is utilized indirectly: in fact the totally defined collapsing functions d and d_σ appear in the analysis, which are also introduced in [8].

As contrasted with it G. Takeuti [20, 21, 22] uses his systems of ordinal diagrams directly for a proof theory of Π^1_1-Comprehension. The definition of ordinal diagrams is closely related to the cut-elimination procedure due to him. Actually Buchholz [7] shows that Takeuti's reduction steps can be explained in terms of the cut elimination

T. Arai
Graduate School of Mathematical Sciences, The University of Tokyo, 3-8-1 Komaba, Meguro-ku, Tokyo 153-8914, Japan. e-mail: tosarai@ms.u-tokyo.ac.jp

R. Kahle and M. Rathjen (eds.), *The Legacy of Kurt Schütte*,
https://doi.org/10.1007/978-3-030-49424-7_15

procedure through the $\Omega_{\mu+1}$-rule. But unfortunately Takeuti's systems of ordinal diagrams are equipped with many order relations and are too small to handle such a strong theory Π^1_2-Sep + BI.

Turning to the problem of cut-elimination in second order, and higher order logic calculi (known as Takeuti's Fundamental Conjecture), W. Tait [19] proves cut-eliminability (Hauptsatz) for the classical second order (full impredicative) logic calculus based on Schütte's [16] reformulation of it in terms of a semantical notion, *semivaluation*. For an alternative proof of the Hauptsatz, see S. Maehara [13] and its reformulation in [3]. Although these proofs show the Hauptsatz, a terminating cut-eliminating procedure is required in a solution of the Takeuti's problem. Such a procedure is given by Takeuti [20, 21] for logic calculi for which the Hauptsatz yields (actually is equivalent to) the 1-consistency of Π^1_1-CA$_0$.

On the other side G. Jäger [11] shows wellfoundedness up to each ordinal< $\bar{\theta}^0(\bar{\theta}^1\varepsilon_{I+1}0)0$ in the constructive theory T_0 due to S. Feferman [9], which is interpretable in Δ^1_2-CA + BI. Jäger's proof is based on Ausgezeichnete Klassen introduced by Buchholz [4], and the ordinal $\bar{\theta}^0(\bar{\theta}^1\varepsilon_{I+1}0)0$ seems to be equal to the ordinal $\psi_0 I$.

Given these advances in the 1980's, we had introduced a system $(O(I), <)$ of ordinal diagrams and proved a cut-elimination theorem for a logic calculus SBL in the style of Gentzen-Takeuti [10, 22] by transfinite induction on the system. SBL corresponds to the system Π^1_2-Sep + BI in the sense that the Hauptsatz (normal form theorem) for SBL is equivalent to the 1-consistency of Π^1_2-Sep + BI over a weak theory, e.g., over $I\Sigma_1$. By giving a terminating cut-elimination procedure for the logic calculus SBL, we aimed to contribute the Takeuti's problem.

This was done in the original version of this paper written in 1988. The system $(O(I), <)$ of ordinal diagrams was obtained as a kind of mixture of totally defined collapsing functions d_σ in [8] and Takeuti's ordinal diagrams. Specifically d_σ is a primitive constructor of ordinal terms in $O(I)$, whereas it is a derived term in [8]. In the original version of this paper it was also shown that each initial segment determined by $\alpha < \Omega_1 \in O(I)$ is well-founded. The proof is formalizable in T_0 as Jäger's proof in [11]. This was a starting point for us to construct larger notation systems of ordinals, e.g., in [1].

The proof of cut-elimination in the original version was inspired from Schütte's proof in [17]. In the present version let us update the original proof via the collapsing functions ψ_σ and the operator controlled derivations both due to Buchholz [6].

In section 15.2 let us recall the collapsing functions ψ_σ up to $\sigma \leq I$, where I is the least weakly inaccessible cardinal. A wellfoundedness proof in T_0 is omitted in the present version since it should not be hard. In subsection 15.2.1 we define an essentially less than relation $\alpha \ll \beta\,\{\eta\}$ for ordinals α, β, η in terms of Skolem hulls $\mathcal{H}_\gamma(\psi_\sigma\gamma)$. In section 15.3 a second order logic calculus SBL is introduced. In section 15.4 we introduce a stratified logic calculus SBL$'$ inspired by Schütte [17]. SBL is then embedded in SBL$'$, and a cut-free proof in SBL$'$ denotes essentially a cut-free proof in SBL. For each proof P in SBL$'$ we assign an ordinal $o(P) < \psi_{\Omega_1}\varepsilon_{I+1}$ in such a way that if P contains a cut rule, then we can construct another proof P' of the same end sequent in SBL$'$ such that $o(P') < o(P)$ (Main Lemma 15.27). It turns out that each proof appearing in the cut-elimination procedure enjoys some conditions on assigned ordinals, which are spelled out in Definition 15.24.3. Restrictions similar to these conditions are found in the proof through operator controlled derivations due to Buchholz [6]. It is not easy for us to expound on technicalities here. In Notes 15.18 and 15.25, a proof schema is given, and let us try to explain how our proof is derived from Takeuti's reduction [20], Buchholz' $\Omega_{\mu+1}$-rule [5], Pohlers' method of local predicativity in [14], and operator controlled derivations in [6].

The final section 15.5 is devoted to a proof of Main Lemma 15.27.

15.2 Collapsing Functions ψ_σ

Let I denote the least weakly inaccessible cardinal, and $\Omega_\alpha := \omega_\alpha$ for $0 < \alpha < I$. Put $\Omega_0 := 0$ and $R = \{\Omega_{\alpha+1} : \alpha < I\} \cup \{I\} = \{\sigma \leq I : \omega < \sigma \text{ is regular}\}$. σ, τ, κ range over elements in R.

In this section let us recall the collapsing functions ψ_σ $(\sigma \in R)$ due to W. Buchholz [6].

Definition 15.1. $\mathcal{H}_\alpha(X)$ denote the *Skolem hull* of the set $X \cup \{0, I\}$ of ordinals under the functions $+, \beta \mapsto \omega^\beta, \beta \mapsto \Omega_\beta$ and $(\sigma, \beta) \mapsto \psi_\sigma\beta$ $(\beta < \alpha)$.

$$\psi_\sigma\alpha = \min(\{\beta < \sigma : \sigma \in \mathcal{H}_\alpha(\beta) \,\&\, \mathcal{H}_\alpha(\beta) \cap \sigma \subset \beta\} \cup \{\sigma\}).$$

The following facts are shown in Lemma 4.5 of [6]. We see that $\psi_\sigma\alpha < \sigma$ from the regularity of σ and $\alpha + 1 < \Omega_{\alpha+1}$ for $\alpha < I$. When $\sigma = \Omega_{\mu+1}$, we have $\mu \leq \Omega_\mu < \psi_\sigma\alpha < \Omega_{\mu+1}$. Hence $\sigma \in \mathcal{H}_0(\psi_\sigma\alpha)$. If $\alpha_0 < \alpha_1$, then $\mathcal{H}_{\alpha_0}(\beta) \subset \mathcal{H}_{\alpha_1}(\beta)$, and $\psi_\sigma\alpha_0 \leq \psi_\sigma\alpha_1$. Also $\mathcal{H}_\alpha(\beta)$ is closed under the *natural sum* $\gamma\#\delta$ and the functions $\gamma \mapsto \omega^\gamma$ and $\gamma \mapsto \Omega_\gamma$ in the reverse direction, i.e., $\gamma\#\delta \in \mathcal{H}_\alpha(\beta) \Rightarrow \{\gamma, \delta\} \subset \mathcal{H}_\alpha(\beta)$, $\omega^\gamma \in \mathcal{H}_\alpha(\beta) \Rightarrow \gamma \in \mathcal{H}_\alpha(\beta)$ and $\Omega_\gamma \in \mathcal{H}_\alpha(\beta) \Rightarrow \gamma \in \mathcal{H}_\alpha(\beta)$.

Let ε_{I+1} denote the next epsilon number above I. Each ordinal $\alpha \in \mathcal{H}_{\varepsilon_{I+1}}(0)$ is seen to be denoted uniquely by an ordinal term in normal form as follows. A set OT of ordinal terms is generated from $\{0, I\}$ by $+, \beta \mapsto \omega^\beta, \beta \mapsto \Omega_\beta$ and $(\sigma, \beta) \mapsto \psi_\sigma\beta$. The set OT together with finite sets $G_\beta(\gamma)$ for $\gamma \in OT$ is defined recursively as follows. $\alpha_0 + \cdots + \alpha_n$ is in OT iff each additively closed α_i is in OT, and $\alpha_0 \geq \cdots \geq \alpha_n$. $\omega^\alpha \in OT$ iff $\alpha \in OT$ is not an epsilon number. $\Omega_\beta \in OT$ iff $\beta \in OT$ is a successor ordinal below I. $\beta = \psi_\sigma\alpha \in OT$ iff $\sigma, \alpha \in OT$ and $G_\beta(\alpha) \cup G_\beta(\sigma) < \alpha$, where either $\sigma = I$ or $\sigma = \Omega_{\mu+1}$ for a $\mu < I$. $G_\beta(0) = G_\beta(I) = \emptyset$, $G_\beta(\alpha_0 + \cdots + \alpha_n) = \bigcup\{G_\beta(\alpha_i) : i \leq n\}$, $G_\beta(\omega^\alpha) = G_\beta(\Omega_\alpha) = G_\beta(\alpha)$, and

$$G_\beta(\psi_\sigma\alpha) = \begin{cases} \emptyset & \text{if } \psi_\sigma\alpha < \beta, \\ G_\beta(\sigma) \cup G_\beta(\alpha) \cup \{\alpha\} & \text{otherwise.} \end{cases}$$

We see then that $\gamma \in \mathcal{H}_\alpha(\beta) \Leftrightarrow G_\beta(\gamma) < \alpha$ for $\gamma, \alpha \in OT$, and the relation $\alpha < \beta$ $(\alpha, \beta \in OT)$ is computable. For example we see that in OT, $\Omega_\alpha < \psi_{\Omega_{\alpha+1}}\beta < \Omega_{\alpha+1}$, $\psi_{\Omega_{\alpha+1}}\beta < \psi_I\gamma \Leftrightarrow \alpha < \psi_I\gamma$, and $\psi_I\alpha < \psi_I\beta \Leftrightarrow \alpha < \beta$.

Moreover each ordinal term in OT denotes an ordinal in $\mathcal{H}_{\varepsilon_{I+1}}(0)$, and conversely each ordinal in $\mathcal{H}_{\varepsilon_{I+1}}(0)$ is denoted by an ordinal term in OT. In what follows by ordinals we mean ordinal terms in the set OT, and $\mathcal{H}_{\varepsilon_{I+1}}(0)$ is identified with OT. $\alpha, \beta, \gamma, \delta, \ldots$ range over ordinal terms in $\mathcal{H}_{\varepsilon_{I+1}}(0)$, and $\sigma, \tau, \ldots$ over elements in the set $R = \{I\} \cup \{\Omega_{\mu+1} : \mu \in \mathcal{H}_{\varepsilon_{I+1}}(0)\}$. Let

$$Fx := \{\alpha \in \mathcal{H}_{\varepsilon_{I+1}}(0) : \Omega_\alpha = \alpha > 0\} = \{\psi_I \alpha : \alpha \in \mathcal{H}_{\varepsilon_{I+1}}(0), \alpha \in \mathcal{H}_\alpha(\psi_I \alpha)\} \cup \{I\}.$$

15.2.1 Relations $\alpha \ll \beta \{\eta\}$

In this subsection an essentially less than relation $\alpha \ll \beta \{\eta\}$ is defined through Skolem hulls $\mathcal{H}_\gamma(\psi_\sigma \gamma)$.

Definition 15.2. For ordinal terms $\delta_0, \delta_1, \eta \in OT = \mathcal{H}_{\varepsilon_{I+1}}(0)$,

$$\delta_0 \ll \delta_1 \{\eta\} :\Leftrightarrow \delta_0 < \delta_1 \wedge \forall \sigma \forall \alpha [\{\delta_1, \eta\} \subset \mathcal{H}_\alpha(\psi_\sigma \alpha) \Rightarrow \delta_0 \in \mathcal{H}_\alpha(\psi_\sigma \alpha)],$$
$$\delta_0 \underline{\ll} \delta_1 \{\eta\} :\Leftrightarrow \delta_0 = \delta_1 \vee (\delta_0 \ll \delta_1 \{\eta\}),$$
$$\delta_0 \ll \delta_1 :\Leftrightarrow \delta_0 \ll \delta_1 \{0\}.$$

Proposition 15.3. 1. $\delta_0 \ll \delta_1 \{\eta\} \Rightarrow \delta_0 \# \alpha \ll \delta_1 \# \alpha \{\eta\}$.
2. $\delta_0 \ll \delta_1 \{\eta\} \Rightarrow \omega^{\delta_0} \ll \omega^{\delta_1} \{\eta\}$.
3. *Assume* $\{\gamma, \delta_1, \eta\} \subset \mathcal{H}_\gamma(\psi_\sigma(\gamma \# \delta_1))$ *and* $\delta_0 \ll \delta_1 \{\eta\}$. *Then* $\psi_\sigma(\gamma \# \delta_0) \ll \psi_\sigma(\gamma \# \delta_1) \{\eta\}$.
4. *Assume* $\gamma \in \mathcal{H}_\gamma(\psi_{\Omega_{\mu+1}} \gamma)$, $\alpha \ll \gamma$, *and* $\alpha \le \Omega_\mu$. *Then* $\alpha \ll \psi_{\Omega_{\mu+1}} \gamma$.
5. *Assume* $\alpha \ll \psi_\tau \gamma \{\eta\}$, $\gamma \in \mathcal{H}_\gamma(\psi_\tau \gamma)$ *and* $\forall \beta > \gamma \forall \sigma [\{\gamma, \tau\} \subset \mathcal{H}_\beta(\psi_\sigma \beta) \Rightarrow \eta \in \mathcal{H}_\beta(\psi_\sigma \beta)]$. *Then* $\alpha \ll \psi_\tau \gamma$.
6. *Assume* $\delta_0 \ll \delta_1 \{\eta\}$, $\psi_\tau(\gamma \# \delta_0) < \psi_\tau(\gamma \# \delta_1)$, *and* $\{\psi_\tau(\gamma \# \delta_1), \eta\} \subset \mathcal{H}_\alpha(\psi_\sigma \alpha)$. *Then* $\psi_\tau(\gamma \# \delta_0) \in \mathcal{H}_\alpha(\psi_\sigma \alpha)$.

Proof. 15.3.3.. Assume $\{\gamma, \delta_1, \eta\} \subset \mathcal{H}_\gamma(\psi_\sigma(\gamma \# \delta_1))$ and $\delta_0 \ll \delta_1 \{\eta\}$. Then $\delta_0 \in \mathcal{H}_\gamma(\psi_\sigma(\gamma \# \delta_1))$, and $\{\gamma, \delta_0, \sigma\} \subset \mathcal{H}_{\gamma \# \delta_1}(\psi_\sigma(\gamma \# \delta_1))$. Hence $\psi_\sigma(\gamma \# \delta_0) \in \mathcal{H}_{\gamma \# \delta_1}(\psi_\sigma(\gamma \# \delta_1)) \cap \sigma = \psi_\sigma(\gamma \# \delta_1)$.

Next suppose $\{\psi_\sigma(\gamma \# \delta_1), \eta\} \subset \mathcal{H}_\alpha(\psi_\tau \alpha)$. We show $\psi_\sigma(\gamma \# \delta_0) \in \mathcal{H}_\alpha(\psi_\tau \alpha)$. We can assume $\psi_\sigma(\gamma \# \delta_1) > \psi_\tau \alpha$. Then $\{\sigma, \gamma, \delta_1, \eta\} \subset \mathcal{H}_\alpha(\psi_\tau \alpha)$ and $\gamma \# \delta_1 < \alpha$. Therefore $\{\sigma, \gamma, \delta_0\} \subset \mathcal{H}_\alpha(\psi_\tau \alpha)$ and $\gamma \# \delta_0 < \alpha$. These yield $\psi_\sigma(\gamma \# \delta_0) \in \mathcal{H}_\alpha(\psi_\tau \alpha)$.
15.3.6. is seen as in Proposition 15.3.3.. □

15.3 The Logic Calculus SBL

In this section a second-order logic calculus SBL is introduced. $\mathcal{L}$ denotes a second-order language consisting of logical symbols $\vee, \wedge, \exists, \forall$, individual constants $c, \ldots$, function symbols $f, \ldots$, first-order free variables $a, b, \ldots$, first-order bound variables

$x, y, \ldots$, relation symbols $R, \ldots$, second-order free variables $U, V, \ldots$, and second-order bound variables $X, Y, \ldots$. Let us assume that each relation symbol and each second-order (free or bound) variable is *unary* for simplicity, and $\mathcal{L}$ contains an individual constant c and a (unary) relation symbol R, cf. pure variable condition in Definition 15.17.

T stands for either a (unary) second-order free variable or a relation symbol. Tt and $\neg Tt$ are *literals* for terms t. Formulas are generated from literals by means of $\vee, \wedge$ and first-order and second-order quantifications $\exists, \forall$ as usual. Negations $\neg A$ of formulas A are defined recursively through de Morgan's law and the elimination of double negations $\neg(\neg Tt) :\equiv (Tt)$.

For formal expressions E, s, t such as terms and proofs $E[s/t]$ denotes the expression obtained from E by replacing some occurrences of the expression t in E by the expression s. Let F be a formula[1] with a second-order bound variable X, and A a formula with a variable x. Then $F[A/X]$ denotes the formula obtained from F by replacing each occurrence of Xt by $A[t/x]$ and each occurrence of $\neg Xt$ by $\neg A[t/x]$.

Definition 15.4. 1. For formulas A, $VT(A)$ denotes the set of second-order free variables occurring in a scope of a second-order quantifier in A. $U \in VT(A)$ iff U is tied by a second-order quantifier in A in the sense of [22].

2. Two classes Π^1_2 and Σ^1_2 of formulas are defined as follows. Each of the classes Π^1_2 and Σ^1_2 is defined to contain literals, and be closed under positive propositional connectives $\vee, \wedge$ and first-order quantifications $\exists x, \forall x$.

$$\begin{aligned}
\forall X\, A[X/U] \in \Pi^1_2 &\Leftrightarrow A \in \Pi^1_2,\\
\exists X\, A[X/U] \in \Sigma^1_2 &\Leftrightarrow A \in \Sigma^1_2,\\
\forall X\, A[X/U] \in \Sigma^1_2 &\Leftrightarrow A \in \Pi^1_2 \cap \Sigma^1_2 \ \&\ U \notin VT(A),\\
\exists X\, A[X/U] \in \Pi^1_2 &\Leftrightarrow A \in \Pi^1_2 \cap \Sigma^1_2 \ \&\ U \notin VT(A).
\end{aligned}$$

$A \in \Pi^1_2 \cap \Sigma^1_2$ iff A is isolated in the sense of [22].

3. An occurrence of a second-order quantifier QX in a formula $QXA[X/U]$ is said to be *distinguished* if either $Q = \forall$, $\forall XA[X/U] \in \Pi^1_2$ and $U \notin VT(A)$, or $Q = \exists$, $\exists XA[X/U] \in \Sigma^1_2$ and $U \notin VT(A)$.

For example every second-order quantifier occurring in a formula is distinguished if the formula is in $\Pi^1_2 \cap \Sigma^1_2$. Moreover the outermost quantifier $\exists X$ in the formula $\exists X(A \subset X \subset B)$ is distinguished if $A \in \Pi^1_2$ and $B \in \Sigma^1_2$, where $(A \subset B) :\Leftrightarrow (\forall x(A(x) \to B(x)))$ and $(A \subset X \subset B) :\Leftrightarrow (A \subset X \wedge X \subset B)$.

Definition 15.5. The logic calculus SBL.

Axioms or *initial sequents* are

$$(Ax) \qquad \Gamma, \neg L, L \qquad \text{for literals } L.$$

[1] Strictly speaking, we should say that F is a semi-formula as in [22], but for simplicity let us call semi-formulas and semi-terms with bound variables as formulas and terms, resp.

Inference rules are the following.

$$\frac{\Gamma, A_i}{\Gamma}\,(\vee) \qquad \frac{\Gamma, A_0 \quad \Gamma, A_1}{\Gamma}\,(\wedge) \qquad \frac{\Gamma, F[t/x]}{\Gamma}\,(\exists_1) \qquad \frac{\Gamma, F[a/x]}{\Gamma}\,(\forall_1)$$

where in the rule $(\vee)$, $A_0 \vee A_1$ is the *main formula* of $(\vee)$ and is in the *lower sequent* Γ. The formula A_i $(i = 0, 1)$ in the *upper sequent* is the *minor formula* of the $(\vee)$. In the rule $(\wedge)$, $A_0 \wedge A_1$ is the main formula, and is in the lower sequent Γ. Formulas $A_i, (i = 0, 1)$ in the upper sequents are the minor formulas of the $(\wedge)$. In the rule $(\exists_1)$, $\exists x\, F$ is the main formula of $(\exists_1)$ and is in the lower sequent Γ. The formula $F[t/x]$ in the upper sequent is the minor formula of the $(\exists_1)$. In the rule $(\forall_1)$, $\forall x\, F$ is the main formula, and is in the lower sequent Γ. The $F[a/x]$ in the upper sequent is the minor formula of the $(\forall_1)$, and the free variable a is the *eigenvariable* of the $(\forall_1)$, which does not occur in the lower sequent Γ.

$$\frac{\Gamma, \neg C \quad C, \Delta}{\Gamma, \Delta}\,(cut)$$

C is the *cut formula* of the (cut).

$$\frac{\Gamma, F[T/X]}{\Gamma}\,(\exists_2) \qquad \frac{\Gamma, F[U/X]}{\Gamma}\,(\forall_2)$$

where in the rule $(\exists_2)$, $\exists X\, F$ is the main formula of $(\exists_2)$ and is in the lower sequent Γ, and T is either a variable or a relation symbol. The formula $F[T/X]$ in the upper sequent is the minor formula of the $(\exists_2)$. In the rule $(\forall_2)$, $\forall X\, F$ is the main formula, and is in the lower sequent Γ. The $F[U/X]$ in the upper sequent is the minor formula of the $(\forall_2)$, and the free variable U is the *eigenvariable* of the $(\forall_2)$, which does not occur in the lower sequent Γ.

$$\frac{\Gamma, F[A/X]}{\Gamma}\,(BI)$$

where $\exists X F$ is the main formula of the (BI) and is in the lower sequent Γ, $F[A/X]$ is the minor formula, and either

$(BI)_1$ $\exists X F \in \Pi^1_2 \cap \Sigma^1_2$, or
$(BI)_2$ $A \in \Pi^1_2 \cap \Sigma^1_2$.

This means that when $\exists X\, F \in \Pi^1_2 \cap \Sigma^1_2$, A is an arbitrary formula, while $\exists X\, F$ is arbitrary if $A \in \Pi^1_2 \cap \Sigma^1_2$.

$$\frac{\Gamma, A \subset B}{\Gamma}\,(\Pi^1_2\text{-Sep})$$

where $\exists X(A \subset X \subset B)$ is the main formula of the $(\Pi^1_2\text{-Sep})$, and is in the lower sequent Γ. $A \subset B$ is the minor formula, $A \in \Pi^1_2$ and $B \in \Sigma^1_2$.

Our aim of this paper is to show the following theorem.

Theorem 15.6. Cut-elimination theorem for SBL.

There is a rewriting procedure r on derivations in SBL *such that for any* SBL-*derivation P of a sequent, $r(P)$ is an* SBL-*derivation of the same sequent, and there is an n such that its n-th iterate $r^{(n)}(P)$ is cut-free.*

Let us first reduce the problem of cut-eliminability to derivations ending with first-order sequents.

Definition 15.7. 1. A formula is said to be *first-order* if no second-order quantifier occurs in it.
2. A sequent is *first-order* if every formula in it is first-order.
3. We say that the *cut-elimination theorem holds for derivations in* SBL *ending with first-order sequents* if there is a rewriting procedure r on derivations of first-order sequents for which Theorem 15.6 holds.

Proposition 15.8. *If the cut-elimination theorem holds for derivations in* SBL *ending with first-order sequents, then the cut-elimination theorem 15.6 holds for* SBL.

Proof. This is seen by cut-elimination by absorption combined with the joker translation due to P. Päppinghaus [15]. Note that

$$\frac{\Gamma, \neg C \quad \Gamma, C}{\Gamma, \neg B_0} \qquad \frac{\Gamma, F(\{x : B_0\})}{\Gamma, \exists Y F(Y)}$$

are admissible rules in the presence of the inference rules $(BI)_1$ and $(BI)_2$ for $B_0 \equiv (\forall X \forall y[X(y) \to X(y)]) \in \Pi^1_2 \cap \Sigma^1_2$, cf. Theorem 1.3 and Lemma 1.5(ii) in [15], resp. □

Remark 15.9. Let SBL_1 denote temporarily the calculus SBL without the rule $(BI)_2$. Namely in SBL_1, the rule (BI) is restricted to the case when the main formula $\exists X F \in \Pi^1_2 \cap \Sigma^1_2$. SBL_1 is equivalent to SBL with respect to derivable sequents since $\exists X \forall y[X(y) \leftrightarrow A(y)] \in \Pi^1_2 \cap \Sigma^1_2$ for $A \in \Pi^1_2 \cap \Sigma^1_2$. The reason why we introduce the superfluous $(BI)_2$ in SBL is as follows: when we operate our cut-elimination procedure to an SBL_1-derivation, then we obtain a cut-free SBL-derivation with rules $(BI)_2$ since we need the inference 'infer $\exists Y F(Y)$ from $F(\{x : B_0\})$' in replacing the joker J_0 by B_0 in Lemma 1.5(ii) of [15]. It is open for us whether or not the 'inner' cut-elimination theorem for SBL_1 holds besides its intrinsic interests.

15.4 The Stratified Logic Calculus SBL′

A stratified (in German: geschichtet) calculus SBL′ is introduced.

Definition 15.10. The stratified language $\mathcal{L}'$ is obtained from the second-order language $\mathcal{L}$ by modifying relation symbols and second-order variables as follows.

1. (unary) relation symbols with *index* 0: R^0.

2. (unary second-order) *unstratified bound* variables: $X, Y, \ldots$.
3. *stratified* variables :

   a. free variables with *index* $\alpha \leq I$: for each $\alpha \in \mathcal{H}_{\varepsilon_{I+1}}(0) \cap (I+1)$ and each free variable U in $\mathcal{L}$, U^α is a stratified free variable with index α.
   b. bound variables with *index* η : for each ordinal $\eta \in Fx$ and each bound variable X in $\mathcal{L}$, X^η is a stratified bound variable with index η.

When T denotes a relation symbol R, $T^\alpha :\equiv R^0$, i.e., $\alpha = 0$.

Definition 15.11. An $\mathcal{L}'$-formula A' is obtained from an $\mathcal{L}$-formula A by attaching indices as follows.

1. Attach the index 0 to each relation symbol R in A.
2. Attach an index $\alpha \leq I$ to every occurrence of each free variable U in A. Every occurrence of the same variable U receives the same index, while the indices may depend on free variables.
3. Attach an index $\eta \in Fx$ to *all undistinguished* quantifiers. In a formula each undistinguished quantifier receives the same index. Also leave distinguished quantifiers without indices.

$A^{\mathcal{L}}$ denotes the (unstratified) $\mathcal{L}$-formula obtained from an $\mathcal{L}'$-formula A by erasing all indices. Conversely A' denotes ambiguously an $\mathcal{L}'$-formula obtained from an $\mathcal{L}$-formula A by attaching some indices.

$A' \in \Pi^1_2[\Sigma^1_2]$ iff $A \in \Pi^1_2[\Sigma^1_2]$, resp. $\forall^\eta, \exists^\eta$ denote stratified quantifiers $\forall X^\eta, \exists X^\eta$ for a bound variable X.

Definition 15.12. $Gr(A), gr(A) < \omega$ for $\mathcal{L}'$-formulas A.

1. a. $Gr(A) = 0$ if neither $\forall^I$ nor $\exists^I$ occurs in A.
   In what follows assume that either $\forall^I$ or $\exists^I$ occurs in A.
   b. $Gr(A) = \max\{Gr(A_0), Gr(A_1)\} + 1$ if $A \in \{A_0 \vee A_1, A_0 \wedge A_1\}$.
   c. $Gr(A) = Gr(B) + 1$ if $A \in \{\forall x\, B[x/u], \exists x\, B[x/u]\}$.
   d. $Gr(A) = 1$ if $A \in \{\forall X F, \exists X F\}$.
   e. $Gr(A) = \max\{2, Gr(F[R^0/X]) + 1\}$ if $A \in \{\forall X^I F, \exists X^I F\}$.
2. a. $gr(A) = 0$ if A is either a literal or of the form QXF.
   b. $gr(A_0 \vee A_1) = gr(A_0 \wedge A_1) = \max\{gr(A_0), gr(A_1)\} + 1$.
   c. $gr(\exists x\, B[x/u]) = gr(\forall x\, B[x/u]) = gr(B) + 1$.
   d. $gr(QX^\eta F) = gr(F[R^0/X]) + 1$.

Definition 15.13. Let A be an $\mathcal{L}'$-formula.

1. $A \in \Sigma^I$ (in Schütte's terminology 'A ist klein') if $\forall^I$ does not occur in A.
2. A is said to be *stratified* if $\alpha < I$ holds for all indices α of free variables U^α occurring in A.

Note that each $\mathcal{L}'$-formula of the form $\exists X\, F$ is in the class Σ^I since the outermost quantifier $\exists X$ is distinguished and $F \in \Sigma^1_2$, i.e., no undistinguished universal quantifier $\forall Y^\eta$ occurs in F.

Definition 15.14. For $\mathcal{L}'$-formulas A, the ordinals $st_{\Pi}(A) < I + \omega$ if $A \in \Pi^1_2$, and ordinals $st_{\Sigma}(A) < I + \omega$ if $A \in \Sigma^1_2$ are defined. Let $\Lambda \in \{\Pi, \Sigma\}$. $st_{\Lambda}(T^{\alpha}t) = st_{\Lambda}(\neg T^{\alpha}t) = \alpha$. $st_{\Lambda}(A_0 \circ A_1) = \max(st_{\Lambda}(A_0), st_{\Lambda}(A_1))$ for $\circ \in \{\wedge, \vee\}$. $st_{\Lambda}(Qx\, B[x/u]) = st_{\Lambda}(B)$ for $Q \in \{\forall, \exists\}$.

In what follows let $F_0 \equiv F[R^0/X]$.

$$st_{\Pi}(\forall XF) = st_{\Pi}(F_0) \;\; ; \;\; st_{\Sigma}(\forall XF) = st_{\Pi}(F_0) + 1,$$

$$st_{\Sigma}(\exists XF) = st_{\Sigma}(F_0) \;\; ; \;\; st_{\Pi}(\exists XF) = st_{\Sigma}(F_0) + 1. \tag{15.1}$$

$$st_{\Pi}(\forall X^{\eta}F) = \max(\eta, st_{\Pi}(F_0)),$$

$$st_{\Sigma}(\exists X^{\eta}F) = \max(\eta, st_{\Sigma}(F_0)). \tag{15.2}$$

Proposition 15.15. Let A be an $\mathcal{L}'$-formula such that $A \in \Pi^1_2$. $I(A)$ denotes the set of indices α for which a free variable U^{α} occurs in A.

1. $st_{\Sigma}(\neg A) = st_{\Pi}(A)$.
2. Let η denote the index of an undistinguished quantifier in A if such a quantifier occurs. Otherwise let $\eta = 0$. Also let $\nu = \max I(A)$. Then there is a $k < \omega$ such that $st_{\Pi}(A) = \max\{\eta, \nu + k\}$, cf. (15.2).
3. Let $A \equiv (\forall XF)$, and U a variable not occurring in A. Then $st_{\Pi}(F[U^{\alpha}/X]) = st_{\Pi}(\forall XF)$ if $\alpha = st_{\Pi}(\forall XF) < I$.

Next we define axioms and inference rules in SBL$'$. $\Gamma, \neg A, A$ is defined to be an axiom, not just for literals A, cf. *Case 4* in section 15.5.

Definition 15.16. Axioms and inference rules in SBL$'$.

$$(Ax) \qquad \Gamma, \neg A, A \text{ with } Gr(A) = 0.$$

Inference rules $(\wedge), (\vee), (\forall_1), (\exists_1), (cut)$ are the same as in SBL.

1. $$\frac{\Gamma}{\Gamma, \Delta}\ (th)$$
   is the *thinning*.
2. *Critical rule.*
   $$\frac{\Gamma, \exists X^{\eta}F, F(T^{\alpha})}{\Gamma, \exists X^{\eta}F}\ (c)$$
   where $\eta \neq I \Rightarrow \alpha < \eta$. α is the *index*, and η the *type* of the inference.
3. *Distinguished rules.*
   a. $$\frac{\Gamma, \exists X(A \subset X \subset B), A \subset B}{\Gamma, \exists X(A \subset X \subset B)}\ (d1)$$
   where $Gr(\exists X(A \subset X \subset B)) \neq 0$.

b.

$$\frac{\Gamma, \exists XF, F(T^{\alpha})}{\Gamma, \exists XF}\ (d2)$$

where $Gr(\exists XF) \neq 0$.

4. *Bar-Induction.*

$$\frac{\Gamma, \exists XF, F(A)}{\Gamma, \exists XF}\ (BI)$$

where

a. $st_{\Sigma}(\exists XF) \geq I$ only when $\exists XF \in \Pi^1_2$, and
b. $Gr(\exists XF) = 0$.

5. *Strong rules.*

a.

$$\frac{\Gamma, \forall XF, F(U^I)}{\Gamma, \forall XF}\ (s1)$$

where $Gr(\forall XF) \neq 0$, U does not occur in the lower sequent. $(s1)$ is of *type* I.

b.

$$\frac{\Gamma, \forall X^{\eta}F, F(U^{\eta})}{\Gamma, \forall X^{\eta}F}\ (s2)$$

where U does not occur in the lower sequent. $(s2)$ is of *type* η.

6. *Weak rule.*

$$\frac{\Gamma, \forall XF, F(U^{\alpha})}{\Gamma, \forall XF}\ (w)$$

where

a. $Gr(\forall XF) = 0$.
b. $\alpha = st_{\Pi}(\forall XF) < I$.
c. U does not occur in the lower sequent.

α is the *index* of the (w).

7. *Substitution of level* α.

$$\frac{\Gamma}{\Gamma[A/U^{\alpha}]}\ (sub)^{\alpha}$$

where

a. $\alpha < I$.
b. A is a stratified formula such that $C[A/U^{\alpha}]$ is an $\mathcal{L}'$-formula for each $C \in \Gamma$.
c. The *eigenvariable* U does not occur in the lower sequent.
d. Each $C \in \Gamma$ enjoys the following.
   i. C is a Π^1_2-formula such that $st_{\Pi}(C) \leq \alpha$.
   ii. $U \notin VT(C^{\mathcal{L}})$.

8. $\forall^I$*-reduction of type* $\eta < I$.

$$\frac{\Delta_0, \Delta_1}{\Delta_0[\forall^\eta/\forall^I], \Delta_1}\ (\forall^I\text{-red})^\eta$$

where each $A \in \Delta_0 \cup \Delta_1$ is either $A \in \Sigma^I$ or $\neg A \in \Sigma^I$, and $A[\forall^\eta/\forall^I]$ denotes the $\mathcal{L}'$-formula obtained from A by replacing $\forall X^I$ by $\forall X^\eta$.

9. $\exists^I$*-reduction of type* $\eta < I$.

$$\frac{\Delta_0, \Delta_1}{\Delta_0[\exists^\eta/\exists^I], \Delta_1}\ (\exists^I\text{-red})^\eta$$

where $\Delta_0 \cup \Delta_1 \subset \Sigma^I$ and $A[\exists^\eta/\exists^I]$ denotes the $\mathcal{L}'$-formula obtained from A by replacing $\exists X^I$ by $\exists X^\eta$.

Definition 15.17. A *preproof* is a finite tree with axioms (Ax) and inference rules in SBL′.

A preproof P enjoys the *pure variable condition* if all eigenvariables are distinct from each other, each eigenvariable does not occur in the end-sequent of P and if a free variable occurs in an upper sequent of a rule, but not in the lower sequent, then the variable is the eigenvariable of the rule.

Note 15.18. Let us give some comments on Definition 15.16.

Inference rules without main formulas are (cut), (th), (sub), $(\forall^I\text{-red})$ and $(\exists^I\text{-red})$.

In Definition 15.16.3a of the inference $(d1)$, the existential quantifier $\exists X$ in $\exists X(A \subset X \subset B)$ is distinguished, and hence $(\exists X(A \subset X \subset B)) \in \Sigma^1_2$. In particular $A \in \Pi^1_2$ and $B \in \Sigma^1_2$. Moreover $Gr(\exists X(A \subset X \subset B)) \neq 0$ iff $Gr(\exists X(A \subset X \subset B)) = 1$.

A distinguished existential quantifier can be introduced by a distinguished rule (d) or by a (BI) possibly with a complex witness; infer $\exists X F(X)$ from $F(A)$. On the other side we can infer undistinguished existential quantifiers $\exists X^\eta F(X^\eta)$ only when we have a simple witness, i.e., either a relation symbol or a variable T^α with a smaller index $\alpha < \eta$ such that $F(T^\alpha)$.

A pair of inferences are $[(w), (BI)]$, $[(s2), (c)]$ and $[(s1), (d)]$. First consider the second pair of a critical rule (c) with a strong rule $(s2)$:

$$\frac{\dfrac{\Gamma, \neg F(U^\eta)}{\Gamma, \forall X^\eta \neg F}\ (s2) \qquad \dfrac{F(T^\alpha), \Delta}{\exists X^\eta F, \Delta}\ (c)}{\Gamma, \Delta}\ (cut)$$

Such a (cut) with the cut formula $\exists X^\eta F$ is transformed to another (cut) with the cut formula $F(T^\alpha)$ when $\alpha < \eta$. In the transformation, the eigenvariable U^η is replaced by T^α.

The first pair of a (BI) with a weak rule (w) is transformed to a substitution rule (sub) followed by a (cut) as in [22, 5], cf. *Case 10* in section 15.5. Here it is crucial for us to have following facts.

Let $\forall X \neg F(X)$ be the major formula of the weak rule. Then we see from (15.1) in Definition 15.14 and Proposition 15.15.3 that $st_{\Pi}(\exists XF) > \alpha = st_{\Pi}(\neg F(U^{\alpha}))$ for $\alpha = st_{\Pi}(\forall X \neg F(X)) < I$. We obtain from Definition 15.16.7(d)i that $\alpha < \beta$ as long as the major formula $\exists XF$ is in the upper sequent of a $(sub)^{\beta}$ of level β. Moreover we see from Definition 15.16.7(d)ii that the eigenvariable of the $(sub)^{\beta}$ does not occur in the formula $\exists XF$. In the same vein the eigenvariable of any (sub) does not occur in the formula $\forall X \neg F$ provided that the formula $\forall X \neg F$ is in the upper sequent of the (sub). If $st_{\Pi}(\neg F) < \gamma$, then the eigenvariable V^{γ} of a substitution of level γ does not occur in the formula $\neg F$. Otherwise we would have $st_{\Pi}(\neg F) \geq \gamma$.

Conversely the transformation with the substitution rules does work as far as formulas are graded in such a way that the above facts hold. Specifically if a variable V^{γ} with index γ occurs in a formula F, then F is graded not lower than γ, $st_{\Lambda}(F) \geq \gamma$ for $\Lambda \in \{\Pi, \Sigma\}$. Given that $Gr(\forall X \neg F) = 0$ and $\alpha = st_{\Pi}(\forall X \neg F) < st_{\Pi}(\exists XF)$ for the major formula $\forall X \neg F$ of weak rules (w), we obtain $st_{\Lambda}(G) \geq \alpha \geq \max\{\gamma, \eta\}$ when the eigenvariable U^{α} of the rule (w) occurs in a formula for G, a variable V^{γ} occurs in F, and an undistinguished universal quantifier $\forall Y^{\eta}$ occurs in F.

$$
\dfrac{\dfrac{\Delta_0, \forall X \neg F, \neg F(U^{\alpha})}{\Delta_0, \forall X \neg F}\,(w) \quad \vdots \quad \Delta, \forall X \neg F
\qquad
\dfrac{F(A), \exists XF, \Gamma_0}{\exists XF, \Gamma_0}\,(BI) \quad \vdots \quad \dfrac{\exists XF, \Gamma_1}{\exists XF, \Gamma_1'}\,(sub)^{\beta} \quad \vdots \quad \exists XF, \Gamma}{\Delta, \Gamma}
\quad \vdots \quad \dfrac{\Pi}{\Pi'}\,(sub)^{\gamma}
$$

Finally consider the third pair of a distinguished rule (d) with a strong rule $(s1)$.

$$
\dfrac{\dfrac{\Gamma, F(U^{I})}{\Gamma, \forall X \neg F}\,(s1) \qquad \dfrac{G, \Delta}{\exists XF, \Delta}\,(d)}{\Gamma, \Delta}\,(cut)
$$

where $Gr(\exists XF) \neq 0$, and G is the minor formula such that either $G \equiv (A \subset B)$ and $F \equiv (A \subset X \subset B)$, or $G \equiv F(T^{\alpha_0})$. From $Gr(\exists XF) \neq 0$ we see that there occurs an undistinguished existential quantifier $\exists Y^{I}$ in the Σ^1_2-formula G. The fact that the undistinguished quantifier receives the ordinal I means that the quantifier has not yet been stratified, i.e., not yet been graded. The grading is done through the reduction rules $(\exists^{I}\text{-red})^{\eta}$ and $(\forall^{I}\text{-red})^{\eta}$ for an ordinal η, cf. *Case 9* in section 15.5. And then the rules (d) and $(\exists^{I}\text{-red})^{\eta}$ are interchanged to a combination of a rule $(\exists^{I}\text{-red})^{\eta}$ and a rule (BI), cf. *Cases 4* and *5*. Moreover the rules $(s1)$ and $(\forall^{I}\text{-red})^{\eta}$ are interchanged to a combination of a rule $(\forall^{I}\text{-red})^{\eta}$ and a rule (w), cf. *Cases 7*. This results in a pair $[(w), (BI)]$ of the first kind.

$$\dfrac{\dfrac{\dfrac{\Gamma, \neg F(U^{\alpha'})}{\Gamma, \neg F'(U^{\alpha'})}\ (\forall^I\text{-red})^\eta}{\Gamma, \forall X \neg F'}\ (w) \qquad \dfrac{\begin{array}{c}\vdots\ P_0\\ A' \subset A'\end{array} \quad \dfrac{G, \Delta}{G', \Delta}\ (\exists^I\text{-red})^\eta}{\exists X F', \Delta}\ (BI)}{\Gamma, \Delta}$$

where the subproof P_0 of $A' \subset A'$ is appended only when $G \equiv (A \subset B)$ with $G' \equiv (A' \subset B')$.

The type $\eta < I$ of rules $(\exists^I\text{-red})^\eta$ (and of $(\forall^I\text{-red})^\eta$) is chosen so that $\alpha < \eta$ for every index α which may occur above the rule $(\exists^I\text{-red})^\eta$, cf. (15.5) in Definition 15.24. We can assume that the undistinguished existential quantifier $\exists Y^I$ comes from the major formula $\exists Y^I H$ of a critical rule (c). Let us assume that $H \in \Sigma^1_2 \cap \Pi^1_2$, and $H(T^\alpha)$ be its minor formula. The critical rule (c) is interchanged with the rule $(\exists^I\text{-red})^\eta$, cf. *Case 6* in section 15.5. This results in another critical rule (c) with major formula $\exists Y^\eta H$ and the minor formula $H(T^\alpha)$. Hence we need the fact $\alpha < \eta$.

Here we consider the indices not only the existing ones when the rule $(\exists^I\text{-red})^\eta$ is introduced, but also ones which may be created in the future. Therefore we need to collapse ordinals to get the type η as in the Pohlers' local predicativity method in [14]. In the method it is crucial for the minor formula G of the rules (d) to be in Σ^I. Then each formula in the upper sequents of rules $(\exists^I\text{-red})$ can be restricted to be in Σ^I, cf. *Case 7*. Consider the index $\alpha' = st_\Pi(\forall X \neg F')$ of the eigenvariable $U^{\alpha'}$ of the weak rule (w), where the formula $\neg F'$ is obtained from $\neg F$ by replacing the undistinguished universal quantifiers $\forall Y^I$ by $\forall Y^\eta$. From Proposition 15.15.2 we see that either $\alpha' = \eta$ or $\alpha' = \beta + k$ for an index β of a free variable occurring in F and a $k < \omega$. Since the transformation from the pair $[(s1), (d)]$ to $[(w), (BI)]$ could be iterated, there may be another rule $(\exists^I\text{-red})^\nu$ of type ν below Γ, Δ. Thus the new index η should be chosen so that $\eta < \nu$, cf. (15.6) in Definition 15.24.

Definition 15.19. 1. The *degree* $\text{dg}(A) < \omega + \omega$.

$$\text{dg}(A) = \begin{cases} gr(A) & \text{if } Gr(A) = 0, \\ \omega + (Gr(A) - 1) & \text{otherwise.} \end{cases}$$

2. The *height* $h(\Gamma) = h(\Gamma; P)$ of a sequent Γ in a preproof P is defined by bottom-up recursion as follows.

   a. $h(\Gamma) = 0$ if Γ is either the end-sequent of P or an upper sequent of a (sub).
   b. $h(\Gamma) = \omega$ if Γ is an upper sequent of a $(Q^I\text{-red})$.
      In what follows assume that Γ is an upper sequent of a rule J other than $(sub), (Q^I\text{-red})$ with the lower sequent Δ.
   c. $h(\Gamma) = \max\{h(\Delta), \text{dg}(A)\}$ if J is either a (cut) with the cut formula A, or a (BI) with the minor formula A.
   d. $h(\Gamma) = h(\Delta)$ in other cases.

Observe that the degrees $\text{dg}(A)$ and the Grades $Gr(A)$ in Definition 15.12 are defined non-monotonically. This means that $\text{dg}(B) > \text{dg}(A)$ can happen for a subformula B of a formula A. For example let $\exists Y^I B$ be a subformula of a formula $\exists X F$. Then $Gr(\exists X F) = 1$, while $Gr(\exists Y^I B) = 2$, and $\text{dg}(\exists X F) = \omega$ with $\text{dg}(\exists Y^I B) = \omega + 1$.

This causes no problem in transformations since the (cut) with the cut formula $\exists XF$ is transformed to a (cut) with a cut formula $\exists XF'$ with $Gr(\exists XF') = 0$ through a rule ($\exists^I$-red), cf. *Case 9* in in section 15.5.

Relations between occurrences A, B of formulas in a preproof such as 'A is a *descendant* of B' or equivalently 'B is an *ancestor* of A', and 'an occurrence of inference rule is *implicit* or *explicit*' are defined as in [22, 2]. Any of inferences $(cut), (th), (sub), (\forall^I\text{-red})$ and $(\exists^I\text{-red})$ is neither implicit nor explicit since these have no main formulas.

Definition 15.20. Let P be a preproof.

1. Let Δ be a sequent in P.
   a. Δ is in the *explicit part* of P if every rule below Δ is either an explicit rule or a thinning (th), and Δ is either an (Ax) or a lower sequent of an explicit rule or a (th).
   b. Δ is a *bar sequent* of P if Δ is not in the explicit part of P, and either Δ is the end-sequent or an upper sequent of an explicit rule or a (th) whose lower sequent is in the explicit part of P.
2. Let Δ_0 be a bar sequent of P. The *end-piece* of Δ_0 consists of the following sequents in P: Δ_0 is in the end-piece. If a lower sequent of a rule other than implicit rule is in the end-piece, then its upper sequents are in the end-piece.
3. An implicit rule is a *boundary rule* if its lower sequent is in the end-piece of a bar sequent of P.
4. A triple (J_1, J_2, J) of rules in P is a *suitable triangle* if J_i is a boundary rule with its main formula C_i for $i = 1, 2$, and J is a (cut):

$$\frac{\Gamma, \neg C \quad C, \Delta}{\Gamma, \Delta}\ (cut)\ J$$

   where $\neg C$ is a descendant of C_1, C is a descendant of C_2 and $\{\neg C, C\} \cap (\Gamma \cup \Delta) = \emptyset$. C is said to be a *suitable cut formula*.

Proposition 15.21. *For a preproof P, P contains no bar sequent iff P consists solely of explicit rules and (th)'s.*

Definition 15.22. Let P be a preproof enjoying the pure variable condition. A *stack function stk* for P assigns an ordinal $stk(J) \in OT$ (the *stack* of J) to each occurrence J of rules ($\exists^I$-red) and (sub) in P.

Given a stack function stk, we assign ordinals $o(\Delta) = o(\Delta; P, stk), o(J) = o(J; P, stk)$ to each sequent Δ and each line of a rule J by top-down recursion as follows.

1. $o(\Delta) = 1$ if Δ is an (Ax).
   In what follows let Δ be the lower sequent of a rule J with upper sequents Γ (and Γ').
2. $o(J) = o(\Gamma)$ if J is one of the rules (Q-red), $(sub), (th)$.

3. $o(J) = o(\Gamma) + 1$ if J is one of the rules $(\vee), (\forall_1), (\exists_1), (c), (d), (s)$ and (w).
4. $o(J) = o(\Gamma)\#o(\Gamma')$ if J is either a $(\wedge)$ or a (cut).
5. Let J be a (BI) with its main formula $\exists XF$.

$$o(J) = \begin{cases} \Omega_{\alpha+1}\#o(\Gamma) & \text{if } \alpha = st_\Sigma(\exists XF) < I, \\ I\#o(\Gamma) & \text{otherwise.} \end{cases}$$

6. If J is a $(sub)^\alpha$ of level α with $\gamma = stk(J)$ and $\beta = o(J) = o(\Gamma)$, then

$$o(\Delta) = \psi_{\Omega_{\alpha+1}}(\gamma\#\omega^\beta).$$

7. If J is an $(\exists^I\text{-red})$ with $\gamma = stk(J)$ and $\beta = o(J) = o(\Gamma)$, then

$$o(\Delta) = \begin{cases} \psi_I(\gamma\#\omega^\beta) & \text{if } h(\Delta) < \omega, \\ \beta & \text{otherwise.} \end{cases}$$

8. Let J be a rule other than $(sub), (\exists^I\text{-red})$.
   a. $o(\Delta) = \omega_m(o(J))$ where $h(\Gamma) = h(\Delta) + m$ for an $m < \omega$.
   b. $o(\Delta) = 0$ if $h(\Delta) < \omega \le h(\Gamma)$, cf. the condition in Definition 15.24.2.

   For ordinals α and $m < \omega$, $\omega_0(\alpha) = \alpha$ and $\omega_{m+1}(\alpha) = \omega^{\omega_m(\alpha)}$.

   Finally $o(P) = o(P, stk) = o(\Gamma_{end}; P, stk)$ for the end sequent Γ_{end} of P.

Note that we have $Gr(\exists XF) = 0$ for the main formula $\exists XF$ of a (BI), and $st_\Sigma(\exists XF) \ge I$ only when $\exists XF \in \Pi^1_2$, cf. Definition 15.16.4a, where $st_\Sigma(\exists XF) \ge I$ iff $\exists XF \in \Pi^1_2$ and a free variable U^I occurs in the formula $\exists XF$. Suppose that the rule (BI) occurs in a preproof with the pure variable condition and the condition (15.3) in Proposition 15.23 is fulfilled for the preproof. Then the free variable U^I is the eigenvariable of an $(s2)$ with type I, which occurs below the (BI).

$$\dfrac{}{\Delta, \exists XF(U^I)}\ (BI)$$
$$\vdots$$
$$\dfrac{\Gamma, G(U^I)}{\Gamma, \forall Y^I G}\ (s2)$$

Proposition 15.23. Let P be a preproof enjoying the pure variable condition. Assume that P satisfies the following condition:

(15.3) The end-sequent of P is a first-order sequent Γ_{end} such that any $A \in \Gamma_{end}$ is stratified and $st_\Pi(A) = 0$.

Let Δ be a sequent in P.

1. If $h(\Delta; P) < \omega$, then $\text{dg}(A) < \omega$, i.e., $Gr(A) = 0$ for any $A \in \Delta$.
2. Let U^α be a variable occurring in Δ. Assume that there occurs no strong rule below the sequent Δ.Then $\alpha < I$, and each formula in Δ is stratified.
3. For any upper sequent Δ of a (sub), $o(\Delta) < I$.

Proof. Proposition 15.23.1 is seen inductively from below to above. If Δ is an upper sequent of a $(sub)^\alpha$ of level $\alpha < I$ and $A \in \Delta$, then $st_\Pi(A) \leq \alpha$, and hence $Gr(A) = 0$.

Proposition 15.23.2 is shown inductively from below to above. □

Definition 15.24. Let P be a preproof enjoying the pure variable condition and stk a stack function for P. P together with stk is said to be a *proof* (in SBL′) if the following conditions are satisfied:

1. The end-sequent of P is a first-order sequent Γ_{end} such that $st_\Pi(A) = 0$ for any $A \in \Gamma_{end}$, cf. (15.3).
2. Let J be a rule with its lower sequent Δ and an upper sequent Γ such that $h(\Delta) < \omega \leq h(\Gamma)$. Then the rule J is a vacuous ($\exists^I$-red).
   For the lower sequents Δ of rules (Q^I-red) ($Q \in \{\exists, \forall\}$), $h(\Delta) \leq \omega$ holds.
3. Let J be either an ($\exists^I$-red) or a $(sub)^\mu$, Δ the upper sequent of J, $\alpha = o(\Delta; P, stk)$ and $\gamma = stk(J)$, the stack of J with respect to the stack function stk. Let $\sigma = I$ when J is an ($\exists^I$-red), and $\sigma = \Omega_{\mu+1}$ when it is a rule $(sub)^\mu$. Then
$$\{\gamma, \alpha\} \subset \mathcal{H}_\gamma(\psi_\sigma \gamma) \tag{15.4}$$
   and for any index ξ occurring above J
$$\xi \in \mathcal{H}_\gamma(\psi_\sigma \gamma) \tag{15.5}$$
   where by an index ξ occurring above J we mean
   a. T^ξ occurs above J, or
   b. X^ξ occurs above J, or
   c. there is an ($\exists^I$-red)$^\xi$ rule occurring above J, or
   d. there is an ($\exists^I$-red) rule J_0 occurring above J such that $\xi = \psi_I(\gamma_0 \# \omega^{\alpha_0})$ with $\gamma_0 = stk(J_0)$ and $\alpha_0 = o(J_0; P, stk)$.
4. Let J be an ($\exists^I$-red) of type η with the stack $\gamma = stk(J)$, Δ the upper sequent of J with $\alpha = o(\Delta; P, stk)$. Then
$$\eta \geq \psi_I(\gamma \# \omega^\alpha). \tag{15.6}$$
5. a. Every (sub) is in an end-piece of a bar sequent.
   b. The eigenvariable of a (sub) does not occur in any explicit formula in the upper sequent of the (sub).
   c. Each bar sequent Γ is the lower sequent of a vacuous $(sub)^0$ of level 0. The vacuous (sub) is of the form
$$\frac{\Gamma}{\Gamma}\ (sub)^0$$
   with an eigenvariable U^0 not occurring in Γ.

Note 15.25. In a proof, both of the ordinal term $\delta = \psi_{\Omega_{\alpha+1}}(\gamma \# \omega^\beta)$ attached to the lower sequent of a rule $(sub)^\alpha$, and the ordinal term $\delta = \psi_I(\gamma \# \omega^\beta)$ attached to the lower sequent of a rule ($\exists^I$-red) are seen to be in OT as follows. Let $\sigma = \Omega_{\alpha+1}$ when the rule is a $(sub)^\alpha$, and $\sigma = I$ when the rule is an $(\exists^I\text{-red})$. The fact $G_\delta(\gamma,\beta,\sigma) < \gamma\#\omega^\beta$, i.e., $\{\sigma,\gamma\#\omega^\beta\} \subset \mathcal{H}_{\gamma\#\omega^\beta}(\delta)$ follows from $\{\sigma,\gamma,\beta\} \subset \mathcal{H}_\gamma(\psi_\sigma\gamma)$, cf. $\alpha \leq \Omega_\alpha < \psi_\sigma\gamma$ and (15.4) in Definition 15.24. Similarly we see that both of the ordinal term $\xi = \psi_I(\gamma_0\#\omega^{\alpha_0})$ in Definition 15.24.3d, and the ordinal term $\psi_I(\gamma\#\omega^\alpha)$ in (15.6) are in OT. Therefore for any proof P, $o(P) \in \mathcal{H}_{\varepsilon_{I+1}}(0) \cap \Omega_1 = OT \cap \Omega_1$ by vacuous $(sub)^0$'s in Definition 15.24.5c. Roughly a proof is depicted as follows.

$$\begin{array}{l} \vdots \\ \dfrac{\Delta_0}{\Delta}\ \text{boundary } J \\ \vdots\ (th),(sub),(cut),(red),\text{explicit rules} \\ \dfrac{\Gamma}{\text{bar }\Gamma}\ (sub)^0 \\ \vdots\ (th),\text{explicit rules} \\ \text{1st order end-sequent} \end{array}$$

where each sequent from a bar sequent Γ to the lower sequent Δ of a boundary rule is in the end-piece. A boundary rule J is an implicit rule which we encounter first on a path from the end-sequent to an axiom.

The stack function stk is inspired from the ordinals γ in the assumption $\mathcal{A}(\Theta;\gamma,\kappa,\mu)$ of Theorem 4.8(Collapsing and impredicative cutelimination) in [6]. Roughly a stack γ of a rule is a basis for bounding witnesses, i.e., indices, cf. (15.4) and (15.5) in Definition 15.24. For the rules $(\exists^I\text{-red})^\eta$, we intend to bound witnesses by the type η, cf. (15.6). The type remains the same, and the stack is unchanged in almost cases. Initially each stack of rules $(\exists^I\text{-red})$ is defined to be 0, and each stack of rules (sub) is defined to be higher than the ordinals attached to rules $(\exists^I\text{-red})$, cf. Proposition 15.26.1. Let J be an $(\exists^I\text{-red})$. A stack of the rule J is piled up when a transformation introduces a new rule $(\exists^I\text{-red})$ in *Case 9*, and the same happens to substitution rules (sub) in *Case 10*. The stack of the new rules is defined to be the one of J, whereas the stack of J is increased to keep the condition (15.5). When we transform proofs, the attached ordinals decrease. To keep the conditions (15.4), (15.5), (15.6), and $o(P) \in OT$, the stack γ and the attached ordinal α are separated each other, and the additively closed ordinal ω^α is introduced, cf. (15.8) in *Case 9*.

For a first-order sequent Γ in the language $\mathcal{L}$, let Γ^0 denote the sequent in $\mathcal{L}'$ obtained from Γ by attaching the index 0 to every second-order free variable and relation symbol occurring in Γ.

Proposition 15.26. Let Γ be a first-order sequent in $\mathcal{L}$.

1. If Γ is derivable in SBL, then so is Γ^0 in SBL′.
2. If there is a proof in SBL′ ending with Γ^0 and containing no bar sequent, then Γ is (cut-free) derivable in the first-order sequent calculus LK.

Proof. Proposition 15.26.1. Let P be an SBL-derivation of the first-order sequent Γ. We can assume that P enjoys the pure variable condition, P contains no rule $(BI)_2$, cf. Remark 15.9, and any main formula $\exists X(A \subset X \subset B)$ of a $(\Pi^1_2\text{-Sep})$ is not Π^1_2, for otherwise $\exists X(A \subset X \subset B) \in \Pi^1_2 \cap \Sigma^1_2$, and it is derivable by the rule $(BI)_1$ and $A \subset B$, i.e., from, e.g., $A \subset A \subset B$.

Then construct a proof P^0 of Γ^0 from P as follows: attach the index I to every undistinguished quantifiers, attach the index 0 to every relation symbol, attach suitable indices to every second-order free variable from below to above. In particular attach the indices $\alpha = st_\Pi(\forall XF) \in \{0, I\}$ to the eigenvariables of weak rules (w), and the index I to the eigenvariables of strong rules (s). Clearly the condition (15.3) in Definition 15.24.1 is enjoyed, and $st_\Pi(A) = 0$ for any A in the end-sequent Γ^0, which is first-order.

In the resulting preproof P_0, insert vacuous $(\exists^I\text{-red})$ immediately below a (cut) such that $h(\Delta_0, \Delta_1; P_0) < \omega \leq h(\Delta_0, \neg C; P_0)$ for the lower sequent Δ_0, Δ_1 and an upper sequent $\Delta_0, \neg C$ of the (cut). Namely change

$$\frac{\Delta_0, \neg C \quad C, \Delta_1}{\Delta_0, \Delta_1}\ (cut)$$

to

$$\cfrac{\cfrac{\Delta_0, \neg C \quad C, \Delta_1}{\Delta_0, \Delta_1}\ (cut)}{\Delta_0, \Delta_1}\ (\exists^I\text{-red})$$

Then the condition in Definition 15.24.2 is fulfilled. Note that $\mathrm{dg}(A) < \omega$, i.e., $Gr(A) = 0$ for any $A \in \Delta_0 \cup \Delta_1$ since the end-sequent Γ^0 is first-order. In particular no undistinguished quantifier Q^I occurs in $\Delta_0 \cup \Delta_1$.

Moreover insert vacuous $(sub)^0$ at bar sequents. Note that any formula B in any bar sequent is first-order, and hence $st_\Pi(B) = 0$.

The resulting preproof is denoted P^0. Any main formula of rules (BI) and (w) in P^0 is in $\Pi^1_2 \cap \Sigma^1_2$, and $\eta = I$ for any main formula $\exists X^\eta F$ of rules (c) in P^0.

A stack function stk^0 together with types of vacuous $(\exists^I\text{-red})$ is defined as follows. First put $stk^0(J_0) = 0$ for any $(\exists^I\text{-red})$ J_0. Then the condition (15.5) is fulfilled since any index α occurring in P^0 is in $\{n, I+n : n < \omega\}$, and there is no rules $(\exists^I\text{-red})$ nor (sub) above J_0 in P^0. Next the type η of $(\exists^I\text{-red})$ J_0 is defined to be $\delta_0 = \psi_I(0\#\alpha_0)$ for $\alpha_0 = o(J_0; P^0, stk^0)$. Obviously $\alpha_0 \in \mathcal{H}_0(0) \subset \mathcal{H}_0(\psi_I 0)$, and the normality condition (15.4) is fulfilled for J_0. Then assign ordinals up to upper sequents of $(sub)^0$ J_1. Let $\alpha_1 = o(J_1; P^0, stk^0)$, and pick an $n < \omega$ so that $\alpha_0 < \omega_n(I+1)$ for any $(\exists^I\text{-red})$ J_0 occurring above J_1 with $\alpha_0 = o(J_0; P^0, stk^0)$. Then let $stk^0(J_1) = \omega_n(I+1)$, and $\delta_1 = o(\Delta; P^0, stk^0) = \psi_{\Omega_1}(\omega_n(I+1)\#\alpha_1)$ for the lower (bar) sequent Δ of the $(sub)^0$ J_1. Then $\psi_I(0\#\alpha_0), \omega_n(I+1)\#\alpha_1 \in \mathcal{H}_{\omega_n(I+1)}(\psi_{\Omega_1}(\omega_n(I+1)))$. Hence the conditions (15.5) and (15.4) are fulfilled for J_1. Thus (P^0, stk^0) is a proof in SBL'.

Proposition 15.26.2 is seen from Proposition 15.21. Namely erase all the indices 0 from the proof of Γ^0 without bar sequent. It results in a cut-free LK-derivation of Γ. □

By Propositions 15.8 and 15.26, and the well-foundedness of $(\mathcal{H}_{\varepsilon_{I+1}}(0) \cap \Omega_1, <)$, for a proof of Theorem 15.6 it suffices to show the following.

Main Lemma 15.27. *For any proof* (P, stk) *in* SBL′ *with a bar sequent, we can construct another proof* (P', stk') *with the same end-sequent such that* $o(P', stk') < o(P, stk)$.

Main Lemma is proved in the next section 15.5.

15.5 Proof of Main Lemma 15.27

Throughout this section P together with a stack function stk denotes a proof with a bar sequent. For simplicity let us suppress stack functions in ordinals attached to sequents and rules. Namely $o(\Gamma; P, stk)$ $[o(J; P, stk)]$ is denoted by $o(\Gamma; P)$ $[o(J; P)]$, resp.

Each reduction, i.e., rewriting step is performed within the end-piece of a bar sequent Δ_b. By Definition 15.24.5c the bar sequent Δ_b is the lower sequent of a vacuous $(sub)^0$ with its stack γ,

$$\frac{\Delta_b; \alpha_0}{\Delta_b; \psi_{\Omega_1}(\gamma \# \omega^{\alpha_0})} \; (sub)^0; \alpha_0$$

where and everywhere in this section, $\Gamma; \alpha$ designates that $o(\Gamma; P) = \alpha$ for sequents Γ in P, and $J; \alpha$ that $o(J; P) = \alpha$ for rules J in P. Also we see from (15.3) in Definition 15.24.1 and the pure variable condition that each formula in an end-piece is stratified.

When P is rewritten to another P' below, a stack function stk' for P' is defined in an obvious way except otherwise stated explicitly. Namely a rule J' in P' receives the same stack as one for the corresponding rule J in P in most cases except in *Case 9* and *Case 10*. In each step we need to verify that P' is a proof and $o(P') < o(P)$. In most cases this amounts to show that P' together with a stack function stk' fulfills the conditions in (15.4), (15.5) and (15.6).

Case 1. An explicit rule is in an end-piece of a bar sequent Δ_b in P.

Let J_0 be one of the lowest explicit rule in the end-piece of Δ_b. By (15.3) in Definition 15.24.1 the end-sequent of P is a first-order sequent, and hence J_0 is one of rules $(\wedge), (\vee), (\forall_1), (\exists_1)$. Consider the case when J_0 is a rule $(\forall_1)$, and let the end-piece of P be the following:

$$\cfrac{\cfrac{\Gamma, B(a); \beta}{\Gamma; \beta + 1} \; (\forall_1)\, J_0}{\begin{array}{c}\vdots\\ \cfrac{\Delta_b; \alpha}{\Delta_b; \psi_{\Omega_1}(\gamma \# \omega^{\alpha})} \; J\end{array}}$$

where $\forall x\, B(x) \in \Gamma \cap \Delta_b$ and $\Delta_b; \alpha$ for the upper sequent Δ_b of the vacuous $(sub)^0$ J with its lower sequent Δ_b and its stack $\gamma = stk(J)$. Note that by Definition 15.24.5b no (sub) change explicit formulas, and the end-piece ends with a vacuous (sub) by Definition 15.24.5c.

Let P' be obtained from P by replacing the end-piece by the following.

$$\cfrac{\cfrac{\begin{array}{c}\Gamma, B(a); \beta \\ \vdots \\ \Delta_b, B(a); \alpha'\end{array}}{\Delta_b, B(a); \psi_{\Omega_1}(\gamma\#\omega^{\alpha'})}\ J'}{\Delta_b; \psi_{\Omega_1}(\gamma\#\omega^{\alpha'})+1}\ (\forall_1)$$

We see from $\beta \ll \beta + 1$ and Proposition 15.3 that $\alpha' \ll \alpha$. From this we see $o(P') < o(P)$. Let us verify that P' is a proof. The condition (15.6) on rules ($\exists^I$-red) in P' is fulfilled by $\beta \ll \beta + 1$. We have $\alpha \in \mathcal{H}_\gamma(\psi_{\Omega_1}\gamma)$ by (15.4) for J. Hence $\alpha' \in \mathcal{H}_\gamma(\psi_{\Omega_1}\gamma)$ for the stack $\gamma = stk'(J')$ of the vacuous $(sub)^0$ J' in P'. Similarly we see that the conditions (15.5) and (15.4) on rules ($\exists^I$-red), (sub) in P' are fulfilled. Therefore P' is a proof.

Case 2. The end-piece of a bar sequent Δ_b contains a (cut) of the following form:

$$\begin{array}{c}\cfrac{\Gamma, \neg A; \alpha \quad A, \Delta; \beta}{\Gamma, \Delta; \gamma}\ (cut); \alpha\#\beta \\ \vdots \\ \Delta_b\end{array}$$

where $\{\neg A, A\} \cap (\Gamma \cup \Delta) \neq \emptyset$. Suppose $\neg A \in \Gamma \cup \Delta$. By Proposition 15.23.1 we have $h(\Gamma, \Delta) < \omega \Rightarrow h(\Gamma, \neg A) < \omega$. In other words, $h(\Gamma, \neg A) = h(\Gamma, \Delta) + m$ for an $m < \omega$. Thus $\gamma = \omega_m(\alpha\#\beta)$. Rewrite the subproof of Δ_b as follows.

$$\begin{array}{c}\cfrac{\Gamma, \neg A; \alpha'}{\Gamma, \Delta; \alpha'}\ (th) \\ \vdots \\ \Delta_b\end{array}$$

Note that the height of an upper sequent of a (sub) is defined to be 0 in Definition 15.19.2a, and the height of an upper sequent of an ($\exists^I$-red) is equal to ω by Definition 15.19.2b. Hence there is no (sub) nor ($\exists^I$-red) in the height lowering part in P'. Thus we see that $\alpha' \underline{\ll} \omega_m(\alpha) \ll \gamma$, and P' is a proof such that $o(P') < o(P)$.

Case 3. $\{\neg A, A\} \subset \Delta_b$ for a formula A and a bar sequent Δ_b.

By (15.3) in Definition 15.24.1 A is a first-order formula, and $Gr(A) = 0$.

$$P = \begin{array}{c}\vdots \\ \Delta_b, \neg A, A; \psi_{\Omega_1}(\gamma\#\omega^\alpha) \\ \vdots\end{array} \qquad P' = \begin{array}{c}\overline{\Delta_b, \neg A, A; 1}\ (Ax) \\ \vdots\end{array}$$

By virtue of *Case 1-3* we can assume that any end-piece of P contains no explicit rule nor axiom. Then we see as in Sublemma 12.9 of [22] that each end-piece of P contains a suitable triangle as follows.

Proposition 15.28. *Suppose that an end-piece P_0 of P contains no explicit rule nor axiom. Then P_0 contains a suitable triangle.*

Proof. There exists a (cut) in P_0, one of whose cut formulas is a descendant of a major formula of a boundary rule. Let us call such a (cut) *linked*. Pick one of the lowest linked (cut) in P_0. Then its lower sequent contains no descendant of a major formula of a boundary rule. Call such a linked (cut) *exhausted*. We then see that each of the topmost exhausted (cut) in P_0 is a suitable one, i.e., each of its cut formulas is a descendant of a major formula of a boundary rule. □

Before reducing suitable triangles, let us consider the following cases. *Cases 4–6* when a descendant of the main formula of a boundary rule $J_0 = (d), (c)$ is changed by a $J_1 = (\exists^I\text{-red})$, and *Cases 7–8* when a descendant of the main formula of a boundary rule $J_0 = (s)$ is changed by a $J_1 = (\forall^I\text{-red})$. In each of these cases, J_0 and J_1 are exchanged. When $J_0 = (d)$, the distinguished rule (d) is changed to a (BI). When $J_0 = (s1)$, the strong rule $(s1)$ is changed to a weak rule (w).

Case 4. A descendant of the main formula of a boundary rule $(d1)$ is changed by an $(\exists^I\text{-red})$. Let P be the following.

$$P = \begin{array}{c} \dfrac{\Delta_0, \exists XF, A \subset B; \gamma}{\Delta_0, \exists XF; \gamma + 1}\,(d1) \\ \vdots \\ \dfrac{\Delta, \exists XF; \delta}{\Delta', \exists XF'}\,(\exists^I\text{-red})^\eta \\ \vdots \\ \dfrac{\Gamma_0, \exists XF'; \delta}{\Gamma_1, \exists XF'}\,(\exists^I\text{-red}) \end{array}$$

where the lower rule $(\exists^I\text{-red})$ is a vacuous one such that $h(\Gamma_1, \exists XF') < \omega = h(\Gamma_0, \exists XF')$, cf. Definition 15.24.2. $F \equiv (A \subset X \subset B)$ with $(\exists XF) \in \Sigma^I$ and $Gr(\exists XF) = 1$. Also $F' \equiv F[\exists^\eta/\exists^I] \equiv (A' \subset X \subset B') \equiv (A[\forall^\eta/\forall^I] \subset X \subset B[\exists^\eta/\exists^I])$. Hence $Gr(A') = Gr(B') = Gr(\exists XF') = 0$. From Propositions 15.23.2 and 15.15.2 we see that $st_\Sigma(\exists XF') < I$. Note that there is no $(sub)^s$ between the boundary $(d1)$ and the $(\exists^I\text{-red})^\eta$ since the formula $\exists XF \notin \Pi^1_2$ with $Gr(\exists XF) \neq 0$ is not in the upper sequent of a (sub), cf. Definition 15.16.7(d)i.

Let P' be the following.

$$P' = \dfrac{\begin{array}{c}\vdots \\ \Gamma_1, \exists XF', A' \subset A'; 4\end{array} \quad \begin{array}{c} \Delta_0, \exists XF, A \subset B; \gamma \\ \vdots \\ \dfrac{\Delta, \exists XF, A \subset B; \delta'}{\Delta', \exists XF', A' \subset B'}\,(\exists^I\text{-red})^\eta \\ \vdots \\ \dfrac{\Gamma_0, \exists XF', A' \subset B'; \delta'}{\Gamma_1, \exists XF', A' \subset B'}\,(\exists^I\text{-red}) \end{array}}{\dfrac{\Gamma_1, \exists XF', F'(A')}{\Gamma_1, \exists XF'}\,(BI)}\,(\wedge)$$

where $(A' \subset A') \equiv (\forall x[\neg A'(x) \vee A'(x)])$ is derived from the axiom $\neg A'(u), A'(u)$ with $Gr(A'(u)) = 0$ by two $(\vee)$'s followed by a $(\forall_1)$.

It is easy to see that $o(\Gamma, \exists XF, A \subset B; P') = \gamma$ since $Gr(F'(A')) = 0$ and $dg(F'(A')) = gr(F'(A')) < \omega = h(\Gamma_0, \exists XF', A' \subset B'; P') = h(\Gamma_0, \exists XF'; P)$. δ' is an ordinal such that $\delta' \ll \delta$ by Proposition 15.3. In particular $\psi_I(\alpha\#\omega^{\delta'}) < \psi_I(\alpha\#\omega^{\delta}) \leq \eta$ for the stack α of the rules $(\exists^I\text{-red})^\eta$ by the condition (15.6).

Let ν be an index occurring in the formula $\exists XF$. Then by the condition (15.5) we obtain $\nu \in \mathcal{H}_\alpha(\psi_I\alpha)$, and $\nu \in \mathcal{H}_{\alpha\#\omega^\delta}(\psi_I(\alpha\#\omega^\delta)) \cap I = \psi_I(\alpha\#\omega^\delta) \leq \eta$. Hence we obtain $st_\Sigma(\exists XF') = \eta$ by Proposition 15.15.2.

Let us show $o(\Gamma_1, \exists XF'; P') \lll o(\Gamma_1, \exists XF'; P)\,\{\eta\}$. Let β be the stack of the lower vacuous rule $(\exists^I\text{-red})$. Then $o(\Gamma_1, \exists XF'; P) = \psi_I(\beta\#\omega^\delta)$, while $o(\Gamma_1, \exists XF'; P') = \omega_m(4\#\psi_I(\beta\#\omega^{\delta'})\#\Omega_{\eta+1})$ for an $m < \omega$.

We see $o(\Gamma_1, \exists XF'; P') < o(\Gamma_1, \exists XF'; P)$ from $\psi_I(\beta\#\omega^{\delta'}) \lll \psi_I(\beta\#\omega^\delta)$ and $\eta < \psi_I(\beta\#\omega^\delta)$. The latter follows from (15.5), i.e., from $\eta \in \mathcal{H}_\beta(\psi_I\beta) \cap I \subset \mathcal{H}_{\beta\#\omega^\delta}(\psi_I(\beta\#\omega^\delta)) \cap I = \psi_I(\beta\#\omega^\delta)$. This yields $\Omega_{\eta+1} \lll \psi_I(\beta\#\omega^\delta)\,\{\eta\}$, and $o(\Gamma_1, \exists XF'; P') \lll o(\Gamma_1, \exists XF'; P)\,\{\eta\}$. We see that the conditions (15.4) and (15.6) is fulfilled for rules $(\exists^I\text{-red})$ in P'.

Consider the condition (15.5) for rules $(\exists^I\text{-red})$, e.g., for the lower vacuous rule $(\exists^I\text{-red})$. There occur new indices, e.g., $\psi_I(\alpha\#\omega^{\delta'})$ for the rule $(\exists^I\text{-red})^\eta$ in P'. We need to show $\psi_I(\alpha\#\omega^{\delta'}) \in \mathcal{H}_\beta(\psi_I\beta)$. We have $\psi_I(\alpha\#\omega^\delta) \in \mathcal{H}_\beta(\psi_I\beta)$ for the stack β of the vacuous rule. From $\psi_I(\alpha\#\omega^{\delta'}) < \psi_I(\alpha\#\omega^\delta)$, $\delta' \lll \delta$ and Proposition 15.3.6. we see that $\psi_I(\alpha\#\omega^{\delta'}) \lll \psi_I(\alpha\#\omega^\delta)$. In particular we obtain $\psi_I(\alpha\#\omega^{\delta'}) \in \mathcal{H}_\beta(\psi_I\beta)$.

Let Π_0 be an upper sequent of a $(sub)^\mu\, J$ occurring below $\Gamma_1, \exists XF'$ in P with its lower sequent Π_1 and its stack $\gamma = stk(J)$. Assume $\alpha'_0 \lll \alpha_0\,\{\eta\}$ for $\alpha'_0 = o(\Pi_0; P')$ and $\alpha_0 = o(\Pi_0; P)$. Let the stack $\gamma = stck'(J')$ of the corresponding rule $(sub)^\mu\, J'$ in P'. We see that the condition (15.5) on $(sub)^\mu\, J'$ is fulfilled as above from Proposition 15.3.6.. For $\sigma = \Omega_{\mu+1}$, let $\alpha_1 = o(\Pi_1; P) = \psi_\sigma(\gamma\#\omega^{\alpha_0})$ and $\alpha'_1 = o(\Pi_1; P') = \psi_\sigma(\gamma\#\omega^{\alpha'_0})$. Then $\alpha'_1 \lll \alpha_1\,\{\eta\}$ follows from Proposition 15.3.3., (15.4), $\gamma, \alpha_0 \in \mathcal{H}_\gamma(\psi_\sigma\gamma)$, and (15.5), $\eta \in \mathcal{H}_\gamma(\psi_\sigma\gamma)$. Hence $o(P') \lll o(P)\,\{\eta\}$, and $o(P') < o(P)$.

Case 5. A descendant of the main formula of a boundary rule $(d2)$ is changed by an $(\exists^I\text{-red})$. Let P be the following.

$$P = \begin{array}{c} \dfrac{\Delta_0, \exists XF, F(T^\nu); \gamma}{\Delta_0, \exists XF}\,(d2) \\ \vdots \\ \dfrac{\Delta, \exists XF; \delta}{\Delta', \exists XF'}\,(\exists^I\text{-red})^\eta \\ \vdots \\ \dfrac{\Gamma_0, \exists XF'}{\Gamma_1, \exists XF'}\,(\exists^I\text{-red}) \end{array}$$

where the lower rule $(\exists^I\text{-red})$ is a vacuous one such that $h(\Gamma_1, \exists XF') < \omega = h(\Gamma_0, \exists XF')$, $F' \equiv F[\exists^\eta/\exists^I]$, $Gr(\exists XF) \neq 0$, and T^ν is either a relation symbol R^0 or a stratified free variable U^ν. Proposition 15.23.2 yields $\nu < I$. Also $o(\Delta_0, \exists XF; P) =$

$\gamma + 1$ and $Gr(\exists XF') = 0$ with stratified $\exists XF'$. Similarly as in *Case 4* we see that $st_\Sigma(\exists XF') = \eta$, and the following P' is a proof such that $o(P') < o(P)$.

$$P' = \begin{array}{c} \Delta_0, \exists XF, F(T^\nu) \\ \vdots \\ \dfrac{\Delta, \exists XF, F(T^\nu)}{\Delta', \exists XF', F'(T^\nu)}\ (\exists^I\text{-red})^\eta \\ \vdots \\ \dfrac{\dfrac{\Gamma_0, \exists XF', F'(T^\nu)}{\Gamma_1, \exists XF', F'(T^\nu)}\ (\exists^I\text{-red})}{\Gamma_1, \exists XF'}\ (BI) \end{array}$$

Case 6. A descendant of the main formula of a boundary rule (c) is changed by an ($\exists^I$-red). Let P be the following.

$$P = \begin{array}{c} \dfrac{\Delta_0, \exists X^I F, F(T^\nu); \gamma}{\Delta_0, \exists X^I F; \gamma + 1}\ (c) \\ \vdots \\ \dfrac{\Delta, \exists X^I F}{\Delta', \exists X^\eta F'}\ (\exists^I\text{-red})^\eta \\ \vdots \\ \dfrac{\Gamma_0, \exists X^\eta F'; \delta}{\Gamma_1, \exists X^\eta F'}\ (\exists^I\text{-red}) \end{array}$$

where the lower rule ($\exists^I$-red) is a vacuous one such that $h(\Gamma_1, \exists XF') < \omega = h(\Gamma_0, \exists XF')$, $F' \equiv F[\exists^\eta/\exists^I]$. We see as in *Case 4* that there is no (sub) between the boundary (c) and the ($\exists^I$-red) since $Gr(\exists X^I F) \neq 0$. By (15.5) and (15.6) we have $\nu \in \mathcal{H}_\gamma(\psi_I \gamma)$ and $\psi_I \gamma \leq \eta$ with the stack γ of the rule $(\exists^I\text{-red})^\eta$. Moreover $\nu < I$ by Proposition 15.23.2. Hence $\nu < \eta$, and the rule (c) in the following P' is a legitimate one.

$$P' = \begin{array}{c} \Delta_0, \exists X^I F, F(T^\nu); \gamma \\ \vdots \\ \dfrac{\Delta, \exists X^I F, F(T^\nu)}{\Delta', \exists X^\eta F', F'(T^\nu)}\ (\exists^I\text{-red})^\eta \\ \vdots \\ \dfrac{\dfrac{\Gamma_0, \exists X^\eta F', F'(T^\nu); \delta'}{\Gamma_1, \exists X^\eta F', F'(T^\nu)}\ (\exists^I\text{-red})}{\Gamma_1, \exists X^\eta F'}\ (c) \end{array}$$

We have $o(\Gamma_1, \exists X^\eta F'; P) = \psi_I(\alpha \# \omega^\delta)$ and $o(\Gamma_1, \exists X^\eta F'; P') = \psi_I(\alpha \# \omega^{\delta'}) + 1$ for the stack α of lower vacuous rules ($\exists^I$-red). From $\delta' \ll \delta$ we see $\psi_I(\alpha \# \omega^{\delta'}) + 1 \ll \psi_I(\alpha \# \omega^\delta)$. We see easily that P' is a proof such that $o(P') < o(P)$.

Case 7. A descendant of the main formula of a boundary rule $(s1)$ is changed by an $(\forall^I\text{-red})$: Let P be the following.

$$P = \begin{array}{c} \vdots\ P_0 \\ \dfrac{\Delta_0, \forall XF, F(U^I); \gamma}{\Delta_0, \forall XF; \gamma+1}\ (s1) \\ \vdots \\ \dfrac{\Delta, \forall XF; \delta}{\Delta', \forall XF'}\ (\forall^I\text{-red})^\eta \\ \vdots \\ \dfrac{\Gamma_0, \forall XF'; \delta}{\Gamma_1, \forall XF'}\ (\exists^I\text{-red}) \end{array}$$

where the lower rule $(\exists^I\text{-red})$ is a vacuous one such that $h(\Gamma_1, \forall XF') < \omega = h(\Gamma_0, \forall XF')$, and $F' \equiv F[\forall^\eta/\forall^I]$ with $Gr(\forall XF) \neq 0$. Since each formula occurring in lower sequents or in upper sequents of rules $(\exists^I\text{-red})$ is in Σ^I by Definition 15.16.9, there occurs no rule $(\exists^I\text{-red})$ in P_0 with $\forall XF \notin \Sigma^I$, and the rule $(\forall^I\text{-red})^\eta$ occurs above the vacuous $(\exists^I\text{-red})$. Since $\forall XF \in \Pi^1_2$ and $Gr(\forall XF) \neq 0$, $\forall^I$ occurs in $\forall XF$, i.e., $\forall XF \notin \Sigma^I$, and $\forall Z^\eta$ occurs in F' for a bound variable Z^η. Therefore there occurs no (sub) between the boundary $(s1)$ and the $(\forall^I\text{-red})$. On the other, $Gr(\forall XF') = 0$ and $\forall XF'$ is stratified. Hence the rule (w) in the following P' is a legitimate one with $\beta = st_\Pi(\forall XF')$.

$$P' = \begin{array}{c} \vdots\ P'_0 \\ \Delta_0, \forall XF, F(U^\beta); \gamma' \\ \vdots \\ \dfrac{\Delta, \forall XF, F(U^\beta); \delta'}{\Delta', \forall XF', F'(U^\beta)}\ (\forall^I\text{-red})^\eta \\ \vdots \\ \dfrac{\dfrac{\Gamma_0, \forall XF', F'(U^\beta); \delta'}{\Gamma_1, \forall XF', F'(U^\beta)}\ (\exists^I\text{-red})}{\Gamma_1, \forall XF'}\ (w) \end{array}$$

P'_0 is obtained from P_0 by replacing the variable U^I by U^β.

Suppose that in P_0, there is a rule (BI) with a main formula $\exists YB$ such that the eigenvariable U^I of the boundary $(s1)$ occurs in $\exists YB$. Assume that the result $(\exists YB)[U^\beta/U^I]$ of replacing U^I by U^β is stratified. At such a rule (BI), I is added in the ordinal attached in P, while $\Omega_{\mu+1}$ is added in P' for $\mu = st_\Sigma((\exists YB)[U^\beta/U^I]) < I$.

$$P = \cfrac{\cfrac{\cfrac{\vdots}{\Delta_1, \exists Y B(Y, U^I), B(A, U^I); \gamma_0}}{\Delta_1, \exists Y B(Y, U^I); I\#\gamma_0}\,(BI) \quad \vdots\, P_0}{\cfrac{\Delta_0, \forall X F(X, \forall Z^I G), F(U^I, \forall Z^I G); \gamma}{\Delta_0, \forall X F(X, \forall Z^I G); \gamma + 1}\,(s1)} \quad \vdots$$

$$P' = \cfrac{\cfrac{\vdots}{\Delta_1, \exists Y B(Y, U^\beta), B(A, U^\beta); \gamma'_0}}{\Delta_1, \exists Y B(Y, U^\beta); \Omega_{\mu+1}\#\gamma'_0}\,(BI) \quad \vdots\, P'_0 \quad \Delta_0, \forall X F(X, \forall Z^I G), F(U^\beta, \forall Z^I G); \gamma' \quad \vdots$$

Let $\mathcal{S}_{P_0}$ be the set of all indices α_0 such that either a free variable V^{α_0} or a bound variable Y^{α_0} occurs either in F or in a main formula of a (BI) in P_0. Let $\eta' = \eta\#\{\alpha_0 : \alpha_0 \in \mathcal{S}_{P_0}\}$. Then we obtain $\Omega_{\mu+1} \ll I\,\{\eta'\}$ for each such (BI) since $\mu \in \{\beta + k, \alpha_0 + k : \alpha_0 \in \mathcal{S}_{P_0}, k < \omega\} \subset \{\eta + k, \alpha_0 + k : \alpha_0 \in \mathcal{S}_{P_0}, k < \omega\}$ by Proposition 15.15.2 for $\beta = st_\Pi(\forall X F')$. Hence $\gamma' \ll \gamma + 1\,\{\eta'\}$. This yields $\delta' \ll \delta\,\{\eta'\}$.

On the other hand we have $\{\alpha, \delta, \eta'\} \subset \mathcal{H}_\alpha(\psi_I \alpha)$ for the stack α of the lower vacuous rules $(\exists^I\text{-red})$ by (15.4) and Definitions 15.24.3a, 15.24.3b. Hence by Proposition 15.3.3. we obtain $\psi_I(\alpha\#\omega^{\delta'}) \ll \psi_I(\alpha\#\omega^\delta)\,\{\eta'\}$ for $o(\Gamma_1, \forall X F'; P) = \psi_I(\alpha\#\omega^\delta)$ and $o(\Gamma_1, \forall X F'; P') = \psi_I(\alpha\#\omega^{\delta'}) + 1$. Thus $o(\Gamma_1, \forall X F'; P') \ll o(\Gamma_1, \forall X F'; P)\,\{\eta'\}$, and we obtain $o(P') \ll o(P)\,\{\eta'\}$, and $o(P') < o(P)$.

Let us verify that P' is a proof. Although η may be a new index in the upper part of $\Delta, \forall X F, F(U^\beta)$ in P'_0, there is no rule $(\exists^I\text{-red})$ nor (sub) in the part since there is no (sub) above the boundary $(s1)$ by Definition 15.24.5a. Hence the conditions (15.5) and (15.4) are enjoyed for the upper part. The condition in (15.6) is fulfilled as we saw above. We see that the conditions (15.5) and (15.4) are fulfilled below the $(\forall^I\text{-red})^\eta$ in P' by Proposition 15.3.6. and (15.5) for P.

Case 8. A descendant of the main formula of a boundary rule $(s2)$ is changed by an $(\forall^I\text{-red})$. Let P be the following.

$$P = \cfrac{\cfrac{\cfrac{\Delta_0, \forall X^I F, F(U^I); \gamma}{\Delta_0, \forall X^I F; \gamma + 1}\,(s2) \quad \vdots \quad \Delta, \forall X^I F}{\Delta', \forall X^\eta F'}\,(\forall^I\text{-red})^\eta \quad \vdots \quad \Gamma_0, \forall X^\eta F'}{\Gamma_1, \forall X^\eta F'}\,(\exists^I\text{-red})$$

where the lower rule ($\exists^I$-red) is a vacuous one such that $h(\Gamma_1, \forall X^\eta F') < \omega = h(\Gamma_0, \forall X^\eta F')$, and $F' \equiv F[\forall^\eta/\forall^I]$. Let P' be the following.

$$P' = \begin{array}{c}
\Delta_0, \forall X^I F, F(U^\eta); \gamma' \\
\vdots \\
\dfrac{\Delta, \forall X^I F, F(U^\eta)}{\Delta', \forall X^\eta F', F'(U^\eta)}\ (\forall^I\text{-red})^\eta \\
\vdots \\
\dfrac{\dfrac{\Gamma_0, \forall X^\eta F', F'(U^\eta)}{\Gamma_1, \forall X^\eta F', F'(U^\eta)}\ (\exists^I\text{-red})}{\Gamma_1, \forall X^\eta F'}\ (s2)
\end{array}$$

In P', the eigenvariable U^I is replaced by the variable U^η. As in *Case 7* we see that P' is a proof such that $o(P') < o(P)$.

In the following cases let us reduce suitable triangles (J_1, J_2, J), where descendants of main formulas of J_1 and J_2 are not changed by any rules (Q^I-red) by virtue of *Cases 4–8*.

Case 9. J_1 is an $(s1)$ and J_2 is a (d). Let P be the following.

$$P = \begin{array}{c}
\dfrac{\begin{array}{cc}
\dfrac{\Delta_0, \forall X \neg F(U^I)}{\Delta_0, \forall X \neg F}\ (s1)\, J_1 & \dfrac{G, \exists X F, \Gamma_0}{\exists X F, \Gamma_0}\ (d)\, J_2 \\
\vdots & \vdots \\
\Delta, \forall X \neg F; \alpha & \exists X F, \Gamma; \beta
\end{array}}{\Delta, \Gamma; \alpha \# \beta}\ J \\
\vdots \\
\dfrac{\Pi; \delta}{}\ J_3 \\
\vdots \\
\dfrac{}{\Phi; \psi_I(\alpha_4 \# \omega^\delta)}\ J_4
\end{array}$$

where $Gr(\exists X F) = 1$, i.e., $\mathrm{dg}(\exists X F) = \omega = h(\Pi)$, J_2 is either a $(d1)$ with $G \equiv (A \subset B)$ and $F \equiv (A \subset X \subset B)$, or a $(d2)$ with $G \equiv F(T^\xi)$. Π denotes the upper sequent of the uppermost ($\exists^I$-red) J_3 below J, cf. Definition 15.24.2. Φ denotes the lower sequent of the lowest vacuous rule ($\exists^I$-red) J_4. In other words Φ is the uppermost sequent below the $(cut)\, J$ such that $h(\Phi) < \omega$, cf. Definition 15.24.2. Let $\alpha_n = stk(J_n)$ be the stack of the rule J_n for $n = 3, 4$.

Note that no (sub) occurs between J and Φ since the height of the upper sequent of a (sub) is defined to be 0, cf. Definition 15.19.2a. Furthermore there is no (sub) between the $(s1)\, J_1$ and $(cut)\, J$, and no (sub) between the $(d)\, J_2$ and J since $Gr(\exists X F) \neq 0$. Let P' be the following.

$$
P' = \dfrac{\dfrac{\dfrac{\dfrac{\dfrac{\vdots}{\Delta, \forall X \neg F; \alpha}}{\Delta, \Gamma, \forall X \neg F}(th) \quad \vdots \quad \dfrac{\Pi, \forall X \neg F; \delta_1}{\Pi, \forall X \neg F'}(\forall^I\text{-red})^\eta}{} J_{31} \quad \vdots}{\Phi, \forall X \neg F'; \psi_I(\alpha'_4 \# \omega^{\delta_1})} J_{41} \qquad \dfrac{\dfrac{\dfrac{\dfrac{\vdots}{\exists X F, \Gamma; \beta}}{\exists X F, \Delta, \Gamma}(th) \quad \vdots \quad \dfrac{\exists X F, \Pi; \delta_2}{\exists X F', \Pi}(\exists^I\text{-red})^\eta J_{new}}{} J_{32} \quad \vdots}{\exists X F', \Phi; \psi_I(\alpha'_4 \# \omega^{\delta_2})} J_{42}}{\Phi}(cut)
$$

where $F' \equiv F[\exists^\eta/\exists^I]$. Hence $\mathrm{dg}(\exists X F') = gr(\exists X F') < \omega = h(\Pi; P)$. Then $o(\Delta, \forall X \neg F; P) = \alpha = o(\Delta, \forall X \neg F; P')$ and $o(\exists X F, \Gamma; P) = \beta = o(\exists X F, \Gamma; P')$.

From $\alpha, \beta \ll \alpha \# \beta$ we see that

$$\delta_1, \delta_2 \ll \delta. \tag{15.7}$$

A new rule $(\forall^I\text{-red})^\eta$ and a new $(\exists^I\text{-red})^\eta$ are introduced in P'.

The stack of the new rule $(\exists^I\text{-red})^\eta$ is defined to be equal to the one of J_3, and the stacks of the rules $(\exists^I\text{-red})\, J_{31}$ and of J_{32} are increased by $\omega^{\delta_2} + 1$.

$$\alpha_3 = stk'(J_{new}) = stk(J_3)\, \&\, \alpha'_3 = stk'(J_{31}) = stk'(J_{32}) = \alpha_3 \# \omega^{\delta_2} \# 1. \tag{15.8}$$

The type η of the new $(\forall^I\text{-red})^\eta$ and of $(\exists^I\text{-red})^\eta$ is defined to be $\eta = \psi_I(\alpha_3 \# \omega^{\delta_2})$ with $\delta_2 = o(\exists X F, \Pi; P')$ for the upper sequent $\exists X F, \Pi$ of the new $(\exists^I\text{-red})^\eta$. Let us verify the conditions (15.6), (15.5) and (15.4) for the new rule $(\exists^I\text{-red})^\eta$. (15.6) is obvious. (15.5) inherits from one for J_3 in P since the stack of the new $(\exists^I\text{-red})^\eta$ is equals to the one of the rule J_3. We obtain

$$\alpha_3, \delta \in \mathcal{H}_{\alpha_3}(\psi_I \alpha_3) \tag{15.9}$$

by (15.4) for J_3. The condition (15.4), $\alpha_3, \delta_2 \in \mathcal{H}_{\alpha_3}(\psi_I \alpha_3)$ follows from this and (15.7).

Next stacks of the rules $(\exists^I\text{-red})\, J_{3i}$ are increased by $\omega^{\delta_2} + 1$. We see that the conditions (15.5), (15.4) and (15.6) are fulfilled for J_{3i} with $i = 1, 2$ as follows. The new index $\eta \in \mathcal{H}_{\alpha'_3}(\psi_I \alpha'_3)$ for (15.5). This is seen from (15.9), (15.7) and $\alpha_3 \# \omega^{\delta_2} < \alpha'_3$. For $i = 1, 2$, we see $\alpha'_3, \delta_i \in \mathcal{H}_{\alpha'_3}(\psi_I \alpha'_3)$ and $\psi_I(\alpha'_3 \# \omega^{\delta_i}) \ll \psi_I(\alpha_3 \# \omega^{\delta})$ from (15.7) and (15.9). Thus the conditions (15.4) and (15.6) are enjoyed for rules J_{3i}.

Suppose that there exists a rule $(\exists^I\text{-red})\, K$ occurring below J_3 in P, and $\gamma = stk(K)$ its stack. Then the stack of the corresponding rules K' in P' is defined to be $stk'(K') = stk(K) = \gamma$, and let $\delta' = o(K'; P') \in \{\delta_1, \delta_2\}$. In particular the stack $stk'(J_{41}) = stk'(J_{42}) = stk(J_4) = \alpha_4$ of the rules $(\exists^I\text{-red})\, J_{41}$ and of J_{42}. We obtain $\gamma, \omega^{\delta'} \in \mathcal{H}_\gamma(\psi_I \gamma)$ and $\psi_I(\gamma \# \omega^{\delta'}) \ll \psi_I(\gamma \# \omega^{\delta})$ from (15.7) and $\{\gamma, \delta\} \subset \mathcal{H}_\gamma(\psi_I \gamma)$. Thus the conditions (15.4) and (15.6) are enjoyed for rules K'.

Consider (15.5) for K'. Let μ be the type of J_3. Then we have $\mu \in \mathcal{H}_\gamma(\psi_I\gamma)$ by Definition 15.24.3c for K. On the other hand we have $\eta < \psi_I\alpha_3' < \psi_I(\alpha_3\#\omega^\delta) \leq \mu$ by (15.6) for J_3. Hence $\eta < \mu \in \mathcal{H}_\gamma(\psi_I\gamma) \cap I = \psi_I\gamma$ and $\eta \in \mathcal{H}_\gamma(\psi_I\gamma)$. Moreover we have $\psi_I(\alpha_3'\#\omega^{\delta'}) \in \mathcal{H}_\gamma(\psi_I\gamma)$ by $\psi_I(\alpha_3'\#\omega^{\delta'}) < \psi_I(\alpha_3\#\omega^\delta)$, (15.7), Proposition 15.3.6. and $\psi_I(\alpha_3\#\omega^\delta) \in \mathcal{H}_\gamma(\psi_I\gamma)$. Thus (15.5) is fulfilled for K'.

For $i = 1, 2$, we obtain $\psi_I(\alpha_4\#\omega^{\delta_i}) \ll \psi_I(\alpha_4\#\omega^\delta)$. Hence $o(\Phi; P') = \omega_m(\psi_I(\alpha_4\#\omega^{\delta_1})\#\psi_I(\alpha_4\#\omega^{\delta_2})) \ll o(\Phi; P)$ for an $m < \omega$.

Finally let S be a $(sub)^\nu$ occurring below Φ in P, and S' be the corresponding rule in P'. Let $\gamma = stk'(S') = stk(S)$, and $\sigma = \Omega_{\nu+1}$. We have $\psi_I(\alpha_3\#\omega^\delta) \in \mathcal{H}_\gamma(\psi_\sigma\gamma)$ by Definition 15.24.3d for S. Proposition 15.3.6. yields $\psi_I(\alpha_3'\#\omega^{\delta_i}) \in \mathcal{H}_\gamma(\psi_\sigma\gamma)$ for $i = 1, 2$, and $\eta \in \mathcal{H}_\gamma(\psi_\sigma\gamma)$ by (15.7) and $\eta < \psi_I(\alpha_3'\#\omega^{\delta_i}) < \psi_I(\alpha_3\#\omega^\delta)$.

Thus the conditions (15.5) and (15.4) for S' are seen from $o(S'; P') \ll o(S; P)$ using Proposition 15.3.6. as above.

Case 10. J_1 is a (w) and J_2 is a (BI). Let P be the following.

$$
\begin{array}{c}
\dfrac{\Delta_0,\forall X\neg F,\neg F(U^{\xi});\beta}{\Delta_0,\forall X\neg F;\beta+1}\,(w)\,J_1 \qquad \dfrac{F(A),\exists XF,\Gamma_0;\alpha}{\exists XF,\Gamma_0;\alpha_1}\,(BI)\,J_2 \\
\vdots \qquad\qquad\qquad\qquad \vdots \\
P = \quad \dfrac{\Delta,\forall X\neg F \qquad\qquad\qquad \exists XF,\Gamma}{\Delta,\Gamma}\,J \\
\vdots \\
\dfrac{\Pi;\delta}{\Phi}\,(sub)\,J_3
\end{array}
$$

where $Gr(\exists XF) = 0$, $\xi = st_\Pi(\forall X\neg F) = st_\Sigma(\exists XF)$, and let $\sigma = \Omega_{\xi+1}$. Then $o(J_2; P) = \sigma\#\alpha$ and $\alpha_1 = \omega_m(\sigma\#\alpha)$ for m such that $h(F(A), \exists XF, \Gamma_0) = \max\{h(\exists XF, \Delta_0), \mathrm{dg}(F(A))\} = h(\exists XF, \Delta_0) + m$. We see $\xi < I$ from Proposition 15.23.2. Also Π denotes the upper sequent of the uppermost $(sub)\,J_3$ of level$\leq \xi$ below J, cf. Definition 15.24.5c.

Note that no (sub) changes the descendants of $\forall X\neg F$ nor of $\exists XF$ by the condition in Definition 15.16.7(d)ii.

From Proposition 15.15.3 and (15.1) in Definition 15.14 we see that $\xi = st_\Pi(\neg F(U^\xi))$, and $st_\Pi(\exists XF) = \xi + 1$. Hence from the Definition 15.16.7(d)i of the rule (sub), we see that the level ν of any $(sub)^\nu$ occurring between $\exists XF, \Gamma_0$ and Π is larger than ξ, $\nu > \xi$. In particular no eigenvariable of a (sub) occurring between Δ, Γ and Π occurs in $\neg F$, cf. Note 15.18. Let P' be the following.

$$
\begin{array}{c}
\vdots \\
\Delta_0, \forall X \neg F, \neg F(U^{\xi}); \beta \\
\vdots \qquad\qquad\qquad \vdots \\
\dfrac{\Delta, \neg F(U^{\xi}), \forall X \neg F \qquad \exists X F, \Gamma}{\Delta, \Gamma, \neg F(U^{\xi})} \\
\vdots
\end{array}
$$

$$
\dfrac{\dfrac{\Pi, \neg F(U^{\xi}); \delta'}{\Pi, \neg F(A); \psi_{\sigma}(\gamma \# \omega^{\delta'})}\,(sub)^{\xi} \qquad F(A), \exists X F, \Gamma_0; \alpha}{\exists X F, \Gamma_0, \Pi; \alpha'_1}\,(cut)\, J'_2
$$

$$
\begin{array}{c}
\vdots \qquad\qquad\qquad\qquad \vdots \\
\dfrac{\Delta, \forall X \neg F \qquad\qquad \exists X F, \Gamma, \Pi}{\Delta, \Gamma, \Pi} \\
\vdots \\
\dfrac{\Pi; \delta''}{\Phi}\; J'_3
\end{array}
$$

where $h(\Pi, \neg F(U^{\xi}); P') = h(\Pi; P) = 0$ for the upper sequent $\Pi, \neg F(U^{\xi})$ of the new $(sub)^{\xi}$, $h(F(A), \exists X F, \Gamma_0; P') = h(F(A), \exists X F, \Gamma_0; P)$. Note that the new $(sub)^{\xi}$ is in an end-piece, cf. Definition 15.24.5a. The rules occurring above $\Pi, \neg F(U^{\xi})$ in P' receives the same stack of the corresponding rule in P. From $\beta = o(\Delta_0, \forall X \neg F, \neg F(U^{\xi}); P') = o(\Delta_0, \forall X \neg F, \neg F(U^{\xi}); P) \ll o(\Delta_0, \forall X \neg F; P) = \beta + 1$, we see for $\delta' = o(\Pi, \neg F(U^{\xi}); P')$ that

$$
\delta' \ll \delta. \tag{15.10}
$$

The stack γ of the new $(sub)^{\xi}$ is defined to be the stack $\gamma = stk(J_3)$ of the (sub) J_3 in P. Then the conditions (15.4) and (15.5) in Definition 15.24.3 are enjoyed for the rule $(sub)^{\xi}$, $\{\gamma, \delta'\} \subset \mathcal{H}_{\gamma}(\psi_{\sigma}\gamma)$, $o(J'_2; P') = \psi_{\sigma}(\gamma \# \omega^{\delta'}) \# \alpha$, and for $\alpha'_1 = \omega_m(\psi_{\sigma}(\gamma \# \omega^{\delta'}) \# \alpha)$

$$
\alpha'_1 \ll \alpha_1 \, \{\psi_{\sigma}(\gamma \# \omega^{\delta'})\}. \tag{15.11}
$$

Rules occurring between J'_2 and J'_3 in P' receive the same stacks of the corresponding rules in P. Then the condition (15.4) is enjoyed for these (sub)'s by (15.11). Note that the level ν of any $(sub)^{\nu}$ between J'_2 and J'_3 is higher than ξ, $\nu > \xi$. Then $\psi_{\sigma}(\gamma \# \omega^{\delta'}) < \sigma \leq \Omega_{\nu}$ and for the stack λ of such a rule $(sub)^{\nu}$,

$$
\psi_{\sigma}(\gamma \# \omega^{\delta'}) \in \mathcal{H}_{\lambda}(\psi_{\Omega_{\nu+1}} \lambda). \tag{15.12}
$$

Since no eigenvariable of a (sub) between Δ, Γ and Π occurs in $\neg F$, such a (sub) does not change the descendants of $\neg F(U^{\xi})$.

The stack of the (sub) J'_3 is increased by $\omega^{\delta'} \# 1$, i.e., $\gamma' = stk'(J'_3) = \gamma \# \omega^{\delta'} \# 1$. Let $\tau = \Omega_{\mu+1}$ with the level μ of J_3. Then we have $\{\gamma, \delta\} \subset \mathcal{H}_{\gamma}(\psi_{\tau}\gamma)$, and $\delta' \in \mathcal{H}_{\gamma}(\psi_{\tau}\gamma)$ by (15.10). Hence $\gamma' \in \mathcal{H}_{\gamma'}(\psi_{\tau}\gamma)$. Next from (15.11), (15.12) and Proposition 15.3.3. we obtain

$$\delta'' \ll \delta\, \{\psi_\sigma(\gamma \# \omega^{\delta'})\}. \tag{15.13}$$

On the other hand we have $\sigma \ll \alpha_1$. Consider a rule $(sub)^\nu$ occurring between J'_2 and J'_3. We have $\sigma \leq \Omega_\nu$ with $\nu > \xi$, $\{\lambda, \alpha_2\} \subset \mathcal{H}_\lambda(\psi_{\Omega_{\nu+1}}\lambda)$, and $\sigma \ll \lambda \# \omega^{\alpha_2}$ for the stack λ of the rule $(sub)^\nu$ and the ordinal α_2 attached to the upper sequent of the rule $(sub)^\nu$. We obtain $\sigma \ll \delta$ by Proposition 15.3.4.. Hence

$$\psi_\sigma(\gamma \# \omega^{\delta'}) \in \mathcal{H}_{\gamma'}(\psi_\tau \gamma) \tag{15.14}$$

and $\delta'' \in \mathcal{H}_{\gamma'}(\psi_\tau\gamma)$ by (15.13). Therefore we obtain $\{\gamma', \delta''\} \subset \mathcal{H}_{\gamma'}(\psi_\tau\gamma)$. Thus the condition (15.4) is enjoyed for J'_3.

The condition (15.5) for J'_3 is enjoyed since no essentially new index occurs above J'_3.

Let us show $o(\Phi; P') = \psi_\tau(\gamma' \# \omega^{\delta''}) \ll \psi_\tau(\gamma \# \omega^{\delta}) = o(\Phi; P)$. We have $\psi_\tau(\gamma' \# \omega^{\delta''}) \ll \psi_\tau(\gamma \# \omega^{\delta})\, \{\psi_\sigma(\gamma \# \omega^{\delta'})\}$ by (15.13), Proposition 15.3.3. and (15.14). Moreover we see from (15.10) that for any $\alpha > \gamma \# \omega^\delta$ and any ρ, if $\{\gamma, \delta, \sigma\} \subset \mathcal{H}_\alpha(\psi_\rho \alpha)$, then $\psi_\sigma(\gamma \# \omega^{\delta'}) \in \mathcal{H}_\alpha(\psi_\rho\alpha)$. $\psi_\tau(\gamma' \# \omega^{\delta''}) \ll \psi_\tau(\gamma \# \omega^\delta)$ is seen from Proposition 15.3.5..

The stacks of (sub)'s below J'_3 remain the same. (15.4) and (15.5) are fulfilled for these (sub)'s by $\psi_\tau(\gamma' \# \omega^{\delta''}) \ll \psi_\tau(\gamma \# \omega^\delta)$.

Case 11. J_1 is an $(s2)$ and J_2 is a (c) with a main formula $\exists X^I F$. Let P be the following.

$$P = \begin{array}{c}
\begin{array}{cc}
\dfrac{\begin{array}{c}\vdots\, P_1 \\ \Delta_0, \forall X^I \neg F, \neg F(U^I); \delta\end{array}}{\Delta_0, \forall X^I \neg F; \delta + 1}\,(s2)\,J_1
&
\dfrac{F(T^\nu), \exists X^I F, \Gamma_0; \xi}{\exists X^I F, \Gamma_0; \xi + 1}\,(c)\,J_2 \\
\vdots & \vdots \\
\Delta, \forall X^I \neg F & \exists X^I F, \Gamma
\end{array} \\
\hline
\Delta, \Gamma
\end{array}\, J; \gamma$$

$$\dfrac{\begin{array}{c}\vdots\end{array}}{\Phi; \beta}\, J_0; \alpha$$

where $\beta = o(\Phi)$, $\alpha = o(J_0)$, $\gamma = o(J)$, $\delta = o(\Delta_0, \forall X^I \neg F, \neg F(U^I))$, $\xi = o(F(T^\nu), \exists X^I F, \Gamma_0)$, and the lower sequent Φ of the rule J_0 denotes the uppermost sequent below $\Delta, \forall X^I \neg F$ such that $h(\Phi) < h(\Delta, \forall X^I \neg F)$. $\nu < I$ follows from Proposition 15.23.2. By Definition 15.12, $Gr(\forall X^I \neg F) > 1$, and hence $h(\Delta, \forall X^I \neg F; P) > \omega$. Since the height of upper sequents of $(Q^I$-red) is defined to be ω, we see that there is no $(Q^I$-red) between J and J_0 in P.

Note that no (sub) changes the descendants of $\forall X^I \neg F$ nor of $\exists X^I F$ by the condition in Definition 15.16.7(d)ii. We see that there is no (sub) between J_2 and J, nor between J_1 and J in P since $\exists X^I F \notin \Pi^1_2$ and $st_\Pi(\forall X^I \neg F) \geq I$. Moreover there is no (sub) between J and J_0 since the height of the upper sequents of any (sub) is defined to be 0. Let P' be the following.

$$
\begin{array}{cc}
\begin{array}{c} \vdots\, P'_1 \\ \Delta_0, \forall X^I \neg F, \neg F(T^\nu); \delta' \end{array} & F(T^\nu), \exists X^I F, \Gamma_0; \xi \\
\vdots & \vdots
\end{array}
$$

$$
\dfrac{\Delta, \neg F(T^\nu), \forall X^I \neg F \qquad \exists X^I F, \Gamma}{\Delta, \Gamma, \neg F(T^\nu)}\; J'; \gamma' \qquad \dfrac{\Delta, \forall X^I \neg F \qquad \exists X^I F, F(T^\nu), \Gamma}{F(T^\nu), \Delta, \Gamma}
$$

$$
\dfrac{\dfrac{\vdots}{\Phi, \neg F(T^\nu); \beta_1}\; J_{01}; \alpha_1 \qquad \dfrac{\vdots}{F(T^\nu), \Phi; \beta_2}\; J_{02}; \alpha_2}{\Phi; \beta'}\; (cut)
$$

where P'_1 is obtained from P_1 by replacing the eigenvariable U^I by T^ν. Let $\beta' = o(\Phi; P')$, $\beta_1 = o(\Phi, \neg F(T^\nu); P')$, $\beta_2 = o(F(T^\nu), \Phi; P')$, $\alpha_1 = o(J_{01}; P')$, $\alpha_2 = o(J_{02}; P')$. Also $\delta' = o(\Delta_0, \forall X^I \neg F, \neg F(T^\nu); P')$ and $\gamma' = o(J'; P')$.

For P' to be a proof, we need to verify the condition on rules (sub) in Definition 15.16.7(d)i, the condition (15.6) on rules ($\exists^I$-red), and the conditions (15.4) and (15.5) on rules (sub), ($\exists^I$-red).

We saw that a rule occurring between J_1 and J_0 is neither a (sub) nor an ($\exists^I$-red). Let K be a rule occurring in P_1, which is either a (sub) or an ($\exists^I$-red). Since each (sub) is in an end-piece by Definition 15.24.5a, K is not a (sub). Also from $h(\Delta, \forall X^I \neg F) > \omega$ and $h(\Delta_0, \forall X^I \neg F, \neg F(U^I)) > \omega$, we see that K is not an ($\exists^I$-red) by Definition 15.24.2. Therefore there is no such rule K.

Let $\mathcal{S}_{P_1}$ be the set of all indices ν_1 such that either a free variable W^{ν_1} or a bound variable Y^{ν_1} occurs in a main formula of a (BI) in P_1. Let $\nu' = \nu \# \{\nu_1 : \nu_1 \in \mathcal{S}_{P_1}\}$.

As in *Case 7* we see $\delta' \ll \delta\{\nu'\}$, and $\gamma' \ll \gamma\{\nu'\}$. Hence $\alpha_1 \ll \alpha\{\nu'\}$ and $\alpha_2 \ll \alpha$. Let Π denote an upper sequent of J_0, and let $h = h(\Pi; P)$. Then $h = h(\Phi) + m$ for an $m < \omega$. From $\mathrm{dg}(F(T^\nu)) < \mathrm{dg}(\exists X^I F) \leq h$ and $h(F(T^\nu), \Phi; P') < h$, we see $\beta' \ll \beta\{\nu'\}$.

Let K be the uppermost ($\exists$-red) below J_0 in P, and K' the corresponding rule in K' with their stacks $\lambda = stk(K) = stk'(K')$. Consider the conditions (15.4) and (15.6) on K'. We have $\alpha_{K'} = o(K'; P') \ll o(K; P) = \alpha_K\{\nu'\}$, and $\nu' \in \mathcal{H}_\lambda(\psi_I \lambda)$ since the indices ν, ν_1 occur above K. From $\{\alpha_K, \lambda\} \subset \mathcal{H}_\lambda(\psi_I \lambda)$ we obtain $\alpha_{K'} \in \mathcal{H}_\lambda(\psi_I \lambda)$, and $\psi_I(\lambda \# \omega^{\alpha_{K'}}) < \psi_I(\lambda \# \omega^{\alpha_K})$. Similarly we see that rules ($\exists^I$-red) below J_{01} enjoy the conditions (15.4), (15.5) and (15.6).

Next assume that K is a $(sub)^\mu$ occurring below J_0, and consider the conditions (15.4) and (15.5) on K'. Let $\alpha_K = o(K; P)$ and $\alpha_{K'} = o(K'; P')$. We need to show that $\alpha_{K'} \in \mathcal{H}_\lambda(\psi_\sigma \lambda)$, where $\lambda = stk(K) = stk'(K')$ and $\sigma = \Omega_{\mu+1}$. We have $\alpha_{K'} \ll \alpha_K\{\nu'\}$, and $\nu' \in \mathcal{H}_\lambda(\psi_\sigma \lambda)$ by (15.5) for K. Then $\alpha_{K'} \in \mathcal{H}_\lambda(\psi_\sigma \lambda)$.

Case 12. J_1 is an $(s2)$ and J_2 is a (c) with a main formula $\exists X^\eta F$ for an $\eta < I$. Let P be the following.

$$
P = \dfrac{\dfrac{\begin{array}{c}\vdots\ P_1\\ \dfrac{\Delta_0, \forall X^{\eta}\neg F, \neg F(U^{\eta}); \delta}{\Delta_0, \forall X^{\eta}\neg F; \delta+1}\,(s2)\,J_1 \\ \vdots \\ \Delta, \forall X^{\eta}\neg F\end{array} \qquad \begin{array}{c}\dfrac{F(T^{\nu}), \exists X^{\eta} F, \Gamma_0; \xi}{\exists X^{\eta} F, \Gamma_0; \xi+1}\,(c)\,J_2 \\ \vdots \\ \exists X^{\eta} F, \Gamma\end{array}}{\begin{array}{c}\Delta, \Gamma \\ \vdots\end{array}}\,J;\gamma}{\Phi; \beta}\,J_0; \alpha
$$

where $\beta = o(\Phi)$, $\alpha = o(J_0)$, $\gamma = o(J)$, and $\delta = o(\Delta_0, \forall X^{\eta}\neg F, \neg F(U^{\eta}))$, $\xi = o(F(T^{\nu}), \exists X^{I} F, \Gamma_0)$, and the lower sequent Φ of the rule J_0 denotes the uppermost sequent below $\Delta, \forall X^{\eta}\neg F$ such that $h(\Phi) < h(\Delta, \forall X^{\eta}\neg F)$.

Note that no (sub) changes the descendants of $\forall X^{\eta}\neg F$ nor of $\exists X^{\eta} F$ by the condition in Definition 15.16.7(d)ii.

From $\exists X^{\eta} F \notin \Pi^1_2$, we see that there is no (sub) between J_2 and J_0 in P since the height of the upper sequents of any (sub) is defined to be 0.

By Definition 15.16.2 we obtain $\nu < \eta$. Then Proposition 15.15.2 with a limit η yields $st_{\Pi}(\neg F(T^{\nu})) \le st_{\Pi}(\forall X^{\eta}\neg F)$ when $\forall X^{\eta}\neg F \in \Pi^1_2$. Let P' be the following.

$$
\dfrac{\dfrac{\dfrac{\begin{array}{c}\vdots\ P'_1\\ \Delta_0, \forall X^{\eta}\neg F, \neg F(T^{\nu}); \delta' \\ \vdots \\ \Delta, \neg F(T^{\nu}), \forall X^{\eta}\neg F\end{array} \quad \begin{array}{c}\vdots\\ \exists X^{\eta} F, \Gamma\end{array}}{\begin{array}{c}\Delta, \Gamma, \neg F(T^{\nu})\\ \vdots\end{array}}\,J';\gamma'}{\Phi, \neg F(T^{\nu}); \beta_1}\,J_{01}; \alpha_1 \qquad \dfrac{\dfrac{\begin{array}{c}\vdots\\ \Delta, \forall X^{\eta}\neg F\end{array} \quad \begin{array}{c}F(T^{\nu}), \exists X F, \Gamma_0; \xi \\ \vdots \\ \exists X^{\eta} F, F(T^{\nu}), \Gamma\end{array}}{\begin{array}{c}F(T^{\nu}), \Delta, \Gamma\\ \vdots\end{array}}}{F(T^{\nu}), \Phi; \beta_2}\,J_{02}; \alpha_2}{\Phi; \beta'}
$$

where P'_1 is obtained from P_1 by replacing the eigenvariable U^{η} by T^{ν}. Let $\beta' = o(\Phi; P')$, $\beta_1 = o(\Phi, \neg F(T^{\nu}); P')$, $\beta_2 = o(F(T^{\nu}), \Phi; P')$, $\alpha_1 = o(J_{01}; P')$, $\alpha_2 = o(J_{02}; P')$. Also $\delta' = o(\Delta_0, \forall X^{\eta}\neg F, \neg F(T^{\nu}); P')$ and $\gamma' = o(J'; P')$.

For P' to be a proof, we need to verify the condition on rules (sub) in Definition 15.16.7(d)i, the condition (15.6) on rules ($\exists^I$-red), and the conditions (15.4) and (15.5) on rules (sub), ($\exists^I$-red).

First consider the condition on rules (sub) in Definition 15.16.7(d)i. It suffices to examine a (sub) occurring between $\Delta_0, \forall X^{\eta}\neg F$ and $\Delta, \forall X^{\eta}\neg F$ with the added formula $\neg F(T^{\nu})$ in P'. From the same condition for the (sub) in P we see that $\forall X^{\eta}\neg F \in \Pi^1_2$, and hence $st_{\Pi}(\neg F(T^{\nu})) \le st_{\Pi}(\forall X^{\eta}\neg F)$ by $\nu < \eta$.

Next consider the conditions on rules (sub), ($\exists^I$-red) in P'. Let K be a rule in P, which is either a (sub) or an ($\exists^I$-red). Assume that K occurs either in P_1 or between J_1 and J_0. Let K' be the corresponding rule occurring in the left part of J' in P'. If the eigenvariable U^{η} does not occur above K, then the new index ν does not occur above K' except it occurs already above K, and the ordinal remains the same. In this case there is nothing to prove. Assume that U^{η} occurs above K.

Let $\sigma = I$ when K is an ($\exists^I$-red), and $\sigma = \Omega_{\mu+1}$ when K is a $(sub)^{\mu}$. Then $\eta < \sigma$ is seen from $\eta < I$ when K is an ($\exists^I$-red), which is in P_1. Also $\eta < \sigma$ is seen from $\eta \leq st_{\Pi}(\forall X^{\eta} \neg F) \leq \mu < \sigma$ when K is a $(sub)^{\mu}$, which is between J_1 and J, and the formula $\forall X^{\eta} \neg F$ is in the upper sequent of K, cf. Definition 15.16.7(d)i.

Let $\lambda = stk(K) = stk'(K')$ be the stack of the rule K in P, and of the rule K' in P'. Let $\mathcal{S}_{P_1}$ be the set of all indices ν_1 such that either a free variable W^{ν_1} or a bound variable Y^{ν_1} occurs in a main formula of a (BI) in P_1. Let $\nu' = \nu\#\{\nu_1 : \nu_1 \in \mathcal{S}_{P_1}\}$. Since the variable U^{η} as well as indices ν_1 in P_1 occurs above K, we have $\{\eta\} \cup \mathcal{S}_{P_1} \subset \mathcal{H}_{\lambda}(\psi_{\sigma}\lambda)$ by Definition 15.24.3a. Then $\nu < \eta \in \mathcal{H}_{\lambda}(\psi_{\sigma}\lambda) \cap \sigma = \psi_{\sigma}\lambda$, and

$$\nu' \in \mathcal{H}_{\lambda}(\psi_{\sigma}\lambda) \tag{15.15}$$

Thus (15.5) is fulfilled for K.

Next let $\alpha_K = o(K; P)$ and $\alpha_{K'} = o(K'; P')$. We have $\alpha_{K'} \ll \alpha_K \{\nu'\}$. By (15.4) we obtain $\alpha_K \in \mathcal{H}_{\lambda}(\psi_{\sigma}\lambda)$. Hence $\alpha_{K'} \in \mathcal{H}_{\lambda}(\psi_{\sigma}\lambda)$ by (15.15). Thus (15.4) is fulfilled for K'. (15.6) follows from $\alpha_{K'} \in \mathcal{H}_{\lambda}(\psi_I\lambda)$ and $\alpha_{K'} < \alpha_K$.

Finally let us show $o(P') < o(P)$. We have $\delta' \underline{\ll} \delta \{\nu'\}$. Consider a $(sub)^{\mu}$ K occurring between J_1 and J. Then $\nu < \eta \leq st_{\Pi}(\forall X^{\eta} \neg F) \leq \mu < \Omega_{\mu+1} = \sigma$, and $\eta \in \mathcal{H}_{\lambda}(\psi_{\sigma}\lambda) \cap \sigma$ with the stack $\lambda = stk(K)$. Hence $\nu' \in \mathcal{H}_{\lambda}(\psi_{\sigma}\lambda)$, and this yields $\gamma' \ll \gamma \{\nu'\}$. We see that $\beta' \ll \beta \{\nu'\}$ as in *Case 11*.

Case 13. The case when the suitable cut formula is a disjunction $A \vee B$.

Case 14. The case when the suitable cut formula is an existential formula $\exists x A[x/u]$.

These cases are seen as in *Case 11*. This completes a proof of Main Lemma 15.27.

References

1. T. Arai. Ordinal diagrams for recursively Mahlo universes. *Arch. Math. Logic*, 39:353–391, 2000.
2. T. Arai. Proof theory for theories of ordinals I: recursively Mahlo ordinals. *Ann. Pure Appl. Logic*, 122:1–85, 2003.
3. T. Arai. Cut-eliminability in second order calculus. *Ann. Japan Asso. Phil. Sci.*, 27:45–60, 2018.
4. W. Buchholz. Normalfunktionen und konstruktive Systeme von Ordinalzahlen. In J. Diller and G.H Müller, editors, *Proof Theory Symposion Kiel, 1974*, volume 500 of *Lecture Notes in Mathematics*, pages 4–25. Springer, Berlin, 1975.
5. W. Buchholz. *Eine Erweiterung der Schnitteliminationsmethode*. Habilitationsschrift, LMU München, 1977.
6. W. Buchholz. A simplified version of local predicativity. In P. Aczel, H. Simmons, and S. S. Wainer, editors, *Proof Theory*, pages 115–147. Cambridge UP, 1992.
7. W. Buchholz. Explaining the Gentzen-Takeuti reduction steps: a second-order system. *Arch. Math. Logic*, 40:255–272, 2001.

8. W. Buchholz and K. Schütte. Ein Ordinalzahlensystem für die Abgrenzung der Π^1_2-Separation und Bar-Induktion. *Sitzungsber. d. Bayer. Akad. d. Wiss., Math.-Nat. Kl.*, pages 99–132, 1983.
9. S. Feferman. Constructive theories of functions and classes. In M. Boffa, D. van Dalen, and K. McAloon, editors, *Logic Colloquium*, volume 78, pages 159–224. North-Holland, Amsterdam, 1979.
10. G. Gentzen. Neue Fassung des Widerspruchsfreiheitsbeweises für die reine Zahlentheorie. *Forschungen zur Logik und zur Grundlegung der exakter Wissenschaften*, Neue Folge 4:19–44, 1938.
11. G. Jäger. A well-ordering proof for Feferman's theory t_0. *Arch. Math. Logic*, 23:65–77, 1983.
12. G. Jäger and W. Pohlers. Eine beweistheoretische Untersuchung von (Δ^1_2-CA) + (BI) und verwandter Systeme. *Sitzungsber. d. Bayer. Akad. d. Wiss., Math.-Nat. Kl.*, pages 1–28, 1982.
13. S. Maehara. Lattice-valued representation of the cut-elimination theorem. *Tsukuba J. Math*, pages 509–521, 1991.
14. W. Pohlers. Proof-theoretic analysis of ID_ν by the method of local predicativity. In W. Sieg W. Buchholz, S. Feferman and W. Pohlers, editors, *Iterated Inductive Definitions and Subsystems of Analysis: Recent Proof-Theoretical Studies*, volume 897 of *Lect. Notes Math.* Springer, 1981.
15. P. Päppinghaus. Completeness properties of classical theories of finite type and the normal form theorem. *Dissertations Mathematicae*, 207:1–66, 1983.
16. K. Schütte. Syntactical and semantical properties of simple type theory. *Jour. Symb. Logic*, 25:305–326, 1960.
17. K. Schütte. Eine beweistheoretische Abgrenzung des Teilsystems der Analysis mit Π^1_2-Separation und Bar-Induktion. *Sitzungsber. d. Bayer. Akad. d. Wiss., Math.-Nat. Kl.*, pages 11–41, 1987.
18. S. G. Simpson. *Subsystems of second order arithmetic*. Cambridge UP, Cambridge, 2nd edition edition, 2009.
19. W. Tait. A non-constructive proof of Gentzen's Hauptsatz for second order predicate logic. *Bull. AMS*, 72:980–983, 1966.
20. G. Takeuti. On the fundamental conjecture of GLC V. *J. Math Soc. Japan*, 10:121–134, 1958.
21. G. Takeuti. On the fundamental conjecture of GLC VI. *Proc. Japan Acad.*, 37:440–443, 1961.
22. G. Takeuti. *Proof Theory*, volume 81 of *Studies in Logic*. North-Holland, Amsterdam, second edition edition, 1987. reprinted from Dover, 2013.

Chapter 16
An Upper Bound for the Proof-Theoretic Strength of Martin-Löf Type Theory with W-type and One Universe

Anton Setzer

Abstract We present an upper bound for the proof theoretic strength of Martin-Löf's type theory with W-type and one universe. This proof, together with the well ordering proof carried out in [12] shows that the proof theoretic strength of this theory is precisely $\psi_{\Omega_1}\Omega_{I+\omega}$, which is slightly more than the strength of Feferman's theory T_0, the classical set theory KPI, and the subsystem of analysis $(\Delta^1_2-CA)+(BI)$. The strength of the intensional and extensional version, and of the version à la Tarski and à la Russell are shown to be the same. The proof is carried out by interpreting the type theories in question in an extension of Kripke-Platek set theory KPI^+. We show that validity of Π^1_1-sentences is preserved in this interpretation.

16.1 Introduction

16.1.1 Proof Theory and Type Theory

Proof theory and type theory are two answers of mathematical logic to the crisis of the foundations of mathematics at the beginning of the 20th century. Proof theory was established by Hilbert in order to prove the consistency of theories by using finitary methods. When Gödel showed that Hilbert's program cannot be carried out as originally intended, the focus of proof theory changed towards analysing theories and determining the minimum ordinal in a natural ordinal notation system such that transfinite induction up to this ordinal proves the consistency of the theory. That ordinal is called the proof theoretic strength, which turned out to be an excellent measure for the strength of theories.

A. Setzer
Department of Computer Science, Swansea University Bay Campus, Fabian Way, Swansea SA1 8EN, UK, e-mail: a.g.setzer@swansea.ac.uk

R. Kahle and M. Rathjen (eds.), *The Legacy of Kurt Schütte*,
https://doi.org/10.1007/978-3-030-49424-7_16

On the other hand, type theories were designed to provide a new framework for mathematics, the consistency of which can be justified, as far as possible, by itself (see [18] for a discussion about why a justification of the consistency of mathematical theories is needed, what can be achieved, and what the limitations are).

Both directions of mathematical logic have become quite important because of their applicability to computer science. Proof theoretic methods can be used for instance to extract programs from proofs, to analyse term rewriting systems and for theoretical questions the area of logic programming.

On the other hand many machine assisted theorem provers, such as Agda, Coq, Idris, Lego, and Epigram, are based on type theory. One reason why type theory is an excellent basis theory is that in type theory algorithms and proofs are essentially the same. Therefore proving becomes very similar to programming, and techniques from software engineering can be used to develop proofs.

Therefore it seems to be interesting to apply proof theory to type theory. In particular, the question mainly answered in this article is: what is the precise proof theoretic strength of Martin-Löf's type theory? This is interesting because the answer determines the exact place of Martin-Löf's type theory on the proof theoretic scale. This allows to compare it with other theories, the strength of which is already known.

More precisely, in this article we are dealing with the strength of Martin-Löf's type theory with one universe and W-type. This work was first presented in our PhD thesis [10][1]. There are two directions to be proved. One is to determine a lower bound, which is presented in [12]. This was done by carrying out a well ordering proof directly in Martin-Löf Type Theory. In this article we present a refined version of the upper bound. We embed type theory in a Kripke-Platek style set theory, KPI^+, the strength of which can then be determined easily using standard proof theoretic techniques. We show that validity of Π^1_1-sentences is preserved in this interpretation.

16.1.2 Related Work

An overview over the state of the art of proof theory of Martin-Löf Type Theory can be found in [14] and [15]. As mentioned before [12] contains the lower bound for the proof theoretic strength of the theories discussed in this article. A more easy readable introduction to well ordering proofs in Martin-Löf type theory can be found in [11].

In [3] Griffor and Rathjen were, independent of the author and in parallel, following another approach towards determining the proof theoretic strength of

[1] In that thesis we were using the formulation of [19] of intensional type theory, which trivially embeds into extensional type theory for which an upper bound is obtained in the current article. In [12] we transferred the well-ordering proof from [10] to the formulation of intensional type theory as it is used in the Martin-Löf Type Theory community. We introduce that theory and embed it into extensional type theory in the current article in Sect. 16.3. The transfer of the well-ordering proof was possible since the equality type was used in the well-ordering proof in [12] only for natural numbers.

Martin-Löf's type theory by embedding constructive set theory into type theory. [3] contains an excellent review of all the research carried out in the past in this area. We refer the interested reader to that article and only mention the main new results concerning type theory obtained in [3]. Griffor and Rathjen showed, that the theory $\mathrm{ML}_1\mathrm{V}$, Martin-Löf's type theory with one universe and Aczel's iterative set V or elimination rules for the universe or both has the strength of Kripke-Platek set theory $\mathrm{KP}\omega$. They showed, that type theory with one universe and the W-type restricted to elements of the universe only, which they called $\mathrm{ML}_1\mathrm{W}$, has strength $(\Delta^1_2 - \mathrm{CA}) + (\mathrm{BI})$. Adding elimination rules for the universe and/or Aczel's iterative set V is shown to yield the same strength. In [3] the obvious generalisation of these results to n universes and ω universes together with their strength is mentioned as well (no detailed proof is given).

In [13] the author has introduced the Mahlo universe, which is significantly stronger than the type theory with W and finitely many universes, and determined a lower bound for its proof theoretic strength. In [16] the basic model construction for type theory with one universe and W-type and with one Mahlo universe and the W-type was given. That article concentrated mainly on the universe constructions and presented the main constructions, whereas the current article contains full details for the upper bound of Martin-Löf Type Theory with W-type and one universe. In the current article we show as well that all Π^1_1-sentences provable in the type theory in question can be shown in KPI^+.

The author has developed an autonomous Mahlo universe [17], a Π_3-reflecting universe, and a Π_n-reflecting universe, but these constructions have not been published yet.

16.1.3 Overview

The article is organised as follows: In Sect. 16.2 we introduce the versions of extensional Martin-Löf's type theory for which we prove the upper bound. To make a precise definition of the substitution, we introduce sets of b-objects, g-terms, g-types, R-terms and R-types, which should contain all the terms and types occurring in Martin-Löf's type theory (g-terms and g-types correspond to the Tarski-formalisation, R-terms and R-types to the Russel-formalisation). These concepts will be needed afterwards for the interpretation of the theory $\mathrm{ML}^{\mathrm{e}}_1\mathrm{W}_{\mathrm{T,U}}$ in KPI^+. Then we define the rules for the extensional Martin-Löf Type Theories with a universe à la Tarski ($\mathrm{ML}^{\mathrm{e}}_1\mathrm{W}_{\mathrm{T}}$, $\mathrm{ML}^{\mathrm{e}}_1\mathrm{W}_{\mathrm{T,U}}$) and à la Russell ($\mathrm{ML}^{\mathrm{e}}_1\mathrm{W}_{\mathrm{R}}$, $\mathrm{ML}^{\mathrm{e}}_1\mathrm{W}_{\mathrm{R,U}}$) considered in this paper. In Sect. 16.3 we introduce intensional variants $\mathrm{ML}^{\mathrm{i}}_1\mathrm{W}_{\mathrm{T}}$, $\mathrm{ML}^{\mathrm{i}}_1\mathrm{W}_{\mathrm{T,U}}$, $\mathrm{ML}^{\mathrm{i}}_1\mathrm{W}_{\mathrm{R}}$, $\mathrm{ML}^{\mathrm{i}}_1\mathrm{W}_{\mathrm{R,U}}$ of those theories. We carry out the well-known result that they embed in a straightforward way into the extensional versions of those theories. In Sect. 16.4 we compare the Tarski and the Russel-versions and show that all of them can be embedded into the strongest Tarski version $\mathrm{ML} := \mathrm{ML}^{\mathrm{e}}_1\mathrm{W}_{\mathrm{T,U}}$. In Sect. 16.5 we introduce Kripke-Platek set theory and the extension KPI^+ used for modelling the type theories in question in this article.

Having introduced the relevant theories we now develop an interpretation of ML in the extension of Kripke-Platek style called KPI^+. This embedding is a quite general and flexible method, which can be adopted to variations of Martin-Löf's type theory. In Sect. 16.6, we show, how to interpret terms and types in KPI^+. Types will be introduced as sets of pairs of terms, which are considered to be equal in that type. We see, how much strength is needed in order to interpret the W-type and the universe, which correspond to the proof theoretic strength added by those constructions. In Sect. 16.7 we prove basic properties of this interpretation, such as monotonicity of operators, that one obtains equivalence relations, and substitution lemmas.

In Sect. 16.8 we prove that the interpretation is correct, that is

$$\text{If } \mathrm{ML} \vdash r : A \text{ then } \mathrm{KPI}^+ \vdash \langle r, r\rangle \in A^* .$$

Furthermore, we conclude, that the extended version $|\mathrm{ML}_1^{\mathrm{e}}\mathrm{W}_{\mathrm{T,U}}|$ can be interpreted as well.

In Sect. 16.9 we interpret sentences of Arithmetic A as $\widehat{A}$ in $\mathrm{ML}_1^{\mathrm{e}}\mathrm{W}_{\mathrm{T}}$ and prove, that for Π_1^1-sentences A we get that $\mathrm{ML}_1^{\mathrm{e}}\mathrm{W}_{\mathrm{T}} \vdash a \in \widehat{A}$ implies $\mathrm{KPI}^+ \vdash A$. We conclude in Sect. 16.10 that $|\mathrm{ML}| \leq |\mathrm{KPI}^+|$. To complete the proof, we show $|\mathrm{KPI}^+| \leq \psi_{\Omega_1}(\Omega_{\mathrm{I}+\omega})$, which implies because of the embedding of the Martin-Löf type theories used into ML, that $|\mathrm{ML}_1^{\mathrm{e}}\mathrm{W}_{\mathrm{T,U}}|$, $|\mathrm{ML}_1^{\mathrm{e}}\mathrm{W}_{\mathrm{T}}|$, $|\mathrm{ML}_1^{\mathrm{e}}\mathrm{W}_{\mathrm{R,U}}|$, $|\mathrm{ML}_1^{\mathrm{e}}\mathrm{W}_{\mathrm{R}}|$, $|\mathrm{ML}_1^{\mathrm{i}}\mathrm{W}_{\mathrm{T,U}}|$, $|\mathrm{ML}_1^{\mathrm{i}}\mathrm{W}_{\mathrm{T}}|$, $|\mathrm{ML}_1^{\mathrm{i}}\mathrm{W}_{\mathrm{R,U}}|$, $|\mathrm{ML}_1^{\mathrm{i}}\mathrm{W}_{\mathrm{R}}| \leq \psi_{\Omega_1}(\Omega_{\mathrm{I}+\omega})$. By [12], this bound is sharp.

16.2 Definition of the Formal System of Extensional Martin-Löf's Type Theory

In this section we will introduce the formulations of extensional Martin-Löf Type Theory, both a là Tarski and a là Russell, which we will embed into KPI^+. In Sect. 16.3 we will introduce intensional type theory and show that it embeds into extensional type theory. Therefore the upper bound for extensional type theory is an upper bound for intensional type theory as well.

Definition 16.1. (a) The symbols of extensional Martin-Löf's type theory, are infinitely many variables z_i ($i \in \omega$); the symbols $\Rightarrow$, :, „ (,), =, $\in$, λ; the term constructors (with their arity in parenthesis) 0 (0), r (0), $\widehat{\mathbb{N}}$ (0), A_n^k (for each $n < k$, with arity 0), $\widehat{\mathbb{N}}_k$ (for each $k \in \omega$, with arity 0), S (1), i (1), j (1), p_0 (1), p_1 (1), p (2), sup (2), R (2), Ap (2), $\widehat{+}$ (2), $\widehat{\Pi}$ (2), $\widehat{\Sigma}$ (2), $\widehat{\mathrm{W}}$ (2), D (3), P (3), $\widehat{\mathrm{I}}$ (3), C_n ($n \in \omega$, arity $n+1$); the type constructors with their arity $\mathbb{N}_k$ (for each $k \in \omega$, arity 0), $\mathbb{N}$ (0), U (0), T (1), + (2), Π (2), Σ (2), W (2) and I (3).

To make it easier to remember the meaning of the symbols, we give the following hints: r is the (unique) element of an identity type I; A_n^k is the nth element of the finite type $\mathbb{N}_k$ with k elements, C_k the **C**ase distinction for $\mathbb{N}_k$; O is the zero,

S the **S**uccessor, P **P**rimitive recursion or induction over the natural numbers $\mathbb{N}$; i stands for left **i**nclusion, j for right inclusion, D is the case distinction for a type $A + B$, which is the disjoint union of A and B; p_0 and p_1 are the projections, p the pairing for the Σ-type; R the **R**ecursion operator over a W-type; Ap the application of a function (as an element of a Π-type) to an argument; $\widehat{\mathbb{N}}$, $\widehat{\mathbb{N}}_k$, $\widehat{\mathrm{I}}$, $\widehat{+}$, $\widehat{\Sigma}$, $\widehat{\Pi}$, $\widehat{\mathrm{W}}$ are codes for the types $\mathbb{N}$, $\mathbb{N}_k$, I, +, Σ, Π, W, respectively, as elements of the universe U, which become, if the **T**arski-operator T is applied to them, the corresponding type.

(b) We usually write "symbols of Martin-Löf Type Theory" for "symbols of extensional Martin-Löf Type Theory", when there is no confusion.

(c) The b-objects are variables, $\lambda x.b$ and $C(b_1, \ldots, b_n)$, if C is an n-ary term or type constructor and $b, b_1, \ldots, b_n$ are b-objects.
The set of free variables of a b-object $\mathrm{FV}(b)$ are defined as usual. We write +, $\widehat{+}$ infix (that is $(a + b)$ for $+(a, b)$) $(x)t$ for $\lambda x.t$, $(x, y)t$ for $(x)(y)t$, $(x, y, z)t$ for $(x)(y, z)t$, and if $S \in \{\Sigma, \Pi, \mathrm{W}, \widehat{\Sigma}, \widehat{\Pi}, \widehat{\mathrm{W}}\}$, $Sx \in s.t := S(s, (x)t)$. Furthermore, we write sometimes $r\ s$ or rs for $\mathrm{Ap}(r, s)$.
We have the usual conventions about omitting brackets, especially the scope of $\lambda x.$ is as long as possible, for instance $\lambda x.s\ t$ should be read as $\lambda x.(s\ t)$. We define for b-objects $b_1, \ldots, b_n, b$ and variables $x_1, \ldots, x_n$ the simultaneous substitution $b[x_1/b_1, \ldots, x_n/b_n]$ as usual (using the convention, that if $x_i = x_j$ then the first substitution applies) and "$b[x_1/b_1, \ldots, x_n/b_n]$ is an allowed substitution".
α-equality ($=_\alpha$) is defined as usual.

(d) The set of g-terms (for generalised terms) is inductively defined as: variables x are g-terms; if $n < k$, $n, k \in \mathbb{N}$, then A_n^k is a g-term; and if $k \in \mathbb{N}$, then $\widehat{\mathbb{N}}_k$ is a g-term; if r, s, t are g-terms, $x, y, z, x' \in \mathrm{Var}_{\mathrm{ML}}$, $x \neq y \neq z \neq x$, then 0, r, $\widehat{\mathbb{N}}$, $\mathrm{S}(r)$, $\lambda x.r$, $\mathrm{p}(r, s)$, $\mathrm{sup}(r, s)$, $\mathrm{i}(r)$, $\mathrm{j}(r)$, $\mathrm{P}(r, s, (x, y)t)$, $\mathrm{Ap}(r, s)$, $\mathrm{p}_0(r)$, $\mathrm{p}_1(r)$, $\mathrm{R}(r, (x, y, z)s)$, $\mathrm{D}(r, (x)s, (x')t)$, $\widehat{\Pi}x \in r.s$, $\widehat{\Sigma}x \in r.s$, $\widehat{\mathrm{W}}x \in r.s$, $r \widehat{+} s$, $\widehat{\mathrm{I}}(r, s, t)$ are g-terms; if $n \in \mathbb{N}$ and $r, s_1, \ldots, s_n$ are g-terms, then $\mathrm{C}_n(r, s_1, \ldots, s_n)$ is a g-term. Let $\mathrm{Term}_{\mathrm{Cl}}$ be the set of closed g-terms.

(e) The g-types are $\mathbb{N}_k$ ($k \in \omega$), $\mathbb{N}$, U, and, if A, B are g-types, $x \in \mathrm{Var}_{\mathrm{ML}}$, r, s g-terms, then $\Pi x \in A.B$, $\Sigma x \in A.B$, $\mathrm{W}x \in A.B$, $A + B$, $\mathrm{I}(A, r, s)$, $\mathrm{T}(r)$ are g-types.

(f) A g-context-piece is a string $x_1 : A_1, \ldots, x_n : A_n$, where $n \geq 0$, x_i different variables, A_i g-types.
A g-context is a g-context-piece $x_1 : A_1, \ldots, x_n : A_n$, s.t. $\mathrm{FV}(A_i) \subset \{x_1, \ldots, x_{n-1}\}$.
A g-judgement is $A = B$: type or $s = t : A$ where A, B are g-types, s, t are g-terms. A g-dependent judgements is $\Gamma \Rightarrow \theta$ where Γ is a g-context and θ a g-judgement.

(g) R-terms (for Russel-terms) are defined as the g-terms, except, that we replace $\widehat{\Pi}$, $\widehat{\Sigma}$, $\widehat{\mathrm{W}}$, $\widehat{+}$, $\widehat{\mathrm{I}}$ by $\Pi, \Sigma, \mathrm{W}, +, \mathrm{I}$ respectively.
R-types are defined by the same definition as the g-types, but referring to R-terms instead of g-terms, and replacing that "$\mathrm{T}(r)$ is a g-type" by "r is an R-type for r an R-term". R-context-pieces, -contexts, -judgements, and -dependent judgements are defined as the corresponding g-constructions, but referring to R-terms and -types instead of the g-terms and -types.

(h) We treat the usual judgements A : type and $s : A$ as abbreviations:
$(A : \text{type}) :\equiv (A = A : \text{type})$, $(s : A) :\equiv (s = s : A)$.

(i) We abbreviate $[\vec{x}/\vec{t}] := [x_1/t_1, \ldots, x_n/t_n]$ if $\vec{x} = x_1, \ldots, x_n$ and $\vec{t} = t_1, \ldots, t_n$.
$[x_1/t_1, \ldots, x_n/t_n] \setminus \{y\} := [\vec{x}/\vec{t}] \setminus \{y\}$ is the result of omitting in $[\vec{x}/\vec{t}]$ the x_i/t_i s.t. $x_i = y$, and $[\vec{x}/\vec{t}] \setminus \{y_1, \ldots, y_m\} := (\cdots(([\vec{x}/\vec{t}] \setminus \{y_1\}) \setminus \{y_2\}) \cdots \setminus \{y_m\})$.

Definition 16.2. of extensional Martin-Löf Type Theory with W-type and one Universe. We will define the rules, which are of the form

$$(Rule)\ \frac{\Gamma_1 \Rightarrow \Theta_1 \qquad \cdots \qquad \Gamma_n \Rightarrow \Theta_n}{\Gamma \Rightarrow \Theta}$$

where $\Gamma_1, \ldots, \Gamma_n, \Gamma$ are g-context-pieces, $\Theta_1, \ldots, \Theta_n, \Theta$ are g-judgements ($n = 0$ is allowed) in the version à la Tarski, and R-context-pieces and R-judgements in the version à la Russell.

Then we define for $\mathrm{ML} = \mathrm{ML}_1^{\mathrm{e}}\mathrm{W}_{\mathrm{T}}$ (the extensional version à la Tarski) or $\mathrm{ML} = \mathrm{ML}_1^{\mathrm{e}}\mathrm{W}_{\mathrm{T,U}}$ (the extensional version à la Tarski with additional rules for the universe), $\mathrm{ML} \vdash \Gamma \Rightarrow \Theta$ inductively by:

If (*Rule*) is a rule of ML as above, Δ is a g-context-piece such that $\Delta, \Gamma_1, , \ldots, \Delta, \Gamma_n$, Δ, Γ are g-contexts, and if $\mathrm{ML} \vdash \Delta, \Gamma_i \Rightarrow \Theta_i$ for $i = 1, \ldots, n$, then $\mathrm{ML} \vdash \Delta, \Gamma \Rightarrow \Theta$.

Analogously we define $\mathrm{ML}_1^{\mathrm{e}}\mathrm{W}_{\mathrm{R}} \vdash \Gamma \Rightarrow \Theta$ (the extensional version à la Russell) and $\mathrm{ML}_1^{\mathrm{e}}\mathrm{W}_{\mathrm{R,U}} \vdash \Gamma \Rightarrow \Theta$ $\mathrm{ML}_1^{\mathrm{e}}\mathrm{W}_{\mathrm{R,U}} \vdash \Gamma \Rightarrow \Theta$ (the version à la Russell corresponding to $\mathrm{ML}_1^{\mathrm{e}}\mathrm{W}_{\mathrm{T,U}}$), but referring to R-context-pieces, R-contexts etc.instead of g-context-pieces, g-contexts etc.

We will write Θ for $\Rightarrow \Theta$ as a premise of a rule.

In the following, let A, B, A', B' be g-types (or R-types in the formulation à la Russell), $a, b, r, s, t, r_i, s_i, t_i, a', b', r', s', t', r_i', s_i', t_i', t''$ be g-terms (or R-terms) θ, θ' be g-(or R-)judgements Γ' be a g-(R-)context-pieces.

Furthermore, let $x, y, z, u \in \mathrm{Var}_{\mathrm{ML}}$. Additionally assume for all rules, that all substitution mentioned explicitly are allowed. For instance in the rule $(\mathbb{N}_{\mathrm{S}}^{=})$, assume that $s_1[x/t, y/\mathrm{P}(t, s_0, (x, y)s_1)]$ and $A[z/\mathrm{S}(t)]$ are allowed substitutions.

General Rules

$$(\mathrm{ASS})\ \frac{A : \text{type}}{x : A \Rightarrow x : A} \qquad (\mathrm{THIN})\ \frac{A : \text{type} \qquad \Gamma' \Rightarrow \Theta}{x : A, \Gamma' \Rightarrow \Theta}$$

$$(\mathrm{SYM})\ \frac{t = t' : A}{t' = t : A} \quad \frac{A = B : \text{type}}{B = A : \text{type}} \qquad (\mathrm{SUB})\ \frac{x : A, \Gamma' \Rightarrow \Theta \qquad \Rightarrow t : A}{\Gamma'[x/t] \Rightarrow \Theta[x/t]}$$

$$(\mathrm{TRANS})\ \frac{t = t' : A \qquad t' = t'' : A}{t = t'' : A} \qquad \frac{A = B : \text{type} \qquad B = C : \text{type}}{A = C : \text{type}}$$

$$(\mathrm{REPL1})\ \frac{x : A, \Gamma' \Rightarrow B : \text{type} \qquad \Rightarrow t = t' : A}{\Gamma'[x/t] \Rightarrow B[x/t] = B[x/t'] : \text{type}}$$

$$\text{(REPL2)}\ \frac{x : A, \Gamma' \Rightarrow s : B \qquad \Rightarrow t = t' : A}{\Gamma'[x/t] \Rightarrow s[x/t] = s[x/t'] : B[x/t]}$$

$$\text{(REPL3)}\ \frac{t = t' : A \qquad A = B : \text{type}}{t = t' : B}$$

$$\text{(ALPHA)}\ \frac{x : A, \Gamma' \Rightarrow \theta}{x : A', \Gamma' \Rightarrow \theta} \quad \frac{A : \text{type}}{A = A' : \text{type}} \quad \frac{t : A}{t = t' : A} \quad (\text{if } A =_\alpha A',\ t =_\alpha t')$$

Formation Rules

$$(\mathbb{N}_k^{\text{F}})\quad \mathbb{N}_k : \text{type} \quad (k \in \mathbb{N}) \qquad\qquad (\mathbb{N}^{\text{F}})\quad \mathbb{N} : \text{type}$$

$$(\Pi^{\text{F}})\ \frac{A = A' : \text{type} \qquad x : A \Rightarrow B = B' : \text{type}}{\Pi x \in A.B = \Pi x \in A'.B' : \text{type}} \qquad (\Sigma^{\text{F}})\ \frac{A = A' : \text{type} \qquad x : A \Rightarrow B = B' : \text{type}}{\Sigma x \in A.B = \Sigma x \in A'.B' : \text{type}}$$

$$(\text{W}^{\text{F}})\ \frac{A = A' : \text{type} \qquad x : A \Rightarrow B = B' : \text{type}}{\text{W}x \in A.B = \text{W}x \in A'.B' : \text{type}} \qquad (+^{\text{F}})\ \frac{A = A' : \text{type} \qquad B = B' : \text{type}}{A + B = A' + B' : \text{type}}$$

$$(\text{I}^{\text{F}})\ \frac{A = A' : \text{type} \qquad t = t' : A \qquad s = s' : A}{\text{I}(A, t, s) = \text{I}(A', t', s'') : \text{type}}$$

Introduction Rules

$$(\mathbb{N}_k^{\text{I}})\quad \text{A}_n^k : \mathbb{N}_k \quad (n < k,\ n, k \in \mathbb{N}) \qquad\qquad (\mathbb{N}^{\text{I}})\quad 0 : \mathbb{N} \quad \frac{t = t' : \mathbb{N}}{\text{S}(t) = \text{S}(t') : \mathbb{N}}$$

$$(\Pi^{\text{I}})\ \frac{x : A \Rightarrow B : \text{type} \qquad x : A \Rightarrow t = t' : B}{\lambda x.t = \lambda x.t' : \Pi x \in A.B} \qquad (\Sigma^{\text{I}})\ \frac{x : A \Rightarrow B : \text{type} \qquad s = s' : A \qquad t = t' : B[x/s]}{\text{p}(s, t) = \text{p}(s', t') : \Sigma x \in A.B}$$

$$(\text{W}^{\text{I}})\ \frac{x : A \Rightarrow B : \text{type} \qquad r = r' : A \qquad s = s' : B[x/r] \to \text{W}x \in A.B}{\text{sup}(r, s) = \text{sup}(r', s') : \text{W}x \in A.B}$$

$$(+^{\mathrm{I}}_1)\ \frac{A : \text{type} \quad B : \text{type} \quad s = s' : A}{\mathrm{i}(s) = \mathrm{i}(s') : A + B} \qquad (+^{\mathrm{I}}_2)\ \frac{A : \text{type} \quad B : \text{type} \quad s = s' : B}{\mathrm{j}(s) = \mathrm{j}(s') : A + B}$$

$$(\mathrm{I}^{\mathrm{I}})\ \frac{t = t' : A}{\mathrm{r} : \mathrm{I}(A, t, t')}$$

Elimination Rules

$$(\mathbb{N}^{\mathrm{E}}_k)\ \frac{x : \mathbb{N}_k \Rightarrow A : \text{type} \quad t = t' : \mathbb{N}_k \quad s_i = s_i' : A[x/\mathrm{A}^k_i](i = 0 \ldots k-1)}{\mathrm{C}_k(t, s_0, \ldots, s_{k-1}) = \mathrm{C}_k(t', s_0', \ldots, s_{k-1}') : A[x/t]} \quad (k \in \mathbb{N})$$

$$(\mathbb{N}^{\mathrm{E}})\ \frac{x : \mathbb{N} \Rightarrow A : \text{type} \quad r = r' : \mathbb{N} \quad s = s' : A[x/0] \quad x : \mathbb{N}, y : A \Rightarrow t = t' : A[x/\mathrm{S}(x)]}{\mathrm{P}(r, s, (x, y)t) = \mathrm{P}(r', s', (x, y)t') : A[x/r]}$$

$$(\Pi^{\mathrm{E}})\ \frac{x : A \Rightarrow B : \text{type} \quad s = s' : \Pi x \in A.B \quad r = r' : A}{\mathrm{Ap}(s, r) = \mathrm{Ap}(s', r') : B[x/r]}$$

$$(\Sigma^{\mathrm{E}}_0)\ \frac{x : A \Rightarrow B : \text{type} \quad r = r' : \Sigma x \in A.B}{\mathrm{p}_0(r) = \mathrm{p}_0(r') : A} \qquad (\Sigma^{\mathrm{E}}_1)\ \frac{x : A \Rightarrow B : \text{type} \quad r = r' : \Sigma x \in A.B}{\mathrm{p}_1(r) = \mathrm{p}_1(r') : B[x/\mathrm{p}_0(r)]}$$

$$(\mathrm{W}^{\mathrm{E}})\ \frac{\begin{array}{c} x : A \Rightarrow B : \text{type} \\ u : \mathrm{W}x \in A.B \Rightarrow C : \text{type} \\ r = r' : \mathrm{W}x \in A.B \\ x : A, y : B \to \mathrm{W}x \in A.B, z : \Pi v \in B.C[u/\mathrm{Ap}(y, v)] \\ \qquad \Rightarrow t = t' : C[u/\sup(x, y)] \end{array}}{\mathrm{R}(r, (x, y, z)t) = \mathrm{R}(r', (x, y, z)t') : C[u/r]}$$

$$(+^{\mathrm{E}})\ \frac{A : \text{type} \quad B : \text{type} \quad z : A + B \Rightarrow C : \text{type} \quad r = r' : A + B \quad x : A \Rightarrow s = s' : C[z/\mathrm{i}(x)] \quad y : B \Rightarrow t = t' : C[z/\mathrm{j}(y)]}{\mathrm{D}(r, (x)s, (y)t) = \mathrm{D}(r', (x)s', (y)t') : C[z/r]}$$

$$(\mathrm{I}^{\mathrm{E}})\ \frac{s : A \quad t : A \quad r : \mathrm{I}(A, s, t)}{s = t : A}$$

Equality Rules

$$(\mathbb{N}_k^=)\ \frac{x : \mathbb{N}_k \Rightarrow A : \text{type} \qquad s_i : A[x/\mathrm{A}_i^k](i = 0 \ldots k-1)}{\mathrm{C}_k(\mathrm{A}_n^k, s_0, \ldots, s_{k-1}) = s_n : A[x/\mathrm{A}_n^k]} \quad (n < k,\ n, k \in \mathbb{N})$$

$$(\mathbb{N}_0^=)\ \frac{x : \mathbb{N} \Rightarrow A : \text{type} \qquad s : A[x/0] \qquad x : \mathbb{N}, y : A \Rightarrow t : A[x/\mathrm{S}(x)]}{\mathrm{P}(0, s, (x, y)t) = s : A[x/0]}$$

$$(\mathbb{N}_\mathrm{S}^=)\ \frac{\begin{array}{cc} x : \mathbb{N} \Rightarrow A : \text{type} & r : \mathbb{N} \\ s : A[x/0] & x : \mathbb{N}, y : A \Rightarrow t : A[x/\mathrm{S}(x)] \end{array}}{\mathrm{P}(\mathrm{S}(r), s, (x, y)t) = t[x/r, y/\mathrm{P}(r, s, (x, y)t)] : A[x/\mathrm{S}(r)]}$$

$$(\Pi^=)\ \frac{\begin{array}{c} x : A \Rightarrow B : \text{type} \\ \lambda x.t : \Pi x \in A.B \\ r : A \end{array}}{\mathrm{Ap}(\lambda x.t, r) = t[x/r] : B[x/r]}$$

$$(\Pi^\eta)\ \frac{x : A \Rightarrow B : \text{type} \qquad t : \Pi x \in A.B}{\lambda x.\mathrm{Ap}(t, x) = t : \Pi x \in A.B} \quad (\text{if } x \notin \mathrm{FV}(t))$$

$$(\Sigma_0^=)\ \frac{A : \text{type} \qquad \mathrm{p}(r, s) : \Sigma x \in A.B}{\mathrm{p}_0(\mathrm{p}(r, s)) = r : A}$$

$$(\Sigma_1^=)\ \frac{x \in A \Rightarrow B : \text{type} \qquad \mathrm{p}(r, s) : \Sigma x \in A.B}{\mathrm{p}_1(\mathrm{p}(r, s)) = s : B[x/r]}$$

$$(\Sigma_2^=)\ \frac{x \in A \Rightarrow B : \text{type} \qquad t : \Sigma x \in A.B}{t = \mathrm{p}(\mathrm{p}_0(t), \mathrm{p}_1(t)) : \Sigma x \in A.B}$$

$$(\mathrm{W}^=)\ \frac{\begin{array}{c} x : A \Rightarrow B : \text{type} \\ u : \mathrm{W}x \in A.B \Rightarrow C : \text{type} \\ r : A \\ s : B[x/r] \to \mathrm{W}x \in A.B \\ x : A, y : B \to \mathrm{W}x \in A.B, z : (\Pi v \in B.C[u/\mathrm{Ap}(y, v)]) \Rightarrow t : C[u/\sup(x, y)] \end{array}}{\mathrm{R}(\sup(r, s), (x, y, z)t) = t[x/r, y/s, z/\lambda v.\mathrm{R}(\mathrm{Ap}(s, v), (x, y, z)t)] : C[u/\sup(r, s)]}$$

$$(\text{If } v \notin \mathrm{FV}(s) \cup \mathrm{FV}((x, y, z)t))$$

$$(+_0^=)\ \frac{z \in A + B \Rightarrow C : \text{type} \quad r : A \quad x : A \Rightarrow s : C[z/\mathrm{i}(x)] \quad y : B \Rightarrow t : C[z/\mathrm{j}(y)]}{\mathrm{D}(\mathrm{i}(r), (x)s, (y)t) = t[x/r] : C[z/\mathrm{i}(r)]}$$

$$(+_1^=)\ \frac{z \in A + B \Rightarrow C : \mathrm{type} \quad r : B \quad x : A \Rightarrow s : C[z/\mathrm{i}(x)] \quad y : B \Rightarrow t : C[z/\mathrm{j}(y)]}{\mathrm{D}(\mathrm{j}(r), (x)s, (y)t) = t[y/r] : C[z/\mathrm{j}(r)]}$$

$$(\mathrm{I}^=)\ \frac{A : \mathrm{type} \quad s : A \quad t : A \quad r : \mathrm{I}(A, s, t)}{r = \mathrm{r} : \mathrm{I}(A, s, t)}$$

Rules for the Universe (à la Tarski)

Formation Rules for the Universe

$(\mathrm{U}^{\mathrm{I}})\quad \mathrm{U} : \mathrm{type}$

$$(\mathrm{T}^{\mathrm{I}})\ \frac{a = a' : \mathrm{U}}{\mathrm{T}(a) = \mathrm{T}(a') : \mathrm{type}}$$

Introduction Rules for the Universe

$(\widehat{\mathbb{N}}_k^{\mathrm{I}})\quad \widehat{\mathbb{N}}_k : \mathrm{U} \quad (k \in \omega)$

$(\widehat{\mathbb{N}}^{\mathrm{I}})\quad \widehat{\mathbb{N}} : \mathrm{U}$

$$(\widehat{\Pi}^{\mathrm{I}})\ \frac{a = a' : \mathrm{U} \qquad x : \mathrm{T}(a) \Rightarrow b = b' : \mathrm{U}}{\widehat{\Pi} x \in a.b = \widehat{\Pi} x \in a'.b' : \mathrm{U}}$$

$$(\widehat{\Sigma}^{\mathrm{I}})\ \frac{a = a' : \mathrm{U} \qquad x : \mathrm{T}(a) \Rightarrow b = b' : \mathrm{U}}{\widehat{\Sigma} x \in a.b = \widehat{\Sigma} x \in a'.b' : \mathrm{U}}$$

$$(\widehat{\mathrm{W}}^{\mathrm{I}})\ \frac{a = a' : \mathrm{U} \qquad x : \mathrm{T}(a) \Rightarrow b = b' : \mathrm{U}}{\widehat{\mathrm{W}} x \in a.b = \widehat{\mathrm{W}} x \in a'.b' : \mathrm{U}}$$

$$(\widehat{+}^{\mathrm{I}})\ \frac{a = a' : \mathrm{U} \qquad b = b' : \mathrm{U}}{a \widehat{+} b = a' \widehat{+} b' : \mathrm{U}}$$

$$(\widehat{\mathrm{I}}^{\mathrm{I}})\ \frac{a = a' : \mathrm{U} \qquad s = s' : \mathrm{T}(a) \qquad t = t' : \mathrm{T}(a)}{\widehat{\mathrm{I}}(a, s, t) = \widehat{\mathrm{I}}(a', s', t') : \mathrm{U}}$$

Equality Rules for the Universe

$(\widehat{\mathbb{N}}_k^=)\quad \mathrm{T}(\widehat{\mathbb{N}}_k) = \mathbb{N}_k : \mathrm{type} \quad (k \in \omega)$

$(\widehat{\mathbb{N}}^=)\quad \mathrm{T}(\widehat{\mathbb{N}}) = \mathbb{N} : \mathrm{type}$

$$(\widehat{\Pi}^=)\ \frac{a : \mathrm{U} \qquad x : \mathrm{T}(a) \Rightarrow b : \mathrm{U}}{\mathrm{T}(\widehat{\Pi} x \in a.b) = \Pi x \in \mathrm{T}(a).\mathrm{T}(b) : \mathrm{type}}$$

$$(\widehat{\Sigma}^{=})\ \frac{a : \mathrm{U} \qquad x : \mathrm{T}(a) \Rightarrow b : \mathrm{U}}{\mathrm{T}(\widehat{\Sigma} x \in a.b) = \Sigma x \in \mathrm{T}(a).\mathrm{T}(b) : \mathrm{type}}$$

$$(\widehat{\mathrm{W}}^{=})\ \frac{a : \mathrm{U} \qquad x : \mathrm{T}(a) \Rightarrow b : \mathrm{U}}{\mathrm{T}(\widehat{\mathrm{W}} x \in a.b) = \mathrm{W} x \in \mathrm{T}(a).\mathrm{T}(b) : \mathrm{type}}$$

$$(\widehat{+}^{=})\ \frac{a : \mathrm{U} \qquad b : \mathrm{U}}{\mathrm{T}(a \widehat{+} b) = \mathrm{T}(a) + \mathrm{T}(b) : \mathrm{type}}$$

$$(\widehat{\mathrm{I}}^{=})\ \frac{a : \mathrm{U} \qquad t : \mathrm{T}(a) \qquad s : \mathrm{T}(a)}{\mathrm{T}(\widehat{\mathrm{I}}(a, t, s)) = \mathrm{I}(\mathrm{T}(a), t, s) : \mathrm{type}}$$

The rules of $\mathrm{ML}_1^e\mathrm{W}_\mathrm{T}$ are all rules mentioned above (using g-terms, g-types etc.).

The rules of $\mathrm{ML}_1^e\mathrm{W}_{\mathrm{T,U}}$ are all rules of $\mathrm{ML}_1^e\mathrm{W}_\mathrm{T}$ and additionally the following rules:

$$(\widehat{\Sigma}^{\mathrm{E}})\ \frac{\widehat{\Sigma} x \in a.b : \mathrm{U}}{a : \mathrm{U}} \quad \frac{\widehat{\Sigma} x \in a.b : \mathrm{U}}{x : \mathrm{T}(a) \Rightarrow b : \mathrm{U}} \qquad (\widehat{\Pi}^{\mathrm{E}})\ \frac{\widehat{\Pi} x \in a.b : \mathrm{U}}{a : \mathrm{U}} \quad \frac{\widehat{\Pi} x \in a.b : \mathrm{U}}{x : \mathrm{T}(a) \Rightarrow b : \mathrm{U}}$$

$$(\widehat{\mathrm{W}}^{\mathrm{E}})\ \frac{\widehat{\mathrm{W}} x \in a.b : \mathrm{U}}{a : \mathrm{U}} \quad \frac{\widehat{\mathrm{W}} x \in a.b : \mathrm{U}}{x : \mathrm{T}(a) \Rightarrow b : \mathrm{U}} \qquad (\widehat{+}^{\mathrm{E}})\ \frac{a \widehat{+} b : \mathrm{U}}{a : \mathrm{U}} \quad \frac{a \widehat{+} b : \mathrm{U}}{b : \mathrm{U}}$$

$$(\widehat{\mathrm{I}}^{=})\ \frac{\widehat{\mathrm{I}}(a, s, t) : \mathrm{U}}{a : \mathrm{U}} \quad \frac{\widehat{\mathrm{I}}(a, s, t) : \mathrm{U}}{s : \mathrm{T}(a)} \quad \frac{\widehat{\mathrm{I}}(a, s, t) : \mathrm{U}}{t : \mathrm{T}(a)}$$

The rules of $\mathrm{ML}_1^e\mathrm{W}_\mathrm{R}$ are the same as the rules of $\mathrm{ML}_1^e\mathrm{W}_\mathrm{T}$, but referring to R-terms, R-types etc. and replacing the Rules for the Universe by:

Rules for the Universe à la Russell

Formation Rules for the Universe

$(\mathrm{U}^{\mathrm{I}})\quad \mathrm{U} : \mathrm{type}$

$$(\mathrm{T}^{\mathrm{I}})\ \frac{a = a' : \mathrm{U}}{a = a' : \mathrm{type}}$$

Introduction Rules for the Universe

$(\mathbb{N}_k^{\mathrm{U}})\quad \mathbb{N}_k : \mathrm{U} \quad (k \in \omega)$

$(\mathbb{N}_k^{\mathrm{U}})\quad \mathbb{N} : \mathrm{U}$

$$(\Pi^{\mathrm{U}})\ \frac{a = a' : \mathrm{U} \qquad x : a \Rightarrow b = b' : \mathrm{U}}{\Pi x \in a.b = \Pi x \in a'.b' : \mathrm{U}} \qquad (\Sigma^{\mathrm{U}})\ \frac{a = a' : \mathrm{U} \qquad x : a \Rightarrow b = b' : \mathrm{U}}{\Sigma x \in a.b = \Sigma x \in a'.b' : \mathrm{U}}$$

$$(\mathrm{W}^{\mathrm{U}})\ \frac{a = a' : \mathrm{U} \qquad x : a \Rightarrow b = b' : \mathrm{U}}{\mathrm{W}x \in a.b = \mathrm{W}x \in a'.b' : \mathrm{U}} \qquad (+^{\mathrm{U}})\ \frac{a = a' : \mathrm{U} \qquad b = b' : \mathrm{U}}{a + b = a' + b' : \mathrm{U}}$$

$$(\mathrm{I}^{\mathrm{U}})\ \frac{a = a' : \mathrm{U} \qquad t = t' : a \qquad s = s' : a}{\mathrm{I}(a, t, s) = \mathrm{I}(a', t', s') : \mathrm{U}}$$

The rules of $\mathrm{ML}_1^{\mathrm{e}}\mathrm{W}_{\mathrm{R,U}}$ are all rules of $\mathrm{ML}_1^{\mathrm{e}}\mathrm{W}_{\mathrm{R}}$ and additionally the following rules:

$$(\Sigma^{\mathrm{U,E}})\ \frac{\Sigma x \in a.b : \mathrm{U}}{a : \mathrm{U}} \quad \frac{\Sigma x \in a.b : \mathrm{U}}{x : a \Rightarrow b : \mathrm{U}} \qquad (\Pi^{\mathrm{U,E}})\ \frac{\Pi x \in a.b : \mathrm{U}}{a : \mathrm{U}} \quad \frac{\Pi x \in a.b : \mathrm{U}}{x : a \Rightarrow b : \mathrm{U}}$$

$$(\mathrm{W}^{\mathrm{U,E}})\ \frac{\mathrm{W}x \in a.b : \mathrm{U}}{a : \mathrm{U}} \quad \frac{\mathrm{W}x \in a.b : \mathrm{U}}{x : a \Rightarrow b : \mathrm{U}}$$

$$(+^{\mathrm{U,E}})\ \frac{a + b : \mathrm{U}}{a : \mathrm{U}} \quad \frac{a + b : \mathrm{U}}{b : \mathrm{U}}$$

$$(\mathrm{I}^{\mathrm{U,E}})\ \frac{\mathrm{I}(a, s, t) : \mathrm{U}}{a : \mathrm{U}} \quad \frac{\mathrm{I}(a, s, t) : \mathrm{U}}{s : a} \quad \frac{\mathrm{I}(a, s, t) : \mathrm{U}}{t : a}$$

16.3 Intensional Martin-Löf Type Theory and Its Embedding into Extensional Type Theory

In this section we introduce the intensional version of Martin-Löf Type Theory, and repeat the standard proof how to interpret it in extensional type theory. The well-ordering proof in [12] was carried out in that theory. We note that [19] has a different way of formulating intensional type theory, which we used in [10], and which is trivially a sub theory of extensional type theory.

Definition 16.3. (a) The symbols and constructors of intensional Martin-Löf type theory are the symbols and constructors of extensional Martin-Löf type theory, but replacing r by refl of arity 1 and adding a constructor J of arity 3.

(b) The set of $\mathrm{b}^{\mathrm{int}}$-objects, $\mathrm{g}^{\mathrm{int}}$-terms, $\mathrm{g}^{\mathrm{int}}$-types, $\mathrm{g}^{\mathrm{int}}$-context pieces, $\mathrm{g}^{\mathrm{int}}$-contexts, $\mathrm{g}^{\mathrm{int}}$-judgements, $\mathrm{g}^{\mathrm{int}}$-dependent judgements, $\mathrm{R}^{\mathrm{int}}$-terms, $\mathrm{R}^{\mathrm{int}}$-types, $\mathrm{R}^{\mathrm{int}}$-context pieces, $\mathrm{R}^{\mathrm{int}}$-contexts, $\mathrm{R}^{\mathrm{int}}$-judgements, $\mathrm{R}^{\mathrm{int}}$-dependent judgements, are defined as the corresponding b-/g-/R-constructions, but replacing r by refl, adding J as a constructor for terms, omitting that r is a $\mathrm{g}^{\mathrm{int}}$-term, and adding the clause for refl, J for $\mathrm{g}^{\mathrm{int}}$-terms as follows: If r, s, t, t' are $\mathrm{g}^{\mathrm{int}}$-terms $x \in \mathrm{Var}_{\mathrm{ML}}$, then $\mathrm{refl}(r)$ and $\mathrm{J}(r, s, t, (x)t')$ are $\mathrm{g}^{\mathrm{int}}$-terms.

Definition 16.4. of intensional Martin-Löf Type Theory with W-type and one Universe.

The rules for $\mathrm{ML}_1^{\mathrm{i}}\mathrm{W}_{\mathrm{T}}$ (the intensional version à la Tarski) and $\mathrm{ML}_1^{\mathrm{i}}\mathrm{W}_{\mathrm{R}}$ (the intensional version à la Russell) and their extensions $\mathrm{ML}_1^{\mathrm{i}}\mathrm{W}_{\mathrm{T,U}}$ and $\mathrm{ML}_1^{\mathrm{i}}\mathrm{W}_{\mathrm{R,U}}$ are the rules for $\mathrm{ML}_1^{\mathrm{e}}\mathrm{W}_{\mathrm{T}}$, $\mathrm{ML}_1^{\mathrm{e}}\mathrm{W}_{\mathrm{R}}$, $\mathrm{ML}_1^{\mathrm{e}}\mathrm{W}_{\mathrm{T,U}}$, $\mathrm{ML}_1^{\mathrm{e}}\mathrm{W}_{\mathrm{R,U}}$, respectively, but referring to $\mathrm{g}^{\mathrm{int}}$-objects and $\mathrm{R}^{\mathrm{int}}$-objects instead of g-objects and R-objects, and replacing the rules $(\mathrm{I}^{\mathrm{I}})$, $(\mathrm{I}^{\mathrm{E}})$, and $(\mathrm{I}^{=})$ by the following rules:

Intensional Equality Rules

$$(\mathrm{I}^{\mathrm{I}})\ \frac{t : A}{\mathrm{refl}(t) : \mathrm{I}(A,t,t)}$$

$$(\mathrm{I}^{\mathrm{E}})\ \frac{\begin{array}{c} A : Set \\ x : A, y : A, z : \mathrm{I}(A,x,y) \Rightarrow C : \mathrm{type} \\ r = r' : A \\ s = s' : A \\ t = t' : \mathrm{I}(A,r,s) \\ z' : A \Rightarrow u = u' : C[x/z', y/z', z/\mathrm{refl}(z')] \end{array}}{\mathrm{J}(r,s,t,(z')u) = \mathrm{J}(r',s',t',(z')u') : C[x/r, y/s, z/t]}$$

$$(\mathrm{I}^{=})\ \frac{\begin{array}{c} A : \mathrm{type} \\ x : A, y : A, z : \mathrm{I}(A,x,y) \Rightarrow C : \mathrm{type} \\ r : A \\ z' : A \Rightarrow u : C[x/z', y/z', z/\mathrm{refl}(z')] \end{array}}{\mathrm{J}(r,r,\mathrm{refl}(r),(z')u) = u[z'/r] : C[x/r, y/r, z/\mathrm{refl}(r)]}$$

Definition 16.5. We define a translation φ of $\mathrm{b}^{\mathrm{int}}$-, $\mathrm{g}^{\mathrm{int}}$- and $\mathrm{R}^{\mathrm{int}}$-constructions into the corresponding respective b-, g- and R-constructions by recursively replacing

- $\mathrm{refl}(r)$ by r,
- $\mathrm{J}(r,s,t,(z')u)$ by $u[z'/r]$.

Lemma 16.6. *If* $\mathrm{ML}_1^{\mathrm{i}}\mathrm{W}_{\mathrm{T}}$, $\mathrm{ML}_1^{\mathrm{i}}\mathrm{W}_{\mathrm{R}}$, $\mathrm{ML}_1^{\mathrm{i}}\mathrm{W}_{\mathrm{R,U}}$, *or* $\mathrm{ML}_1^{\mathrm{i}}\mathrm{W}_{\mathrm{T,U}}$ *proves* $\Gamma \Rightarrow \theta$, *then the corresponding respective extensional version* $\mathrm{ML}_1^{\mathrm{e}}\mathrm{W}_{\mathrm{T}}$, $\mathrm{ML}_1^{\mathrm{e}}\mathrm{W}_{\mathrm{R}}$, $\mathrm{ML}_1^{\mathrm{e}}\mathrm{W}_{\mathrm{R,U}}$, $\mathrm{ML}_1^{\mathrm{e}}\mathrm{W}_{\mathrm{T,U}}$, *proves* $\varphi(\Gamma) \Rightarrow \varphi(\theta)$.

Proof. By induction on the derivation. For all unchanged rules this follows by applying the same rule to the IH. All what remains to show is that for the intensional equality rules, the translation of the conclusion is derivable in extensional type theory from the translated assumptions using the extensional equality rules.

In case of $(\mathrm{I}^{\mathrm{I}})$ this is obvious (the translated conclusion is $\mathrm{r} : \mathrm{I}(A,t,t)$). In case of I^{E} one needs to show that the assumptions imply the translated conclusion which is $u[z'/r] = u'[z'/r'] : C[x/r, y/s, z/t]$. From $t = t' : \mathrm{I}(A,r,s)$ one concludes $t : \mathrm{I}(A,r,s)$, therefore $r = s : A$ and therefore $\mathrm{I}(A,r,s) = \mathrm{I}(A,r,r)$. Therefore $t = t' : \mathrm{I}(A,r,r)$, and $t = \mathrm{r} : \mathrm{I}(A,r,s)$. It follows $C[x/r, y/s, z/t] = C[x/r, y/r, z/\mathrm{r}]$.

We have $u[z'/r] = u'[z'/r'] : C[x/r, y/r, z/\mathrm{r}]$ and therefore $u[z'/r] = u'[z'/r'] : C[x/r, y/s, z/t]$.

In case of ($\mathrm{I}^{=}$) the translated conclusion is $u[z'/r] = u[z'/r] : C[x/r, y/r, z/\mathrm{r}]$ which follows from the last translated assumption by substitution. □

16.4 Embedding of the Russell Version of Martin-Löf Universes into the Tarski Version

In this section we prove, that Martin-Löf Type Theory à la Russell $\mathrm{ML}_1^{\mathrm{e}}\mathrm{W}_{\mathrm{R,U}}$ can be embedded into Martin-Löf Type Theory à la Tarski $\mathrm{ML}_1^{\mathrm{e}}\mathrm{W}_{\mathrm{T,U}}$. Therefore, the upper bound proved for $\mathrm{ML}_1^{\mathrm{e}}\mathrm{W}_{\mathrm{T,U}}$ is an upper bound for $\mathrm{ML}_1^{\mathrm{e}}\mathrm{W}_{\mathrm{R}}$ and $\mathrm{ML}_1^{\mathrm{e}}\mathrm{W}_{\mathrm{R,U}}$ as well, and as well for $\mathrm{ML}_1^{\mathrm{e}}\mathrm{W}_{\mathrm{T}}$ (since it is a proper sub theory). By Lemma 16.6 it is as well an upper bound for $\mathrm{ML}_1^{\mathrm{i}}\mathrm{W}_{\mathrm{T}}$, $\mathrm{ML}_1^{\mathrm{i}}\mathrm{W}_{\mathrm{T,U}}$, $\mathrm{ML}_1^{\mathrm{i}}\mathrm{W}_{\mathrm{R}}$ and $\mathrm{ML}_1^{\mathrm{i}}\mathrm{W}_{\mathrm{R,U}}$.

Definition 16.7. (a) Define for C constructors,

$\psi(\mathbb{N}) :\equiv \widehat{\mathbb{N}}$, $\psi(\Sigma) :\equiv \widehat{\Sigma}$, $\psi(\mathrm{I}) :\equiv \widehat{\mathrm{I}}$,
$\psi(\mathbb{N}_k) :\equiv \widehat{\mathbb{N}}_k$, $\psi(\mathrm{W}) :\equiv \widehat{\mathrm{W}}$, $\psi(C) :\equiv C$ otherwise.
$\psi(\Pi) :\equiv \widehat{\Pi}$, $\psi(+) :\equiv \widehat{+}$,

(b) Define ψ : R-term $\longrightarrow$ g-term by recursion on the b-objects:
$\psi(x) :\equiv x$ ($x \in \mathrm{Var_{ML}}$),
$\psi(\lambda x.t) :\equiv \lambda x.\psi(t)$,
$\psi(C(t_1, \ldots, t_n)) :\equiv \psi(C)(\varphi(t_1), \ldots, \varphi(t_n))$ (for constructors C).

(c) Define the function ρ : R-type $\longrightarrow$ g-type by recursion on the g-types:
$\rho(Sx \in r.s) :\equiv Sx \in \rho(r).\rho(s)$ ($S \in \Sigma, \Pi, \mathrm{W}$),
$\rho(r + s) :\equiv \rho(r) + \rho(s)$, $\rho(\mathrm{I}(r, s, t)) :\equiv \mathrm{I}(\rho(r), \psi(s), \psi(t))$,
$\rho(C) :\equiv C$ for $C \in \{\mathbb{N}, \mathbb{N}_k, \mathrm{U}\}$, $\rho(t) :\equiv \mathrm{T}(\psi(t))$, otherwise.

(d) If $\Gamma = x_1 : A_1, \ldots, x_n : A_n$ is a g-context-piece, then
$\rho(\Gamma) :\equiv x_1 : \rho(A_1), \ldots, x_n : \rho(A_n)$.

(e) If r, s are g-terms, A is a g-type, then
$\rho(r = s : A) :\equiv (\psi(r) = \psi(s) : \rho(A))$,
$\rho(A = B : \mathrm{type}) :\equiv (\rho(A) = \rho(B) : \mathrm{type})$.

(f) Define μ : g-type $\longrightarrow$ g-type by recursion on the g-types:
$\mu(\mathrm{T}(sx \in r.s)) :\equiv Sx \in \mu(\mathrm{T}(r)).\mu(\mathrm{T}(s))$ (where $s = \widehat{\Sigma}, \widehat{\Pi}, \widehat{\mathrm{W}}$ and $S = \Sigma, \Pi, \mathrm{W}$ respectively),
$\mu(\mathrm{T}(r \widehat{+} s)) :\equiv \mu(\mathrm{T}(r)) + \mu(\mathrm{T}(s))$, $\mu(\mathrm{T}(\widehat{\mathrm{I}}(r, s, t)))) :\equiv \mathrm{I}(\mu(\mathrm{T}(r)), s, t)$,
$\mu(\mathrm{T}(\widehat{\mathbb{N}})) :\equiv \mathbb{N}$, $\mu(\mathrm{T}(\widehat{\mathbb{N}}_k)) :\equiv \mathbb{N}_k$,
$\mu(r + s) :\equiv \mu(r) + \mu(s)$, $\mu(\mathrm{I}(r, s, t)) :\equiv \mathrm{I}(\mu(r), s, t)$,
$\mu(Sx \in r.s) :\equiv Sx \in \mu(r).\mu(s)$ ($S \in \{\Sigma, \Pi, \mathrm{W}\}$),
$\mu(t) :\equiv t$, otherwise.

Lemma 16.8. *Assume r, s, t, s_i b-objects, $x_i \in \mathrm{Var_{ML}}$.*

(a) $\mathrm{FV}(t) = \mathrm{FV}(\psi(t)) = \mathrm{FV}(\rho(t)) = \mathrm{FV}(\mu(t))$.

(b) *If $t[x_1/s_1, \ldots, x_n/s_n]$ allowed, then $\psi(t)[x_1/\psi(s_1), \ldots, x_n/\psi(s_n)]$, $\mu(t)[x_1/\psi(s_1), \ldots, x_n/\psi(s_n)]$, $\rho(t)[x_1/\psi(s_1), \ldots, x_n/\psi(s_n)]$ are allowed.*

(c) *If t is an R-term, then* $\rho(t) = \mu(\mathrm{T}(\psi(t)))$.
(d) *If* t, s_i *are b-objects, then* $\psi(t[x_1/s_1, \dots, x_n/s_n]) = \psi(t)[x_1/\psi(s_1), \dots, x_n/\psi(s_n)]$.
If t is a g-type, s_i are g-terms, then $\mu(\mu(t)) = \mu(t)$,
$\mu(t[x_1/s_1, \dots, x_n/s_n]) = \mu(\mu(t)[x_1/s_1, \dots, x_n/s_n])$,
If t is an R-type, s_i are R-terms, then
$\rho(t[x_1/s_1, \dots, x_n/s_n]) = \mu(\rho(t)[x_1/\psi(s_1), \dots, x_n/\psi(s_n)])$.
(e) *If t is a g -judgement, -context, -context-piece, or dependent judgement, then $\rho(t)$ is an R-judgement, -context, -context-piece, or dependent judgement, respectively.*
(f) *If* $r{=_\alpha}s$, *then* $\psi(r){=_\alpha}\psi(s)$, $\rho(r){=_\alpha}\rho(s)$, $\mu(r){=_\alpha}\mu(s)$.
(g) *If r is a g-term and b-term, then* $\psi(r) = r$.
If r is a g-type and b-type, then $\rho(r) = r$.

Lemma 16.9. *In all versions of Martin-Löf Type Theory, we have the following useful derived rule:*

$$\frac{\Gamma, x : A, \Gamma' \Rightarrow \theta \quad \Gamma \Rightarrow A = A' : \text{type}}{\Gamma, x : A', \Gamma' \Rightarrow \theta}$$

Proof. Let y be a fresh variable. Then by (THIN) $\Gamma, y : A', x : A, \Gamma' \Rightarrow \theta$, easily we have $\Gamma, y : A' \Rightarrow y : A$, By (SUB) $\Gamma, y : A', \Gamma'[x/y] \Rightarrow \theta[x/y]$, and by change of the variable we obtain the assertion. □

Lemma 16.10. *Let* $\mathrm{ML_T}$ *be* $\mathrm{ML_1^eW_T}$ *or* $\mathrm{ML_1^eW_{T,U}}$ *and* Γ, Γ' *be g-context-pieces, a, b, t g-terms, A, B g-types, x a variable. The following applies:*

(a) *If* $\mathrm{ML_T} \vdash \Gamma \Rightarrow s : A$, *or* $\mathrm{ML_T} \vdash \Gamma \Rightarrow s = t : A$ *then* $\mathrm{ML_T} \vdash \Gamma \Rightarrow A : \text{type}$.
(b) *If* $\mathrm{ML_T} \vdash \Gamma \Rightarrow Sx \in A.B : \text{type}$ $(S \in \{\Sigma, \Pi, \mathrm{W}\})$, *then* $\mathrm{ML_T} \vdash \Gamma \Rightarrow A$, $\mathrm{ML_T} \vdash \Gamma, y : A \Rightarrow B[x/y] : \text{type}$, *for all* $y \in \mathrm{Var_{ML}} \setminus X$ *for some finite set* X.
(c) *If* $\mathrm{ML_T} \vdash \Gamma \Rightarrow A + B : \text{type}$ *then* $\mathrm{ML_T} \vdash \Gamma \Rightarrow A : \text{type}$, $\mathrm{ML_T} \vdash \Gamma \Rightarrow B : \text{type}$.
(d) *If* $\mathrm{ML_T} \vdash \Gamma \Rightarrow \mathrm{I}(A, b, c) : \text{type}$, *then* $\mathrm{ML_T} \vdash \Gamma \Rightarrow b : A$, $\mathrm{ML_T} \vdash \Gamma \Rightarrow c : A$.
(e) *If* $\mathrm{ML_T} \vdash \Gamma \Rightarrow \mathrm{T}(b) : \text{type}$ *then* $\mathrm{ML_T} \vdash \Gamma \Rightarrow b : \mathrm{U}$.
(f) *If* $\mathrm{ML_T} \vdash \Gamma \Rightarrow A = B : \text{type}$, *then* $\mathrm{ML_T} \vdash \Gamma \Rightarrow A : \text{type}$, $\mathrm{ML_T} \vdash \Gamma \Rightarrow B : \text{type}$.
(g) *If* $\mathrm{ML_T} \vdash \Gamma, x : A, \Gamma \Rightarrow \theta$ *then* $\mathrm{ML_T} \vdash \Gamma \Rightarrow A : \text{type}$.

Note that (f) is trivial, but will be needed in the following as an additional premise for the induction.

Proof. We first change the calculus, treating A : type no longer as an abbreviation for $A = A$: type. For every instance of a rule with conclusion $A = A$: type we add a rule with conclusion A : type where any premise $B = B$: type is replaced by B : type, and any premise $s = s : B$ is replaced by $s : B$. Furthermore, add the rules

$$(\text{REPL1})\ \frac{x : A, \Gamma', \Gamma'' \Rightarrow B : \text{type} \qquad \Rightarrow t = t' : A}{\Gamma'[x/t], \Gamma''[x/t'] \Rightarrow B[x/t'] : \text{type}}$$

$$(\text{ALPHA}_1)\ \frac{A : \text{type}}{A' : \text{type}} \quad \text{if } A{=_\alpha}A'$$

$$(\text{REFL})\ \frac{A : \text{type}}{A = A : \text{type}}$$ □

If for this calculus the theorem is provable, then this calculus is equivalent to the original: If we have a proof in the original calculus, then embed it into the calculus, by applying, whenever we need the removed rules the weak inferences. If we have a proof in the new calculus, the result is a proof in the original calculus, since we only added derived rules.

Now the proof follows by induction on the length of the (new) derivation. The only difficult case is (SUB), where the difficulty are the second conclusion in the cases (b): let the conclusion be for instance $\Gamma, \Gamma'[x/t] \Rightarrow (\Sigma x' \in A.B)[x/t] : \text{type}$. By IH $\Gamma, \Gamma', y : A \Rightarrow B[x'/y] : \text{type}$ for $y \notin X$, therefore $\Gamma, \Gamma'[x/t], y : A[x/t] \Rightarrow B[x'/y][x/t]$ for $x \neq y$, $x \notin X$ (the substitution is allowed). Then for $y \notin X \cup \{x\}$, if $x = x'$ or $x \notin \text{FV}(B)$, it follows $(\Sigma x' \in A.B)[x/t] = \Sigma x' \in A[x/t].B$ and we have the assertion, otherwise $x' \notin \text{FV}(t)$ and $B[x'/y][x/t] = B[x/t][x'/y] : \text{type}$.

Similarly we argue in (REPLi), the other rules are easy.

Lemma 16.11. (a) $\text{ML}_1^e\text{W}_{\text{T,U}} \vdash \Gamma \Rightarrow r = s : \text{U}$, *then*
$\text{ML}_1^e\text{W}_{\text{T,U}} \vdash \Gamma \Rightarrow \text{T}(r) = \mu(\text{T}(r)) : \text{type}$ *and*
$\text{ML}_1^e\text{W}_{\text{T,U}} \vdash \Gamma \Rightarrow \text{T}(s) = \mu(\text{T}(s)) : \text{type}$.
(b) $\text{ML}_1^e\text{W}_{\text{T,U}} \vdash \Gamma \Rightarrow A = B : \text{type}$, *then* $\text{ML}_1^e\text{W}_{\text{T,U}} \vdash \Gamma \Rightarrow A = \mu(A) : \text{type}$ *and* $\text{ML}_1^e\text{W}_{\text{T,U}} \vdash \Gamma \Rightarrow B = \mu(B) : \text{type}$.

Proof. (a): Induction on the definition of r being a b-object. If for instance $\text{ML}_1^e\text{W}_{\text{T,U}} \vdash \Gamma \Rightarrow \widehat{\Sigma} x \in a.b : \text{U}$, then by the additional rules of $\text{ML}_1^e\text{W}_{\text{T,U}}$ it follows $\Gamma \Rightarrow a : \text{U}$ and $\Gamma, x : \text{T}(a) \Rightarrow b : \text{U}$, by IH therefore $\Gamma \Rightarrow \text{T}(a) = \mu(\text{T}(a)) : \text{type}$, $\Gamma, x : \text{T}(a) \Rightarrow \text{T}(b) = \mu(\text{T}(b)) : \text{type}$, by type introduction follows the assertion, similarly for the other terms, for which $\mu(\text{T}(t))$ does something.

(b) Similar, using Lemma 16.10 instead of the new rules. □

Lemma 16.12. *If* $\text{ML}_1^e\text{W}_{\text{R,U}} \vdash \Gamma \Rightarrow \theta$ *then* $\text{ML}_1^e\text{W}_{\text{T,U}} \vdash \rho(\Gamma) \Rightarrow \rho(\theta)$.

Especially, if $\Gamma \Rightarrow \theta$ *is a dependent judgement of both* $\text{ML}_1^e\text{W}_{\text{R,U}}$ *and* $\text{ML}_1^e\text{W}_{\text{T,U}}$, *then we have:*

If $\text{ML}_1^e\text{W}_{\text{R,U}} \vdash \Gamma \Rightarrow \theta$ *then* $\text{ML}_1^e\text{W}_{\text{T,U}} \vdash \Gamma \Rightarrow \theta$.

Proof. Induction on the derivation.

In most rules, the assertion follows by the same rules.

Difficult rules: (SUB), (REPLi): Use 16.8 (d), 16.11 and 16.9.

Equality rules and extensional equality rules: use for the substitution part the same argument.

Second and third rule in (U^I): we conclude $\text{T}(\psi(A))$, and using 16.11 and an easy argument follows the assertion.

Universe introduction rules (possibly extensional): easy. □

Notation. In the following we will write ML for $\text{ML}_1^e\text{W}_{\text{T,U}}$.

16.5 Definition of KPI$^+$

We introduce now the Kripke-Platek set theory KPI$^+$, in which we will interpret afterwards ML. For more details on it, the reader should refer to [1], [4], [5], [7], [6], and [8].

Definition 16.13. Definition of Kripke-Platek set theory:

(a) Let $\mathcal{L}_{KP}$be the classical first-order language, with terms being variables, atomic formulas being $u \in v$, $\neg(u \in v)$, $\mathrm{Ad}(u)$, $\neg\mathrm{Ad}(u)$. The set of Variables should be $\mathrm{Var}_{\mathrm{KP}} = \{u_i | i \in \mathbb{N}\}$ (a meta-set), $u_i \neq u_j$ for $i \neq j$.
The formulas are built from atomic formulas by $\wedge$, $\vee$, $\forall$, $\exists$. We define $\neg A$ by the deMorgan's laws. The quantifier in $\forall x.\varphi$ ($\exists x.\varphi$) is bounded, if φ of the form $x \in v \longrightarrow B$ ($x \in v \wedge B$) with $x \neq v$. A Δ_0-formula is a formula with no unbounded quantifier.
We abbreviate
$A \longrightarrow B :\equiv ((\neg A) \vee B)$,
$\forall x \in v.B :\equiv \forall x.x \in v \longrightarrow B$,
$\exists x \in v.B :\equiv \exists x.(x \in v \wedge B)$,
$(u = v) :\equiv ((\forall x \in u.x \in v) \wedge (\forall x \in v.x \in u))$,
$u \notin v :\equiv \neg(u \in v)$,
$\mathrm{tran}(u) :\equiv \forall x \in u.\forall y \in x.y \in u$,
$\mathrm{infinite}(u) :\equiv \exists x \in u.(x = x) \wedge \forall x \in u.\exists y \in u.x \in y$,
$\mathrm{Inacc}(x) :\equiv \mathrm{Ad}(x) \wedge \forall y \in x.\exists z \in x.y \in z \wedge \mathrm{Ad}(z)$,
$$\mathrm{Inacc}_n(x) :\equiv \exists x_0, \dots, x_n.\mathrm{Inacc}(x_0) \wedge \mathrm{Ad}(x_1) \wedge \mathrm{Ad}(x_2) \wedge \cdots \wedge \mathrm{Ad}(x_n) \wedge x_0 \in x_1 \wedge x_1 \in x_2 \wedge \cdots \wedge x_{n-1} \in x_n \wedge x = x_n.$$
ψ a formula, then ψ^u means the replacing of every unbounded quantifier $\forall v$ by $\forall v \in u$ and $\exists v$ by $\exists v \in u$.
Note, that $\mathrm{Inacc}(x)$ expresses, that x is an admissible, closed under admissibles, the ordinal of which is an inaccessible, and $\mathrm{Inacc}_n(x)$, that x is an admissible, which is at least the nth admissible above an x s.t. $\mathrm{Inacc}(x)$.

(b) Definition of axiom schemes:

(Ext) $\forall x.\forall y.\forall z.x = y \longrightarrow (x \in z \longrightarrow y \in z) \wedge (\mathrm{Ad}(x) \longrightarrow \mathrm{Ad}(y))$.

(Found) $\forall \vec{z}.[\forall x.(\forall y \in x.\varphi(y, \vec{z}) \longrightarrow \varphi(x, \vec{z})) \longrightarrow \forall x.\varphi(x, \vec{z})]$ (φ an arbitrary formula).

(Pair) $\forall x.\forall y.\exists z.x \in z \wedge y \in z$.

(Union) $\forall x.\exists z.\forall y \in x.\forall u \in y.u \in z$.

(Δ_0-sep) $\forall \vec{z}.\forall w.\exists y.[(\forall x \in y.(x \in w \wedge \varphi(x, \vec{z}))) \wedge \forall x \in w.\varphi(x, \vec{z}) \longrightarrow x \in y]$ (φ a Δ_0-formula).

(Δ_0-coll) $\forall \vec{z}.\forall w.[(\forall x \in w.\exists y.\varphi(x, y, \vec{z})) \longrightarrow \exists w'.\forall x \in w.\exists y \in w'.\varphi(x, y, \vec{z})]$ (φ a Δ_0-formula).

(Ad.1) $\forall x.\mathrm{Ad}(x) \longrightarrow \mathrm{tran}(x) \wedge \exists w \in x.\mathrm{infinite}(w)$.

(Ad.2) $\forall x.\forall y.\mathrm{Ad}(x) \wedge \mathrm{Ad}(y) \longrightarrow x \in y \vee x = y \vee y \in x$.

(Ad.3)	$\forall x.\mathrm{Ad}(x) \rightarrow \psi^x$ (ψ an instance of (Pair), (Union), (Δ_0-sep), (Δ_0-coll)).
(Lim)	$\forall x.\exists y.\mathrm{Ad}(y) \wedge x \in y$.
(inf)	$\exists x.\mathrm{infinite}(x)$.
$(+_n)$	$\exists x.\mathrm{Inacc}_n(x)$.

(c) KPI^+ is the theory (Ext) + (Found) + (Pair) + (Union) + (Δ_0-sep) + (Δ_0-coll) + (inf) + (Ad.1–3) + $\{(+_{\mathrm{n}})|\mathrm{n} \in \omega\}$.
So KPI^+ is a theory, which guarantees the existence of one recursive inaccessible, and of finitely many admissibles above it.

Definition 16.14. (a) Ord is the class of ordinals.
(b) $\alpha(a) := \bigcup(a \cap \mathrm{Ord})$.
(c) $\mathrm{Ad}_1 := \bigcap\{x|\mathrm{Ad}(x)\}$, $\mathrm{Ad}_2 := \bigcap\{x|\mathrm{Ad}(x) \wedge \mathrm{Ad}_1 \in x\}$, $\mathrm{Ad}_{\mathrm{I}} := \bigcap\{x|\mathrm{Inacc}(x)\}$, $\mathrm{Ad}_{\mathrm{I},n} := \bigcap\{x|\mathrm{Inacc}_n(x)\}$.
Note, that Ad_1, Ad_2, Ad_{I}, $\mathrm{Ad}_{\mathrm{I},n}$ can be defined, since there exists b s.t. $\mathrm{Ad}(b)$ or $\mathrm{Inacc}(b)$ or $\mathrm{Inacc}_n(b)$, and therefore we can replace the class by $\{x \in b \cup \{b\}| \cdots\}$.
(d) $\Omega_1 := \alpha(\mathrm{Ad}_1)$, $\mathrm{I} := \alpha(\mathrm{Ad}_{\mathrm{I}})$, $\mathrm{I}_n := \alpha(\mathrm{Ad}_{\mathrm{I},n})$.
(e) $\mathrm{ad}(u) := \bigcap(\{c \in \mathrm{Ad}_{\mathrm{I}}|\mathrm{Ad}(c) \wedge u \in c\} \cup \{\mathrm{Ad}_{\mathrm{I}}\})$, $\alpha^+(u) := \alpha(\mathrm{ad}(u))$.

Remark 16.15. In KPI^+ we have

(a) $\mathrm{Ad}(\mathrm{Ad}_1)$, $\mathrm{Ad}(\mathrm{Ad}_2)$, $\mathrm{Inacc}(I)$, $\mathrm{Inacc}_n(\mathrm{I}_n)$.
(b) $u \in \mathrm{Ad}_{\mathrm{I}} \rightarrow \mathrm{ad}(u) \in \mathrm{Ad}_{\mathrm{I}} \wedge \mathrm{Ad}(\mathrm{ad}(u)) \wedge u \in \mathrm{ad}(u)$.

16.6 Interpretation of Terms and Types

A type A will be interpreted basically as a set of pairs of closed terms: $\langle t, s\rangle \in A^*$ means that t and s are equal elements of this types. We will define A^* as the set of terms, which are by an introductory rule elements of this type, and close it under the reduction rule. For instance, if A^* and B^* are already defined, then

$$(A + B)^* := +^*(A^*, B^*)$$

where again

$$+^*(u, v) = \mathrm{Compl}(+^{\mathrm{basis}}(u, v)),$$
$$+^{\mathrm{basis}}(u, v) := \{\langle \mathrm{i}(a), \mathrm{i}(a')\rangle|\langle a, a'\rangle \in u\} \cup \{\langle \mathrm{j}(b), \mathrm{j}(b')\rangle|\langle b, b'\rangle \in v\},$$
$$\mathrm{Compl}(u) := \{\langle r, s\rangle|\exists r', s'.\langle r', s'\rangle \in u \wedge r \longrightarrow_{\mathrm{red}} r' \wedge s \longrightarrow_{\mathrm{red}} s'\}.$$

Since we only want to interpret finitely many types, namely those types, which occur in a certain derivation, we interpret dependent types as Σ-functions, the arguments of which are represented by the free variables of the type, in such a way, that $A^*[x_1/t_1, \ldots, x_n/t_n] = (A[x_1/t_1, \ldots, x_n/t_n])^*$.

The Π-type has an introductory rule with a premise, where dependency occurs. The intended meaning of the premise $x : A \Rightarrow t = t' : B$ is

$$\forall r, r'.\langle r, r' \rangle \in A^* \longrightarrow \langle t[x/r], t'[x/r'] \rangle \in B^*.$$

Furthermore, we know

$$\langle r, r' \rangle \in A^* \Rightarrow B^*[x/r] = B'^*[x/r'].$$

Since we have to close it under α-conversion we can therefore define $(\Pi x \in A.B)^* := \Pi^*(A^*, (x)B^*)$, where $\Pi^*(u, f) = \mathrm{Compl}(\Pi^{\mathrm{basis}}(u, f))$ and

$$\Pi^{\mathrm{basis}}(u, f) := \{\langle \lambda y.t, \lambda y'.t' \rangle | \forall \langle r, r' \rangle \in u.\langle t[y/r], t'[y'/r'] \rangle \in f(r) \wedge f(r) = f(r')\}).$$

The condition $f(r) = f(r')$ has been added for technical reasons.

In order to make proofs about the terms easy, we will have deterministic reduction-rules. We will allow e.g. $\mathrm{Ap}(\lambda x.r, s) \longrightarrow_{\mathrm{red}} r[x/s]$ only, if s is in normal form. Furthermore, we do not allow any reductions of $\lambda x.r$, so giving reduction rules generally only for closed terms This simple approach is possible, since, from the definition of $(\Pi x \in A.B)^*$, we see, that, if $\langle \lambda x.t, \lambda x.t \rangle \in (\Pi x \in A.B)^*$, and if $t \longrightarrow t'$ in a general sense for open terms, $t[x/r] \longrightarrow_{\mathrm{red}} t'[x/r]$ for closed terms r. But now, if B^* is closed under $\longrightarrow_{\mathrm{red}}$, we conclude $\langle \lambda x.t, \lambda x.t' \rangle \in (\Pi x \in A.B)^*$.

The interpretation of the W-type, which represents an inductive definition, is done in the usual way: we take some operator on sets and iterate it up to the closure ordinal, which is the next admissible above the components A and B of it. By the introduction rule, we get as the operator F such that

$$\begin{aligned} F_{\mathrm{W}}(u, f)(v) = \mathrm{Compl}(\{\langle \sup(s, \lambda y.t), \sup(s', \lambda y'.t') \rangle | \langle s, s' \rangle \in u \\ \wedge \forall \langle r, r' \rangle \in f(s).\langle t[y/r], t'[y'/r'] \rangle \in v\}). \end{aligned}$$

$(\mathrm{W}x \in A.B)^* := F_{\mathrm{W}}^{\alpha}(A^*, (x)B^*)$. We can choose as α any admissible ordinal s.t. A^* and $(x)B^*$ are elements of L_α. We will take as α I_n, the nth admissible above I, and n is the maximum of $\mathrm{lev}(A)$ and $\mathrm{lev}(B)$, which is the number of nestings of W-types in A and B. Although this ordinal is usually too big, it suffices for our construction. In the introduction rules for the elements of the universe, e.g.

$$(\widehat{\Pi}^{\mathrm{I}})\ \frac{a : \mathrm{U} \qquad x : \mathrm{T}(a) \Rightarrow b : \mathrm{U}}{\widehat{\Pi} x : a.b : \mathrm{U}}$$

we introduce simultaneously the elements a of the universe and its interpretation $\mathrm{T}(a)$ as a type. We will therefore first define a set $\widehat{\mathrm{U}}$ of triples $\langle a, A, b \rangle$, where a and b are terms, considered as equal elements of the universe, such that $\mathrm{T}(a)^* = \mathrm{T}(b)^* = A$. Therefore

$$\mathrm{U}^* = \{\langle a, b \rangle | \exists A.\langle a, A, b \rangle \in \widehat{\mathrm{U}}\}$$

and $\mathrm{T}(a)^* = f(a)$, where

$$f = \{\langle a, A\rangle | \exists b.\langle a, A, b\rangle \in \widehat{\mathrm{U}}\}.$$

$\widehat{\mathrm{U}}$ is again the fixed point of an operator $\widetilde{\mathrm{U}}$, so $\widehat{\mathrm{U}} = \widetilde{\mathrm{U}}^\alpha$ for some admissible α. Since U is closed under the W-type, in the definition of $\widetilde{\mathrm{U}}$ we need to go to the next admissible. Therefore, α must be closed under the step to the next admissible. We can choose $\alpha := \mathrm{I}$, which is a recursive inaccessible, i.e. an admissible closed under the step to the next admissible. Here we see, why we need the theory KPI^+: We need, one admissible, closed under admissibles, and finitely many admissibles above it.

We want to interpret an intuitionistic theory in a classical one, using some realisation. Now if we have a realisation interpretation as indices for recursive functions there is naturally a very easy realisation of $\neg\forall x.\exists y.\varphi(x, y)$, if $\forall x.\exists y.\varphi(x, y)$ is arithmetical formula, such that from x we cannot compute a y such that $\varphi(x, y)$ holds: every term $\lambda x.t$ does it since there is no realisation of $\forall x.\exists y.\varphi(x, y)$. Therefore there are false recursive realisations. But proofs carried out in an intuitionistic theory like Martin-Löf Type Theory should not prove false statements. The reason, why it does not prove any, is, that we can apply the realising term $\lambda u.s$ of $\neg\forall x.\exists y.\varphi(x, y)$, in some sense to non intuitionistic proofs as well. In order to fix this, we allow to add to our model a constructor, which has a non recursive reduction rule, and gives us the y depending on the x. Then we have a realising term t for $\forall x.\exists y.\varphi(x, y)$, and can apply any proof $\lambda u.s$ of $\neg\forall x.\exists y.\varphi(x, y)$ to it to get an element of type $\perp$ which is empty. Therefore, adding non recursive constructors, we can achieve that there is no realising term of a false formula.

We want to extend our result even for Π^1_1-sentences. Here again we have the problem, that the powerset of the natural numbers, $\mathbb{N} \to \mathrm{U}$, does not represent all sets in KPI^+. We will not be able to get a result, where we conclude from $\mathrm{ML} \vdash \forall X \in (\mathbb{N} \to \mathrm{U}).\varphi(X)$, that $\mathrm{KPI}^+ \vdash \forall x.x \subset \omega \to \varphi'(x)$ for the translation φ' of φ, but only, if we have, that x is an element of the first admissible. (We could easily extend it for x being an element of the first recursive inaccessible admissible without any trouble, but the result is enough to get an upper bound for the provable proof theoretic strength).)

Anyway, this text serves only to motivate the introduction of non recursive constructors. We have to quantify over all possible choices of new constructors. We will have either constructors, that give as a natural number (functions $\omega^l \to \omega$ for some l), or functions, that gives us an element of the universe, in order to get elements of the powerset of $\mathbb{N}$, but we only need the elements $\widehat{\mathbb{N}}_0$ (for is not element) and $\widehat{\mathbb{N}}_1$ (for is an element).

Definition 16.16. (a) We assume, that we have chosen some Gödel-numbers $\ulcorner S \urcorner$ for all symbols S of ML.

(b) A triple $\langle \ulcorner C \urcorner, l, f\rangle$ is a constructor definition, if C, l are natural numbers (C is a Gödel number for the constructor), such that C is different from the Gödel-numbers for the symbols, $l > 0$ and f is a function $f : \omega^{l-1} \to \omega$ or $l = 0 \wedge f : \omega \to \{\ulcorner \widehat{\mathbb{N}}_0 \urcorner, \ulcorner \widehat{\mathbb{N}}_1 \urcorner\}$. In this situation we define $\mathrm{arity}(C) := \max\{l - 1, 0\}$.

(c) A constructor extension set is a finite set of constructor definitions, such that the Gödel-numbers for the constructors are different. We write $\mathrm{CES}(a_0)$ for a_0 being a constructor extension set and $a_0 \in \mathrm{Ad}_2$.
(d) The a_0-extended g-terms, g-types, b-objects are defined as the g-terms, g-types, b-objects, but having in addition for each element $\langle C, l, f\rangle$ of a_0 a term constructor Constr_C of arity $\mathrm{arity}(C)$, and allowing to form a g-term $\mathrm{Constr}_C(r_1, \ldots, r_n)$ for g-terms r_i and $n = \mathrm{arity}(C)$. For simplicity we write usually C instead of Constr_c. Let Term_{Cl,a_0} be the set of closed a_0-extended g-terms.
(e) $\forall CES(a_0).\varphi(a_0) :\equiv \forall a_0 \in \mathrm{Ad}_2.\mathrm{CES}(a_0) \to \varphi(a_0)$.

Assumption. As long as there is no other CES mentioned, let a_0 be a CES s.t. $a_0 \in \mathrm{Ad}_2$. Most of the next definitions will depend on a_0, which we will not mention in the following. If we have to mention it, we will add it as a subscript.

Definition 16.17. (a) We choose some Gödelization of a_0-extended b-terms, but will omit the Gödel-brackets.
(b) The introductory term constructors are the term constructors 0, r, $\widehat{\mathbb{N}}$, S, i, j, p, sup, $\widehat{+}$, $\widehat{\Pi}$, $\widehat{\Sigma}$, $\widehat{\mathrm{W}}$, $\widehat{\mathrm{I}}$; $\widehat{\mathbb{N}}_k$ for $k \in \omega$, and A^k_n for $n < k \in \omega$.
(c) Let $\longrightarrow_{\mathrm{red,imm}_{a_0}}$ or short $\longrightarrow_{\mathrm{red,imm}}$ be the relation between closed a_0-extended g-terms, defined by
$\mathrm{p}_0(\mathrm{p}(r, s)) \longrightarrow_{\mathrm{red,imm}} r$, $\mathrm{p}_1(\mathrm{p}(r, s)) \longrightarrow_{\mathrm{red,imm}} s$,
$\mathrm{Ap}(\lambda x.r, s) \longrightarrow_{\mathrm{red,imm}} r'[x/s]$ where $r' =_\alpha r$ s.t. $r'[x/s]$ is allowed and r' is chosen minimal w.r.t. the choice of variables substituted in lexicographic order.
$\mathrm{C}_n(\mathrm{A}^n_i, r_1, \ldots, r_n) \longrightarrow_{\mathrm{red,imm}} r_i$,
$\mathrm{D}(\mathrm{i}(r), s, t) \longrightarrow_{\mathrm{red,imm}} s\ r$, $\mathrm{D}(\mathrm{j}(r), s, t) \longrightarrow_{\mathrm{red,imm}} t\ r$,
(note that we write $r\ s$ for $\mathrm{Ap}(r, s)$).
$\mathrm{P}(0, s, t) \longrightarrow_{\mathrm{red,imm}} s$, $\mathrm{P}(\mathrm{S}(r), s, t) \longrightarrow_{\mathrm{red,imm}} t\ r\ \mathrm{P}(r, s, t)$,
$\mathrm{R}(\mathrm{sup}(r, s), t) \longrightarrow_{\mathrm{red,imm}} (t\ r\ s\ (\lambda z_i.\mathrm{R}(s\ z_i, t)))$, where i is minimal such that $z_i \notin \mathrm{FV}(s) \cup \mathrm{FV}(t)$.
$C(\mathrm{S}^{n_1}(0), \ldots, \mathrm{S}^{n_l}(0)) \longrightarrow_{\mathrm{red,imm}_{a_0}} \mathrm{S}^{f(n_1,\ldots,n_l)}(0)$, (if $\langle C, l+1, f\rangle \in a_0$),
$C(\mathrm{S}^n(0)) \longrightarrow_{\mathrm{red,imm}_{a_0}} f(n)$ (if $\langle C, 0, f\rangle \in a_0$, and $f : \omega \to \{\widehat{\mathbb{N}}_0, \widehat{\mathbb{N}}_1\}$.)
(d) We define inductively a set of (indices for) terms in normal-form $\mathrm{Term}_{\mathrm{nf}}$, a subset of the closed a_0-extended g-terms:
If C is an introductory n-ary term constructor, $t_1, \ldots, t_n \in \mathrm{Term}_{\mathrm{nf}}$, then $C(t_1, \ldots, t_n) \in \mathrm{Term}_{\mathrm{nf}}$.
If C is a n-ary term constructor (possibly an extended term constructor) that is not introductory, $t_1, \ldots, t_n \in \mathrm{Term}_{\mathrm{nf}}$, and there exists no t such that $C(t_1, \ldots, t_n) \longrightarrow_{\mathrm{red,imm}} t$, then $C(t_1, \ldots, t_n) \in \mathrm{Term}_{\mathrm{nf}}$.
If $t \in \mathrm{Term}$, $x \in \mathrm{Var}_{\mathrm{ML}}$, $\mathrm{FV}(t) \subset \{x\}$, then $\lambda x.t \in \mathrm{Term}_{\mathrm{nf}}$.
(e) We define for a_0-extended g-terms t, the next reduced term t^{red}.
For $t \in \mathrm{Term}_{\mathrm{nf}}$. $t^{\mathrm{red}} := t$.
If C is a n-ary (possibly extended) term constructor, $r_i \in \mathrm{Term}_{\mathrm{Cl}}$, $\exists i.r_i \notin \mathrm{Term}_{\mathrm{nf}}$, then $C(r_1, \ldots, r_n)^{\mathrm{red}} := C(r_1^{\mathrm{red}}, \ldots, r_n^{\mathrm{red}})$.
If $t := C(r_1, \ldots, r_n) \notin \mathrm{Term}_{\mathrm{nf}}$, $r_i \in \mathrm{Term}_{\mathrm{nf}}$, then $t \longrightarrow_{\mathrm{red,imm}} t'$ for some t', $t^{\mathrm{red}} := t'$.

We define $r \longrightarrow_{\text{red}} s$ if and only if there exists a sequence $\langle s_0, \ldots, s_n \rangle$ such that $r = s_0$, $s = s_n$ and $\forall i < n. s_{i+1} = (s_i)^{\text{red}}$.

Lemma 16.18.

(a) $\text{KPI}^+ \vdash \forall r, s, s' \in \text{Term}_{\text{Cl}}.(r \longrightarrow_{\text{red}} s \wedge r \longrightarrow_{\text{red}} s') \rightarrow (s \longrightarrow_{\text{red}} s' \vee s' \longrightarrow_{\text{red}} s)$.
(b) $\text{KPI}^+ \vdash \forall r, s, s' \in \text{Term}_{\text{Cl}}.(r \longrightarrow_{\text{red}} s \wedge r \longrightarrow_{\text{red}} s' \wedge s, s' \in \text{Term}_{\text{nf}}) \rightarrow s = s'$.
(c) *If C is a n-ary constructor, then*

$$\begin{aligned} \text{KPI}^+ \vdash \forall r_1, \ldots, r_n, r_1', \ldots, r_n'.& \\ (r_1 \longrightarrow_{\text{red}} r_1' \in \text{Term}_{\text{nf}} \wedge \cdots \wedge r_n \longrightarrow_{\text{red}} r_n' \in \text{Term}_{\text{nf}})& \\ \rightarrow (C(r_1, \ldots, r_n) \longrightarrow_{\text{red}} C(r_1', \ldots, r_n')).& \end{aligned}$$

(d) $\text{KPI}^+ \vdash \forall t, t', s \in \text{Term}_{\text{Cl}}.(t \longrightarrow_{\text{red}} s \wedge t =_\alpha t') \rightarrow \exists s' \in \text{Term}_{\text{Cl}}.t' \longrightarrow_{\text{red}} s' \wedge s =_\alpha s'$.
(e) $\text{KPI}^+ \vdash \forall t, t' \in \text{Term}_{\text{Cl}}.t =_\alpha t' \rightarrow (t \in \text{Term}_{\text{nf}} \iff t' \in \text{Term}_{\text{nf}})$.

Definition 16.19. If F is a Σ function, we define by recursion on $\alpha \in \text{Ord}$

$$F^\alpha := \begin{cases} \emptyset & \text{if } \alpha = 0, \\ F(F^\beta) & \text{if } \alpha = \beta + 1, \\ \bigcup_{\beta < \alpha, \beta \in \text{Ord}} F^\beta & \text{if } \alpha \in \text{Lim}. \end{cases}$$

Definition 16.20. (a) Let Compl be the Σ-function

$$\begin{aligned} \text{Compl}(u) := \; & \{\langle r, s \rangle \in \text{Term}_{cl} \times \text{Term}_{cl} | \exists r', s' \in \text{Term}_{a,\text{nf}}. \\ & r \longrightarrow_{\text{red}} r' \wedge s \longrightarrow_{\text{red}a} s' \wedge \langle r', s' \rangle \in u\}. \end{aligned}$$

(b) $\mathbb{N}_k^{\text{basis}} := \{\langle \text{A}_n^k, \text{A}_n^k \rangle | n < k\}$, $\mathbb{N}_k^{**} := \text{Compl}(\mathbb{N}_k^{\text{basis}})$, which are Σ-functions, depending on the parameter k.
(c) $\mathbb{N}^{\text{basis}} := \{\langle \text{S}^n(0), \text{S}^n(0) \rangle | n < \omega\}$, $\mathbb{N}_k^{**} := \text{Compl}(\mathbb{N}^{\text{basis}})$.
(d)
$$\begin{aligned} \Pi^{\text{basis}}(u, f) := \; & \{\langle \lambda x.s, \lambda x'.s' \rangle \in \text{Term}_{\text{nf}} \times \text{Term}_{\text{nf}} | \\ & \forall \langle r, r' \rangle \in u. \langle s[x/r], s'[x'/r'] \rangle \in f(r) \wedge f(r) = f(r')\} \end{aligned}$$

(more precisely we have to write:

$$\begin{aligned} \Pi^*(u, f) := \; & \{\langle t, t' \rangle \in \text{Term}_{\text{nf}} \times \text{Term}_{\text{nf}} | \\ & \exists x, x' \in \text{Var}_{\text{ML}}, r, r' \in \text{Term}.t = \lambda x.s \wedge t' = \lambda x'.s' \wedge \\ & \forall r, r' \in \text{Term}_{\text{Cl}}. \langle r, r' \rangle \in u \rightarrow \\ & \quad \langle s[x/r], s[x'/r'] \rangle \in f(r) \wedge f(r) = f(r')\} \end{aligned}$$

similarly in the following definitions).
$\Pi^*(u, f) := \text{Compl}(\Pi^{\text{basis}}(u, f))$.
(e)
$$\begin{aligned} \Sigma^{\text{basis}}(u, f) := \; & \{\langle \text{p}(r, s), \text{p}(r', s') \rangle \in \text{Term}_{\text{nf}} \times \text{Term}_{\text{nf}} | \\ & \langle r, r' \rangle \in u \wedge \langle s, s' \rangle \in f(r) \wedge f(r) = f(r')\}. \end{aligned}$$

$\Sigma^*(u, f) := \text{Compl}(\Sigma^{\text{basis}}(u, f))$.

(f) Let $\lambda^*(u) := \{\langle t, u\rangle | t \in \mathrm{Term}_{\mathrm{Cl}}\}$.

(g) $\mathrm{W}^*(u, f, \alpha) := F_{\mathrm{W}}(u, f)^{\alpha}$,
where $F_{\mathrm{W}}(u, f)(v) = \mathrm{Compl}(F_{\mathrm{W}}^{\mathrm{basis}}(u, f)(v))$, and

$$F_{\mathrm{W}}^{\mathrm{basis}}(u, f)(v) := \{\langle \mathrm{sup}(r, s), \mathrm{sup}(r', s')\rangle \in \mathrm{Term}_{\mathrm{nf}} \times \mathrm{Term}_{\mathrm{nf}} | \\ \langle r, r'\rangle \in u \wedge \langle s, s'\rangle \in \Pi^{\mathrm{basis}}(f(r), \lambda^*(v)) \wedge f(r) = f(r')\}.$$

(h)

$$+^{\mathrm{basis}}(u, v) := \{\langle \mathrm{i}(r), \mathrm{i}(r')\rangle \in \mathrm{Term}_{\mathrm{nf}} \times \mathrm{Term}_{\mathrm{nf}} | \langle r, r'\rangle \in u\} \\ \cup \{\langle \mathrm{j}(r), \mathrm{j}(r')\rangle \in \mathrm{Term}_{\mathrm{nf}} \times \mathrm{Term}_{\mathrm{nf}} | \langle r, r'\rangle \in v\}.$$

$+^*(u, v) := \mathrm{Compl}(+^{\mathrm{basis}}(u, v))$.

(i)

$$\mathrm{I}^{\mathrm{basis}}(u, s, t) := \{\langle \mathrm{r}, \mathrm{r}\rangle \in \mathrm{Term}_{\mathrm{nf}} \times \mathrm{Term}_{\mathrm{nf}} | \langle s, t\rangle \in u\}.$$

$\mathrm{I}^*(u, s, t,) := \mathrm{Compl}(\mathrm{I}^{\mathrm{basis}}(u, s, t))$.

We will interpret each g-type occurring in a proof of Martin-Löf's type theory as a Σ-function, with arguments represented by the free variables of the type. More precisely, if $\mathrm{FV}(A) = \{z_1, \ldots, z_n\}$, ($z_i$ as in the definition 16.1 of $\mathrm{Var}_{\mathrm{ML}}$) the arguments of the interpretation A^* will have arguments given by the variables $\{u_1, \ldots, u_n\}$ (u_i as in definition 16.13 (a) of $\mathrm{Var}_{\mathrm{KP}}$). We introduce the following abbreviation:

Definition 16.21. (a) If A is a Σ function in KPI^+, u_i as in the definition 16.13 (a) of $\mathrm{Var}_{\mathrm{KP}}$ z_i as in the definition 16.1 of $\mathrm{Var}_{\mathrm{ML}}$, $r_1, \ldots, r_m$ extended b-objects,

$$A[z_{i_1}/r_1, \ldots, z_{i_n}/r_n] := A[u_{i_1}/r_{j_1}, \ldots, u_{i_n}/r_{j_n}],$$

where on the right-hand side we have the real substitution in KPI^+ in such a way, that, if a variable occurs more than once, only the first occurrence is carried out. If we introducing symbols for Σ-functions, this substitution is the application of the Σ-function to the arguments in the ordering as specified by the definition of the function.
We will write $A[\vec{x}/\vec{n}]$ for $A[x_1/n_1, \ldots, x_n/n_n]$.

(b) In the situation as above let $(z_i)A$ be the Σ-function with the same arguments as A except u_i s.t. $((z_i)A)[\vec{x}/\vec{r}] = \{\langle u, A[z_i/u, \vec{x}/\vec{r}]\rangle | u \in \mathrm{Term}_{\mathrm{Cl}}\}$.

Definition 16.22. We define for every g-type A the Σ-function A^* together with $\mathrm{lev}(A) \in \omega$.

If $\mathrm{FV}(A) = \{z_1, \ldots, z_n\}$, ($z_i$ as in definition 16.1 of $\mathrm{Var}_{\mathrm{ML}}$), A^* will have arguments given by the variables $\{u_1, \ldots, u_n\}$ (u_i as in definition 16.13 (a) of $\mathrm{Var}_{\mathrm{KP}}$). We will define it by giving the values $A^*[\vec{x}/\vec{s}]$.

Let for $A \subseteq \mathrm{Term}_{\mathrm{nf}} \times \mathrm{Term}_{\mathrm{nf}}$, t possibly depending on r $\lambda\langle r, \cdot\rangle \in A.t := \{\langle r, t\rangle \mid \langle r, r\rangle \in A\}$.

For $k \in \omega$, $\mathbb{N}_k^*[\,] := \mathbb{N}_k^{**}$, $\mathrm{lev}(\mathbb{N}_k) := 0$.

$\mathbb{N}^*[\,] := \mathbb{N}^{**}$, $\mathrm{lev}(\mathbb{N}) := 0$.

Let A, B be g-types, $m := \max\{\mathrm{lev}(A), \mathrm{lev}(B)\}$.
$\mathrm{lev}(\Pi x \in A.B) := m$,
$(\Pi x \in A.B)^*[\vec{x}/\vec{s}] := \Pi^*(A^*[\vec{x}/\vec{s}], \lambda\langle r, \cdot\rangle \in A^*[\vec{x}/\vec{s}].B^*[\vec{x}/\vec{s}, x/r])$,
$\mathrm{lev}(\Sigma x \in A.B) := m$,
$(\Sigma x \in A.B)^*[\vec{x}/\vec{s}] := \Sigma^*(A^*[\vec{x}/\vec{s}], \lambda\langle r, \cdot\rangle \in A^*[\vec{x}/\vec{s}].B^*[\vec{x}/\vec{s}, x/r])$,
$\mathrm{lev}(\mathrm{W} x \in A.B) := m + 1$,
$(\mathrm{W} x \in A.B)^*[\vec{x}/\vec{s}] := \mathrm{W}^*(A^*[\vec{x}/\vec{s}], \lambda\langle r, \cdot\rangle \in A^*[\vec{x}/\vec{s}].B^*[\vec{x}/\vec{s}, x/r], \mathrm{I}_m)$,
$\mathrm{lev}(A + B) := m$, $(A + B)^*[\vec{x}/\vec{s}] := +^*(A^*[\vec{x}/\vec{s}], B^*[\vec{x}/\vec{s}])$,
$\mathrm{lev}(\mathrm{I}(A, s, t)) := \mathrm{lev}(A)$, $(\mathrm{I}(A, s, t))^*[\vec{x}/\vec{s}] := \mathrm{I}^*(A^*[\vec{x}/\vec{s}], s[\vec{x}/\vec{s}], t[\vec{x}/\vec{s}])$,
$\mathrm{lev}(\mathrm{U}) := 1$, $\mathrm{U}^*[\,] := \sim(\widehat{\mathrm{U}})$,
$\mathrm{lev}(\mathrm{T}(t)) := 0$, $(\mathrm{T}(t))^*[\vec{x}/\vec{s}] := \mathrm{func}(\widehat{\mathrm{U}})(t[\vec{x}/\vec{s}])$,
where $\widehat{\mathrm{U}}$, $\sim(u)$, $\mathrm{func}(u)$ will be defined in the next definition.

Definition 16.23.

(a) $\sim(u) := \{\langle s, s'\rangle \in \mathrm{Term}_{\mathrm{Cl}} \times \mathrm{Term}_{\mathrm{Cl}} | \exists v \in \mathrm{TC}(u).\langle s, v, s'\rangle \in u\}$.
(b) $\mathrm{func}(u) := \{\langle s, v\rangle \in \mathrm{Term}_{\mathrm{Cl}} \times \mathrm{TC}(u) | \exists s' \in \mathrm{Term}_{\mathrm{Cl}}.\langle s, v, s'\rangle \in u\}$.
(c) $\mathrm{Compl}_{\mathrm{U}}(u) := \{\langle r, b, r'\rangle \in \mathrm{Term}_{\mathrm{Cl}} \times \mathrm{TC}(u) \times \mathrm{Term}_{\mathrm{Cl}} | \exists s, s' \in \mathrm{Term}_{\mathrm{nf}}.$
$r \longrightarrow_{\mathrm{red}} s \wedge r' \longrightarrow_{\mathrm{red}} s' \wedge \langle s, b, s'\rangle \in u\}$.
(d) $\widetilde{\mathrm{U}}(u) := \mathrm{Compl}_{\mathrm{U}}(\widetilde{\mathrm{U}}^{\mathrm{basis}}(u))$, where

$$
\begin{aligned}
&\widetilde{\mathrm{U}}^{\mathrm{basis}}(u) := \\
&\quad \{\langle \widehat{\mathrm{N}}_k, \mathrm{N}_k^{**}, \widehat{\mathrm{N}}_k\rangle \in \mathrm{ad}(u) | k \in \omega\} \\
&\cup \{\langle \widehat{\mathrm{N}}, \mathrm{N}^{**}, \widehat{\mathrm{N}}\rangle\} \\
&\cup \{\langle \widehat{\Pi} x \in r.s, \Pi^*(b, f), \widehat{\Pi} x' \in r'.s'\rangle | \varphi(r, x, s, r', x', s', b, f, u) \wedge b, f \in \mathrm{ad}(u)\} \\
&\cup \{\langle \widehat{\Sigma} x \in r.s, \Sigma^*(b, f), \widehat{\Sigma} x' \in r'.s'\rangle | \varphi(r, x, s, r', x', s', b, f, u) \wedge b, f \in \mathrm{ad}(u)\} \\
&\cup \{\langle \widehat{\mathrm{W}} x \in r.s, \mathrm{W}^*(b, f, \alpha^+(u)), \widehat{\mathrm{W}} x' \in r'.s'\rangle | \\
&\qquad \varphi(r, x, s, r', x', s', b, f, u) \wedge b, f \in \mathrm{ad}(u)\} \\
&\cup \{\langle r \widehat{+} s, +^*(b, c), r' \widehat{+} s'\rangle \in \mathrm{ad}(u) | \psi_+(r, s, r', s', b, c, u) \wedge b, c \in \mathrm{ad}(u)\} \\
&\cup \{\langle \widehat{\mathrm{I}}(r, s, t), \mathrm{I}^*(b, s, t), \widehat{\mathrm{I}}(r', s', t')\rangle \in \mathrm{ad}(u) | \\
&\qquad \psi_i(r, s, t, r', s', t', b, u) \wedge b \in \mathrm{ad}(u)\}
\end{aligned}
$$

and

$$
\begin{aligned}
&\varphi(r, x, s, r', x', s', b, f, u) \\
&:= r, r' \in \mathrm{Term}_{\mathrm{nf}} \wedge s, s' \in \mathrm{Term} \\
&\qquad \wedge \mathrm{FV}(s) \subset \{x\} \wedge \mathrm{FV}(s') \subset \{x'\} \wedge \langle r, b, r'\rangle \in u \wedge \\
&\qquad (\forall \langle t, t'\rangle \in b.\langle s[x/t], f(t), s'[x'/t']\rangle \in u)
\end{aligned}
$$

(note that $f(t) = \bigcup\{c \in \mathrm{TC}(f) | \langle t, c\rangle \in f\}$).

$$
\begin{aligned}
\psi_+(r, s, r', s', b, c, u) &:= r, s, r', s' \in \mathrm{Term}_{\mathrm{nf}} \wedge \langle r, b, r'\rangle \in u \wedge \langle s, c, s'\rangle \in u, \\
\psi_i(r, s, t, r', s', t', b, u) &:= r, s, t, r', s', t' \in \mathrm{Term}_{\mathrm{nf}} \wedge \langle r, b, r'\rangle \in u \wedge \\
&\qquad \langle s, s'\rangle \in b \wedge \langle t, t'\rangle \in b.
\end{aligned}
$$

(e) $\widehat{\mathrm{U}} := \widetilde{\mathrm{U}}^{\mathrm{I}}$.

16.7 Properties of the Interpretation

Lemma 16.24. (a) $\forall v \subset v'.F^*_{\mathrm{W}}(b,f)(v) \subset F^*_{\mathrm{W}}(b,f)(v')$.
(b) $\forall \gamma < \delta.\mathrm{W}^*(b,f,\gamma) \subset \mathrm{W}^*(b,f,\delta)$.
(c) $(b \in a \wedge f \in a \wedge \mathrm{Ad}(a)) \to \forall \gamma > \alpha(a).\mathrm{W}^*(b,f,\gamma) = \mathrm{W}^*(b,f,\alpha(a))$.

Proof. (a) immediate, (b) follows from (a) by induction on δ.
(c) It is sufficient to show, with $\alpha := \alpha(u)$, $v := \mathrm{W}^*(b,f,\alpha)$, that $F_{\mathrm{W}}(b,f)(v) \subset v$. Since $\mathrm{Compl}(v) \subset v$ it is sufficient to prove $F^{\mathrm{basis}}_{\mathrm{W}}(b,f)(v) \subset v)$.
Now, if $\langle \sup(r,t), \sup(r',t')\rangle \in F^{\mathrm{basis}}_{\mathrm{W}}(b,f)(v)$, then $t = \lambda x.s$, $t' = \lambda x'.s'$, and $\forall u,u' \in \mathrm{Term}_{\mathrm{Cl}}.\langle u,u'\rangle \in f(r) \to \exists \delta \in \mathrm{Ord} \cap \alpha.\langle s[x/u], s'[x'/u']\rangle \in \mathrm{W}^*(b,f,\delta)$.
Since $\mathrm{Ad}(a)$, we have $(\Delta_0\text{-coll})^a$, therefore for some $\rho < \alpha$,
$\forall t,t' \in \mathrm{Term}_{\mathrm{Cl}}.\langle t,t'\rangle \in f(r) \to \exists \delta < \rho.\langle s[x/t], s'[x'/t']\rangle \in \mathrm{W}^*(b,f,\delta)$.
Now it follows $\langle \sup(r,\lambda x.s), \sup(r',\lambda x'.s')\rangle \in \mathrm{W}^*(b,f,\rho) \subset v$ and the assertion. □

Definition 16.25.

(a) $\mathrm{equiv}(u) :\Leftrightarrow \forall r,s,t,r',s' \in \mathrm{Term}_{\mathrm{Cl}}.((\langle r,s\rangle \in u \to \langle s,r\rangle \in u) \wedge$
$((\langle r,s\rangle \in u \wedge \langle s,t\rangle \in u) \to \langle r,t\rangle \in u)$
(note that we do not claim reflexivity).
(b) $\mathrm{equivfun}(f) :\Leftrightarrow \forall x \in \mathrm{dom}(f).\mathrm{equiv}(f(x))$.
(c) $\mathrm{Cor}(u) :\Leftrightarrow$
$\forall r,r',r'' \in \mathrm{Term}_{\mathrm{Cl}}.\forall b,b'.(\langle r,b,r'\rangle \in u \to [\langle r',b,r\rangle \in u \wedge \mathrm{equiv}(b)$
$\wedge[\langle r',b',r''\rangle \in u \to (\langle r,b,r''\rangle \in u \wedge b = b')]])$.

Remark 16.26.

(a) $(\mathrm{Cor}(u) \wedge \langle r,b,r'\rangle \in u \wedge \langle r,b',r''\rangle \in u) \to (b = b' \wedge \langle r,b,r''\rangle \in u)$.
(b) If $\mathrm{Cor}(u)$ then with $\sim := \sim(u)$, $f := \mathrm{func}(u)$ we have $\sim$ is a symmetric and transitive relation, f is a function s.t. $\forall a,b.a \sim b \to f(a) = f(b)$ and $\mathrm{equivfun}(f)$.

Lemma 16.27. (a) $(\mathrm{equiv}(u) \wedge u \subset \mathrm{Term}_{\mathrm{nf}} \times \mathrm{Term}_{\mathrm{nf}}) \to \mathrm{equiv}(\mathrm{Compl}(u))$.
(b) $(\mathrm{equiv}(u) \wedge \mathrm{equiv}(v) \wedge \mathrm{equivfun}(f) \wedge k \in \omega \wedge s,t \in \mathrm{Term}_{\mathrm{Cl}}) \to$
$(\mathrm{equiv}(\mathbb{N}^{\mathrm{basis}}) \wedge \mathrm{equiv}(\mathbb{N}^{\mathrm{basis}}_k) \wedge \mathrm{equiv}(\Pi^{\mathrm{basis}}(u,f)) \wedge \mathrm{equiv}(\Sigma^{\mathrm{basis}}(u,f)) \wedge$
$\mathrm{equiv}(F^{\mathrm{basis}}_{\mathrm{W}}(u,f)(v)) \wedge \mathrm{equiv}(+^{\mathrm{basis}}(u,v)) \wedge \mathrm{equiv}(\mathrm{i}^{\mathrm{basis}}(u,s,t)))$.
(c) $(\mathrm{equiv}(u) \wedge \mathrm{equiv}(v) \wedge \mathrm{equivfun}(f) \wedge \alpha \in \mathrm{Ord} \wedge k \in \omega \wedge s,t \in \mathrm{Term}_{\mathrm{Cl}}) \to$
$(\mathrm{equiv}(\mathbb{N}^*) \wedge \mathrm{equiv}(\mathbb{N}^*_k) \wedge \mathrm{equiv}(\Pi^*(u,f)) \wedge \mathrm{equiv}(\Sigma^*(u,f))$
$\wedge \mathrm{equiv}(F^*_{\mathrm{W}}(u,f)(v)) \wedge \mathrm{equiv}(\mathrm{W}^*(u,f,\alpha))$
$\wedge \mathrm{equiv}(+^*(u,v)) \wedge \mathrm{equiv}(\mathrm{i}^*(u,s,t)))$.

Lemma 16.28. *Assume* $r,s,t,r',s',t' \in \mathrm{Term}$, $x,x' \in \mathrm{Var}_{\mathrm{ML}}$, b,f,u *sets.*

(a) $(\varphi(r,x,s,r',x',s',b,f,u) \wedge \mathrm{Cor}(u)) \to$
$(b \in \mathrm{ad}(u) \wedge \exists f \in \mathrm{ad}(u).\forall \langle t,t'\rangle \in b.f(t) = f(t') = f'(t) = f'(t'))$.
(b) $\psi_+(r,s,r',s',b,c,u) \to b,c \in \mathrm{ad}(u)$.
(c) $\psi_i(r,s,t,r',s',t',b,u) \to b \in \mathrm{ad}(u)$.
(d) $\mathrm{Cor}(u) \to \widetilde{\mathrm{U}}(u) \in \mathrm{ad}(\mathrm{ad}(u))$.
(e) $\forall \gamma \in \mathrm{Ord} \cap \mathrm{I}.\widetilde{\mathrm{U}}^{\gamma} \in \mathrm{Ad}_{\mathrm{I}}$.

Proof. (a) $b \in \mathrm{TC}(u) \in \mathrm{ad}(u)$.
Let $f' := \{\langle t, c\rangle \in \mathrm{Term}_{\mathrm{Cl}} \times \mathrm{TC}(u) | \langle t, t\rangle \in b \wedge \langle s[x/t], c, s[x/t]\rangle \in u\}$.
$f' \in \mathrm{ad}(u)$. Furthermore, if $\langle t, t'\rangle \in b$, it follows $\langle s[x/t], f(t), s'[x/t']\rangle \in u$, by $\mathrm{Cor}(u)$ $\langle s[x/t], f(t), s[x/t]\rangle \in u$, $f(t) = f'(t)$, and, since
$\langle s'[x/t], f(t), s[x/t]\rangle \in u \wedge \langle s'[x/t], f(t'), s[x/t']\rangle \in u$, it follows $f(t) = f(t')$.
(b), (c),(d): easy.
(e): Induction on γ, using (d) and 16.15 (b). □

Lemma 16.29. *Assume* $r, s, t, r', s', t', r'', s'', t'' \in \mathrm{Term}$, $x, x', x'' \in \mathrm{Var}_{\mathrm{ML}}$, b, b', f, f', u, u' *sets.*

(a) $\varphi(r, x, s, r', x', s', b, f, u) \wedge \mathrm{Cor}(u) \to \varphi(r', x', s', r, x, s, b, f, u)$.
(b) $(\varphi(r, x, s, r', x', s', b, f, u) \wedge \varphi(r', x', s', r'', x'', s'', b', f', u') \wedge \mathrm{Cor}(u \cup u')) \to$
$\varphi(r, x, s, r'', x'', s'', b, f, u \cup u') \wedge b = b' \wedge$
$\forall \langle t, t'\rangle \in b. f(t) = f'(t) = f(t') = f'(t')$.
(c) $\psi_+(r, s, r', s', b, c, u) \wedge \mathrm{Cor}(u) \to \psi_+(r', s', r, s, b, c, u)$.
(d) $(\psi_+(r, s, r', s', b, c, u) \wedge \psi_+(r', s', r'', s'', b', c', u') \wedge \mathrm{Cor}(u \cup u') \to$
$(\psi_+(r, s, r'', s'', b, c, u \cup u') \wedge b = b' \wedge c = c'))$.
(e) $\psi_i(r, s, t, r', s', t', b, u) \wedge \mathrm{Cor}(u) \to \psi_i(r', s', t', r, s, t, b, u)$.
(f) $(\psi_i(r, s, t, r', s', t', b, u) \wedge \psi_i(r', s', t', r'', s'', t'', b', u') \wedge \mathrm{Cor}(u \cup u') \to$
$(\psi_i(r, s, t, r'', s'', t'', b, u \cup u') \wedge b = b'))$.

Lemma 16.30. (a) $(\mathrm{Cor}(u) \wedge \sim(u) \subset \mathrm{Term}_{\mathrm{nf}} \times \mathrm{Term}_{\mathrm{nf}}) \to \mathrm{Cor}(\mathrm{Compl}_{\mathrm{U}}(u))$.
(b) $\mathrm{Cor}(u) \to \mathrm{Cor}(\widetilde{\mathrm{U}}(u))$.
(c) $u \subset u' \wedge \mathrm{Cor}(u') \to \widetilde{\mathrm{U}}(u) \subset \widetilde{\mathrm{U}}(u')$.
(d) $\mathrm{Cor}(\widehat{\mathrm{U}})$.

Lemma 16.31. *If A g-type, then* $\mathrm{KPI}^+ \vdash \forall s_1, \ldots, s_n \in \mathrm{Term}_{\mathrm{Cl}}.\mathrm{equiv}(A^*[\vec{x}/\vec{s}])$.

Proof. Induction on the definition of types. □

Definition 16.32. Let A, B g-types, s, t g-terms, $\mathrm{FV}(A)$, $\mathrm{FV}(B)$, $\mathrm{FV}(s)$, $\mathrm{FV}(t) \subset \{x_1, \ldots, x_n\}$, $r_1, \ldots, r_n$, $s_1, \ldots, s_n$ be extended g-terms.

(a) $(A = B : \mathrm{type})^*[\vec{x}/\vec{r}; \vec{s}] :\Leftrightarrow (A = B : \mathrm{type})^*[x_1/r_1; s_1, \ldots, x_n/r_n; s_n] :\Leftrightarrow$ $(A^*[\vec{x}/\vec{r}] = B^*[\vec{x}/\vec{s}])$.
(b) $(t = t' : A)^*[\vec{x}/\vec{r}; \vec{s}] :\Leftrightarrow (t = t' : A)^*[x_1/r_1; s_1, \ldots, x_n/r_n; s_n] :\Leftrightarrow$ $\langle t[\vec{x}/\vec{r}], t'[\vec{x}/\vec{s}]\rangle \in A^*[\vec{x}/\vec{r}]$.

We will not mention the variables $x_1, \ldots, x_n$ explicitly, if they are the variables, mentioned in the context, writing $(A = B : \mathrm{type})^*[\vec{r}; \vec{s}]$ and $(t = t' : A)^*[\vec{r}; \vec{s}]$.

Note that $s : A$, $A : \mathrm{type}$ abbreviate $s = s : A$, $A = A : \mathrm{type}$, therefore $(s : A)^*[\vec{r}; \vec{s}]$, $(A : \mathrm{type})^*[\vec{r}; \vec{s}]$ are defined as well.

Lemma 16.33 (Substitution Lemma). *Let C, D be g-types, r, s, t_i, t_i' g-terms, $x_i, y_i \in \mathrm{Var_{ML}}$. Then:*

(a) *If $r[\vec{x}/\vec{t}]$ is an allowed substitution, $\mathrm{FV}(r[\vec{x}/\vec{t}]) \subset \{y_1, \ldots, y_n\}$, then*
$\mathrm{KPI}^+ \vdash \forall \vec{r} \in \mathrm{Term_{Cl}}.r[\vec{x}/\vec{t}][\vec{y}/\vec{r}] = r[x_1/t_1[\vec{y}/\vec{r}], \ldots, x_n/t_n[\vec{y}/\vec{r}], \vec{y}/\vec{r}]$.
(Note that, if variables occur more than once in $[\vec{y}/\vec{r}]$, only the first substitution is relevant.)

(b) *If $C[\vec{x}/\vec{t}]$ is an allowed substitution, $\mathrm{FV}(C[\vec{x}/\vec{t}]) \subset \{y_1, \ldots, y_n\}$, then*
$\mathrm{KPI}^+ \vdash \forall \vec{r}, r' \in \mathrm{Term_{Cl}}.C[\vec{x}/\vec{t}]^*[\vec{y}/\vec{r}] = C^*[x_1/t_1[\vec{y}/\vec{r}], \ldots, x_n/t_n[\vec{y}/\vec{r}], \vec{y}/\vec{r}]$.

(c) *If $A[\vec{x}/\vec{t}]$, $B[\vec{x}/\vec{t}']$ are allowed substitutions,*
$\mathrm{FV}(A[\vec{x}/\vec{t}]), \mathrm{FV}(B[\vec{x}/\vec{t}']) \subset \{y_1, \ldots, y_n\}$, *then*
$\mathrm{KPI}^+ \vdash \forall \vec{r}, \vec{s} \in \mathrm{Term_{Cl}}.(A = B : \mathrm{type})^*[\vec{x}/(\vec{t}[\vec{y}/\vec{r}]); (\vec{t}'[\vec{y}/\vec{s}]), \vec{y}/\vec{r}; \vec{s}] \iff$
$(A[\vec{x}/\vec{t}] = B[\vec{x}/\vec{t}'] : \mathrm{type})^*[\vec{y}/\vec{r}; \vec{s}]$.

(d) *If $A[\vec{x}/\vec{t}]$, $r[\vec{x}/\vec{t}]$ are allowed substitutions,*
$\mathrm{FV}(A[\vec{x}/\vec{t}]), \mathrm{FV}(r[\vec{x}/\vec{t}]) \subset \{y_1, \ldots, y_n\}$, *then*
$\mathrm{KPI}^+ \vdash \forall \vec{r}, \vec{s} \in \mathrm{Term_{Cl}}.(r : A)^*[\vec{x}/(\vec{t}[\vec{x}/\vec{r}]); (\vec{t}[\vec{x}/\vec{s}]), \vec{x}/\vec{r}; \vec{s}] \iff$
$(r[x/t] : A[x/t])^*[\vec{x}/\vec{r}; \vec{s}]$.

(e) *If $A[\vec{x}/\vec{t}]$, $r[\vec{x}/\vec{t}], s[\vec{x}/\vec{t}']$ are allowed substitutions, $\mathrm{FV}(A[\vec{x}/\vec{t}])$, $\mathrm{FV}(r[\vec{x}/\vec{t}])$, $\mathrm{FV}(s[\vec{x}/\vec{t}']) \subset \{y_1, \ldots, y_n\}$, then*
$\mathrm{KPI}^+ \vdash \forall \vec{r}, \vec{s} \in \mathrm{Term_{Cl}}.(r = s : A)^*[\vec{x}/(\vec{t}[\vec{x}/\vec{r}]); (\vec{t}'[\vec{x}/\vec{s}]), \vec{x}/\vec{r}; \vec{s}] \iff$
$(r[\vec{x}/\vec{t}] = s[x/\vec{t}'] : A[x/\vec{t}])^*[\vec{x}/\vec{r}; \vec{s}].r$

Proof. By induction on the definition of the terms and types. □

Lemma 16.34. *For every g-type A $\mathrm{FV}(A) \subset \{x_1, \ldots, x_n\}$, it follows*

(a) $\forall \vec{r}, r, r', s, s' \in \mathrm{Term_{Cl}}.(r \longrightarrow_{\mathrm{red}} r') \to (s \longrightarrow_{\mathrm{red}} s')$
$\to \langle r, s \rangle \in A^*[\vec{x}/\vec{r}] \to \langle r', s' \rangle \in A^*[\vec{x}/\vec{r}]$.

(b) $\forall \vec{r}, r, r' \in \mathrm{Term_{Cl}}.\langle r, r' \rangle \in A^*[\vec{x}/\vec{r}] \to \exists s, s' \in \mathrm{Term_{nf}}.r \longrightarrow_{\mathrm{red}} s \wedge r' \longrightarrow_{\mathrm{red}} s'$.

Proof. Easy, since for each type, Compl was applied to some set. □

Definition 16.35.

(a) $\mathrm{Stable}(a) := \forall r, s, r', s' \in \mathrm{Term_{Cl}}.\langle r, s \rangle \in a \to r =_\alpha r' \to s =_\alpha s' \to \langle r', s' \rangle \in a$.

(b) For every g-type A with $\mathrm{FV}(A) = \{x_1, \ldots, x_n\}$ we define
$\mathrm{Flex}(A) := \forall r_1, \ldots, r_n, s_1, \ldots, s_n \in \mathrm{Term_{Cl}}.v(r_1 =_\alpha s_1 \wedge \cdots \wedge r_n =_\alpha s_n) \to$
$A^*[\vec{x}/\vec{r}] = A^*[\vec{x}/\vec{s}]$.

Lemma 16.36. *For every g-type C, D with $\mathrm{FV}(C) = \{x_1, \ldots, x_n\}$ and $C =_\alpha D$ we have*

(a) $\mathrm{KPI}^+ \vdash \mathrm{Flex}(C)$.

(b) $\mathrm{KPI}^+ \vdash \forall r_1, \ldots, r_n \in \mathrm{Term_{Cl}}.\mathrm{Stable}(C^*[\vec{x}/\vec{r}])$.

(c) $\mathrm{KPI}^+ \vdash \forall r_1, \ldots, r_n \in \mathrm{Term_{Cl}}.C^*[\vec{x}/\vec{r}] = D^*[\vec{x}/\vec{r}]$.

Proof. Easy, simultaneously by induction on the definition of g-types. In the case $C \equiv \mathrm{U}, \mathrm{T}(t)$ we define

$$\begin{aligned}\mathrm{Stable_U}(u) := \ & \forall s,s',t,t' \in \mathrm{Term_{Cl}}.\forall b \in \mathrm{TC}(u).\\ & s{=_\alpha}s' \wedge t{=_\alpha}t' \wedge \mathrm{pair}(s,b,t) \in u\\ & \to (\mathrm{pair}(s',b,t') \in u \wedge \mathrm{Stable}(b)).\end{aligned}$$

We conclude $\mathrm{Stable_U}(u) \to \mathrm{Stable_U}(\widetilde{\mathrm{U}}(u))$ and therefore $\mathrm{Stable_U}(\widehat{\mathrm{U}})$, from which we obtain the assertion for $C = \mathrm{U}$ and $C = \mathrm{T}(t)$. The other cases are straightforward. □

In order to state our Main Lemma, we need to express, that, if we assume elements of the types of the context, the interpretation of the conclusion Θ of a judgement of Martin-Löf is valid. Since we need, that this is independent of the choice of equal elements of A_i, we will introduce the following abbreviation:

Definition 16.37. Let $\Gamma \equiv x_1 : A_1, \ldots, x_k : A_k$ be a g-context.

$$\begin{aligned}\forall\Gamma^{=}(\vec{r};\vec{s}).\varphi :\equiv \ & \forall r_1,\ldots,r_k,s_1,\ldots,s_k \in \mathrm{Term_{Cl}}.(\langle r_1,s_1\rangle \in A_1^*[\,] \wedge\\ & \langle r_2,s_2\rangle \in A_2^*[x_1/r_1] \wedge \cdots \wedge \langle r_k,s_k\rangle \in A_k^*[x_1/r_1,\ldots,x_{k-1}/r_{k-1}]) \to \varphi.\end{aligned}$$

"Assume $\Gamma^{=}(\vec{r};\vec{s})$" means:
"Assume $r_1,\ldots,r_k,s_1,\ldots,s_k \in \mathrm{Term_{Cl}}$ such that $\langle r_1,s_1\rangle \in A_1^*[\,]\wedge$
$\langle r_2,s_2\rangle \in A_2^*[x_1/r_1] \wedge \cdots \wedge \langle r_k,s_k\rangle \in A_k^*[x_1/r_1,\ldots,x_{k-1}/r_{k-1}]$".

16.8 Main Lemma

In this section we prove the Main Lemma, which expresses that if $\mathrm{ML} \vdash r : A$, then $\mathrm{KPI}^+ \vdash \langle r,r\rangle \in A^*$. We have to go through all judgements.

Lemma 16.38 (Main Lemma). *Let* Γ,Δ *be g-context-pieces,* $x, x_i \in \mathrm{Var_{ML}}$, A_i, A, B *g-types,* t, t' *g-terms,* θ *a g-judgement. Assume* $\Gamma = x_1 : A_1, \ldots, x_n : A_n$.

(a) *If* $\mathrm{ML} \vdash \Gamma \Rightarrow t = t' : A$, *then*

  (i) $\mathrm{KPI}^+ \vdash \forall\Gamma^{=}(\vec{r};\vec{s}).(t = t' : A)^*[\vec{x}/\vec{r};\vec{s}]$,
  (ii) $\mathrm{KPI}^+ \vdash \forall\Gamma^{=}(\vec{r};\vec{s}).(A : \mathrm{type})^*[\vec{x}/\vec{r};\vec{s}]$.

(b) *If* $\mathrm{ML} \vdash \Gamma \Rightarrow A = A' : \mathrm{type}$, *then*

$$\mathrm{KPI}^+ \vdash \forall\Gamma^{=}(\vec{r};\vec{s}).(A = A' : \mathrm{type})^*[\vec{x}/\vec{r};\vec{s}].$$

(c) *If* $\mathrm{ML} \vdash \Gamma, x : A, \Delta \Rightarrow \theta$, *then*

$$\mathrm{KPI}^+ \vdash \forall\Gamma^{=}(\vec{r};\vec{s}).(A : \mathrm{type})^*[\vec{x}/\vec{r};\vec{s}].$$

Proof of the Main Lemma. We prove simultaneously (a) – (c) by induction on the derivation. We write IH 3 for the Induction-hypothesis for the 3rd premise, etc. IH 3(c) for the Induction-hypothesis (c) for the 3rd premise of the rule etc.

If there is more than one rule of one category (as in the case of (TRANS)), we refer to them by $(\mathrm{TRANS})_1$, $(\mathrm{TRANS})_2$, etc.

Let $\Gamma = x_1 : A_1, \ldots, x_n : A_n$, $\Gamma' = y_1 : B_1, \ldots, y_m : B_m$.

If $\vec{r} = r_1, \ldots, r_n$, $i \leq n$, then $\hat{r}_i := r_1, \ldots, r_{i-1}$ ($\hat{r}_1$ is empty).

If θ is $t = t' : A$ or $A = B$: type, let $\theta' = (A : \text{type})$ (the judgement treated in the cases (i) of (a),(b), or which follows from the assertion in (c).

Case distinction by the last rule applied.

Assume that lemma is proved for the premises of a rule, as stated in the definition, weakened by the context Γ.

We treat only some examples of the rules, covering the more complicated ones.

Case (SYM)$_1$ Assume $\Gamma^{=}(\vec{r}; \vec{s})$. From $\langle r_i, s_i \rangle \in A_i^*[\hat{r}_i]$ it follows $\langle s_i, r_i \rangle \in A_i^*[\hat{r}_i]$ and by IH (a,ii) $\langle s_i, r_i \rangle \in A_i^*[\hat{s}_i]$. By IH (a,i) it follows $\langle t[\vec{s}], t'[\vec{r}] \rangle \in A^*[\vec{s}]$, and by IH (a,ii) $A^*[\vec{r}] = A^*[\vec{s}]$, and by 16.31 it follows $(t' = t : A)^*[\vec{r}; \vec{s}]$.

(a,ii) follows from IH (a,ii).

Case (SYM)$_2$ Assume $\Gamma^{=}(\vec{r}; \vec{s})$. As for (SYM)$_1$ we have $\langle s_i, r_i \rangle \in A_i^*[\hat{s}_i]$, by IH $A^*[\vec{s}] = B^*[\vec{r}]$ and therefore the assertion.

Case (SUB) Assume $\Gamma^{=}(\vec{r}; \vec{r}')$, $\langle s_i, s_i' \rangle \in B_i[x/t]^*[\vec{r}, \hat{s}_i]$. Now by Lemma 16.33

$$B_i[x/t]^*[\vec{r}, \hat{s}_i] = B_i^*[x/t[\vec{r}, \hat{s}_i], \vec{r}, \hat{s}_i] = B_i^*[x/t[\vec{r}], \vec{r}, \hat{s}_i].$$

By IH 2 (a,i) $\langle t[\vec{r}], t'[\vec{r}'] \rangle \in A^*[\vec{r}]$, therefore

$$\theta^*[\vec{x}/\vec{r}; \vec{r}', x/t[\vec{r}]; t'[\vec{r}'], \vec{y}/\vec{s}; \vec{s}'],$$

and by Lemma 16.33 $\theta[x/t]^*[\vec{x}/\vec{r}; \vec{r}', \vec{s}; \vec{s}']$, similarly for θ'.

Proof for (c): If "$y : B$" in Γ, the assertion follows by IH.

If "$y : B$" in $\Gamma'[x/t]$, it follows by IH $B^*[\vec{r}, x/\vec{t}[\vec{r}], \hat{s}_i] = B^*[\vec{r}', x/t[\vec{r}'], \hat{s}_i']$, and by 16.33 the assertion.

Case (REPL1) Assume $\Gamma^{=}(\vec{r}; \vec{r}')$, $\langle s_i, s_i' \rangle \in B_i[x/t]^*[\vec{r}, \hat{s}_i]$. We have $\langle t[\vec{r}], t'[\vec{r}'] \rangle \in A^*[\vec{r}]$. By 16.33, it follows $B_i^*[\vec{x}/\vec{r}, x/t[\vec{r}], \hat{y}_i/\hat{s}_i] = B_i[x/t]^*[\vec{r}, \hat{s}_i]$. Therefore we have $\langle s_i, s_i' \rangle \in B_i^*[\vec{x}/\vec{r}, x/t[\vec{r}], \hat{y}_i/\hat{s}_i]$. Then by IH 1 $B^*[\vec{x}/\vec{r}, x/t[\vec{r}], \vec{y}/\vec{s}] = B^*[\vec{x}/\vec{r}', x/t'[\vec{r}'], \vec{y}/\vec{s}']$, and by 16.33 follows the assertion.

Proof for (c): From IH 2 it follows, arguing as for the rule (SYM), the assertion for $\Gamma \Rightarrow t' = t : A$ and further, arguing as for (TRANS) the assertion for $\Gamma \Rightarrow t = t : A$, which is the same as for $\Gamma \Rightarrow t : A$ and now the proof follows as in (SUB).

Case (REPL2) Assume $\Gamma^{=}(\vec{r}, \vec{r}')$, $\langle s_i, s_i' \rangle \in B_i[x/t]^*[\vec{r}, \hat{s}_i]$. Then $\langle t[\vec{r}], t'[\vec{r}'] \rangle \in A^*[\vec{r}]$, and by IH 1(a,i)

$$(s = s : B)^*[\vec{x}/\vec{r}; \vec{r}', x/t[\vec{r}]; t'[\vec{r}'], \vec{y}/\vec{s}; \vec{s}']$$

and by 16.33 follows the assertion for (a,i). (a,ii) follows as in (REPL1), using that we get the assertion for $\Gamma \Rightarrow t = t : A$, and (c) follows exactly as in (REPL1).

Case (ALPHA): Immediate by the IH since if $A {=_\alpha} A'$, $t {=_\alpha} t'$, $A[\vec{s}] = A'[\vec{s}]$, $t[\vec{r}'] {=_\alpha} t'[\vec{r}']$ and $\langle t[\vec{r}], t[\vec{r}'] \rangle \in A^*[\vec{r}] \iff \langle t[\vec{r}], t'[\vec{r}'] \rangle \in A^*[\vec{r}]$.

Case ($\Pi^{T,=}$) Assume $\Gamma^{=}(\vec{r}; \vec{r}')$. By IH (c) $A^*[\vec{r}] = A'^*[\vec{r}']$, and, if $\langle r, s \rangle \in A^*[\vec{r}]$, it follows $\langle r, r \rangle, \langle s, s \rangle \in A^*[\vec{r}]$, therefore by IH $B^*[\vec{r}, x/r] = B'^*[\vec{r}', x/r]$, $B^*[\vec{r}, x/s] = B'^*[\vec{r}', x/s]$, $\Pi x \in A.B^*[\vec{r}] = (\Pi x \in A'.B')^*[\vec{r}']$.

Case $(\mathbb{N}^{\mathrm{I},=})_2$: Assume $\Gamma^=(\vec{r};\vec{r}')$. By IH we have for some $k \in \omega$ $t[\vec{r}]\longrightarrow_{\mathrm{red}}\mathrm{S}^k(0)$ and $t'[\vec{r}']\longrightarrow_{\mathrm{red}}\mathrm{S}^k(0)$, therefore $\mathrm{S}(t)[\vec{r}]\longrightarrow_{\mathrm{red}}\mathrm{S}^{k+1}(0)$, $\mathrm{S}(t')[\vec{r}']\longrightarrow_{\mathrm{red}}\mathrm{S}^{k+1}(0)$, and we have the assertion.

Case $(\Pi^{\mathrm{I},=})$: Assume $\Gamma^=(\vec{r};\vec{r}')$, $\langle r,r'\rangle \in A^*[\vec{r}]$. Then by IH (a,i)
$\langle t[x/r,\vec{r}],t'[x/r',\vec{r}']\rangle \in B^*[x/r,\vec{r}] = B^*[\vec{r},x/r]$,
$\langle(\lambda x.t)[\vec{x}/\vec{r}],(\lambda x.t')[\vec{x}/\vec{r}']\rangle \in (\Pi x \in A.B)^*[\vec{r}]$.

(a,ii) follows as in $(\Pi_1^{T,=})$, since from IH (a,ii) follows (b) for $x : A \Rightarrow B$: type.

Case $(\mathrm{W}^{\mathrm{I},=})$: Let $n := \max\{\mathrm{lev}(A),\mathrm{lev}(B)\}$, $\mathrm{W}x \in A.B^*$.

Assume $\Gamma^=(\vec{r};\vec{r}')$. Let $F := F_{\mathrm{W}}(A^*[\vec{x}/\vec{r}],(x)B^*[\vec{x}/\vec{r}])$. Then by IH $\langle r[\vec{r}],r'[\vec{r}']\rangle \in A^*[\vec{r}]$, $s[\vec{r}]\longrightarrow_{\mathrm{red}}\lambda x.t$, $s'[\vec{r}']\longrightarrow_{\mathrm{red}}\lambda x'.t'$, $B^*[x/r[\vec{r}],\vec{r}] = B^*[x/r'[\vec{r}'],\vec{r}']$, and

$$\forall\langle u,u'\rangle \in B[x/t]^*[\vec{x}/\vec{r}](= B^*[x/t[\vec{r}],\vec{x}/\vec{r}])\exists\gamma < \mathrm{I}_n.\langle t[x/u],t'[x/u']\rangle \in F^\gamma.$$

By (Δ_0-coll) and $\mathrm{Ad}(\mathrm{L}_{\mathrm{I}_n})$ there exist a $\delta < \mathrm{I}_n$ such that the γ can be chosen to be $< \delta$. Then $\langle\sup(r,s)[\vec{r}],\sup(r,s)[\vec{r}']\rangle \in F^{\gamma+1} \subset \mathrm{W}x \in A.B^*[\vec{r}]$.

(a,ii) follows as in $(\mathrm{W}^{T,=})$.

Case$(\mathbb{N}^{E,=})$: Assume $\Gamma^=(\vec{r};\vec{r}')$. Then by IH 1 $\langle r[\vec{r}],r'[\vec{r}']\rangle \in \mathbb{N}^*$, therefore $r[\vec{r}]\longrightarrow_{\mathrm{red}}\mathrm{S}^n(0)$, $r'[\vec{r}]\longrightarrow_{\mathrm{red}}\mathrm{S}^n(0)$ for some $n < \omega$. Furthermore, by IH 2 and 16.34 (b) exist $\widetilde{s}$, $\widetilde{s}' \in \mathrm{Term}_{\mathrm{nf}}$ such that $s[\vec{r}]\longrightarrow_{\mathrm{red}}\widetilde{s}$, $t[\vec{r}]\longrightarrow_{\mathrm{red}}\widetilde{s}'$, $\langle\widetilde{s},\widetilde{s}'\rangle \in A[x/0]^*[\vec{r}] = A^*[x/u,\vec{r}]$.

Let $[\vec{x}'/\vec{s}] := [\vec{x}/\vec{r}] \setminus \{x,y\}$, $[\vec{x}'/\vec{s}''] := [\vec{x}/\vec{r}'] \setminus \{x,y\}$.

Let $P_0(r) := \mathrm{P}(r,\widetilde{s},\lambda x.\lambda y.(t[\vec{x}'/\vec{s}]))$. $P_1(r) := \mathrm{P}(r,\widetilde{s}',\lambda x.\lambda y.(t'[\vec{x}'/\vec{s}']))$. Then

$$\mathrm{P}(r,s,(x,y)t)[\vec{r}]\longrightarrow_{\mathrm{red}}P_0(\mathrm{S}^n(0)),\ \mathrm{P}(r',s',(x,y)t')[\vec{r}']\longrightarrow_{\mathrm{red}}P_1(\mathrm{S}^n(0)).$$

We have $A[z/r]^*[\vec{r}] = A^*[z/r[\vec{r}],\vec{r}] = A^*[z/\mathrm{S}^n(0),\vec{r}]$, and therefore assertion (a,i).

We show: $\forall m \in \omega.\langle P_0(\mathrm{S}^m(0)),P_1(\mathrm{S}^m(0))\rangle \in A^*[z/\mathrm{S}^m(0),\vec{r}]$.

If $m = 0$, $P_0(\mathrm{S}^m(0))\longrightarrow_{\mathrm{red}}\widetilde{s}$, $P_1(\mathrm{S}^m(0))\longrightarrow_{\mathrm{red}}\widetilde{s}'$, $\langle\widetilde{s},\widetilde{s}'\rangle \in A^*[z/0,\vec{r}]$.

If $m = k+1$, it follows by IH $P_0(\mathrm{S}^k(0))\longrightarrow_{\mathrm{red}}\widetilde{s}$, $P_1(\mathrm{S}^k(0))\longrightarrow_{\mathrm{red}}\widetilde{s}'$,
$\widetilde{s},\widetilde{s}' \in \mathrm{Term}_{\mathrm{nf}}$, $\langle\widetilde{s},\widetilde{s}'\rangle \in A^*[z/\mathrm{S}^k(0),\vec{r}] = A[z/x]^*[x/\mathrm{S}^k(0),\vec{r}]$.
$P_0(\mathrm{S}^m(0))\longrightarrow_{\mathrm{red}}t[\vec{x}'/\vec{s},x/\mathrm{S}^k(0),y/\widetilde{s}]$, $P_1(\mathrm{S}^m(0))\longrightarrow_{\mathrm{red}}t'[\vec{x}'/\vec{s}'',x/\mathrm{S}^k(0),y/\widetilde{s}']$.
Now $\langle\mathrm{S}^k(0),\mathrm{S}^k(0)\rangle \in \mathbb{N}^*$, therefore by IH 3 it follows

$$\langle t[\vec{x}'/\vec{s},x/\mathrm{S}^k(0),y/\widetilde{s}],t'[\vec{x}'/\vec{s}'',x/\mathrm{S}^k(0),y/\widetilde{s}']\rangle \in A[z/\mathrm{S}(x)]^*[\vec{r}] = A^*[z/\mathrm{S}^m(0),\vec{r}],$$

and the side induction is finished.

(a,ii) is easy.

Case$(\Pi^{E,=})$: Assume $\Gamma^=(\vec{r};\vec{r}')$. By IH 1,2 there exist $\widetilde{r},\widetilde{r}' \in \mathrm{Term}_{\mathrm{nf}}$ such that $r[\vec{r}]\longrightarrow_{\mathrm{red}}\widetilde{r}_1$, $r'[\vec{r}']\longrightarrow_{\mathrm{red}}\widetilde{r}'$ $\langle\widetilde{r}_1,\widetilde{r}'\rangle \in A^*[\vec{r}]$, and there are $t,t' \in \mathrm{Term}$ and Variables $x,x' \in \mathrm{Var}_{\mathrm{ML}}$ such that

$$s[\vec{r}]\longrightarrow_{\mathrm{red}}\lambda x.r,\ s'[\vec{s}]\longrightarrow_{\mathrm{red}}\lambda x'.r',\ \langle\lambda x.r,\lambda x'.r'\rangle \in (\Pi x \in A.B)^{\mathrm{basis}}[\vec{r}].$$

Therefore

$$\mathrm{Ap}(s,r)[\vec{r}]\longrightarrow_{\mathrm{red}}\mathrm{Ap}(\lambda x.t,\widetilde{r})\longrightarrow_{\mathrm{red}}t[x/\widetilde{r},\vec{r}],$$
$$\mathrm{Ap}(s',r')[\vec{r}']\longrightarrow_{\mathrm{red}}t'[x'/\widetilde{r},\vec{r}'],$$
$$\langle t[x/\widetilde{r},\vec{r}],t'[x'/\widetilde{r}',\vec{r}']\rangle \in B^*[x/\widetilde{r},\vec{r}].$$

As before we conclude

$$\langle r[\vec{r}],r[\vec{r}]\rangle \in A^*[\vec{r}],$$
$$\langle \widetilde{r},r_1[\vec{r}]\rangle \in A^*[\vec{r}],$$
$$B^*[x/\widetilde{r},\vec{r}] = B^*[x/r[\vec{r}],\vec{r}] = B[x/r]^*[\vec{r}],$$

and we have IH (a,i).

(a,ii) follows as before.

Case($\Sigma^{E,=}$): Assume $\Gamma^=(\vec{r};\vec{r}')$. By IH 1 exist $s,s',t,t' \in \mathrm{Term}_{\mathrm{nf}}$ such that

$$r[\vec{r}]\longrightarrow_{\mathrm{red}}\mathrm{p}(s,t),\ r'[\vec{r}']\longrightarrow_{\mathrm{red}}\mathrm{p}(s',t'),\ \langle s,s'\rangle \in A^*[\vec{r}],\ \langle t,t'\rangle \in B^*[x/s,\vec{r}].$$

Then $\mathrm{p}_0(r[\vec{r}])\longrightarrow_{\mathrm{red}}s$, $\mathrm{p}_0(r'[\vec{r}'])\longrightarrow_{\mathrm{red}}s'$, and we are done for the first rule, and $\mathrm{p}_1(r[\vec{r}])\longrightarrow_{\mathrm{red}}t$, $\mathrm{p}_1(r'[\vec{r}'])\longrightarrow_{\mathrm{red}}t'$, and since from $\langle s,s'\rangle \in A[\vec{r}]$, it follows

$$\langle s,s'\rangle \in A[\vec{r}],\ \langle \mathrm{p}_0(r)[\vec{r}],s\rangle \in A[\vec{r}],$$

therefore by IH 2

$$B^*[x/s,\vec{r}] = B^*[x/\mathrm{p}_0(r)[\vec{r}],\vec{r}] = B[x/\mathrm{p}_0(r)]^*[\vec{r}]$$

follows (a,i) for the second rule.

(a,ii) is in $(\Sigma^{E,=})_1$ trivial, in $(\Sigma^{E,=})_2$ we use the proof of $(\Sigma^{E,=})_1$ and argue as before.

Case($\mathrm{W}^{E,=}$): Assume $\Gamma^=(\vec{r};\vec{r}')$, $n := \max\{\mathrm{lev}(A),\mathrm{lev}(B)\}$, $F := F_{\mathrm{W}}(A^*[\vec{r}],(x)B^*[\vec{r}])$. By IH $r[\vec{r}]\longrightarrow_{\mathrm{red}}\widetilde{r}$, $r'[\vec{r}']\longrightarrow_{\mathrm{red}}\widetilde{r}'$, $\langle \widetilde{r},\widetilde{r}'\rangle \in F^{\delta}(\vec{r},\cdot)$. for some $\delta < \alpha$. Let

$$[\vec{x}'/\vec{s}] := [\vec{x}/\vec{r}] \setminus \{x,y,z\},$$
$$[\vec{x}'/\vec{s}'] := [\vec{x}/\vec{r}'] \setminus \{x,y,z\},$$
$$R_0(r) := \mathrm{R}(r,(x,y,z)t)[\vec{r}](= \mathrm{R}(r,\lambda x.\lambda y.\lambda z.(t[\vec{s}]))),$$
$$R_1(r) := \mathrm{R}(r,(x,y,z)t')[\vec{r}'].$$

We show by induction on γ,

$$\forall\gamma < \alpha.\forall\langle \widetilde{s},\widetilde{s}'\rangle \in F^{\gamma}.\langle R_0(\widetilde{s}),R_1(\widetilde{s}')\rangle \in C^*[u/\widetilde{s},\vec{r}]. \qquad (+)$$

Since $C[u/t]^*[\vec{r}] = C^*[u/r[\vec{r}],\vec{r}] = C^*[u/\widetilde{r},\vec{r}] = C^*[u/\widetilde{r}',\vec{r}']$ (using arguments as before), follows the assertion.

The case $\gamma = 0$ is trivial, and if $\gamma \in \mathrm{Lim}$ follows the assertion by IH.

Let now $\gamma = \gamma' + 1$, $u' := F^{\gamma'}$, $\langle \widetilde{s}, \widetilde{s}' \rangle \in F(u')$.

If $\widetilde{s} \longrightarrow_{\mathrm{red}} s$, $\widetilde{s}' \longrightarrow_{\mathrm{red}} s'$, $\langle s, s' \rangle \in F^{\mathrm{basis}}(\vec{r}', \cdot)$, $\langle R_0(s), R_1(s') \rangle \in C^*[u/s, \vec{r}]$, it follows $\langle R_0(\widetilde{s}), R_1(\widetilde{s}') \rangle \in C^*[u/s, \vec{r}]$, further, like similar arguments before,

$$C^*[u/s, \vec{r}] = C^*[u/\widetilde{s}, \vec{r}] = C^*[u/\widetilde{s}'].$$

We therefore assume $\langle \widetilde{s}, \widetilde{s}' \rangle \in F^{\mathrm{basis}}(\vec{r}, u')$.

Let $\langle \widetilde{s}, \widetilde{s}' \rangle = \langle \mathrm{sup}(a, \lambda x.s), \mathrm{sup}(a', \lambda x'.s') \rangle$, $\langle a, a' \rangle \in A^*[\vec{r}]$. Let $\langle r'', r''' \rangle \in B^*[x/a, \vec{r}]$. Then $r'' \longrightarrow_{\mathrm{red}} b$, $r''' \longrightarrow_{\mathrm{red}} b'$ for $\langle b, b' \rangle \in B^*[x/a, \vec{r}]$, $b, b' \in \mathrm{Term}_{\mathrm{nf}}$, and we have $\langle s[x/r''], s'[x'/r'''] \rangle \in u'$ and

$$\langle s[x/b], s'[x'/b'] \rangle \in u'. \tag{$*$}$$

Since $u' \subset (\mathrm{W}x \in A.B)^*[\vec{r}]$ it follows from the first of these assertions

$$\langle \lambda x.s, \lambda x'.s' \rangle \in (B \to \mathrm{W}x \in A.B)^*[\vec{r}].$$

Furthermore, for $\langle b, b' \rangle \in B^*[x/r, \vec{r}]$,

$$(R_0((\lambda x.s)v))[v/r''] \longrightarrow_{\mathrm{red}} R_0(s[x/b]) \qquad (v \notin \mathrm{FV}(\lambda x.s)),$$
$$(R_1((\lambda x.s')v'))[v'/r'''] \longrightarrow_{\mathrm{red}} R_1(s'[x/b']) \qquad (v' \notin \mathrm{FV}(\lambda x.s)).$$

and by side IH, it follows

$$\begin{aligned}\langle (R_0((\lambda x.s)v))[v/r''], (R_1((\lambda x'.s')v))[v'/r'''] \rangle &\in C^*[u/(s[x/b]), \vec{r}] \\ &= C^*[u/(s[x/r'']), \vec{r}].\end{aligned}$$

Now we have $\langle r_i, r_i \rangle \in A_i[\hat{r}_i]$, $\mathrm{Ap}(\lambda x.s, r'') \longrightarrow_{\mathrm{red}} s[x/b]$, and by $(*)$, $u' \subset (\mathrm{W}x \in A.B)^*[\vec{r}]$, equiv$((\mathrm{W}x \in A.B)^*[\vec{r}])$ and 16.34 it follows

$$\langle s[x/b], \mathrm{Ap}(\lambda x.s, r'') \rangle \in (\mathrm{W}x \in A.B)^*[\vec{r}]$$

therefore

$$C[u/\mathrm{Ap}(y, v)]^*[v/r'', y/\lambda x.s, \vec{x}/\vec{r}] = C^*[u/\mathrm{Ap}(\lambda x.s, r''), \vec{r}] = C^*[u/(s[x/b]), \vec{r}]$$

further

$$C[u/\mathrm{Ap}(y, v)]^*[v/r'', y/\lambda x.s, \vec{x}/\vec{r}] = C[u/\mathrm{Ap}(y, v)]^*[v/r''', y/\lambda x.s, \vec{x}/\vec{r}],$$

and we have

$$\langle \lambda v.R_0((\lambda x.s)v), \lambda v'.R_1((\lambda x'.s')v') \rangle \in (\Pi v \in B.C[u/\mathrm{Ap}(y, v)])^*[y/\lambda x.s, \vec{x}/\vec{r}].$$

Now by IH 2 it follows

$$\langle t[x/r, y/\lambda x.s, z/\lambda v.R_0((\lambda x.s)v), \vec{r}], t'[x/r', y/\lambda x'.s', z/\lambda v'.R_1((\lambda x.s')v'), \vec{r}']\rangle$$
$$\in C[u/\mathrm{sup}(x,y)]^*[x/r, y/\lambda x.s, \vec{r}].$$

Since

$$C[u/\mathrm{sup}(x,y)]^*[x/r, y/\lambda x.s, \vec{r}] = C^*[u/\mathrm{sup}(r, \lambda x.s), \vec{r}] = C^*[u/s, s],$$

and

$$R_0(s) \longrightarrow_{\mathrm{red}} (\lambda x.\lambda y.\lambda z.t[\vec{x}'/\vec{s}])r(\lambda x.s)(\lambda v.R_0((\lambda x.s)v))$$
$$\longrightarrow_{\mathrm{red}} t[x/r, y/\lambda x.s, z/(\lambda v.R_0((\lambda x.s)v))],$$

$$R_1(s') \longrightarrow_{\mathrm{red}} t'[x/r', y/\lambda x.s', z/(\lambda v'.R_1((\lambda x'.s')v'))]$$

follows (+), and we are done. (a,ii) follows as in the case ($\mathbb{N}_k^{E,=}$).

Case ($+^{E,=}$): Assume $\Gamma^=(\vec{r};\vec{r}')$. By IH $r[\vec{r}] \longrightarrow_{\mathrm{red}} \mathrm{i}(\widetilde{r}) \in \mathrm{Term}_{\mathrm{nf}}$, $r'[\vec{r}'] \longrightarrow_{\mathrm{red}} \mathrm{i}(\widetilde{r}') \in \mathrm{Term}_{\mathrm{nf}}$ and $\langle \widetilde{r}, \widetilde{r}'\rangle \in A^*[\vec{r}]$ or $r[\vec{r}] \longrightarrow_{\mathrm{red}} \mathrm{j}(\widetilde{r}) \in \mathrm{Term}_{\mathrm{nf}}$, $r'[\vec{r}'] \longrightarrow_{\mathrm{red}} \mathrm{j}(\widetilde{r}') \in \mathrm{Term}_{\mathrm{nf}}$ and $\langle \widetilde{r}, \widetilde{r}'\rangle \in B^*[\vec{r}]$. Let $[\vec{x}'/\vec{s}] := [\vec{x}/\vec{r}] \setminus \{x\}$. In the first case we have

$$\mathrm{D}(r,(x)s,(y)t)[\vec{r}] \longrightarrow_{\mathrm{red}} (\lambda x.(s[\vec{s}]))\widetilde{r} \longrightarrow_{\mathrm{red}} s[x/\widetilde{r}, \vec{x}/\vec{r}],$$

$$\mathrm{D}(r',(x)s',(y)t')[\vec{r}] \longrightarrow_{\mathrm{red}} s'[x/\widetilde{r}', \vec{x}/\vec{r}'],$$

$$\langle s[x/\widetilde{r}, \vec{r}], s'[x/\widetilde{r}', \vec{r}']\rangle \in C[z/\mathrm{i}(x)]^*[x/\widetilde{r}, \vec{r}] = C^*[z/\mathrm{i}(\widetilde{r}), \vec{r}]$$

and using arguments as before

$$C^*[z/\mathrm{i}(\widetilde{r}), \vec{r}] = C^*[z/r[\vec{r}], \vec{r}] = C[z/r]^*[\vec{r}]$$

and we are done. The second assertion follows in the same way.

(a,ii) follows as before.

Case (I^E): Assume $\Gamma^=(\vec{r};\vec{r}')$. By IH 1 follows $(\mathrm{I}(A,s,t))^*[\vec{r}] \neq \emptyset$, $\langle s[\vec{r}], t[\vec{r}]\rangle \in A^*[\vec{r}]$. Furthermore, by IH 3 $\langle t[\vec{r}], t[\vec{r}']\rangle \in A^*[\vec{r}]$, and by $\mathrm{equiv}(A^*[\vec{r}])$ follows (a,i). (a,ii) is trivial.

Case ($\Pi^=$), ($\Sigma_0^=$), ($\Sigma_1^=$): By using the proof for the elimination rules we see, that if the conclusion is $r = s : C$, we conclude assuming $\Gamma^=(\vec{r};\vec{r}')$, that $(r = r : C)[\vec{r};\vec{r}']$, further $(r[\vec{r}'] \longrightarrow_{\mathrm{red}} t \in \mathrm{Term}_{\mathrm{nf}}) \to (s[\vec{r}'] \longrightarrow_{\mathrm{red}} t)$, therefore follows $(r = s : C)[\vec{r};\vec{r}']$.

Case (Π^η):Assume $\Gamma^=(\vec{r},\vec{r}')$ By IH we have

$$\langle t[\vec{r}], t[\vec{r}']\rangle \in (\Pi x \in A.B)^*[\vec{r}],$$

therefore $t[\vec{r}] \longrightarrow_{\mathrm{red}} \lambda x.s$, $t[\vec{r}'] \longrightarrow_{\mathrm{red}} \lambda x'.s'$,

$$\langle \lambda x.s, \lambda x'.s'\rangle \in (\Pi x \in A.B)^{\mathrm{basis}}[\vec{r}].$$

Assume $\langle r, r'\rangle \in A^*[\vec{r}]$. Then $r \longrightarrow_{\mathrm{red}} \widetilde{r}$, $r' \longrightarrow_{\mathrm{red}} \widetilde{r}'$, $\langle \widetilde{r}, \widetilde{r}'\rangle \in A^*[\vec{r}]$, $\widetilde{r}, \widetilde{r}' \in \mathrm{Term}_{\mathrm{nf}}$.

$$\mathrm{Ap}(t,x))[\vec{r}][x/r] = \mathrm{Ap}(t[\vec{r}], r) \longrightarrow_{\mathrm{red}} \mathrm{Ap}(\lambda x.s, \widetilde{r}) \longrightarrow_{\mathrm{red}} s[x/\widetilde{r}],$$

and since

$$\langle s[x/\widetilde{r}], s'[x'/\vec{r}']\rangle \in B^*[x/\widetilde{r},\vec{r}] = B^*[x/r,\vec{r}],$$

follows

$$\langle \mathrm{Ap}(t,x)[\vec{r}][x/r], s'[x'/\vec{r}']\rangle \in B^*[x/r,\vec{r}],$$

therefore

$$\langle \lambda x.\mathrm{Ap}(t,x)[\vec{r}], \lambda x'.s'\rangle \in (\Pi x \in A.B)^*[\vec{r}],$$

$$\langle \lambda x.\mathrm{Ap}(t,x)[\vec{r}], t[\vec{r}]\rangle \in (\Pi x \in A.B)^*[\vec{r}].$$

Case ($\Sigma_2^=$): Assume $\Gamma^=(\vec{r};\vec{r}')$. By IH $t[\vec{r}]\longrightarrow_{\mathrm{red}} r$, $t[\vec{r}']\longrightarrow_{\mathrm{red}} r'$ for some $\langle r,r'\rangle \in \Sigma x \in A.B^*[\vec{r}] \cap (\mathrm{Term}_{\mathrm{nf}} \times \mathrm{Term}_{\mathrm{nf}})$, $\mathrm{p}(\mathrm{p}_0(t),\mathrm{p}_1(t))[\vec{r}]\longrightarrow_{\mathrm{red}} r$ and we are done.

Case ($\mathrm{I}^=$): Assume $\Gamma^=(\vec{r};\vec{r}')$. By IH we conclude $\langle t_0[\vec{r}], t_0[\vec{r}']\rangle \in \mathrm{I}(A,t_1,t_2)^*[\vec{r}]$, therefore $t_0[\vec{r}']\longrightarrow_{\mathrm{red}}\mathrm{r}$, $\langle \mathrm{r},\mathrm{r}\rangle \in \mathrm{I}(A,t_1,t_2)^*[\vec{r}]$, $\langle t_0[\vec{r}],\mathrm{r}\rangle \in \mathrm{I}(A,t_1,t_2)^*[\vec{r}]$. (a,ii) is trivial.

Case other equality rules: Let $\widetilde{r} = \widetilde{s} : A$ be the conclusion of the rules. By using several times the general rules, elimination rules and in case $\mathrm{W}^=$ the introduction rules we can conclude $\widetilde{r} = \widetilde{r} : A$, and $\widetilde{s} = \widetilde{s} : A$.
(For ($\mathrm{W}^=$) we argue that $\Gamma, v : B[x/t_0] \Rightarrow \mathrm{Ap}(t_1,v) : \mathrm{W}x \in A.B$,
by ($\mathrm{W}^{E,=}$) $\Gamma, v : B[x/t_0] \Rightarrow \mathrm{R}(\mathrm{Ap}(s',v),(x,y,z)t') : C[u/\mathrm{Ap}(s',v)]$,
by $\Pi^{\mathrm{I},=}$ $\Gamma \Rightarrow \lambda v.\mathrm{R}(\mathrm{Ap}(s',v),(x,y,z)t') : \Pi v \in B.C[u/\mathrm{Ap}(s',v)]$,
by (ALPHA) for the z_i, that we need it follows
$\Gamma \Rightarrow \lambda z_i.\mathrm{R}(\mathrm{Ap}(s',z_i),(x,y,z)t') : \Pi v \in B.C[u/\mathrm{Ap}(s',v)]$,
and now by (SUB) follows the assertion.)
Now, assuming $\Gamma^=(\vec{r},\vec{r}')$, and using the proofs above we can conclude $\langle r[\vec{r}], r[\vec{r}']\rangle \in A^*[\vec{r}]$ and $A^*[\vec{r}] = A^*[\vec{r}']$, so (a,i). In all the cases, we have, if the right side is written as $t[x_1/r_1,\ldots,x_n/t_n]$, if x_i corresponds to the type B_i (read off from the rule) follows easily by IH and using the proofs of several rules handled before the assertion for $\Gamma \Rightarrow r_i : B_i$, therefore $r_i[\vec{r}]\longrightarrow_{\mathrm{red}}\widetilde{r_i} \in \mathrm{Term}_{\mathrm{nf}}$ for some $\widetilde{r_i}$, $\langle \widetilde{r_i}, r_i[\vec{r}]\rangle \in B_i[\vec{r}]$, further $\widetilde{r}[\vec{r}]\longrightarrow_{\mathrm{red}} t[x_1/\widetilde{r_1},\cdots x_n\widetilde{r_n},\vec{r}]$. We conclude

$$\langle t[x_1/\widetilde{r_1},\ldots,x_n/\widetilde{r_n},\vec{r}], t[x_1/r_1[\vec{r}],\ldots,x_n/r_n[\vec{r}],\vec{r}]\rangle \in A^*[\vec{r}].$$

Now using $\mathrm{equiv}(A^*[\vec{r}])$ and Lemma 16.34 we conclude

$$\langle \widetilde{r}[\vec{r}], \widetilde{s}[\vec{r}]\rangle \in A^*[\vec{r}], \langle \widetilde{s}[\vec{r}], \widetilde{s}[\vec{r}']\rangle \in A^*[\vec{r}],$$

and have (a,i).

Case (U^{I}): trivial.

($\mathrm{T}^{\mathrm{I},=}$) we have by IH, assuming $\Gamma^=(\vec{r};\vec{r}')$,

$$\langle a[\vec{r}], a'[\vec{r}']\rangle \in \mathrm{U}^*$$

therefore,

$$\langle a[\vec{r}], b, a'[\vec{r}']\rangle \in \widehat{\mathrm{U}}$$

for some b, by $\mathrm{Cor}_{\mathrm{U}}(\widehat{\mathrm{U}})$,

$$\langle a[\vec{r}], b, a[\vec{r}]\rangle \in \widehat{\mathrm{U}}, \langle a[\vec{r}'], b, a[\vec{r}']\rangle \in \widehat{\mathrm{U}},$$

and

$$\mathrm{T}(a)^*[\vec{r}] = b = \mathrm{T}(a')^*[\vec{r}'].$$

Case $(\widehat{\Pi}^{\mathrm{I},=})$: Assume $\Gamma^=(\vec{r};\vec{r}')$. By IH $a[\vec{r}]\longrightarrow_{\mathrm{red}}\widetilde{a}$, $a'[\vec{r}']\longrightarrow_{\mathrm{red}}\widetilde{a}'$,

$$\exists\gamma < \mathrm{I}\exists b' \in \mathrm{TC}(\widetilde{\mathrm{U}}^\gamma)(\langle\widetilde{a}, b', \widetilde{a}\rangle, \langle\widetilde{a}', b', \widetilde{a}\rangle \in \widetilde{\mathrm{U}}^\gamma),$$

and

$$\forall\langle t, t'\rangle \in b' \to \exists\delta < \mathrm{I}.\exists c \in \mathrm{TC}(\widetilde{\mathrm{U}}^\delta).$$
$$(\langle b[x/t, \vec{r}], c, b'[x/t', \vec{r}']\rangle \in \widetilde{\mathrm{U}}^\delta).$$

Since $\mathrm{Ad}(\mathrm{L_I})$ (here is the central point where we need (Δ_0-coll) and an admissible a which is closed under the step to the next admissible), and $\mathrm{TC}(\widetilde{\mathrm{U}}^\beta) \in \mathrm{L_I}$ $(\beta < \mathrm{I})$, there is a $\rho < \mathrm{I}$, such that $\gamma < \rho$ and δ can be chosen $< \rho$. There are now b, f such that $([\vec{x}'/\vec{s}] := [\vec{x}/\vec{r}] \setminus \{x\}, [\vec{x}'/\vec{s}'] := [\vec{x}/\vec{r}'] \setminus \{x\})$ $\varphi(\widetilde{a}, x, b[\vec{s}], \widetilde{a}', x, b[\vec{s}'], b, f, \widetilde{\mathrm{U}}^\rho)$, (note that the c we used above is correct by $\mathrm{Cor}(\widehat{\mathrm{U}})$) and by 16.28 (a) follows

$$\langle\widehat{\Pi}x \in a.b, \Pi^*(b, f), \widehat{\Pi}x \in a'.b'\rangle \in \widetilde{\mathrm{U}}^{\rho+1}.$$

Case $(\widehat{\Pi}^=)$: Assume $\Gamma^=(\vec{r};\vec{r}')$ and chose b', f, ρ as in $(\widehat{\Pi}^{\mathrm{I},=})$. Then $\mathrm{T}(a)^*[\vec{r}'] = b'$, and if $\langle t, t'\rangle \in b'$, $\mathrm{T}(b)^*[x/t, \vec{r}] = f(t) = f(t') = \mathrm{T}(b)^*[x/t', \vec{r}']$. Since we have $\mathrm{Cor}(\widehat{\mathrm{U}})$ (by Lemma 16.28 (d)) it follows

$$\mathrm{T}((\widehat{\Pi}x \in a.b))^*[\vec{r}] = \Pi^*(\mathrm{T}(a)[\vec{r}], (x)\mathrm{T}(b)[\vec{r}]) = (\Pi x \in \mathrm{T}(a).\mathrm{T}(b))^*[\vec{r}].$$

In the case of $(\widehat{\mathrm{W}}^=)$ we conclude as before, that

$$F_{\mathrm{W}}(\mathrm{T}(a)^*[\vec{r}], (x)\mathrm{T}(b)^*[\vec{r}]) = F_{\mathrm{W}}(\mathrm{T}(a)^*[\vec{r}'], (x)\mathrm{T}(b)^*[\vec{r}'])$$

and, since $\alpha^+(\widetilde{\mathrm{U}}^\rho) < \mathrm{I}$, ($\rho$ chosen as in $(\widehat{\Pi}^{\mathrm{I},=})$) it follows by 16.24 $\mathrm{T}(\widehat{\mathrm{W}}x \in a.b)^*[\vec{r}] = F_{\mathrm{W}}^{\alpha^+(\widetilde{\mathrm{U}}^\rho)}(\mathrm{T}(a)^*[\vec{r}], (x)(\mathrm{T}(b))^*[\vec{r}]) = F_{\mathrm{W}}^{\mathrm{I}}(\mathrm{T}(a)^*[\vec{r}], (x)(\mathrm{T}(b))^*[\vec{r}]) = (\mathrm{W}x \in \mathrm{T}(a).\mathrm{T}(b))^*[\vec{r}'].$

Case $(\widehat{\Sigma}^{\mathrm{E}})$: Assume $\Gamma^=(\vec{r};\vec{r}')$. By IH exists c, $\alpha < \mathrm{I}$ such that

$$\langle(\widehat{\Sigma}x \in a.b)[\vec{r}], c, (\widehat{\Sigma}x \in a.b)[\vec{r}']\rangle \in \widetilde{\mathrm{U}}^\alpha.$$

Let α be chosen minimal. Then, $\alpha = \alpha' + 1$, and with $u := \widetilde{\mathrm{U}}^{\alpha'}$ there exist $r, r' \in \mathrm{Term_{nf}}$, c, c', f, f' such that $a[\vec{r}]\longrightarrow_{\mathrm{red}}r$, $a[\vec{r}']\longrightarrow_{\mathrm{red}}r'$, $\langle r, c, r'\rangle \in u$ and (with $[\vec{x}'/\vec{s}] := [\vec{x}/\vec{r}] \setminus \{x\}$, $[\vec{x}'/\vec{s}'] := [\vec{x}/\vec{r}'] \setminus \{x\}$) $\forall\langle t, t'\rangle \in c.\langle s[\vec{s}][x/t], f(t), s[\vec{s}'][x/t']\rangle \in u$. Therefore $\mathrm{T}(a[\vec{r}])^* = c = \mathrm{T}(a[\vec{r}'])^*$, and for $\langle t, t'\rangle \in \mathrm{T}(a[\vec{r}])^*$, $\langle s[x/t, \vec{r}], s[x/t', \vec{s}']\rangle \in \mathrm{U}^*[\ ]$.

$(\widehat{\Pi}^{\mathrm{E}})$, $(\widehat{\mathrm{W}}^{\mathrm{E}})$, $(\widehat{+}^{\mathrm{E}})$ are checked in the same way. For $(\widehat{\mathrm{I}}^{\mathrm{E}})$ we observe, that $a[\vec{r}]\longrightarrow_{\mathrm{red}}\widetilde{a}$, $a[\vec{r}']\longrightarrow_{\mathrm{red}}\widetilde{a}'$, $s[\vec{r}]\longrightarrow_{\mathrm{red}}\widetilde{s}$, $s[\vec{r}']\longrightarrow_{\mathrm{red}}\widetilde{s}'$, $t[\vec{r}]\longrightarrow_{\mathrm{red}}\widetilde{t}$, $t[\vec{r}']\longrightarrow_{\mathrm{red}}\widetilde{t}'$, and $\langle\widetilde{a}, c, \widetilde{a}'\rangle \in u$, for some u as before, $\mathrm{T}(a)^*[\vec{r}] = c$, $\langle\widetilde{s}, \widetilde{s}'\rangle \in c$, $\langle\widetilde{t}, \widetilde{t}'\rangle \in c$, and since c is closed under $\longrightarrow_{\mathrm{red}}$ follows the assertion. □

16.9 Π^1_1-soundness of the Interpretation of Martin-Löf Type Theory into KPI$^+$

In this section we want to evaluate the results we have found out to get the proof theoretic strength of Martin-Löf's type theory. We will interpret the language of analysis ($\mathcal{L}_{\text{analysis}}$, introduced in 16.39) in $\mathcal{L}_{ML}$ and $\mathcal{L}_{KP}$ (definition16.43) and prove that it permutes with the interpretation of Martin-Löf's type theory in KPI$^+$ (Lemmata 16.45 and 16.46). Next we observe that every proof of ML can be interpreted in KPI$^+$ (Lemma 16.47). This preserves Π^1_1-sentences, where second order quantifiers in Kripke Platek set theory refer to elements of Ad_2. In the next section we will analyse the strength of KPI$^+$ and obtain the desired upper bound for the type theories in question.

Definition 16.39. Definition of the language of Peano Arithmetic $\mathcal{L}_{\text{analysis}}$: we have first order variables v_i $(i \in \omega$, $\text{var}_{\text{analysis}} := \{v_i | i \in \omega\})$; second order variables V_i $(i \in \omega$, $\text{VAR}_{\text{analysis}} := \{V_i | i \in \omega\})$; further we have symbols for each primitive recursive function, $=, \wedge, \vee, \rightarrow, \forall, \exists, \bot$, and .,,,(,).

Terms are first-order variables and $f(t_1, \ldots, t_n)$ if t_i are terms and f is a symbol for a n-ary primitive recursive function.

Prime formulas are $\bot$, equations $r = s$, and $r \in X$ for r, s terms, $X \in \text{VAR}_{\text{analysis}}$.

Formulas are prime formulas and $A \rightarrow B$, $A \wedge B$, $A \vee B$, $\forall x.A$, $\exists x.A$, if A, B formulas, $x \in \text{var}_{\text{analysis}} \cup \text{VAR}_{\text{analysis}}$.

A Δ^1_0 formula is a formula, not containing bounded second-order quantifiers, and a Π^1_1-formula is $\forall X.\varphi$, where φ is a Δ^1_0-formula.

Remark 16.40. We could omit that $A \wedge B$, $A \vee B$, $\exists x.A$ are formulas in the above definition and simplify the correctness theorem below. We keep those sets since the proof of Lemma 16.46 gives some insights about how the interpretation works.

Assumption. After renaming all variables, we assume, we have additional new variables U_i of KPI$^+$ $(i \in \omega)$ and Z_i of ML $(i \in \omega)$, s.t. in the step from a g-type to A^*, Z_i becomes U_i and in 16.21 (a), if $x_i = Z_j$, then on the right side we put U_j.

Definition 16.41. (a) Let $\mathrm{C}^{\text{set}}_i$ be new Gödelnumbers for new constructors $\text{Constr}_{\mathrm{C}^{\text{set}}_i}$ (for which we write as before for simplicity $\mathrm{C}^{\text{set}}_i$).

(b) Let for a set b, $\text{Embset}_{\text{ML}}(f) := \{\langle n, \widehat{\mathbb{N}}_0\rangle | n \in \omega \setminus b\} \cup \{\langle n, \widehat{\mathbb{N}}_1\rangle | n \in \omega \cap b\}$.

(c) If $\text{CES}(b)$, the Gödel numbers of the constructors in b are $\neq \mathrm{C}^{\text{set}}_{i_j}$ $(j = 1, \ldots, m)$, $b_1, \ldots, b_m \in \text{Ad}_2$, $b_i \subset \omega$, $Z_{i_j} \neq Z_{i_k}$, $(j \neq k)$, then

$$\text{CES}^+(b, Z_{i_1}/b_1, \ldots, Z_{i_m}/b_m) := b \cup \{\langle \mathrm{C}^{\text{set}}_{i_1}, 0, f_1\rangle, \ldots, \langle \ulcorner \mathrm{C}^{\text{set}}_{i_m} \urcorner, 0, f_m\rangle\},$$

where $f_i := \text{Embset}_{\text{ML}}(b_i)$.

Definition 16.42. Let $\mathcal{P}(\mathbb{N}) := \mathbb{N} \rightarrow \mathrm{U}$, $\text{ML}^{\text{i}}_1\text{W}_{\text{T}} \vdash \mathbb{N} \rightarrow \mathrm{U}$: type.

Definition 16.43. (a) For each primitive recursive $g : \mathbb{N}^k \to \mathbb{N}$ we define a closed g-term $\mathrm{int}_{\mathrm{PA,ML}}(g)$, (we abbreviate this as $\hat{g} := \mathrm{int}_{\mathrm{PA,ML}}(g)$) such that

$$\mathrm{ML} \vdash \hat{g} : \underbrace{\mathbb{N} \to \cdots \mathbb{N}}_{k \text{ times}} \to \mathbb{N},$$

and we define a set $\mathrm{int}_{\mathrm{PA,KP}}(g)$ short $\widetilde{g}$ in $\mathcal{L}_{KP}$ such that $\mathrm{KPI}^+ \vdash \mathrm{fun}(\widetilde{g}) \wedge \mathrm{dom}(\widetilde{g}) = \mathbb{N}^k \wedge \forall x \in \mathbb{N}^k.\widetilde{g}(x) \in \mathbb{N}$.
Case $g = \mathrm{S}$: $\hat{g} := \lambda x.\mathrm{S}(x)$, $\widetilde{g} := \{\langle x, x+1\rangle | x \in \mathbb{N}\}$.
Case $g = \mathrm{Proj}^n_i$ $\hat{g} := \lambda x_1, \ldots, x_n.x_i$, $\widetilde{g} := \{\langle\langle x_1, \ldots, x_n\rangle, x_i\rangle | x_1, \ldots, x_n \in \mathbb{N}\}$.
Case $g = \mathrm{Cons}^n_c$:
$\hat{g} := \lambda x_1, \ldots, x_n.\mathrm{S}^c(0)$, $\widetilde{g} := \{\langle\langle x_1, \ldots, x_n\rangle, c\rangle | x_1, \ldots, x_n \in \mathbb{N}\}$.
Case $g(x_1, \ldots, x_n) = h(g_1(x_1, \ldots, x_n), \ldots, g_m(x_1, \ldots, x_n))$:
$\hat{g} := \lambda x_1, \ldots, x_n.\hat{h}\ (\hat{g}_1\ x_1 \cdots x_n)\ \cdots\ (\hat{g}_m\ x_1 \cdots x_n)$,
$\widetilde{g} := \{\langle\langle x_1, \ldots, x_n\rangle, h(g_1(x_1, \ldots, x_n), \ldots, g_m(x_1, \ldots, x_n))\rangle$
$| x_1, \ldots, x_n \in \mathbb{N}\}$.
Case $g(x_1, \ldots, x_n, 0) = h(x_1, \ldots, x_n)$,
$g(x_1, \ldots, x_n, y+1) = k(x_1, \ldots, x_n, y, g(x_1, \ldots, x_n, y))$:
$\hat{g} := \lambda x_1, \ldots, x_n, y.\mathrm{P}(y, \hat{h}\ x_1 \cdots x_n, (u,v)\hat{k}\ x_1 \cdots x_n\ u\ v)$,
define $l(x_1, \ldots, x_n, 0) := \widetilde{h}(x_1, \ldots, x_n)$,
$l(x_1, \ldots, x_n, \mathrm{S}(y)) := \widetilde{k}(x_1, \ldots, x_n, y, l(x_1 \cdots x_n, y))$,
then
$\widetilde{g} := \{\langle\langle x_1, \ldots, x_n, y\rangle, l(x_1, \ldots, x_n, y)\rangle | x_1, \ldots, x_n, y \in \mathbb{N}\}$.

(b) For each term t of analysis we define a g-term $\mathrm{int}_{\mathrm{PA,ML}}(t)$, short $\hat{t}$ and a term of $\mathcal{L}_{KP}$ $\mathrm{int}_{\mathrm{PA,KP}}(t)$, short $\widetilde{t}$, such that, if $\mathrm{FV}(t) = \{v_{i_1}, \ldots, v_{i_n}\}$ $(i_1 < \cdots < i_n)$ (v_i as in definition 16.39 of $\mathrm{var}_{\mathrm{analysis}}$) then $\mathrm{FV}(\hat{t}) \subset \{z_{i_1}, \ldots, z_{i_n}\}$, $\mathrm{FV}(\widetilde{t}) \subset \{u_{i_1}, \ldots, u_{i_n}\}$ and $\mathrm{ML} \vdash z_{i_1} : \mathbb{N}, \ldots, z_{i_n} : \mathbb{N} \Rightarrow \hat{t} : \mathbb{N}$, and $\mathrm{KPI}^+ \vdash \forall u_{i_1}, \ldots, u_{i_n} \in \omega.(\widetilde{t} \in \omega)$.
Case $t = v_i$: $\hat{t} := z_i$, $\widetilde{t} := u_i$.
Case $t = 0$: $\hat{t} := 0$, $\widetilde{t} := 0$.
Case $t = g(t_1, \ldots, t_n)$: $\hat{t} := \hat{g}\ \hat{t}_1\ \cdots\ \hat{t}_n$, $\widetilde{t} := \widetilde{g}(\widetilde{t}_1, \ldots, \widetilde{t}_n)$.

(c) For each formula A of analysis we define a g-type $\mathrm{int}_{\mathrm{PA,ML}}(A)$ (short $\widehat{A}$), and a formula of $\mathcal{L}_{KP}$ $\mathrm{int}_{\mathrm{PA,KP}}(A)$, short $\widetilde{A}$, such that in KPI^+ $\widetilde{A}$ is equivalent to a Δ_0-formula, and if $\mathrm{FV}(A) = \{v_{i_1}, \ldots, v_{i_n}, V_{j_1}, \ldots, V_{j_m}\}$, $i_k \neq i_l$, $j_k \neq j_l$ for $k \neq l$, then $\mathrm{FV}(\widehat{A}) \subset \{z_{i_1}, \ldots, z_{i_n}, Z_{j_1}, \ldots, Z_{j_m}\}$, $\mathrm{FV}(\widetilde{A}) \subset \{u_{i_1}, \ldots, u_{i_n}, U_{j_1}, \ldots, U_{j_m}\}$ and for all versions ML of Martin-Löf Type Theory considered in this article we get $\mathrm{ML} \vdash z_{i_1} : \mathbb{N}, \ldots, z_{i_n} : \mathbb{N}, Z_{j_1} : \mathcal{P}(\mathbb{N}), \ldots, Z_{j_m} : \mathcal{P}(\mathbb{N}) \Rightarrow \widehat{A} : \mathrm{type}$.
Case $A = (s = t)$: $\widehat{A} := I(\mathbb{N}, \hat{s}, \hat{t})$, $\widetilde{A} := (\widetilde{s} = \widetilde{t})$.
Case $A = (t \in V_i)$: $\widehat{A} := \mathrm{T}(Z_i\ \hat{t})$ for the Tarski-version $\widehat{A} := Z_i\ \hat{t}$ for the Russell-version, $\widetilde{A} := \widetilde{t} \in U_i$.
Case $A = (B \wedge C)$: $\widehat{A} := (\widehat{B} \times \widehat{C})$, $\widetilde{A} := \widetilde{B} \wedge \widetilde{C}$.
Case $A = (B \vee C)$: $\widehat{A} := (\widehat{B} + \widehat{C})$, $\widetilde{A} := (\widetilde{B} \vee \widetilde{C})$.
Case $A = (B \to C)$: $\widehat{A} := (\widehat{B} \to \widehat{C})$, $\widetilde{A} := (\widetilde{B} \to \widetilde{C})$.
Case $A = \forall v_i.B$: $\widehat{A} := \Pi z_i \in \mathbb{N}.\widehat{B}$, $\widetilde{A} := \forall u_i \in \omega.\widetilde{B}$.
Case $A = \exists v_i.B$: $\widehat{A} := \Sigma z_i \in \mathbb{N}.\widehat{B}$, $\widetilde{A} := \exists u_i \in \omega.\widetilde{B}$.

Case $A = \forall V_i.B$: $\widehat{A} := \Pi Z_i \in \mathcal{P}(\mathbb{N}).\widehat{B}$, $\widetilde{A} := \forall U_i \in \mathrm{Ad}_2.U_i \subset \omega \to \widetilde{B}$.
Case $A = \exists V_i.B$: $\widehat{A} := \Sigma Z_i \in \mathcal{P}(\mathbb{N}).\widehat{B}$, $\widetilde{A} := \exists U_i \in \mathrm{Ad}_2.U_i \subset \omega \wedge \widetilde{B}$.
Case $A = \perp$: $\widehat{A} := \mathbb{N}_0$, $\widetilde{A} := (0 \neq 0)$.

Definition 16.44. (a) We define emb : $\omega \to \omega$, $\mathrm{emb}(n) := \mathrm{S}^n(0)(=: \hat{n})$ (or more precisely $\ulcorner \mathrm{S}^n(0) \urcorner$), a function definable in KPI^+.
(b) $\langle a, \cdot \rangle := \langle a, a \rangle$.

Lemma 16.45. (a) *If* $g : \mathbb{N}^k \to \mathbb{N}$ *is primitive recursive, then*

$$\mathrm{KPI}^+ \vdash \forall t_1, \ldots, t_k \in \mathrm{Term}_{\mathrm{Cl}}.\forall n_1, \ldots, n_k. \\ (r_1 \longrightarrow_{\mathrm{red}} \hat{\hat{n}}_1 \wedge \cdots \wedge r_k \longrightarrow_{\mathrm{red}} \hat{\hat{n}}_k) \to \hat{g}\ r_1\ \cdots\ r_k \longrightarrow_{\mathrm{red}} \mathrm{emb}(\widetilde{g}(n_1, \ldots, n_k)).$$

(b) *If t is a term of analysis,* $\mathrm{FV}(t) \subset \{v_1, \ldots, v_n\}$, *then*

$$\mathrm{KPI}^+ \vdash \forall r_1, \ldots, r_n \in \mathrm{Term}_{\mathrm{Cl}}.\forall n_1, \ldots, n_k.(r_1 \longrightarrow_{\mathrm{red}} \hat{\hat{n}}_1 \wedge \cdots \wedge r_k \longrightarrow_{\mathrm{red}} \hat{\hat{n}}_k) \to \\ \hat{t}[z_1/r_1, \ldots, z_n/r_n] \longrightarrow_{\mathrm{red}} \mathrm{emb}(\widetilde{t}[u_1/n_1, \ldots, u_n/n_n]).$$

Proof. (a) Case $g = \mathrm{S}$: $\hat{g}\ r_1 \longrightarrow_{\mathrm{red}} (\lambda x.\mathrm{S}(x))\ \hat{n}_1 \longrightarrow_{\mathrm{red}} \mathrm{S}(\hat{n}_1) = \mathrm{emb}(\mathrm{S}(n_1))$.
Case $g = \mathrm{Proj}^n_i$: $\hat{g}\ r_1\ \cdots\ r_n \longrightarrow_{\mathrm{red}} \hat{\hat{n}}_i = \mathrm{emb}(\widetilde{g}(n_1, \ldots, n_k))$.
Case $g = \mathrm{Cons}^n_c$: trivial.
Case $g(\vec{x}) = h(g_1(\vec{x}), \ldots, g_m(\vec{x}))$, $l_i := \widetilde{g}_i(n_1, \ldots, n_n)$:

$$\hat{g}_i\ r_1\ \cdots\ r_n \longrightarrow_{\mathrm{red}} \mathrm{emb}(\widetilde{g}_i(n_1, \ldots, n_n)) =: \mathrm{emb}(l_i),$$

therefore

$$\begin{aligned} &\hat{g}\ r_1\ \cdots\ r_n \\ &\qquad = \hat{h}\ (\hat{g}_1\ r_1\ \cdots\ r_n))\ \cdots\ (\hat{g}_m\ r_1\ \cdots\ r_n)) \\ &\quad \longrightarrow_{\mathrm{red}} \hat{h}\ \hat{l}_1\ \cdots\ \hat{l}_n \\ &\quad \longrightarrow_{\mathrm{red}} \mathrm{emb}(\widetilde{h}(l_1, \ldots, l_m)) \\ &\qquad = \mathrm{emb}(g(n_1, \ldots, n_n)). \end{aligned}$$

Case $g(x_1, \ldots, x_n, 0) = h(x_1, \ldots, x_n)$, $g(\vec{x}, x_{n+1}) = k(\vec{x}, y, g(\vec{x}, x_{n+1})))$.
Let $\hat{h}\ r_1\ \cdots\ r_n \longrightarrow_{\mathrm{red}} \mathrm{emb}(\widetilde{h}(n_1, \ldots, n_n)) =: l_0$. We show by induction on l,

$$\mathrm{P}(\hat{\hat{l}}, l_0, (\lambda u, v.\hat{k}\ r_1\ \ldots\ r_n\ u\ v)) \longrightarrow_{\mathrm{red}} \mathrm{emb}(\widetilde{g}(n_1, \ldots, n_n, l)).$$

Then follows the assertion, since

$$\begin{aligned} &g(r_1, \ldots, r_n, r_{n+1}) \\ &\longrightarrow_{\mathrm{red}} \mathrm{P}(\hat{\hat{n}}_{n+1}, l_0, (\lambda u, v.\hat{k}\ r_1\ \cdots\ r_n\ u\ v)) \\ &\longrightarrow_{\mathrm{red}} \mathrm{emb}(\widetilde{g}(n_1, \ldots, n_n, n_{n+1})). \end{aligned}$$

Proof of the statement: If $l = 0$,

$$
\begin{aligned}
&(\lambda u, v.\hat{k}\ r_1\ \cdots\ r_n\ u\ v))\\
&\qquad\qquad \longrightarrow_{\mathrm{red}} l_0\\
&\qquad\qquad\qquad = \mathrm{emb}(\widetilde{g}(n_1,\ldots,n_n,l)).
\end{aligned}
$$

If $l = m + 1$,

$$
\begin{array}{ll}
& \mathrm{P}(\hat{\hat{l}}, l_0, (\lambda u, v.\hat{k}\ r_1,\ \cdots\ r_n\ u\ v))\\
= & \mathrm{P}(\mathrm{S}(\hat{\hat{m}}), l_0, (\lambda u, v.\hat{k}\ r_1\ \cdots\ r_n\ u\ v))\\
\longrightarrow_{\mathrm{red}} & (\lambda u, v.\hat{k}\ r_1\ \cdots\ r_n\ u\ v)\ \hat{\hat{m}}\ \mathrm{P}(\hat{\hat{m}}, l_0, (\lambda u, v.\hat{k}\ r_1\ \ldots\ r_n\ u\ v))\\
\longrightarrow_{\mathrm{red}} & (\lambda u, v.\hat{k}\ r_1\ \cdots\ r_n\ u\ v)\ \hat{\hat{m}}\ \mathrm{emb}(g(n_1,\ldots,n_n,m))\\
\longrightarrow_{\mathrm{red}} & \hat{k}\ r_1\ \cdots\ r_n\ \hat{\hat{m}}\ \mathrm{emb}(g(n_1,\ldots,n_n,m))\\
\longrightarrow_{\mathrm{red}} & \hat{k}\ \hat{\hat{n}}_1,\cdots\ \hat{\hat{n}}_n\ \hat{\hat{m}}\ \mathrm{emb}(g(n_1,\ldots,n_n,m)))\\
\longrightarrow_{\mathrm{red}} & \mathrm{emb}(\widetilde{k}(n_1,\ldots,n_n,m,g(n_1,\ldots,n_n,m)))\\
= & \mathrm{emb}(\widetilde{g}(n_1,\ldots,n_n,l)).
\end{array}
$$

(b): If $t = v_i, 0$, this is trivial, and if $t = g(t_1,\ldots,t_n)$ it follows by IH

$$\hat{t}_i[\vec{r}] \longrightarrow_{\mathrm{red}} \mathrm{emb}(\widetilde{t}_i[\vec{n}]),$$

by (a) therefore

$$\hat{t}[\vec{r}] = \hat{g}\ \hat{t}_1[\vec{r}]\ \cdots\ \hat{t}_n[\vec{r}] \longrightarrow_{\mathrm{red}} \widetilde{g}(\mathrm{emb}(\widetilde{t}_1[\vec{n}]),\ldots,\mathrm{emb}(\widetilde{t}_n[\vec{n}])) = \widetilde{t}[\vec{n}].$$

Next task would now be to prove, that, when we first interpret a formula of $\mathcal{L}_{\mathrm{analysis}}$ in $\mathcal{L}_{ML}$ and then use the interpretation, as we have done in section 16.6, we get an equivalent formula to the one, we get by directly interpreting $\mathcal{L}_{\mathrm{analysis}}$ in $\mathcal{L}_{KP}$. But in this formulation, this is not correct, here is the place, where we need to extend the set of term constructors by non constructive constructors. In order to interpret a true Π^0_2 formula $A = \forall x.\exists y.\varphi$ in such a way, that for the false formula $\neg A$ we have $(\neg A)^* = \emptyset$ we need an element of A^*, which gives for the x a witness y of φ. But this might be non constructive, so we add here a (possibly) non constructive new constructor. So for every formula we need certain new constructors. Furthermore, we want that Π^1_1-formulas are interpreted correctly as well, that is, we want, that if we have a free set variable V_i, we can replace it in the KPI^+-interpretation by arbitrary subsets $U_i \subset \omega$, $U_i \in \mathrm{Ad}_2$. We achieve this by allowing here arbitrary interpretations for the constructor $\mathrm{C}^{\mathrm{set}}_i$. □

Lemma 16.46. *For every Δ^1_0-formula A with*

$$\mathrm{FV}(A) \subset \{v_{i_1},\ldots,v_{i_l},V_{j_1},\ldots,V_{j_m}\}$$

with $i_k \neq i_l$, $j_k \neq j_l$, $(k \neq l)$ there is a CES *c not referring to the constructors $\mathrm{C}^{\mathrm{set}}_i$ $(i \in \omega)$ and a g-term $h \in \mathrm{Term}_{\mathrm{Cl}}$ s.t. $\mathrm{FV}(h) \subset \{z_{i_1},\ldots,z_{i_l}\}$, and with $\vec{z} := z_{i_1},\ldots,z_{i_l}$, $\vec{Z} := Z_{j_1},\ldots,Z_{j_m}$, $\vec{u} := u_{i_1},\ldots,u_{i_l}$, $\vec{U} := U_{j_1},\ldots,U_{j_m}$, $\vec{C^Z} := \mathrm{C}^{\mathrm{set}}_{j_1},\ldots,\mathrm{C}^{\mathrm{set}}_{j_m}$, $\vec{n} := n_1,\ldots,n_l$, we have*

$$\begin{aligned}
\mathrm{KPI}^+ \vdash\ & \forall n_1,\dots,n_l \in \omega.\forall r_1,\dots,r_l \in \mathrm{Term}_{\mathrm{Cl}}.\\
& \forall b_1,\dots,b_m \in \mathrm{Ad}_2.\forall a_0 \in \mathrm{Ad}_2.\\
& (b_1 \subset \omega \wedge \cdots \wedge b_m \subset \omega\\
& \ \wedge \mathrm{CES}(a_0) \wedge \mathrm{CES}^+(c, Z_{i_1}/b_1,\dots,Z_{i_m}/b_m) \subset a_0\\
& \ \wedge (r_1 \longrightarrow_{\mathrm{red}} \hat{\hat{n}}_1 \wedge \cdots \wedge r_l \longrightarrow_{\mathrm{red}} \hat{\hat{n}}_l))\\
& \rightarrow ((\exists r \in \mathrm{Term}_{\mathrm{nf}}.h[\vec{z}/\vec{r}] \longrightarrow_{\mathrm{red}} r)\\
& \quad \wedge (\widetilde{A}[\vec{u}/\vec{n}, \vec{U}/\vec{b}] \iff \langle h[\vec{z}/\vec{r}], \cdot\rangle \in \widehat{A}^*[\vec{z}/\vec{r}]_{a_0})\\
& \quad \wedge (\widetilde{A}[\vec{u}/\vec{n}, \vec{U}/\vec{b}] \iff \widehat{A}^*[\vec{z}/\vec{r}]_{a_0} \neq \emptyset)).
\end{aligned}$$

Proof. By induction on the definition of the formulas.

Note that by Remark 16.40 we could have omitted the formulas $A \wedge B$, $A \vee B$, and $\exists x.A$ and the corresponding cases in the current proof. We kept them because the proofs give some interesting insights how this method works.

As before we will not mention explicitly Variables, that occur in subterms, or do not occur at all.

Case $A = \perp$: Choose $c := \emptyset$, $h := 0$. We have $\neg\widetilde{A}[\vec{n}]$, $\widehat{A}^*[\vec{r}] = \emptyset$.

Case $A = (s = t)$: Choose as $c := \emptyset$, $h := \mathrm{r} \in \mathrm{Term}_{\mathrm{nf}}$. We have, using that for $r \longrightarrow_{\mathrm{red}} s \in \mathrm{Term}_{\mathrm{nf}}$ s is unique, and by Lemma 16.45

$$\begin{aligned}
\widetilde{A}[\vec{n}] &\iff \widetilde{s}[\vec{n}] = \widetilde{t}[\vec{n}]\\
&\iff (\exists n \in \omega.\hat{s}[\vec{r}] \longrightarrow_{\mathrm{red}} \mathrm{S}^n(0) \wedge \hat{t}[\vec{r}] \longrightarrow_{\mathrm{red}} \mathrm{S}^n(0))\\
&\iff \langle \hat{s}[\vec{r}], \hat{t}[\vec{r}]\rangle \in \mathbb{N}^*\\
&\iff \langle \mathrm{r}, \mathrm{r}\rangle \in \widehat{A}^*[\vec{r}]\\
&\iff \widehat{A}^*[\vec{r}] \neq \emptyset.
\end{aligned}$$

Case $A = (s \in V_i)$: Let $c := \emptyset$, $h := \mathrm{A}^1_0$. Assume $\vec{r}$, $\vec{n}$, b_i, a_0 as in the assumption. Then $\hat{s}[\vec{z}/\vec{r}] \longrightarrow_{\mathrm{red}} \mathrm{emb}(\widetilde{s}[\vec{u}/\vec{n}]) = \mathrm{S}^k(0)$ for some k.

$$\begin{aligned}
\widehat{A}[\vec{z}/\vec{r}, \vec{Z}/\vec{C^Z}]^*_{a_0} &= \mathrm{T}(\mathrm{C}^{\mathrm{set}}_i(\widehat{s}[\vec{z}/\vec{r}]))_{a_0}\\
&= \begin{cases} \mathrm{T}(\widehat{\mathbb{N}}_0) & \text{if } k \notin b_i,\\ \mathrm{T}(\widehat{\mathbb{N}}_1) & \text{if } k \in b_i \end{cases}\\
&= \begin{cases} \emptyset & \text{if } k \notin b_i,\\ \mathrm{Compl}(\{\mathrm{A}^1_0\}) & \text{if } k \in b_i. \end{cases}
\end{aligned}$$

$\widetilde{A}[\vec{u}/\vec{n}, \vec{U}/\vec{b}] = \widetilde{s}[\vec{u}/\vec{n}] \in b_i$. This implies the assertion.

Case $A = (A_1 \wedge A_2)$: Let c_i, h_i for A_i chosen, $c := c_1 \cup c_2$, $h := \mathrm{p}(h_1, h_2)$. Then for $\vec{n}, \vec{r}, \vec{b}, a_0$ as in the assumption of the assertion there exist $s_1, s_2 \in \mathrm{Term}_{\mathrm{nf}}$, such that $r_i[\vec{r}] \longrightarrow_{\mathrm{red}} s_i$, $h[\vec{r}] \longrightarrow_{\mathrm{red}} \mathrm{p}(s_1, s_2) \in \mathrm{Term}_{\mathrm{nf}}$.

$$\begin{aligned}\widetilde{A}[\vec{n}] &\iff \widetilde{B}_1[\vec{n}] \wedge \widetilde{B}_2[\vec{n}]\\ &\iff \langle s_1, \cdot\rangle \in \widehat{B}_1^*[\vec{n}] \wedge \langle s_2, \cdot\rangle \in \widehat{B}_2^*[\vec{n}]\\ &\iff \langle h[\vec{r}], \cdot\rangle \in \widehat{A}^*[\vec{n}]\\ &\iff \widehat{B}_1^*[\vec{r}] \neq \emptyset \wedge \widehat{B}_2^*[\vec{r}] \neq \emptyset\\ &\iff \widehat{A}^*[\vec{n}] \neq \emptyset.\end{aligned}$$

Case $A = (B_0 \vee B_1)$: Let c_i, h_i for B_i chosen. Let

$$f := \{\langle\langle n_1, \ldots, n_l\rangle, i\rangle | (i = 0 \wedge \widetilde{B}_0[\vec{n}]) \vee (i = 1 \wedge \neg\widetilde{B}_0[\vec{n}])\}$$

(note that $\widetilde{B}_0$ is a Δ_0-formula).

Let $\ulcorner C\urcorner$ be a Gödel-number for a new constructor, different from all $\ulcorner C^{\text{set}}\urcorner$, $c := c_1 \cup c_2 \cup \{\langle\ulcorner C\urcorner, l+1, f\rangle\}$, $h := \mathrm{P}(C\ \vec{z}, h_1, (u, v)h_2)$ (u, v new variables). Assume $\vec{n}$, $\vec{r}$, $\vec{b}$, a_0 as in the assumption. Then there exist s_i such that $h_i[\vec{r}] \longrightarrow_{\text{red}} s_i \in \text{Term}_{\text{nf}}$

$$C(r_1, \ldots, r_n) \longrightarrow_{\text{red}} C(\hat{\hat{n}}_1, \ldots, \hat{\hat{n}}_i) \longrightarrow_{\text{red}} \mathrm{S}^i(0)$$

for $i = f(\vec{n}) \in \{0, 1\}$.

We get, if $i = 0$,

$$h[\vec{r}] \longrightarrow_{\text{red}} \mathrm{P}(0, s_1, (u, v)h_2[\vec{r}]) \longrightarrow_{\text{red}} s_1 \in \text{Term}_{\text{nf}},$$

and if $i = 1$,

$$h[\vec{r}] \longrightarrow_{\text{red}} s_2.$$

We have

$$\begin{aligned}\widetilde{A}[\vec{n}] &\iff \widetilde{B}_1[\vec{n}] \vee (\neg\widetilde{B}_1[\vec{n}] \wedge \widetilde{B}_2[\vec{n}])\\ &\iff (f(\vec{n}) = 0 \wedge \langle s_1, \cdot\rangle \in \widehat{B}_1^*[\vec{r}]) \vee (f(\vec{n}) = 1 \wedge \langle s_2, \cdot\rangle \in \widehat{B}_2^*[\vec{r}])\\ &\iff \langle h[\vec{r}], \cdot\rangle \in \widehat{A}^*[\vec{r}]\\ &\iff \widehat{B}_1^*[\vec{r}] \neq \emptyset \vee \widehat{B}_2^*[\vec{r}] \neq \emptyset\\ &\iff \widehat{A}^*[\vec{r}] \neq \emptyset.\end{aligned}$$

Case $A = (B_1 \to B_2)$. Let c_i, h_i for B_i chosen, $c := c_1 \cup c_2$, $h := \lambda x.h_2$. Assume $\vec{r}$, $\vec{n}$, $\vec{b}$, a_0 as in the assumption. Then $h_2[\vec{r}] \longrightarrow_{\text{red}} s_2$ for some $s_2 \in \text{Term}_{\text{nf}}$.

Subcase $\widetilde{A}[\vec{n}]$. If $\widetilde{B}_1[\vec{n}]$ is false, then by IH $\widehat{B}_1^*[\vec{r}] = \emptyset$, therefore

$$\forall\langle r, r'\rangle \in \widehat{B}_1^*[\vec{r}].\langle h_2[x/r, \vec{r}], h_2[x/r', \vec{r}]\rangle \in \widehat{B}_2^*[\vec{r}]$$

therefore $\langle h[\vec{r}], \cdot\rangle \in \widehat{A}^*[\vec{r}]$.

If $\widetilde{B}_1[\vec{n}]$ is true, then $\widetilde{B}_2[\vec{r}]$ is true, therefore $\langle s_2, \cdot\rangle \in \widehat{B}_2^*[\vec{r}]$,

$$\forall\langle r, r'\rangle \in \widehat{B}_1^*[\vec{r}].h_2[x/r, \vec{r}] \longrightarrow_{\text{red}} s_2 \wedge h_2[x/r', \vec{r}] \longrightarrow_{\text{red}} s_2 \wedge \langle s_2, \cdot\rangle \in \widehat{B}_2^*[\vec{n}],$$

$h[\vec{r}] \in \widehat{A}^*[\vec{n}]$.

Subcase $\neg\widetilde{A}[\vec{n}]$. Then by IH exists s_1 such that $h_1[\vec{r}]\longrightarrow_{\text{red}} s_1 \in \text{Term}_{\text{nf}}$ and we have $\langle s_1, \cdot\rangle \in \widehat{B}_1^*[\vec{r}]$ and, if we had $\langle s, s'\rangle \in \widehat{A}^*[\vec{r}]$, then $\langle s, \cdot\rangle \in \widehat{A}^*[\vec{r}]$, $s\longrightarrow_{\text{red}}\lambda x.t$ for some t, $\langle t[x/s_1], \cdot\rangle \in \widehat{B}_2^*[\vec{r}] = \emptyset$, a contradiction, therefore $\widehat{A}^*[\vec{r}] = \emptyset$.

Case $A = \forall v_i.B$: Let c_1, h_1 for B be chosen, $c := c_1$, $h := \lambda v_i.h_1$. Assume $\vec{n}$, $\vec{r}$, $\vec{b}$, a_0 as in the assertion, $h[\vec{r}] \in \text{Term}_{\text{nf}}$.

Assume $\langle r, r'\rangle \in \widehat{A}^*[\vec{r}]$, then $\langle r, r\rangle \in \widehat{A}^*[\vec{r}]$, $r\longrightarrow_{\text{red}}\lambda x.t$ and

$$\forall k \in \omega\langle t[x/\hat{k}, \vec{r}], \cdot\rangle. \in \widehat{B}^*[z_i/\hat{k}, \vec{r}],$$

by IH it follows $\forall k \in \omega.\widetilde{B}[u_i/k, \vec{n}]$, therefore $\widetilde{A}[\vec{n}]$.

Assume $\widetilde{A}[\vec{n}]$. Then for all $k \in \omega$ $\widetilde{B}[v_i/k, \vec{r}]$, therefore by IH $\langle h[v_i/r, \vec{r}], \cdot\rangle \in \widehat{B}^*[z_i/r, \vec{r}]$, whenever $r\longrightarrow_{\text{red}}\text{S}^k(0)$, therefore $\langle h[\vec{r}], \cdot\rangle \in \widehat{A}[\vec{r}]$.

Case $A = \exists v_i.B$, c_1, h_1 be chosen for B,

$$\begin{aligned} f := \{\langle\langle\vec{n}\rangle, k\rangle|\ & (\widetilde{B}[u_i/k, \vec{n}] \wedge \forall k' < k.\neg(\widetilde{B}[u_i/k, \vec{n}])) \vee \\ & (k = 0 \wedge \forall k \in \omega.\neg(\widetilde{B}[u_i/k, \vec{n}])). \end{aligned}$$

Let $\ulcorner C\urcorner$ be a new name for a constructor $\neq \text{C}_i^{\text{set}}$, $c := c_1 \cup \{\langle\ulcorner C\urcorner, l+1, f\rangle\}$. $h := \text{p}(C(\vec{z}), h_1[z_i/C(\vec{z})])$ Assume $\vec{n}$, $\vec{r}$ $\vec{b}$, a_0 as in the assertion, $k := f(\vec{n})$.

$$C(\vec{r})\longrightarrow_{\text{red}}C(\hat{n}_1, \ldots, \hat{n}_n)\longrightarrow_{\text{red}}\text{S}^k(0).$$

By IH we have $h_1[z_i/C(\vec{z})][\vec{r}] = h_1[z_i/C(\vec{r}), \vec{r}]\longrightarrow_{\text{red}}t_1$ for some $t_1 \in \text{Term}_{\text{nf}}$, therefore $h[\vec{r}]\longrightarrow_{\text{red}}\text{p}(\text{S}^k(0), t_1)$.

Assume $\langle r, r'\rangle \in \widehat{A}^*[\vec{r}]$. Then $\langle r, r\rangle \in \widehat{A}^*[\vec{r}]$, $r\longrightarrow_{\text{red}}\text{p}(\text{S}^l(0), r'') \in \text{Term}_{\text{nf}}$. Then $\langle r'', \cdot\rangle \in \widehat{B}^*[z_i/\text{S}^l(0), \vec{r}]$, by IH $\widetilde{B}[u_i/l, \vec{n}]$, therefore $\widetilde{A}[\vec{n}]$.

Assume $\widetilde{A}[\vec{n}]$. Then by definition $\widetilde{B}[u_i/k, \vec{n}]$ and by IH

$$\langle t_1, \cdot\rangle \in \widehat{B}^*[z_i/\hat{\hat{k}}, \vec{r}] = \widehat{B}^*[z_i/C(\vec{z})[\vec{r}], \vec{r}] = \widehat{B}[z_i/C(\vec{z})]^*[\vec{z}/\vec{r}],$$

therefore $\langle h[\vec{r}], \cdot\rangle \in \widehat{A}^*[\vec{r}]$. □

Lemma 16.47. *If φ is a Π_1^1-formula,* $\text{ML} \vdash s : \widehat{\varphi}$, *then* $\text{KPI}^+ \vdash \widetilde{\varphi}$.

Proof. Let $\varphi = \forall V_i.B$. $\text{ML} \vdash s : \widehat{\varphi}$. By Lemma 16.38 it follows $\text{KPI}^+ \vdash \forall\text{CES}(b).\langle\hat{s}_b^*, \cdot\rangle \in \widehat{\varphi}_b^*$. $\widehat{\varphi}^*[\vec{r}] = \Pi^*(\widehat{\mathbb{N} \to \text{U}}^*, \lambda\langle r, \cdot\rangle \in \widehat{\mathbb{N} \to \text{U}}^*.\widehat{B}[\vec{r}, r])$. Assume $b_i \in \text{Ad}_2$, x a variable, c the CES chosen for B as in Lemma 16.46, $a_0 := b_0 \cup \{\langle\text{C}_i^{\text{set}}, 0, \text{Embset}(b_i)\rangle\}$. We have $\text{C}_i \in (\mathbb{N} \to \text{U})_{a_0}^*$. Therefore, under the assumption $b_i \in \text{Ad}_2$, $b_i \subset \omega$, $\text{KPI}^+ \vdash \langle\hat{s}^*\ \text{C}_i, \cdot\rangle \in \widehat{B}_{a_0}^*[V_i/\text{C}_i^{\text{set}}]$, by Lemma 16.46 $\text{KPI}^+ \vdash \widetilde{B}[U_i/b_i]$, and we have $\text{KPI}^+ \vdash \forall V_i \in \text{Ad}_2.V_i \subset \omega \to \widetilde{B}$ which is $\widetilde{\varphi}$. □

16.10 Main Theorem

In this final section we will prove the result about the proof theoretic strength of the type theories used in this article (Theorem 16.52). We will show that the result about the embedding is sufficient to show that we have an upper bound for the proof-theoretic strength. We have to overcome the fact, that we did only prove, that if $\mathrm{ML} \vdash \mathrm{TI}(\prec)$ ($\mathrm{TI}(\prec)$ for transfinite induction over a primitive recursive relation $\prec$), we get $\mathrm{KPI}^+ \vdash \mathrm{TI}^{\mathrm{Ad}_2}(\prec)$, where $\mathrm{TI}^{\mathrm{Ad}_2}$ means transfinite induction, with the quantifier over subsets of ω, which are elements of Ad_2. But we will see, that this will be sufficient to obtain the result.

Definition 16.48. We define some formulas in $\mathcal{L}_{KP}$:

(a) In the following, $(a, \prec)$ will be a pair where a is a set, and $\prec \subset a \times a$. In this context $s \prec t := \langle s,t\rangle \in \prec$, $\forall x \prec t.\varphi := \forall x \in a.x \prec t \to \varphi$, and $\exists x \prec t.\varphi := \exists x \in a.x \prec t \wedge \varphi$. Furthermore, $s \preceq t := s \prec t \vee s = t$.
(b) $\mathrm{Wf}^d(a, \prec) := \prec \subset a \times a \wedge \forall x \in d.x \subset a \to x \neq \emptyset \to \exists y \in x.\forall z \prec y.z \notin x$. ($\prec$ is a relation on a which is well-founded, restricted to d).
(c) $\mathrm{Collaps}(a, \prec, f) := Fun(f) \wedge \mathrm{dom}(f) = a \wedge \forall x \in a.f(x) = \{f(y) | y \prec x\}$. ($f$ is a collapsing function on $(a, \prec)$.

Lemma 16.49. *If $\varphi(y, y_1, \dots, y_n)$ is a Δ_0-formula with only the free variables mentioned, then* $\mathrm{KPI}^+ \vdash \mathrm{Wf}^d(a, \prec) \to \mathrm{Ad}(c) \to \forall y_1, \dots, y_n \in c.(\forall x \in a.(\forall y \prec x.\varphi(y, y_1, \dots, y_n) \to \varphi(x, y_1, \dots, y_n))) \to \forall x \in a.\varphi(x, y_1, \dots, y_n)$. *The formula after the second arrow is called principle of restricted induction over* $(a, \prec)$.

Proof. Immediate. □

Lemma 16.50. $\mathrm{KPI}^+ \vdash \mathrm{Ad}(c) \to \mathrm{Ad}(d) \to c \in d \to \mathrm{Wf}^d(a, \prec) \to c, d \in a \to$
$\exists f \in c.\mathrm{Collaps}(a, \prec, f)$.

Proof. As in [6], Theorem 4.6, but replacing Δ_0-induction by d-induction. □

Lemma 16.51. *If* $\mathrm{KPI}^+ \vdash \forall x.\mathrm{Ad}(x) \to \varphi(x)$ *for a Σ_1-formula φ^x, then $\mathrm{L}_\nu \models \varphi$, where φ^x is the restriction of all unrestricted quantifiers to x, and $\nu := \psi_{\Omega_1}(\Omega_{\mathrm{I}+\omega})$.*

Proof. We follow the lines of [2]. First observe, that we can prove as in Theorem 2.9 there, using several applications of $\exists_\kappa$, $(\bigwedge)$, and $\vdash^* \mathrm{Ad}(\mathrm{L}_\kappa)$ for $\kappa \in \mathrm{R}$, and if we have $\lambda \in \mathrm{Lim}$, $(\kappa_i)_{i \in \omega}$ a sequence, s.t. $\kappa_0 \in \mathrm{R}$, $\forall \alpha \in \kappa.\exists \rho \in \kappa.\alpha \in \rho \in \mathrm{R}$ and $\kappa_i \in \kappa_{i+1} \in \mathrm{R} \cap \lambda$ $(i \in \omega)$, and if we extend X^* by including as well κ_i $(i \in \omega)$ it follows

$$\vdash^*_\lambda (\mathrm{KPI}^+)^\lambda.$$

($\vdash^*_\lambda \varphi^\lambda$ for every axiom φ of KPI^+). We can adjust Theorem 3.12 of [2] to obtain, if we have in this situation, if $\lambda \in \mathcal{H}$, $\kappa_i \in \mathcal{H}$ $(i \in \omega)$ and $\mathcal{H}$ closed under $\xi \mapsto \xi^{\mathrm{R}}$, then:

For each theorem φ of KPI^+ exists $k \in \mathbb{N}$ such that $\mathcal{H} \vdash^{\omega^{\lambda+k}}_{\lambda+k} \varphi^\lambda$.

Now observe, that $\mathcal{H}_\gamma$ in [2] has the desired properties (with $\lambda := \Omega_{\mathrm{I}+\omega}, \kappa_i := \Omega_{\mathrm{I}+i}$) and we conclude that if $\mathrm{KPI}^+ \vdash \forall x.\mathrm{Ad}(x) \to \varphi^x$, where φ is a Σ-sentence, then $\mathrm{L}_\nu \models \varphi$ for $\nu := \psi_{\Omega_1}(\Omega_{\mathrm{I}+\omega})$. □

Theorem 16.52. $|\mathrm{ML}_1^{\mathrm{e}}\mathrm{W}_{\mathrm{T,U}}|$, $|\mathrm{ML}_1^{\mathrm{e}}\mathrm{W}_{\mathrm{T}}|$, $|\mathrm{ML}_1^{\mathrm{e}}\mathrm{W}_{\mathrm{R,U}}|$, $|\mathrm{ML}_1^{\mathrm{e}}\mathrm{W}_{\mathrm{R}}|$, $|\mathrm{ML}_1^{\mathrm{i}}\mathrm{W}_{\mathrm{T,U}}|$, $|\mathrm{ML}_1^{\mathrm{i}}\mathrm{W}_{\mathrm{T}}|$, $|\mathrm{ML}_1^{\mathrm{i}}\mathrm{W}_{\mathrm{R}}|$, $|\mathrm{ML}_1^{\mathrm{i}}\mathrm{W}_{\mathrm{R,U}}| = \psi_{\Omega_1}(\Omega_{\mathrm{I}+\omega})$, *where the ordinal denotation is as in* [2].

Remark 16.53. Since intensional type theory using the formulation of [19] which was used in [10] is a proper subtheory of $|\mathrm{ML}_1^{\mathrm{e}}\mathrm{W}_{\mathrm{T}}|$, $\psi_{\Omega_1}(\Omega_{\mathrm{I}+\omega})$ is an upper bound for the proof theoretic strength of the Tarski and Russell formulations of intensional type theory used in [10].

Proof of Theorem 16.52. Let $\nu := \psi_{\Omega_1}(\Omega_{\mathrm{I}+\omega})$. By [12] we have $|\mathrm{ML}_1^{\mathrm{i}}\mathrm{W}_{\mathrm{T}}| \geq= \nu$. Since the Russell formulations embed trivially into the Tarski Formulations (replace $\mathrm{T}(r)$ by r and $\widehat{\mathbb{N}}_k, \widehat{\mathbb{N}}, \widehat{\Pi}, \widehat{\Sigma}, \widehat{\mathrm{W}}, \widehat{+}, \widehat{\mathrm{I}}$ by $\mathbb{N}_k, \mathbb{N}, \Pi, \Sigma, \mathrm{W}, +, \mathrm{I}$, respectively), we obtain as well $|\mathrm{ML}_1^{\mathrm{i}}\mathrm{W}_{\mathrm{R}}| \geq \nu$. Since $\mathrm{ML}_1^{\mathrm{i}}\mathrm{W}_{\mathrm{T,U}}$ is an extension of $\mathrm{ML}_1^{\mathrm{i}}\mathrm{W}_{\mathrm{T}}$, which both embed into $\mathrm{ML}_1^{\mathrm{e}}\mathrm{W}_{\mathrm{T,U}}$ and $\mathrm{ML}_1^{\mathrm{e}}\mathrm{W}_{\mathrm{T}}$, respectively, and $\mathrm{ML}_1^{\mathrm{i}}\mathrm{W}_{\mathrm{R,U}}$ is an extension of $\mathrm{ML}_1^{\mathrm{i}}\mathrm{W}_{\mathrm{R}}$, which both embed into $\mathrm{ML}_1^{\mathrm{e}}\mathrm{W}_{\mathrm{R,U}}$ and $\mathrm{ML}_1^{\mathrm{e}}\mathrm{W}_{\mathrm{R}}$, respectively, ν is a lower bound for all theories in question.

Regarding the upper bound, we show $|\mathrm{ML}| \leq \nu$. Since all the theories can be embedded into $\mathrm{ML} := \mathrm{ML}_1^{\mathrm{e}}\mathrm{W}_{\mathrm{T,U}}$ in such a way, that the principle of transfinite induction remains unchanged (except, that $\mathrm{T}(U_i(t))$ becomes $U_i(t)$ in the version à là Russell), we obtain that ν is an upper bound for the proof theoretic strength of all other theories in the theorem as well.

Proof of $|\mathrm{ML}| \leq \nu$: Assume $\prec$ is a primitive recursive linear ordering on the primitive recursive subset T of ω,

$$\varphi := \forall X.(\forall y.y \in T \to (\forall z.z \prec y \to z \in X) \to y \in X) \to \forall y.y \in T \to y \in X$$

and $\mathrm{ML} \vdash \widehat{\varphi}$. Then by Lemma 16.47 $\mathrm{KPI}^+ \vdash \widetilde{\varphi}$. We follow the proof of [9] Theorem 7.14. Let $a := \{x \in \omega | x \in T\}$, $\prec' := \{\langle x, y\rangle \in \omega \times \omega | x \prec y\}$. Then $\mathrm{KPI}^+ \vdash \mathrm{Ad}(\mathrm{Ad}_1) \wedge \mathrm{Ad}(\mathrm{Ad}_2) \wedge \mathrm{Ad}_1 \in \mathrm{Ad}_2 \wedge \mathrm{Wf}^{\mathrm{Ad}_2}(a, \prec') \wedge a, \prec' \in \mathrm{Ad}_1$, and by 16.50 $\mathrm{KPI}^+ \vdash \exists f \in \mathrm{Ad}_1.\mathrm{Collaps}(a, \prec', f)$, $\mathrm{KPI}^+ \vdash \forall x.\mathrm{Ad}(x) \to \exists f \in x.\mathrm{Collaps}(a, \prec', f)$. Therefore $\mathrm{L}_\nu \models \exists f.\mathrm{Collaps}(a, \prec'\ o, f)$. Since $\prec$ is linear ordering, it follows that $\mathrm{Image}(f)$ is an ordinal, and, because $\nu \in \mathrm{Lim}$ we have $\mathrm{Image}(f) \in \mathrm{L}_\nu$, $\mathrm{ordertype}(\prec) = \mathrm{Image}(f) < \nu$. □

Acknowledgements Supported by CORCON (Correctness by Construction, FP7 Marie Curie International Research Project, PIRSES-GA-2013-612638), COMPUTAL (Computable Analysis, FP7 Marie Curie International Research Project, PIRSES-GA-2011-294962), CID (Computing with Infinite Data, Marie Curie RISE project, H2020-MSCA-RISE-2016-731143), CA COST Action CA15123 European research network on types for programming and verification (EUTYPES), EPSRC grant GR/S30450/01 and EPSRC grant EP/G033374/1; part of this research was done while the author was visiting the university of Leeds as part of the EC Twinning Project "Proofs and Computation".

References

1. J. Barwise. *Admissible Sets and Structures. An Approach to Definability Theory*. Omega-series. Springer, 1975.
2. W. Buchholz. A simplified version of local predicativity. In P. Aczel, H. Simmons, and S. S. Wainer, editors, *Proof Theory. A selection of papers from the Leeds Proof Theory Programme 1990*, pages 115 – 147, Cambridge, 1992. Cambridge University Press.
3. E. Griffor and M. Rathjen. The strength of some Martin-Löf type theories. *Arch. math. Log.*, 33:347 – 385, 1994.
4. G. Jäger. *Die konstruktible Hierarchie als Hilfsmittel zur beweistheoretischen Untersuchung von Teilsystemen der Analysis*. Dissertation, Universität München, 1979.
5. G. Jäger. A well-ordering proof for Feferman's theory T_0. *Arch. math. Log.*, 23:65 – 77, 1983.
6. G. Jäger. *Theories for Admissible Sets: A Unifying Approach to Proof Theory*. Bibliopolis, Naples, 1986.
7. G. Jäger and W. Pohlers. Eine beweistheoretische Untersuchung von $(\Delta^1_2 - CA) + BI$ und verwanter Systeme. *Sitzungsberichte der Bayer. Akad. d. Wiss., Math.-Nat. Kl.*, pages 1 – 28, 1982.
8. W. Pohlers. Admissibility in proof theory, a survey. In *Logic, Methodology and Philosphy of Science VI, Hannover 1979*, pages 123 – 139, Amsterdam, 1982. North-Holland.
9. M. Rathjen. Proof-theoretical analysis of KPM. *Arch. math. Log.*, 30:377 – 403, 1991.
10. A. Setzer. *Proof theoretical strength of Martin-Löf Type Theory with W-type and one universe*. PhD thesis, Mathematisches Institut, Universität München, Munich, Germany, 1993. Available from http://www.cs.swan.ac.uk/ csetzer/articles/weor0.pdf.
11. A. Setzer. An introduction to well-ordering proofs in Martin-Löf's type theory. In G. Sambin and J. Smith, editors, *Twenty-five years of constructive type theory*, pages 245 – 263, Oxford, 1998. Clarendon Press.
12. A. Setzer. Well-ordering proofs for Martin-Löf type theory. *Annals of Pure and Applied Logic*, 92:113 – 159, 1998.
13. A. Setzer. Extending Martin-Löf Type Theory by one Mahlo-universe. *Archive for Mathematical Logic*, 39:155–181, 2000.
14. A. Setzer. Proof theory of Martin-Löf Type Theory – An overview. *Mathematiques et Sciences Humaines*, 42 année, n° 165:59 – 99, 2004.
15. A. Setzer. Proof theory and Martin-Löf Type Theory. In P. Atten, M. v.; Boldini and G. Bourdeau, M.; Heinzmann, editors, *One Hundred Years of Intuitionism (1907 – 2007)*, Publications des Archives Henri-Poincaré Publications of the Henri Poincaré Archives, pages 257 – 279. Birkhäuser, 2008.
16. A. Setzer. Universes in type theory part I – Inaccessibles and Mahlo. In A. Andretta, K. Kearnes, and D. Zambella, editors, *Logic Colloquium '04*, pages 123 – 156. Association of Symbolic Logic, Lecture Notes in Logic 29, Cambridge University Press, 2008.
17. A. Setzer. Universes in Type Theory Part II – Autonomous Mahlo. Submitted. Available from http://www.cs.swan.ac.uk/ csetzer/articles/modelautomahlomain.pdf, 2011.
18. A. Setzer. The use of trustworthy principles in a revised Hilbert's program. In Reinhard Kahle and Michael Rathjen, editors, *Gentzen's Centenary*, pages 45–60. Springer International Publishing, 2015.
19. A. Troelstra and D. van Dalen. *Constructivism in Mathematics. An Introduction, Vol. II*. North-Holland, Amsterdam, 1988.

Chapter 17
Normalization Proof for Derivations in PA after P. Cohen

Grigori Mints

Abstract P. Cohen (unpublished) gave a new normalization proof for derivations in PA. We present it here as a simplification of Gentzen's second consistency proof.

17.1 Introduction

Theorem 17.1 (Gentzen). *Every derivation of a quantifier-free sentence in PA can be reduced by a finite number of standard cut-elimination transformations to an axiom.*

Consider antecedent Tait-style formalization of PA as a calculus of contradictions. Sequents are finite sets of prenex formulas

$$\Gamma = A_1, \ldots, A_n \text{ interpreted as } \neg \bigwedge_{i \leq n} A_i .$$

Axioms: closure under weakening of any reasonable set of quantifier-free sentences including F, Γ where F is a false quantifier-free sentence.

"Reasonable" should ensure derivability of Peano axioms other than induction.

Inference Rules:

$$\exists \frac{\Gamma, A[c]}{\Gamma, \exists x A} \quad \forall \frac{\Gamma, A[t]}{\Gamma, \forall x A} \quad Ind \frac{\bar{A}[0], \Gamma \quad A[c], \bar{A}[c+1], \Delta \quad \forall x A, \Sigma}{\Gamma, \Delta, \Sigma}$$

with a standard proviso for eigenvariable c.

Branching (=Gentzen's *Kettenschluss*) is a series of cuts

$$C \frac{\Gamma, A \quad \Delta, \bar{A}}{\Gamma, \Delta}$$

G. Mints (†)
Stanford University, USA

R. Kahle and M. Rathjen (eds.), *The Legacy of Kurt Schütte*,
https://doi.org/10.1007/978-3-030-49424-7_17

with some formulas possibly deleted, for example:

$$\frac{\Gamma_0, \bar{A}[0] \quad \Gamma_1, A[0], \bar{A}[1] \quad \ldots \Gamma_n, A[n-1], \bar{A}[n] \quad A[n], \Gamma}{\Gamma_0, \ldots, \Gamma_n, \Gamma}$$

Note. If a derivation does not contain any branchings or inductions, then it is an axiom. Hence it is sufficient to eliminate branchings and inductions.

Consider an additional repetition rule which does not change anything:

$$Rep\ \frac{\Gamma}{\Gamma}$$

Definition 17.2. The *order* of an induction inference Ind is the number of quantifiers in the formula $\forall x A$; the *order* of a branching is the maximal number of quantifiers in cut formulas A.

Derivation is a finite tree beginning with axioms and proceeding down by inference rules (hence the last sequent is the lowermost sequent). From now on all derivations should satisfy the following additional restriction.

AR *Every branching or induction of the order r is immediately followed by at least r inferences each of which is Rep or branching.*

AR can be achieved by inserting *Rep* inferences.

Definition 17.3. A variable is redundant if it occurs in a premise of an inference, but not in its conclusion, and is different from the eigenvariable of that inference (if any).

Preliminary reductions.

1. Elimination of redundant (occurrences of) variables: they are replaced by any numeral, say 0.
2. Elimination of quantifier-free cuts. One of two cut-formulas in such a cut C, say A is true. Delete it from the whole branch Γ, A [one should check that axioms are transformed into axioms] and delete the branc ending in $\Gamma, \bar{A}$.

In the following we always assume there are no redundant variables or quantifier-free cuts.

Definition 17.4. The *principal branch* of a derivation contains its endsequent, the rightmost premise of induction and the premise of branching containing only cut-formulas beginning with $\forall$.

Note. The lowermost $\forall$-inference in the principal branch has a constant term t. Indeed, there are no eigenvariables below that inference, hence any variable would be redundant.

Proper reductions.

Case 1. The principal branch does not contain ∀-inferences. Then all ∀-formulas are redundant and can be deleted from the branch, which is reduced to an axiom as required.

Case 2. The principal branch contains a ∀-inference. Take the lowermost one and call it $\mathcal{A}$. By the Note its term t is a constant, and using provable equality $t = n$ for some numeral n we can assume that t is n. Note that there are only Ind and branching inferences in the principal branch.

Case 2.1. $\mathcal{A}$ is traceable to a cut (= branching).

$$\dfrac{\alpha_1\ \Gamma,\Sigma_1\ \ldots\ \alpha_k\ \Gamma,\Sigma_k,\exists x\bar{A}[x] \quad \alpha\ \Gamma,\Delta,\forall x A[x]}{\beta \qquad \ldots\ \Gamma \qquad \sum_{i\le k}\omega^{\alpha_i}+\omega^{\alpha}} \quad \text{where} \quad \dfrac{c:\ \ldots\ \Pi,\bar{A}[c]}{d_c:\ \Pi,\exists x\bar{A}[x]} \ \ldots \qquad \dfrac{\ldots\ \Phi, A[n]}{d_n:\ \Phi,\forall x A[x]}\ \mathcal{A}\ \ldots$$

$$\overline{d:\ \Theta\ \ldots}$$

$$\ldots$$

$$ord(d) = \omega^{\beta} + \omega^{\sum_{i\le k}\omega^{\alpha_i}+\omega^{\alpha}}$$

$$c := n\ \ldots \qquad d_c^-:\ \Pi,\bar{A}[n] \qquad \ldots$$

$$\sum_{i<k}\omega^{\alpha_i}+\omega^{\alpha_k^-}\quad \dfrac{\alpha_1\ \Gamma,\Sigma_1,\bar{A}[n]\ldots\alpha_k^-\ \Gamma,\Sigma_k,\bar{A}[n]}{\beta\ \ldots\Gamma,\bar{A}[n]}$$

$$\ldots \qquad d_n^-:\ \Phi,\forall x A[x],A[n] \qquad \ldots$$

$$\dfrac{\alpha_1\ \Gamma,\Sigma_1,A[n]\ldots\alpha_k\ \Gamma,\exists x\bar{A},\Sigma_k,A[n] \quad \alpha^-\ \Gamma,\Delta,\forall xA,A[n]}{\Gamma,A[n] \qquad \sum_{i\le k}\omega^{\alpha_i}+\omega^{\alpha^-}}$$

$$\overline{d':\ \Theta\ldots}$$

$$ord(d') = \omega^{\beta} + \omega^{\sum_{i<k}\omega^{\alpha_i}+\omega^{\alpha_k^-}} + \omega^{\sum_{i\le k}\omega^{\alpha_i}+\omega^{\alpha^-}} <$$
$$\omega^{\beta} + \omega^{\sum_{i\le k}\omega^{\alpha_i}+\omega^{\alpha^-}}\cdot 2 < \omega^{\beta} + \omega^{\sum_{i\le k}\omega^{\alpha_i}+\omega^{\alpha^-}+1} \le ord(d).$$

Case 2.2. $\mathcal{A}$ is traceable to induction. Transform the derivation as follows.
Original derivation

$$
\dfrac{\dfrac{\alpha\ \bar{A}[0],\Gamma \quad \beta\ A[c],\bar{A}[c+1],\Gamma \quad \gamma\ \begin{array}{c} \dots \\ \mathcal{A}\ \dfrac{A[n],\Phi}{\forall x A[x],\Phi} \\ \dots \\ \forall x A[x],\Gamma \end{array}}{\delta \dots \qquad \Gamma \qquad \omega^{\alpha}+\omega^{\beta}+\omega^{\gamma}}}{d:\ \Theta}
$$

$$\dots$$

with $ord(d) = \omega^{\delta} + \omega^{\omega^{\alpha}+\omega^{\beta}+\omega^{\gamma}}$ reduces to

$$
\dfrac{\begin{array}{cccc} & c := k\dots & c := n-1\dots & \\ \alpha\ \bar{A}[0],\Gamma & \beta\ A[k],\bar{A}[k+1],\Gamma & \beta\ A[n-1],\bar{A}[n],\Gamma & \dfrac{\bar{A}[0],\Gamma \quad A[c],A[c+1],\Gamma \quad \begin{array}{c} \dots \\ A[n],\forall x A[x],\Phi \\ \dots \\ \forall x A[x],\Gamma,A[n] \end{array}}{\delta \qquad \dots\Gamma,A[n]} \end{array}}{d':\ \Theta}
$$

$$\dots$$

with $ord(d') = \omega^{\delta} + \omega^{\alpha} + \omega^{\beta}\cdot(n-1) + \omega^{\omega^{\alpha}+\omega^{\beta}+\omega^{\gamma^{-}}} < \omega^{\delta} + \omega^{\alpha}\cdot(n+2) + \omega^{\beta}\cdot(n+2) + \omega^{\omega^{\alpha}+\omega^{\beta}+\omega^{\gamma^{-}}} < \omega^{\delta} + \omega^{\omega^{\alpha}+\omega^{\beta}+\omega^{\gamma^{-}}+1} \leq ord(d)$.

17.2 Finite Trees as Ordinals. Termination of Reduction Sequence

Let's define lexicographic ordering of finite trees with root by induction on the height.

1. One-point tree is $<$ any other tree.
2. Let subtrees in

$$
\begin{array}{c} T_1\ \dots\ T_n \\ \nwarrow\ \nearrow \\ T \end{array} \qquad\qquad \begin{array}{c} U_1\ \dots\ U_m \\ \nwarrow\ \nearrow \\ U \end{array}
$$

be ordered: $T_1 \geq \dots \geq T_n$, $U_1 \geq \dots \geq U_m$.

Then $T > U$ iff $T_i > U_i$ for the first $i:\ T_i \neq U_i$ or $[(\forall i \leq m)T_i = U_i$ and $n > M]$. It is easy to prove that $>$ is a linear well-ordering of the order-type ε_0:

define $ord(.) = 0$; and for the tree T with immediate subtrees $T_1, \ldots, T_n$ as above, $ord(T) = \omega^{ord(T_1)} + \cdots + \omega^{ord(T_n)}$ where + is the natural (commutative) addition of ordinals.

Lemma 17.5. (a) *Replacing a subtree by a strictly smaller tree results in a strictly smaller tree:*

$$U \prec U' \vdash T[U] \prec T[U'];$$

the same for replacing by any finite number of smaller trees;
(b) *every reduction step strictly decreases the tree*

Proof. (a) Easy by the definition of $\prec$.
(b) Check all reductions. Ordinals in the proof-figures indicate necessary steps. □

Proof. This implies Gentzen's consistency theorem. It is necessary to check that AR is preserved, so that a derivation containing a cut or induction can be reduced. □

17.3 Comparison with Gentzen's Second Consistency Proof

1. Reductions."I"n Gentzen's proof new cuts are placed separately at a proper place down the derivation (the height line). In Cohen's proof new cuts are added to existing cuts or *Rep*.
2. Ordinal assignment."I"n Gentzen's proof an ordinal is assigned to any derivation d taking into consideration mutual positions of cuts and inductions in d. In Cohen's proof there is no special ordinal assignment: derivations = ordinals.

1 + 2 combined simplify ordinal assignment to induction and eliminate several combinatorial complications from Gentzen's proof: induction is treated in the same way as cut.

Refinements. This proof provides the bound ω_{k+2}, where k is the height of the original derivation. I can modify it to obtain (known) optimal bound ω_{n+2} where n is the maximal complexity of induction. Here $\omega_0 = 0, \omega_{n+1} = \omega^{\omega_n}$.

References

1. G. Gentzen. Neue Fassung des Widerspruchsfreiheitsbeweises für die reine Zahlentheorie. *Forschungen zur Logik und zur Grundlegung der exacten Wissenschaften, NS*, 4:19–44, 1938.
2. S. Kleene. *Mathematical Logic*. John Wiley and sons, New York, 1967.
3. G. Mints. A normal form theorem for logical derivations implying one for arithmetic. *Annals of Pure and Applied Logic*, 62:65–79, 1993.
4. G. Takeuti. *Proof Theory*. North-Holland, 1987.

Part III
Further Legacy

Chapter 18
From Probability Measures to Each Lévy Triplet and Back

Horst Osswald

Abstract It will be shown that all finite-dimensional Lévy processes are completely determined by Borel probability measures on a fixed finite-dimensional Euclidian space in a certain sense. Each Lévy process is infinitely close to a fixed multilinear process, living on that space, depending only on the dimension. Lévy processes can be identified, if they satisfy the same Lévy triplet (Lévy-Khintchine formula). The components of this triplet appear in the Fourier transformation of the current probability measure. Lindstrøms work "Hyperfinite Lévy processes, *Stochastics* **76** (6) (2004)" is used to show that each Lévy triplet is satisfied by an appropriate probability measure on the fixed Euclidian space.

Key words: Lévy processes, Lévy-Khintchine formula, Simple Type Logic

Mathematics subject classification: Primary 60G51, 60G15 Secondary 03B15

18.1 Introduction

The purpose of this paper is to show that all finite-dimensional Lévy processes are completely determined by Borel probability measures on a fixed finite-dimensional Euclidian space in a certain sense. Each Lévy process is infinitely close to a fixed multilinear process, living on that space, depending only on the dimension. Lévy processes are stochastic processes, depending on chance and time, starting in 0 and having independent and identically distributed increments. The most important Lévy processes are Brownian motion and Poisson processes. We refer to the book of Bertain [2] for the standard theory of Lévy processes.

H. Osswald
Mathematische Institut, LMU München, Germany, e-mail: osswald@mathematik.uni-muenchen.de

R. Kahle and M. Rathjen (eds.), *The Legacy of Kurt Schütte*,
https://doi.org/10.1007/978-3-030-49424-7_18

We are working in a highly saturated nonstandard model $\mathfrak{W}$ of a standard model $\mathfrak{V}$ of Russel's simple type logic, studied for example by Kurt Schütte [26] and Abraham Robinson [24]. Models of simple type logic can be seen as models of mathematics.

For each mathematical entity M within the standard model $\mathfrak{V}$ there exists an extension ${}^*M$ within the nonstandard model $\mathfrak{W}$. This extension is strict if and only if M is infinite.

For example, the set $\mathbb{R}^d$ of real d-tuples, $d \in \mathbb{N}$, extends to ${}^*\mathbb{R}^d$, $\mathbb{N}$ extends to ${}^*\mathbb{N}$ and $<$ extends to ${}^*<$. However, ${}^*\{1,\ldots,d\} = \{1,\ldots,d\}$. If there is no risk of confusion, we omit the index $*$.

It is perhaps surprising that both models $\mathfrak{V}$ and $\mathfrak{W}$ have the same formal properties, although they are quite different. The distinguishing feature cannot be told by usual mathematical formalism (see Section 2.1.).

Fix $H \in {}^*\mathbb{N} \setminus \mathbb{N}$. Since there is no integer between n and $n \,{}^*\!+ 1$ with $n \in \mathbb{N}$, H is infinitely large (Huge), i.e., $H \,{}^*\!> n$ for all $n \in \mathbb{N}$. It follows that $0 \,{}^*\!< \frac{1}{H} \,{}^*\!< \frac{1}{n}$ for all $n \in \mathbb{N}$, thus $\frac{1}{H} > 0$ is infinitely small. We also fix an infinitely fine partition T of the continuous time line ${}^*[0,\infty[$, where

$$T := \left\{\frac{1}{H}, \frac{2}{H}, \ldots, H\right\}.$$

Note that T is infinite, but finite in the sense of $\mathfrak{W}$. This set T serves as time line for discrete processes, which are very close to processes on the continuous time line $[0,\infty[$.

All d-dimensional Lévy processes are living on the $d \cdot H^2$-dimensional Euclidian space $\left({}^*\mathbb{R}^d\right)^T$ within the model $\mathfrak{W}$. It is the set of all functions from T into ${}^*\mathbb{R}^d$ in $\mathfrak{W}$. The elements of $\left({}^*\mathbb{R}^d\right)^T$ represent chance. Note that the elements of $\left({}^*\mathbb{R}^d\right)^T$ are H^2-tuples $X = (X_s)_{s\in T}$ with $X_s \in {}^*\mathbb{R}^d$. These X_s are the increments $B(\cdot,s) - B\left(\cdot, s - \frac{1}{H}\right)$ (with $B(\cdot,0) := 0$) of the discrete and multilinear stochastic process B within $\mathfrak{W}$, thus:

$$B : \left({}^*\mathbb{R}^d\right)^T \times T \to {}^*\mathbb{R}^d, (X,t) \longmapsto \sum_{s\in T, s\leq t} X_s = \sum_{s\in T, s\leq t} B(X,s) - B(X, s - \frac{1}{H}).$$

The value $B(X,t)$ can be seen as win or loss of a game at time t or as the place, where a particle arrives at time t, under the random event $X \in \left({}^*\mathbb{R}^d\right)^T$. However, the characteristic feature of the process B depends on a Borel probability measure $\overline{\mu}$ (in $\mathfrak{W}$) on $\left({}^*\mathbb{R}^d\right)^T$. In our context $\overline{\mu}$ is the H^2-fold product of a Borel probability measure μ (in $\mathfrak{W}$) on ${}^*\mathbb{R}^d$. It follows that the sequence $(X \longmapsto X_s)_{s\in T}$ of random variables $X \longmapsto X_s$ on $\left({}^*\mathbb{R}^d\right)^T$ with values in ${}^*\mathbb{R}^d$, is independent in $\mathfrak{W}$ and the expected values $\mathbb{E}_{\overline{\mu}}(X_s)$ and $\mathbb{E}_{\mu}(x)$ coincide (see Proposition 3.3.1 in [21]).

In order to construct from B a standard Lévy process ${}^\circ B : \left({}^*\mathbb{R}^d\right)^T \times [0,\infty[\to \mathbb{R}^d$, it is necessary to restrict the set of probability measures μ on ${}^*\mathbb{R}^d$ to the set of Lévy-limited measures μ on ${}^*\mathbb{R}^d$, defined later. Let $\widehat{\overline{\mu}}$ be the Loeb measure over $\overline{\mu}$, defined below. Then ${}^\circ B$ is, roughly speaking, infinitely close to B $\widehat{\overline{\mu}}$-almost surely.

The standard process $^\circ B$ is called the *standard part* of B and depends only on the probability measure μ. We will see that each d-dimensional Lévy process can be obtained as the standard part of B, where we identify two Lévy processes if they satisfy the same Lévy triplet, because then they have the same laws. The Fourier transformations of Lévy-limited probability measures μ produce Lévy triplets. The results in this article are slight modifications of Tom Lindstrøm's work [12] applied to the setting here and perhaps may simplify parts of his work.

Now it is time to present some examples:

(i) Fix a standard $\beta > 0$. Let π be the internal Borel probability measure on $^*\mathbb{R}$, concentrated on $\{0, 1\}$, where $\pi(\{0\}) = 1 - \frac{\beta}{H}$ and $\pi(\{1\}) = \frac{\beta}{H}$. Then $^\circ B$ is Loeb's construction of the Poisson process [10] of rate β.

(ii) Let α be the internal Borel probability measure on $^*\mathbb{R}$, concentrated on $\left\{-\frac{1}{\sqrt{H}}, \frac{1}{\sqrt{H}}\right\}$, setting $\alpha\left(\left\{\frac{1}{\sqrt{H}}\right\}\right) = \frac{1}{2} = \alpha\left(\left\{-\frac{1}{\sqrt{H}}\right\}\right)$. Then $^\circ B$ is Anderson's construction of Brownian motion [1].

(iii) Let c be the internal centered Gaußian distribution on $^*\mathbb{R}$ of infinitely small variance $\frac{1}{H}$, i.e., $c(A) := \int_A e^{-\frac{H}{2}x^2} dx \sqrt{\frac{H}{2\pi}}$ for all Borel sets A in $\mathfrak{W}$ on $^*\mathbb{R}$. Then $^\circ B$ is Cutland's construction of Brownian motion [4].

(iv) Fix $\beta \in \mathbb{R}$. Let δ be concentrated on $\left\{\frac{\beta}{H}\right\}$ with $\delta(\left\{\frac{\beta}{H}\right\}) := 1$. Then $^\circ B(\cdot, r) = \beta r$ is a deterministic process.

(v) Let σ be concentrated on $\{0, 1, -1\}$ with $\sigma(\{1\}) = \frac{\beta}{2H} = \sigma(\{-1\})$, $\sigma(\{0\}) = 1 - \frac{\beta}{H}$. Then $^\circ B$ is the symmetric Poisson process with rate β.

(vi) Fix a strictly monotone decreasing mapping f from $\mathbb{N}$ into the set of positive real numbers such that $\sum_{i \in \mathbb{N}} f(i)$ converges. Let μ be concentrated on $\{^*f(i) \mid i \leq H\}$ and define $\mu(^*f(i)) := \frac{1}{H}$ for all $i \leq H$. Later we will see that the Lévy measure, constructed from μ, is infinite on any neighborhood of 0.

(vii) The product measure $\mu := \alpha \otimes \pi$ of α and π (see (i) and (ii)) and also the measure in Example (viii) may be seen as prototypes for this paper.

(viii) Here we have a Borel probability measure on $^*\mathbb{R}^2$, which is not a product.measure. Let μ be the measure in (vii) and define $F : {}^*\mathbb{R}^2 \to {}^*\mathbb{R}^2, (a, b) \vdash (a - b, a + b)$. Let μ^F be the *image measure of μ under F*, which means that for all Borel sets $E \in \mathfrak{W}$ in $^*\mathbb{R}^2$,

$$\mu^F(E) := \mu\left(\left\{x \in {}^*\mathbb{R}^2 \mid F(x) \in E\right\}\right).$$

Let us now sketch the nonstandard approach to infinite-dimensional Brownian motion. Fix an infinite-dimensional but separable real Hilbert space $\mathbb{H} \in \mathfrak{V}$. By Theorem 9.8.1 in [21] there exists an $\omega \in {}^*\mathbb{N} \setminus \mathbb{N}$ and an ω-dimensional Euclidian space $\mathbb{F}$ with $\mathbb{H} \subseteq \mathbb{F} \subseteq {}^*\mathbb{H}$, such that for each $n \in \mathbb{N}$ any element f of the n-fold tensor product $\mathbb{H}^{\otimes n}$ of $\mathbb{H}$ is infinitely close to $^*f$ restricted to the n-fold tensor product of $\mathbb{F}$. The range of the integrands of the n-fold iterated integrals belongs to $\mathbb{H}^{\otimes n}$.

Now we have $B : \mathbb{F}^T \times T \to \mathbb{F}, (X, t) \longmapsto \sum_{s \leq t} X_s$. Let μ be the ω-fold product of Cutland's measure c in Example (iii) on $\mathbb{F}$. Then $\overline{\mu}$ is the $\omega \cdot H^2$ fold product of c on $\mathbb{F}^T$. Unfortunately, the standard part $^\circ B$ of B, defined on $\mathbb{F}^T \times [0, \infty[$, does not exist in the Hilbert space $\mathbb{H}$ itself, but it exists in the completion $\mathbb{B}$ of $\mathbb{H}$, endowed with a separating countable sequence $(|\cdot|_i)_{i\in\mathbb{N}}$ of Gross measurable semi-norms on $\mathbb{H}$. Then $\mathbb{B}$ is a Fréchet space and $^{\circ}B$ is a Brownian motion.

What is a Brownian motion with values in the Fréchet space $\mathbb{B}$? Recall first that a process b with values in a d-dimensional Hilbert space $\mathbb{G}$, $d \in \mathbb{N}$, is a d-dimensional Brownian motion if each component of b is a Brownian motion and these components are running independently on orthogonal axes. To be more precise, fix an orthonormal basis $(\mathfrak{e}_i)_{i\leq d}$ of $\mathbb{G}$. Then b is a Brownian motion if each $\langle b, \mathfrak{e}_i\rangle$ is a one-dimensional Brownian motion and the sequence $\langle b, \mathfrak{e}_i\rangle_{i=1,\ldots,d}$ is independent.

Now, what about $\mathbb{B}$-valued Brownian motion? Not only $\mathbb{H}$ is dense in $\mathbb{B}$ under $(|\cdot|_i)_{i\in\mathbb{N}}$, the topological dual $\mathbb{B}'$of $\mathbb{B}$ is dense in $\mathbb{H}' = \mathbb{H}$ under the original norm $\|\cdot\|$ on $\mathbb{H}$ (see Proposition 4.3.6 in [21]). So we have orthogonality in $\mathbb{B}'$. Now a process b with values in $\mathbb{B}$ is a Brownian motion if $\varphi \circ b$ is a one dimensional Brownian motion for each $\varphi \in \mathbb{B}'$ with $\|\varphi\| = 1$ and $\varphi \circ b$ and $\psi \circ b$ are independent for $\varphi \perp \psi$. Note that this notion of infinite-dimensional Brownian motion entirely extends the finite-dimensional case. For more details see Chapter 11 in [21].

In a subsequent article we will study infinite-dimensional Lévy processes in general, using results of this paper. Here are some ideas to Poisson processes: Let $\mu := \pi^H$ be the H-fold product of the measure π in Example (i), μ is concentrated on the set $\{0,1\}^H$ of all H-tuples of 0 and 1 in $\mathfrak{W}$. Since we need that μ is Lévy-limited (see Section 4.1), we introduce a new norm $\|\cdot\|$ on $\{0,1\}^H$ by setting

$$\|x_1, \ldots, x_H\|^2 := \sum_{i=1}^{H} \frac{1}{2^i} x_i^2.$$

Then μ fulfills (LL 1),. . . ,(LL3) with $\|\cdot\|$ instead of $|\cdot|$. In the finite-dimensional case both norms are equivalent. Now

$$^{\circ}B : \left(\{0,1\}^H\right)^T \times [0,\infty[\to \mathbb{N}_0^{\mathbb{N}} \; \widehat{\overline{\mu}}\text{-a.s.}$$

Using nonstandard models of type logic, stochastic analysis including Malliavin calculus for a large class of one-dimensional Lévy processes and for the infinite-dimensional Brownian has been studied in [14, 15, 16, 13, 23, 17, 20, 18, 19, 22], extending Cutland's [4] and Cutland and Ng's [5] approach to one-dimensional Brownian motion.

18.2 Models of Type Logic

Kurt Schütte has made fundamental contributions to syntax and semantics of simple type logic (see Remark 18.4 below). Models of simple type logic has been applied to probability theory in many articles. For simplicity and without any loss of generality, we only have two types, the type 0 of urelements and the type 1 of sets of large complexity.

Let $\mathbb{N}_0$ denote the nonnegative integers. A sequence $V := (V_n)_{n\in\mathbb{N}_0}$ is called a *model* if $V_0 \neq \emptyset$ and $V_{n+1} \subseteq \mathcal{P}(V_0 \cup \cdots \cup V_n)$, where $\mathcal{P}(V_0 \cup \cdots \cup V_n)$ denotes the set of all subset of $V_0 \cup \cdots \cup V_n$. Let us call a model $V := (V_n)_{n\in\mathbb{N}_0}$ a *model over* X, if $V_0 = X$.

The entities in V_0 are called the *urelements* in V, the entities in V_{n+1} are called the *sets* in V. A subset S of $V_0 \cup \cdots \cup V_n$ is called *internal in* V, if S is a set in V, otherwise S is called *external in* V. The urelements are internal entities. Since V_{n+1} may be a strict subset of $\mathcal{P}(V_0 \cup \cdots \cup V_n)$, external sets are admitted. A model is called a *standard model*, if it does not contain external sets.

Instead of adding the type index 0 to the urelements and the type index 1 to the sets, according to the work of Schütte [26] and Robinson [24], we code the urelements in such a way that they never appear as sets in the model. For the proof we refer to Proposition A.2.1 in [21].

Proposition 18.1. (O. [13]) *For each set $Y \neq \emptyset$ there exists a bijection c from Y onto a set X such that the elements of X never appear as sets in any model over X.*

It depends on the theory we have in mind, which objects we take either as urelements or as sets. If we are not interested in the shape of elements of a mathematical entity, then this entity can be chosen as an urelement. For example, in arbitrary Banach spaces the elements may be urelements, but in the Lebesgue L^p-spaces the elements are functions, thus sets. In general, the real numbers are urelements. However, if we want to study real numbers as equivalence classes of Cauchy-sequences of rationals, then real numbers become sets and now the rationals may be chosen to be urelements. However, if we want to study the rationals as equivalence classes of pairs of integers, then ... and so on!

18.2.1 The Formal Language of Mathematics

Syntax and semantics for a model $V := (V_n)_{n\in\mathbb{N}_0}$ is similar to first order logic. The language $\mathcal{L}_V$ contains *logical symbols* $\wedge, \vee, \neg, \exists, \forall, =, \in$ and *variables* $x_1, x_2, \ldots$. The elements of $V_{<\infty} := \bigcup_{n\in\mathbb{N}_0} V_n$ are the *constants* in $\mathcal{L}_V$. We also need *auxiliary symbols* like parentheses " (", ") ", point "·" and comma ",". We assume that all these symbols are distinct.

The *sentences* in $\mathcal{L}_V$ are built up inductively from the prime formulas $(a \in b)$ and $(a = b)$, using the logical symbols, where a, b are constants in $\mathcal{L}_V$. By Proposition 18.2, k-placed relations are definable, if V is closed under definition, in particular, if V is a standard model. The quantifiers $\exists$ and $\forall$ are restricted to the sets of V. For example, fix a sentence A in $\mathcal{L}_V$ and let a, b be constants in $\mathcal{L}_V$, where a is a set in V. If x is a variable, not occurring in A, then $(\exists x \in aA_b(x))$ is a sentence in $\mathcal{L}_V$. Here $A_b(x)$ is the string of signs of the alphabet of $\mathcal{L}_V$ that results from A by replacing each b with x, where b occurs in A. Note that the elements of sets in V may be also sets in V.

A *formula* in $\mathcal{L}_V$ results from a sentence in $\mathcal{L}_V$ by replacing some constant with a variable x, not occurring in the sentence. We then say that x is *free* in the formula. We shall write $A(x)$, to indicate that x is free in the formula A. If there is no risk of confusion, we omit the parentheses "(","')".

Fix a model $V := (V_n)_{n\in\mathbb{N}_0}$. The truth predicate for sentences in $\mathcal{L}_V$ is defined in the usual way. Following a common practice in model theory, we will write $V \models A$ if A is true in V.

18.2.2 Special Properties of Models

We will now study models, where the internal sets have nice closure properties: A model $V := (V_n)_{n\in\mathbb{N}_0}$ is called *closed under definition*, if for each formula $A(x)$ in $\mathcal{L}_V$ and each $n \in \mathbb{N}_0$

$$\{a \in V_0 \cup \cdots \cup V_n \mid V \models A(x)(a)\} \in V_{n+1},$$

where the sentence $A(x)(a)$ results from $A(x)$ by replacing x by a. Obviously, standard models are closed under definition. For the proof of the following result we refer to the proof of Proposition A.5.1 in [21]. Let us use the shorthand $V_{\leq n} := V_0 \cup \cdots \cup V_n$.

Proposition 18.2. *Suppose that V is closed under definition and that the positive integers belong to the urelements in V, coded according to* Proposition 18.1. *Fix* $n \geq 1$. *Then*

(1) $V_1 \subseteq \cdots \subseteq V_n \subseteq \cdots$, *thus* $V_{\leq n} = V_0 \cup V_n$.
(2) *If F is a finite subset of* $V_{\leq n-1}$, *then* $F \in V_n$.
(3) *If* $a, b \in V_{\leq(n-1)}$, *then* $\langle a, b\rangle := \{\{a, b\}, \{b\}\} \in V_{n+1}$.
(4) *Fix* $a_1, \ldots, a_k \in V_{\leq(n-1)}$. *Then* $(a_1, \ldots, a_k) := \{\langle 1, a_1\rangle, \ldots, \langle k, a_k\rangle\} \in V_{n+2}$.
(5) *Fix* $a_1, \ldots, a_k \in V_n$. *Then* $a_1 \cup \cdots \cup a_k \in V_n$ *and* $a_1 \cap \cdots \cap a_k \in V_n$.
(6) *If* $a, b \in V_n$, *then* $a \setminus b \in V_n$.
(7) $\emptyset, V_{\leq(n-1)}, V_{n-1} \in V_n$.
(8) *Fix* $a_1, \ldots, a_k \in V_n$. *Then*

$$a_1 \times \cdots \times a_k = \{(\alpha_1, \ldots, \alpha_k) \mid \alpha_1 \in a_1 \wedge \cdots \wedge \alpha_k \in a_k\} \in V_{n+3}.$$

Fix a model $V = (V_n)_{n\in\mathbb{N}_0}$, which is closed under definition and a second model $W = (W_n)_{n\in\mathbb{N}_0}$. A mapping $* : \bigcup_{n\in\mathbb{N}_0} V_n \to \bigcup_{n\in\mathbb{N}_0} W_n$ is called an *elementary embedding* from V into W, if (E 1) and (E 2) are true, where (E 2) means that V and W have the same formal properties.

(E 1) For each $n \in \mathbb{N}_0$, $^*(V_n) = W_n$.
(E 2) (*Transfer Principle*) For each sentence $A(a_1, \ldots, a_k)$ in $\mathcal{L}_V$ with constants $a_1, \ldots, a_k \in \bigcup_{n\in\mathbb{N}_0} V_n$,

$A(a_1, \ldots, a_k)$ is true in V if and only if $A(^*a_1, \ldots, ^*a_k)$ is true in W, where $^*a$ is a shorthand of $^*(a)$. The Transfer Principle means that the models V and W have the same formal properties.

For the proof of the properties of elementary embeddings we refer to the proof of Proposition A.6.1 in [21]:

Proposition 18.3. *Let $*$ be an elementary embedding from V into W and assume that the positive integers belong to the urlements in V, coded according to* Proposition 18.1. *We obtain for each $n \in \mathbb{N}_0$:*

(1) *The restriction $* \upharpoonright V_n$ maps V_n into W_n, and W_n is a set in W.*
(2) *The map $*$ is injective and a* homomorphism, *that is, for $a, b \in V_{<\infty}$,*

$$a \in b \Leftrightarrow {}^*a \in {}^*b.$$

(3) $*(V_{\leq n}) = W_{\leq n} \in W_{n+1}$.
(4) *W is closed under definition. This is Keisler's Internal Definition Principle.*
(5) *Let $A(x)$ be a formula in $\mathcal{L}_V$. Then*

$$^*\{a \in V_{\leq n} \mid V \models A(x)(a)\} = \{b \in W_{\leq n} \mid W \models (^*(A(x)))(b)\}.$$

(6) *If $\{a_1, \ldots, a_k\}$ is a finite subset of $V_{<\infty}$, then $^*\{a_1, \ldots, a_k\} = \{^*a_1, \ldots, {}^*a_k\}$, in particular, $^*\emptyset = \emptyset$. It follows that $^*(a_1, \ldots, a_k) = (^*a_1, \ldots, {}^*a_k)$ for all $a_1, \ldots, a_k \in V_{<\infty}$.*
(7) *Let $a_1, \ldots, a_k, a$ be sets in V. Let f be a set in V and a mapping from $a_1 \times \cdots \times a_k$ into a. Then $^*f$ is a mapping from $^*a_1 \times \cdots \times {}^*a_k$ into $^*a$ such that for all $(b_1, \ldots, b_k) \in a_1 \times \cdots \times a_k$,*

$$^*(f(b_1, \ldots, b_k)) = {}^*f(^*b_1, \ldots, {}^*b_k)).$$

If, in addition, f is surjective onto a or injective or bijective, then $^*f$ is surjective onto $^*a$, injective or bijective, respectively.

Remark 18.4. Here is the connection to Schütte's work [26]. Takeuti's conjecture, in particular the consistency of analysis, follows from the fact that Schütte's semi valuation can be extended to a total valuation.

Let $* : V \to W$ be an elementary embedding, where V is a standard model. Recall that W has been called standard, if the internal sets are "total". We may call W STANDARD, if $*$ is surjective, thus W does not contain any nonstandard numbers. The following question arises: are "standard" and "STANDARD" equivalent, in particular, does "standard" imply "STANDARD"? In a joint paper together with Christian Hort [7] we have proved that this implication is not true if and only if there exists a large cardinal, larger than the first inaccessible cardinal. Therefore, if "standard" does not imply "STANDARD", then Zermelo-Fraenkel set theory is consistent.

In the following let us fix a standard model $\mathfrak{V} := (\mathfrak{V}_n)_{n \in \mathbb{N}_0}$ such that the real numbers are urelements at least. Let us call a set *small* if its cardinality is not larger than the cardinality κ of $\bigcup_{n\in\mathbb{N}_0} \mathfrak{V}_n$. A non-empty family $\mathcal{B}$ of sets is called *deep* if $\mathcal{B}$ has the *finite intersection property*, which means that the intersection of all finitely many sets in $\mathcal{B}$ is nonempty. Let κ^+ be the smallest cardinal number larger than κ. A model $\mathfrak{W} = (\mathfrak{W}_n)_{n\in\mathbb{N}_0}$, which is closed under definition, is called κ^+*-saturated* if for any $n \geq 1$, and for any deep and small family $(B_i)_{i\in I}$ of sets $B_i \in \mathfrak{W}_n$ the intersection $\bigcap_{i\in I} B_i$ of all the sets B_i is nonempty. It follows that the internal sets in a κ^+-saturated model can be treated as though they were κ^+-compact:

Proposition 18.5. *Assume that $\mathfrak{W}$ is κ^+-saturated. Fix $n > 0$ in $\mathbb{N}$ and an $A \in \mathfrak{W}_n$. Let $(B_i)_{i\in I}$ be a small family of sets $B_i \in \mathfrak{W}_n$ such that $A \subseteq \bigcup_{i\in I} B_i$. Then there exist finitely many $i_1,\ldots,i_k \in I$ with $A \subseteq B_{i_1} \cup \cdots \cup B_{i_k}$. Therefore, internal sets in $\mathfrak{W}$ are finite or of cardinality larger than κ.*

Proof. Assume that the assertion fails. Then the family $(A \setminus B_i)_{i\in I}$ is deep and small. By saturation, there exists an $a \in \bigcap_{i\in I} (A \setminus B_i)$, which contradicts $A \subseteq \bigcup_{i\in I} B_i$. □

18.2.3 The Main Result on Saturation

Here is the main result about κ^+-saturated models. For the proof we refer to the Appendix B in the book [21].

Theorem 18.6. *For each standard model $\mathfrak{V}$ of cardinality κ there exists a κ^+-saturated model $\mathfrak{W}$ and an elementary embedding $*$ from $\mathfrak{V}$ into $\mathfrak{W}$, thus $\mathfrak{W}$ and $\mathfrak{V}$ have the same formal properties.*

The proof of Theorem 18.6 in [21] uses ideas of Sacks' work [25] and is quite elementary. This means that the ambitious existence of a countably incomplete κ^+ good ultrafilter is not used as for example in the proof of Theorem 6.1.8 in the book of Chang and Keisler [3].

The idea of the proof is similar to the proof in algebra that each field k has an algebraically closed extension K. In the algebraic proof one has to find roots of polynomials; here we have to find elements in the intersection of families of internal sets. Roughly speaking, in algebra one constructs a suitable transfinite increasing chain $(k_\alpha)_{\alpha<\gamma}$ of fields k_α with $k_0 = k$ such that, if α is a successor ordinal, then k_α contains one root of each irreducible polynomial over $k_{\alpha-1}$. If α is a limit number, then k_α is the union of the preceding fields. The union of all the k_α provides an algebraically closed extension of k. Here a transfinite elementary chain is constructed; in the successor step we apply the compactness theorem, which says that for each model $\mathfrak{A}$ and each deep family $\mathcal{I}$ of internal sets in $\mathfrak{A}$ there exists a second model $\mathfrak{B}$ and an elementary embedding # from $\mathfrak{A}$ into $\mathfrak{B}$ such that the whole intersection $\bigcap \left\{{}^{\#}I \mid I \in \mathcal{I}\right\}$ is nonempty. In the limit case we take the so called elementary limit of the preceding models.

18.3 Some Properties of the Nonstandard Model

The proofs of the results in this section can be found in almost each book on Nonstandard Analysis. We refer to the book [21], Chapters 8 and 9, because the terminology there corresponds to the terminology here.

We fix $\mathfrak{V}$, $\mathfrak{W}$ and $*$ according to Theorem 18.6 and assume that the real numbers are urelements in $\mathfrak{V}$. Fix $d \in \mathbb{N}$, the dimension of the finite-dimensional Lévy process, we have in mind. If there is no risk of confusion, we omit *, in particular we identify $r \in \mathbb{R}^d$ with $^*r \in {}^*\mathbb{R}^d$.

Since $^*\mathbb{N}$ is an internal set, and thus $\mathbb{N}$, as a countable infinite set, is a strict subset of $^*\mathbb{N}$, there exists an $H \in {}^*\mathbb{N} \setminus \mathbb{N}$. Then $n < H$ for all standard $n \in \mathbb{N}$ and therefore $\frac{1}{H} < \frac{1}{n}$ for all standard $n \in \mathbb{N}$. We have dropped $*$ at $\div$, at $<$ and at n. The number $\frac{1}{H}$ is an example of an infinitesimal number, different from 0. The discrete set

$$T := \left\{\frac{1}{H}, \frac{2}{H}, \ldots, H\right\}$$

can be seen as infinitely fine partition of the continuous interval $[0, \infty[$ and is a finite set in the sense of the nonstandard model $\mathfrak{W}$, T contains H^2 many elements. However, T is an infinite set in the sense of naive set theory, in which we have constructed the models $\mathfrak{V}$ and $\mathfrak{W}$. The elements of T represent time for stochastic processes within the model $\mathfrak{W}$.

Fix $\alpha \in {}^*\mathbb{R}^d$. Then α is called *infinitesimal*, if $|\alpha| := \sqrt{\sum_{i=1}^d \alpha_i^2} < r$ for all standard strictly positive real numbers r; and α is called *limited*, if there exists a standard $n \in \mathbb{N}$ with $|\alpha| < n$. Otherwise, α is called *unlimited*.

Theorem 18.7. (Robinson [24]) *Fix $\alpha \in {}^*\mathbb{R}^d$. Then α is limited if and only if there exists a standard $a \in \mathbb{R}^d$ such that $\alpha \approx a$, i.e., $\alpha - a$ is infinitesimal. Since this a is uniquely determined by α, we call a the* standard part *of α, written*

$$a := {}^\circ\alpha.$$

We need the nonstandard characterization of "open" and "closed": and of "limits".

Proposition 18.8. (Robinson [24])

(a) *A subset $G \subseteq \mathbb{R}^d$ is open iff ${}^\circ\alpha \in G$ implies $\alpha \in {}^*G$ for all limited $\alpha \in {}^*\mathbb{R}^d$.*
(b) *A subset $F \subseteq \mathbb{R}^d$ is closed iff $\alpha \in {}^*F$ implies ${}^\circ\alpha \in F$ for all limited $\alpha \in {}^*\mathbb{R}^d$.*
(c) *Fix $f : \mathbb{N} \to \mathbb{R}^d$ and $a \in \mathbb{R}^d$. Then $\lim_{n\to\infty} f(n) = a$ iff ${}^*f(K) \approx a$ for all unlimited $K \in {}^*\mathbb{N}$.*

The *fixed sample space* for d-dimensional stochastic processes within $\mathfrak{W}$ is $({}^*\mathbb{R}^d)^T$, the set of all internal functions from T into ${}^*\mathbb{R}^d$. The elements of $({}^*\mathbb{R}^d)^T$ represent chance and are internal H^2-tuples of the form $\left(X_{\frac{1}{H}}, \ldots, X_H\right)$ with $X_s \in {}^*\mathbb{R}^d$. It turns out that each standard d-dimensional Lévy process, defined on the standard time line $[0, \infty[$ lives on $({}^*\mathbb{R}^d)^T$, endowed with a suitable probability measure.

The following, perhaps surprising but simple, observation shows how rich the space $\left({}^*\mathbb{R}^d\right)^T$ is. Let $\mathfrak{R}$ be the set of all $X \in \left({}^*\mathbb{R}^d\right)^T$ such that

$$^\circ B(X,\cdot) : [0,\infty[\to \mathbb{R}^d, r \mapsto \lim_{^\circ t \downarrow r, t \in T} {}^\circ(B(X,t) \text{ exists,}$$

where $\lim_{^\circ t\downarrow r}$ denotes the right hand limit at r. Recall from the introduction that $B(X,t) := \sum_{s\leq t, s\in T} X_s$.

Proposition 18.9. *Suppose $f : [0,\infty[\to \mathbb{R}^d$ is right continuous. Then there exists an $X \in \mathfrak{R}$ such that $(^\circ B)(X,\cdot) = f$.*

Proof. Define $X_{\frac{1}{H}} := {}^*f\left(\frac{1}{H}\right)$ and $X_{\frac{i}{H}} := {}^*f\left(\frac{i}{H}\right) - {}^*f\left(\frac{i-1}{H}\right)$ if $i > 1$ and $\frac{i}{H} \in T$. Then $B(X,t) = {}^*f(t)$ for all $t \in T$. Fix $r \in [0,\infty[$. Then,

$$f(r) = \lim_{^\circ t\downarrow r} {}^\circ({}^*f(t)) = \lim_{^\circ t\downarrow r} {}^\circ(B(X,t)) = {}^\circ B(X,r).$$

because for all $t \in T$, ${}^*f(t) \approx f(r)$, if $r \leq t \approx r$. □

The well known *Underspill-* and *Overspill-Principles* (see Proposition 8.4.1 in [21]) show the equivalence of the philosophical terms "potentially infinite" and "actually infinite". More precisely:

Proposition 18.10. *Suppose that A is an internal subset of ${}^*\mathbb{N}$.*

(a) (The Underspill Principle) *Assume that for each unlimited $G \in {}^*\mathbb{N}$ there exists an unlimited $K \in A$ with $K \leq G$. Then A contains arbitrarily large standard positive integers.*
(b) (The Overspill Principle) *Assume that $A \cap \mathbb{N}$ is infinite. Then for each unlimited number $M \in {}^*\mathbb{N}$ there is an unlimited $K \in A$ with $K < M$.*

Now we will present a short introduction to Loeb measures [10] (for details see [21]). Let Λ be an internal nonempty set in $\mathfrak{W}$ and let C be an internal *algebra* on Λ, which means, C is a set of subsets of Λ, closed under finitely many unions and intersections, $\Lambda \in C$ and $A \setminus B \in C$ for $A, B \in C$. Assume that μ is a finitely additive measure in $\mathfrak{W}$, defined on C (i. e. $\mu(\emptyset) = 0$ and $\mu(A \cup B) = \mu(A) + \mu(B)$ for disjoint $A, B \in C$) with values in the limited part of ${}^*[0,\infty[$. By Theorem 18.7, ${}^\circ\mu : C \ni A \mapsto {}^\circ(\mu(A))$ exists and is a finitely additive measure on C. If $(A_n)_{n\in\mathbb{N}}$ is a countable sequence in C such that $\bigcup_{n\in\mathbb{N}} A_n \in C$, then, by Proposition 18.5, there exist finitely many $n_1, .., n_k \in \mathbb{N}$ with $\bigcup_{n\in\mathbb{N}} A_n = A_{n_1} \cup \cdots \cup A_{n_k}$. Therefore, ${}^\circ\mu$ is *σ-additive on C*, which means, if the sets $A_n \in C$ are pairwise disjoint and $A := \bigcup_{n\in\mathbb{N}} A_n \in C$, then ${}^\circ\mu(A) = \sum_{n=1}^{\infty} {}^\circ\mu(A_n)$. By Caratheodory's Extension Theorem, ${}^\circ\mu$ can be extended to a σ-additive measure $\widehat{\mu}$ on the standard σ-algebra $\sigma(C)$, generated by C. The measure on the completion $L_\mu(C)$ of $\sigma(C)$ is also denoted by $\widehat{\mu}$. Note that in $L_\mu(C)$ there exist many external sets. However, each set in $L_\mu(C)$ is equivalent to an internal set, in details:

Theorem 18.11. (P. Loeb [10]) *A set B belongs to $L_\mu(C)$ if and only if there exists a set $A \in C$ such that*

$$\widehat{\mu}(B \,\Delta\, A) = 0,$$

where $B \Delta A$ denotes the symmetric difference $(A \setminus B) \cup (B \setminus A)$ *of A and B. It follows that $\widehat{\mu}(B) = \widehat{\mu}(A) = {}^\circ\mu(A)$.*

The measure space $(\Lambda, L_\mu(C), \widehat{\mu})$ is called the *Loeb space over* (Λ, C, μ). Although Loeb spaces are measure spaces in the usual sense, they neither belong to the standard model nor to the nonstandard model. Loeb spaces are elements of naive set theory, in which we have defined standard and nonstandard models.

Recall from Measure Theory that functions f from Λ into $\mathbb{R}^d$ are called *measurable*, if $f^{-1}[A] := \{x \in \Lambda \mid f(x) \in A\} \in L_\mu(C)$ for all Borel sets $A \subseteq \mathbb{R}^d$. An internal C-measurable function $F : \Lambda \to {}^*\mathbb{R}^d$ is called a *lifting* of $f : \Lambda \to \mathbb{R}^d$ if $\widehat{\mu}\{x \in \Lambda \mid f(x) \not\approx F(x)\} = 0$, which is equivalent to the fact that for all standard $\varepsilon > 0$,

$$\widehat{\mu}\{x \in \Lambda \mid |f(x) - F(x)| \geq \varepsilon\} < \varepsilon.$$

Theorem 18.12. (Anderson [1], Loeb [10]) *The function f is measurable if and only if F has a lifting.*

Liftings are used to convert functions with values in $\mathfrak{V}$ to internal functions with values in $\mathfrak{W}$. In general it is not too difficult to find liftings at least in the context of separable spaces. In [8] there is an example, where liftings do not exist. In general, it is more subtle to convert results from the nonstandard world $\mathfrak{W}$ back to standard mathematics (so called *pushing down* results). The construction of Loeb measures from internal measures and the construction of standard Lévy processes as standard parts of the fixed process in $\mathfrak{W}$ are examples of pushing down results.

There is also a close relationship between nonstandard and standard integrals. An internal C-measurable function $F : \Lambda \to {}^*\mathbb{R}^d$ is called *S-integrable* if for all unlimited $K \in {}^*\mathbb{N}$,

$$\int_{\{|x| \geq K\}} |F| \, d\mu \approx 0.$$

For example, limited C-measurable functions F are S-integrable.

Theorem 18.13. (Anderson [1], Loeb [10]) *The $L_\mu(C)$-measurable function $f : \Lambda \to \mathbb{R}^d$ is integrable if and only if f has an S-integrable lifting $F : \Lambda \to {}^*\mathbb{R}^d$, in which case*

$$\int_\Lambda F d\mu \approx \int_\Lambda f d\widehat{\mu}.$$

18.4 Finite-Dimensional Lévy Processes

We start with Lévy-limited measures in order to produce all standard finite-dimensional *Lévy processes*. These are processes $L : \Lambda \times [0, \infty[\to \mathbb{R}^d$ with $L(\cdot, 0) = 0$

and with independent and identically distributed increments. The most important Lévy processes are Brownian motion and Poisson processes. Let us often use the common shorthand

$$\{\varphi(x)\} := \left\{x \in {}^*\mathbb{R}^d \mid \varphi(x)\right\}.$$

18.4.1 Lévy-Limited Measures on $^*\mathbb{R}^d$

Following the introduction, we fix $d \in \mathbb{N}$ and the space $^*\mathbb{R}^d$. An internal probability measure μ on the internal Borel sets of $^*\mathbb{R}^d$ is called *Lévy-limited*, if

(LL 1) $H \int_{\{|x|\leq\delta\}} |x|^2 \, d\mu$ is limited for all limited $\delta \geq 0$,

(LL 2) $H \left| \int_{\{|x|\leq\delta\}} x d\mu \right|$ is limited for all limited noninfinitesimal $\delta > 0$,

(LL 3) $\lim_{k\to\infty} {}^\circ H \cdot \mu\{k \leq |x|\} = 0$, where ${}^\circ a := \infty$ for unlimited a.

Lindstrøm [12] has introduced these conditions in order to characterize limitedness of internal Lévy processes. Note that all measures defined under (i),. . . ,(viii) in the introduction are Lévy-limited.

Fix an internal Borel set $A \subseteq {}^*\mathbb{R}^d$. Let $\mathbf{1}_A$ be the *indicator function* of A, which means, $\mathbf{1}_A(x) := 1$ if $x \in A$ and $\mathbf{1}_A(x) = 0$, if $x \notin A$. The measure μ^A on $^*\mathbb{R}^d$ is defined as the image measure of μ under the mapping $x \mapsto \mathbf{1}_A(x) \cdot x$ (see Example (viii) in the introduction). Note that

$$\mu^A(D) = \begin{cases} \mu(A \cap D) & \text{if } 0 \notin D, \\ \mu\left((A \cap D) \cup \left({}^*\mathbb{R}^d \setminus A\right)\right) & \text{if } 0 \in D. \end{cases}$$

Here are two simple facts about the measure μ^A.

Proposition 18.14. *If μ is Lévy-limited, then $\mu^{\{\sigma<|x|\}}$ and $\mu^{\{|x|\leq\sigma\}}$ are both Lévy-limited for all limited, noninfinitesimal $\sigma > 0$.*

Proof. The proof is straightforward, using the well known fact that for all Borel measurable functions f in $\mathfrak{W}$, defined on $^*\mathbb{R}^d$,

$$\int_{^*\mathbb{R}^d} f d\mu^A = \int_{^*\mathbb{R}^d} f(\mathbf{1}_A(x) \cdot x) d\mu,$$

if at least one of the integrals exists. □

Now and later on we will use the well known fact that two Borel probability measures μ and ν on $\mathbb{R}^d$ coincide if their *Fourier transformations* are identical, which means that

$$\int_{\mathbb{R}^d} e^{i\langle a,x\rangle} d\mu = \int_{\mathbb{R}^d} e^{i\langle a,x\rangle} d\nu \text{ for all } a \in \mathbb{R}^d.$$

By the Transfer Principle, this result is also true for internal Borel probability measures on ${}^*\mathbb{R}^d$.

Proposition 18.15. *The processes B and $B^A : \left({}^*\mathbb{R}^d\right)^T \times T \to {}^*\mathbb{R}^d$, $(X,t) \mapsto \sum_{s\le t} \mathbf{1}_A(X_s) \cdot X_s$ are the same under $\overline{\mu^A}$ and $\overline{\mu}$, respectively.*

Proof. Fix $t \in T$. It suffices to prove that the image measures $\left(\overline{\mu^A}\right)^{B(\cdot,t)}$ of $\overline{\mu^A}$ under $B(\cdot,t)$ and $(\overline{\mu})^{B^A(\cdot,t)}$ under $B^A(\cdot,t)$ coincide, which means, we have to prove that for all $\mathfrak{a} \in {}^*\mathbb{R}^d$

$$\int_{{}^*\mathbb{R}^d} e^{i\langle a,x\rangle} d(\overline{\mu^A})^{B(\cdot,t)} = \mathbb{E}_{\overline{\mu^A}} e^{i\langle \mathfrak{a},B(\cdot,t)\rangle} = \mathbb{E}_{\overline{\mu}} e^{i\left\langle \mathfrak{a},B^A(\cdot,t)\right\rangle} = \int_{{}^*\mathbb{R}^d} e^{i\langle a,x\rangle} d\,(\overline{\mu})^{B^A(\cdot,t)}.$$

The proof of the second equality uses the independence of both $(X \mapsto X_s)_{s\le t}$ and $(X \mapsto \mathbf{1}_A(X_s) \cdot X_s)_{s\le t}$ under $\overline{\mu^A}$ and $\overline{\mu}$, respectively. □

For example, if $A := \{|x| \le \sigma\}$ for a limited noninfinitesimal $\sigma \in {}^*\mathbb{R}$, then B^A has *limited increments under* $\overline{\mu}$. The next result shows that processes B under Lévy-limited measures can be approximated by processes with limited increments in the following sense.

Theorem 18.16. (Lindstrøm [12]) *Let μ be Lévy-limited. Then for each standard $n \in \mathbb{N}$,*

$$\lim_{r\to\infty} {}^\circ\overline{\mu}\left\{\forall t \in T_n \left(B_t = B_t^{\{|x|\le r\}}\right)\right\} = 1.$$

Proof. Define $q_r := H\mu\{r < |x|\}$. Then

$$\overline{\mu}\left\{\forall t \in T_n \left(B_t = B_t^{\{|x|\le r\}}\right)\right\} =$$

$$\overline{\mu}\{\forall t \in T_n (|X_t| \le r)\} = \mu\{|x| \le r\}^{Hn} = \left(1 - \frac{q_r}{H}\right)^{Hn} \approx e^{-n\cdot q_r},$$

where we have used the nonstandard characterization of limits. By (LL 3), we can choose r such that ${}^\circ e^{-n\cdot q_r}$ is arbitrarily close to 1. □

Theorem 18.17. *Let μ be Lévy-limited. Fix a limited noninfinitesimal $r > 0$ and set $A := \left\{x \in {}^*\mathbb{R}^d \mid |x| \le r\right\}$. Then $\mathbb{E}_{\overline{\mu}}\left(B^A(X,t) - Ht\mathbb{E}_{\mu^A} x\right)^2 d\overline{\mu}$ is limited for all limited $t \in T$.*

Proof. Using the independence of $\left(\mathbf{1}_A(X_s) \cdot X_s - \mathbb{E}_{\mu^A} x\right)_{s\le t}$, we compute

$$\mathbb{E}_{\overline{\mu}}\left|B^A(X,t) - Ht\mathbb{E}_{\mu^A} x\right|^2 = \mathbb{E}_{\overline{\mu}}\left|\sum_{s\le t}\left(\mathbf{1}_A(X_s) \cdot X_s - \mathbb{E}_{\mu^A} x\right)\right|^2 =$$

$$\mathbb{E}_{\overline{\mu}}\left\langle\sum_{s\le t}\left(\mathbf{1}_A(X_s) \cdot X_s - \mathbb{E}_{\mu^A} x\right), \sum_{s\le t}\left(\mathbf{1}_A(X_s) \cdot X_s - \mathbb{E}_{\mu^A} x\right)\right\rangle = \alpha + \beta$$

with

$$\alpha = 2\mathbb{E}_{\overline{\mu}} \sum_{s<\sigma\leq t} \left\langle \mathbf{1}_A(X_s)\cdot X_s - \mathbb{E}_{\mu^A} x, \mathbf{1}_A(X_\sigma)\cdot X_\sigma - \mathbb{E}_{\mu^A} x \right\rangle =$$

$$2\sum_{s<\sigma\leq t} \mathbb{E}_{\overline{\mu}} \left\langle \mathbf{1}_A(X_s)\cdot X_s - \mathbb{E}_{\mu^A} x, \mathbf{1}_A(X_\sigma)\cdot X_\sigma - \mathbb{E}_{\mu^A} x \right\rangle =$$

$$2\sum_{s<\sigma\leq t} \left\langle \mathbb{E}_{\overline{\mu}}\left(\mathbf{1}_A(X_s)\cdot X_s - \mathbb{E}_{\mu^A} x\right), \mathbb{E}_{\overline{\mu}}\left(\mathbf{1}_A(X_\sigma)\cdot X_\sigma - \mathbb{E}_{\mu^A} x\right)\right\rangle = 0,$$

and

$$\beta = \mathbb{E}_{\overline{\mu}} \sum_{s\leq t} \left|\mathbf{1}_A(X_s)\cdot X_s - \mathbb{E}_{\mu^A} x\right|^2 =$$

$$\sum_{s\leq t} \mathbb{E}_{\overline{\mu}} \left|\mathbf{1}_A(X_s)\cdot X_s - \mathbb{E}_{\mu^A} x\right|^2 = Ht\cdot \mathbb{E}_{\mu}\left|\mathbf{1}_A(x)\cdot x - \mathbb{E}_{\mu^A} x\right|^2 =$$

$$Ht\cdot \mathbb{E}_{\mu}\left\langle \mathbf{1}_A(x)\cdot x - \mathbb{E}_{\mu^A} x, \mathbf{1}_A(x)\cdot x - \mathbb{E}_{\mu^A} x\right\rangle^2 =$$

$$Ht\cdot\left(\int_{{}^*\mathbb{R}^d} |x|^2\, d\mu^A - \left|\int_{{}^*\mathbb{R}^d} x d\mu^A\right|^2\right) = Ht\cdot\left(\int_A |x|^2\, d\mu - \left|\int_A x d\mu\right|^2\right)$$

is limited by Proposition 18.14. □

Proposition 18.18. *Let μ be Lévy-limited. Then $H\cdot\mu\{\sigma\leq|x|\}$ is limited for all limited noninfinitesimal $\sigma > 0$.*

Proof. By (LL 3) there exists $k\in\mathbb{N}$ with $k>\sigma$ such that $H\cdot\mu\{k<|x|\}$ is limited. Then

$$H\cdot\mu\{\sigma\leq|x|\leq k\} = H\cdot\int_{\left\{1\leq\frac{1}{\sigma^2}|x|^2\right\}\cap\{|x|\leq k\}} 1 d\mu \leq \frac{H}{\sigma^2}\cdot\int_{\{|x\leq k|\}} |x|^2\, d\mu$$

is limited, thus, $H\cdot\mu\{\sigma\leq|x|\} = H\cdot(\mu\{\sigma\leq|x|\leq k\}+\mu\{k<|x|\})$ is limited. □

18.4.2 Splitting Infinitesimals

Following Lindstrøm [12], we will use splitting infinitesimals in order to separate the continuous part of a Lévy process from its pure jump part.

Proposition 18.19. *Fix a Lévy-limited μ. Then there exists an unlimited $N_\infty\leq\sqrt[4]{H}$ in ${}^*\mathbb{N}$ such that*

(I) $\lim_{n\to\infty} {}^\circ H\int_{\left\{\frac{1}{N_\infty}<|x|\leq\frac{1}{n}\right\}} |x|^2\, d\mu = 0$,

(II) $H\mu\left\{\frac{1}{N_\infty}<|x|\leq N_\infty\right\}<\sqrt[4]{H}$.

Proof. (I) Set $\iota := \lim_{n\to\infty} {}^\circ H\int_{\{|x|\leq\frac{1}{n}\}} |x|^2\, d\mu$. By (LL 1) in 4.1, $i<\infty$. Using the Overspill Principle, there exists an unlimited $N_\infty\leq\sqrt[4]{H}$ in ${}^*\mathbb{N}$ such that $i\approx$

$H \int_{\left\{|x| \leq \frac{1}{N_\infty}\right\}} |x|^2 \, d\mu$. Then

$$^\circ H \int_{\{\frac{1}{N_\infty} < |x| \leq \frac{1}{n}\}} |x|^2 \, d\mu = {}^\circ H \left(\int_{\{|x| \leq \frac{1}{n}\}} |x|^2 \, d\mu - \int_{\{|x| \leq \frac{1}{N_\infty}\}} |x|^2 \, d\mu \right) \to_{n \to \infty} 0.$$

By Proposition 18.18, $H\mu \left\{\frac{1}{n} < |x| \leq n\right\} < \sqrt[4]{H}$ for all $n \in \mathbb{N}$. By the Overspill Principle, we my choose N_∞ such that (II) is also true. □

Lindstrøm has called $\eta := \frac{1}{N_\infty}$ a *splitting infinitesimal for* μ. If

$$\lim_{n \to \infty} {}^\circ H \int_{\{|x| \leq \frac{1}{n}\}} |x|^2 \, d\mu = 0,$$

then μ is called a *pure jump measure*, in which case $^\circ B$ is called a *pure jump process under* μ.

Note that the measures in example (i), (iv), (v), (vi) are pure jump measures, the others not. In the case of Anderson's Brownian motion, $\frac{1}{\sqrt[4]{H}}$ is a splitting infinitesimal. This is also true for Cutland's Brownian motion, because, by Proposition 18.8 (c),

$$\lim_{n \to \infty} {}^\circ H \int_{\left\{H^{-\frac{1}{4}} < |x| \leq \frac{1}{n}\right\}} x^2 \left(\exp(-\frac{H}{2} x^2)\right) dx \sqrt{\frac{H}{2\pi}} =$$

$$\lim_{n \to \infty} \left(\lim_{k \to \infty} k \int_{\left\{k^{-\frac{1}{4}} < |x| \leq \frac{1}{n}\right\}} x^2 \left(\exp(-\frac{k}{2} x^2)\right) dx \sqrt{\frac{k}{2\pi}} \right) = 0.$$

18.4.3 Lévy Measure

Using Lindstrøm's work [12], we construct from the Lévy-limited measure μ a so called *Lévy measure* $^\infty\mu$ by setting for all Borel sets $A \subseteq \mathbb{R}^d$

$$^\infty\mu(A) := \lim_{\varepsilon \downarrow 0} \widehat{H\mu} \left(\left\{ x \in {}^*\mathbb{R}^d \mid \varepsilon < |x| \quad \text{and} \quad {}^\circ x \in A \right\} \right).$$

By Proposition 18.18, the Loeb measure $\widehat{H\mu}$ above is limited on $\{\varepsilon < |x|\}$. In case of Loeb's Poisson process $^\infty\mu(A) = \beta \cdot \mathbf{1}_A(1)$, and $^\infty\mu$ is identically 0 in case of Brownian motion. Note that for μ in Example (vi) in the introduction we have

$$^\infty\mu(A) = |\{i \in \mathbb{N} \mid f(i) \in A\}|,$$

where $|\cdot|$ denotes the number of elements in the set $\cdot$. It follows that the Lévy measure is infinite on any neighborhood of 0. The next result shows that $^\infty\mu$ satisfies the usual properties of Lévy measures.

Theorem 18.20.

(α) ${}^{\infty}\mu\{0\} = 0$.
(β) $\int_{\mathbb{R}^d} \left(|x|^2 \wedge 1\right) d\,{}^{\infty}\mu < \infty$, *where* $a \wedge b := \min\{a, b\}$.

Proof. (α) is obvious. To prove (β), note that for standard $n \in \mathbb{N}$,

$$\int_{\{|x|\leq 1\}} |x|^2 \, d\,{}^{\infty}\mu = \lim_{n\to\infty} \int_{\{\frac{1}{n}\leq|x|\leq 1\}} |x|^2 \, d\,{}^{\infty}\mu \leq$$

$$\lim_{n\to\infty} \int_{{}^*\{\frac{1}{2n}\leq|x|\leq 1+\frac{1}{n}\}} |x|^2 \, d\,\widehat{H\mu} \leq {}^{\circ}\int_{\{x\in {}^*\mathbb{R}^d \mid |x|\leq 2\}} |x|^2 \, d\,H\mu < \infty,$$

by (LL 1). Moreover,

$$\int_{\{1<|x|\}} 1 d\,{}^{\infty}\mu = {}^{\infty}\mu\{1 < |x|\} \leq {}^{\circ}H\mu\left({}^*\left\{\frac{1}{2} < |x|\right\}\right) < \infty,$$

by Proposition 18.18. This proves (β). □

We have the following properties of Lévy measures, in particular, Property (e) is convenient for the following. Fix a *Lévy measure* ρ, i.e., ρ is a Borel measure on $\mathbb{R}^d$, having the properties (α) and (β) with ${}^{\infty}\mu = \rho$. Define

$$\mathfrak{N}(\rho) := \left\{r \in \mathbb{R}, r > 0 \mid \rho\left\{x \in \mathbb{R}^d \mid |x| = r\right\} = 0\right\}.$$

Let us call a number in $\mathfrak{N}(\rho)$ a *nonatom for* ρ. Note that we do not put 0 to $\mathfrak{N}(\rho)$.

Theorem 18.21. *Let* ρ *be a Lévy measure on* $\mathbb{R}^d$. *Then*

(a) $\rho\left\{\frac{1}{n} < |x|\right\} < \infty$ *for all standard* $n \in \mathbb{N}$.
(b) $\int_{\{|x|\leq n\}} |x|^2 \, d\rho(x) < \infty$ *for all* $n \in \mathbb{N}$.
(c) $\int_{\{\frac{1}{n}<|x|\leq n\}} |x| \, d\rho(x) < \infty$ *for all* $n \in \mathbb{N}$.
(d) $\lim_{k\to\infty} \rho(\{k < |x|\}) = 0$.
(e) *For each open nonempty interval* $I :=]a, b[$ *in* $\mathbb{R}$ *with* $0 < a$ *there exists a nonatom* $r \in I$ *for* ρ, *thus,* $\widehat{H\mu}(\{|x| \approx r\})) = 0$ *if* $\rho = {}^{\infty}\mu$.
(f) *There exists an unlimited* $N_\infty \leq \sqrt[4]{H}$ *such that*

$${}^*\rho\left\{\frac{1}{N_\infty} < |x|\right\} < \sqrt[4]{H},$$

in particular, $\frac{1}{H} \cdot {}^*\rho\left\{\frac{1}{N_\infty} < |x|\right\} \approx 0$.

Proof. (a) By Theorem 18.20 Part (β),

$$\rho\left\{\frac{1}{n} < |x|\right\} = \rho\left\{\frac{1}{n} < |x| \leq 1\right\} + \rho\{1 < |x|\} =$$

$$\rho\left\{1 < n^2 |x|^2, |x| \leq 1\right\} + \int_{\{1<|x|\}} 1 d\rho \leq$$

$$n^2 \int_{\{|x|\leq 1\}} |x|^2 \, d\rho + \int_{\{1<|x|\}} 1 d\rho < \infty.$$

(b) By (a),

$$\int_{\{|x|\leq n\}} |x|^2 \, d\rho(x) = \int_{\{|x|\leq 1\}} |x|^2 \, d\rho(x) + \int_{\{1<|x|\leq n\}} |x|^2 \, d\rho(x) \leq$$

$$\int_{\{|x|\leq 1\}} |x|^2 \, d\rho(x) + n^2 \cdot \rho\{1 < |x|\} < \infty.$$

(c) By (a),

$$\int_{\{\frac{1}{n}<|x|\leq n\}} |x| \, d\rho(x) \leq n\rho\left\{\frac{1}{n} < |x|\right\} < \infty.$$

(d) By (a),

$$0 = \lim_{k\to\infty} \left(\rho\{1 < |x|\} - \rho\{1 < |x| \leq k\}\right) = \lim_{k\to\infty} \rho\{k < |x|\}.$$

(e) By (a), each set $A_n = \left\{r \in I \mid \rho(\{|x| = r\}) > \frac{1}{n}\right\}$, $n \in \mathbb{N}$, is finite. Therefore, $\bigcup_{n\in\mathbb{N}} A_n$ is countable. Each $r \in I \setminus \bigcup_{n\in\mathbb{N}} A_n$ satisfies (e).

(f) follows from (a) by applying the Overspill Principle to ${}^*\rho$. □

Here is an application of Part (e) of the preceding result: fix a nonatom r for ${}^\infty\mu$ and for each $k \in \mathbb{N}$ a nonatom $n_k \in]\frac{1}{k+1}, \frac{1}{k}[$ for ${}^\infty\mu$. Define for fixed $a \in \mathbb{R}^d$,

$$f_r : \mathbb{R}^d \to \mathbb{C}, x \mapsto e^{i\langle a,x\rangle} - 1 - \mathbf{1}_{\{x\in\mathbb{R}^d \mid |x|\leq r\}}(x) \cdot i\langle a, x\rangle,$$
$$F_r : {}^*\mathbb{R}^d \to {}^*\mathbb{C}, x \mapsto e^{i\langle a,x\rangle} - 1 - \mathbf{1}_{\{x\in {}^*\mathbb{R}^d \mid |x|\leq r\}}(x) \cdot i\langle a, x\rangle.$$

Using the Taylor formula at 0 up to the second summand we can see that

$$|F_r|(x), |f_r|(x) \leq \frac{1}{2}\langle a, x\rangle^2 \leq \frac{1}{2}|a|^2 |x|^2, \text{ if } |x| \leq r. \tag{18.1}$$

Proposition 18.22. *Let η be a splitting infinitesimal for μ. Fix r in $\mathfrak{N}({}^\infty\mu)$. Then*

$$\int_{\mathbb{R}^d} f_r d\, {}^\infty\mu = {}^\circ\int_{\{x\in {}^*\mathbb{R}^d \mid \eta<|x|\}} F_r dH\mu. \tag{18.2}$$

Proof. Note that F_r is a lifting of f_r on each $A_k := \left\{x \in {}^*\mathbb{R}^d \mid n_k < |x|\right\}$, because

$$\{x \in A_k \mid F_r(x) \not\approx f_r({}^\circ x)\} \subseteq \{x \text{ is unlimited}\} \cup \{|x| \approx r\} \cup \{|x| \approx n_k\},$$

which is a $\widehat{H\mu}$-nullset. Since F_r is limited, F_r is $S_{H\mu}$-integrable on each A_k. We obtain

$$\int_{\mathbb{R}^d} f_r d\,{}^{\infty}\mu = \lim_{k\to\infty} \int_{\{n_k<|x|\}} f_r d\,{}^{\infty}\mu =$$

$$\lim_{k\to\infty} {}^{\circ}\int_{{}^*\{n_k<|x|\}} f_r({}^{\circ}x) d\,\widehat{H\mu} = \lim_{k\to\infty} {}^{\circ}\int_{{}^*\{n_k<|x|\}} F_r(x) d\,H\mu \approx$$

$$\int_{\{n_K<|x|\}} F_r(x) d\,H\mu$$

for some unlimited $K \in {}^*\mathbb{N}$. Since $r > 0$ and standard, we can choose $n \in \mathbb{N}$ with $\frac{1}{n} < r$. Then

$$\left|{}^{\circ}\int_{\{\eta<|x|\le n_K\}} F_r d\,H\mu\right| \le \frac{1}{2}|a|^2\,{}^{\circ}\int_{{}^*\{\eta<|x|\le\frac{1}{n}\}} |x|^2 d\,H\mu \to_{n\to\infty} 0.$$

This proves the desired result. □

18.4.4 The Lévy-Triplets

Using the proof of Theorem 8.1 in Lindstrøm's work [12], we study the Fourier transformation of a Lévy-limited probability measure μ on ${}^*\mathbb{R}^d$. We assume that d-dimensional vectors are columns. Let $\pi_i(\mathfrak{x})$ denote the projection of the vector $\mathfrak{x}$ to the ith component, $i = 1, \ldots, d$. Fix a splitting infinitesimal η for μ and let us use the terminology in the preceding section. Define the $d \times d$-matrix C_η, setting

$$C_\eta := \left(H\int_{\{|\mathfrak{x}|\le\eta\}} \pi_i(\mathfrak{x}) \cdot \pi_j(\mathfrak{x})\, d\mu\right)_{i,j\in\{1,\ldots,d\}}. \tag{18.3}$$

For each $r \in \mathfrak{R}({}^{\infty}\mu)$ set

$$\gamma_r := H\int_{\{|\mathfrak{x}|\le r\}} \mathfrak{x}\, d\mu. \tag{18.4}$$

Note that γ_r and C_η are limited and C_η is nonnegative definite. Now for all $\mathfrak{a} \in {}^*\mathbb{R}^d$ and $r \in \mathfrak{R}({}^{\infty}\mu)$

$$\int_{{}^*\mathbb{R}^d} e^{i\langle\mathfrak{a},\mathfrak{x}\rangle} d\mu = 1 + \int_{{}^*\mathbb{R}^d} \left(e^{i\langle\mathfrak{a},\mathfrak{x}\rangle} - 1 - \mathbf{1}_{\{|\mathfrak{x}|\le r\}}(\mathfrak{x}) \cdot i\langle\mathfrak{a},\mathfrak{x}\rangle + i\left\langle\mathfrak{a},\frac{\gamma_r}{H}\right\rangle\right) d\mu =$$

$$1 + i\left\langle\mathfrak{a},\frac{\gamma_r}{H}\right\rangle + \int_{{}^*\mathbb{R}^d} \left(e^{i\langle\mathfrak{a},\mathfrak{x}\rangle} - 1 - \mathbf{1}_{\{|\mathfrak{x}|\le r\}}(\mathfrak{x}) \cdot i\langle\mathfrak{a},\mathfrak{x}\rangle\right) d\mu = u + v + w + \alpha$$

with $u := 1 + i\left\langle\mathfrak{a},\frac{\gamma_r}{H}\right\rangle$,

$$v := \int_{\{|\mathfrak{x}|\le\eta\}} \left(e^{i\langle\mathfrak{a},\mathfrak{x}\rangle} - 1 - i\langle\mathfrak{a},\mathfrak{x}\rangle\right) d\mu,\quad w := \int_{\{\eta<|\mathfrak{x}|\le r\}} \left(e^{i\langle\mathfrak{a},\mathfrak{x}\rangle} - 1 - i\langle\mathfrak{a},\mathfrak{x}\rangle\right) d\mu,$$

$$\alpha := \int_{\{r<|\mathfrak{x}|\}} \left(e^{i\langle \mathfrak{a},\mathfrak{x}\rangle} - 1\right) d\mu.$$

We expand the integrand of v into its Taylor formula up to the third summand and obtain

$$e^{i\langle \mathfrak{a},\mathfrak{x}\rangle} - 1 - i\langle \mathfrak{a},\mathfrak{x}\rangle = -\frac{\langle \mathfrak{a},\mathfrak{x}\rangle^2}{2} + j(\mathfrak{x}),$$

where $\int_{\{|\mathfrak{x}|\le\eta\}} j dH\mu \approx 0$. Note that

$$\int_{\{|\mathfrak{x}|\le\eta\}} \langle \mathfrak{a},\mathfrak{x}\rangle^2 d\mu = \frac{1}{H}\left\langle C_\eta \cdot \mathfrak{a},\mathfrak{a}\right\rangle.$$

Therefore, we have the following result:

$$\int_{^*\mathbb{R}^d} e^{i\langle \mathfrak{a},\mathfrak{x}\rangle} d\mu \approx 1 + \frac{1}{H}\left(i\langle \mathfrak{a},\gamma_r\rangle - \frac{1}{2}\left\langle C_\eta \cdot \mathfrak{a},\mathfrak{a}\right\rangle + \int_{\{\eta<|\mathfrak{x}|\}} F_r dH\mu\right). \tag{18.5}$$

By Proposition 18.22, $\int_{\mathbb{R}^d} f_r d\,{}^{\infty}\mu \approx \int_{\{\eta<|\mathfrak{x}|\}} F_r dH\mu$. Therefore, in the Fourier transformation of the measure μ there exist three constants in standard terms, a Lévy measure ${}^{\infty}\mu$, a symmetric nonnegative definite matrix ${}^{\circ}C_\eta$ and a sequence $({}^{\circ}\gamma_r)_{r\in\mathfrak{N}({}^{\infty}\mu)}$ of vectors in $\mathbb{R}^d$. These three constants together are called *Lévy-triplets*.

Examples 18.23. The set $\{|x| \le \eta\}$ and the matrix ${}^{\circ}C_\eta$ represent the diffusion part of μ, the set $\{\eta < |x|\}$ and the Lévy measure ${}^{\infty}\mu \equiv \widehat{H\mu}$ represent the jump part of μ. In Example (ii) in the introduction, the Lévy measure is constant 0 and in Example (i) there the 1×1 matrix C_η is 0.

Take the d-fold product of the measures in Examples (ii) or (iii) in the introduction. Then C_η is the $d\times d$ unit matrix, infinitely close to the $d\times d$ unit matrix, respectively.

Now we proceed the other way round, again using Lindstrøm's work [12]. We start with a Lévy triplet, i.e., with a symmetric nonnegative definite matrix C, with a Lévy measure ρ on the Borel sets of $\mathbb{R}^d$ and with a sequence of vectors $(\gamma_r)_{r\in\mathfrak{N}(\rho)}$ in $\mathbb{R}^d$. We will construct an internal Borel probability measure μ on ${}^*\mathbb{R}^d$ and a splitting infinitesimal η, such that

$$C \approx C_\eta,\ \rho = {}^{\infty}\mu \quad \text{and} \quad \gamma_r = H\int_{\{|\mathfrak{x}|\le r\}} \mathfrak{x}\, d\mu \ \text{ for each } r \in \mathfrak{N}(\rho),$$

thus each Lévy triplet can be reproduced by a suitable measure μ.

Since C is symmetric and nonnegative definite, there exists a matrix D with $D \cdot D^T = C$. By Theorem 18.21 (a) and (c) and Overspill, there exists an unlimited $E \le \sqrt[4]{H}$ such that

$${}^*\rho\left\{\frac{1}{E} < |\mathfrak{x}| \le E\right\} \le \sqrt[4]{H} \quad \text{and} \quad \int_{\{\frac{1}{E}<|\mathfrak{x}|\le E\}} |\mathfrak{x}|\, d\,{}^*\rho \le \sqrt[4]{H}. \tag{18.6}$$

The number $\eta := \frac{1}{E}$ plays the role of a splitting infinitesimal for μ. Define

$$A_1 := \left\{ \mathfrak{x} \in {}^*\mathbb{R}^d \mid \eta < |\mathfrak{x}| \leq \frac{1}{\eta} \right\},$$

and define for the internal Borel set $A \subseteq A_1$,

$$\mu(A) := \frac{{}^*\rho(A)}{H} \approx 0.$$

Let $(\mathfrak{e}_i)_{i\in\{1,\ldots,d\}}$ be the canonical basis of $\mathbb{R}^d$. Fix an arbitrary $r \in \mathfrak{R}(\rho)$. In order to extend the measure μ to

$$A_2 := \{a^{+,1}, a^{-,1}, \ldots, a^{+,d}, a^{-,d}\} \subseteq \left\{ \mathfrak{x} \in {}^*\mathbb{R}^d \mid |\mathfrak{x}| \leq \eta \right\},$$

set $\delta := 1 - \mu(A_1) \approx 1$, $\sigma_r := \gamma_r - H \int_{\{\eta<|\mathfrak{x}|\leq r\}} \mathfrak{x} d\mu$ and for $i \in \{1, \ldots, d\}$,

$$a^{+,i} := \sqrt{\frac{d}{\delta H}} \cdot D \cdot \mathfrak{e}_i + \frac{\sigma_r}{\delta H}, \; a^{-,i} := -\sqrt{\frac{d}{\delta H}} \cdot D \cdot \mathfrak{e}_i + \frac{\sigma_r}{\delta H}.$$

Extend the measure μ, first defined on A_1 above, to A_2, setting

$$\mu\left(\left\{a^{+,i}\right\}\right) := \frac{\delta}{2d} =: \mu\left(\left\{a^{-,i}\right\}\right).$$

The number $|A_2|$ of different elements in A_2 is at least 1 and at most $2d$. If, for example, $a^{+,1} = a^{-,1}$ and $|A_2| = 2d - 1$, then $\mu\left(\left\{a^{+,1}\right\}\right) = \frac{\delta}{d}$ and $\mu\left(\left\{a^{+,j}\right\}\right) = \mu\left(\left\{a^{-,j}\right\}\right) = \frac{\delta}{2d}$ for $j > 1$. In any case, μ is a probability measure on $A_1 \cup A_2$.

Theorem 18.24. *The given Lévy-triplet ρ, $C, (\gamma_r)_{r\in\mathfrak{R}(\rho)}$ is reproduced by μ, which means,*

(1) $\rho = {}^{\infty}\mu$ *on the Borel sets on* $\mathbb{R}^d$.
(2) $C \approx C_\eta$.
(3) $\gamma_r = H \int_{\{|\mathfrak{x}|\leq r\}} \mathfrak{x} d\mu$ *for all* $r \in \mathfrak{R}(\rho)$.
(4) μ *is Lévy limited.*

Proof. (1) Fix an open set $G \subseteq \mathbb{R}^d$. In order to prove that $\rho(G) = {}^{\infty}\mu(G)$ it suffices to prove that $\rho(G_n) = {}^{\infty}\mu(G_n)$ for all $n \in \mathbb{N}$, where $G_n = \left\{x \in G \mid \frac{1}{n} < |x| < n\right\}$ There exists an increasing sequence $(U_k)_{k\in\mathbb{N}}$ of open sets U_k such that $G_n = \bigcup U_k = \bigcup \overline{U_k}$, where $\overline{U_k}$ is the closed hull of U_k. By Proposition 18.8 (a), (b),

$$^{\infty}\mu(U_k) = \widehat{H\mu}\left\{x \in {}^*\mathbb{R}^d \mid {}^{\circ}x \in U_k\right\} \leq {}^{\circ}H\mu\,({}^*U_k) =$$

$$^{\circ}\,({}^*\rho)\,({}^*U_k) = \rho\,(U_k),$$

$$^{\infty}\mu(\overline{U_k}) = \widehat{H\mu}\left\{{}^{\circ}x \in \overline{U_k}\right\} \geq {}^{\circ}H\mu\left({}^*\overline{U_k}\right) = {}^{\circ}({}^*\rho)\left({}^*\overline{U_k}\right) = \rho\left(\overline{U_k}\right).$$

It follows that ${}^{\infty}\mu(G_n) = \rho\,(G_n)$.

(2) For the reader's and my own convenience, choose $d = 2$ and let $D = \begin{pmatrix} s & t \\ u & v \end{pmatrix}$.
Recall that $D \cdot D^T = C = \begin{pmatrix} s^2+t^2 & su+tv \\ su+tv & u^2+v^2 \end{pmatrix}$. Then

$$C_\eta = \begin{pmatrix} H\int_{A_2} \pi_1^2 d\mu & H\int_{A_2} \pi_1 \cdot \pi_2 d\mu \\ H\int_{A_2} \pi_1 \cdot \pi_2 d\mu & H\int_{A_2} \pi_2^2 d\mu \end{pmatrix} =$$

$$\begin{pmatrix} s^2+t^2+\frac{(\pi_1(\sigma_r))^2}{Hq} & su+tv+\frac{\pi_1(\sigma_r)\cdot\pi_2(\sigma_r)}{Hq} \\ su+tv+\frac{\pi_1(\sigma_r)\cdot\pi_2(\sigma_r)}{Hq} & u^2+v^2+\frac{(\pi_2(\sigma_r))^2}{Hq} \end{pmatrix} \approx C, \text{ using (18.6).}$$

(3) Note that $H\int_{\{|x|\leq r\}} x d\mu = H\left(\int_{\{|x|\leq\eta\}} x d\mu + \int_{\{\eta<|x|\leq r\}} x d\mu\right) = \gamma_r$.

(4) (LL 1) and (LL 3) follow from Theorem 18.21 Part (b) and (d), respectively. To prove (LL 2), fix a noninfinitesimal $\sigma > 0$. Then

$$H\int_{\{|x|\leq s\}} x d\mu = H\int_{\{|x|\leq r\}} x d\mu \begin{cases} +H\int_{\{r<|x|\leq s\}} x d\mu, & \text{if } r \leq s, \\ -H\int_{\{s<|x|\leq r\}} x d\mu, & \text{if } s \leq r. \end{cases}$$

The second summands are limited by Theorem 18.21 Part (c). By (18.3), the first summand is equal to the limited γ_r. □

18.4.5 Standard Lévy Processes

Fix a Lévy-limited measure μ. We sketch the proof of the existence of the standard part ${}^\circ B : \left({}^*\mathbb{R}^d\right)^T \times [0,\infty[\to \mathbb{R}^d$ of B under μ. The techniques for finding standard parts of internal processes are well established by the work of Loeb [10], Anderson [1], Keisler [9], Lindstrøm [11], Hoover and Perkins [6]. First assume that B has limited increments. Note that $M := \left(B(X,t) - Ht\mathbb{E}_\mu x\right)_{t\in T}$ is a martingale. By Theorem 18.17, the standard part ${}^\circ M : \left({}^*\mathbb{R}^d\right)^T \times [0,\infty[\to \mathbb{R}^d$ of M exists (see Corollary 10.16.3 in [21]). We have for $\widehat{\overline{\mu}}$-almost all $X \in \left({}^*\mathbb{R}^d\right)^T$,

$${}^\circ M(X,r) := \lim_{{}^\circ s \downarrow r} {}^\circ(M(X,s)) \text{ for all } r \in [0,\infty[.$$

It is easy to see that the standard ${}^\circ B$ of B also exists. Since arbitrary B can be approximated by processes with limited increments, according to Theorem 18.16, the standard part of B exists. We obtain:

Theorem 18.25. *The standard ${}^\circ B$ exists $\widehat{\overline{\mu}}$-a.s. Moreover, it can be shown that ${}^\circ B$ is right continuous and has left hand limits $\widehat{\overline{\mu}}$-a.s and is a Lévy process.*

In order to compute the Fourier transformation of ${}^\circ B$ at each $r \in [0,\infty[$, we need

Lemma 18.26. *Fix $r \in [0,\infty[$ and $t \in T$ with $t \approx r$. Then $B(\cdot,t)$ is a lifting of $({}^{\circ}B)(\cdot,r)$.*

Proof. Let $(t_k)_{k\in\mathbb{N}}$ be a sequence of limited elements in T with ${}^{\circ}t_k > r$ and $\lim_{k\to\infty} {}^{\circ}t_k = r$. It follows that $\lim_{k\to\infty} {}^{\circ}(B(\cdot,t_k)) = ({}^{\circ}B)(\cdot,r)$ $\widehat{\overline{\mu}}$-a.s. Therefore, for all standard $\varepsilon > 0$,

$$\lim_{k\to\infty} \widehat{\overline{\mu}}\{X \mid {}^{\circ}(B(X,t_k)) - ({}^{\circ}B)(X,r)| \geq \varepsilon\} = 0.$$

Note that it suffices to prove that for all standard $\varepsilon > 0$,

$$\overline{\mu}\{X \mid B(X,t_k) - B(X,t)| \geq \varepsilon\} < \varepsilon$$

for small $t_k - t$. First assume that B has limited increments, in other words, $\mu\{|x| > S\} = 0$ for some noninfinitesimal limited S. Then, similar to the computation in the proof of Theorem 18.17,

$$\overline{\mu}\{|B(X,t_k) - B(X,t)| \geq \varepsilon\} = \overline{\mu}\left\{\frac{1}{\varepsilon^2}|B(X,t_k) - B(X,t)|^2 \geq 1\right\} \leq$$
$$\frac{1}{\varepsilon^2}\mathbb{E}_{\overline{\mu}}|B(X,t_k) - B(X,t)|^2 =$$
$$\frac{1}{\varepsilon^2}\left(H(t_k - t)\,\mathbb{E}_{\mu}|x|^2 + H(t_k - t)(H(t_k - t) - 1)\left|\mathbb{E}_{\mu}x\right|^2\right) < \varepsilon$$

for small $t_k - t$. For general B fix $S \in \mathbb{N}$ and set $A := \{|x| \leq S\}$ and $B^S := B^A$. Then by Theorem 18.16,

$$\overline{\mu}\{|B_{t_k} - B_t)| \geq \varepsilon\} \leq$$
$$\overline{\mu}\left\{|B_{t_k} - B^S_{t_k})| \geq \frac{\varepsilon}{3}\right\} + \overline{\mu}\left\{|B^S_{t_k} - B^S_t)| \geq \frac{\varepsilon}{3}\right\} + \overline{\mu}\left\{|B^S_t - B_t)| \geq \frac{\varepsilon}{3}\right\}$$

is smaller than ε for small $t_k - t$ and large S. □

Using the Fourier transformation of the Lévy-limited measure μ, in particular equation (18.5), and the preceding lemma, we compute the Fourier transformation of B at $r \in [0,\infty[$, following again Lindstrøm's work [12]. The result is the standard Lévy-Khintchine formula.

Theorem 18.27. *Fix $\mathfrak{a} \in \mathbb{R}^d$ and $u \in \mathfrak{N}({}^{\infty}\mu)$. Then*

$$\mathbb{E}_{\widehat{\overline{\mu}}}e^{i\langle a,({}^{\circ}B)_r\rangle} = \exp\left(ir\langle \mathfrak{a}, {}^{\circ}\gamma_u\rangle - \frac{r}{2}\left\langle {}^{\circ}C_{\eta}\mathfrak{a},\mathfrak{a}\right\rangle + r\int_{\mathbb{R}^d} f_u d\,{}^{\infty}\mu\right).$$

Proof. Again using the independence of $(X \mapsto X_s)_{s\leq t}$,

$$\mathbb{E}_{\widehat{\overline{\mu}}}e^{i\langle a,({}^{\circ}B)_r\rangle} \approx \mathbb{E}_{\overline{\mu}}e^{i\langle a,B_t\rangle} = \mathbb{E}_{\overline{\mu}}e^{i\left\langle a,\Sigma_{s\leq t}X_s\right\rangle} = \mathbb{E}_{\overline{\mu}}\prod_{s\leq t}e^{i\langle a,X_s\rangle} =$$

$$\prod_{s\leq t}\mathbb{E}_{\overline{\mu}}e^{i\langle a,X_s\rangle} = \prod_{s\leq t}\mathbb{E}_{\mu}e^{i\langle a,x\rangle} = (\mathbb{E}_{\mu}e^{i\langle a,x\rangle})^{Ht} =$$

$$\left(1+\frac{1}{H}\left(i\left\langle \mathfrak{a},\gamma_u\right\rangle-\frac{1}{2}\left\langle C_\eta\cdot\mathfrak{a},\mathfrak{a}\right\rangle+\int_{\{\eta<|x|\}}F_u dH\mu\right)\right)^{Ht}\approx$$

$$\exp\left(ir\left\langle \mathfrak{a},{}^\circ\gamma_u\right\rangle-\frac{r}{2}\left\langle {}^\circ C_\eta\mathfrak{a},\mathfrak{a}\right\rangle+r\int_{\mathbb{R}^d}f_u d\,{}^\infty\mu\right).$$

□

Acknowledgements I am grateful to Gregor Svindland for many helpful discussions.

References

1. R. M. Anderson. A non-standard representation for Brownian motion and Itô integration. *Israel J. Math.*, 25:15–46, 1976.
2. J. Bertoin. *Lévy processes*. Cambridge University Press, Cambridge, 1996.
3. C. C. Chang and H. J. Keisler. *Model theory*. North-Holland, Amsterdam, 1973.
4. N. Cutland. Infinitesimals in action. *J. Lond. Math. Soc.*, 35:202–216, 1987.
5. N. Cutland and N. Siu-Ah. A nonstandard approach to the Malliavin calculus. In S. Albeverio, W. A. J. Luxemburg, and M. P. H. Wolff, editors, *Advances in Analysis, Probability and Mathematical Physics — Contributions of Nonstandard Analysis*, pages 149–170. Kluwer Acad. Publ., Dordrecht, 1995.
6. D. L. Hoover and E. A. Perkins. Nonstandard construction of the stochastic integral and applications to stochastic differential equations I and II. *Trans. Amer. Math. Soc.*, 275, 1983.
7. C. Hort and H. Osswald. On nonstandard models in higher order logic. *J. of Symbolic Logic*, 49, 1984.
8. H.Osswald. A note on liftings of linear continuous functionals. *Proc. Amer. Math. Soc.*, 120, 1994.
9. H. J. Keisler. An infinitesimal approach to stochastic analysis. *Memoirs Amer. Math. Soc.*, 48, 1984.
10. P. A. Loeb. Conversion from nonstandard to standard measure spaces and applications in probability theory. *Trans. Amer. Math. Soc.*, 211:113–122, 1975.
11. T. Lindstrøm. Hyper-finite stochastic integration I, II, III, and addendum. *Math. Scand.*, 46:265–333, 1980.
12. T. Lindstrøm. Hyperfinite Lévy processes. *Stochastics*, 76(6):517–548, 2004.
13. H. Osswald. The existence of poly-saturated models. In M. Wolff P. Loeb, editor, *Nonstandard Analysis for the working mathematician*. Kluwer Scientific, 2000.
14. H. Osswald. Infinitesimals in abstract Wiener spaces. In F. Gesztesy et al., editor, *Stochastic Processes, Physics and Geometry: New Interplays, II (Leipzig, 1999): A Volume in Honor of Sergio Albeverio*, volume 29 of *CMS Conf. Proc.*, pages 539–546, Providence, RI, 2000. Amer. Math. Soc.
15. H. Osswald. Malliavin calculus in abstract Wiener spaces using infinitesimals. *Advances in Mathematics*, 176:1–37, 2003.
16. H. Osswald. On the Clark-Ocone formula for the abstract Wiener space. *Advances in Mathematics*, 176:38–52, 2003.
17. H. Osswald. Malliavin calculus on product measures of $\mathbb{R}^\varkappa$ based on chaos. *Stochastics*, 77(6):501–514, 2005.
18. H. Osswald. Malliavin calculus on extensions of abstract Wiener spaces. *Journal of Mathematics of the Kyoto University*, 8(2):239–262, 2008.
19. H. Osswald. On anicipative Girsanov transformations. *Journal of Theoretical Probability*, 22:474–481, 2009.
20. H. Osswald. A smooth approach to Malliavin calculus for Lévy processes. *Journal of Theoretical Probability*, 22:441–473, 2009.

21. H. Osswald. Malliavin calculus for Lévy processes and infinite dimensional Brownian motion. *Cambridge Tracts in Mathematics*, 191, 2012.
22. H. Osswald. Computation of the kernels of Lévy functionals and applications. *Illinois Journal of Mathematics*, 55(3), Fall 2011.
23. H. Osswald and J.-L. Wu. On infinite-dimensional continuous Ornstein-Uhlenbeck processes. *Acta Appl. Math.*, 83:289–312, 2004.
24. A. Robinson. *Non-standard Analysis*. North-Holland, 1966.
25. J. Sacks. *Saturated Model theory*. W.A. Benjamin, 1972.
26. K. Schütte. Syntactical and semantical properties of simple type logic. *J. of Symbolic Logic*, 25, 1960.

Chapter 19
On the Strength of the Uniform Fixed Point Principle in Intuitionistic Explicit Mathematics

Michael Rathjen and Sergei Tupailo

Abstract The paper is concerned with a line of research that plumbs the scope of constructive theories. The object of investigation here is Feferman's intuitionistic theory of explicit mathematics augmented by the monotone fixed point principle which asserts that every monotone operation on classifications (Feferman's notion of set) possesses a least fixed point. To be more precise, the new axiom not merely postulates the existence of a least solution, but, by adjoining a new functional constant to the language, it is ensured that a fixed point is uniformly presentable as a function of the monotone operation. The strength of the classical non-uniform version, **MID**, was investigated in [6] whereas that of the uniform version was determined in [16, 17] and shown to be that of subsystems of second order arithmetic based on $\mathbf{\Pi}^1_2$-comprehension. This involved a rendering of $\mathbf{\Pi}^1_2$-comprehension in terms of fixed points of non-monotonic Π^1_1-operators and a proof-theoretic interpretation of the latter in specific operator theories that can be interpreted in explicit mathematics with the uniform monotone fixed point principle. The intent of the current paper is to show that the same strength obtains when the underlying logic is taken to be intuitionistic logic.

19.1 Introduction

This paper continues research (cf. [5, 2, 20, 6, 15, 16, 17, 18, 22]) addressing the status of monotone inductive definitions in the general constructive setting of Feferman's explicit mathematics [3, 4], called $\mathbf{T_0}$. It has a strong bearing on the problem of determining the limits of what is constructively justifiable that was of great interest to logicians ever since the 1960s (cf. [10]).

M. Rathjen · S. Tupailo
School of Mathematics, University of Leeds, Leeds, LS2 9JT, UK, e-mail: M.Rathjen@leeds.ac.uk, sergei@cs.ioc.ee

R. Kahle and M. Rathjen (eds.), *The Legacy of Kurt Schütte*,
https://doi.org/10.1007/978-3-030-49424-7_19

The question of the strength of systems of explicit mathematics with fixed point principles **MID** and **UMID** was raised by Feferman in [5]; we quote:

> *What is the strength of* $\mathbf{T_0}$ + **MID**? *[...] I have tried, but did not succeed, to extend my interpretation of* $\mathbf{T_0}$ *in* $\Sigma^1_2 - AC + BI$ *to include the statement* **MID**. *The theory* $\mathbf{T_0}$ + **MID** *includes all constructive formulations of iteration of monotone inductive definitions of which I am aware, while* $\mathbf{T_0}$ *(in its* IG *axiom) is based squarely on the general iteration of accessibility inductive definitions. Thus it would be of great interest for the present subject to settle the relationship between these theories.* (p. 88)

We are particularly interested in the intuitionistic strength of the axiom $\mathbf{UMID}_{\mathbb{N}}$ which postulates the existence of a least fixed point for any monotone operation f on subsets of the natural numbers, where a least solution $\mathbf{lfp}(f)$ is presented as a function of the operation by adjoining a new constant **lfp** to the language of $\mathbf{T_0}$. To relate the state of the art in these matters we shall need some terminology. Below we shall distinguish between the classical and the intuitionistic version of a theory by appending the superscript c and i, respectively. For a system S of explicit mathematics we denote by $S\restriction$ the version wherein the induction principles for the natural numbers and for inductive generation are restricted to sets. $\mathbf{IND}_{\mathbb{N}}$ stands for the schema of induction on natural numbers for arbitrary formulas of the language of explicit mathematics. $(\mathbf{\Pi}^1_2\text{-}\mathbf{CA})_0$ denotes the subsystem of second order arithmetic (based on classical logic) with $\mathbf{\Pi}^1_2$-comprehension but with induction restricted to sets, whereas $(\mathbf{\Pi}^1_2\text{-}\mathbf{CA})$ also contains the full schema of induction on $\mathbb{N}$.

The papers [16, 17] yielded the following results:

Theorem 19.1. (i) $(\mathbf{\Pi}^1_2\text{-}\mathbf{CA})_0$ *and* $\mathbf{T}^c_0\restriction + \mathbf{UMID}_{\mathbb{N}}$ *have the same proof-theoretic strength.*

(ii) $(\mathbf{\Pi}^1_2\text{-}\mathbf{CA})$ *and* $\mathbf{T}^c_0\restriction + \mathbf{IND}_{\mathbb{N}} + \mathbf{UMID}_{\mathbb{N}}$ *have the same proof-theoretic strength.*

The first result about $\mathbf{UMID}_{\mathbb{N}}$ on the basis of intuitionistic explicit mathematics was obtained by the second author in [22].

Theorem 19.2. $(\mathbf{\Pi}^1_2\text{-}\mathbf{CA})_0$ *and* $\mathbf{T}^i_0\restriction + \mathbf{UMID}_{\mathbb{N}}$ *have the same proof-theoretic strength.*

[22] uses a characterization of $(\mathbf{\Pi}^1_2\text{-}\mathbf{CA})_0$ via a classical μ-calculus (a theory which extends the concept of an inductive definition), dubbed $\mathbf{ACA}_0(\mathcal{L}^\mu)$, given by Möllerfeld [12] and then proceeds to show that $\mathbf{ACA}_0(\mathcal{L}^\mu)$ can be interpreted in its intuitionistic version, $\mathbf{ACA}^i_0(\mathcal{L}^\mu)$, by means of a double negation translation. Finally, as the latter theory is readily interpretable in $\mathbf{T}^i_0\restriction + \mathbf{UMID}_{\mathbb{N}}$, the proof-theoretic equivalence stated in Theorem 19.2 follows in view of Theorem 19.1.

The proof of [22], however, does not readily generalize to $\mathbf{T}^i_0\restriction + \mathbf{IND}_{\mathbb{N}} + \mathbf{UMID}_{\mathbb{N}}$ and extensions by further induction principles. The main reason for this is that adding induction principles such as induction on natural numbers for all formulas to $\mathbf{ACA}_0(\mathcal{L}^\mu)$ only slightly increases the strength of the theory. It is obvious from the results of [16] that in order to arrive at a μ-calculus of the strength of $(\mathbf{\Pi}^1_2\text{-}\mathbf{CA})$ one has to allow for transfinite nestings of the μ-operator for any ordinal $\alpha < \varepsilon_0$. By engineering a double negation translation in a similar vein as in [22], we will be able to conclude the following result.

Theorem 19.3. (i) $(\mathbf{\Pi^1_2\text{-CA}})_0$ *and* $\mathbf{T}_0^i \restriction +\mathbf{UMID}_{\mathbb{N}}$ *have the same proof-theoretic strength.*

(ii) $(\mathbf{\Pi^1_2\text{-CA}})$ *and* $\mathbf{T}_0^i \restriction +\mathbf{IND}_{\mathbb{N}} + \mathbf{UMID}_{\mathbb{N}}$ *have the same proof-theoretic strength.*

Through Theorem 19.3 we get another proof of Theorem 19.2 (which also does not hinge upon [12]).

Finally, it's worth mentioning that the same results could be obtained by subjecting the operator theories $\mathbf{T}^{\mathbf{OP}}_{<\omega}$ and $\mathbf{T}^{\mathbf{OP}}_{<\varepsilon_0}$ to a double negation interpretation. Moreover, this translation works for extensions of Theorem 19.3(ii) where one allows for transfinite nestings of the μ-operator as long as the ordinals come from a primitive recursive ordinal representation system.

19.2 Fixed Point Theories

We consider different frameworks for expressing the existence of fixed point of operators.

19.2.1 The μ-calculus

The μ-calculus extends the concept of an inductive definition. It is basically an algebra of monotone functions over the power class of the domain of a first order structure (or over a complete lattice), whose basic constructors are first order definable operators, functional composition and least and greatest fixed point operators. The μ-calculus arose from numerous works of logicians and computer scientists. It originated with Scott and DeBakker [19] and was developed by Hitchcock and Park [8], Park [13], Kozen [9], Pratt [14], and others (see [1]). The μ-calculus is used in verification of computer programs and provides a tool box for modelling a variety of phenomena, from finite automata to alternating automata on infinite trees and infinite games with finitely presentable winning conditions. Here we will be interested in the μ-calculus over the natural numbers. The μ-definable sets over the natural numbers were first described by Lubarsky [11]. He determined their complexity in the constructible hierarchy and showed that their ordinal ranks in that hierarchy can reach rather large countable ordinals. In the following we denote by $\mathbf{ACA}_0(\mathcal{L}^\mu)$ an axiomatic theory whose language is an extension of that of the classical μ-calculus over $\mathbb{N}$, $\mathcal{L}^\mu$ (see [11]), by set quantifiers and comprehension for first-order properties. This version was formalized in [12]. The letters "*ACA*" stand for "arithmetic comprehension axiom" and the subscript 0 indicates that the induction principle on natural numbers holds for sets rather than arbitrary classes. The name "$\mathbf{ACA}_0(\mathcal{L}^\mu)$" for this theory is somewhat misleading as its comprehension axioms allow for the formation of non-arithmetic sets. However, we will stick to this notation for 'historical' reasons.

Definition 19.4. *The language of* $\mathbf{ACA}_0(\mathcal{L}^\mu)$ *builds on the language of Peano arithmetic,* **PA**. *It has variables* $x, y, z, \ldots, X, Y, Z, \ldots$ *ranging over numbers and sets of numbers, respectively. The terms of* **PA** *will be referred to as number terms.* Number terms, set terms *and* formulas *of the language* $\mathcal{L}^\mu$ *are defined as follows.*

1. *The terms of* **PA** *are* number terms *of* $\mathcal{L}^\mu$.
2. *Set variables are* set terms.
3. $\bot$ *is a* formula.
4. *If* s *and* t *are* number terms *then* $s = t$ *is a* formula.
5. *If* s *is a* number term *and* S *is a* set term *then* $s \in S$ *is a* formula.
6. *If* φ_0 *and* φ_1 *are* formulas *then* $\varphi_0 \wedge \varphi_1$, $\varphi_0 \vee \varphi_1$ *and* $\varphi_0 \to \varphi_1$ *are* formulas.
7. *If* ψ *is a* formula *then* $\forall x \psi$ *and* $\exists x \psi$ *are* formulas.
8. *If* ψ *is a* formula *then* $\forall X \psi$ *and* $\exists X \psi$ *are* formulas.
9. *If* φ *is an* X*-positive first-order* formula *then* $\mu x X.\varphi$ *is a* set term.

In the definition above we call a formula *first-order* if it does not contain set quantifiers $\exists X, \forall X$. For X a set variable an expression $\mathfrak{E}$ is said to be X*-positive* (X*-negative*) if every occurrence of X in $\mathfrak{E}$ is positive (negative). In classical logic we can restrict ourselves to the connectives $\neg, \wedge, \vee$ and then X is positive in a formula φ if every occurrence of X in φ is in the scope of an even number of negations. But as we shall also be concerned with the intuitionistic μ-calculus, we define this notion inductively as follows:

(1) X is X*-positive*; (2) Y is both X*-positive* and X*-negative* if Y is a set variable different from X; (3) $\bot$ and $s = t$ are also both X*-positive* and X*-negative*; (4) $s \in S$ is X*-positive (-negative)* iff S is; (5) polarity does not change with $\wedge$, $\vee$, quantifiers and the μ-symbol; (6) and, finally, $\varphi_0 \to \varphi_1$ is X*-positive (-negative)* iff φ_0 is X*-negative (-positive)* and φ_1 is X*-positive (-negative)*.

For set terms S, T, $S \subseteq T$ is the formula $\forall x(x \in S \to x \in T)$.

Definition 19.5. *The axioms of* $\mathbf{ACA}_0(\mathcal{L}^\mu)$ *are the following:*

1. *The axioms of* **PA**.
2. *(Induction)* $\forall X \big(0 \in X \ \wedge \ \forall u(u \in X \to u + 1 \in X) \ \to \ \forall u \, u \in X\big)$.
3. *(First-order comprehension)* $\exists Z \forall x[x \in Z \leftrightarrow \varphi(x)]$ *for every first-order formula* φ *in which the set variable* Z *does not appear free.*
4. *(Least fixed point axiom)*

$$\forall x[x \in P \leftrightarrow \varphi(x, P)] \ \wedge \ \forall Y\big[\forall x\big(\varphi(x, Y) \to x \in Y\big) \to P \subseteq Y\big] \tag{19.1}$$

where P *stands for the set term* $\mu x X.\varphi$.

$\mathbf{ACA}_0(\mathcal{L}^\mu)$ *is based on classical logic. The system with the underlying logic changed to intuitionistic logic will be denoted by* $\mathbf{ACA}_0^i(\mathcal{L}^\mu)$.

The theories with the full induction scheme **IND** *will be denoted by* $\mathbf{ACA}(\mathcal{L}^\mu)$ *and* $\mathbf{ACA}^i(\mathcal{L}^\mu)$*, respectively.* **IND** *is the scheme*

$$\psi(0) \ \wedge \ \forall x[\psi(x) \to \psi(x + 1)] \ \to \ \forall x \psi(x)$$

for all formulas ψ.

That X is positive (negative) in ψ will be notated by $\psi(X^+)$ ($\psi(X^-)$). Positivity is a guarantor of monotonicity, while negativity guarantees anti-monotonicity.

Lemma 19.6. *For every X-positive formulas $\psi(X^+)$ and and every X-negative formula $\theta(X^-)$ of $\mathbf{ACA}_0(\mathcal{L}^\mu)$ we have:*

(i) $\mathbf{ACA}_0^i(\mathcal{L}^\mu) \vdash \forall X \forall Y\, [X \subseteq Y \ \wedge\ \psi(X) \ \rightarrow\ \psi(Y)]$.
(ii) $\mathbf{ACA}_0^i(\mathcal{L}^\mu) \vdash \forall X \forall Y\, [X \subseteq Y \ \wedge\ \theta(Y) \ \rightarrow\ \theta(X)]$.

Proof. Use induction on the complexity of the formulas. □

At first blush, the μ-calculus appears to be innocent enough. Though a first order formula $\varphi(X^+, x)$ may contain complicated μ-terms, it might seem that these act solely as parameters and therefore one could obtain $\mu x X.\varphi(X^+, x)$ via an ordinary first order arithmetic inductive definition in these parameters, so that all the μ-definable sets would turn out to be sets recursive in finite iterations of the hyperjump. But this is far from being true. The μ-calculus allows for nestings of least fixed point operators. Better yet, there can be feedback. This provides the major difficulty in understanding the expressive power of $\mathcal{L}^\mu$. To illustrate the complexity of nested set terms in $\mathcal{L}^\mu$, let $\theta(X^+, Y^-, Z^+, W^-)$ be a first order formula of $\mathcal{L}^\mu$. Then the following are set terms: $\mu z Z.\theta$, $\mu y Y.\, w \notin \mu z Z.\theta$, $\mu x X.\mu y Y.\, w \notin \mu z Z.\theta$, $\mu w W.\mu x X.\mu y Y.\, w \notin \mu z Z.\theta$.

In the μ-calculus one can also define the *greatest fixed point* constructor ν: If $\varphi(X^+, x)$ is first order, $\nu x X.\varphi(X^+, x)$ is $\{u \mid u \notin \mu x X.\neg\varphi(\neg X, x)\}$. The appropriate measure for the complexity of μ-terms was determined by Lubarsky [11]. μ and ν can be viewed as higher order quantifiers giving rise to complexity classes Σ_n^μ and Π_n^μ of $\mathcal{L}^\mu$ formulas which measure the alternations of μ and ν.

The pivotal proof-theoretic connection between $\mathbf{ACA}_0(\mathcal{L}^\mu)$ and $\mathbf{ACA}_0^i(\mathcal{L}^\mu)$ was established by Tupailo.

Theorem 19.7. (Tupailo) *$\mathbf{ACA}_0(\mathcal{L}^\mu)$ can be interpreted in $\mathbf{ACA}_0^i(\mathcal{L}^\mu)$ via a double negation translation.*

Proof. [22]. □

19.2.2 Fragments of Second Order Arithmetic

The proof-theoretic strength of theories is commonly calibrated using standard theories and their canonical fragments. In classical set theory this linear line of consistency strengths is couched in terms of large cardinal axioms while for weaker theories the line of reference systems traditionally consist in second order arithmetic and its fragments, owing to Hilbert's and Bernays' [7] observation that large chunks of mathematics can already be formalized in second order arithmetic.

Definition 19.8. *The language $\mathcal{L}_2$ of second-order arithmetic contains number variables $x, y, z, u, \ldots$, set variables $X, Y, Z, U, \ldots$ (ranging over subsets of $\mathbb{N}$), the constant 0, function symbols $Suc, +, \cdot$, and relation symbols $=, <, \in$. Suc stands for the* *successor function.* Terms *are built up as usual. For* $n \in \mathbb{N}$*, let* $\bar{n}$ *be the canonical term denoting n. Formulae are built from the prime formulae* $s = t$, $s < t$, *and* $s \in X$ *using* $\wedge, \vee, \neg, \forall x, \exists x, \forall X$ *and* $\exists X$ *where* s, t *are terms. Note that equality in* $\mathcal{L}_2$ *is only a relation on numbers. However, equality of sets will be considered a defined notion, namely* $X = Y$ *if and only if* $\forall x[x \in X \leftrightarrow x \in Y]$. *As per usual, number quantifiers are called bounded if they occur in the context* $\forall x(x < s \rightarrow \ldots)$ *or* $\exists x(x < s \wedge \ldots)$ *for a term s which does not contain x. The* Σ^0_0*-formulae are those formulae in which all quantifiers are bounded number quantifiers. For* $k > 0$, Σ^0_k*-formulae are formulae of the form* $\exists x_1 \forall x_2 \ldots Q x_k \varphi$, *where* φ *is* Σ^0_0; $\mathbf{\Pi}^0_k$*-formulae are those of the form* $\forall x_1 \exists x_2 \ldots Q x_k \varphi$. *The union of all* $\mathbf{\Pi}^0_k$*- and* Σ^0_k*-formulae for all* $k \in \mathbb{N}$ *is the class of* arithmetical *or* $\mathbf{\Pi}^0_\infty$-formulae. *The* Σ^1_k*-formulae* ($\mathbf{\Pi}^1_k$*-formulae) are the formulae* $\exists X_1 \forall X_2 \ldots Q X_k \varphi$ *(resp.* $\forall X_1 \exists X_2 \ldots Q x_k \varphi$*) for arithmetical* φ.

The basic axioms in all theories of second-order arithmetic are the defining axioms of $0, 1, +, \cdot, <$ *and the* induction axiom

$$\forall X(0 \in X \wedge \forall x(x \in X \rightarrow x + 1 \in X) \rightarrow \forall x(x \in X)),$$

respectively the scheme of induction

$$\mathbf{IND} \qquad \varphi(0) \wedge \forall x(\varphi(x) \rightarrow \varphi(x+1)) \rightarrow \forall x \varphi(x),$$

where φ *is an arbitrary* $\mathcal{L}_2$*-formula. We consider the axiom scheme of C-* comprehension *for formula classes C which is given by*

$$C\text{-}\mathbf{CA} \qquad \exists X \forall u(u \in X \leftrightarrow \varphi(u))$$

for all formulae $\varphi \in C$ *in which X does not occur.*

For each axiom scheme **Ax** *we denote by* (**Ax**) *the theory consisting of the basic arithmetical axioms, the scheme* $\mathbf{\Pi}^0_\infty$-**CA**, *the scheme of induction and the scheme* **Ax**. *If we replace the scheme of induction by the induction axiom, we denote the resulting theory by* $(\mathbf{Ax})_0$. *An example for these notations is the theory* ($\mathbf{\Pi}^1_1$-**CA**) *which contains the induction scheme, whereas* $(\mathbf{\Pi}^1_1\text{-}\mathbf{CA})_0$ *only contains the induction axiom in addition to the comprehension scheme for* $\mathbf{\Pi}^1_1$*-formulae.*

In the framework of these theories one can introduce defined symbols for all primitive recursive functions. Especially, let $\langle , \rangle : \mathbb{N} \times \mathbb{N} \longrightarrow \mathbb{N}$ *be a primitive recursive and bijective pairing function. The* x^{th} *section of U is defined by* $U_x := \{y : \langle x, y \rangle \in U\}$. *Observe that a set U is uniquely determined by its sections on account of* $\langle , \rangle$*'s bijectivity. Any set R gives rise to a binary relation* $<_R$ *defined by* $y <_R x := \langle y, x \rangle \in R$. *Using the foregoing coding, we can formulate the schema of* Bar induction

$$\mathbf{BI} \qquad \forall X \big[\mathbf{WF}(<_X) \ \wedge \ \forall u \big(\forall v <_X u \varphi(v) \rightarrow \varphi(u) \big) \ \rightarrow \ \forall u \varphi(u) \big]$$

for all formulae φ, *where* $\mathbf{WF}(<_X)$ *expresses that* $<_X$ *is well-founded, i.e.,* $\mathbf{WF}(<_X)$ *stands for the formula* $\forall Y \ \big[\forall u \big[(\forall v <_X u \ v \in Y) \ \rightarrow \ u \in Y \big] \ \rightarrow \ \forall u \ u \in Y \big]$.

For a collection of formulas, $\mathcal{F}$, we also formulate the axiom of choice for these formulas:

$$\mathcal{F}\text{-}\mathbf{AC} \qquad \forall x \exists Y\, F(x,Y) \rightarrow \exists Y \forall x\, F(x, Y_x),$$

where $F(x,X)$ belongs to $\mathcal{F}$.

Definition 19.9. *A binary relation $<$ on $\mathbb{N}$ is said to be a prewellordering if $<$ is well-founded and transitive and satisfies*

$$\forall x,y\, [\, x < y \,\vee\, y < x \,\vee\, x \equiv_< y\,],$$

where $x \equiv_< y$ signifies $\forall u\, ([u < x \leftrightarrow u < y] \wedge [x < u \leftrightarrow y < u])$.

Given two prewellorderings $\lhd$ and $<$, we say that a function $f : \mathbb{N} \rightarrow \mathbb{N}$ embeds *$\lhd$* into *$<$ if $\forall xy\,(\,y \lhd x \;\rightarrow\; f(y) < f(x))$ and $\forall xz\,[\,z < f(x) \;\rightarrow\; \exists y(\,y \lhd x \;\wedge\; f(y) \equiv_< z\,)]$.*

Note that $(\mathbf{\Pi}^1_1 - \mathbf{CA})_0$ suffices to prove that there exists a function f such that f embeds $\lhd$ into $<$ or f embeds $<$ into $\lhd$.

We use the abbreviations $\mathbf{PWO}(<)$ to express that $<$ is a prewellordering. By field($<$) we mean the set $\{x : \exists y\,(x < y \;\vee\; y < x)\}$. For a set V, let field(V) $=$ field($<_V$).

$<$ is a wellordering *(written $\mathbf{WO}(<)$) if $\mathbf{PWO}(<)$ and $\forall xy \in$ field($<$) $[x < y \;\vee\; y < x \;\vee\; x = y]$. $<$ is a* wellordering of $\mathbb{N}$ *if $\mathbf{WO}(<)$ and field($<$) $= \mathbb{N}$.*

Definition 19.10. *Let $\mathcal{F}$ be a collection of $\mathcal{L}_2$-formulae. The principle that any operator which is describable via an $\mathcal{F}$-formula inductively defines a set, $\mathcal{F}$-**Fix**, is expressed by the schema*

$$\forall X \exists! Y\, \varphi(X,Y) \rightarrow$$
$$\exists V \exists Z \exists U \left[\mathbf{PWO}(V) \;\wedge\; \forall x\, \varphi(Z_{Vx}, Z_x) \;\wedge\; \varphi(\bigcup\nolimits_x Z_x, U) \;\wedge\; U \subseteq \bigcup\nolimits_x Z_x\,\right]$$

where φ belongs to $\mathcal{F}$.

Note that $\bigcup_x Z_x$ is uniquely determined by φ, that is if

$$\text{(19.2)} \qquad \mathbf{PWO}(\bar{V}) \;\wedge\; \forall x\, \varphi(\bar{Z}_{\bar{V}x}, \bar{Z}_x) \;\wedge\; \varphi(\bigcup\nolimits_x \bar{Z}_x, \bar{U}) \;\wedge\; \bar{U} \subseteq \bigcup\nolimits_x \bar{Z}_x,$$

then $\bigcup_x Z_x = \bigcup_x \bar{Z}_x$. To see this assume that f embeds $<_V$ into $<_{\bar{V}}$. By induction on $<_V$ one then verifies that $\forall x Z_x = \bar{Z}_{f(x)}$. The latter implies the assertion. As to a fragment of second order arithmetic in which the previous proof can be carried out, one needs provability of comparability of prewellorderings; thus, e.g. $(\mathbf{\Pi}^1_1\text{-}\mathbf{CA})_0$ suffices.

We shall denote $\bigcup_x Z_x$ by $\mathbf{I}^\infty_\varphi$. By $<_\varphi$ we shall refer to an arbitrary choice of prewellordering $<_{\bar{V}}$ satisfying (19.2).

The crucial result linking Π^1_2-comprehension and and the schema $\mathcal{F}$-**Fix** is the following.

Theorem 19.11. (i) $(\Sigma^1_2\text{-}\mathbf{AC})_0 + \mathbf{\Pi}^1_1\text{-}\mathbf{Fix} = (\mathbf{\Pi}^1_2\text{-}\mathbf{CA})_0$.
(ii) $(\Sigma^1_2\text{-}\mathbf{AC}) + \mathbf{\Pi}^1_1\text{-}\mathbf{Fix} = (\mathbf{\Pi}^1_2\text{-}\mathbf{CA})$.

Proof. [16] Theorem 3.15 and Corollary 3.16. □

19.2.3 The Theories $\mathcal{M}_{<\gamma}$

To begin with, we fix an ordinal notation system OT. For this paper it will be sufficient to assume that OT is a standard notation system for ε_0. $<$ will refer to the primitive recursive "less than" relation which comes with OT, and $\alpha < \varepsilon_0$ will mean $\gamma \in$ OT.

The language $\mathcal{L}_\mu$, extending the language $\mathcal{L}_2$ of 2nd order arithmetic, is described below. It is a generalization of the language of μ-calculus as presented in [12, Section 1b] and [22, Section 1]. *Number terms*, *set terms* and *formulas* of the language $\mathcal{L}_\mu$ are defined as follows.

Definition 19.12. 1. Number terms of $\mathcal{L}_2$ are *number terms* of $\mathcal{L}_\mu$.
2. Set variables are *set terms*.
3. $\bot$ is a *formula*.
4. If s and t are *number terms* then $s = t$ is a *formula*.
5. If s is a *number term* and S is a *set term* then $s \in S$ is a *formula*.
6. If φ_0 and φ_1 are *formulas* then $\varphi_0 \wedge \varphi_1$, $\varphi_0 \vee \varphi_1$ and $\varphi_0 \rightarrow \varphi_1$ are *formulas*.
7. If ψ is a *formula* then $\forall x\psi$ and $\exists x\psi$ are *formulas*.
8. If ψ is a *formula* then $\forall X\psi$ and $\exists X\psi$ are *formulas*.
9. If φ is a first-order (i.e. not containing set quantifiers) *formula* then $\mu xyXY.\varphi$ is a *set term*.

Fix a *limit ordinal* $\gamma \leq \varepsilon_0$ for the remainder of this article. $\mathcal{M}_{<\gamma}$ is based on classical logic. The basic axioms of $\mathcal{M}_{<\gamma}$ are those of $\mathbf{ACA_0}(\mathcal{L}_\mu)$, i.e. $\mathbf{ACA_0}$ extended to the language $\mathcal{L}_\mu$. The main additional axiom scheme $\mathbf{LFP}_\gamma[\varphi[x,y,X,Y],T]$ will govern the μ-term $T := \mu xyXY.\varphi$; its formulation requires some preparation. The basic idea is to state it as

$$\forall y{<}\alpha \forall Y \text{ "}\varphi[x,y,X,Y] \text{ is monotone in } X\text{"} \rightarrow \forall y{<}\alpha\, \mathrm{LFP}_{x,X}[\varphi[x,y,X,T_{<y}],T_y] \tag{19.3}$$

(for abbreviations see below), for every first-order formula φ and every $\alpha < \gamma$, but for technical reasons we formulate it in a slightly different way. In our formulation, in order to claim $\forall y{<}\alpha\, \mathrm{LFP}_{x,X}[\varphi[x,y,X,T_{<y}],T_y]$, we require monotonicity not only of the formula φ, but of all formulas χ s.t. T "depends on" a μ-term $\mu xyXY.\chi$. Exact definitions are given next.

Definition 19.13. For every formula ψ and every μ-term $T := \mu xyXY.\varphi$, assuming that all bound variables in them are renamed so as to avoid collisions, we define finite sets $\mathcal{M}(\psi)$ and $\mathcal{M}(T)$ of the form $\{\langle \ell_i, \mu x_i y_i X_i Y_i.\varphi_i\rangle \mid \ldots\}$, where ℓ_i is a finite list of variables, inductively as follows:

$$\mathcal{M}(\psi) := \begin{cases} \emptyset & \text{if } \psi \text{ is } \bot,\ s = t \text{ or } s \in X; \\ \mathcal{M}(\psi_0) \cup \mathcal{M}(\psi_1) & \text{if } \psi \text{ is } \psi_0 \circ \psi_1 \text{ and } \circ \in \{\wedge, \vee, \rightarrow\}; \\ \{\langle(\ell_i, x), T_i\rangle \mid \langle \ell_i, T_i\rangle \in \mathcal{M}(\chi[x])\} & \text{if } \psi \text{ is } Qx\chi[x] \text{ and } Q \in \{\forall, \exists\}; \\ \mathcal{M}(\chi[X]) & \text{if } \psi \text{ is } QX\chi[X] \text{ and } Q \in \{\forall, \exists\}; \\ \{\langle \emptyset, \mu xyXY.\varphi\rangle\} \cup \{\langle(\ell_i, x, y, X, Y), T_i\rangle \mid \langle \ell_i, T_i\rangle \in \mathcal{M}(\varphi)\} & \\ & \text{if } \psi \text{ is } s \in \mu xyXY.\varphi. \end{cases}$$

Finally we define $\mathcal{M}(T) := \mathcal{M}(0 \in T)$.

Now $\mathbf{LFP}_\alpha[\varphi, T]$ is defined as an axiom scheme

$$\mathrm{Mon}(\mathcal{M}(T), \alpha) \rightarrow \forall y < \alpha\, \mathrm{LFP}_{x,X}[\varphi[x, y, X, T_{<y}], T_y], \tag{19.4}$$

for every first-order formula φ and every $\alpha < \gamma$, where we adopt the following abbreviations:

$\mathrm{Mon}(\mathcal{M}(T), \alpha)$ stands for the conjunction of all formulas

$$\forall \vec{z}_i \forall \vec{Z}_i \forall y_i < \alpha \forall Y_i \text{ “}\varphi_i[x_i, y_i, X_i, Y_i] \text{ is monotone in } X_i\text{”},$$

where $\langle(\vec{z}_i, \vec{Z}_i), \mu x_i y_i X_i Y_i.\varphi_i\rangle \in \mathcal{M}(T)$.

$$\begin{aligned} X \subseteq Y \quad & \text{for } \forall x\,(x \in X \rightarrow x \in Y), \\ \mathrm{LFP}_{x,X}[\psi[x, X], Z] \quad & \text{for } \forall x\,(x \in Z \leftrightarrow \psi[x, Z]) \wedge \\ & \qquad \forall U(\forall z(\psi[z, U] \rightarrow z \in U) \rightarrow Z \subseteq U), \\ x \in Z_y \quad & \text{for } (y, x) \in Z, \\ x \in Z_{<y} \quad & \text{for } (x)_0 < y \wedge x \in Z, \\ (\cdot,\cdot), (\cdot)_0, (\cdot)_1 \quad & \text{for the usual pairing and unpairing operations} \\ & \qquad \text{on natural numbers.} \end{aligned}$$

Spelling out the formulas the abbreviations $\mathrm{Mon}(\mathcal{M}(T), \alpha)$ and $\mathbf{LFP}_\alpha[\varphi, T]$ stand for, we have that $\mathrm{Mon}(\mathcal{M}(T), \alpha)$ is the conjunction of the formulas

$$\forall \vec{z}_i \forall \vec{Z}_i \forall y_i < \alpha \forall Y_i \forall X_i' \forall X_i''\ (X_i' \subseteq X_i'' \rightarrow (\varphi_i[x_i, y_i, X', Y_i] \rightarrow \varphi_i[x_i, y_i, X_i'', Y_i])) \tag{19.5}$$

where $\langle(\vec{z}_i, \vec{Z}_i), \mu x_i y_i X_i Y_i.\varphi_i\rangle \in \mathcal{M}(T)$; and $\mathbf{LFP}_\alpha[\varphi, T]$ is the formula

$$\begin{aligned} (19.6) \quad \mathrm{Mon}(\mathcal{M}(T), \alpha) \rightarrow \forall y < \alpha\, (\forall x\, (x \in T_y \leftrightarrow \varphi[x, y, T_y, T_{<y}]) \\ \wedge \forall U(\forall z(\varphi[z, y, U, T_{<y}] \rightarrow z \in U) \rightarrow T_y \subseteq U)). \end{aligned}$$

In addition to the above, $\mathcal{M}_{<\gamma}$ contains the scheme of transfinite induction $\mathbf{TI}_\alpha$, for every formula $\psi[x] \in \mathcal{L}_\mu$ and every $\alpha < \gamma$:

$$\forall x(\forall y < x\, \psi[y] \rightarrow \psi[x]) \rightarrow \forall x < \alpha\, \psi[x]. \tag{19.7}$$

Lemma 19.14. *Let $\varphi(x, X)$ be a first order formula of $\mathcal{M}_{<\gamma}$ (which usually contains other free variables) such that $\varphi(x, X)$ is $\mathcal{M}_{<\gamma}$-provably monotone in x, X, i.e.,*

$$\mathcal{M}_{<\gamma} \vdash \forall x \forall X \forall Z [X \subseteq Z \wedge \varphi(x,X) \to \varphi(x,Z)].$$

Let the first-order formula $\varphi_{st}(x,X)$ be defined by

$$\varphi_{st}(x,X) := \exists u,v \left[x = (u,v) \wedge \varphi(u,\{z : (z,u) \in X\}) \wedge \neg\varphi(v,\{z : \neg\varphi(u,\{w : (w,z) \in X\})\})\right]. \tag{19.8}$$

Then $\varphi_{st}(x,X)$ is provably monotone in $\mathcal{M}_{<\gamma}$ with respect to x,X. Letting $x <_\varphi y$ stand for $(x,y) \in \mu xX.\varphi_{st}(x,X)$, we get

$$\mathcal{M}_{<\gamma} \vdash \mathrm{WF}(<_\varphi) \wedge \forall u\,[u \in \mu xX.\varphi(x,X) \leftrightarrow \varphi(u,\{v : v <_\varphi u\})]. \tag{19.9}$$

Proof. The monotonicity of $\varphi_{st}(x,X)$ follows from that of $\varphi(x,X)$. (19.9) is proved in [15], section 3 and stated in [15], Corollary 3.3. The proof presented in [15], though, is formally carried out in a system of explicit mathematics with an extra axiom asserting that every monotone operation on sets has a least fixed point. However, one easily checks that that proof carries over to $\mathcal{M}_{<\gamma}$. □

Proposition 19.15. *To every first-order formula $\theta(x,X)$ of $\mathcal{M}_{<\gamma}$ and variables x,X we can effectively assign a first-order formula $\Upsilon(x,X)$ of $\mathcal{M}_{<\gamma}$ with the same free variables such that $\Upsilon(x,X)$ is $\mathcal{M}_{<\gamma}$-provably monotone with respect to x,X, i.e.,*

$$\mathcal{M}_{<\gamma} \vdash \forall x \forall X \forall Z [X \subseteq Z \wedge \Upsilon(x,X) \to \Upsilon(x,Z)]. \tag{19.10}$$

Moreover, setting

$$\begin{aligned}
\Upsilon^u &:= \{y : \Upsilon(y,\{v : v <_\Upsilon u\})\},\\
\Theta^u &:= \{a : (0,a,a) \in \Upsilon^u\},\\
\Theta^{<u} &:= \bigcup_{y <_\Upsilon u} \Theta^y := \{b : \exists y <_\Upsilon u\; b \in \Theta^y\},\\
\Theta^\infty &:= \bigcup_u \Theta^u := \{z : \exists u\, z \in \Theta^u\},\\
\Gamma_\theta(X) &:= \{x : \theta(x,X)\},
\end{aligned}$$

$\mathcal{M}_{<\gamma}$ proves that

$$\Theta^u = \Gamma_\theta(\Theta^{<u}) \cup \Theta^{<u}, \tag{19.11}$$

$$\Gamma_\theta(\Theta^\infty) \subseteq \Theta^\infty. \tag{19.12}$$

In particular, Θ^∞ is first-order definable in the language of $\mathcal{M}_{<\gamma}$ and $\mathcal{M}_{<\gamma}$ proves that Θ^∞ is a set and that Θ^∞ arises by iterating the operator Γ_θ along the stage comparison prewellordering $<_\Upsilon$ of $\mu xX.\Upsilon(x,X)$. Moreover, Θ^∞ is closed under Γ_θ. In other words, Θ^∞ is the set inductively defined by the operator Γ_θ.

Proof. The details of the definition of Υ can be found in [15] Definition 4.3. More precisely, one has to substitute Γ_θ for the operator Θ in [15] Definition 4.3 and then define $\Upsilon(x, X)$ by $x \in \Upsilon(X)$, where the latter Υ denotes the operator defined in [15] Definition 4.3. The statement we want to prove is [15] Theorem 4.1 except that we have to replace the theory $\mathbf{T_0} \restriction + \mathbf{MID}$ by $\mathcal{M}_{<\gamma}$. Upon nearer inspection of the proof of [15] Theorem 4.1, one sees that in works in $\mathcal{M}_{<\gamma}$ as well. □

Definition 19.16. *For every first order formula $\theta(x, X)$ of $\mathcal{M}_{<\gamma}$ and variables x, X we notate the first order definable set Θ^∞ of Proposition 19.15 by $\mu_\nu xX.\theta(x, X)$.*

Proposition 19.17. *To every first order formula $\theta(x, y, X, Y)$ of $\mathcal{M}_{<\gamma}$, $\delta < \gamma$, and variables x, y, X, Y we can assign a first order definable set S such that*

$$\mathcal{M}_{<\gamma} \vdash \forall \alpha \leq \delta \; S_\alpha \; = \; \nu_\mu xX.\,\theta(x, \alpha, X, S_{<\alpha}) \tag{19.13}$$

where $S_\alpha := \{u : \; (\alpha, u) \in S\}$ and $S_{<\alpha} := \{(\beta, v) \in S : \; \beta < \alpha\}$.

Proof. Let $\Upsilon(x, y, X, Y)$ be the formula of Proposition 19.15 assigned to θ and the variables x, X. Then $\Upsilon(x, y, X, Y)$ is $\mathcal{M}_{<\gamma}$-provably monotone with respect to x, X. Let $\Upsilon_{st}(x, y, X, Y)$ be the formula introduced in Lemma 19.14 that inductively defines the stage comparison relation on $\mu xX.\Upsilon(x, y, X, Y)$. This formula is also $\mathcal{M}_{<\gamma}$-provably monotone with respect to x, X. Let

$$\begin{aligned}
\tilde{Y}(\alpha) &:= \{(\beta, (z)_1) : \; z \in Y_\beta \;\wedge\; (z)_0 = 1 \;\wedge\; \beta < \alpha\} \\
\tilde{\Upsilon}(x, \alpha, X, Y) &:= \exists u, v \left(x = (u, v) \;\wedge\; \left([u = 0 \;\wedge\; \Upsilon_{st}(v, \alpha, \{y : \; (0, y) \in X\}, \tilde{Y}(\alpha))] \right.\right. \\
&\qquad \left.\left. \vee\; [u = 1 \;\wedge\; \exists w \, \Upsilon((0, v, v), \alpha, \{z : \; (0, (z, w)) \in X\}, \tilde{Y}(\alpha))] \right) \right).
\end{aligned}$$

Note that $\tilde{\Upsilon}(x, \alpha, X, Y)$ is monotone (provably so in $\mathcal{M}_{<\gamma}$) with respect to x, X. Thus $T := \mu xyXY.\tilde{\Upsilon}$ is a term of $\mathcal{M}_{<\gamma}$. Finally put

$$S := \{(\alpha, w) : \; (\alpha, (1, w)) \in T\}.$$

We shall now prove (19.13). Letting $Y = T_{<\alpha}$ we have

$$\begin{aligned}
\tilde{Y}(\alpha) &= \{(\beta, (z)_1) : \; z \in T_\beta \;\wedge\; (z)_0 = 1 \;\wedge\; \beta < \alpha\} \\
&= S_{<\alpha}.
\end{aligned}$$

Let

$$\begin{aligned}
A &:= \mu xX.\Upsilon_{st}(x, \alpha, X, S_{<\alpha}), \\
B &:= \{v : \; \exists w \, \Upsilon((0, v, v), \alpha, \{z : \; (z, w) \in A\}, S_{<\alpha})\}, \\
C &:= \{(0, y) : \; y \in A\} \;\cup\; \{(1, z) : \; z \in B\}.
\end{aligned}$$

As $\forall v \, [v \in A \leftrightarrow \Upsilon_{st}(v, \alpha, A, S_{<\alpha})]$ we then have

$$\begin{aligned}
\tilde{\Upsilon}(x, \alpha, C, T_{<\alpha}) &\leftrightarrow \exists u, v \left(x = (u, v) \;\wedge\; \left([u = 0 \;\wedge\; \Upsilon_{st}(v, \alpha, A, S_{<\alpha})] \right.\right. \\
&\qquad \left.\left. \vee\; [u = 1 \;\wedge\; \exists w \, \Upsilon((0, v, v), \alpha, \{z : \; (z, w) \in A\}, S_{<\alpha})] \right) \right) \\
&\leftrightarrow \exists u, v \left(x = (u, v) \;\wedge\; ([u = 0 \wedge v \in A] \;\vee\; [u = 1 \wedge v \in B]) \right) \\
&\leftrightarrow x \in C,
\end{aligned} \tag{19.14}$$

so that

$$(19.15) \qquad T_\alpha \subseteq C.$$

From $\forall x\,[x \in T_\alpha \leftrightarrow \tilde{\Upsilon}(x, \alpha, T_\alpha, T_{<\alpha})]$ it follows that

$$\forall v\,[(0, v) \in T_\alpha \leftrightarrow \Upsilon_{st}(v, \alpha, \{y : (0, y) \in T_\alpha\}, S_{<\alpha})]$$

and hence

$$(19.16) \qquad A \subseteq \{y : (0, y) \in T_\alpha\}.$$

(19.15) and (19.16) together yield $\{y : (0, y) \in T_\alpha\} = A$. Moreover, as

$$\forall v\,[(1, v) \in T_\alpha \leftrightarrow \exists w\,\Upsilon((0, v, v), \alpha, \{z : (z, w) \in \{y : (0, y) \in T_\alpha\}, S_{<\alpha})]$$

we also get $B = \{v : (1, v) \in T_\alpha\}$ and consequently arrive at

$$(19.17) \qquad T_\alpha = C,$$

$$(19.18) \qquad S_\alpha = B.$$

In view of Proposition 19.15 this entails that

$$S_\alpha = \mu_V x X.\,\theta(x, \alpha, X, S_{<\alpha}).$$

□

Next, we want to show that $(\mathbf{\Pi}^1_2\text{-}\mathbf{CA})_0$ and $(\mathbf{\Pi}^1_2\text{-}\mathbf{CA})$ can be reduced to $\mathcal{M}_{<\omega}$ and $\mathcal{M}_{<\varepsilon_0}$, respectively. Here we shall draw on [16]. It follows from [16] Theorem 3.15, Corollary 3.16, Corollary 4.25 and Corollary 4.29 that $(\mathbf{\Pi}^1_2\text{-}\mathbf{CA})_0$ and $(\mathbf{\Pi}^1_2\text{-}\mathbf{CA})$ can be reduced to certain operator theories $\mathbf{T}^{\mathbf{OP}}_{<\omega}$ and $\mathbf{T}^{\mathbf{OP}}_{<\varepsilon_0}$, respectively. More precisely, we have

Theorem 19.18. (i) $(\mathbf{\Pi}^1_2\text{-}\mathbf{CA})_0$ *and* $\mathbf{T}^{\mathbf{OP}}_{<\omega}$ *prove the same* Π^1_3 *sentences.*
(ii) $(\mathbf{\Pi}^1_2\text{-}\mathbf{CA})$ *and* $\mathbf{T}^{\mathbf{OP}}_{<\varepsilon_0}$ *prove the same* Π^1_3 *sentences.*

Thus it suffices to show that $\mathbf{T}^{\mathbf{OP}}_{<\omega}$ and $\mathbf{T}^{\mathbf{OP}}_{<\varepsilon_0}$ can be interpreted in $\mathcal{M}_{<\omega}$ and $\mathcal{M}_{<\varepsilon_0}$, respectively. The foregoing theories are based on axioms for finite and transfinite iterations of certain operators. $\mathbf{T}^{\mathbf{OP}}_{<\omega}$ is the the theory $(\mathbf{\Pi}^0_\infty - \mathbf{CA})_0$ augmented by the axioms

$$\mathbf{OP}_n \qquad \forall \vec{X} \forall Y \exists Z\, \Phi^{\vec{X}}_n(Y) = Z$$

for all n.

$\mathbf{T}^{\mathbf{OP}}_{<\varepsilon_0}$ be the theory $(\mathbf{\Pi}^0_\infty - \mathbf{CA})_0$ augmented by the axioms

$$\mathbf{OP}_\alpha \qquad \forall \vec{X} \forall Y \exists Z\, \Phi^{\vec{X}}_\alpha(Y) = Z$$

for all $\alpha < \varepsilon_0$.

For a detailed account of the syntax and axioms of $\mathbf{T}^{\mathbf{OP}}_{<\omega}$ and $\mathbf{T}^{\mathbf{OP}}_{<\varepsilon_0}$ we refer to [16]. The crucial observation is that the operators $\Phi^{\vec{X}}_n$ can be defined in $\mathcal{M}_{<\omega}$ and the operators $\Phi^{\vec{X}}_\alpha$ can be defined in $\mathcal{M}_{<\varepsilon_0}$, using Proposition 19.17. As a result we can conclude the following theorem.

Theorem 19.19. (i) *$\mathcal{M}_{<\omega}$ has the same proof-theoretic strength as* $(\mathbf{\Pi}^1_2\text{-}\mathbf{CA})_0$.
(ii) *$\mathcal{M}_{<\varepsilon_0}$ has the same proof-theoretic strength as* $(\mathbf{\Pi}^1_2\text{-}\mathbf{CA})$.

Proof. In light of the foregoing remarks we only need to show that $(\mathbf{\Pi}^1_2\text{-}\mathbf{CA})_0$ and $(\mathbf{\Pi}^1_2\text{-}\mathbf{CA})$ can accommodate $\mathcal{M}_\omega$ and $\mathcal{M}_{\varepsilon_0}$, respectively. This follows from Theorem 19.11, i.e., [16] Theorem 3.15 and Corollary 3.16. □

19.3 Double-Negation Translation

Fix $\gamma \leq \varepsilon_0$. Let $\mathcal{M}^i_{<\gamma}$ result from $\mathcal{M}_{<\gamma}$ by changing the logic from classical to intuitionistic. In this section, following the method of [22, Section 1], we will prove that $\mathcal{M}^i_{<\gamma}$ has the same strength as $\mathcal{M}_{<\gamma}$.

Definition 19.20. *(Negative, completely negative, φ^N)* A formula φ is *negative* iff occurrences of every atom, $\vee$, $\exists x$ or $\exists X$ not in the scope of the μ-symbol in φ are negated.

For any φ by φ^N we define the formula obtained from φ by putting $\neg\neg$ in front of every atom, $\vee$, $\exists x$ or $\exists X$ in φ not in the scope of μ.

An expression is *completely negative* iff all occurrences of atoms, $\vee$, $\exists x$ and $\exists X$ in it, including those in the scope of μ, are negated.

Note that φ^N is negative and $\mathcal{M}(\varphi) = \mathcal{M}(\varphi^N)$ for every formula φ.

Definition 19.21. *(Complete negation operation $N(e)$)* For any expression e we define $N(e)$ recursively as follows:

1. $N(e) := e$ if e is $\perp$, a number term or a set variable.
2. N commutes with $=$, $\in$ and logical connectives.
3. $N(\mu xyXY.\varphi) := \mu xyXY.(N(\varphi))^N$.

Note that if e is a μ-term then $N(e)$ is completely negative.

Lemma 19.22. *For any μ-term T we have*

$$\mathcal{M}(N(T)) = \{\langle \ell_i, \mu x_i y_i X_i Y_i.(N(\varphi_i))^N\rangle \mid \langle \ell_i, \mu x_i y_i X_i Y_i.\varphi_i\rangle \in \mathcal{M}(T)\}.$$

Proof. By induction on T. □

The calculus $\mathcal{M}^N_{<\gamma}$ is the same as $\mathcal{M}_{<\gamma}$, with the only difference that $\mathbf{LFP}_\alpha[\varphi[x,y,X,Y],T]$ is replaced by an axiom $\mathbf{LFP}^N_\alpha[\varphi[x,y,X,Y],T]$, where T is $\mu xyXY.\varphi$, which is

$$\mathrm{Mon}(\mathcal{M}(N(T),\alpha) \rightarrow \forall x\,(x \in (N(T))_y \leftrightarrow N(\varphi)[x,y,(N(T))_y,(N(T))_{<y}]) \\ \wedge \forall U(\forall z(N(\varphi)[z,y,U,(N(T))_{<y}] \rightarrow z \in U) \rightarrow (N(T))_y \subseteq U), \tag{19.19}$$

where $\mathrm{Mon}(\mathcal{M}(N(T),\alpha)$ signifies the conjunction of all formulas

$$\forall \vec{z}_i \forall \vec{Z}_i \forall y_i < \alpha \forall Y_i \forall X_i' \forall X_i''\,(X_i' \subseteq X_i'' \rightarrow (N(\varphi_i[x_i,y_i,X',Y_i]) \rightarrow \\ N(\varphi_i[x_i,y_i,X_i'',Y_i])))$$

with $\langle(\vec{z}_i,\vec{Z}_i),\mu x_i y_i X_i Y_i.\varphi_i\rangle \in \mathcal{M}(T)$.

Lemma 19.23. *$\mathcal{M}_{<\gamma}$ can be interpreted in $\mathcal{M}^N_{<\gamma}$.*

Proof. Given a derivation d in $\mathcal{M}_{<\gamma}$, replace every formula φ in d by $N(\varphi)$ in order to obtain a derivation in $\mathcal{M}^N_{<\gamma}$. The only little thing to check is that $N\,(\mathbf{LFP}_\gamma[\varphi[x,y,X,Y],\mu xyXY.\varphi])$ is of the form (19.19), but this is straightforward. □

For a set Z by $\overline{Z}$ we denote the set $\{x \mid \neg x \in Z\}$, which exists by Arithmetical Comprehension. Below we use the standard notation $Y \doteq Z$ to mean $\forall x(x \in Y \leftrightarrow x \in Z)$.

Lemma 19.24. (Extensionality Lemma) *For every $\alpha < \gamma$ $\mathcal{M}^i_{<\gamma}$ proves that if $Z_1 \doteq Z_2$ then:*

(a) *if* $\mathrm{Mon}(\mathcal{M}(\varphi[Z_1]),\alpha) \wedge \mathrm{Mon}(\mathcal{M}(\varphi[Z_2]),\alpha)$ *then* $\varphi[Z_1] \leftrightarrow \varphi[Z_2]$, *for every formula φ;*
(b) *for a term* $T[Z] := \mu xyXY.\varphi[x,y,X,Y,Z]$ *if*

$$\mathrm{Mon}(\mathcal{M}(T[Z_1]),\alpha) \wedge \mathrm{Mon}(\mathcal{M}(T[Z_2]),\alpha)$$

then $\forall y < \alpha(\mu xyXY.\varphi[Z_1])_y \doteq (\mu xyXY.\varphi[Z_2])_y$, *for every first-order formula φ.*

Proof. The proof proceeds by induction on the buildup of an expression $e[Z]$. Below we use IH as an abbreviation for "induction hypothesis". The assertion is obvious when e is an elementary atom, i.e. an atom not of the form $s \in S$ where S is a μ-term. The induction step for logical connectives is also straightforward, we consider $\rightarrow$ and $\exists z$ for illustration.

Assume φ is $\varphi_0 \rightarrow \varphi_1$. From $\mathrm{Mon}(\mathcal{M}(\varphi[Z_1]),\alpha) \wedge \mathrm{Mon}(\mathcal{M}(\varphi[Z_2]),\alpha)$ we get $\mathrm{Mon}(\mathcal{M}(\varphi_i[Z_1]),\alpha) \wedge \mathrm{Mon}(\mathcal{M}(\varphi_i[Z_2]),\alpha)$ for both $i = 0,1$. By IH $\varphi_i[Z_1] \leftrightarrow \varphi_i[Z_2]$ for both $i = 0,1$, thus yielding $\varphi[Z_1] \leftrightarrow \varphi[Z_2]$.

Assume φ is $\exists z \psi[z]$. From $\mathrm{Mon}(\mathcal{M}(\varphi[Z_1]),\alpha) \wedge \mathrm{Mon}(\mathcal{M}(\varphi[Z_2]),\alpha)$ we get $\mathrm{Mon}(\mathcal{M}(\psi[z,Z_1]),\alpha) \wedge \mathrm{Mon}(\mathcal{M}(\psi[z,Z_2]),\alpha)$, for any z. By IH $\psi[z,Z_1] \leftrightarrow \psi[z,Z_2]$, for any z, thus yielding $\varphi[Z_1] \leftrightarrow \varphi[Z_2]$. The induction step for (b) requires subsidiary transfinite induction on y. So assuming $\mathrm{Mon}(\mathcal{M}(T[Z_1]),\alpha) \wedge \mathrm{Mon}(\mathcal{M}(T[Z_2]),\alpha)$, $\varphi[x,y,X,Y,Z_1] \leftrightarrow \varphi[x,y,X,Y,Z_2]$, $y < \alpha$, $(T[Z_1])_z \doteq (T[Z_2])_z$ for all $z < y$ and φ be first-order, it remains to show $(T[Z_1])_y := (\mu xyXY.\varphi[x,y,X,Y,Z_1])_y \doteq (\mu xyXY.\varphi[x,y,X,Y,Z_2])_y =: (T[Z_2])_y$. From (19.6) we have

$$\forall x\,(x \in (T[Z_1])_y \leftrightarrow \varphi[x, y, (T[Z_1])_y, (T[Z_1])_{<y}, Z_1]) \wedge$$
$$\forall U(\forall z(\varphi[z, y, U, (T[Z_1])_{<y}, Z_1] \to z \in U) \to (T[Z_1])_y \subseteq U) \quad (19.20)$$

and

$$\forall x\,(x \in (T[Z_2])_y \leftrightarrow \varphi[x, y, (T[Z_2])_y, (T[Z_2])_{<y}, Z_2]) \wedge$$
$$\forall U(\forall z(\varphi[z, y, U, (T[Z_2])_{<y}, Z_2] \to z \in U) \to (T[Z_2])_y \subseteq U). \quad (19.21)$$

From the first conjunct of (19.20), by the main IH, we have

$$\forall x\,(x \in (T[Z_1])_y \leftrightarrow \varphi[x, y, (T[Z_1])_y, (T[Z_1])_{<y}, Z_2]);$$

applying the subsidiary IH and then the main IH again we arrive at

$$\forall x\,(x \in (T[Z_1])_y \leftrightarrow \varphi[x, y, (T[Z_1])_y, (T[Z_2])_{<y}, Z_2]).$$

From the second conjunct of (19.21), taking $U := (T[Z_1])_y$, we obtain $(T[Z_2])_y \subseteq (T[Z_1])_y$. Similarly we get $(T[Z_1])_y \subseteq (T[Z_2])_y$. Together this gives $(T[Z_1])_y \doteq (T[Z_2])_y$. □

Lemma 19.25. *$\mathcal{M}^N_{<\gamma}$ can be interpreted in $\mathcal{M}^i_{<\gamma}$.*

Proof. Apply the double-negation translation $(\cdot)^N$. Classical logic goes into intuitionistic. It's easily checked (and well-known) that translations of all axioms of $\mathbf{ACA_0}(\mathcal{L}_\mu)$ are derivable intuitionistically from the axioms of $\mathbf{ACA_0}(\mathcal{L}_\mu)$. Transfinite induction (19.7) is also no problem, since $<$ is a recursive relation and $\neg\neg z_1 < z_2$ can be equivalently replaced by $z_1 < z_2$. So we need only to derive in $\mathcal{M}^i_{<\gamma}$ the formula $\left(\mathbf{LFP}^N_\gamma[\varphi[x, y, X, Y], T]\right)^N$ ($T := \mu xyXY.\varphi$), i.e.

$$\bigwedge_{\langle(\vec{z}_i, \vec{Z}_i), \mu x_i y_i X_i Y_i.\varphi_i\rangle \in \mathcal{M}(T)} \forall \vec{z}_i \forall \vec{Z}_i \forall y_i \overline{\overline{<}} \alpha \forall Y_i \forall X'_i \forall X''_i$$
$$\left(\overline{\overline{X'_i}} \subseteq \overline{\overline{X''_i}} \to \left(\left(N(\varphi_i[x_i, y_i, X', Y_i])\right)^N \to \left(N(\varphi_i[x_i, y_i, X''_i, Y_i])\right)^N\right)\right) \to$$
$$\forall y \overline{\overline{<}} \alpha \left(\forall x\,(x \in \overline{\overline{N(T)_y}} \leftrightarrow (N(\varphi))^N[x, y, (N(T))_y, (N(T))_{<y}]) \wedge\right.$$
$$\left.\forall U(\forall z((N(\varphi))^N[z, y, U, (N(T))_{<y}] \to z \in \overline{\overline{U}}) \to \overline{\overline{N(T)_y}} \subseteq \overline{\overline{U}})\right), \quad (19.22)$$

where by $z_1 \overline{\overline{<}} z_2$ we denote $\neg\neg z_1 < z_2$. As remarked above, $\overline{\overline{<}}$ can be equivalently replaced by $<$; also, by Lemma 19.22 the premise of the implication (19.22) can be equivalently replaced by

$$\bigwedge_{\langle(\vec{z}_i,\vec{Z}_i),\mu x_i y_i X_i Y_i.\varphi_i\rangle\in\mathcal{M}(N(T))} \forall\vec{z}_i\forall\vec{Z}_i\forall y_i<\alpha\forall Y_i\forall X_i'\forall X_i'' \left(\overline{\overline{X_i'}} \subseteq \overline{\overline{X_i''}} \to (\varphi_i[x_i,y_i,X',Y_i] \to \varphi_i[x_i,y_i,X_i'',Y_i])\right).$$

We will derive now in $\mathcal{M}^i_{<\gamma}$ the formula

$$\bigwedge_{\langle(\vec{z}_i,\vec{Z}_i),\mu x_i y_i X_i Y_i.\varphi_i\rangle\in\mathcal{M}(T)} \forall\vec{z}_i\forall\vec{Z}_i\forall y_i<\alpha\forall Y_i\forall X_i'\forall X_i''$$
$$\left(X_i' \subseteq X_i'' \to \left(\left(N(\varphi_i[x_i,y_i,X',Y_i])\right)^N \to \left(N(\varphi_i[x_i,y_i,X_i'',Y_i])\right)^N\right)\right) \to$$
$$\forall y<\alpha\left(\forall x\,(x\in\overline{\overline{N(T)_y}} \leftrightarrow (N(\varphi))^N[x,y,(N(T))_y,(N(T))_{<y}]) \wedge\right.$$
$$\left.\forall U(\forall z((N(\varphi))^N[z,y,U,(N(T))_{<y}] \to z\in\overline{\overline{U}}) \to \overline{\overline{N(T)_y}} \subseteq \overline{\overline{U}})\right), \quad (19.23)$$

which is stronger than (19.22) and will imply the latter by intuitionistic logic. So assume

$$\bigwedge_{\langle(\vec{z}_i,\vec{Z}_i),\mu x_i y_i X_i Y_i.\varphi_i\rangle\in\mathcal{M}(N(T))} \forall\vec{z}_i\forall\vec{Z}_i\forall y_i<\alpha\forall Y_i\forall X_i'\forall X_i'' \left(X_i' \subseteq X_i'' \to (\varphi_i[x_i,y_i,X',Y_i] \to \varphi_i[x_i,y_i,X_i'',Y_i])\right);$$

this means $\mathrm{Mon}(\mathcal{M}(N(T)),\alpha)$. By the $\mathbf{LFP}_\gamma[(N(\varphi))^N,N(T)]$ axiom of $\mathcal{M}^i_{<\gamma}$ we get now $\forall y<\alpha\,\mathrm{LFP}_{x,X}[(N(\varphi))^N[x,y,X,(N(T))_{<y}],(N(T))_y]$, i.e.

$$\forall y<\alpha\left(\forall x\,(x\in N(T)_y \leftrightarrow (N(\varphi))^N[x,y,(N(T))_y,(N(T))_{<y}]) \wedge\right.$$
$$\left.\forall U(\forall z((N(\varphi))^N[z,y,U,(N(T))_{<y}] \to z\in U) \to N(T)_y \subseteq U)\right). \quad (19.24)$$

Fixing $y < \alpha$, the first conjunct in the conclusion of (19.23) is derived by prenexing $\neg\neg$ to the first conjunct in (19.24). For the second conjunct, given U, assume $\forall z\left((N(\varphi))^N[z,y,U,(N(T))_{<y}] \to z\in\overline{\overline{U}}\right)$. For any z, we have $\mathrm{Mon}(\mathcal{M}((N(\varphi))^N[z,y,U,(N(T))_{<y}]),\alpha)$. Since the formula $(N(\varphi))^N$ is completely negative, we can use the fact $\overline{U} \doteq \overline{\overline{\overline{U}}}$ and Lemma 19.24 to prove by induction $\mathrm{Mon}(\mathcal{M}((N(\varphi))^N[z,y,\overline{\overline{U}},(N(T))_{<y}]),\alpha)$. Now using again complete negativeness, the fact $\overline{U} \doteq \overline{\overline{\overline{U}}}$ and Lemma 19.24(a), we obtain $(N(\varphi))^N[z,y,U,(N(T))_{<y}] \leftrightarrow (N(\varphi))^N[z,y,\overline{\overline{U}},(N(T))_{<y}]$, for any z. Therefore we can conclude

$\forall z \left((N(\varphi))^N[z, y, \overline{\overline{U}}, (N(T))_{<y}] \to z \in \overline{\overline{U}}\right)$. By the second conjunct of (19.24) we obtain $(N(T))_y \subseteq \overline{\overline{U}}$, which intuitionistically implies $\overline{\overline{N(T)}}_y \subseteq \overline{\overline{U}}$. □

Theorem 19.26. *$\mathcal{M}^i_{<\gamma}$ has the strength of $\mathcal{M}_{<\gamma}$.*

Proof. This follows now from Lemmata 19.23 and 19.25. □

19.4 Embedding into Intuitionistic Explicit Mathematics

In this section we will give embeddings of systems $\mathcal{M}^i_\omega$ and $\mathcal{M}^i_{\varepsilon_0}$ into theories of intuitionistic Explicit Mathematics **EETJ**↾ + **UMID**$_\mathbf{N}$ and **EETJ** + **UMID**$_\mathbf{N}$, respectively. Together with Theorems 19.19 and 19.26 above and Theorem 6.1 of [17] about classical **EETJ**^{c}↾ +**UMID**$_\mathbf{N}$ and **EETJ**^{c} + **UMID**$_\mathbf{N}$ this will prove the following

Theorem 19.27. (a) $\mathbf{T_0}$↾ + **UMID**$_\mathbf{N}$ *has exactly the strength of* $(\mathbf{\Pi^1_2}\text{-}\mathbf{CA})_0$*;*
(b) $\mathbf{T_0}$↾ + **Ind**$_\mathbf{N}$ + **UMID**$_\mathbf{N}$ *has exactly the strength of* $(\mathbf{\Pi^1_2}\text{-}\mathbf{CA})$.

The proof of part (a) of this theorem is another way to the main result of [22]; part (b) is a new result.

19.4.1 Explicit Mathematics: a Reminder

Language $\mathcal{L}_{\mathrm{EM}}$. All theories of Explicit Mathematics, considered in this paper, are formulated in a two-sorted language, containing variables for operations (individuals) and names, along with operation constants. Names are thought of as a special kind of operations, coding types (sets) of operations. We use **variables** $a, b, c, \ldots$ as ranging over operations, and $A, B, C, \ldots$ as ranging over names. The main **constants** of $\mathcal{L}_{\mathrm{EM}}$ are the following: combinators k, s, pairing p and projections p_0, p_1, zero 0, successor s_N and predecessor p_N, distinction by cases on natural numbers d_N, join j and inductive generation i. Additionally we have the following 9 **constants** called *name generators*: nat, id, inv, emp, and, or, imp, all, ex. **Terms** are built from variables and constants by the following application clause: if s and t are *terms* then $s \cdot t$ (also written as st) is a *term*, so that the *application* function symbol $\cdot$ accepts arguments of both sorts and returns an operation. In writing terms, parentheses are thought of as associated to the left. **Atomic formulas** are $\bot$ (falsity), $s = t$ (s coincides with t) and $s \,\varepsilon\, t$ (s belongs to the type named by t, s is classified under t), where s and t are terms. **Formulas** are built from atomic formulas by $\wedge, \vee, \to$ and two kinds of quantifiers, over operations and over names, e.g. $\forall a, \exists a, \forall A, \exists A$. Finally, **expression** is a term or a formula. We use the following standard abbreviations:

- $\neg F \;:\Leftrightarrow\; F \to \bot$;

- $F_0 \leftrightarrow F_1 \;:\Leftrightarrow\; (F_0 \rightarrow F_1) \wedge (F_1 \rightarrow F_0)$;
- $t\downarrow \,:\Leftrightarrow\; \exists x(t = x)$;
- $\mathcal{N}[t] \;:\Leftrightarrow\; \exists A(t = A)$;
- $t \doteq \{s[x_1,\ldots,x_n] \mid F[x_1,\ldots,x_n]\} \;:\Leftrightarrow$
  $\mathcal{N}[t] \wedge \forall x(x\,\varepsilon\, t \leftrightarrow \exists x_1 \ldots \exists x_n(x = s[x_1,\ldots,x_n] \wedge F[x_1,\ldots,x_n]))$;
- $s \simeq t \;:\Leftrightarrow\; (s\downarrow \vee\, t\downarrow) \rightarrow s = t$;
- $s \subseteq t \;:\Leftrightarrow\; \forall x\,\varepsilon\, s(x\,\varepsilon\, t)$;
- $s \doteq t \;:\Leftrightarrow\; s \dot\subseteq t \wedge t \dot\subseteq s$;
- t' for $\mathsf{s_N} \cdot t$; 1 for $0'$; st for $s \cdot t$; $t(s_1,\ldots,s_n)$ for $(\ldots(ts_1)\ldots s_n)$; $\langle s,t\rangle$ for $\mathsf{p}st$; $s \neq t$ for $\neg s = t$, etc.

Logic. Intuitionistic 2-sorted logic of partial terms with equality.

Axioms. The axioms are divided in several groups, according to their nature.

I. Applicative axioms. These axioms formalise that operations form a partial combinatory algebra, that we have pairing and projections, usual closure conditions on natural numbers, as well as definition by numerical cases: (1) $\mathsf{k}ab = a$; (2) $\mathsf{s}ab\downarrow \wedge\, \mathsf{s}abc \simeq ac(bc)$; (3) $\mathsf{p}ab\downarrow \wedge\, \mathsf{p_0}a\downarrow \wedge\, \mathsf{p_1}a\downarrow \wedge\, \mathsf{p_0}(\mathsf{p}ab) = a \wedge \mathsf{p_1}(\mathsf{p}ab) = b$; (4) $0\,\varepsilon\,\mathsf{nat} \wedge \forall x\,\varepsilon\,\mathsf{nat}(\mathsf{s_N}x\,\varepsilon\,\mathsf{nat})$; (5) $\forall x\,\varepsilon\,\mathsf{nat}(\mathsf{s_N}x \neq 0 \wedge \mathsf{p_N}(\mathsf{s_N}x) = x)$; (6) $\forall x\,\varepsilon\,\mathsf{nat}(x \neq 0 \rightarrow \mathsf{p_N}x\,\varepsilon\,\mathsf{nat} \wedge \mathsf{s_N}(\mathsf{p_N}x) = x)$; (7) $a\,\varepsilon\,\mathsf{nat} \wedge b\,\varepsilon\,\mathsf{nat} \rightarrow (a = b \rightarrow \mathsf{d_N}xyab = x) \wedge (a \neq b \rightarrow \mathsf{d_N}xyab = y)$.

II. Induction on nat. $F[0]\wedge\forall x(F[x] \rightarrow F[\mathsf{s_N}x]) \rightarrow \forall x\,\varepsilon\,\mathsf{nat}F[x]$, for every formula F.

The following lemmata 19.28 and 19.29 are provable using only applicative axioms I (see, for example, [4]).

Lemma 19.28. λ-abstraction. *For every term $t[x]$ there exists a term $\lambda x.t[x]$ such that $\lambda x.t[x]\downarrow$ and for every term s*

$$s\downarrow\rightarrow (\lambda x.t[x])s \simeq t[s]).$$

Lemma 19.29. Recursion Theorem. *There exists a closed term* rec *such that*

$$\mathsf{rec}f\downarrow \wedge\, \mathsf{rec}fx \simeq f(\mathsf{rec}f)x.$$

III. Explicit representation. This axiom states that each name is an operation: $\exists x(x = A)$.

IV. Elementary comprehension (ECA). These axiomatise *name generators*: (1) $\mathcal{N}[\mathsf{nat}]$; (2) $\mathcal{N}[\mathsf{id}] \wedge \forall x(x\,\varepsilon\,\mathsf{id} \leftrightarrow x = \langle \mathsf{p_0}x, \mathsf{p_1}x\rangle \wedge \mathsf{p_0}x = \mathsf{p_1}x)$; (3) $\mathcal{N}[\mathsf{inv}(f,A)] \wedge \forall x(x\,\varepsilon\,\mathsf{inv}(f,A) \leftrightarrow fx\,\varepsilon\,A)$; (4) $\mathcal{N}[\mathsf{emp}] \wedge \forall x(x\,\varepsilon\,\mathsf{emp} \leftrightarrow \bot)$; (5) $\mathcal{N}[\mathsf{and}(A,B)] \wedge \forall x(x\,\varepsilon\,\mathsf{and}(A,B) \leftrightarrow x\,\varepsilon\,A \wedge x\,\varepsilon\,B)$; (6) $\mathcal{N}[\mathsf{or}(A,B)] \wedge \forall x(x\,\varepsilon\,\mathsf{or}(A,B) \leftrightarrow x\,\varepsilon\,A \vee x\,\varepsilon\,B)$; (7) $\mathcal{N}[\mathsf{imp}(A,B)] \wedge \forall x(x\,\varepsilon\,\mathsf{imp}(A,B) \leftrightarrow x\,\varepsilon\,A \rightarrow x\,\varepsilon\,B)$; (8) $\mathcal{N}[\mathsf{all}A] \wedge \forall x(x\,\varepsilon\,\mathsf{all}A \leftrightarrow \forall y(\langle x,y\rangle\,\varepsilon\,A))$; (9) $\mathcal{N}[\mathsf{ex}A] \wedge \forall x(x\,\varepsilon\,\mathsf{ex}A \leftrightarrow \exists y(\langle x,y\rangle\,\varepsilon\,A))$.

Definition 19.30. *Elementary formula.* A formula is *elementary* iff it's constructed from $\bot$, $s = t$ and $t\,\varepsilon\,A$ by means of $\wedge, \vee, \rightarrow, \forall x, \exists x$ only. (No occurrences of $t\,\varepsilon\,s$ with s not a name variable and name quantifiers are allowed.)

The following lemma reduces Elementary Comprehension to a finite number of its instances; its proof requires only axioms I, III and IV (see [21, L.1.4]).

Lemma 19.31. ECA. *If a formula* $F := F[x; \bar{a}; \bar{A}]$ *is elementary then there exists a term* t^x_F *such that* $\mathrm{FV}(\mathsf{t}^x_F) = \mathrm{FV}(F) \setminus \{x\}$ *and*

$$\mathcal{N}[\mathsf{t}^x_F] \wedge \forall x(x \,\varepsilon\, \mathsf{t}^x_F \leftrightarrow F).$$

V. Join (J). This axiom states that if f is an operation from a type named by A, each value of which is a name, then $\mathsf{j}(A, f)$ names a disjoint union of all fx for $x \,\varepsilon\, A$:

$$\forall x \,\varepsilon\, A \mathcal{N}[fx] \rightarrow$$
$$\Big(\mathcal{N}[\mathsf{j}(A, f)] \ \wedge\ \forall z(z \,\varepsilon\, \mathsf{j}(A, f) \leftrightarrow z = \langle \mathsf{p}_0 z, \mathsf{p}_1 z\rangle \wedge \mathsf{p}_0 z \,\varepsilon\, A \wedge \mathsf{p}_1 z \,\varepsilon\, f(\mathsf{p}_0 z))\Big).$$

VI. Inductive Generation (IG). The first part of this axiom states that $\mathsf{i}(A, B)$ names a wellfounded part of a type named by A along an ordering named by B; the second part allows induction over that type for an arbitrary formula:

$$\mathcal{N}[\mathsf{i}(A, B)] \ \wedge\ \forall x \,\varepsilon\, A(\forall y(\langle y, x\rangle \,\varepsilon\, B \rightarrow y \,\varepsilon\, \mathsf{i}(A, B)) \rightarrow x \,\varepsilon\, \mathsf{i}(A, B))$$
$$\wedge\ (\forall x \,\varepsilon\, A(\forall y(\langle y, x\rangle \,\varepsilon\, B \rightarrow F[y]) \rightarrow F[x]) \rightarrow \forall x \,\varepsilon\, \mathsf{i}(A, B) F[x]),$$

where F is an arbitrary formula.

The theory App is the one containing only applicative axioms I; **EON** has axioms I–II. The theory **EONN** has axioms of the groups I–III. **EET** is **EONN**+**ECA**, **EETJ** is **EET** + **J** and $\mathbf{T_0}$ is **EETJ** + **IG**. By $\mathbf{T}\!\restriction$ we mean a version of a theory **T** where both induction on natural numbers II and inductive generation VI are restricted to formulas $F := x \in C$.

By $\mathcal{L}_{\mathrm{EM,lfp}}$ we denote the language $\mathcal{L}_{\mathrm{EM}}$ extended by an operation constant lfp. For the statement of **UMID** and $\mathbf{UMID_N}$ principles see e.g. [18, Section 2.2]. We repeat these definitions here:

$\mathbf{Clop}[f, A]$	means	$\forall X \subseteq A \exists Y \subseteq A\ fX = Y;$
$\mathbf{Mon}[f, A]$	-"-	$\forall X \subseteq A \forall Y \subseteq A\ (X \subseteq Y \rightarrow fX \subseteq fY);$
$\mathbf{Lfp}[Y, f, A]$	-"-	$fY \subseteq Y \wedge Y \subseteq A \wedge \forall X \subseteq A\,(fX \subseteq X \rightarrow Y \subseteq X);$
$\mathbf{UMID}_A$	-"-	$\forall f\,\Big(\mathbf{Clop}[f, A] \wedge \mathbf{Mon}[f, A] \rightarrow \mathbf{Lfp}[\mathsf{lfp} f, f, A]\Big).$

Now, **UMID** is the principle $\mathbf{UMID_V}$, where $\mathbf{V} \doteq \{x \mid x = x\}$ is (a name of) the universal type, and $\mathbf{UMID_N}$ is $\mathbf{UMID}_{\mathsf{nat}}$.

$\mathbf{Mon}[f, A]$ above means that the operation f is *monotone on* A. Plain "*f monotone*" means that f is monotone on **V**.

Lemma 19.32. *Define* $\mathbf{Lfp'}[Y, f, A]$ *to be* $\mathbf{Lfp}[Y, f, A] \wedge Y \subseteq fY$, $\mathbf{UMID}'_A$ *to be* $\forall f\,\Big(\mathbf{Clop}[f, A] \wedge \mathbf{Mon}[f, A] \rightarrow \mathbf{Lfp'}[\mathsf{lfp} f, f, A]\Big)$. *Then, on the basis of intuitionistic logic,* $\mathbf{UMID}_A \leftrightarrow \mathbf{UMID}'_A$.

Proof. The direction $\leftarrow$ is obvious. For $\rightarrow$, assume $\mathbf{UMID}_A$, $\mathbf{Clop}[f,A]$ and $\mathbf{Mon}[f,A]$. This implies $\mathsf{lfp}f \subseteq A$ and $f(\mathsf{lfp}f) \subseteq \mathsf{lfp}f$. By monotonicity of f we have $f(f(\mathsf{lfp}f)) \subseteq f(\mathsf{lfp}f)$. But this yields $\mathsf{lfp}f \subseteq f(\mathsf{lfp}f)$ by the remaining part of the $\mathbf{Lfp}[\mathsf{lfp}f,f,A]$ assertion. □

For every number term t of $\mathcal{L}_\mu$ one defines in the standard way its translation t^{EM} into the language $\mathcal{L}_{\mathrm{EM}}$.

19.4.2 Embeddings of $\mathcal{M}^i_\omega$ and $\mathcal{M}^i_{\varepsilon_0}$

In this subsection we start with proving Theorem 19.27(a), by showing how to translate a derivation in $\mathcal{M}^i_\omega$ into a derivation in $\mathbf{EETJ}\!\upharpoonright + \mathbf{UMID_N}$. Changes necessary to upgrade the argument to $\mathcal{M}^i_{\varepsilon_0}$ (Theorem 19.27(b)) will be indicated in the end.

First we note that every proof in $\mathcal{M}^i_\omega$ is a proof in $\mathcal{M}^i_n$ for some natural number n; from now on we fix this n. In Explicit Mathematics, we reason here in $\mathbf{EETJ}\!\upharpoonright + \mathbf{UMID_N}$.

Lemma 19.33. *There is an operation* J *s.t. if* $\forall y < n \forall Y \subseteq \mathsf{nat}\,(\mathbf{Clop}[fyY,\mathsf{nat}] \wedge \mathbf{Mon}[fyY,\mathsf{nat}])$ *then the following holds:*

(a) $\exists Z \subseteq \mathsf{nat}\,(Z = \mathsf{J}f)$;

(b) *for every* $y < n$ $(\mathsf{J}f)_y \doteq \mathsf{lfp}(fy \bigoplus_{z<y} (\mathsf{J}f)_z)$, *where* $A_i := \{x\,\varepsilon\,\mathsf{nat} \mid (i,x)\,\varepsilon\,A\}$, $\bigoplus_{z<y} A_z := \{(z,x) \mid z < y \wedge (z,x)\,\varepsilon\,A\}$, $(\cdot,\cdot)$ *and* $<$ *denote (translations of) appropriate operations/relations on natural numbers.*

Proof. Assume $\forall y < n \forall Y \subseteq \mathsf{nat}\,(\mathbf{Clop}[fyY,\mathsf{nat}] \wedge \mathbf{Mon}[fyY,\mathsf{nat}])$. By the $\mathbf{UMID_N}$ axiom we can define types J^y, for all $y < n$, to satisfy $\mathsf{J}^y = \mathsf{lfp}(fy \bigoplus_{z<y} \mathsf{J}^z)$. Finally we put $\mathsf{J}f := \bigoplus_{z<n} \mathsf{J}^z$. □

Definition 19.34. (t^x_φ, $f^{x,X}_\varphi$) For every first-order formula φ and a variable x of $\mathcal{L}_\mu$ we define a term t^x_φ of $\mathcal{L}_{\mathrm{EM},\mathsf{lfp}}$ by recursion on φ in the following way:

$$\mathsf{t}^x_\varphi := \begin{cases} \mathsf{emp} & \text{if } \varphi \text{ is } \bot; \\ \mathsf{inv}(\lambda x.\langle s^{\mathrm{EM}}[x], \mathsf{t}^{\mathrm{EM}}[x]\rangle, \mathsf{id}) & \text{if } \varphi \text{ is } s[x] = t[x]; \\ \mathsf{inv}(\lambda x.s^{\mathrm{EM}}[x], X) & \text{if } \varphi \text{ is } s[x] \in X; \\ \mathsf{and}(\mathsf{t}^x_{\varphi_0[x]}, \mathsf{t}^x_{\varphi_1[x]}) & \text{if } \varphi \text{ is } \varphi_0[x] \wedge \varphi_1[x]; \\ \mathsf{or}(\mathsf{t}^x_{\varphi_0[x]}, \mathsf{t}^x_{\varphi_1[x]}) & \text{if } \varphi \text{ is } \varphi_0[x] \vee \varphi_1[x]; \\ \mathsf{imp}(\mathsf{t}^x_{\varphi_0[x]}, \mathsf{t}^x_{\varphi_1[x]}) & \text{if } \varphi \text{ is } \varphi_0[x] \rightarrow \varphi_1[x]; \\ \mathsf{t}^x_{\forall y \varepsilon \mathsf{nat}(\langle y,x\rangle \varepsilon \mathsf{j}(\mathsf{nat},\lambda y.\mathsf{t}^x_{\psi[x,y]}))} & \text{if } \varphi \text{ is } \forall y \psi[x,y]; \\ \mathsf{t}^x_{\exists y \varepsilon \mathsf{nat}(\langle y,x\rangle \varepsilon \mathsf{j}(\mathsf{nat},\lambda y.\mathsf{t}^x_{\psi[x,y]}))} & \text{if } \varphi \text{ is } \exists y \psi[x,y]; \\ \mathsf{inv}(\lambda x.\langle x, s^{\mathrm{EM}}[x]\rangle, \mathsf{j}(\mathsf{nat}, \lambda x.\mathsf{J}(f[x]))), & \\ \quad \text{where } f[x] := \lambda z \lambda Z \lambda Y.\mathsf{and}(\mathsf{t}^y_{\psi[x,y,z,Y,Z]}, \mathsf{nat}) & \\ & \text{if } \varphi \text{ is } s[x] \in \mu yzYZ.\psi[x,y,z,Y,Z], \end{cases}$$

where in the quantifier clauses for an elementary formula $\eta[x]$ of $\mathcal{L}_{\mathrm{EM,lfp}}$ $\mathsf{t}^x_{\eta[x]}$ is the standard term s.t. $\mathsf{t}^x_{\eta[x]} :\doteq \{x \mid \eta[x]\}$ (see Lemma 19.31). The operation $f_\varphi^{x,X}$ is now defined as $\lambda X.\mathsf{and}(\mathsf{t}^x_\varphi, \mathsf{nat})$.

Definition 19.35. (φ^{EM}) For every formula φ of $\mathcal{L}_\mu$ we define its translation φ^{EM} of $\mathcal{L}_{\mathrm{EM,lfp}}$ by recursion on φ in the following way:

$$\varphi^{\mathrm{EM}} := \begin{cases} \bot & \text{if } \varphi \text{ is } \bot; \\ s^{\mathrm{EM}} = t^{\mathrm{EM}} & \text{if } \varphi \text{ is } s = t; \\ s^{\mathrm{EM}} \in X & \text{if } \varphi \text{ is } s \in X; \\ \varphi_0^{\mathrm{EM}} \wedge \varphi_1^{\mathrm{EM}} & \text{if } \varphi \text{ is } \varphi_0 \wedge \varphi_1; \\ \varphi_0^{\mathrm{EM}} \vee \varphi_1^{\mathrm{EM}} & \text{if } \varphi \text{ is } \varphi_0 \vee \varphi_1; \\ \varphi_0^{\mathrm{EM}} \to \varphi_1^{\mathrm{EM}} & \text{if } \varphi \text{ is } \varphi_0 \to \varphi_1; \\ \forall x \,\varepsilon\, \mathsf{nat}\, \psi^{\mathrm{EM}}[x] & \text{if } \varphi \text{ is } \forall x \psi[x]; \\ \exists x \,\varepsilon\, \mathsf{nat}\, \psi^{\mathrm{EM}}[x] & \text{if } \varphi \text{ is } \exists x \psi[x]; \\ \forall X \subseteq \mathsf{nat}\, \psi^{\mathrm{EM}}[X] & \text{if } \varphi \text{ is } \forall X \psi[X]; \\ \exists X \subseteq \mathsf{nat}\, \psi^{\mathrm{EM}}[X] & \text{if } \varphi \text{ is } \exists X \psi[X]; \\ s^{\mathrm{EM}} \,\varepsilon\, \mathsf{J}(\lambda y \lambda Y. f_\psi^{x,X}) & \text{if } \varphi \text{ is } s \in \mu xyXY.\psi[x,y,X,Y]. \end{cases}$$

In short, this translation leaves everything intact, except it relativizes quantifiers to (subtypes of) nat and replaces μ by iterated lfp applied to appropriate operations.

Lemma 19.36. *For a term $T := \mu xyXY.\varphi \in \mathcal{L}_\mu$ if $\left(\mathrm{Mon}(\mathcal{M}(T), n)\right)^{\mathrm{EM}}$ then the following holds:*

(a) t^x_φ *is a name, i.e.* $\exists Z(Z = \mathsf{t}^x_\varphi)$*;*
(b) $\forall y < n \forall Y \subseteq \mathsf{nat}\, (\mathbf{Clop}(f_\varphi^{x,X}, \mathsf{nat}) \wedge \mathbf{Mon}(f_\varphi^{x,X}, \mathsf{nat}))$.

Proof. (a) and (b) are proved simultaneously by induction on T. By IH we may assume that the assertion holds for all μ-terms $\mu x_i y_i X_i Y_i.\chi$ occurring in φ but not inside its μ-terms. Assuming $\left(\mathrm{Mon}(\mathcal{M}(T), n)\right)^{\mathrm{EM}}$, we therefore have $\forall y_i < n \forall Y_i \subseteq \mathsf{nat} \forall X_i \subseteq \mathsf{nat}\, \exists Z_i (Z_i = \mathsf{t}^{x_i}_{\chi_i}) \wedge \forall y_i < n \forall Y_i \subseteq \mathsf{nat}\, (\mathbf{Clop}(f_{\chi_i}^{x_i,X_i}, \mathsf{nat}) \wedge \mathbf{Mon}(f_{\chi_i}^{x_i,X_i}, \mathsf{nat}))$, for all i, and for all $x \varepsilon \mathsf{nat}$, $y < n$, $X \subseteq \mathsf{nat}$ and $Y \subseteq \mathsf{nat}$. By the $\mathbf{LFP}_n[\chi_i, \mu x_i y_i X_i Y_i.\chi_i]$ axioms and Definition 19.34 this proves $\exists Z\,(Z = \mathsf{t}^x_\varphi)$, i.e. (a), for any $y < n$, $X \subseteq \mathsf{nat}$ and $Y \subseteq \mathsf{nat}$. Now another use of $\left(\mathrm{Mon}(\mathcal{M}(T), n)\right)^{\mathrm{EM}}$ and Definition 19.34 gives us (b). □

Lemma 19.37. *If $\varphi[x, y, X, Y]$ is first-order then* $\mathbf{EETJ}{\restriction} + \mathbf{UMID_N}$ *proves* $\forall y < n$ $\forall Y \subseteq \mathsf{nat} \forall x \,\varepsilon\, \mathsf{nat} \forall X \subseteq \mathsf{nat} \left(\varphi^{\mathrm{EM}}[x, y, X, Y] \leftrightarrow x \,\varepsilon\, (\lambda y \lambda Y. f_\varphi^{x,X}) y Y X\right)$.

Proof. This follows immediately from Definitions 19.34 and 19.35. □

Lemma 19.38. *$\mathcal{M}_n^i$ can be interpreted in* $\mathbf{EETJ}{\restriction} + \mathbf{UMID_N}$.

Proof. This is straightforward now. Interpretations of all axioms, except the $\mathbf{LFP}_n$ axiom (19.4) and $\mathbf{TI}_n$, are standard. In particular, Arithmetical Comprehension requires Elementary Comprehension of **EM**, and restricted induction calls for the restricted induction on nat. $(\mathbf{TI}_n)^{\mathrm{EM}}$ is derivable by logic. Now we turn to the $\mathbf{LFP}_n$ axiom. Applying Definition 19.35, in Explicit Mathematics this turns into

$$\Big(\mathrm{Mon}(\mathcal{M}(T),n)\Big)^{\mathrm{EM}} \to \Big(\forall y{<}n\, \mathrm{LFP}_{x,X}[\varphi[x,y,X,T_{<y}],T_y]\Big)^{\mathrm{EM}},$$

where $T := \mu xyXY.\varphi[x,y,X,Y]$. So assume $\Big(\mathrm{Mon}(\mathcal{M}(T),n)\Big)^{\mathrm{EM}}$. We are left with the task to prove

$$\forall y{<}n\Bigg(\Big(\forall x\,(x \in T_y \leftrightarrow \varphi[x,y,T_y,T_{<y}])\Big)^{\mathrm{EM}} \wedge \\ \Big(\forall U(\forall z(\varphi[z,y,U,T_{<y}] \to z \in U) \to T_y \subseteq U)\Big)^{\mathrm{EM}}\Bigg),$$

i.e.

$$\forall y{<}n\Big(\forall x\,\varepsilon\,\mathsf{nat}\,(x\,\varepsilon\,\mathsf{T}_y \leftrightarrow \varphi^{\mathrm{EM}}[x,y,\mathsf{T}_y,\mathsf{T}_{<y}]) \wedge \\ \forall U{\subseteq}\mathsf{nat}(\forall z\,\varepsilon\,\mathsf{nat}(\varphi^{\mathrm{EM}}[z,y,U,\mathsf{T}_{<y}] \to z\,\varepsilon\,U) \to \mathsf{T}_y \subseteq U)\Big),$$

where $\mathsf{T} := \mathsf{J}(\lambda y\lambda Y.f_\varphi^{x,X})$. By Lemma 19.37 this is equivalent to

$$\forall y{<}n\Big(\forall x\,\varepsilon\,\mathsf{nat}\,(x\,\varepsilon\,\mathsf{T}_y \leftrightarrow x\,\varepsilon\,(\lambda y\lambda Y.f_\varphi^{x,X})y\mathsf{T}_{<y}\mathsf{T}_y) \wedge \\ \forall U{\subseteq}\mathsf{nat}(\forall z\,\varepsilon\,\mathsf{nat}(z\,\varepsilon\,(\lambda y\lambda Y.f_\varphi^{x,X})y\mathsf{T}_{<y}U \to z\,\varepsilon\,U) \to \mathsf{T}_y \subseteq U)\Big),$$

which follows from $\mathbf{UMID_N}$ axiom by Lemmata 19.36, 19.33 and 19.32. □

This completes embedding of $\mathcal{M}^i_\omega$ into $\mathbf{EETJ}{\restriction} + \mathbf{UMID_N}$, and thereby the proof of Theorem 19.27(a). For Theorem 19.27(b), we proceed in exactly the same way, with the following small changes: (1) From the very beginning, we observe that a derivation in $\mathcal{M}^i_{\varepsilon_0}$ is a derivation in $\mathcal{M}^i_\beta$ for some $\beta < \varepsilon_0$. We use this β in place of $n < \omega$ from the case of $\mathcal{M}^i_\omega$. (2) In the proof of Lemma 19.33 we need to refer to Recursion Theorem 19.29 and transfinite induction for β. Namely, existence of an operation $\tilde{\mathsf{J}}$ s.t. $(\tilde{\mathsf{J}}f)_y \simeq \mathsf{lfp}(fy\bigoplus_{z<y}(\tilde{\mathsf{J}}f)_z)$ follows from Recursion Theorem, but, taking as before $\mathsf{J}f := \bigoplus_{y<\beta}\tilde{\mathsf{J}}f$, in order to prove $\exists Z \subseteq \mathsf{nat}\,(Z = \mathsf{J}f)$ and $\forall y{<}\beta(\mathsf{J}f)_y \doteq \mathsf{lfp}(fy\bigoplus_{z<y}(\mathsf{J}f)_z)$ we need to use $\mathbf{TI}_\beta$. (3) In the beginning of proof of Lemma 19.38, we observe that $(\mathbf{TI}_\beta)^{\mathrm{EM}}$ is a theorem of **EETJ**.

Acknowledgements The authors were supported in this work by United Kingdom Engineering and Physical Sciences Research Council Grant GR/R 15856/01. The bulk of the research was carried out in 2005 at the Ohio State University. As regards the completion of this paper, the first author is grateful for support by the John Templeton Foundation ("A new dawn of intuitionism: mathematical and philosophical advances," ID 60842).[1]

References

1. A. Arnold and D. Niwinski. *Rudiments of the μ-calculus*. Elsevier, Amsterdam, 2001.
2. W. Buchholz, S. Feferman, W. Pohlers, and W. Sieg. *Iterated Inductive Definitions and Subsystems of Analysis*, volume 897 of *Lecture Notes in Mathematics*. Springer, 1981.
3. S. Feferman. A language and axioms for explicit mathematics. In *Algebra and Logic*, volume 450 of *Lecture Notes in Mathematics*, pages 87–139. Springer, 1975.
4. S. Feferman. Constructive theories of functions and classes. In J.N. Crossley, editor, *Logic Colloquium '78*, pages 159–224. North-Holland, 1979.
5. S. Feferman. Monotone inductive definitions. In A.S. Troelstra and D. van Dalen, editors, *The L.E.J. Brouwer Centenary Symposium*, pages 77–89. North-Holland, 1982.
6. T. Glaß, M. Rathjen, and A. Schlüter. The strength of monotone inductive definitions in explicit mathematics. *Annals of Pure and Applied Logic*, 85:1–46, 1997.
7. D. Hilbert and P. Bernays. *Grundlagen der Mathematik II*. Springer, Berlin, 1938.
8. P. Hitchcock and D. M. R. Park. Induction rules and termination proofs. In *Procceedings of the 1st International Colloquium on Automata, Languages, and Programming*, pages 225–251. North-Holland, Amsterdam, 1973.
9. D. Kozen. Results on the propositional μ-calculus. *Theoretical Computer Science*, 27:333–354, 1983.
10. G. Kreisel. Generalized inductive definitions. In *Stanford Report on the Foundations of Analysis Section III*. Mimeographed, Stanford, 1963.
11. B. Lubarsky. μ-definable sets of integers. *Journal of Symbolic Logic*, 58:291–313, 1993.
12. M. Möllerfeld. *Generalized Inductive Definitions. The μ-calculus and Π^1_2-comprehension*. Phd dissertation, Universität Münster, 2002.
13. D. M. R. Park. Fixpoint induction and proof of program semantics. In Mitchie Meltzer, editor, *Machine Intelligence V*, pages 59–78. Edinburgh University Press, Edinburgh, 1970.
14. V. Pratt. A decidable μ-calculus. In *Proceedings 22nd IEEE Symposium on Foundations of Computer Science*, pages 421–427. 1981.
15. M. Rathjen. Monotone inductive definitions in explicit mathematics. *Journal of Symbolic Logic*, 61:125–146, 1996.
16. M. Rathjen. Explicit mathematics with the monotone fixed point principle. *Journal of Symbolic Logic*, 63:509–542, 1998.
17. M. Rathjen. Explicit mathematics with the monotone fixed point principle II: Models. *Journal of Symbolic Logic*, 64:517–550, 1999.
18. M. Rathjen. Explicit mathematics with monotone inductive definitions: a survey. In W. Sieg et al., editor, *Reflections on the Foundations of Mathematics: Essays in Honor of Solomon Feferman*, volume 15 of *Lecture Notes in Logic*, pages 329–346. ASL and AK Peters, 2002.
19. D. Scott and J.W. DeBakker. A theory of programs. Unpublished manuscript, 1969.
20. S. Takahashi. Monotone inductive definitions in a constructive theory of functions and classes. *Ann. Pure Appl. Logic*, 42:255–279, 1989.
21. S. Tupailo. Realization of constructive set theory into explicit mathematics: a lower bound for impredicative Mahlo universe. *Annals of Pure and Applied Logic*, 120:165–196, 2003.
22. S. Tupailo. On the intuitionistic strength of monotone inductive definitions. *Journal of Symbolic Logic*, 69:790–798, 2004.

[1] The opinions expressed in this publication are those of the authors and do not necessarily reflect the views of the John Templeton Foundation.

Chapter 20
Foundations of Mathematics: an Optimistic Message

Stephen G. Simpson

Abstract Historically, mathematics has often been regarded as a role model for all of science—a paragon of abstraction, logical precision, and objectivity. The 19th and early 20th centuries saw tremendous progress. The great mathematician David Hilbert proposed a sweeping program whereby the entire panorama of higher mathematical abstractions would be justified objectively and logically, in terms of finite processes. But then in 1931 the great logician Kurt Gödel published his famous incompleteness theorems, leading to an era of confusion and skepticism. In this talk I show how modern foundational research has opened a new path toward objectivity and optimism in mathematics.[1]

In this short essay I will try to convey something of the flavor of my research area, mathematical logic and foundations of mathematics.

Throughout history, mathematics and logic have been regarded not only as indispensable components of most other sciences, but also as role models for all of science – the most objective and logically perfect of all the sciences. For these reasons, scientists throughout history have thought that it is particularly important to understand the logical structure of mathematics. And that is the need which my research area seeks to fulfill. *Foundations of mathematics* is the study of the most basic concepts and logical structure of mathematics.

Of course, mathematics is an ancient science which has evolved and continues to evolve over time. And it may be hard to believe, but throughout history there have been a number of *foundational crises* – times when there was *severe doubt* about the logical basis of mathematics. What I want to do here is to explore one of the issues

S. G. Simpson
Vanderbilt University, Nashville TN 34720, USA · Pennsylvania State University, University Park, PA 16802, USA, e-mail: sgslogic@gmail.com

[1] This is the text of a lecture for the general public, delivered at the National University of Singapore in 2016. Figure 20.12 is a publicity poster, and the slides for the lecture are included as Figures 20.1 through 20.11.

R. Kahle and M. Rathjen (eds.), *The Legacy of Kurt Schütte*,
https://doi.org/10.1007/978-3-030-49424-7_20

which have dominated foundational research. The issue in question is centered on the concept of *infinity*.

"The infinite! No other question has ever moved so profoundly the spirit of man."

Fig. 20.1 David Hilbert (1862–1943) Source: Wikipedia, public domain.

I begin with a quotation from the greatest mathematician of the early 20th century, David Hilbert. See Figure 20.1.

> "The infinite! No other question has ever moved so profoundly the spirit of man."

This ringing statement was the beginning of Hilbert's profound essay "On the infinite." See Reference [8]. Hilbert was a great mathematician, but why did he say this? Surely there are many *finite* things that move us profoundly, such as our families and our careers. What is so moving about the infinite? Let me return to this question later, after giving you some background concerning the infinite and its role in mathematics.

Hilbert's essay "On the infinite" was a mature and definitive statement of his plan to place mathematics on a firm, solid, objective, logical foundation. But why was there even a need for such a foundation? After all, mathematics is normally regarded as the most basic, most certain, most logically perfect of all the sciences. Mathematics is the science of measurement and quantity, and *as* the science of measurement and quantity, it is an essential component of all of the hard sciences.

Furthermore, beyond its role in the hard sciences, mathematics has often been viewed as an excellent laboratory for intellectual training. Among the ancient Greek thinkers such as Plato, rigorous training in mathematics was regarded as an indispensable prerequisite for all serious intellectual activity. Above the gateway entrance to Plato's Academy was inscribed the phrase "Let none but geometers enter here."

This widely accepted story is perhaps unverifiable, but at the same time it is certainly true in spirit. The point is that mathematics, perhaps more than other sciences, deals in powerful but precisely defined abstractions. By developing our ability to handle mathematical abstractions correctly, we also develop our ability to handle abstractions in other areas such as philosophy, law, politics, and even ethics.

So again, in an era when mathematics was flourishing and making great strides in many directions, why were great mathematicians such as Hilbert concerned about the foundations of mathematics, i.e., the most basic concepts and logical structure of mathematics? Why did they feel a need to shore up the foundations?

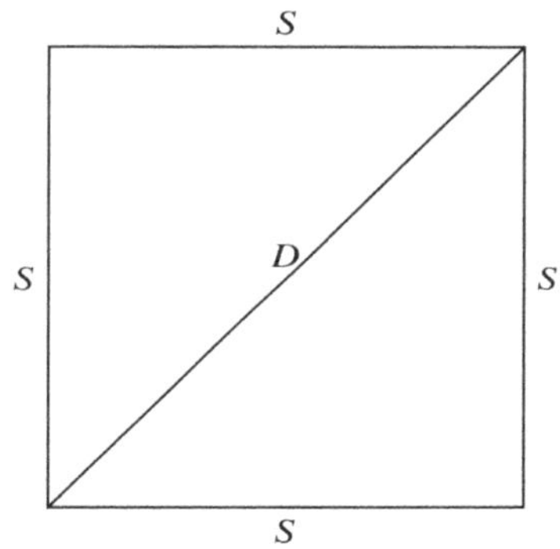

$\sqrt{2}$ = the square root of 2 = $\frac{D}{S}$.

$\sqrt{2}$ is approximately equal to $\frac{99}{70}$.

$\sqrt{2}$ is approximately equal to $\frac{665857}{470832}$.

$\sqrt{2}$ is "exactly" equal to 1.4142135623730950488016887 2 ....

Fig. 20.2 The square root of 2.

The answer is that, even as mathematics was flowering, some of the newer and more abstract mathematical abstractions became, well, a little bit hard to swallow. These new abstractions seemed to be more and more remote from observable reality, and this made mathematical practitioners uneasy. There was a real and growing concern that mathematics may have moved into some sort of "twilight zone."

Actually, this kind of unease had a precedent in ancient Greece. More than 2500 years ago, the philosophical school of Pythagoras was obsessed with *pure numbers*. But consider for example the abstraction of a geometrical square. Clearly, the ratio of the diagonal of the square to any one of the four sides of the square is a pure number, independent of the physical lengths of the diagonal or of the sides. This number is now known as $\sqrt{2}$, the square root of 2. It is the unique positive number which, when multiplied by itself, gives exactly 2. See Figure 20.2. For a long period of time, it was believed that each and every pure number could be described as a ratio of whole numbers, for example the ratio 7 divided by 5, or the ratio 99 divided by 70. However, the Pythagoreans were horrified to discover that this is not the case. In fact, they discovered that the square root of 2 *can never be described exactly* as the ratio of two whole numbers. It is *approximated* rather well as the ratio 99 divided by 70, and it is *approximated* even better as the ratio 665857 divided by 470832. But the Pythagoreans proved that there is no way to express $\sqrt{2}$ *exactly* as a ratio a divided by b where a and b are whole numbers, no matter how large a and b are. In other words, $\sqrt{2}$ was the first example of what is now called an *irrational number*. The Pythagoreans viewed this discovery as horrifying and subversive. For a significant amount of time they tried to keep the irrationality of $\sqrt{2}$ hidden as a deep, dark, esoteric secret, known only among their own exclusive group. Historically we may say that the discovery of the irrationality of $\sqrt{2}$ led to the first major crisis in the foundations of mathematics.

Moving forward through history, there has been a sequence of higher and higher mathematical abstractions, more and more remote from real-world measurements and real-world quantities. Even in the realm of *numbers*, the names chosen for these new classes of numbers give a hint of the kind of tension that has always existed in mathematics and science, between the abstract and the concrete. The very phrase *irrational number* may be viewed as an arrogant challenge to the power of reason and science. And far beyond the irrational numbers, 19th and 20th century mathematicians have introduced ever-more-remote abstractions such as the *transcendental* numbers, the *imaginary* numbers, the *complex* numbers, the *transfinite* ordinal numbers, the *uncountable* cardinal numbers, the *inaccessible* numbers, the *hyperinaccessible* numbers, the *ineffable* numbers, etc. Each of these advances met with considerable resistance when it was first introduced. Note also that each of these classes of numbers involves an appeal to the infinite. For instance, the decimal expansion of $\sqrt{2}$ is an infinite and nonrepeating sequence of digits,

$$1.41421356237309504880168872420969807856967187537694807317667973799 \ldots .$$

And the other kinds of abstract numbers which I just mentioned involve infinities which are even more extreme.

Throughout history, the tension between the abstract versus the concrete, or the infinite versus the finite, has been been pondered deeply. This tension is depicted graphically in a famous fresco, known as "The School of Athens," by the Italian Renaissance artist Raphael. See Figures 20.3 and 20.4. In this 16th century masterpiece, Raphael depicts the two masters of ancient Greek philosophy, Plato and Aristotle, walking among their Athenian colleagues and debating the nature of the infinite and its relationship to the finite. Plato is pointing upward, toward the heavens, while Aristotle is spreading his hand powerfully and benevolently over the earth. This painting by Raphael is often regarded as the perfect embodiment of the classical spirit of the Renaissance. In the context of Raphael's fresco, Hilbert's statement about the spirit of man may be more readily understood.

Fig. 20.3 The School of Athens (about 360 B.C.), by Raphael (1483–1520).
Source: Wikipedia, public domain.

Aristotle is arguably the greatest philosopher of all time. His major work on the philosophy of existence is known as *the Metaphysics*. The Metaphysics is a massive tome, hundreds of pages long. The purpose of the Metaphysics is to answer the most profound questions about *the nature of existence*. A typical question discussed in the Metaphysics is, do abstractions exist in reality, or only in the human mind? As part of the Metaphysics, Aristotle includes a highly nuanced and thorough discussion of the nature of *mathematical existence*. Books M and N of the Metaphysics are the first-ever treatise on the philosophy of mathematics. See Reference [1].

As part of Books M and N of the Metaphysics, Aristotle fashions a *grand compromise* between the infinite and the finite. This grand compromise depends on a crucial distinction between two kinds of infinity: *potential infinity* versus *actual infinity*.

Potential infinity is the kind of infinity associated with a process that can be repeated or continued indefinitely. Many processes of this kind can be observed in reality. For instance, there is the earth turning on its axis, the planets revolving around the sun, the march of time, the motion of a body in outer space, the cycle of generations, the accumulation of wealth, the accumulation of knowledge, etc. Each of these processes is perfectly capable of repeating or continuing, without any forseeable limit. There are many examples of potential infinity in mathematics. The most important example is the potentially infinite sequence of whole numbers, 1, 2, 3, 4, 5, 6, 7, dot dot dot. Another example is a potentially infinite sequence of smaller

Fig. 20.4 Plato and Aristotle (about 360 B.C.) Source: Wikipedia, public domain.

and smaller rational numbers, for example 1, 1/2, 1/4, 1/8, 1/16, 1/32, dot dot dot. Yet another example is the potentially infinite sequence of decimal approximations to the square root of 2, namely 1 point 4 1 4 2 1 3 5 dot dot dot. The process of computing more and more decimal digits of the square root of 2 can be continued indefinitely. All of these potential infinities in mathematics play an important role in our understanding of real-world phenomena.

Actual infinity is another kind of infinity, standing in contrast to potential infinity. An *actual infinity* is a *completed infinite totality*, with emphasis on the word *completed*. As a mathematical example, imagine that we have just now finished enumerating *the entire sequence of all the whole numbers*, and then imagine that we pause to look backward over the entire sequence. This stopping point is sometimes denoted by the symbol ω, because ω (pronounced "omega") is the last letter of the Greek alphabet. In abstract mathematics, the symbol ω denotes an actual infinity, a completed infinite totality. The symbol ω may be viewed, and often *is* viewed, as standing for the collection of *all* whole numbers, all gathered together into one infinite basket.

The philosophical question at issue here is:

Which kinds of infinity have an objective existence?

Do potential infinities exist objectively? Do actual infinities exist objectively? Aristotle comes down on the side of a rather compelling philosophical position. The Aristotelean position may be called *finitism*, but it is really a compromise between the finite and the infinite. According to Aristotelean finitism, potential infinities are part of objective existence, but actual infinities are not.

Over the centuries, many thinkers have agreed with Aristotle's finitism, simply because in observable reality there are many obvious examples of potential infinity, but there are no convincing examples of actual infinity. On the other hand, there have also been many non-finitistic thinkers who profoundly disagreed with Aristotle. Some of them have endorsed a great variety of actual infinities, typically on the grounds that these actual infinities exist as an important component of theological, philosophical, or mathematical thought. As a possible response to these non-finitist thinkers, we "hard-headed realists" could note that existence-in-thought is different from the kind of objective existence which Aristotle was analyzing.

While we are at it, let me note that a great majority of contemporary mathematical practitioners are indifferent to philosophical questions such as potential infinity versus actual infinity. They wish only to carry on with their day-to-day mathematical work, extending current mathematical paradigms and unencumbered by philosophical issues.

We have been dwelling on ancient Greece, and especially Aristotle's finitism. And again I want to stress that Aristotle's finitism does not reject all infinities. Rather, Aristotle's finitism fully accepts the concept of potential infinity, while at the same time rejecting the concept of actual infinity.

But now let us fast-forward to the late 19th and early 20th centuries. During this relatively modern period of history, mathematics was undergoing a great expansion in all directions: differential equations, geometry, number theory, algebra, etc. And as part of this tremendous flowering of mathematics, great strides were being made in the direction of mathematical rigor and logical precision. The work of Cauchy, Weierstrass, Dedekind, Peano, Frege, Cantor, Russell, Whitehead, and Zermelo cleared the way toward a perfect logical framework for all of mathematics. A landmark was the treatise of Hilbert and his assistant Ackermann. Hilbert and Ackermann presented an elegant and completely symbolic formulation of the logical axioms and rules for what is now known as the predicate calculus. Even though digital computers did not yet exist, it was in some sense clear that mathematics and perhaps many other scientific disciplines could in principle be completely *formalized*, i.e., programmed for digital computers.

Later in the 20th century, levels of mathematical abstraction continued to increase rapidly, with a great variety of actual infinities. Consequently, due in part to the non-objective nature of these actual infinities, there were nagging and sometimes massive doubts about the validity of mathematics. The philosopher Russell famously characterized mathematics as "*the science in which we never know what we are talking about, nor whether what we say is true.*" The mathematical physicist Wigner declared that there is no rational explanation for the usefulness of mathematics in the physical sciences. The mathematician Morris Kline published an influential popular book entitled "*Mathematics: the loss of certainty.*" None of these scientists were viewed as enemies of mathematics. They were viewed as merely raising legitimate doubts about the validity of mathematics.

Highlighting a key step in this loss of certainty in mathematics, I want to say something about the ideas of the great mathematician David Hilbert and the great logician Kurt Gödel. See Figures 20.1 and 20.5.

"Objective concepts of mathematics are fundamental to my work in logic."

Fig. 20.5 Kurt Gödel (1906–1978) Source: Wikipedia, public domain.

Fig. 20.6 Hilbert's Program (1926):

Using the tools of mathematical logic, Hilbert proposed to prove that all of mathematics, including the non-finitistic parts of mathematics, is reducible to purely finitistic mathematics. In this way, the objectivity of mathematics would be confirmed.

Fig. 20.7 Gödel's refutation of Hilbert's Program (1931):

Gödel used mathematical logic to prove that some parts of non-finitistic mathematics are not reducible to finitistic mathematics. This includes the "medium" and "strong" levels of the Gödel hierarchy. Thus, the objectivity of mathematics is left in doubt.

Concerning our question about Aristotle's finitism, Hilbert had already made dramatic use of non-finitistic methods early in his career, in his work in a branch of algebra called invariant theory. Other finitistically-minded mathematicians criticized Hilbert's method. For instance, the mathematician Gordan said that Hilbert's work in this area was not mathematics but rather theology. Later, at the peak of his career, Hilbert argued forcefully for the acceptance of actual infinities in mathematics. "*No one shall expel us from the paradise that Cantor has created for us!*" The paradise of which Hilbert was speaking was, of course, a massively non-finitistic paradise, chock full of actual infinities.

However, Hilbert also appreciated the finitistic viewpoint. Indeed, he saw finitistic mathematics as the most indispensable part, the *meaningful core* of mathematics. Attempting to reconcile these two viewpoints, Hilbert developed a sophisticated plan whereby non-finitistic mathematics would be "reduced" to finitistic mathematics, i.e., validated in a fully finitistic way. This would be achieved in three steps.

1. Formalize finitistic and non-finitistic mathematics as two distinct but logically perfect systems of axioms and rules of inference. The key technical tool here would be the predicate calculus of Hilbert and Ackermann.
2. Embed the finitistic system as a subsystem of the non-finitistic system. In essence, these first two steps had already been accomplished by the late 1920s.
3. Argue finitistically that any finitistically meaningful conclusion which is deducible in the non-finitistic system would also be deducible in the finitistic system.

In this way, all of the actual infinities would be seen as "ideal elements" or "harmless but convenient fictions." Because of this reduction, mathematicians would have full license to use the actual infinities freely. The actual infinities would be *guaranteed to be harmless*, once and for all, because we would know in advance that they could never contaminate the finitistically meaningful core of mathematics. See Figure 20.6.

In short, when it came to the question of potential infinity versus actual infinity in mathematics, Hilbert was trying to ride to the rescue. He was proposing a bold, sophisticated, far-reaching plan to compromise between finitism and non-finitism. He wanted mathematicians to have it both ways, i.e., to have our cake and eat it too. This proposal is now known as *Hilbert's program* or, in a more technical vein, *finitistic reductionism*. The full statement of Hilbert's program is spelled out in his 1926 essay entitled "On the infinite." See References [8, 15].

But then, five years later in 1931, Gödel published his famous paper proving that Hilbert's program is impossible to carry out in its entirety. In fact, the work of Gödel revealed a hierarchy of stronger and stronger formal mathematical systems, relying more and more extensively on actual infinities, leading to more and more consequences which are finitistically meaningful but not finitistically verifiable. This hierarchy is known as *the Gödel hierarchy*. See Figure 20.8 and References [6, 11, 15]. The Gödel hierarchy undercuts Hilbert's program, because it shows us that every time we introduce a new kind of actual infinity, it can result in more and more contamination of the finitistically meaningful core. See Figures 20.7 and 20.8.

This discovery by Gödel was viewed as a complete and total defeat for mathematical finitism. The mathematical world seemed to be left with only two options. Either cling to objectivity by rejecting all non-finitistic methods, or abandon objectivity by embracing the panorama of actual infinities. Since mathematicians were by that time quite comfortable with actual infinities, they almost universally chose the second option. Finitism, finitistic reducibility, and objectivity were now apparently out the window. Even today, despite modern advances, the popular wisdom or "settled science" is that Gödel's theorems meant the end of Hilbert's program and the end of objectivity in mathematics.

Fig. 20.8 The Gödel hierarchy

"strong":

- ⋮
- huge cardinal numbers
- ⋮
- ineffable cardinal numbers
- ⋮
- ZFC (Zermelo/Fraenkel set theory)
- ZC (Zermelo set theory)
- simple type theory

"medium":

- Z_2 (second-order arithmetic)
- ⋮
- Π^1_2-CA_0 (Π^1_2 comprehension)
- Π^1_1-CA_0 (Π^1_1 comprehension)
- ATR_0 (arithmetical transfinite recursion)
- ACA_0 (arithmetical comprehension)

"weak":

- WKL_0 (weak König's lemma)
- RCA_0 (recursive comprehension)
- PRA (primitive recursive arithmetic)
- EFA (elementary function arithmetic)
- bounded arithmetic
- ⋮

In the title of this lecture, I promised an optimistic message. So let me now finish by reporting on some results of modern research, dating from the 1970s to the present. The big discovery here is that, despite Gödel, a large and significant part of Hilbert's program can in fact be carried out with brilliant success!

To explain how this can be so despite Gödel, let me return to the Gödel hierarchy. There are two important points to make. See Figure 20.10 and References [9, 10, 11, 12, 13].

Fig. 20.9 Reverse mathematics

Reverse mathematics is a series of case studies to determine which parts of mathematics belong to which levels of the Gödel hierarchy.

Fig. 20.10 Two discoveries

1. Using the methods envisioned by Hilbert, we can prove that "weak" levels of the Gödel hierarchy are finitistically reducible, in the sense of Hilbert's program.
2. Reverse-mathematical case studies provide solid evidence that the "weak" levels cover at least 85 percent of mathematics. This includes most or all of the "applicable" parts of mathematics.

Combining 1 and 2, we conclude that Hilbert's program is largely valid.

The first point is that Hilbert's program can be carried out for the so-called "weak" levels of the Gödel hierarchy. See Figures 20.6 and 20.8 and References [2, 10, 11]. In other words, we can prove finitistically that for any proof of a finitistically meaningful statement in one of these "weak" non-finitistic systems, we can find an alternative proof which is purely finitistic. This is a consequence of the proof-theoretical methods pioneered by Hilbert and Gentzen, as outlined by Hilbert in his 1926 essay.

In order to explain the second point, I need to discuss an ongoing research direction in the foundations of mathematics. Beginning in the 1970s and continuing through the present, mathematical logicians have carried out a series of precise foundational case studies known as *reverse mathematics*. See Figure 20.9 and References [10, 11]. The goal of reverse mathematics is to determine which levels of the Gödel hierarchy are essential for the formalization of which parts of mathematics. This huge and ongoing research program, reverse mathematics, is the subject of an international conference or workshop being held this week and next (January 3–16, 2016) here at NUS, at the Institute for Mathematical Sciences.

As an outcome of reverse mathematics, we now know that the bulk of mathematics, including a great deal of non-finitistic mathematics, can in fact be formalized at the "weak" levels of the Gödel hierarchy. This is of course an ongoing project, but I estimate that it applies to something like 85 to 95 percent of mathematics as it exists today, including not only constructive or computable mathematics but also a great deal of highly nonconstructive mathematics. See References [9] and [10, Chapter IV].

In many cases, some care is needed. For instance, a central feature of the usual proof of Peano's existence theorem for solutions of ordinary differential equations is a sequential compactness argument involving the Ascoli lemma. In order to show that the Peano theorem is finitistically reducible, we need to rework this sequential compactness argument into a covering compactness argument. See [10, §IV.8]. However, the reworking is purely technical and does not affect the mathematical significance.

So what is my optimistic message? My optimistic message is that, by focusing on the "weak" levels of the Gödel hierarchy, we can carry out Hilbert's program for at least 85 percent of mathematics. In other words, we can have our cake and eat at least 85 percent of it. Yes, some of the most blatant non-finitistic moves can result in contamination, but not by any means all of them. Our message is that, if mathematicians exercise some care and work within the strictures of the known "weak" systems or other systems which are finitistically reducible, then all is well. The finitistic core is protected from contamination. Hilbert's program lives! See Figure 20.11 and Reference [9].

Fig. 20.11 My optimistic message:

Most of mathematics has an objective basis!

To summarize, what I am suggesting is that, even though Hilbert's program cannot be carried out in its entirety, there is a less sweeping version of Hilbert's program which can be carried out, and which represents the *proper balance* between the finite and the infinite in mathematics.

Now, if only we could find the proper balance between the finite and the infinite in theology!

Acknowledgements The author's research is supported by Simons Foundation Collaboration Grant 276282.

References

1. Hippocrates G. Apostle. *Aristotle's Philosophy of Mathematics*. University of Chicago Press, 1952.
2. John P. Burgess. On the outside looking in: a caution about conservativeness. In *[4]*, pages 128–141. Association for Symbolic Logic, Cambridge University Press, 2009.
3. C.-T. Chong, Q. Feng, T. A. Slaman, and W. H. Woodin, editors. *Infinity and Truth*, volume 25. Institute for Mathematical Sciences, National University of Singapore. World Scientific, 2014.
4. S. Feferman, C. Parsons, and S. G. Simpson, editors. *Kurt Gödel: Essays for his Centennial*, volume 33 of *Lecture Notes in Logic*. Association for Symbolic Logic, Cambridge University Press, 2010.
5. Kurt Gödel. Über formal unentscheidbare Sätze der Principia Mathematica und verwandter Systeme, I. *Monatshefte für Mathematik und Physik*, 38(1):173–198, 1931.

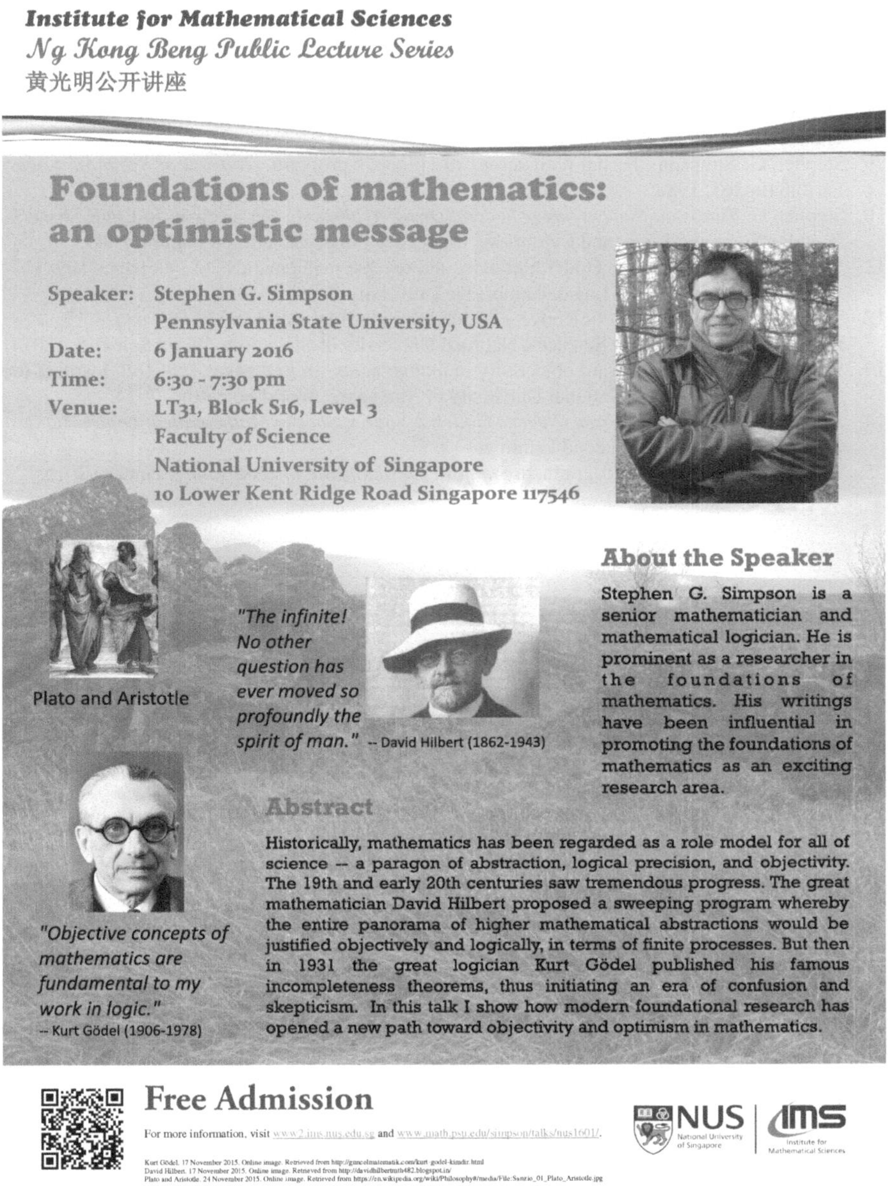

Fig. 20.12 Public lecture at the National University of Singapore.
Source: National Univ. of Singapore, Inst. for Mathematical Sciences. Reprinted with permission.

6. Kurt Gödel. On formally undecidable propositions of Principia Mathematica and related systems. In *[14]*, pages 596–616. Harvard University Press, 1967. Translation of [5] by J. Van Heijenoort.
7. David Hilbert. Über das Unendliche. *Mathematische Annalen*, 95(1):161–190, 1926.
8. David Hilbert. On the infinite. In *[14]*, pages 367–392. Harvard University Press, 1967. Translation of [7] by S. Bauer-Mengelberg.
9. Stephen G. Simpson. Partial realizations of Hilbert's program. *Journal of Symbolic Logic*, 53(2):349–363, 1988.
10. Stephen G. Simpson. *Subsystems of Second Order Arithmetic*. Perspectives in Logic. Association for Symbolic Logic and Cambridge University Press, 2nd edition edition, 2009.
11. Stephen G. Simpson. The Gödel hierarchy and reverse mathematics. In *[4]*, pages 109–127. Association for Symbolic Logic, Cambridge University Press, 2010.
12. Stephen G. Simpson. An objective justification for actual infinity? In *[3]*, pages 225–228. Institute for Mathematical Sciences, National University of Singapore. World Scientific, 2014.
13. Stephen G. Simpson. Toward objectivity in mathematics. In *[3]*, pages 157–169. Institute for Mathematical Sciences, National University of Singapore. World Scientific, 2014.
14. J. van Heijenoort, editor. *From Frege to Gödel: A Source Book in Mathematical Logic*. Harvard University Press, 1971. Second Printing.
15. Richard Zach. Hilbert's Program. In *Stanford Encyclopedia of Philosophy*. January 2015.

Chapter 21
A Glimpse of Σ_3-elementarity

Gunnar Wilken

Abstract We take the first step toward analyzing pure Σ_3-elementary substructures of ordinal numbers.

21.1 Introduction

The question of whether the collection C of objects that do not contain themselves as members, $C = \{x \mid x \notin x\}$, contains itself or not, known as Russell's paradox [52], put an end to the era of so-called naive set theory in the early 20th century. It showed clearly that definitions of mathematical objects need to be made carefully. It further justified the formulation of mathematical theories on the basis of axioms formulated in a formal language from which consequences are derived by inference according to an underlying logical system, and it demonstrated the need to ask for proofs of consistency of mathematical theories. Hilbert's programme, see [29] and Zach [77] for a recent account, to secure consistent parts of mathematics, marks the beginnings of proof theory as a mathematical discipline in its own right, see Pohlers [43] for reference. Hilbert's particular request for a proof of consistency of Peano arithmetic within itself was shown by Gödel's incompleteness theorem [27] to be unsatisfiable through a technical, true, but unprovable statement. This raised the question of how to cope with incompleteness and kindled curiosity to find mathematically interesting theorems that are independent of, i.e. neither provable nor disprovable within, some theory under consideration.

Gentzen [23, 24] proposed a consistency proof for Peano arithmetic using the technique of cut elimination that reduced consistency to the principle of transfinite induction up to the ordinal ε_0, which therefore is called the proof-theoretic ordinal of Peano arithmetic. This established the branch of proof theory called ordinal

G. Wilken
Structural Cellular Biology Unit, Okinawa Institute of Science and Technology, 1919-1 Tancha, Onna-son, 904-0495 Okinawa, Japan. e-mail: wilken@oist.jp

R. Kahle and M. Rathjen (eds.), *The Legacy of Kurt Schütte*,
https://doi.org/10.1007/978-3-030-49424-7_21

analysis, see [40, 41, 48, 50, 43] for reference, which requires increasingly involved recursive notation systems for proof-theoretic ordinals of theories of increasing strength. Thus, this approach establishes a measure for the proof-strength of theories, complementary to comparative or reductive proof theory, where one seeks to embed or emulate theories within one another, and more generally to the large field of reverse mathematics. See Simpson [58] for reference.

An alternative proof of consistency for arithmetic was given by Gödel [28] himself using the technique of functional interpretation (see Diller [18] for reference), which reduces consistency to the computability of primitive-recursive functionals of finite type comprising a term system called Gödel's T, which can be formulated in the setting of (typed) lambda-calculus. See Hindley and Seldin [30] for reference. The latter is studied in many variants in its own right as it is a fundamental paradigm of computability, term rewriting, and hence in foundations of computer science. The computability of all terms of Gödel's T was shown by Howard [31], using the technique of ordinal assignment in the form of a term algebra containing variables and ranging over the ordinals below ε_0. This assignment could be analyzed, including the treatment of fragments of T, and enhanced to yield an assignment of natural numbers to the terms of T through a collapsing technique from ε_0 into natural numbers, developed by Weiermann [60] and first applied by him to Schütte's treatment of Gödel's T formulated in combinatorial logic rather than lambda calculus, see Schütte's book on proof theory [55] and [61, 62], and to provide a detailed derivation lengths classification [76]. We continue with an adapted version of the introduction to [75].

Among the mathematically appealing examples exemplifying mathematical incompleteness are the Paris-Harrington theorem [39], Goodstein sequences [26, 34, 66], Kruskal's theorem [35, 51], its extension by Friedman [57], the graph minor theorem by Robertson and Seymour, see [22] and, for a general reference on so-called concrete mathematical incompleteness, Friedman's book [21]. Concrete incompleteness refers to natural mathematical theorems independent of significantly strong fragments of ZFC, Zermelo-Fraenkel set theory with the axiom of choice. Further developments were initiated by Weiermann, who established results on phase transition from provability to unprovability [63, 64], which were taken up by others, e.g. Lee [36]. Another natural mathematical theorem that can be seen as a generalization of Kruskal's theorem was shown by Carlson [14], using elementary patterns of resemblance [12] as basic structures of nested trees. Theorems regarding the embeddability of tree structures into one another play a central role in the area of well-quasi orderings, cf. [14], maximal order types of which can be measured using ordinal notation systems from proof theory, cf. [54] and the introduction to [65].

Elementary patterns of resemblance (in short: patterns) of order n are finite structures equipped with orderings $(\le_i)_{i\le n}$ such that $\le_0$ is a linear ordering and $\le_1, \ldots, \le_n$ are forests such that 1) $\le_{i+1} \subseteq \le_i$ and 2) $a \le_{i+1} b$ whenever $a \le_i b \le_i c$ and $a \le_{i+1} c$ for all $i < n$ and all a, b, c in the universe of the pattern. Patterns that do not contain (partial, or graphs of) functions are called pure patterns. Whereas according to Kruskal's theorem, the collection of finite trees is well-quasi ordered with respect to inf-preserving embeddings [35], in the context of patterns, the natural embeddings considered are coverings, i.e. $\leq_0$-embeddings that maintain the relations $\leq_i$ for $i = 1, \ldots, n$. Carlson's theorem on pure patterns of order 2 in [14] states that these are well-quasi ordered with respect to coverings. This result can be proved in the extension of the base theory of reverse mathematics, RCA_0, which allows for recursive comprehension, by the uniform Π^1_1-reflection principle for $\mathrm{KP}\ell_0$, a subsystem of set theory axiomatizing a mathematical universe that is a limit of admissible sets, see [3, 14]. We have shown in [74], independently of [15], that Carlson's theorem is unprovable in $\mathrm{KP}\ell_0$, or equivalently, Π^1_1-CA_0. This latter subsystem of second order number theory where induction is restricted to range over sets, and set comprehension is restricted to Π^1_1-formulae, plays a prominent role in reverse mathematics, as it is the strongest of the so-called "big five", cf. Simpson [58] for the relevance of these theories in mathematics. See Pohlers [42] for an extensive exposition of various subsystems of set theory and second order number theory, equivalences and comparisons in strength, and their proof-theoretic ordinals. Note that in [42] $\mathrm{KP}\ell_0$ goes by the name $\mathrm{KP}\ell^r$.

Patterns were discovered by Carlson [11] during model construction establishing the consistency of epistemic arithmetic with the statement *I know I am a Turing machine*, thereby proving a conjecture by Reinhardt called the strong mechanistic thesis (SMT). See the next section for a digression into this discovery. The consistency proof naturally leads to a structure $\mathcal{R}_1 = (\mathrm{Ord}; \leq, \leq_1)$ of ordinals where $\leq$ is the standard ordering on the ordinals and the relation $\leq_1$ is defined recursively in β by

$$\alpha \leq_1 \beta :\Leftrightarrow (\alpha; \leq, \leq_1) \preceq_{\Sigma_1} (\beta; \leq, \leq_1).$$

We therefore have $\alpha \leq_1 \beta$ if and only if $(\alpha; \leq, \leq_1)$ and $(\beta; \leq, \leq_1)$ satisfy the same Σ_1-sentences over the language $(\leq, \leq_1)$ with parameters from $\alpha = \{\gamma \mid \gamma < \alpha\}$, where Σ_1-formulas are quantifier-free formulas preceded by finitely many existential quantifiers. $\mathcal{R}_1$ was analyzed in [10] and shown to be recursive and periodic in multiples of ε_0. It turned out that pure patterns of order 1 comprise the finite isomorphism types of $\mathcal{R}_1$ and that patterns provide ordinal notations denoting their unique pointwise minimal coverings within $\mathcal{R}_1$, an observation Carlson elaborated in [12]. In $\mathcal{R}_1$ the relation $\leq_1$ can still be described by a relatively simple recursion formula, given in terms of Cantor normal form notation, see [10], here quoted by Theorem 21.9. Ordinal notations in Cantor normal form are built up from 0, +, and ω-exponentiation, where ω denotes the least infinite ordinal (see [43, 55] for reference). Note that ε_0 is the least fixed point of ω-exponentiation and that Cantor normal form notations with parameters from an initial segment, say $\tau + 1$, of the ordinals provide notations for the segment of ordinals below the least epsilon number greater than τ, as epsilon numbers are by definition the fixed points of ω-exponentiation.

Gödel's program to solve incompleteness progressively by the introduction of large cardinal axioms, see [20, 33], together with the heuristic correspondence between large cardinals and ordinal notations through reflection properties and embeddings, motivated Carlson to view patterns as a programmatic approach to ordinal notations. As suggested by Carlson, the concept of patterns of resemblance, in which binary relations code the property of elementary substructure, can be generalized within this approach to patterns of embeddings that involve codings of embeddings, to ultimately tie in with inner model theory in form of an ultra-fine structure. It seems plausible that $\mathcal{R}_n$-patterns ($n < \omega$) just suffice to analyze set-theoretic systems of Π_m-reflection ($m < \omega$), see [47, 44].

The extension of $\mathcal{R}_1$ to a relational structure $\mathcal{R}_1^+ = (\mathrm{Ord}; 0, +; \leq, \leq_1)$, containing the graphs of 0 and ordinal addition, gives rise to additive patterns of order 1 and has been the central object of study in [12, 68, 70, 71, 69, 16, 1]. Such extensions are of interest when it comes to applications of patterns to ordinal analysis. [68] gives a sense of the relation $\leq_1$ in $\mathcal{R}_1^+$, as was recommended to acquire in [12]. As pointed out in the introduction to [72], in [68] an easily accessible characterization of the Bachmann-Howard structure, in terms of Σ_1-elementarity is given in full detail. The Bachmann-Howard ordinal characterizes the segment of ordinals below the proof-theoretic ordinal of Kripke-Platek set theory with infinity, $\mathrm{KP}\omega$, which axiomatizes the notion of (infinite) admissible set, cf. Barwise [3], or equivalently ID_1, the theory of non-iterated positive inductive definitions, cf. [43]. This provides illustrative semantics for Skolem hull notations, cf. [40], the type of ordinal notations that have been most-useful in ordinal analysis so far, by the description of ordinals as least solutions to Σ_1-sentences over the language $(0, +, \leq, \leq_1)$. On the other hand the pattern approach to ordinal notations is explained in terms of ordinal arithmetic in the style of [51].

In [69] an ordinal assignment is given that yields minimal instantiations of patterns in the ordinals, and vice versa. Pattern-characterizations are assigned to ordinals given in classical notation. In [16] normal forms are considered, also from the viewpoint of [12]. The elementary recursive isomorphism between Skolem-hull notations and pattern notations for additive patterns of order 1, established in [70, 71, 69], was of great interest to many proof-theorists. It was shown in [71, 69] that notations in terms of $\mathcal{R}_1^+$-patterns characterized the proof-theoretic ordinal of Π_1^1-CA_0. A similar phenomenon for classical notations was shown in [56]: withdrawing basic functions like addition from a notation system causes a collapse, in case of a system for Π_1^1-CA_0 down to ε_0. The analysis of additive patterns of order 1 is easily modified when adding (the graphs of) other basic arithmetic functions for expressive convenience. Such a procedure, however, does not increase the order type of the resulting notation system any further, cf. [12] from Section 9.

In order to analyze $\mathcal{R}_1^+$ and assign minimal ordinal solutions to additive patterns of order 1, a toolkit of ordinal arithmetic based on relativized Skolem hull-based notation systems T^τ was introduced in [70]. Here τ is intended to be either 1 or an arbitrary epsilon number, which we often denote as $\tau \in \mathbb{E}_1 = \mathbb{E} \cup \{1\}$. This type of ordinal notation system builds upon a broad foundation developed mainly by Bachmann, Aczel, Feferman, Bridge, Schütte, Buchholz, Rathjen, and Weiermann (see [2, 4, 5, 6, 45, 51]), and the introduction of [70]. Variants of systems T^τ that can be extended to stronger ordinal notation systems and arithmetic were provided in [65], starting with work by Buchholz and Schütte in [8]. State-of-the-art notation systems of this classical type were introduced by Rathjen, culminating in a notation system for Π_2^1-CA_0 [49], and have recently been elaborated more completely for an analysis of the provably recursive functions of reflection by Pohlers and Stegert [44].

Returning to ordinal arithmetic sufficiently expressive to display pattern notations on the basis of order 1 (additive) or order 2 (pure), let $\Omega^{>\tau}$ be the least regular cardinal greater than $\mathrm{Card}(\tau) \cup \aleph_0$. We also simply write Ω for $\Omega^{>\tau}$ when the dependence of Ω on τ is easily understood from the context. For convenience we set $\mathrm{T}^0 := \mathrm{T}^1 = \mathrm{T}$ for the original notation system that provides notations for the ordinals below the proof theoretic ordinal of theories such as Π^1_1-CA_0, $\mathrm{KP}\ell_0$, and $\mathrm{ID}_{<\omega}$, the latter being the theory of finitely times-iterated inductive definitions, cf. e.g. [42]. In the presence of 0, ordinal addition, and constants for all ordinals less than τ, the stepwise collapsing, total, injective, unary functions $(\vartheta_i)_{i<\omega}$, where $\vartheta_0 = \vartheta^\tau$ is relativized to τ, give rise to unique terms for all ordinals in the notation system T^τ. With ordinal addition being part of the set of functions over which Skolem hulling is performed, the function ϑ^τ enumerates additive principal numbers starting from τ, omitting fixed points. Fixed points are in the range of ϑ^τ, however, as images of multiples of Ω. Letting $\Omega_0 := \tau$, $\Omega_1 := \Omega^{>\tau}$, and generally Ω_{i+1} to be the least (infinite) regular cardinal greater than Ω_i for $i < \omega$, this algorithm of denoting ordinals allows for a powerful mechanism extending the fixed-point free variant of the Veblen functions [59]. The function ϑ_i enumerates additive principal numbers starting with Ω_i in fixed-point free manner, so that fixed-points of increasing level are enumerated by ϑ_i with arguments that are multiples of Ω_{i+1} of appropriate fixed-point level, which in turn are denoted by values of the function ϑ_{i+1}. Recall that the Veblen functions are binary functions in which the first argument denotes the fixed point level. While $\varphi(0,\cdot)$ enumerates the additive principal numbers, i.e. the class of ordinals greater than 0 that are closed under ordinal addition, which we often denote as $\mathbb{P}$, $\varphi(1,\cdot)$ enumerates ε-numbers starting with ε_0. In general, $\alpha \mapsto \varphi(\xi+1,\alpha)$ enumerates the fixed points of $\varphi(\xi,\cdot)$, and $\varphi(\lambda,\cdot)$ enumerates the common fixed points of all $\varphi(\xi,\cdot)$, $\xi < \lambda$. Now, the fixed point-level of Γ-numbers, i.e. ordinals α such that $\alpha = \varphi(\alpha, 0)$, cannot be expressed unless we introduce a ternary Veblen function. This problem of finding names for increasing fixed-point levels is therefore resolved by arithmetization through higher ϑ_j-functions, relying on the regularity properties of the Ω_i, in that they cannot be reached by enumeration functions from below. Examples are $\vartheta_0(0) = 1 = \omega^0 = \varphi(0,0)$, $\vartheta_0(1) = \omega = \varphi(0,1)$, $\vartheta_0(\varepsilon_0) = \omega^{\varepsilon_0+1}$, $\vartheta_0(\vartheta_1(0)) = \vartheta_0(\Omega) = \varepsilon_0 = \varphi(1,0)$, $\vartheta_0(\Omega+\varepsilon_0) = \varepsilon_{\varepsilon_0}$, $\vartheta_0(\Omega+\Omega) = \varphi(2,0)$, $\vartheta_0(\Omega^2) = \Gamma_0$ where $\Omega = \aleph_1$ and $\Omega^2 = \vartheta_1(\vartheta_1(0))$, and $\vartheta_0(\vartheta_1(\vartheta_2(0))) = \vartheta_0(\varepsilon_{\Omega+1})$ denoting the Bachmann-Howard ordinal. Note that we have omitted the superscript 0 in the ϑ-functions above for better readability. Summarizing as in [69], we have

Lemma 21.1 (cf. 4.2 of [70]). *Let $i < \omega$.*

1. $\vartheta_i(0) = \Omega_i$.
2. *For any $\alpha < \Omega_{i+1}$ we have $\vartheta_i(1+\alpha) = \bar{\omega}^{\Omega_i+\alpha}$,*
   where $\xi \mapsto \bar{\omega}^\xi$ denotes the enumeration function of the additive principal numbers that are not epsilon numbers.

Definition 21.2 (Inductive Definition of T^τ, cf. 3.22 of [70]).

1. $\tau \subseteq \mathrm{T}^\tau$.
2. $\xi, \eta \in \mathrm{T}^\tau \Rightarrow \xi + \eta \in \mathrm{T}^\tau$.
3. $\xi \in \mathrm{T}^\tau \cap \Omega_{i+2} \Rightarrow \vartheta_i(\xi) \in \mathrm{T}^\tau$ for all $i < \omega$.

Lemma 21.3 (cf. 3.30 of [70]). *For every $i < \omega$ the function $\vartheta_i \restriction_{\mathrm{T}^\tau \cap \Omega_{i+2}}$ is* 1-1 *and has values in* $\mathbb{P} \cap [\Omega_i, \Omega_{i+1})$.

Thus every ordinal in T^τ can be identified with a unique term built up from parameters below τ using + and the functions ϑ_i ($i < \omega$). The choice of regular cardinals for the Ω_i is just convenient for short proofs. One can choose recursively regular ordinals instead (see [46]), at the cost of more involved proofs to demonstrate recursiveness (see [53]). Patterns collapse these gaps left in Skolem hull notations in a dynamic manner, which is one major reason that explains why, despite the elegance and brevity of their definition, their calculation is quite involved. Another, intrinsically connected, major reason explaining the subtleties in pattern calculation is the necessity to calculate or analyze in terms of the connectivity components of the relations $\leq_i$. We will return to the discussion of connectivity components later.

As explained above, we developed an expressive ordinal arithmetic in [70] on the basis of Skolem hull term systems for initial segments $\tau^\infty := \mathrm{T}^\tau \cap \Omega^{>\tau}$ of Ord, that was further extended in [16], [17], and [74]. The following definition transfinitely iterates the closure under $\tau \mapsto \tau^\infty = \mathrm{T}^\tau \cap \Omega^{>\tau}$ continuously through all of Ord.

Definition 21.4 (modified 9.1 of [70]). Let $(\upsilon_\iota)_{\iota \in \mathrm{Ord}}$ be the sequence defined by

1. $\upsilon_0 := 0$,
2. $\upsilon_{\xi+1} := \upsilon_\xi^\infty = \mathrm{T}^{\upsilon_\xi} \cap \Omega^{>\upsilon_\xi}$,
3. $\upsilon_\lambda := \sup\{\upsilon_\iota \mid \iota < \lambda\}$ for $\lambda \in \mathrm{Lim}$, i.e. λ a limit ordinal.

In Corollary 5.10 of [71] we have shown that the maximal $<_1$-chain in $\mathcal{R}_1^+$ is $\mathrm{Im}(\upsilon) \setminus \{0\}$. The results from [71], [69], and Sections 5 and 6 from [16] were applied in a case study for an ordinal analysis of $\mathrm{ID}_{<\omega}$, [38], following Buchholz' style of operator-controlled derivations [7]. Proof-theoretic analysis based on patterns as originally intended by Carlson uses β-logic [9]. The intrinsic semantic content via the notion of elementary substructure directly on the ordinals in connection with their immediate combinatorial characterization and the concise elegance of their definition on the one hand, and the remarkable intricacies in their calculation, revealing mathematical depth not yet fully understood, on the other hand, create the impression that patterns contribute to the quest for natural well-orderings, cf. [19].

Returning to the discussion of pure patterns of order 2, let

$$\mathcal{R}_2 = (\mathrm{Ord}; \leq, \leq_1, \leq_2)$$

be the structure of ordinals with standard linear ordering $\leq$ and partial orderings $\leq_1$ and $\leq_2$, simultaneously defined by induction on β in

$$\alpha \leq_i \beta :\Leftrightarrow (\alpha; \leq, \leq_1, \leq_2) \preceq_{\Sigma_i} (\beta; \leq, \leq_1, \leq_2)$$

where $\preceq_{\Sigma_i}$ is the usual notion of Σ_i-elementary substructure (without bounded quantification), as analyzed thoroughly below the ordinal $\upsilon_1 = 1^\infty$ of $\mathrm{KP}\ell_0$ in [17, 73, 74].

Core($\mathcal{R}_2$), the *core* of $\mathcal{R}_2$, i.e. the union of pointwise minimal instantiations of all finite isomorphism types of $\mathcal{R}_2$, was analyzed in [74] and shown to coincide (in domain) with the initial segment of the ordinals below 1^∞. Note that any (finite) subset of $\mathcal{R}_2$ gives rise to a substructure of $\mathcal{R}_2$; hence, it represents a (finite) isomorphism type of $\mathcal{R}_2$. It was shown in [74] that collections of pure patterns of order 2 as defined above and of finite isomorphism types of $\mathcal{R}_2$ coincide, and that each pure pattern of order 2 has a *unique isominimal* representative $P \subset \mathcal{R}_2$, where a finite substructure $Q \subset \mathcal{R}_2$ is isominimal if and only if $Q \leq_{\text{pw}} R$ for every $R \subset \mathcal{R}_2$ such that $Q \cong R$. $Q \leq_{\text{pw}} R$ means that Q is pointwise less than or equal to R with respect to increasing enumerations. The notions core and isominimality where introduced by Carlson in [12]. In the case of patterns of order 1 the increased strength resulting from basic arithmetic functions such as addition, is matched by pure patterns of order 2, as we have shown in [17] that any pure pattern of order 2 has a covering below 1^∞, the least such ordinal. According to a conjecture by Carlson, this phenomenon of compensation holds generally for all orders.

Despite the fact that the cores of $\mathcal{R}_1^+$ and $\mathcal{R}_2$ cover the same initial segment of the ordinals, their structures differ considerably. While the core of $\mathcal{R}_1^+$ shows a great deal of uniformity, reminding one of Girard's notion of dilator, cf. [25], and giving rise to Montalbán's Question 27 in [37], the core of $\mathcal{R}_2$ is a structure, the regularity of which is far less obvious, due to the absence of uniformity provided by ordinal addition. In this context the ordinal 1^∞ is obtained as a collapse when weakening additive pattern notations of order 2 on the basis of the structure $\mathcal{R}_2^+ = (\text{Ord}; 0, +; \leq, \leq_1, \leq_2)$ to pure patterns of order 2 arising in $\mathcal{R}_2$. We claim that the segment of countable ordinals denoted by the Skolem-hull notation system derived from the first ω-many weakly inaccessible cardinals covers (the domain of) Core($\mathcal{R}_2^+$). Note that the notation system based on the first weakly inaccessible cardinal matches the proof-theoretic ordinal of the set theory KPI, which axiomatizes an admissible universe that is a limit of admissible sets, and which is equivalent to the system Δ^1_2-CA + BI of second order number theory, first analyzed by Jäger and Pohlers in [32], cf. also [42]. See Buchholz' seminal work [7] for an analysis of KPI via operator controlled derivations. The analysis of $\mathcal{R}_2^+$ is a topic of future work and requires a generalization of ordinal arithmetical methods, the beginning of which is outlined in [65]. Note that [13] discusses patterns of order 2 in a general way over Ehrenfeucht-Mostowski structures; however, on the basis of modified relations $\leq_1$, $\leq_2$, see Section 5 of [13].

Define

$$I := \{\iota \in \text{Ord} \mid \iota > 1 \text{ and not of a form } \iota = \lambda + 1 \text{ where } \lambda \in \text{Lim}\}$$

and let the expression $\iota \dot{-} 1$ for $\iota \in \text{Ord}$ denote ι_0 if $\iota = \iota_0 + 1$ for some ι_0, and simply ι otherwise. We will rely on the following theorem, which in turn is an immediate consequence of Theorem 3.6 of [75].

Theorem 21.5 (Maximal $<_2$-chain in $\mathcal{R}_2$, 1.5 of [75]).

1. *The sequence $(\upsilon_\iota)_{\iota>0}$ is a $<_1$-chain through the ordinals, supporting the maximal $<_2$-chain through the ordinals, which is $(\upsilon_\iota)_{\iota \in I}$. Here maximality means that for any $\iota \in I$ (actually for any ordinal ι), the enumeration of all $<_2$-predecessors of υ_ι is given by $(\upsilon_\xi)_{\xi \in I \cap \iota}$.*
2. *For $\iota \in \text{Ord} \setminus I$ the ordinal υ_ι is $\upsilon_{\iota \dot{-} 1}$-$\leq_1$-minimal, i.e. there does not exist any $\alpha \in (\upsilon_{\iota \dot{-} 1}, \upsilon_\iota)$ such that $\alpha <_1 \upsilon_\iota$, and does not have any $<_2$-successor.*

3. *For every $\iota \in I$ the ordinal υ_ι is a $<_2$-predecessor of every proper multiple of $\upsilon_{\iota \dot{-} 1}$ greater than υ_ι. The ordinals of the form υ_λ where $\lambda \in$ Lim comprise the set of suprema of $<_2$-chains of limit order type.* □

This provides an overview of the structure $\mathcal{R}_2$, and a detailed picture that characterizes the relations $\leq_1$ and $\leq_2$ in all of $\mathcal{R}_2$ arithmetically has been established by Theorem 3.6 of [75], which explicitly describes the $\leq_i$-predecessors, and Corollary 3.10 of [75], which describes the $\leq_i$-successors of a given ordinal, $i = 1, 2$.

In Section 21.4 we are going to apply the results established in [75] to prove that the structures $\mathcal{R}_2$ and $\mathcal{R}_3$ agree on the initial segment υ_{ω^2+2}, where $\mathcal{R}_3 = (\mathrm{Ord}; \leq, \leq_1, \leq_2, \leq_3)$ and

$$\alpha \leq_i \beta :\Leftrightarrow (\alpha; \leq, \leq_1, \leq_2, \leq_3) \preceq_{\Sigma_i} (\beta; \leq, \leq_1, \leq_2, \leq_3)$$

simultaneously for $i = 1, 2, 3$ and recursively in β, while

$$\upsilon_{\omega^2} <_3 \upsilon_{\omega^2+2}$$

is the least occurrence of a $<_3$-pair in $\mathcal{R}_3$. A detailed arithmetical analysis of the structure $\mathcal{R}_3$ is the subject of ongoing work.

21.2 Digression: the Discovery of $\mathcal{R}_1$

As in the introductions to [67, 72] we briefly describe how $\mathcal{R}_1$ naturally arises during Carlson's verification of Reinhardt's conjecture in [11], the consistency of epistemic arithmetic EA with the strong mechanistic thesis SMT, as we mentioned in the introduction. To this end, an extension of the standard model $\mathcal{N}$ of arithmetic is constructed, which satisfies SMT and is a model of EA. The theory EA is an extension of Peano arithmetic, the language $\mathcal{L}^+$ of which contains a modal operator K that formalizes knowledge, the logic of which is a quantified version of modal logic S4, and the axioms of which are all sentences of the form $\mathrm{K}\sigma$ where σ is a Peano axiom for the language of EA. The schema SMT is given by the sentences of a form

$$\mathrm{SMT}_\varphi \; :\equiv \; \mathrm{K}\,\exists e \forall x\, (\mathrm{K}\varphi(x) \leftrightarrow \text{“}x \in \mathrm{W}_e\text{”})$$

where $\varphi(x)$ is a formula of the language of EA and “$x \in \mathrm{W}_e$” expresses (in $\mathcal{N}$) that x is an element of the eth recursively enumerable set. Setting $\mathrm{T} := \mathrm{EA} + \mathrm{SMT}$ we are going to roughly sketch how $\mathcal{N}$ can be extended to a model $\mathcal{N}_\mathrm{T}$ of T.

Since SMT speaks about recursive enumerability, it is reasonable to interpret K as “*it is provable in* T”, which is a Σ_1-notion. We therefore define

$$\mathcal{N}_\mathrm{T} \models \mathrm{K}\varphi[s] \; :\Leftrightarrow \; \mathrm{T} \vdash \varphi^s$$

where φ is a formula of EA and s is an assignment of natural numbers to the free variables that occur in φ. By this interpretation we obtain

$$\mathcal{N}_{\mathrm{T}} \models \mathrm{SMT}_{\varphi} \iff \{x \mid \mathrm{T} \vdash \varphi(x)\} \text{ is r.e.},$$

which is true. We have to show that $\mathcal{N}_{\mathrm{T}} \models \mathrm{T}$. The crucial part is to show that in $\mathcal{N}_{\mathrm{T}}$ the universal closure of any formula of the form $\mathrm{K}\varphi \rightarrow \varphi$ holds. This means that for every EA-formula φ and every assignment s for φ such that φ^s is provable in T we have to show that $\mathcal{N}_{\mathrm{T}} \models \varphi^s$. Here the idea is to proceed by induction on the depth of nestings of the operator K. The depth of nestings of K can be understood as the level of introspection. This motivates the introduction of a concept of stratification along ordinals. Therefore K is replaced by operators K_α and formulas in the stratified language have to satisfy that, whenever some K_α is in the scope of a K_β, we have $\alpha < \beta$. Given a set of ordinals Q (well-ordered by $<:=\in$ as usual) we denote the stratified language referred to by $\mathcal{L}^Q$. An appropriate axiomatization Σ of the stratified analogue S of T over Q can be found, which is r.e. provided the ordering $(Q, <)$ is recursive. This leads us to a refined model construction defining

$$\mathcal{N}^{\Sigma} \models \mathrm{K}_\alpha\varphi[s] \;:\Leftrightarrow\; \Sigma \cap \mathcal{L}^\alpha \vdash \varphi^s$$

where for simplicity we assume Q to be an initial segment of the ordinals and therefore $\mathcal{L}^\alpha$ to be the restriction of $\mathcal{L}^Q$ to levels of introspection below α. In order recover T from S we clearly have to have sufficiently many *collapsing axioms* $\mathrm{K}_\alpha\varphi \leftrightarrow \mathrm{K}_\beta\varphi$ in Σ. First of all we have to require that Q be infinite. The problem we now face is that in general we cannot show

$$\mathcal{N}^{\Sigma} \models \mathrm{K}_\beta\varphi \rightarrow \mathrm{K}_\alpha\varphi$$

where $\alpha < \beta$, since the derivation of φ in $\Sigma \cap \mathcal{L}^\beta$ might use axioms from $\mathcal{L}^\beta \setminus \mathcal{L}^\alpha$. The derivation, however, uses only finitely many axioms. Therefore, there exists a finite subset $Z \subseteq \beta$ such that $\Sigma \cap \mathcal{L}^Z \vdash \varphi$ and $\varphi \in \mathcal{L}^{Z \cap \alpha}$. Let $X := Z \cap \alpha$ and $Y := Z \setminus X$. We would like to find some $\tilde{Y} \subseteq \alpha$ such that $X < \tilde{Y}$, $\tilde{Z} := X \cup \tilde{Y}$ is order-isomorphic to Z, and $\Sigma \cap \mathcal{L}^{\tilde{Z}} \vdash \varphi$. Clearly, this does not in general exist for arbitrary ordinals $\alpha < \beta$, which then leads us to look for a partial suborder $<_1 \subseteq <$ on which this transformation always succeeds. The set of collapsing axioms $\mathrm{K}_\alpha\varphi \leftrightarrow \mathrm{K}_\beta\varphi$ must then be reduced to those collapsing axioms where $\alpha \leq_1 \beta$ holds. To make the exchange of Z by $\tilde{Z}$ work, we need a uniformity property of the axiomatization Σ with respect to $<_1$. We therefore require

$$X \cup \tilde{Y} \;\underset{<,<_1}{\cong}\; X \cup Y$$

to hold. Here, $<_1$-embeddability of $X \cup Y$ into $X \cup \tilde{Y}$ would suffice, since this guarantees that $\Sigma \cap \mathcal{L}^{X \cup \tilde{Y}}$ contains enough collapsing axioms, but isomorphism is the minimal condition. Now we see that the transformation of Z to $\tilde{Z}$ we want to employ is a reflection property. It is a criterion for Σ_1-elementary substructure, see Proposition 21.6 in the next section. We obtain (infinite) initial segments or just substructures of $\mathcal{R}_1$ as candidates along which we can define stratified theories that permit partial collapsing. But we need to be able to recover T from a stratified version. This means that we need a stratified theory which collapses completely.

The analysis of $\mathcal{R}_1$ in [10] yields that $(\varepsilon_0 \cdot \omega, <, <_1)$ is a recursive structure in which $C := \{\varepsilon_0 \cdot (n+1) \mid n < \omega\}$ is the field of an infinite $<_1$-chain, whence $<_1\restriction_C = <\restriction_C$. The crucial property of this subset of $\varepsilon_0 \cdot \omega$ is that we can $<_1$-embed every finite subset Z of $\varepsilon_0 \cdot \omega$ into a $Z' \subseteq C$ which is $<$-isomorphic to Z. This allows us to show that

$$\mathcal{N}^{\Sigma \cap \mathcal{L}^C} = \mathcal{N}^{\Sigma}\restriction_{\mathcal{L}^C}.$$

It turns out that replacing every K_α by K in the set $\Sigma \cap \mathcal{L}^C$ yields an r.e. axiomatization of T. In other words, the stratification of T over C has the r.e. axiomatization $\Sigma \cap \mathcal{L}^C$. We further see that

$$\mathit{collapse}(\mathcal{N}^{\Sigma \cap \mathcal{L}^C}) = \mathcal{N}_{\mathrm{T}}$$

where *collapse* denotes the $\mathcal{L}^+$-structure obtained via taking interpretations of arbitrary $\mathrm{K}_\alpha\varphi$'s, which is well-defined since all collapsing axioms of $\mathcal{L}^C$ hold in $\mathcal{N}^{\Sigma \cap \mathcal{L}^C}$. Finally, we can show

$$\mathcal{N}^{\Sigma} \models \Sigma$$

where universal closures of formulas of a form $\mathrm{K}_\alpha\varphi \rightarrow \varphi$ are treated by induction on α, thus proceeding along the number of nestings of knowledge operators, and collapsing axioms $\mathrm{K}_\alpha\varphi \leftrightarrow \mathrm{K}_\beta\varphi$, where $\alpha \leq_1 \beta$, are treated in the way described above via the finite set criterion for Σ_1-elementary substructure. Altogether we obtain

$$\mathcal{N}_{\mathrm{T}} \models \mathrm{T}$$

and thus the consistency of SMT with EA. □

21.3 A Journey from $\mathcal{R}_1$ via $\mathcal{R}_2$ toward $\mathcal{R}_3$

First, we provide criteria and basic properties for elementary substructurehood. This allows us to avoid dealing with formulas and increases the familiarity with basic notions around patterns. We have the following well-known criterion for Σ_1-elementary substructure for finite relational languages, which already played a central role in [10] and later in [68, 71].

Proposition 21.6 (1.2 of [10]). *Let $\mathcal{A}$ and $\mathcal{B}$ be structures for a finite relational language $\mathcal{L}$. $\mathcal{A}$ is a Σ_1-elementary substructure of $\mathcal{B}$ if and only if $\mathcal{A}$ is a substructure of $\mathcal{B}$ and whenever X is a finite subset of $|\mathcal{A}|$ and Y is a finite subset of $|\mathcal{B}| \setminus |\mathcal{A}|$, there exists a subset $\tilde{Y}$ of $|\mathcal{A}|$ such that*

$$X \cup Y \quad \cong_X \quad X \cup \tilde{Y},$$

i.e. there is an $\mathcal{L}$-isomorphism of $X \cup Y$ and $X \cup \tilde{Y}$ that fixes the elements of X. The criterion implies Σ_1-elementary substructurehood also for infinite relational languages that may or may not contain the identity.

Proof. For reader's convenience we reproduce the proof given in [67]. Note that since $\mathcal{L}$ is relational we may consider arbitrary subsets of the universe of an $\mathcal{L}$-structure as substructures of that structure.

Suppose first that the criterion holds for $\mathcal{L}$-structures $\mathcal{A}$ and $\mathcal{B}$. We show $\mathcal{A} \leq_{\Sigma_1} \mathcal{B}$ by induction on the build-up of Σ_1-sentences with parameters from $A := |\mathcal{A}|$. We will see that this direction of the equivalence stated by the proposition also holds for infinite relational languages that may not contain the identity. For relations $P \in \mathcal{L}$ the equivalence $\mathcal{A} \models P(\boldsymbol{a})$ iff $\mathcal{B} \models P(\boldsymbol{a})$ holds for arbitrary $\boldsymbol{a} \in A^n$ where n is the arity of P, since $\mathcal{A}$ is a substructure of $\mathcal{B}$. The equivalence then also follows for arbitrary quantifier-free sentences with parameters from A. Now suppose $F(\boldsymbol{x}, \boldsymbol{y})$ is a Σ_1-formula of $\mathcal{L}$ with exactly the free variables shown. By i.h. we have $\mathcal{A} \models F(\boldsymbol{a}, \boldsymbol{p})$ iff $\mathcal{B} \models F(\boldsymbol{a}, \boldsymbol{p})$, where the components of $\boldsymbol{a}$ and $\boldsymbol{p}$ range over A. We show

$$\mathcal{A} \models \exists \boldsymbol{x} F(\boldsymbol{x}, \boldsymbol{p}) \quad \Leftrightarrow \quad \mathcal{B} \models \exists \boldsymbol{x} F(\boldsymbol{x}, \boldsymbol{p})$$

where again the components of $\boldsymbol{p}$ range over A. The direction from left to right is obvious. In order to show the direction from right to left, let $\boldsymbol{b} \subseteq B := |\mathcal{B}|$ be such that $\mathcal{B} \models F(\boldsymbol{b}, \boldsymbol{p})$. Suppose without loss of generality that $\boldsymbol{b} \subseteq B \setminus A$. Set $X := \boldsymbol{p}$ and $Y := \boldsymbol{b}$. By the criterion there is a set $\tilde{Y} \subseteq A$ such that $X \cup \tilde{Y} \cong_X X \cup Y$, where the subscript means that the isomorphism fixes X. Let $\tilde{Y} =: \boldsymbol{a}$ be such that a_i is the isomorphic image of b_i for each component b_i of $\boldsymbol{b}$. Then considering $X \cup Y$ a substructure of $\mathcal{B}$ we have $X \cup Y \models F(\boldsymbol{b}, \boldsymbol{p})$, which by the isomorphism is equivalent to $X \cup \tilde{Y} \models F(\boldsymbol{a}, \boldsymbol{p})$. But this implies $\mathcal{A} \models F(\boldsymbol{a}, \boldsymbol{p})$ (since $X \cup \tilde{Y}$ is substructure of $\mathcal{A}$), whence $\mathcal{A} \models \exists \boldsymbol{x} F(\boldsymbol{x}, \boldsymbol{p})$.

Now suppose $\mathcal{A} \leq_{\Sigma_1} \mathcal{B}$. We show that the criterion holds. Let finite sets $X \subseteq A$ and $Y \subseteq B \setminus A$ be given. By the finiteness of $\mathcal{L}$, X, and Y there is a finite conjunction Φ of all prime sentences and negated prime sentences of $\mathcal{L}$ with parameters from $X \cup Y$ that are true in $X \cup Y$. Since we have a language with identity this sentence Φ also contains statements expressing that all elements of Y are pairwise distinct and different from the elements of X. Let $Y =: \boldsymbol{b}$ and $\boldsymbol{x}$ be a vector of pairwise distinct variables of the same length. Then $\Psi := \exists \boldsymbol{x} \Phi[\boldsymbol{b} := \boldsymbol{x}]$ is a Σ_1-sentence of $\mathcal{L}$ with parameters only from X. By $X \cup Y \models \Psi$ we get $\mathcal{B} \models \Psi$, whence also $\mathcal{A} \models \Psi$. Let $\boldsymbol{a}$ be a vector of witnesses of Ψ in $\mathcal{A}$. Then $\mathcal{A} \models \Phi[\boldsymbol{b} := \boldsymbol{a}]$, whence also $X \cup \tilde{Y} \models \Phi[\boldsymbol{b} := \boldsymbol{a}]$ where $\tilde{Y} := \boldsymbol{a}$. But this means that we have $X \cup \tilde{Y} \cong_X X \cup Y$. □

Immediate consequences of the notion of Σ_i-elementary substructure are summarized in the following lemma, which generally applies to generalizations of the structures from the introduction to capture Σ_n-elementary substructures, i.e. to accommodate patterns of order n. Let $\mathcal{R}_n := (\mathrm{Ord}; (\leq_i)_{i \leq n})$ and $\mathcal{R}_\omega = (\mathrm{Ord}; (\leq_i)_{i<\omega})$ where $\leq_0 = \leq$.

Lemma 21.7 (1.2 of [17]). *For the first four assertions below, fix any of the $\mathcal{R}$-structures introduced above.*

1. *$\leq_i$ is a forest, i.e. a partial ordering in which the sets of $\leq_i$-predecessors of an element are linearly ordered by $\leq_i$.*

2. *$\le_{i+1}$ respects $\le_i$:*
   - $\le_{i+1} \subseteq \le_i$ *and*
   - $\alpha \le_i \beta \le_i \gamma \,\&\, \alpha \le_{i+1} \gamma \quad \Rightarrow \quad \alpha \le_{i+1} \beta.$
3. *$\{\beta \mid \alpha \le_1 \beta\}$ is a closed interval.*
4. *$\{\beta \mid \alpha \le_2 \beta\}$ is closed.*
5. *For $n < m \le \omega$ and the least β such that there exists an α with $\mathcal{R}_m \models \alpha <_n \beta$ we have*

$$\mathcal{R}_n \cap \beta = \mathcal{R}_m \restriction_{(\le_i)_{i \le n}} \cap \beta.$$

Proof. Straightforward. □

Remark. Note that the above lemma also holds for the additive counterparts of the above structures containing the graphs of 0 and ordinal addition. Closedness means closedness under suprema. We refrain from pursuing further generalization here.

The next lemma gives a sense of the weakest nontrivial $\le_1$-relations on the ordinals, in $\mathcal{R}_1$ and $\mathcal{R}_2$. For the reader's convenience we give the proofs in full detail.

Lemma 21.8 (7.2 of [17])**.** 1. *In $\mathcal{R}_1$ we have (see* [10]*)*

$$\alpha \le_1 \alpha + 1 \quad \Leftrightarrow \quad \alpha \in \mathrm{Lim}.$$

2. *In $\mathcal{R}_2$ we have*

$$\alpha \le_1 \alpha + 1 \quad \Leftrightarrow \quad \alpha \in \mathrm{Lim} \;\&\; \forall \beta (\beta <_2 \alpha \Rightarrow \alpha = \sup\{\gamma < \alpha \mid \beta \le_2 \gamma\}).$$

Proof. For the direction from left to right of part 1, observe that for any parameter $\xi < \alpha$ we have

$$\alpha + 1 \models \exists x\; x > \xi,$$

namely α. Thus, by assumption $\alpha \models \exists x\; x > \xi$, whence α cannot be of a form $\eta + 1$. $\alpha = 0$ is trivially ruled out by the sentence $\exists x\; x \le x$. For the direction from right to left we use the finite set criterion, Proposition 21.6. Let X be a finite subset of α and assume without loss of generality that $Y = \{\alpha\}$. We need to find an ordinal $\tilde{\alpha}$ such that $X < \tilde{\alpha} < \alpha$ and $\xi \le_1 \tilde{\alpha}$ if and only if $\xi \le_1 \alpha$ for all $\xi \in X$. Since α is a limit ordinal by assumption, $\max X < \alpha$, and by property 3. of Lemma 21.7, for each $\xi \in X$ such that $\xi \not\le_1 \alpha$ there is an $\eta_\xi \in (\max X, \alpha)$ such that $\xi \not\le_1 \eta$ for all $\eta \in [\eta_\xi, \alpha)$. Let $\tilde{\alpha}$ be the maximum of these finitely many η_ξ. Clearly, for each $\xi \in X$ such that $\xi \le_1 \alpha$ we also have $\xi \le_1 \tilde{\alpha}$.

We turn to part 2. Let us first assume that $\alpha <_1 \alpha + 1$. As in part 1 this implies $\alpha \in \mathrm{Lim}$. If there were any β such that $\beta <_2 \alpha$ and $\delta := \sup\{\gamma < \alpha \mid \beta \le_2 \gamma\} < \alpha$ then we would have $\beta \le_2 \delta$ according to property 4. of Lemma 21.7, and $\alpha + 1 \models \exists x > \delta\; \beta <_2 x$ while $\alpha \not\models \exists x > \delta\; \beta <_2 x$. Hence the right hand side of the equivalence holds whenever $\alpha <_1 \alpha + 1$.

Now suppose the right hand side of the equivalence holds. We use Proposition 21.6 to show that $\alpha <_1 \alpha + 1$. Let $X \subseteq_{\mathrm{fin}} \alpha$, i.e. a finite subset of α, be given and let

$Y := \{\alpha\}$. We set $X'_i := \{x \in X \mid x \not<_i \alpha\}$ for $i = 1,2$. As shown similarly in part 1, now also using property 4. of Lemma 21.7, there is $\mu < \alpha$ such that $X < \mu$ and

$$\forall x \in X'_i \;\forall \xi \in (\mu,\alpha)\; x \not<_i \xi,\; i = 1,2.$$

In the case $X'_2 = X$ we may choose any ordinal $\tilde{\alpha} \in (\mu,\alpha)$ and set $\tilde{Y} := \{\tilde{\alpha}\}$, while otherwise there exists $\beta := \max(X \setminus X'_2)$, so that we have $\gamma \leq_2 \beta$ for all $\gamma \in X \setminus X'_2$. Then we choose some $\delta \in (\mu,\alpha)$ such that $\beta <_2 \delta$ and set $\tilde{Y} := \{\delta\}$. It is now easy to see that we have $X \cup Y \cong X \cup \tilde{Y}$. □

For readers' convenience we recall results shown in [10]. Let $\mathrm{lh}_1^{\mathcal{R}}(\alpha)$ be $\max\{\beta \mid \alpha \leq_1^{\mathcal{R}} \beta\}$ if that exists and ∞ otherwise. Note that by the reflection principle of set theory we know that the class of ordinals α such that $\mathrm{lh}_1^{\mathcal{R}}(\alpha) = \infty$ is unbounded. Representations of ordinals in additive normal form, i.e. as a sum of weakly decreasing additive principal summands, are sometimes explicitly marked by ANF, and the more explicit representations in Cantor normal form are sometimes marked by CNF. The following theorem characterizes the structure $\mathcal{R}_1$ completely.

Theorem 21.9 ([10]). *Let* $\mathrm{lh}_1 = \mathrm{lh}_1^{\mathcal{R}_1}$.

1. $\mathcal{R}_1 \cong \mathcal{R}_1 \cap [\alpha + 1, \infty)$ *for all* α.
2. $\mathrm{lh}_1(\varepsilon_0 \cdot (1 + \eta)) = \infty$ *for all* η.
3. *For* $\alpha =_{\mathrm{CNF}} \omega^{\alpha_1} + \cdots + \omega^{\alpha_n}$ $(n > 0)$ *with* $\alpha_n =_{\mathrm{ANF}} \rho_1 + \cdots + \rho_m < \alpha$ *we have*

$$\mathrm{lh}_1(\alpha) = \alpha + \mathrm{lh}_1(\rho_1) + \cdots + \mathrm{lh}_1(\rho_m).$$

The approach to obtain the above result is to consider the connectivity components of $\leq_1$ in $\mathcal{R}_1$ and to compute the enumeration function $\alpha \mapsto \kappa_\alpha$ of the $\leq_1$-minimal ordinals. In this computation the translation invariance of $\mathcal{R}_1$ (see the first claim of the theorem) plays an essential role. It directly follows from the above theorem that there do not exist any $<_1$-connections from within any interval $(\varepsilon_0 \cdot \iota, \varepsilon_0 \cdot (\iota + 1))$, $\iota \in \mathrm{Ord}$, to any ordinal $\geq \varepsilon_0 \cdot (\iota + 1)$. Note that by the above theorem the least ordinal below which we can find a covering of any given pure pattern of order 1 is ε_0. Moreover, the class of ordinals α below which we can find *cofinally* coverings of any pure pattern of order 1, i.e. coverings the minimal element of which can be arbitrarily large below α, is the class of proper multiples of ε_0.

We now come to the elementary observation that in structures containing the relation $\leq_2$, any ordinal that has a proper $<_2$-successor must be the supremum of an infinite $<_1$-chain:

Lemma 21.10 (7.5 of [17]). *If* $\alpha <_2 \beta$ *then* α *is the* sup *of an infinite* $<_1$*-chain.*

Proof. For any $\rho < \alpha$ we have $\beta \models \exists x \,\forall y > x\, (\rho < x <_1 y)$. Hence the same holds true in α. We obtain $\rho_1 <_1 \rho_2 <_1 \rho_3 <_1 \cdots <_1 \alpha$. □

A criterion for the relation $\leq_2$ that actually characterizes $\leq_2$ in $\mathcal{R}_2$, as follows from [75], is provided by the next proposition. Let us assume $\mathcal{L}$ to be a relational language containing $(\leq_i)_{i\leq 2}$ and consider an $\mathcal{L}$-class structure with domain Ord, where $\leq_0$ is the standard linear ordering on the ordinals and $\le_i$, $i = 1,2$, are interpreted as in the introduction by the notion of Σ_i-elementary substructure, however, with respect to the entire language $\mathcal{L}$. Isomorphism indicated by $\cong$ therefore means $\mathcal{L}$-isomorphism. This generality regarding the underlying language $\mathcal{L}$ covers structures like $\mathcal{R}_n, \mathcal{R}_n^+$ for $n \ge 2$ and $\mathcal{R}_\omega, \mathcal{R}_\omega^+$ that were mentioned earlier.

Proposition 21.11 (cf. 7.4 of [17]). *Let α,β be such that $\alpha < \beta$ and $X,\tilde{Y},Y$ be finite sets of ordinals such that $X,\tilde{Y} \subseteq \alpha$ and $Y \subseteq [\alpha,\beta)$. Consider the following properties:*

1. *$X < \tilde{Y} < \alpha$ and there exists an isomorphism $h : X \cup \tilde{Y} \xrightarrow{\cong} X \cup Y$.*
2. *For all finite $\tilde{Y}^+$, where $\tilde{Y} \subseteq \tilde{Y}^+ \subseteq \alpha$, h can be extended to an isomorphism h^+ such that*
$$h^+ : X \cup \tilde{Y}^+ \xrightarrow{\cong} X \cup Y^+$$
for a suitable superset $Y^+ \subseteq \beta$ of Y.

If for all such X and Y there exists a set $\tilde{Y}$ that satisfies property 1, then we have $\alpha <_1 \beta$. If $\tilde{Y}$ can be chosen so that additionally property 2 holds, then we even have $\alpha <_2 \beta$.

Proof. Let ordinals α,β such that $\alpha < \beta$ be given and assume the criterion of the lemma holds. Using 21.6 we obtain $\alpha <_1 \beta$. Let $\varphi(\boldsymbol{x},\boldsymbol{y},\boldsymbol{z})$ be a quantifier free formula of $\mathcal{L}$ with all free variables shown and let $\boldsymbol{\xi} \subseteq \alpha$ be a list of parameters matching the length of $\boldsymbol{x}$. Assume first that

$$\alpha \models \exists \boldsymbol{y}\ \forall \boldsymbol{z}\ \varphi(\boldsymbol{\xi},\boldsymbol{y},\boldsymbol{z}).$$

Let $\boldsymbol{\eta} \subseteq \alpha$ be a list of witnesses for $\boldsymbol{y}$ so that $\alpha \models \forall \boldsymbol{z}\ \varphi(\boldsymbol{\xi},\boldsymbol{\eta},\boldsymbol{z})$. Thanks to $\alpha <_1 \beta$ we have $\beta \models \forall \boldsymbol{z}\ \varphi(\boldsymbol{\xi},\boldsymbol{\eta},\boldsymbol{z})$, whence $\beta \models \exists \boldsymbol{y}\ \forall \boldsymbol{z}\ \varphi(\boldsymbol{\xi},\boldsymbol{y},\boldsymbol{z})$. Now let us assume that

$$\beta \models \exists \boldsymbol{y}\ \forall \boldsymbol{z}\ \varphi(\boldsymbol{\xi},\boldsymbol{y},\boldsymbol{z})$$

and let $\boldsymbol{\eta} \subseteq \beta$ be witnesses for $\boldsymbol{y}$. Without loss of generality we may assume that $\boldsymbol{\eta} \subseteq [\alpha,\beta)$ since we could regard witnesses below α as parameters in the list $\boldsymbol{\xi}$. Set $X := \{\xi \mid \xi \in \boldsymbol{\xi}\}$, $Y := \{\eta \mid \eta \in \boldsymbol{\eta}\}$. By the criterion there exists a set $\tilde{Y}$ such that conditions 1 and 2 of the criterion hold. Let $h : X \cup \tilde{Y} \xrightarrow{\cong} X \cup Y$ be according to condition 2 and let $\tilde{\boldsymbol{\eta}} := h^{-1}[\boldsymbol{\eta}]$. We claim

$$\alpha \models \forall \boldsymbol{z}\ \varphi(\boldsymbol{\xi},\tilde{\boldsymbol{\eta}},\boldsymbol{z}).$$

In order to show this claim let $\tilde{\boldsymbol{\zeta}} \subseteq \alpha$ matching $\boldsymbol{z}$ be given. Set $\tilde{Y}^+ := \tilde{Y} \cup \tilde{\boldsymbol{\zeta}}$ and let h^+ and Y^+ be according to the criterion so that for $\boldsymbol{\zeta} := h^+[\tilde{\boldsymbol{\zeta}}]$, since $\beta \models \varphi(\boldsymbol{\xi},\boldsymbol{\eta},\boldsymbol{\zeta})$ and $X \cup \tilde{Y}^+ \cong X \cup Y^+$, we have $\alpha \models \varphi(\boldsymbol{\xi},\tilde{\boldsymbol{\eta}},\tilde{\boldsymbol{\zeta}})$, whence $\alpha \models \exists \boldsymbol{y}\ \forall \boldsymbol{z}\ \varphi(\boldsymbol{\xi},\boldsymbol{y},\boldsymbol{z})$, concluding the proof of the criterion. □

As mentioned in [17], the above type of criterion can be generalized to the higher levels of elementarity. See Proposition 21.14 for $<_3$. Notice that if the criterion holds for pairs of ordinals α,β and β,γ, showing that $\alpha <_2 \beta <_2 \gamma$ then the criterion also holds for α,γ. Also, if the criterion holds for pairs α,γ_ι, where $\iota \in I$ for some nonempty set I, showing that $\alpha <_2 \gamma_\iota$ for all $\iota \in I$, then the criterion also holds for $\alpha, \sup\{\gamma_\iota \mid \iota \in I\}$.

Another useful elementary observation is the following lemma, where we restrict the underlying language $\mathcal{L}$ to be finite.

Lemma 21.12 (cf. 7.6 of [17]). *Suppose $\alpha <_2 \beta$, $X \subseteq_{\text{fin}} \alpha$, and $\emptyset \neq Y \subseteq_{\text{fin}} [\alpha,\beta)$.*

1. *There exist cofinally many $\tilde{Y} \subseteq \beta$ such that $X \cup \tilde{Y} \cong X \cup Y$. More generally, for any $Z \subseteq_{\text{fin}} \alpha$ with $X < Z$, if $\alpha \models \forall x \exists \tilde{Z}\ (x < \tilde{Z} \wedge$ "$X \cup Z \cong X \cup \tilde{Z}$") then this also holds in β.*
2. *Cofinally in α, copies $\tilde{Y} \subseteq \alpha$ of Y can be chosen which besides $X < \tilde{Y}$ and $X \cup \tilde{Y} \cong X \cup Y$ also maintain $\leq_1$-connections to β, i.e. for any $y \in Y$ such that $y <_1 \beta$ the corresponding $\tilde{y}$ satisfies $\tilde{y} <_1 \alpha$.*

Proof. By $\alpha <_1 \beta$ we have $\alpha \models \forall r\ \exists \tilde{Y}\ (r < \tilde{Y} \wedge$ "$X \cup \tilde{Y} \cong X \cup Y$") where "$X \cup \tilde{Y} \cong X \cup Y$" means that the diagram of $X \cup Y$ in the language $\mathcal{L}$ holds accordingly for $X \cup \tilde{Y}$. By $\alpha <_2 \beta$ this also holds in β. This shows the first part of the lemma. The second part follows from $\alpha <_2 \beta$ by letting $\{y \in Y \mid y <_1 \beta\} = \{y_1,\ldots,y_k\}$ and noting that for any parameter $\xi < \alpha$

$$\beta \models \exists \tilde{Y} > \xi\ \forall r > \tilde{Y} \left(\text{“}X \cup \tilde{Y} \cong X \cup Y\text{”} \wedge \bigwedge_{i=1}^{k} \tilde{y}_i <_1 r \right)$$

which then also holds in α. Notice that we used $\tilde{y}_i$ to indicate the element of $\tilde{Y}$ corresponding to $y_i \in Y$, which can be done by using e.g. increasing enumerations of the elements in Y and $\tilde{Y}$. □

The following theorem follows from Theorem 7.9 of [17] as the least nontrivial example. We give a proof that only makes use of the results given in this section to demonstrate a basic application of the criterion for $<_2$-relations, Proposition 21.11.

Theorem 21.13. *In $\mathcal{R}_n$, $n \geq 2$, we have*

$$\varepsilon_0 \cdot \omega <_2 \varepsilon_0 \cdot (\omega + 1).$$

This is the least such pair of ordinals in $\mathcal{R}_n$. The least $\leq_1$-predecessor of $\varepsilon_0 \cdot \omega$ is ε_0, which is the least element of the $<_1$-chain of the multiples of ε_0 up to $\varepsilon_0 \cdot (\omega + 1)$.

Proof. We apply our knowledge of the structure $\mathcal{R}_1$ formulated in Theorem 21.9. It follows from Lemma 21.10 that $\varepsilon_0 \cdot \omega$ is the least candidate for an ordinal having a nontrivial $<_2$-successor since it is the least supremum of an infinite $<_1$-chain. Clearly, ε_0 is the least $\leq_1$-predecessor of $\varepsilon_0 \cdot \omega$. By part 1 of Lemma 21.12 the least candidate for an ordinal β such that $\varepsilon_0 \cdot \omega <_2 \beta$ is $\varepsilon_0 \cdot (\omega + 1)$. We now use Proposition 21.11 to verify that $\varepsilon_0 \cdot \omega <_2 \varepsilon_0 \cdot (\omega + 1)$ actually holds. To this end, let finite sets $X \subseteq \varepsilon_0 \cdot \omega$ and $Y \subseteq [\varepsilon_0 \cdot \omega, \varepsilon_0 \cdot (\omega + 1))$ be given. Choose $n \in (0, \omega)$ such that $X \subseteq \varepsilon_0 \cdot n$. It follows from Theorem 21.9 that the order-preserving map

$$\pi : \varepsilon_0 \cdot n \cup [\varepsilon_0 \cdot \omega, \varepsilon_0 \cdot (\omega + 1)) \longrightarrow \varepsilon_0 \cdot (n + 1),$$

which simply shifts the interval $[\varepsilon_0 \cdot \omega, \varepsilon_0 \cdot (\omega + 1))$ down to $[\varepsilon_0 \cdot n, \varepsilon_0 \cdot (n + 1))$, is actually an isomorphism in $\mathcal{R}_1$, hence in $\mathcal{R}_n$. We now define $\tilde{Y} := \pi[Y]$ and let h be the order-preserving mapping $h : X \cup \tilde{Y} \to X \cup Y$, which, as we already know, is an isomorphism. Now let a finite set $\tilde{Y}^+$ such that $\tilde{Y} \subseteq \tilde{Y}^+ \subseteq \varepsilon_0 \cdot \omega$ be given. We partition $\tilde{Y}^+$ into three sets $\tilde{Y}_1^+ \subseteq \varepsilon_0 \cdot n$, $\tilde{Y}_2^+ \subseteq [\varepsilon_0 \cdot n, \varepsilon_0 \cdot (n+1))$, and $\tilde{Y}_3^+ \subseteq [\varepsilon_0 \cdot (n+1), \varepsilon_0 \cdot \omega)$. Since $\varepsilon_0 \cdot (n+1) <_1 \varepsilon_0 \cdot \omega$, there exists a copy $\tilde{Z}$ of $\tilde{Y}_3^+$ such that $\tilde{Y}_2^+ < \tilde{Z} \subseteq \varepsilon_0 \cdot (n+1)$ and

$$X \cup \tilde{Y}_1^+ \cup \tilde{Y}_2^+ \cup \tilde{Z} \cong X \cup \tilde{Y}^+.$$

We now define

$$Y^+ := \tilde{Y}_1^+ \cup \pi^{-1}[\tilde{Y}_2^+ \cup \tilde{Z}],$$

which gives rise to the isomorphism

$$h^+ : X \cup \tilde{Y}^+ \stackrel{\cong}{\longrightarrow} X \cup Y^+,$$

which extends h since $\pi^{-1}[\tilde{Y}] = Y$. This settles our verification of the relationship $\varepsilon_0 \cdot \omega <_2 \varepsilon_0 \cdot (\omega + 1)$. □

Let us now assume $\mathcal{L}$ to be a relational language containing $(\le_i)_{i \le 3}$ and consider an $\mathcal{L}$-class structure with domain Ord, where (Ord, $\le_0$) is the standard linear ordering and $\le_i, i = 1, 2, 3$, are interpreted as in the introduction by the notion of Σ_i-elementary substructure with respect to the entire language $\mathcal{L}$. Isomorphism indicated by $\cong$ means $\mathcal{L}$-isomorphism. This covers structures such as $\mathcal{R}_n, \mathcal{R}_n^+$ for $n \ge 3$ and $\mathcal{R}_\omega, \mathcal{R}_\omega^+$.

Proposition 21.14. *In the setting of Proposition 21.11, if for all X and Y the set* $\tilde{Y}$ *can be chosen so that the isomorphism* h^+ *of property 2 additionally satisfies the following property*

3. For all finite sets $\widehat{Y}$ *such that* $Y^+ \subseteq \widehat{Y} \subseteq \beta$, h^+ *can be extended to*

$$\widehat{h} : X \cup \widehat{\tilde{Y}} \stackrel{\cong}{\longrightarrow} X \cup \widehat{Y}$$

for a suitable superset $\widehat{\tilde{Y}} \subseteq \alpha$ *of* $\tilde{Y}^+$,

then we have $\alpha <_3 \beta$.

Proof. Assuming that the criterion holds, we have $\alpha <_2 \beta$ by Proposition 21.11. Let $\varphi(\boldsymbol{w}, \boldsymbol{x}, \boldsymbol{y}, \boldsymbol{z})$ be a quantifier-free formula of the language $\mathcal{L}$, containing $(\le, \le_1, \le_2, \le_3)$. Let $\boldsymbol{\gamma} \subseteq \alpha$ be a list of parameters instantiating $\boldsymbol{w}$ and first assume

$$\alpha \models \exists \boldsymbol{x}\ \forall \boldsymbol{y}\ \exists \boldsymbol{z}\ \varphi(\boldsymbol{\gamma}, \boldsymbol{x}, \boldsymbol{y}, \boldsymbol{z}). \tag{21.1}$$

Let $\boldsymbol{\xi} \subseteq \alpha$ be a list of witnesses for $\boldsymbol{x}$, so that $\alpha \models \forall \boldsymbol{y}\ \exists \boldsymbol{z}\ \varphi(\boldsymbol{\gamma}, \boldsymbol{\xi}, \boldsymbol{y}, \boldsymbol{z})$. Since $\alpha <_2 \beta$ this sentence also holds in β, whence

$$\beta \models \exists \boldsymbol{x}\ \forall \boldsymbol{y}\ \exists \boldsymbol{z}\ \varphi(\boldsymbol{\gamma}, \boldsymbol{x}, \boldsymbol{y}, \boldsymbol{z}),$$

which we now, in turn, assume to prove the converse. Let $\boldsymbol{\xi} \subseteq \beta$ be a list of witnesses for $\boldsymbol{x}$ so that $\beta \models \forall \boldsymbol{y}\, \exists \boldsymbol{z}\, \varphi(\boldsymbol{\gamma}, \boldsymbol{\xi}, \boldsymbol{y}, \boldsymbol{z})$. Without loss of generality we may assume that $\boldsymbol{\xi} \subseteq [\alpha, \beta)$ since elements below α can be considered part of the parameter list $\boldsymbol{\gamma}$. Define

$$X := \{\gamma \mid \gamma \in \boldsymbol{\gamma}\}, \quad Y := \{\xi \mid \xi \in \boldsymbol{\xi}\}.$$

Let $\tilde{Y} \subseteq \alpha$ be according to the criterion and let $\tilde{\boldsymbol{\xi}}$ consist of the elements of $\tilde{Y}$ in the appropriate order. We claim

$$\alpha \models \forall \boldsymbol{y}\, \exists \boldsymbol{z}\, \varphi(\boldsymbol{\gamma}, \tilde{\boldsymbol{\xi}}, \boldsymbol{y}, \boldsymbol{z}). \tag{21.2}$$

In order to prove the claim, let an instantiation $\tilde{\boldsymbol{\zeta}} \subseteq \alpha$ of $\boldsymbol{y}$ be given. We have to show that $\alpha \models \exists \boldsymbol{z}\, \varphi(\boldsymbol{\gamma}, \tilde{\boldsymbol{\xi}}, \tilde{\boldsymbol{\zeta}}, \boldsymbol{z})$. Define

$$\tilde{Y}^+ := \tilde{Y} \cup \tilde{\boldsymbol{\zeta}},$$

let $Y^+ \subseteq \beta$ and

$$h^+ : X \cup \tilde{Y}^+ \xrightarrow{\cong} X \cup Y^+$$

be according to property 2 of the criterion, and set $\boldsymbol{\zeta} := h^+[\tilde{\boldsymbol{\zeta}}]$. By assumption we have $\beta \models \exists \boldsymbol{z}\, \varphi(\boldsymbol{\gamma}, \boldsymbol{\xi}, \boldsymbol{\zeta}, \boldsymbol{z})$. Let $\boldsymbol{\delta} \subseteq \beta$ be witnesses for $\boldsymbol{z}$ so that $\beta \models \varphi(\boldsymbol{\gamma}, \boldsymbol{\xi}, \boldsymbol{\zeta}, \boldsymbol{\delta})$, and define

$$\widehat{Y} := Y^+ \cup \boldsymbol{\delta}.$$

By property 3 of the criterion there exists a superset $\widehat{\tilde{Y}} \subseteq \alpha$ of $\tilde{Y}^+$ and an extension

$$\widehat{h} : X \cup \widehat{\tilde{Y}} \xrightarrow{\cong} X \cup \widehat{Y}$$

of h^+, so that for $\tilde{\boldsymbol{\delta}} := \widehat{h}^{-1}[\boldsymbol{\delta}]$

$$\alpha \models \varphi(\boldsymbol{\gamma}, \tilde{\boldsymbol{\xi}}, \tilde{\boldsymbol{\zeta}}, \tilde{\boldsymbol{\delta}}),$$

which implies the claim 21.2 and hence equation 21.1. □

Lemma 21.15. *Whenever $\alpha <_3 \beta$, the ordinal α is the proper supremum of its $<_2$-predecessors.*

Proof. Let any $\varrho < \alpha$ be given. Since $\alpha <_2 \beta$, α is the proper supremum of an infinite $<_1$-chain:

$$\alpha \models \forall x\, \exists y > x\, \forall z > y\, (y <_1 z),$$

which therefore also holds in β. Hence, β is the proper supremum of its $<_1$-predecessors. Since $\leq_2$ respects $\leq_1$, for every $\eta \in (\alpha, \beta)$ such that $\eta <_1 \beta$ we also have $\alpha <_2 \eta$. Therefore,

$$\beta \models \exists x > \varrho\, \forall y > x\, \exists z > y\, (x <_2 z),$$

namely $x = \alpha$. This sentence in turn holds in α, which produces an infinite $<_2$-chain $\xi_1 <_2 \xi_2 <_2 \cdots <_2 \alpha$. □

Let us continue to assume that $\alpha <_3 \beta$ and consider a structure $\mathcal{R}_n$ where $n \geq 3$. By Theorem 21.5 we know that υ_ω is the least ordinal in $\mathcal{R}_2$ which is the proper supremum of its $<_2$-predecessors. For arbitrary parameter $\varrho < \alpha$,

$$\beta \models \exists x > \varrho\, \forall y < x\, \exists z > y\, (z <_2 x),$$

namely $x = \alpha$, which in turn holds in α, whence α is, moreover, the supremum of infinite $<_2$-chains. The least such, as we know by Theorem 21.5, is the ordinal υ_{ω^2}. Now, for arbitrary $n < \omega$ we have

$$\alpha \models \forall x\, \exists \mathbf{y} > x\, \forall z > \mathbf{y}\, (y_1 <_2 \cdots <_2 y_n <_1 z),$$

which therefore also holds in β. We conclude that β must be some υ_ξ for $\xi > \omega^2$, since we find arbitrarily long finite $<_2$-chains that are $<_1$-connected to β, cofinally below β. Now, since υ_{ω^2+1} is υ_{ω^2}-$\leq_1$-minimal, the least such β is the ordinal υ_{ω^2+2}. Below this ordinal $\mathcal{R}_n$ must therefore agree with $\mathcal{R}_2$.

21.4 A Foretaste of $\mathcal{R}_3$

We are now going to prove the relationship $\upsilon_{\omega^2} <_3 \upsilon_{\omega^2+2}$ in order to give a flavor of the structure $\mathcal{R}_3$ with a hint of its subtleties and new lines of argumentation. We will rely on the analysis of $\mathcal{R}_2$ up to the ordinal υ_{ω^2+2}, which amounts to the necessity of understanding $\mathcal{R}_2$ in full generality, not just up to its core. Our aim is to make the proof understandable to some satisfactory degree by keeping the technical intricacies of [75] "under the hood". To this end we remind the reader of Theorem 21.5 in the introduction and give a high level explanation of the notion of *tracking chain*.

As pointed out in [75], the notions of *tracking sequences and chains*, which were introduced in [17] and generalized in [75], are motivated easily, despite the relatively involved technical preparations and definitions required to formulate them. The approach is natural when analyzing well-partial orderings and was already chosen by Carlson to calculate the structure $\mathcal{R}_1$ in [10]: as explained in the previous section, one considers enumeration functions of connectivity components. In the case of $\mathcal{R}_1$, there is only one such function, called κ, the domain of which is $\varepsilon_0 + 1$, since the ε_0-th component of $\mathcal{R}_1$ was shown in [10] to be $[\varepsilon_0, \infty)$, i.e. $\kappa_{\varepsilon_0} = \varepsilon_0 <_1 \alpha$ for all $\alpha > \varepsilon_0$, which is sometimes written as $\varepsilon_0 <_1 \infty$. Within connectivity components one can consider components relative to the root of the component. In $\mathcal{R}_1$ this is again very easy: $[\varepsilon_0 + 1, \varepsilon_0 \cdot 2) \cong [0, \varepsilon_0)$ as in general $\mathrm{Ord} \cong [\alpha + 1, \infty)$ for all α. Considering the nesting of connectivity components now allows us to locate any ordinal in the segment $\varepsilon_0 \cdot \omega$ by a finite sequence of κ-indices. This means that we can *track down* an ordinal in terms of nested connectivity components. As we already know, the ordinal $\varepsilon_0 \cdot \omega$ is the least supremum of an infinite $<_1$-chain, and it is therefore not surprising that the least $<_2$-pair of $\mathcal{R}_2$ is $\varepsilon_0 \cdot \omega <_2 \varepsilon_0 \cdot (\omega + 1)$, see Theorem 21.13. As shown in the previous section, the structures $\mathcal{R}_1$ and $\mathcal{R}_2$ therefore agree on the initial segment $\varepsilon_0 \cdot (\omega + 1)$.

In $\mathcal{R}_1^+$ the situation is dynamized, and the role of nonzero multiples of ε_0 is taken by $\mathrm{Im}(\upsilon) \setminus \{0\}$, as was shown in [71]. There, the enumeration function κ is therefore relativized to κ^τ where $\tau \in \mathbb{E}_1$, so that κ^τ enumerates the τ-$\leq_1$-minimal ordinals, ordinals that do not have $<_1$-predecessors greater than τ. The index of the maximum τ-$\leq_1$-minimal component that is $\leq_1$-connected to τ is defined to be λ_τ, and thereby the domain of κ^τ is $\lambda_\tau + 1$.

In $\mathcal{R}_2$ any additive principal number $\alpha < \upsilon_\omega$ is the maximum of a maximal finite chain

$$\alpha_0 <_1 \alpha_1 <_2 \cdots <_2 \alpha_n = \alpha,$$

for suitable $n < \omega$, such that α_{i-1} is the greatest $<_2$-predecessor of α_i for $i = 2,\ldots,n$, and α_0 is $\leq_1$-minimal. This was shown in [17] for $\alpha \leq 1^\infty = \upsilon_1$. The *tracking sequence* for such α characterizes the sequence $\alpha_0,\ldots,\alpha_n$ in terms of nested connectivity components by providing their indices. In order to track down additively decomposable ordinals also, the process is iterated, leading to a *tracking chain*: a finite sequence of tracking sequences, see Definition 2.14 of [75]. It is important to note that for additively decomposable ordinals the iterated descent along greatest $<_2$-predecessors does not in general suffice to locate the ordinal in terms of nested $\leq_i$-components, $i = 1,2$. An evaluation function o is defined in [17, 75] in order to recover the actual ordinal from its description by indices. For a preview of how to generalize the concept of tracking chains in the context of structures $\mathcal{R}_n$, $n > 2$, see Section 2.3 of [75].

In Definition 2.31 of Subsection 2.4 of [75] we introduced a generalized notion of *closed set of tracking chains*. Closedness was used in [74] in order not just to locate an ordinal α within the core of $\mathcal{R}_2$, but to collect the tracking chains of ordinals needed to specify a pattern of minimal cardinality, the isominimal realization of which contains and therefore denotes α. The corresponding result for $\mathcal{R}_1^+$ was established in [69] and completed in Sections 5 and 6 of [16]. The generalization of closedness introduced in [75] guarantees isominimality relative to a given setting of relativization in form of fixed interval boundaries υ_ι and $\upsilon_{\iota+1}$, the latter to express possible $<_1$-connections up to $\upsilon_{\iota+1}$ if $\iota \notin \mathrm{Lim}_0$, and can be useful when verifying $<_3$-connections in that it closes under greatest $\leq_i$-predecessors, $i = 1,2$, within the interval $[\upsilon_\iota, \upsilon_{\iota+1})$ of concern. This notion of closedness of a finite set, say $Y \subseteq [\upsilon_\iota, \upsilon_{\iota+1})$, moreover, allows one to detect all parameters from $\mathrm{Im}(\upsilon)$ that occur in the terms of elements of Y and $\mathrm{tc}[Y]$, but misses to include the corresponding greatest $<_2$-predecessors strictly below υ_ι, which therefore requires a separate closure condition. As an illustrative example, cf. the paragraph at the beginning of Subsection 2.4 of [75], consider the closure of the singleton $\{\upsilon_\omega \cdot \upsilon_{17} + 1\} \subseteq [\upsilon_\omega, \upsilon_{\omega+1})$. It includes the ordinals $\upsilon_\omega \cdot \upsilon_{17}$, the greatest $<_2$-predecessor of which is υ_ω, and $\mathrm{lh}_1(\upsilon_\omega \cdot \upsilon_{17}) = \upsilon_\omega \cdot \upsilon_{17} + \upsilon_{17}$, the greatest $<_2$-predecessor of which is υ_{18}, into the closed set, while the ordinal $\upsilon_\omega \cdot \upsilon_{17} + 1$ itself is $\leq_2$-minimal.

Note that it follows from the maximality of the $<_2$-chain $(\upsilon_\iota)_{\iota \in I}$, cf. Theorem 21.5 in the introduction, that from within any interval $(\upsilon_\iota, \upsilon_{\iota+1})$ where $\iota \in \mathrm{Ord}$ there do not exist outgoing $<_2$-connections to ordinals $\geq \upsilon_{\iota+1}$. However, there are incoming $<_2$-connections from ordinals υ_ι, where $\iota \in I$, to proper multiples of ordinals $\upsilon_{\iota\dot{-}1}$ that are greater than υ_ι. If υ_ι is the greatest $<_2$-predecessor of such a multiple, the relevant setting of relativization (in the contexts of enumerations of nested connectivity components), also called its *base* in Definition 2.14 of [75], is $\upsilon_{\iota\dot{-}1}$.

We are now going to introduce mappings of ordinals that enable uniform isomorphic transitions between finite subsets of $\mathcal{R}_2$. Such mappings take and generalize the role of the simple translation mappings in $\mathcal{R}_1$, cf. Theorem 21.13. Recall Definition 21.2 in the introduction. For $\tau = \upsilon_\iota$ the set

$$\mathrm{T}_\tau := \mathrm{T}^\tau \cap \Omega^{>\tau}$$

provides terms for the segment of ordinals below $\upsilon_{\iota+1}$ where the (additive principal) ordinals below υ_ι are considered parameters given as constants. For additive principal $\sigma \leq \tau$ let $\mathrm{T}^\tau[\sigma]$ and $\mathrm{T}_\tau[\sigma]$ denote the subsets of T^τ and T_τ, respectively, in which the occurrence of parameters is restricted to ordinals strictly below σ, cf. [70]. Clearly, $\mathrm{T}^\tau[\tau] = \mathrm{T}^\tau$ and $\mathrm{T}_\tau[\tau] = \mathrm{T}_\tau$.

Consider limit ordinals $\lambda, \mu \in \mathrm{Lim}$, $\lambda < \mu$, and the isomorphism

$$\mathrm{T}_{\upsilon_\lambda} \cong \mathrm{T}_{\upsilon_\mu}[\upsilon_\lambda],$$

a direct consequence of Theorem 3.6 of [75], which moves the base υ_λ to υ_μ, i.e. ϑ-terms relativized to υ_λ simply become ϑ-terms relativized to υ_μ. The restriction of parameters to an initial segment of ordinals can be generalized to a fragmented set of ordinals, in term notation with respect to appropriate settings of relativization. For any $n < \omega$ we obtain the isomorphism

$$\mathrm{T}_{\upsilon_{\lambda+n}}[\mathrm{T}_{\upsilon_{\lambda+n-1}}[\ldots\mathrm{T}_{\upsilon_{\lambda+1}}[\mathrm{T}_{\upsilon_\lambda}]\ldots]] \cong \mathrm{T}_{\upsilon_{\mu+n}}[\mathrm{T}_{\upsilon_{\mu+n-1}}[\ldots\mathrm{T}_{\upsilon_{\mu+1}}[\mathrm{T}_{\upsilon_\mu}[\upsilon_\lambda]]\ldots]].$$

Note that the set on the left side is obtained from $\mathrm{T}_{\upsilon_{\lambda+n}}$ by successive (hereditary) replacement of parameters from the intervals $[\upsilon_{\lambda+i}, \upsilon_{\lambda+i+1})$ by elements from $\mathrm{T}_{\upsilon_{\lambda+i}}$ for $i = n-1$ down to $i = 0$, cf. Definition 2.11 of [75]. Both sides therefore represent sets of notations using parameters below the ordinal υ_λ. The isomorphism moves the sequence of bases $(\upsilon_\lambda, \ldots, \upsilon_{\lambda+n})$ to $(\upsilon_\mu, \ldots, \upsilon_{\mu+n})$.

For $\iota \in \mathrm{Ord} \setminus \mathrm{Lim}_0$ and $\lambda \in \mathrm{Lim}$, $\iota < \lambda$, we have a slightly less obvious isomorphism that follows from Theorem 3.6 of [75] and is made easier to grasp via Theorem 21.5:

$$\upsilon_\iota \cup (\upsilon_{\iota+1} + \mathrm{T}_{\upsilon_\iota}) \cong \mathrm{T}_{\upsilon_\lambda}[\upsilon_\iota],$$

where addition is to be understood elementwise, so that $\upsilon_{\iota+1} + \mathrm{T}_{\upsilon_\iota} = [\upsilon_{\iota+1}, \upsilon_{\iota+1} \cdot 2)$. This isomorphism moves the base υ_ι to υ_λ, for instance the connection $\upsilon_{\iota+1} <_2 \upsilon_{\iota+1} + \upsilon_\iota$ becomes the connection $\upsilon_\lambda <_2 \upsilon_\lambda \cdot 2$, and extends for $n < \omega$ to

$$\upsilon_\iota \cup (\upsilon_{\iota+1} + \mathrm{T}_{\upsilon_{\iota+n}}[\mathrm{T}_{\upsilon_{\iota+n-1}}[\ldots\mathrm{T}_{\upsilon_{\iota+1}}[\mathrm{T}_{\upsilon_\iota}]\ldots]]) \cong \mathrm{T}_{\upsilon_{\lambda+n}}[\mathrm{T}_{\upsilon_{\lambda+n-1}}[\ldots\mathrm{T}_{\upsilon_\lambda}[\upsilon_\iota]]],$$

moving the sequence of bases $(\upsilon_\iota, \ldots, \upsilon_{\iota+n})$ to $(\upsilon_\lambda, \ldots, \upsilon_{\lambda+n})$.

For arbitrary $\alpha \in \mathrm{T}_\iota$, $\iota \in \mathrm{Ord}$, there exist finitely many $\iota_1 < \cdots < \iota_n < \iota$ such that all parameters below υ_ι used in the term representing α are terms of a $\mathrm{T}_{\upsilon_{\iota_i}}$ for some

$i \in \{1, \ldots, n\}$, so that we eventually obtain

$$\alpha \in \mathrm{T}_{\upsilon_\iota}[\mathrm{T}_{\upsilon_{\iota_n}}[\ldots[\mathrm{T}_{\upsilon_{\iota_1}}[1]]\ldots]].$$

For convenience we may assume that the finite set of indices $\{\iota_1, \ldots, \iota_n, \iota\}$ is closed under the operation $\dot{-}$. Clearly, the process of resolving parameters can be truncated at any stage, leaving an initial segment of ordinals below some υ_ν as parameter range.

It is now easy to see that isomorphisms moving υ_λ to υ_μ or a sequence of bases $(\upsilon_\lambda, \ldots, \upsilon_{\lambda+n})$ to $(\upsilon_\mu, \ldots, \upsilon_{\mu+n})$ can be extended to move an increasing sequence $(\upsilon_{\lambda_1}, \ldots, \upsilon_{\lambda_m})$ to another increasing sequence $(\upsilon_{\mu_1}, \ldots, \upsilon_{\mu_m})$ where $\lambda_i, \mu_i \in \mathrm{Lim}$, $i = 1, \ldots, m$, and $\lambda_1 < \mu_1$, with (unresolved) parameters restricted to υ_{λ_1}. This in turn can be extended to mapping sequences with indices $\lambda_1, \lambda_1+1, \ldots, \lambda_1+n_1, \lambda_2, \ldots, \lambda_m+n_m$ to sequences with indices $\mu_1, \mu_1+1, \ldots, \mu_1+n_1, \mu_2, \ldots, \mu_m+n_m$. Fragmentation simply continues to replace the parameter segment υ_{λ_m} by

$$\mathrm{T}_{\upsilon_{\lambda_m+n_m}}[\mathrm{T}_{\upsilon_{\lambda_m+n_m-1}}[\ldots \mathrm{T}_{\upsilon_{\lambda_m+1}}[\mathrm{T}_{\upsilon_{\lambda_m}}[\mathrm{T}_{\upsilon_{\lambda_{m-1}+n_{m-1}}}[\ldots[\mathrm{T}_{\upsilon_{\lambda_1}}]\ldots]]]\ldots]] \tag{21.3}$$

on the domain side and by

$$\mathrm{T}_{\upsilon_{\mu_m+n_m}}[\mathrm{T}_{\upsilon_{\mu_m+n_m-1}}[\ldots \mathrm{T}_{\upsilon_{\mu_m+1}}[\mathrm{T}_{\upsilon_{\mu_m}}[\mathrm{T}_{\upsilon_{\mu_{m-1}+n_{m-1}}}[\ldots[\mathrm{T}_{\upsilon_{\mu_1}}[\upsilon_{\lambda_1}]]\ldots]]]\ldots]] \tag{21.4}$$

on the range side of the isomorphism.

We conclude the explicit display of uniform isomorphic transformations with (the inverse of) a more densely compressing mapping. Let $\iota \in \mathrm{Ord}\backslash\mathrm{Lim}_0$, $n_1, \ldots, n_m < \omega$, and $\lambda_1, \ldots, \lambda_m \in \mathrm{Lim}$ such that $\iota < \lambda_1 < \cdots < \lambda_m$. Define for better readability $s_j := \sum_{l=1}^{j}(n_l + 1)$, $j = 1, \ldots, m$, set $n_0 = s_0 := 0$, and consider the fragmented set

$$C := \upsilon_{\iota+s_m} \setminus \bigcup_{j=0}^{m-1}[\upsilon_{\iota+s_j}, \upsilon_{\iota+s_j+1})$$

in term representation from resolving $\mathrm{T}_{\upsilon_{\iota+s_m \dot{-} 1}}$ over the non-fragmented parameter interval υ_ι. Introducing the abbreviation $\mathrm{T}_{(\nu,\nu+1,\ldots,\nu+k)}$ for $\mathrm{T}_{\nu+k}[\mathrm{T}_{\nu+k-1}[\ldots[\mathrm{T}_\nu]\ldots]]$ we then have

$$C = \upsilon_\iota \cup \bigcup_{j=0}^{m-1}\left(\upsilon_{\iota+s_j+1} + \mathrm{T}_{\left(\upsilon_{\iota+s_j}, \ldots, \upsilon_{\iota+s_{j+1}\dot{-}1}\right)}\right).$$

The set C isomorphically maps onto the set

$$D := \mathrm{T}_{\upsilon_{\lambda_m+n_m}}[\mathrm{T}_{\upsilon_{\lambda_m+n_m-1}}[\ldots \mathrm{T}_{\upsilon_{\lambda_m+1}}[\mathrm{T}_{\upsilon_{\lambda_m}}[\mathrm{T}_{\upsilon_{\lambda_{m-1}+n_{m-1}}}[\ldots[\mathrm{T}_{\upsilon_{\lambda_1}}[\upsilon_\iota]]\ldots]]]\ldots]].$$

We define sets of parameter offsets as a preparation for simultaneous base transformations of bases from D down to bases from C, needed to move $<_2$-predecessors of larger elements in $\mathrm{Im}(\upsilon)$ from D down to C for isomorphic compression in the proof of Theorem 21.16.

$$C' := \mathrm{T}_{(\upsilon_{\iota+s_{m-1}}, \ldots, \upsilon_{\iota+s_{m-1}+(n_m \dot{-} 1)})}[\ldots \mathrm{T}_{(\upsilon_{\iota+s_1}, \ldots, \upsilon_{\iota+s_1+(n_2 \dot{-} 1)})}[\mathrm{T}_{(\upsilon_\iota, \ldots, \upsilon_{\iota+(n_1 \dot{-} 1)})}]\ldots],$$

where the increments of υ-indices in the lists have step size 1. Note that the sets C and C' are not closed under greatest $<_2$-predecessors in $\mathcal{R}_2$.

$$D' := \mathrm{T}_{(\upsilon_{\lambda_m},\ldots,\upsilon_{\lambda_m+(n_m\dot{-}1)})}[\ldots \mathrm{T}_{(\upsilon_{\lambda_1},\ldots,\upsilon_{\lambda_1+(n_1\dot{-}1)})}[\upsilon_\iota]\ldots],$$

again with step size 1. The sets D and D' are closed under greatest $<_2$-predecessors in $\mathcal{R}_2$.

Now consider an increasing sequence of limit ordinals $\mu_1 < \cdots < \mu_r$, $\mu_1 > \lambda_m$, and integers $k_1,\ldots,k_r < \omega$. The simultaneous base transformation of bases $D' \cap \mathrm{Im}(\upsilon)$ down to bases $C' \cap \mathrm{Im}(\upsilon)$

$$\pi_{C,D} : D \cup (\mathrm{T}_{\upsilon_{\mu_r+k_r}}[\ldots \mathrm{T}_{\upsilon_{\mu_1}}[D']\ldots] \setminus \upsilon_{\mu_1}) \cong C \cup (\mathrm{T}_{\upsilon_{\mu_r+k_r}}[\ldots \mathrm{T}_{\upsilon_{\mu_1}}[C']\ldots] \setminus \upsilon_{\mu_1}) \tag{21.5}$$

is obtained by the isomorphic transition from D to C and, for ordinals $\geq \upsilon_{\mu_1}$, the replacement of ϑ-terms relative to

$$\upsilon_{\lambda_1},\ldots,\upsilon_{\lambda_1+(n_1\dot{-}1)},\ldots,\upsilon_{\lambda_m},\ldots,\upsilon_{\lambda_m+(n_m\dot{-}1)}$$

by the ϑ-terms relative to

$$\upsilon_\iota,\ldots,\upsilon_{\iota+(n_1\dot{-}1)},\upsilon_{\iota+s_1},\ldots,\upsilon_{\iota+s_1+(n_2\dot{-}2)},\ldots,\upsilon_{\iota+s_{m-1}},\ldots,\upsilon_{\iota+s_{m-1}+(n_m\dot{-}1)},$$

respectively.

Theorem 21.16. *The least $<_3$-relationship $\alpha <_3 \beta$ in $\mathcal{R}_3$ is*

$$\upsilon_{\omega^2} <_3 \upsilon_{\omega^2+2}.$$

Proof. As we have seen in the previous section, it remains to verify this $<_3$-relationship. In order to show that the criterion given in Proposition 21.14 is fulfilled, let finite sets $X \subseteq \alpha := \upsilon_{\omega^2}$ and $Y \subseteq [\alpha,\beta)$ where $\beta := \upsilon_{\omega^2+2}$ be given. Without loss of generality we may assume that, setting $\gamma := \upsilon_{\omega^2+1}$,

1. $\alpha,\gamma \in Y$,
2. $\mathrm{tc}[X \cup Y]$ is a closed set of tracking chains according to Definition 2.31 of [75], and
3. $X \cup Y$ is closed under greatest $<_2$-predecessors insofar as such exist,

in such a way that the partitioning $X \subseteq \alpha$ and $Y \subseteq [\alpha,\beta)$ is maintained.

Now choose $n \in (0,\omega)$ large enough so that $X \subseteq \upsilon_{\omega\cdot n}$, and note that the closure properties listed above make sure that all parameters from $\mathrm{Im}(\upsilon)$ that occur in the terms of elements of Y and $\mathrm{tc}[Y]$ are strictly less than $\upsilon_{\omega\cdot n}$. Let $\tilde{Y}$ be obtained from Y by collapsing via the "double" base transformation π that moves $\alpha = \upsilon_{\omega^2}$ to $\tilde{\alpha} := \upsilon_{\omega\cdot n}$ and $\gamma = \upsilon_{\omega^2+1}$ to $\tilde{\gamma} := \upsilon_{\omega\cdot n+1}$,

$$\pi := \mathrm{T}_\gamma[\mathrm{T}_\alpha[\tilde{\alpha}]] \stackrel{\cong}{\longrightarrow} \mathrm{T}_{(\tilde{\alpha},\tilde{\gamma})}.$$

It is easy to see that we obtain an isomorphism

$$h : X \cup \tilde{Y} \xrightarrow{\cong} X \cup Y.$$

This takes care of property 1 of Proposition 21.11. Now let a finite superset $\tilde{Y}^+ \subseteq \alpha$ of $\tilde{Y}$ be given, and let us again assume that $\mathrm{tc}[X \cup \tilde{Y}^+]$ is closed and that $X \cup \tilde{Y}^+$ is closed under greatest $<_2$-predecessors insofar as such exist. Clearly, we cannot assume $X < \tilde{Y}^+$ any longer and may therefore assume that $X \subseteq \tilde{Y}^+$. Define the abbreviation $\tilde{\beta} := \upsilon_{\omega \cdot n+2}$ and the partitioning

$$\tilde{Y}_0^+ := \tilde{Y}^+ \cap \tilde{\alpha}, \quad \tilde{Y}_1^+ := \tilde{Y}^+ \cap [\tilde{\alpha}, \tilde{\beta}), \quad \tilde{Y}_2^+ := \tilde{Y}^+ \setminus \tilde{\beta},$$

so that $\tilde{Y} \subseteq \tilde{Y}_1^+$. Term representations of the elements in $\tilde{Y}_1^+ \cup \tilde{Y}_2^+$ can be written in the form (21.3), where we may assume $\lambda_1, \lambda_2, \ldots$ to be $\tilde{\alpha} = \upsilon_{\omega \cdot n}, \upsilon_{\omega \cdot (n+1)}, \ldots$ and where $n_1, n_2, \ldots < \omega$ are chosen sufficiently large. We now apply an $\mathcal{R}_2$-isomorphism of the type that takes (21.3) to (21.4), where $\mu_1, \mu_2, \ldots$ is $\alpha = \upsilon_{\omega^2}, \upsilon_{\omega^2+\omega}, \ldots$. Since this isomorphism extends the mapping π^{-1}, we call the extension π^{-1} as well and define

$$Y_0^+ := \tilde{Y}_0^+, \quad Y_1^+ := \pi^{-1}[\tilde{Y}_1^+], \quad Y_2^+ := \pi^{-1}[\tilde{Y}_2^+],$$

so that $\tilde{Y}^+ = \pi[Y_0^+ \cup Y_1^+ \cup Y_2^+]$. Then we have $X \subseteq Y_0^+ \subseteq \tilde{\alpha}$, $Y \subseteq Y_1^+ \subseteq [\alpha, \beta)$, and $Y_2^+ \subseteq [\beta, \delta)$ where $\delta := \upsilon_{\omega^2 \cdot 2}$. In $\mathcal{R}_2$ we have $\beta <_2 \delta$, so that by the converse of Proposition 21.11, cf. the paragraph preceding the proposition, there exists a copy Z of Y_2^+ such that $Y_1^+ < Z < \beta$ and

$$g : Y_0^+ \cup Y_1^+ \cup Z \xrightarrow{\cong} Y_0^+ \cup Y_1^+ \cup Y_2^+$$

in $\mathcal{R}_2$, satisfying the property of faithful extendibility, property 2 of the criterion. Note that defining

$$Y^+ := Y_0^+ \cup Y_1^+ \cup Z$$

we have $Y \subseteq Y^+ \subseteq \beta$, and the isomorphism h extends to

$$h^+ : X \cup \tilde{Y}^+ \xrightarrow{\cong} X \cup Y^+,$$

witnessing $\alpha <_2 \beta$ in $\mathcal{R}_3$. Keep in mind that h^+ fixes $\tilde{Y}_0^+ = Y_0^+$, $X \subseteq Y_0^+$, and that $h^+[\tilde{Y}_2^+] = Z$.

It remains to show that Y^+ and h^+ are suitable to verify Proposition 21.14, the criterion for $<_3$-relations. To this end let a finite superset $\widehat{Y} \subseteq \beta$ of Y^+ be given. Let us again assume that $\mathrm{tc}[\widehat{Y}]$ is a closed set of tracking chains and that $\widehat{Y}$ is closed under greatest $<_2$-predecessors. The converse of Proposition 21.11 for $\beta <_2 \delta$ in $\mathcal{R}_2$ yields a copy $\widehat{Y}' \subseteq \delta$ of $\widehat{Y}$ containing Y_2^+ such that g extends to

$$g^+ : Y_0^+ \cup Y_1^+ \cup \widehat{Y} \xrightarrow{\cong} Y_0^+ \cup Y_1^+ \cup \widehat{Y}',$$

fixing $Y_0^+ \cup Y_1^+$ and mapping Z onto Y_2^+, i.e. g^+ witnesses $\widehat{Y} \cong_{Y_0^+ \cup Y_1^+} \widehat{Y}'$ since $Y_0^+ \cup Y_1^+ \subseteq Y^+ \subseteq \widehat{Y}$.

Let $\widehat{Y}^\dagger \subseteq \delta$ be the least superset of $\widehat{Y}'$ such that $\mathrm{tc}[\widehat{Y}^\dagger]$ is closed and $\widehat{Y}^\dagger$ is closed under greatest $<_2$-predecessors in $\mathcal{R}_2$. Consider the partitioning

$$\widehat{Y}^\dagger = \widehat{Y}^\dagger_{0,1} \cup \widehat{Y}^\dagger_{0,2} \cup \widehat{Y}^\dagger_1 \cup \widehat{Y}^\dagger_2$$

such that $\widehat{Y}^\dagger_{0,1} \subseteq \tilde{\alpha}$, $\widehat{Y}^\dagger_{0,2} \subseteq [\tilde{\alpha}, \alpha)$, $\widehat{Y}^\dagger_1 \subseteq [\alpha, \beta)$, and $\widehat{Y}^\dagger_2 \subseteq [\beta, \delta)$. We have $Y^+_0 \subseteq \widehat{Y}^\dagger_{0,1}$, $Y^+_1 \subseteq \widehat{Y}^\dagger_1$, and $Y^+_2 \subseteq \widehat{Y}^\dagger_2$. It is important to note that, being the π-preimage of $\tilde{Y}^+_1 \cup \tilde{Y}^+_2$, elements of $Y^+_1 \cup Y^+_2$ do not have greatest $<_2$-predecessors in the interval $[\tilde{\alpha}, \alpha)$, nor do the terms representing these elements make use of parameters from that interval.

We now apply the compression isomorphism $\pi_{C,D}$, introduced in (21.5), with $\iota = \omega \cdot (n-1) + l + 1$ for some $l < \omega$ large enough so that $\widehat{Y}^\dagger_{0,1} \subseteq \upsilon_\iota$, D chosen to cover the set $\widehat{Y}^\dagger_{0,2}$ where $\lambda_1 = \omega \cdot n$, C chosen to match D according to the construction of $\pi_{C,D}$, $\upsilon_{\mu_1} = \alpha$, and so that the domain of $\pi_{C,D}$ covers the set $\widehat{Y}^\dagger_1 \cup \widehat{Y}^\dagger_2$. Let

$$\widehat{Y}^\star := \pi_{C,D}[\widehat{Y}^\dagger]$$

with the partitioning inherited from $\widehat{Y}^\dagger$, so that $\widehat{Y}^\star_{0,1} = \widehat{Y}^\dagger_{0,1}$, $\widehat{Y}^\star_{0,2} \subseteq \tilde{\alpha}$, $\widehat{Y}^\star_1 \subseteq [\alpha, \beta)$, and $\widehat{Y}^\star_2 \subseteq [\beta, \delta)$, whence $\widehat{Y}^\star \cong_{\widehat{Y}^\dagger_{0,1}} \widehat{Y}^\dagger$. Note that $\pi_{C,D}$ acts as the identity on $Y^+_0 \cup Y^+_1 \cup Y^+_2$. We now have $\widehat{Y}^\star \cap [\tilde{\alpha}, \alpha) = \emptyset$, no parameters from the interval $[\tilde{\alpha}, \alpha)$ occur in terms representing elements of $\widehat{Y}^\star_1 \cup \widehat{Y}^\star_2$; thus, we may apply a collapsing isomorphism of the type that takes (21.4) down to (21.3), extending π from above. Let us call this extension π again. Then we obtain

$$\widehat{\tilde{Y}} := \pi[\widehat{Y}^\star] \cong \widehat{Y}^\star,$$

which fixes $\widehat{Y}^\star_{0,1} \cup \widehat{Y}^\star_{0,2}$ and is in turn isomorphic to $\widehat{Y}^\dagger$ with $\widehat{Y}^\dagger_{0,1}$ fixed. We may assume that $\widehat{Y}^\dagger = \widehat{Y}'$, stripping off the auxiliary elements from $\widehat{Y}^\dagger \setminus \widehat{Y}'$ if necessary. We therefore have

$$\widehat{\tilde{Y}} = \pi \circ \pi_{C,D} \circ g^+[\widehat{Y}],$$

which collapses below α, fixing elements below $\tilde{\alpha}$. Since $\pi \circ \pi_{C,D} \circ g[Y^+] = \tilde{Y}^+ \subseteq \widehat{\tilde{Y}}$, the restriction of the isomorphism $(\pi \circ \pi_{C,D} \circ g^+)^{-1}$ to $\widehat{\tilde{Y}}$ is the desired extension $\widehat{h}$ of h^+, which concludes our prototypical verification of a $<_3$-relationship. □

In reviewing our result in the present article, we may conclude with the observation that the so-called *core* of $\mathcal{R}_3$, the union of isominimal realizations of all finite isomorphism types of $\mathcal{R}_3$, cf. [12] and [74], is not an initial segment of the ordinals, as was the case with $\mathcal{R}_1$ (see [12]), $\mathcal{R}^+_1$ (see [71, 69]), $\mathcal{R}_2$ (see [74]), and as will be shown for $\mathcal{R}^+_2$ (work in preparation). This observation, communicated to the author by Carlson some time ago, is confirmed by our results here (Theorem 21.16), as the interval $[\upsilon_\omega, \upsilon_{\omega^2})$ and $\mathrm{Core}(\mathcal{R}_3)$ are disjoint.

Acknowledgements I express my gratitude to Professor Ulf Skoglund for encouragement and support of my research, to Professor Wolfram Pohlers for helpful comments on an earlier draft, and thank Dr. Steven D. Aird for editing the manuscript.

References

1. S. A. Alexander. Arithmetical algorithms for elementary patterns. *Archive for Mathematical Logic*, 54:113–132, 2015.
2. H. Bachmann. Transfinite Zahlen. *Ergebnisse der Mathematik und ihrer Grenzgebiete*, 1, 1955.
3. J. Barwise. *Admissible Sets and Structures*. Springer, Berlin, 1975.
4. J. Bridge. A simplification of the Bachmann method for generating large countable ordinals. *The Journal of Symbolic Logic*, 40:171–185, 1975.
5. W. Buchholz. Normalfunktionen und konstruktive Systeme von Ordinalzahlen. In J. Diller and G. H. Müller, editors, *Proof Theory Symposion Kiel 1974*, volume 500 of *Springer Lecture Notes in Mathematics*, pages 4–25. Springer, 1975.
6. W. Buchholz. Ordinal analysis of ID_ν. In W. Buchholz, S. Feferman, W. Pohlers, and W. Sieg, editors, *Iterated Inductive Definitions and Subsystems of Analysis*, volume 897 of *Lecture Notes in Mathematics*, pages 243–260. Springer, 1981.
7. W. Buchholz. A simplified version of local predicativity. In P. H. G. Aczel, H. Simmons, and S. S. Wainer, editors, *Proof Theory*, pages 115–147. Cambridge University Press, 1992.
8. W. Buchholz and K. Schütte. Die Beziehungen zwischen den Ordinalzahlsystemen Σ und $\overline{\Theta}(\omega)$. *Archiv für mathematische Logik und Grundlagenforschung*, 172:179–189, 1976.
9. T. J. Carlson. Ranked β-calculi. Draft.
10. T. J. Carlson. Ordinal arithmetic and Σ_1-elementarity. *Archive for Mathematical Logic*, 38:449–460, 1999.
11. T. J. Carlson. Knowledge, machines, and the consistency of Reinhardt's strong mechanistic thesis. *Annals of Pure and Applied Logic*, 105:51–82, 2000.
12. T. J. Carlson. Elementary patterns of resemblance. *Annals of Pure and Applied Logic*, 108:19–77, 2001.
13. T. J. Carlson. Patterns of resemblance of order 2. *Annals of Pure and Applied Logic*, 158:90–124, 2009.
14. T. J. Carlson. Generalizing Kruskal's theorem to pairs of cohabitating trees. *Archive for Mathematical Logic*, 55:37–48, 2016.
15. T. J. Carlson. Analysis of a double Kruskal theorem. *Transactions of the American Mathematical Society*, 369:2897–2916, 2017.
16. T. J. Carlson and G. Wilken. Normal forms for elementary patterns. *The Journal of Symbolic Logic*, 77:174–194, 2012.
17. T. J. Carlson and G. Wilken. Tracking chains of Σ_2-elementarity. *Annals of Pure and Applied Logic*, 163:23–67, 2012.
18. J. Diller. Functional interpretations. From the Dialectica interpretation to interpretations of classical and constructive set theory. *World Scientific*. 2019.
19. S. Feferman. Three conceptual problems that bug me. Unpublished lecture text for 7th Scandinavian Logic Symposium, Uppsala, 1996.
20. S. Feferman. Gödel's program for new axioms: Why, where, how and what? In P. Hajek, editor, *Lecture Notes in Logic*, volume 6, pages 3–22. Springer, 1996.
21. H. Friedman. Boolean relation theory and incompleteness, 2011. Available online.
22. H. Friedman, N. Robertson, and P. Seymour. The metamathematics of the graph minor theorem. *Contemporary Mathematics Series of the American Mathematical Society*, 65:229–261, 1987.
23. G. Gentzen. Die Widerspruchsfreiheit der reinen Zahlentheorie. *Mathematische Annalen*, 112:493–565, 1936.

24. G. Gentzen. Neue Fassung des Widerspruchsfreiheitsbeweises für die reine Zahlentheorie. *Forschung in Logik und Grundlegung der exakten Wissenschaften*, 4:19–44, 1938.
25. J.-Y. Girard. Π^1_2-logic, Part I: Dilators. *Annals of Mathematical Logic*, 212:75–219, 1981.
26. R. L. Goodstein. On the restricted ordinal theorem. *The Journal of Symbolic Logic*, 9:33–41, 1944.
27. K. Gödel. Über formal unentscheidbare Sätze der Principia Mathematica und verwandter Systeme I. *Monatshefte für Mathematik und Physik*, 38:173–198, 1931.
28. K. Gödel. Über eine bisher noch nicht benützte Erweiterung des finiten Standpunktes. *Dialectica*, 12:280–287, 1958.
29. D. Hilbert. Die Grundlegung der elementaren Zahlenlehre. *Mathematische Annalen*, 104:485–94, 1931.
30. J. Roger Hindley and Jonathan P. Seldin. *Introduction to Combinators and λ-Calculus*. London Mathematical Society. Cambridge University Press, 1986.
31. W. A. Howard. Assignment of ordinals to terms for primitive recursive functionals of finite type. In J. Myhill, R. E. Vesley, and A. Kino, editors, *Intuitionism and Proof Theory, Proceedings of the Summer Conference at Buffalo N.Y. 1968*, pages 443–458. North Holland, 1970.
32. G. Jäger and W. Pohlers. Eine beweistheoretische Untersuchung von Δ^1_2–CA + BI und verwandter Systeme. *Sitzungsberichte der Bayerischen Akademie der Wissenschaften*, pages 1–28, 1982.
33. A. Kanamori. *The Higher Infinite*. Springer, 2009.
34. L. Kirby and J. Paris. Accessible independence results for Peano arithmetic. *Bulletin of the London Mathematical Society*, 14:285–293, 1982.
35. J. B. Kruskal. Well-quasi-ordering, the tree theorem, and Vazsonyi's conjecture. *Transactions of the American Mathematical Society*, 95:210–225, 1960.
36. G. Lee. Friedman-Weiermann style independence results beyond Peano arithmetic. *Journal of the Korean Mathematical Society*, 51:383–402, 2014.
37. A. Montalbán. Open questions in reverse mathematics. *The Bulletin of Symbolic Logic*, 17:431–454, 2011.
38. T. Paetz. *A Π^1_1-analysis of* $\mathrm{ID}_{<\omega}$ *using the relation* $\le_1$. Thesis (Diplom), University of Münster, 2007.
39. J. Paris and L. Harrington. A mathematical incompleteness in Peano arithmetic. In J. Barwise, editor, *Handbook of Mathematical Logic*, pages 1133–1142. North Holland, 1977.
40. W. Pohlers. Proof theory and ordinal analysis. *Archive for Mathematical Logic*, 30:311–376, 1991.
41. W. Pohlers. A short course in ordinal analysis. In P. H. G. Aczel, H. Simmons, and S. S. Wainer, editors, *Proof Theory*, pages 27–78. Cambridge University Press, Cambridge, 1992.
42. W. Pohlers. Subsystems of set theory and second order number theory. In S. R. Buss, editor, *Handbook of Proof Theory*. Elsevier Science B.V., 1998. Chapter IV.
43. W. Pohlers. *Proof Theory. The First Step into Impredicativity*. Springer, Berlin, 2009.
44. W. Pohlers and J.-C. Stegert. Provably recursive functions of reflection. In U. Berger, H. Diener, P. Schuster, and M. Seisenberger, editors, *Logic, Construction, Computation*, volume 3 of *Ontos Mathematical Logic*, pages 381–474. De Gruyter, 2012.
45. M. Rathjen. Ordinal notations based on a weakly Mahlo cardinal. *Archive for Mathematical Logic*, 29:249–263, 1990.
46. M. Rathjen. How to develop proof-theoretic ordinal functions on the basis of admissible sets. *Mathematical Logic Quarterly*, 39:47–54, 1993.
47. M. Rathjen. Proof theory of reflection. *Annals of Pure and Applied Logic*, 68:181–224, 1994.
48. M. Rathjen. The realm of ordinal analysis. In S. B. Cooper and J. K. Truss, editors, *Sets and Proofs*, volume 258 of *London Mathematical Society Lecture Notes*, pages 219–279. Cambridge University Press, 1999.
49. M. Rathjen. An ordinal analysis of parameter free Π^1_2-comprehension. *Archive for Mathematical Logic*, 48:263–362, 2005.
50. M. Rathjen. The art of ordinal analysis. In *Proceedings of the International Congress of Mathematicians*, 2006.

51. M. Rathjen and A. Weiermann. Proof-theoretic investigations on Kruskal's theorem. *Annals of Pure and Applied Logic*, 60:49–88, 1993.
52. B. Russell. *The Principles of Mathematics*. Merchant Books, 2008.
53. A. Schlüter. *Zur Mengenexistenz in formalen Theorien der Mengenlehre*. Thesis (Dissertation), University of Münster, 1993.
54. D. Schmidt. *Well-Partial Orderings and their Maximal Order Types*. Habilitationsschrift, Heidelberg, 1979.
55. K. Schütte. *Proof Theory*. Springer, Berlin, 1977.
56. K. Schütte and S. G. Simpson. Ein in der reinen Zahlentheorie unbeweisbarer Satz über endliche Folgen von natürlichen Zahlen. *Archiv für mathematische Logik und Grundlagenforschung*, 25:75–89, 1985.
57. S. G. Simpson. Nonprovability of certain combinatorial properties of finite trees. In L. A. Harrington et al., editors, *Harvey Friedman's Research on the Foundations of Mathematics*, pages 87–117. Elsevier Science (North Holland), 1985.
58. S. G. Simpson. *Subsystems of Second Order Arithmetic*. Association of Symbolic Logic, Cambridge University Press, second edition, 2009.
59. O. Veblen. Continuous increasing functions of finite and transfinite ordinals. *Transactions of the American Mathematical Society*, 9:280–292, 1908.
60. A. Weiermann. How to characterize provably total functions by local predicativity. *The Journal of Symbolic Logic*, 61:52–69, 1996.
61. A. Weiermann. A proof of strongly uniform termination for Gödel's T by methods from local predicativity. *Archive for Mathematical Logic*, 36:445–460, 1997.
62. A. Weiermann. How is it that infinitary methods can be applied to finitary mathematics? Gödel's T: A case study. *The Journal of Symbolic Logic*, 63:1348–1370, 1998.
63. A. Weiermann. An application of graphical enumeration to PA. *The Journal of Symbolic Logic*, 68:5–16, 2003.
64. A. Weiermann. Phase transitions for Gödel incompleteness. *Annals of Pure and Applied Logic*, 157:281–296, 2009.
65. A. Weiermann and G. Wilken. Ordinal arithmetic with simultaneously defined Theta-functions. *Mathematical Logic Quarterly*, 57:116–132, 2011.
66. A. Weiermann and G. Wilken. Goodstein sequences for prominent ordinals up to the ordinal of Π^1_1-CA_0. *Annals of Pure and Applied Logic*, 164:1493–1506, 2013.
67. G. Wilken. *Σ_1-Elementarity and Skolem Hull Operators*. Thesis (Dissertation), University of Münster, 2004.
68. G. Wilken. The Bachmann-Howard structure in terms of Σ_1-elementarity. *Archive for Mathematical Logic*, 45:807–829, 2006.
69. G. Wilken. Assignment of ordinals to elementary patterns of resemblance. *The Journal of Symbolic Logic*, 72:704–720, 2007.
70. G. Wilken. Ordinal arithmetic based on Skolem hulling. *Annals of Pure and Applied Logic*, 145:130–161, 2007.
71. G. Wilken. Σ_1-elementarity and Skolem hull operators. *Annals of Pure and Applied Logic*, 145:162–175, 2007.
72. G. Wilken. *Arithmetic Analysis of Elementary Patterns of Order 1 and 2*. Habilitationsschrift (kumulativ), University of Münster, 2011.
73. G. Wilken. Tracking chains revisited. In S.-D. Friedman, D. Raghavan, and Y. Yang, editors, *Sets and Computations*, volume 33 of *Lecture Notes Series*, pages 183–220. Institute for Mathematical Sciences, National University of Singapore, World Scientific Publishing Company, 2017.
74. G. Wilken. Pure patterns of order 2. *Annals of Pure and Applied Logic*, 169:54–82, 2018.
75. G. Wilken. Pure Σ_2-elementarity beyond the core. Submitted. Available on http://arxiv.org/abs/1710.01870, 20xx.
76. G. Wilken and A. Weiermann. Derivation lengths classification of Gödel's T extending Howard's assignment. *Logical Methods in Computer Science*, 8:1–44, 2012.
77. R. Zach. Hilbert's program then and now. *Philosophy of Logic*, 5:411–447, 2006.

Part IV
Kurt Schüttes Spätwerk

Kapitel 22
Ein Wohlordnungsbeweis mit Δ^1_2-Komprehension und Bar-Induktion

Kurt Schütte

Abstract In this paper from 1990, Schütte shows that the ordinal representation system T(I) developed in [2] furnishes a lower bound for the proof-theoretic ordinal of the subsystem of second order arithmetic based on Δ^1_2-comprehension and bar induction. More precisely, it is shown that for every ordinal $<d_0 d\varepsilon_{\mathrm{I}+1}$ in the representation system the wellfoundness of its segment can be demonstrated in the latter theory. This wellordering proof, based on Buchholz' notion of distinguished set, resembles to a certain extent one for a weaker system carried it out in [3] whereas other aspects of it resemble the wellordering proof for the stronger system in [4]. The wellfoundededness up to every ordinal $<d_0 d\varepsilon_{\mathrm{I}+1}$ was first established by W. Pohlers in unpublished work.

Wir beziehen uns auf das in [2] entwickelte Ordinalzahlensystem T(I) und beweisen in einem formalen System der Arithmetik 2. Ordnung mit Δ^1_2-Komprehension und Bar-Induktion die Wohlordnung zu jeder Ordinalzahl $< d_0 d\varepsilon_{\mathrm{I}+1}$. Dieser Beweis, dem der von W. Buchholz [1] eingeführte Begriff der *ausgezeichneten Mengen* zugrunde liegt, hat gewisse Ähnlichkeiten mit dem in [3] für ein schwächeres System und mit dem in [4] für ein stärkeres System durchgeführten Wohlordnungsbeweis. Die Wohlordnung bis zu jeder Ordinalzahl $<d_0 d\varepsilon_{\mathrm{I}+1}$ wurde in entsprechender Weise zuerst von W. Pohlers in einer bisher nicht veröffentlichten Untersuchung nachgewiesen.

22.1 Grundbegriffe

Im folgenden bezeichnen $\alpha, \beta, \gamma, \delta, \eta, \mu, \pi, \sigma$ (auch mit Indizes) stets Ordinalterme des Systems T(I), wobei π, σ Ordinalterme der Menge $\{0\} \cup \{\Psi\alpha : \alpha \in \mathrm{C}(\alpha)\}$ sind.

K. Schütte (†)
Mathematisches Institut, LMU München, Germany

R. Kahle and M. Rathjen (eds.), *The Legacy of Kurt Schütte*,
https://doi.org/10.1007/978-3-030-49424-7_22

Die *Stufe* $\mathrm{S}\gamma$ eines Ordinalterms $\gamma < \mathrm{I}$ ist so definiert, daß $\mathrm{S}\gamma = \sigma \leq \gamma < \sigma^+$ ist, wobei $0^+ = \Psi 0$ und $(\Psi\alpha)^- = \Psi(\alpha+1)$ ist.
Wir legen ein formales System der Arithmetik 2. Ordnung zugrunde, dessen Grundbereich die Ordinalzahlenmenge T(I) ist (bezüglich einer geeigneten Kodierung durch natürliche Zahlen). In diesem System wird sowohl über die Elemente des Grundbereichs (Ordinalterme des Systems T(I)) als auch über Teilmengen des Grundbereichs (Teilmengen von T(I)) quantifiziert. Das formale System habe außer den Axiomen und Schlußregeln über der rekursiven Arithmetik (einschließlich des Axiomenschemas der vollständigen Induktion) die Axiomenschemata der Δ^1_2-Komprehension und der Bar-Induktion.
Mit U, V bezeichnen wir freie Mengenvariablen, mit X eine gebundene Mengenvariable (für Teilmengen von T(I)) und mir A, B Klassenterme (für Teilklassen von T(I)).

Definitionen.

$A \subseteq B := \forall x(x \in A \rightarrow x \in B)$.
$A = B := \forall x(x \in A \leftrightarrow x \in B)$.
$A \cup B := \{\eta : \eta \in A \vee \eta \in B\}$.
$A \cap B := \{\eta : \eta \in A \wedge \eta \in B\}$.
$A < \alpha := \forall x(x \in A \rightarrow x < \alpha)$.
$\alpha \leq A := \exists x(\alpha \leq x \wedge x \in A)$.
$A/\alpha := \{\eta : \eta \in A \wedge \eta < \alpha\}$.
$\mathrm{Prg}[A, B] := \forall x(A/x \subseteq B \wedge x \in A \rightarrow x \in B)$. ($B$ ist *progressiv bezüglich* A.)
$\mathrm{W}A := \{\eta : \eta \in A \wedge \forall X(\mathrm{Prg}[A, X] \rightarrow A/\eta \subseteq X)\}$.
(WA ist der *wohlgeordnete Abschnitt* von A.)
$\mathrm{Wo}[A] := A \subseteq \mathrm{W}A$. ($A$ ist *wohlgeordnet.*)

Die folgenden Lemmata und Korollare geben Formeln an, die in unserem formalen System herleitbar sind.

Lemma 22.1.1.

a) $\mathrm{Prg}[A, B] \rightarrow \mathrm{Prg}[A/\alpha, B]$.
b) $A \subseteq B \wedge \mathrm{Wo}[B] \rightarrow \mathrm{Wo}[A]$.
c) $\mathrm{Wo}[A] \wedge \alpha \in A \rightarrow \forall X(\mathrm{Prg}[A, X] \rightarrow \alpha \in X)$.

Beweis. Dies folgt unmittelbar aus den Definitionen. □

Lemma 22.1.2. $\mathrm{Wo}[A] \wedge \mathrm{Wo}[B] \rightarrow \mathrm{Wo}[A \cup B]$.

Beweis. Man hat

$$\mathrm{Prg}[A \cup B, U] \wedge A/\alpha \subseteq U \rightarrow \mathrm{Prg}[B/\alpha, U].$$

Nach Lemma 22.1.1c) folgt

$$\mathrm{Wo}[B] \wedge \mathrm{Prg}[A \cup B, U] \wedge A/\alpha \subseteq U \rightarrow B/\alpha \subseteq U. \tag{1}$$

Man hat auch

(2) $$\text{Prg}[A \cup B, U] \wedge A/\alpha \subseteq U \wedge B/\alpha \subseteq U\ to(\alpha \in A \rightarrow \alpha \in U).$$

Aus (1) und (2) folgt

$$\text{Wo}[B] \wedge \text{Prg}[A \cup B, U] \rightarrow \text{Prg}[A, U].$$

Nach Lemma 22.1.1c) folgt

(3) $$\text{Wo}[A] \wedge \text{Wo}[B] \wedge \text{Prg}[A \cup B, U] \wedge \alpha \in A \rightarrow \alpha \in U.$$

Ebenso erhält man

(4) $$\text{Wo}[A] \wedge \text{Wo}[B] \wedge \text{Prg}[A \cup B, U] \wedge \alpha \in B \rightarrow \alpha \in U.$$

Aus (3) und (4) folgt

$$\text{Wo}[A] \wedge \text{Wo}[B] \wedge \text{Prg}[A \cup B, U] \rightarrow (A \cup B) \subseteq U.$$

Hiermit ergibt sich die Behauptung. □

Definitionen.

1. $\text{M}^A_\sigma := \{\eta : \eta < \sigma^+ \wedge \forall x(x < \sigma \wedge x \in A \rightarrow \text{K}_{\text{S}x}\eta \subseteq A)\}$.
2. $\text{W}^A_\sigma := \text{WM}^A_\sigma$.

Lemma 22.1.3.

a) $\text{M}^A_0 = \{\eta : \eta < 0^+\}$.
b) $\beta < \omega^\alpha + \beta \rightarrow (\omega^\alpha + \beta \in \text{M}^A_\sigma \leftrightarrow \alpha \in \text{M}^A_\sigma \wedge b \in \text{M}^A_\sigma)$.
c) $A/\sigma = B/\sigma \rightarrow \text{M}^A_\sigma = \text{M}^B_\sigma$.
d) $0 \in \text{W}^A_0$.
e) $\alpha \in \text{W}^A_\sigma \leftrightarrow \text{M}^A_\sigma/\alpha \subseteq \text{W}^A_\sigma \wedge \alpha \in \text{M}^A_\sigma$.

Beweis. Dies folgt unmittelbar aus den Definitionen. □

Lemma 22.1.4. $\gamma \in \text{M}^A_{\text{S}\gamma} \rightarrow \text{S}\gamma \in \text{M}^A_{\text{S}\gamma}$.

Beweis durch Induktion nach $\text{gr}(\gamma)$. Ist $\gamma = \text{S}\gamma$, so gilt die Behauptung. Andernfalls liegt einer der folgenden 2 Fälle vor.

1. $\gamma =_{\text{NF}} \omega^\alpha + \beta$ mit $\text{gr}(\alpha) < \text{gr}(\gamma)$ und $\text{S}\alpha = \text{S}\gamma$. Dann folgt aus $\gamma \in \text{M}^A_{\text{S}\gamma}$ nach Lemma 22.1.3b) $\alpha \in \text{M}^A_{\text{S}\gamma}$ und nach I.V. (Induktions-Voraussetzung) $\text{S}\alpha = \text{S}\gamma \in \text{M}^A_{\text{S}\gamma}$.

2. $\gamma = \psi(\text{S}\gamma)\alpha$ mit $\alpha \in \text{C}_{\text{S}\gamma}(\alpha)$. Für alle $\pi < \text{S}\gamma$ gilt dann $\text{K}_\pi\text{S}\gamma \subseteq \text{K}_\pi\gamma$, so daß $\text{K}_\pi\text{S}\gamma \subseteq A$ aus $\text{K}_\pi\gamma \subseteq A$ folgt. Nach der Definition von $\text{M}^A_{\text{S}\gamma}$ folgt $\text{S}\gamma \in \text{M}^A_{\text{S}\gamma}$. □

Definitionen.

1. $\mathrm{D}[A] := A < \mathrm{I} \wedge \forall x(x \leq A \rightarrow A/(\mathrm{S}x)^{+} = \mathrm{W}^{A}_{\mathrm{S}x})$.
(*A* ist eine *ausgezeichnete Klasse.*)
2. $\underline{\mathrm{W}} := \{\eta : \exists X(\mathrm{D}[X] \wedge \eta \in X)\}$.

Lemma 22.1.5. $\mathrm{D}[A] \rightarrow \mathrm{Wo}[A]$.

Beweis. $\mathrm{D}[A] \wedge \alpha \in A$ impliziert
1) $A/(\mathrm{S}\alpha)^{+} = \mathrm{W}^{A}_{\mathrm{S}\alpha}$ nach der Definition von $\mathrm{D}[A]$,
2) $\alpha \in \mathrm{W}^{A}_{\mathrm{S}\alpha}$ nach 1),
3) $\alpha \in \mathrm{W}A$ nach 1) und 2).
Hiermit ergibt sich die Behauptung. □

Lemma 22.1.6. $\mathrm{D}[A] \wedge \alpha \in A \wedge \gamma < \alpha \wedge \gamma \in \mathrm{M}^{A}_{\mathrm{S}\alpha} \rightarrow \gamma \in A$.

Beweis. $\mathrm{D}[A] \wedge \alpha \in A \wedge \gamma < \alpha \wedge \gamma \in \mathrm{M}^{A}_{\mathrm{S}\alpha}$ impliziert
1) $A/(\mathrm{S}\alpha)^{+} = \mathrm{W}^{A}_{\mathrm{S}\alpha}$ nach der Definition von $\mathrm{D}[A]$,
2) $\mathrm{M}^{A}_{\mathrm{S}\alpha}/\gamma \subseteq \mathrm{W}^{A}_{\mathrm{S}\alpha}$ nach 1),
3) $\gamma \in \mathrm{W}^{A}_{\mathrm{S}\alpha}$ nach 2) und Lemma 22.1.3e),
4) $\gamma \in A$ nach 1) und 3). □

Korollar 22.1.7. $\mathrm{D}[A] \wedge \alpha \in A \rightarrow \mathrm{S}\alpha \in A$.

Beweis. Dies folgt aus den Lemmata 22.1.4 und 22.1.6. □

Lemma 22.1.8. $\mathrm{D}[A] \wedge \alpha \in A \wedge \mathrm{M}^{A}_{\mathrm{S}\alpha}/\gamma \subseteq A \wedge \gamma \in \mathrm{M}^{A}_{\mathrm{S}\alpha} \rightarrow \gamma \in A$.

Beweis. $\mathrm{D}[A] \wedge \alpha \in A \wedge \mathrm{M}^{A}_{\mathrm{S}\alpha}/\gamma \subseteq A \wedge \gamma \in \mathrm{M}^{A}_{\mathrm{S}\alpha}$ impliziert
1) $A/(\mathrm{S}\alpha)^{+} = \mathrm{W}^{A}_{\mathrm{S}\alpha}$ nach der Definition von $\mathrm{D}[A]$,
2) $\mathrm{M}^{A}_{\mathrm{S}\alpha}/\gamma \subseteq \mathrm{W}^{A}_{\mathrm{S}\alpha}$ nach 1),
3) $\gamma \in \mathrm{W}^{A}_{\mathrm{S}\alpha}$ nach 2 und Lemma 22.1.3e),
4) $\gamma \in A$ nach 1) und 3). □

Lemma 22.1.9. $\mathrm{D}[A] \wedge \alpha \in A \wedge \beta \in A \rightarrow \mathrm{K}_{\mathrm{S}\beta}\alpha \subseteq A$.

Beweis durch Induktion nach $\mathrm{gr}(\alpha)$.
$\mathrm{D}[A] \wedge \alpha \in A$ impliziert $A/(\mathrm{S}\alpha)^{+} = \mathrm{W}^{A}_{\mathrm{S}\alpha}$, also $\alpha \in \mathrm{W}^{A}_{\mathrm{S}\alpha}$ und $\alpha \in \mathrm{M}^{A}_{\mathrm{S}\alpha}$.
Für $\beta \in A$ mit $\beta < \mathrm{S}\alpha$ folgt $\mathrm{K}_{\mathrm{S}\beta}\alpha \subseteq A$.
Für $\beta \geq \mathrm{S}\alpha$ hat man folgende drei Fälle.

1. $\alpha = 0$. Dann ist $\mathrm{K}_{\mathrm{S}\beta}\alpha$ leer.
2. $\alpha = \omega^{\alpha_1} + \alpha_2$ mit $\mathrm{gr}(\alpha_i) < \mathrm{gr}(\alpha)$ $(i = 1,2)$. Aus $\alpha \in A$ folgt dann nach den Lemmata 22.1.3b) und 22.1.6 $\alpha_i \in A$ und nach I.V. $\mathrm{K}_{\mathrm{S}\beta}\alpha_i \subseteq A$ $(i = 1,2)$. Es folgt $\mathrm{K}_{\mathrm{S}\beta}\alpha = \mathrm{K}_{\mathrm{S}\beta}\alpha_1 \cup \mathrm{K}_{\mathrm{S}\beta}\alpha_2 \subseteq A$.
3. $\mathrm{K}_{\mathrm{S}\beta}\alpha = \{\alpha\}$, also $\mathrm{K}_{\mathrm{S}\beta}\alpha \subseteq A$ aufgrund von $\alpha \in A$. □

22.2 Herleitungen mit arithmetischer Komprehension

Lemma 22.2.1. $\mathrm{D}[U] \wedge \mathrm{D}[V] \wedge \gamma \leq U \wedge \gamma \leq V \to U/(S\gamma)^+ = V/(S\gamma)^+$.

Beweis. $\mathrm{D}[U] \wedge \mathrm{D}[V] \wedge \gamma \leq U \wedge \gamma \leq V \wedge \alpha < (\mathrm{S}\gamma)^+ \wedge (U \cup V)/\alpha \subseteq (U \cap V)$ impliziert
1) $U/\mathrm{S}\alpha = V/\mathrm{S}\alpha$,
2) $\mathrm{M}^U_{\mathrm{S}\alpha} = \mathrm{M}^V_{\mathrm{S}\alpha}$ nach 1) und Lemma 22.1.3c),
3) $\mathrm{W}^U_{\mathrm{S}\alpha} = \mathrm{W}^V_{\mathrm{S}\alpha}$ nach 2),
4) $U/(\mathrm{S}\alpha)^+ = V/(\mathrm{S}\alpha)^+$ nach 3) und der Definition von $\mathrm{D}[A]$,
5) $\alpha \in (U \cup V) \to \alpha \in (U \cap V)$.
Hiermit ergibt sich

$$(1) \qquad \mathrm{D}[U] \wedge \mathrm{D}[V] \wedge \gamma \leq U \wedge \gamma \leq V \to \mathrm{Prg}[(U \cup V)/(S\gamma)^+, U \cap V].$$

Aus den Lemmata 22.1.2 und 22.1.5 folgt

$$(2) \qquad \mathrm{D}[U] \wedge \mathrm{D}[V] \to \mathrm{Wo}[(U \cup V)/(\mathrm{S}\gamma)^+].$$

Aus (1) und (2) folgt nach Lemma 22.1.1c) mit arithmetischer Komprehension

$$\mathrm{D}[U] \wedge \mathrm{D}[V] \wedge \gamma \leq U \wedge \gamma \leq V \to (U \cup V)/(S\gamma)^+ \subseteq (U \cap V).$$

Hiermit ergibt sich die Behauptung. □

Lemma 22.2.2. $\mathrm{D}[U] \wedge \gamma \leq U \to U/(\mathrm{S}\gamma)^+ = \underline{\mathrm{W}}/(\mathrm{S}\gamma)^+$.

Beweis. Aus Lemma 22.2.1 folgt

$$\mathrm{D}[U] \wedge \gamma \leq U \wedge \alpha \leq (\mathrm{S}\gamma)^+ \wedge \mathrm{D}[V] \wedge \alpha \in V \to \alpha \in U.$$

Nach der Definition von $\underline{\mathrm{W}}$ folgt

$$(1) \qquad \mathrm{D}[U] \wedge \gamma \leq U \wedge \alpha < (\mathrm{S}\gamma)^+ \wedge \alpha \in \underline{\mathrm{W}} \to \alpha \in U.$$

Nach der Definition von $\underline{\mathrm{W}}$ hat man auch

$$(2) \qquad \mathrm{D}[U] \wedge \alpha \in U \to \alpha \in \underline{\mathrm{W}}.$$

Aus (1) und (2) folgt die Behauptung. □

Lemma 22.2.3. $\mathrm{D}[\underline{\mathrm{W}}]$.

Beweis. $\mathrm{D}[U] \wedge \gamma \leq U$ impliziert
1) $U/(\mathrm{S}\gamma)^+ = \mathrm{W}^U_{\mathrm{S}\gamma}$ nach der Definition von $\mathrm{D}[U]$,
2) $U/(\mathrm{S}\gamma)^+ = \underline{\mathrm{W}}/(\mathrm{S}\gamma)^+$ nach Lemma 22.2.2,
3) $\underline{\mathrm{W}}/(\mathrm{S}\gamma)^+ = \mathrm{W}^{\underline{\mathrm{W}}}_{\mathrm{S}\gamma}$ nach 1) und 2).

Man erhält also

$$\mathrm{D}[U] \wedge \gamma \leq U \rightarrow \underline{\mathrm{W}}/(\mathrm{S}\gamma)^+ = \mathrm{W}^{\underline{\mathrm{W}}}_{\mathrm{S}\gamma}.$$

Nach der Definition von $\underline{\mathrm{W}}$ folgt

$$\gamma \leq \underline{\mathrm{W}} \rightarrow \underline{\mathrm{W}}/(\mathrm{S}\gamma)^+ = \mathrm{W}^{\underline{\mathrm{W}}}_{\mathrm{S}\gamma}.$$

Hiermit ergibt sich die Behauptung. □

Korollar 22.2.4. $\mathrm{Wo}[\underline{\mathrm{W}}]$.

Beweis. Dies folgt aus den Lemmata 22.1.5 und 22.2.3. □

Lemma 22.2.5. $\mathrm{D}[U] \wedge \alpha \in U \wedge \beta \in U \rightarrow \omega^\alpha + \beta \in U.$

Beweis. Wir können $\beta < \omega^\alpha + \beta$ annehmen, weil die Behauptung sonst trivial ist. Dann ist $\mathrm{S}(\omega^\alpha + \beta) = \mathrm{S}\alpha$. Wir setzen

$$\begin{aligned}
A_\alpha &:= \{\eta : \eta < \omega^\alpha + \eta \rightarrow \omega^\alpha + \eta \in U\},\\
B &:= \{\eta : U \subseteq A_\eta\},\\
F[\alpha] &:= \mathrm{D}[U] \wedge U/\alpha \subseteq B \wedge \alpha \in U,\\
G[\alpha,\beta] &:= F[\alpha] \wedge U/\beta \subseteq A_\alpha \wedge \beta \in U \wedge \beta < \omega^\alpha + \beta,\\
H[\alpha,\beta,\gamma] &:= G[\alpha,\beta] \wedge \gamma < \omega^\alpha + \beta \wedge \gamma \in \mathrm{M}^U_{\mathrm{S}\alpha}
\end{aligned}$$

und beweisen zunächst durch Induktion nach $\mathrm{gr}(\gamma)$

$$(1) \qquad H[\alpha,\beta,\gamma] \rightarrow \gamma \in U.$$

Für $\gamma < \omega^\alpha + \beta$ hat man folgende drei Fälle.

1. $\gamma \leq \alpha$. Dann folgt $\gamma \in U$ nach Lemma 22.1.6 aus

$$\mathrm{D}[U] \wedge \alpha \in U \wedge \gamma \in \mathrm{M}^U_{\mathrm{S}\alpha}.$$

2. $\alpha < \gamma < \omega^\alpha$. Dann hat man $\gamma =_{\mathrm{NF}} \omega^{\gamma_1} + \gamma_2$ mit $\gamma_1 < \alpha$, $\gamma_2 < \gamma$ und $\mathrm{gr}(\gamma_i) < \mathrm{gr}(\gamma)$ $(i = 1,2)$. Aus $H[\alpha,\beta,\gamma]$ folgt in diesem Fall $H[\alpha,\beta,\gamma_i]$ und nach I.V. $\gamma_i \in U$ $(i = 1,2)$. Mit $U/\alpha \subseteq B$ folgt $\gamma_1 \in B$ und $\gamma \in U$.
3. $\gamma = \omega^\alpha + \delta$ mit $\delta < \beta$ und $\mathrm{gr}(\delta) < \mathrm{gr}(\gamma)$. Aus $H[\alpha,\beta,\gamma]$ folgt in diesem Fall $H[\alpha,\beta,\delta]$ und nach I.V. $\delta \in U$. Mit $U/\beta \subseteq A_\alpha$ folgt $\delta \in A_\alpha$ und $\gamma \in U$.

Hiermit ist (1) bewiesen. Aus (1) folgt

$$(2) \qquad G[\alpha,\beta] \rightarrow \mathrm{M}^U_{\mathrm{S}\alpha}/(\omega^\alpha + \beta) \subseteq U.$$

Man hat auch

$$(3) \qquad G[\alpha,\beta] \rightarrow \mathrm{S}(\omega^\alpha + \beta) = \mathrm{S}\alpha \wedge \alpha \in U \wedge \omega^\alpha + \beta \in \mathrm{M}^U_{\mathrm{S}\alpha}.$$

Aus (2) und (3) folgt nach Lemma 22.1.8

$$G[\alpha,\beta] \to \omega^\alpha+\beta \in U.$$

Nach der Definition von $G[\alpha,\beta]$ folgt

$$F[\alpha] \to \mathrm{Prg}[U, A_\alpha].$$

Nach den Lemmata 22.1.5 und 22.1.1c) folgt mit arithmetischer Komprehension

$$F[\alpha] \to U \subseteq A_\alpha.$$

Nach den Definitionen von B und $F[\alpha]$ folgt

$$\mathrm{D}[U] \to \mathrm{Prg}[U, B].$$

Nach den Lemmata 22.1.5 und 22.1.1c) folgt mit arithmetischer Komprehension

$$\mathrm{D}[U] \wedge \alpha \in U \to \alpha \in B.$$

Hiermit ergibt sich die Behauptung

$$\mathrm{D}[U] \wedge \alpha \in U \wedge \beta \in U \to \beta \in A_\alpha. \qquad \square$$

Korollar 22.2.6. $\alpha \in \underline{\mathrm{W}} \wedge \beta \in \underline{\mathrm{W}} \to \omega^\alpha+\beta \in \underline{\mathrm{W}}$.

Beweis. Wir können $\beta < \omega^\alpha+\beta$ annehmen, weil die Behauptung sonst trivial ist. Dann folgt aus Lemma 22.2.1

$$\mathrm{D}[U] \wedge \alpha \in U \wedge \mathrm{D}[V] \wedge \beta \in V \to \alpha \in U \wedge \beta \in U.$$

Nach Lemma 22.2.5 folgt

$$\mathrm{D}[U] \wedge \alpha \in U \wedge \mathrm{D}[V] \wedge \beta \in V \to \omega^\alpha+\beta \in U.$$

Nach der Definition von $\underline{\mathrm{W}}$ folgt zunächst

$$\mathrm{D}[U] \wedge \alpha \in U \wedge \mathrm{D}[V] \wedge \beta \in V \to \omega^\alpha+\beta \in \underline{\mathrm{W}}$$

und weiterhin

$$\alpha \in \underline{\mathrm{W}} \wedge \beta \in \underline{\mathrm{W}} \to \omega^\alpha+\beta \in \underline{\mathrm{W}}. \qquad \square$$

Lemma 22.2.7. $\mathrm{D}[U] \wedge \alpha \in U \wedge \sigma \in U \wedge \alpha \in \mathrm{C}_\sigma(\alpha) \to \psi\sigma\alpha \in U$.

Beweis. Wir setzen

$$A := \{\eta : \forall\pi(\pi \in U \wedge \eta \in \mathrm{C}_\pi(\eta) \to \psi\pi\eta \in U)\},$$
$$F[\alpha] := \mathrm{D}[U] \wedge U/\alpha \subseteq A \wedge \alpha \in U,$$

$$G[\alpha,\sigma] := F[\alpha] \wedge \sigma \in U \wedge \alpha \in \mathrm{C}_\sigma(\alpha),$$
$$H[\alpha,\sigma,\gamma] := G[\alpha,\sigma] \wedge \gamma < \psi\sigma\alpha \wedge \gamma \in \mathrm{M}_\sigma^U$$

und beweisen zunächst durch Induktion nach $\mathrm{gr}(\gamma)$

(1) $$H[\alpha,\sigma,\gamma] \to \gamma \in U.$$

Für $\gamma < \psi\sigma\alpha$ hat man folgende drei Fälle.

1. $\gamma \leq \sigma$. Dann folgt $\gamma \in U$ nach Lemma 22.1.6 aus

$$\mathrm{D}[U] \wedge \sigma \in U \wedge \gamma \in \mathrm{M}_\sigma^U.$$

2. $\sigma < \gamma =_{\mathrm{NF}} \omega^{\gamma_1}+\gamma_2$ mit $\mathrm{gr}(\gamma_i) \wedge \mathrm{gr}(\gamma)$ $(i = 1,2)$. Aus $H[\alpha,\sigma,\gamma]$ folgt in diesem Fall $H[\alpha,\sigma,\gamma_i]$ und nach I.V. $\gamma_i \in U$ $(i = 1,2)$. Nach Lemma 22.2.5 folgt $\gamma \in U$.
3. $\gamma = \psi\sigma\beta$ mit $\beta \in \mathrm{C}_\sigma(\beta)$, $\beta < \alpha$ und $\mathrm{gr}(\beta) < \mathrm{gr}(\gamma)$.
   In diesem Fall beweisen wir

   (1') $$H[\alpha,\sigma,\gamma] \wedge U/\delta \subseteq B_\beta \wedge \delta \in U \to \delta \in B_\beta$$

   für $B_\beta := \{\eta : \mathrm{K}_{\mathrm{S}\eta}\beta \subseteq U\}$.
   Ist $S\delta < \sigma$, so folgt $\mathrm{K}_{\mathrm{S}\delta}\beta \subseteq U$ aus $\mathrm{D}[U] \wedge \delta \in U \wedge \psi\sigma\beta \in \mathrm{M}_\sigma^U$.
   Sei nun $\sigma \leq \mathrm{S}\delta$ und $\eta \in \mathrm{K}_{\mathrm{S}\delta}\beta$. Aus $\beta \in \mathrm{C}_\sigma(\beta) \wedge \alpha \in \mathrm{C}_{\mathrm{S}\delta}(\alpha)$ folgt dann $\beta \in \mathrm{C}_{\mathrm{S}\delta}(\beta) \wedge \alpha \in \mathrm{C}_{\mathrm{S}\delta}(\alpha)$ und $\eta < \psi(\mathrm{S}\delta)\beta < \psi(\mathrm{S}\delta)\alpha$. Aus $U/\delta \subseteq B_\beta$ und $\eta \in \mathrm{K}_{\mathrm{S}\delta}\beta$ folgt $\mathrm{S}\delta \in U \wedge \eta \in \mathrm{M}_{\mathrm{S}\delta}^U$. Hiermit ergibt sich $H[\alpha,\mathrm{S}\delta,\eta]$. Da $\mathrm{gr}(\eta) \leq \mathrm{gr}(\beta) < \mathrm{gr}(\gamma)$ ist, folgt nach I.V. $\eta \in U$, also $\mathrm{K}_{\mathrm{S}\delta}\beta \subseteq U$.
   Hiermit ist (1') bewiesen. Aus (1') folgt

   $$H[\alpha,\sigma,\gamma] \to \mathrm{Prg}[U, B_\beta].$$

   Nach den Lemmata 22.1.5 und 22.1.1c) folgt mit arithmetischer Komprehension

   $$H[\alpha,\sigma,\gamma] \to \beta \in \mathrm{M}_{\mathrm{S}\alpha}^U.$$

   $\mathrm{D}[U] \wedge \alpha \in U \wedge \beta \in \mathrm{M}_{\mathrm{S}\alpha}^U \wedge \beta < \alpha$ impliziert $\beta \in U$.
   Mit $U/\alpha \subseteq A \wedge \sigma \in U \wedge \beta \in C_\sigma(\beta)$ folgt $\gamma = \psi\sigma\beta \in U$.

Hiermit ist (1) bewiesen. Aus (1) folgt

(2) $$G[\alpha,\sigma] \to \mathrm{M}_\sigma^U/\psi\sigma\alpha \subseteq U.$$

Man hat auch

(3) $$G[\alpha,\sigma] \to \sigma \in U \wedge \psi\sigma\alpha \in \mathrm{M}_\sigma^U.$$

Aus (2) und (3) folgt nach Lemma 22.1.8

$$G[\alpha,\sigma] \to \psi\sigma\alpha \in U.$$

Nach der Definition von $G[\alpha,\sigma]$ folgt

$$F[\alpha] \rightarrow \forall\pi(\pi \in U \wedge \alpha \in \mathrm{C}_\pi(\alpha) \rightarrow \psi\pi\alpha \in U).$$

Nach der Definition von A und $F[\alpha]$ folgt

$$\mathrm{D}[U] \rightarrow \mathrm{Prg}[\dot{U}, A].$$

Nach den Lemmata 22.1.5 und 22.1.1c) folgt mit arithmetischer Komprehension

$$\mathrm{D}[U] \wedge \alpha \in U \rightarrow \alpha \in A.$$

Hiermit ergibt sich die Behauptung

$$\mathrm{D}[U] \wedge \alpha \in U \wedge \sigma \in U \wedge \alpha \in C_\sigma(\alpha) \rightarrow \psi\sigma\alpha \in U. \qquad \square$$

Korollar 22.2.8. $\alpha \in \underline{\mathrm{W}} \wedge \sigma \in \underline{\mathrm{W}} \wedge \alpha \in \mathrm{C}_\sigma(\alpha) \rightarrow \psi\sigma\alpha \in \underline{\mathrm{W}}$.

Beweis. Aus Lemma 22.2.1 folgt

$$\mathrm{D}[U] \wedge \alpha \in U \wedge \mathrm{D}[V] \wedge \sigma \in V \rightarrow (\alpha \in U \wedge \sigma \in U) \vee (\alpha \in V \wedge \sigma \in V).$$

Nach Lemma 22.2.7 folgt

$$\mathrm{D}[U] \wedge \alpha \in U \wedge \mathrm{D}[V] \wedge \sigma \in V \wedge \alpha \in \mathrm{C}_\sigma(\alpha) \rightarrow \psi\sigma\alpha \in U \vee \psi\sigma\alpha \in V.$$

Nach der Definition von $\underline{\mathrm{W}}$ folgt zunächst

$$\mathrm{D}[U] \wedge \alpha \in U \wedge \mathrm{D}[V] \wedge \sigma \in V \wedge \alpha \in \mathrm{C}_\sigma(\alpha) \rightarrow \psi\sigma\alpha \in \underline{\mathrm{W}}$$

und weiterhin

$$\alpha \in \underline{\mathrm{W}} \wedge \sigma \in \underline{\mathrm{W}} \wedge \alpha \in \mathrm{C}_\sigma(\alpha) \rightarrow \psi\sigma\alpha \in \underline{\mathrm{W}}. \qquad \square$$

Lemma 22.2.9. $\mathrm{D}[U] \wedge \alpha \in U \wedge \alpha \leq \sigma^+ \wedge \mathrm{D}[V] \rightarrow \alpha \in \mathrm{M}^V_\sigma$.

Beweis. $\mathrm{D}[U] \wedge \alpha \in U \wedge \alpha < \sigma^+ \wedge \mathrm{D}[V]$ impliziert
1) $U \subseteq V \vee V \subseteq U$ nach Lemma 22.2.1,
2) $U \subseteq V \rightarrow \alpha \in V$,
3) $U \subseteq V \rightarrow \alpha \in \mathrm{M}^V_\sigma$ nach 2) und Lemma 22.1.9,
4) $V \subseteq u \wedge \beta \in V \rightarrow \beta \in U$,
5) $V \subseteq U \wedge \beta \in V \rightarrow \mathrm{K}_{S\beta}\alpha \subseteq U$ nach 4) und Lemma 22.1.9,
6) $V \subseteq U \wedge \beta \in V \rightarrow \mathrm{K}_{S\beta}\alpha \subseteq B$ nach 5) und Lemma 22.2.1,
7) $V \subseteq U \rightarrow \alpha \in \mathrm{M}^V_\sigma$ nach 6),
8) $\alpha \in \mathrm{M}^V_\sigma$ nach 1), 3) und 7). $\square$

22.3 Herleitungen mit Π^1_1-Komprehension

Lemma 22.3.1. $0 \in \underline{\mathrm{W}}$.

Beweis. Für $T := \{\eta : \eta < 0^+\}$ gilt nach Lemma 22.1.3a) und d) $\mathrm{M}_0^{\mathrm{W}T} = T$ und $0 \in \mathrm{W}T$. Es folgt $\mathrm{W}_0^{\mathrm{W}T} = \mathrm{W}T$. Hiermit ergibt sich $\mathrm{D}[\mathrm{W}T]$. $\mathrm{W}T$ ist ein Π^1_1-Term. Mit Π^1_1-Komprehension folgt daher $\exists X(\mathrm{D}[X] \wedge 0 \in X)$, also $0 \in \underline{\mathrm{W}}$. □

Lemma 22.3.2. $\mathrm{S}\alpha \leq \sigma \wedge \mathrm{K}_\sigma\alpha \subseteq \underline{\mathrm{W}} \to \alpha \in \underline{\mathrm{W}}$.

Beweis durch Induktion nach $\mathrm{gr}(\alpha)$. Man hat folgende drei Fälle.

1. $\alpha = 0$. Dann gilt die Behauptung nach Lemma 22.3.1.
2. $\alpha =_{\mathrm{NF}} \omega^{\alpha_1} + \alpha_2$ mit $\mathrm{gr}(\alpha_i) < \mathrm{gr}(\alpha)$ und $\mathrm{S}\alpha_i \leq \sigma$ $(i = 1, 2)$. Aus $\mathrm{K}_\sigma\alpha \subseteq \underline{\mathrm{W}}$ folgt dann $\mathrm{K}_\sigma\alpha_i \subseteq \underline{\mathrm{W}}$ und nach I.V. $\alpha_i \in \underline{\mathrm{W}}$ $(i = 1, 2)$. Nach Korollar 22.2.6 folgt $\alpha \in WW$.
3. $\mathrm{K}_\sigma\alpha = \{\alpha\}$. Dann gilt die Behauptung nach Voraussetzung. □

Lemma 22.3.3. $\alpha < \mathrm{I} \wedge \mathrm{K}\alpha \subseteq \underline{\mathrm{W}} \to \alpha \in \underline{\mathrm{W}}$.

Beweis. Durch Induktion nach $\mathrm{gr}(\alpha)$ entsprechend wie für Lemma 22.3.2. □

Lemma 22.3.4. $\mathrm{D}[U] \wedge U/\sigma = \mathrm{W}_\sigma^U \to \mathrm{D}[\mathrm{W}_\sigma^U]$.

Beweis. Wir setzen $A := \mathrm{W}_\sigma^U$.
$\mathrm{D}[U] \wedge U/\sigma = \mathrm{W}_\sigma^U$ impliziert
1) $\beta < \sigma \wedge \beta \in A \to \beta \in U$,
2) $\beta < \sigma \wedge \beta \in A \wedge \alpha \leq \beta \to U/(\mathrm{S}\alpha)^+ = \mathrm{W}_{\mathrm{S}\alpha}^U$ nach 1),
3) $\beta < \sigma \wedge \beta \in A \wedge \alpha \leq \beta \to A/(\mathrm{S}\alpha)^+ = \mathrm{W}_{\mathrm{S}\alpha}^A$ nach 2),
4) $A/\sigma^+ = \mathrm{W}_\sigma^U$,
5) $A/\sigma^+ = \mathrm{W}_\sigma^A$ nach 4),
6) $\mathrm{D}[A]$ nach 3) und 5). □

Lemma 22.3.5. $\mathrm{D}[U] \wedge \mathrm{M}_\sigma^U/\sigma \subseteq U \wedge \sigma \in \mathrm{M}_\sigma^U \to \sigma \in \underline{\mathrm{W}}$.

Beweis. $\mathrm{D}[U] \wedge \mathrm{M}_\sigma^U/\sigma \subseteq U \wedge \sigma \in \mathrm{M}_\sigma^U$ impliziert
1) $U\sigma = \mathrm{M}_\sigma^U/\sigma$ aufgrund von $U/\sigma \subseteq \mathrm{M}_\sigma^U$ nach Lemma 22.1.9,
2) $U\sigma = \mathrm{W}_\sigma^U/\sigma$ nach 1) und Lemma 22.1.5,
3) $\mathrm{D}[\mathrm{W}_\sigma^U]$ nach 2) und Lemme 22.3.4,
4) $\mathrm{M}_\sigma^U/\sigma \subseteq \mathrm{W}_\sigma^U$ nach 1) und 2),
5) $\sigma \in \mathrm{W}_\sigma^U$ nach 4) und Lemma 22.1.3e),
6) $\sigma \in \underline{\mathrm{W}}$ nach 3) und 5) mit Π^1_1-Komprehension, da W_σ^U ein Π^1_1-Term ist. □

22.4 Herleitungen mit Δ^1_2-Komprehension

Lemma 22.4.1. $\forall X(\mathrm{D}[X] \to \alpha \in \mathrm{M}^X_\sigma) \to \alpha \in \mathrm{M}^{\underline{\mathrm{W}}}_\sigma$.

Beweis. $\forall X(\mathrm{D}[X] \to \alpha \in \mathrm{M}^X_\sigma)$ impliziert
1) $\alpha < \sigma^+$ aufgrund von $\exists X \mathrm{D}[X]$,
2) $\mathrm{D}[U] \to \alpha \in \mathrm{M}^U_\sigma$,
3) $\mathrm{D}[U] \wedge \beta < \sigma \wedge \beta \in U \to \mathrm{K}_{\mathrm{S}\beta}\alpha \subseteq U$ nach 2),
4) $\beta < \sigma \wedge \beta \in \underline{\mathrm{W}} \to \mathrm{K}_{\mathrm{S}\beta}\alpha \subseteq \underline{\mathrm{W}}$ nach 3),
5) $\alpha \in \mathrm{M}^{\underline{\mathrm{W}}}_\sigma$ nach 1) und 4). □

Lemma 22.4.2. $\mathrm{M}^{\underline{\mathrm{W}}}_\sigma/\sigma \subseteq \underline{\mathrm{W}} \wedge \sigma \in \mathrm{M}^{\underline{\mathrm{W}}}_\sigma \to \sigma \in \underline{\mathrm{W}}$.

Beweis. Wir setzen $\underline{\mathrm{V}} := \{\eta : \forall X(\mathrm{D}[X] \to \eta \in \mathrm{M}^{\underline{\mathrm{W}}}_\sigma)\}$.
Nach Lemma 22.2.9 hat man

$$\mathrm{D}[U] \wedge \alpha \in U \wedge \alpha < \sigma^+ \wedge \mathrm{D}[V] \to \alpha \in \mathrm{M}^V_\sigma.$$

Hieraus folgt

$$\underline{\mathrm{W}}/\sigma^+ \subseteq \underline{\mathrm{V}}. \tag{1}$$

Nach Lemma 22.4.1 hat man

$$\alpha \in \underline{\mathrm{V}} \to \alpha \in \mathrm{M}^{\underline{\mathrm{W}}}_\sigma.$$

Hieraus folgt

$$\mathrm{M}^{\underline{\mathrm{W}}}_\sigma/\sigma \subseteq \underline{\mathrm{W}} \to \underline{\mathrm{V}}/\sigma \subseteq \underline{\mathrm{W}}. \tag{2}$$

Aus (1) und (2) folgt

$$\mathrm{M}^{\underline{\mathrm{W}}}_\sigma/\sigma \subseteq \underline{\mathrm{W}} \to \underline{\mathrm{V}}/\sigma = \underline{\mathrm{W}}/\sigma. \tag{3}$$

$\underline{\mathrm{V}}/\sigma$ ist ein Π^1_2-Term und $\underline{\mathrm{W}}/\sigma$ ein Σ^1_2-Term. Aus (3) folgt daher mit Δ^1_2-Komprehension

$$\mathrm{M}^{\underline{\mathrm{W}}}_\sigma/\sigma \subseteq \underline{\mathrm{W}} \to \exists X(X = \underline{\mathrm{W}}/\sigma). \tag{4}$$

Nach Lemma 22.3.5 hat man

$$\mathrm{D}[U] \wedge \mathrm{M}^U_\sigma/\sigma \subseteq U \wedge \sigma \in \mathrm{M}^U_\sigma \to \sigma \in \underline{\mathrm{W}}. \tag{5}$$

Aus (4) und (5) folgt die Behauptung

$$\mathrm{M}^{\underline{\mathrm{W}}}_\sigma/\sigma \subseteq \underline{\mathrm{W}} \wedge \sigma \in \mathrm{M}^{\underline{\mathrm{W}}}_\sigma \to \sigma \in \underline{\mathrm{W}}.$$ □

Definitionen.

1. $\underline{M} := \{\eta : (\eta < I \wedge \eta \in \underline{W}) \vee (I \leq \eta \wedge K\eta \subseteq W)\}$.
2. $\underline{W}_\Psi := \{\eta : \eta \in C(\eta) \to \Psi\eta \in \underline{W}\}$.

Lemma 22.4.3. $\alpha \in \underline{M} \wedge \alpha \in C(\alpha) \to \alpha \in M^{\underline{W}}_{\Psi\alpha}$.

Beweis. $\alpha \in \underline{M} \wedge \alpha \in C(\alpha) \wedge \beta < \Psi\alpha \wedge \beta \in \underline{W}$ impliziert
1) $K_{S\beta}\Psi\alpha = K_{S\beta}\alpha \wedge K_{S\beta}\Psi\alpha = K_{S\beta}K\alpha$,
2) $\alpha \in \underline{W} \vee K\alpha \subseteq \underline{W}$,
3) $K_{S\beta}\Psi\alpha \subseteq \underline{W}$ nach 1), 2) und Lemma 22.1.9.
Hiermit ergibt sich die Behauptung. □

Lemma 22.4.4. $\mathrm{Prg}[\underline{M}, \underline{W}_\Psi]$.

Beweis. Wir setzen

$$F[\alpha] := \underline{M}/\alpha \subseteq \underline{W}_\Psi \wedge \alpha \in \underline{M} \wedge \alpha \in C(\alpha),$$
$$G[\alpha,\gamma] := F[\alpha] \wedge \gamma < \Psi\alpha \wedge \gamma \in M^{\underline{W}}_{\Psi\alpha}$$

und beweisen zunächst durch Induktion nach $\mathrm{gr}(\gamma)$

$$G[\alpha,\gamma] \to \gamma \in \underline{W}. \tag{1}$$

Für $\gamma < \Psi\alpha$ hat man folgende drei Fälle.
1. $\gamma = 0$. Dann gilt die Behauptung nach Lemma 22.3.1.
2. $S\gamma < \gamma$. Dann hat man $\mathrm{gr}(S\gamma) < \mathrm{gr}(\gamma)$ und mit $G[\alpha,\gamma]$ auch $G[\alpha, S\gamma]$. Nach I.V. folgt $S\gamma \in \underline{W}$. Aufgrund von $\gamma \in M^{\underline{W}}_{\Psi\alpha}$ folgt $K_{S\gamma}\gamma \subseteq \underline{W}$. Nach Lemma 22.3.2 folgt $\gamma \in \underline{W}$.
3. $\gamma = \Psi\beta$ mit $\beta \in C(\beta)$, $\beta < \alpha$ und $\mathrm{gr}(\beta) < \mathrm{gr}(\gamma)$.
   3.1. Es gebe $\delta \in \underline{W}$ mit $\gamma \leq \delta < \Psi\alpha$. Aus $\gamma \in M^{\underline{W}}_{\Psi\alpha}$ folgt dann $K_{S\delta}\gamma \subseteq \underline{W}$. Nach Lemma 22.3.2 folgt $\gamma \in \underline{W}$.
   3.2. Es gebe kein $\delta \in \underline{W}$ mit $\gamma \leq \delta < \Psi\alpha$. Aus $\gamma \in M^{\underline{W}}_{\Psi\alpha}$ folgt dann $K\beta \subseteq M^{\underline{W}}_{\Psi\alpha}$. Für alle $\eta \in K\beta$ hat man dann $\mathrm{gr}(\eta) < \mathrm{gr}(\gamma)$ und mit $G[\alpha,\gamma]$ auch $G[\alpha,\eta]$. Nach I.V. folgt $\eta \in \underline{W}$, also $K\beta \subseteq \underline{W}$. Ist $\beta < I$, so folgt nach Lemma 22.3.3 $\beta \in \underline{W}$. Man erhält daher in jedem Fall $\beta \in \underline{M}$. Mit $\underline{M}/\alpha \subseteq \underline{W}_\Psi$ und $\beta \in C(\beta)$ folgt $\gamma = \Psi\beta \in \underline{W}$.

Hiermit ist (1) bewiesen. Aus (1) folgt

$$F[\alpha] \to M^{\underline{W}}_{\Psi\alpha}/\Psi\alpha \subseteq \underline{W}. \tag{2}$$

Nach Lemma 22.4.3 hat man auch

$$F[\alpha] \to \Psi\alpha \in M^{\underline{W}}_{\Psi\alpha}. \tag{3}$$

Aus (2) und (3) folgt nach Lemma 22.4.2

$$F[\alpha] \to \Psi\alpha \in \underline{W}.$$

Hiermit ergibt sich die Behauptung $\mathrm{Prg}[\underline{M}, \underline{W}_\Psi]$. □

22.5 Herleitungen mit Δ^1_2-Komprehension und Bar-Induktion

Lemma 22.5.1. $\mathrm{Prg}[M, A] \to \mathrm{I}{+}1 \in A$.

Beweis. Aus den Lemmata 22.1.5 und 22.1.1c) folgt mit Bar-Induktion

$$\mathrm{D}[U] \wedge \alpha \in U \wedge \mathrm{Prg}[U/(\mathrm{S}\alpha)^+, A].$$

Nach Lemma 22.2.2 folgt

$$\mathrm{Prg}[\underline{W}/(\mathrm{S}\alpha)^+, A] \wedge \alpha \in \underline{W} \to \alpha \in A.$$

Da $\underline{W}/(\mathrm{S}\alpha)^+ \subseteq \underline{M}$ gilt, folgt nach Lemma 22.1.1a)

(1) $$\mathrm{Prg}[\underline{M}, A] \to \underline{W} \subseteq A.$$

Man hat

(2) $$\mathrm{Prg}[\underline{M}, A] \wedge \underline{M}/\mathrm{I} \subseteq A \wedge \mathrm{I} \in \underline{M} \to \mathrm{I} \in A.$$

Da $\underline{M}/I = \underline{W}$ und $\mathrm{I} \in \underline{M}$ gilt,folgt aus (1) und (2)

(3) $$\mathrm{Prg}[\underline{M}, A] \to \mathrm{I} \in A.$$

Man hat auch

(4) $$\mathrm{Prg}[\underline{M}, A] \wedge \underline{M}/\mathrm{I} \subseteq A \wedge \mathrm{I} \in A \wedge \wedge \mathrm{I}{+}1 \in \underline{M} \to \mathrm{I}{+}1 \in A.$$

Da $\mathrm{I}{+}1 \in \underline{M}$ gilt, folgt aus (1), (3) und (4) die Behauptung

$$\mathrm{Prg}[\underline{M}, A] \to \mathrm{I}{+}1 \in A.$$ □

Lemma 22.5.2. *Für* $A' := \{\eta : \forall x(\underline{M}/x \subseteq A \to \underline{M}/(x{+}\omega^\eta) \subseteq A)\}$ *gilt*

$$\mathrm{Prg}[\underline{M}, A] \to \mathrm{Prg}[\underline{M}, A'].$$

Beweis. Wir setzen

$$\begin{aligned} F[\alpha] &:= \mathrm{Prg}[\underline{M}, A] \wedge \underline{M}/\alpha \subseteq A', \\ G[\alpha, \beta] &:= F[\alpha] \wedge \underline{M}/\beta \subseteq A, \\ H[\alpha, \beta, \gamma] &:= G[\alpha, \beta] \wedge \gamma < \beta{+}\omega^\alpha \wedge \gamma \in \underline{M} \end{aligned}$$

und beweisen zunächst

(1) $$H[\alpha\beta,\gamma] \to \gamma \in A.$$

Für $\gamma < \beta+\omega^\alpha$ hat man folgende drei Fälle.

1. $\gamma < \beta$. Dann folgt $\gamma \in A$ aus $\gamma \in \underline{M}$ und $\underline{M}/\beta \subseteq A$.
2. $\gamma = \beta$. Dann folgt $\gamma \in A$ aus $\underline{M}/\beta \subseteq A$, $\mathrm{Prg}[\underline{M}, A]$ und $\gamma \in \underline{M}$.
3. $\gamma = \beta + \omega^{\delta_1}+\ldots+\omega^{\delta_m}$ $(m \geq 1)$ mit $\delta_i < \alpha$ $(i = 1,\ldots,m)$.
   Aus $\gamma \in \underline{M}$ folgt dann $\delta_i \in \underline{M}$, und aus $\underline{M}/\alpha \subseteq A'$ folgt $\delta_i \in A'$ $(i = 1,\ldots,m)$.
   Hiermit ergibt sich
   $$\underline{M}/\beta \subseteq A \to \underline{M}/\gamma \subseteq A.$$
   Mit $\underline{M}/\beta \subseteq A$, $\gamma \in \underline{M}$ und $\mathrm{Prg}[\underline{M}, A]$ folgt $\gamma \in A$.

Hiermit ist (1) bewiesen. Aus (1) folgt

$$G[\alpha,\beta] \to \underline{M}/(\beta+\omega^\alpha) \subseteq A.$$

Nach den Definitionen von $G[\alpha,\beta]$ und A' folgt

$$F[\alpha] \to \alpha \in A'.$$

Hiermit ergibt sich die Behauptung $\mathrm{Prg}[\underline{M}, A] \to \mathrm{Prg}[\underline{M}, A']$. □

Definition. $\omega_0(\alpha) := \alpha$, $\omega_{n+1}(\alpha) := \omega^{\omega_n(\alpha)}$.

Lemma 22.5.3. $\mathrm{Prg}[\underline{M}, A] \to \omega_n(\mathrm{I}+1) \in A$.

Beweis durch Induktion nach n. Für $n = 0$ gilt die Behauptung nach Lemma 22.5.1. Wir beweisen sie nun für $n+1$ unter der I.V., daß sie für n gilt. A' sei gemäß Lemma 22.5.2 definiert. Nach I.V. hat man

$$\mathrm{Prg}[M, A'] \to \omega_n(\mathrm{I}+1) \in A'.$$

Nach der Definition von A' folgt

(1) $$\mathrm{Prg}[\underline{M}, A'] \to \underline{M}/(\omega_{n+1}(\mathrm{I}+1)) \subseteq A.$$

Nach Lemma 22.5.2 hat man

(2) $$\mathrm{Prg}[\underline{M}, A] \to \mathrm{Prg}[\underline{M}, A'].$$

Aus (1) und (2) folgt

$$\mathrm{Prg}[\underline{M}, A] \to \underline{M}/(\omega_{n+1}(I+1)) \subseteq A.$$

Hiermit und mit $\omega_{n+1}(I+1) \in \underline{M}$ ergibt sich die Behauptung

$$\mathrm{Prg}[\underline{M}, A] \to \omega_{n+1}(I+1) \in A.$$ □

Lemma 22.5.4. $\psi 0(\Psi(\omega_n(I+1))) \in \underline{W}$.

Beweis. Nach Lemma 22.4.4 hat man $\mathrm{Prg}[\underline{M}, \underline{W}_\Psi]$. Nach Lemma 22.5.3 folgt $\omega_n(\mathrm{I}+1) \in \underline{W}_\Psi$. Da $\omega_n(I+1) \in C(\omega_n(I+1))$ ist, folgt $\Psi(\omega_n(I+1)) \in \underline{W}$. Nach Lemma 22.3.1 hat man $0 \in \underline{W}$. Da $\Psi(\omega_n(I+1)) \in C_0(\Psi(\omega_n(I+1)))$ ist, folgt nach Korollar 22.2.8 $\psi 0(\Psi(\omega_n(I+1))) \in \underline{W}$. □

Anmerkung. Aus Lemma 22.5.4 folgt die Wohlordnung aller Ordinalzahlen $<\psi 0(\Psi(\omega_n(I+1))) \in \underline{W}$, denn für $\mathrm{T} := \{\eta : \eta < 0^+\}$ gilt $\mathrm{M}_0^{\underline{W}} = \mathrm{T}$, folglich $\underline{W}/0^+ = \mathrm{W}_0^{\underline{W}} = \mathrm{WT}$.

Literatur

1. Wilfried Buchholz. Normalfunktionen und konstruktive Systeme von Ordinalzahlen. In *Proof Theory Symposion Kiel 1974*, volume 500 of *Lecture Notes in Mathematics*, pages 4–25. Springer, 1975.
2. Wilfried Buchholz and Kurt Schütte. Ein Ordinalzahlensystem für die beweistheoretische Abgrenzung der Π^1_2-Separation und Bar-Induktion. *Sitzungsberichte der Bayerischen Akademie der Wissenschaften, Mathematisch-Naturwissenschaftliche Klasse*, pages 99–132, 1983.
3. Wilfried Buchholz and Kurt Schütte. *Proof Theory of Impredicative Subsystems of Analysis*. Studies in Proof Theory. Bibliopolis, Napoli, 1988.
4. Kurt Schütte. Ein Wohlordnungsbeweis für das Ordinalzahlensystem $\mathrm{T}(J)$. *Archive for Mathematical Logic*, 27:5–20, 1988.

Kapitel 23
Beziehungen des Ordinalzahlensystems OT(ϑ) zur Veblen-Hierarchie

Kurt Schütte

Abstract This paper of Schütte's is from 1992. In it Schütte compares two ordinal representation systems, namely, OT(ϑ) from [2] which was utilized by Rathjen and Weiermann in [3] in their proof-theoretic investigations of Kruskal's theorem, and the Veblen hierarchy as presented via Schütte's Klammersymbole in [4]. Schütte then uses this comparison to locate several signature ordinals within OT(ϑ), e.g., Γ_0, Ackermann's ordinal from [1], the small Veblen number, and the big Veblen number. For the proof of the crucial technical Lemma 23.4.2, however, Schütte provides only a hint. As discussions between Wilfried Buchholz and Michael Rathjen could not exactly clarify which proof Schütte had envisaged, it was felt that it was a good idea to supply a more complete one. Thankfully, Wilfried Buchholz devised and wrote up such a proof which is now incorporated in the paper. Of course, we cannot ascertain that this is what Schütte had in mind.

23.1 Grundbegriffe

Ω sei die kleinste überabzählbare Ordinalzahl und E die Klasse der ε-Zahlen.
Im folgenden bezeichnen $\alpha, \beta, \gamma, \delta, \eta, \mu$ stets Ordinalzahlen $< \varepsilon_{\Omega+1}$ und i, k, m, n Ordinalzahlen $< \omega$.
Für eine Ordinalzahlenklasse X sei $X/\alpha := \{\eta \in X : \eta < \alpha\}$.

Definition. $\gamma =_{\mathrm{NF}} \omega^\alpha + \beta$ (γ hat die Normalform $\omega^\alpha + \beta$) bedeute, daß $\gamma = \omega^\alpha + \beta$ ist mit $\alpha < \gamma$ und $\beta < \gamma$.

Lemma 23.1.1.
a) *Gilt* $\gamma =_{\mathrm{NF}} \omega^\alpha + \beta$, *so sind* α *und* β *eindeutig durch* γ *bestimmt.*
b) *Zu* γ *gibt es* α *und* β *mit* $\gamma =_{\mathrm{NF}} \omega^\alpha + \beta$ *genau dann, wenn* $\gamma \neq 0$ *und* $\gamma \notin$ E *ist.*

K. Schütte (†)
Mathematisches Institut, LMU München, Germany

R. Kahle and M. Rathjen (eds.), *The Legacy of Kurt Schütte*,
https://doi.org/10.1007/978-3-030-49424-7_23

Induktive Definition von γ^*.

1. $0^* = \Omega^* := 0$.
2. Für $\gamma =_{\mathrm{NF}} \omega^\alpha + \beta$ sei $\gamma^* := \max\{\alpha^*, \beta^*\}$.
3. Für $\gamma \in \mathrm{E}/\Omega$ sei $\gamma^* := \gamma$.

Lemma 23.1.2.

a) $\gamma^* \leq \gamma$ *und* $\gamma^* < \Omega$.
b) $\gamma^* = 0$ *oder* $\gamma^* \in \mathrm{E}$.

23.2 Das Ordinalzahlensystem OT(ϑ)

Induktive Definition der Ordinalzahlenmengen $\mathrm{C}_n(\alpha)$, $\mathrm{C}(\alpha)$ und Ordinalzahlen $\vartheta\alpha$.

(C1) $\Omega \in \mathrm{C}_n(\alpha)$ und $\eta \in \mathrm{C}_n(\alpha)$ für all $\eta \leq \alpha^*$.
(C2) Gilt $\delta =_{\mathrm{NF}} \omega^\beta + \gamma$ und $\beta, \gamma \in \mathrm{C}_n(\alpha)$, so sei $\delta \in \mathrm{C}_{n+1}(\alpha)$.
(C3) Ist $\beta < \alpha$ und $\beta \in \mathrm{C}_n(\alpha)$, so sei $\vartheta\beta \in \mathrm{C}_{n+1}(\alpha)$.
(C4) $\mathrm{C}(\alpha) := \bigcup\{\mathrm{C}_n(\alpha) : n < \omega\}$
(C5) $\vartheta\alpha := \min\{\eta : \eta \notin \mathrm{C}(\alpha)\}$

Lemma 23.2.1.

a) $\mathrm{C}(\alpha)$ *ist eine abzählbare Menge.*
b) $\alpha^* < \vartheta\alpha < \Omega$.
c) $\vartheta\alpha \in \mathrm{E}$.

Beweis. Trivial. □

Lemma 23.2.2.

a) *Aus* $\alpha^* < \vartheta\beta$ *folgt* $\alpha \in \mathrm{C}(\beta)$.
b) $\alpha \in \mathrm{C}(\alpha)$.
c) *Aus* $\alpha < \beta$, $\alpha^* < \vartheta\beta$ *und* $\gamma \in \mathrm{C}_n(\alpha)$ *folgt* $\gamma \in \mathrm{C}(\beta)$.
d) *Aus* $\alpha < \beta$ *und* $\alpha^* < \vartheta\beta$ *folgt* $\vartheta\alpha < \vartheta\beta$.
e) *Aus* $\vartheta\alpha = \vartheta\beta$ *folgt* $\alpha = \beta$.

Beweis. a) ergibt sich durch Induktion nach α.
b) folgt aus a) aufgrund von $\alpha^* < \vartheta\alpha$.
c) ergibt sich durch Induktion nach n.
d) Aus $\alpha < \beta$ und $\alpha^* < \vartheta\beta$ folgt nach a) und c) $\alpha \in \mathrm{C}(\beta)$ und $\mathrm{C}(\alpha) \subseteq \mathrm{C}(\beta)$. Es folgt $\vartheta\alpha \in \mathrm{C}(\beta)$ und $\vartheta\alpha \leq \vartheta\beta$. Aufgrund von $\vartheta\beta \notin \mathrm{C}(\beta)$ folgt $\vartheta\alpha < \vartheta\beta$.
e) Aus $\vartheta\alpha = \vartheta\beta$ folgt nach Lemma 23.2.1b) $\alpha^* < \vartheta\beta$ und $\beta^* < \vartheta\alpha$. Nach d) folgt $\alpha = \beta$. □

Lemma 23.2.3. *Aus* $\alpha \in \mathrm{C}_n(\beta)$ *folgt* $\alpha^* < \vartheta\beta$.

Beweis durch Induktion nach n.

1. $\alpha \in \mathrm{C}_n(\beta)$ gelte nach (C1). Dann hat man $\alpha^* \leq \beta^* < \vartheta\beta$.

2. $\alpha \in \mathrm{C}_n(\beta)$ gelte nach (C2). Dann folgt $\alpha^* < \vartheta\beta$ aus der I.V.
3. $\alpha \in \mathrm{C}_n(\beta)$ gelte nach (C3). Dann hat man $\alpha = \vartheta\gamma$ mit $\gamma < \beta$ und nach I.V. $\gamma^* < \vartheta\beta$. Nach Lemma 23.2.2d) folgt $\alpha^* = \vartheta\gamma < \vartheta\beta$. □

Korollar 23.2.1. *$\alpha \in \mathrm{C}(\beta)$ gilt genau dann, wenn $\alpha^* < \vartheta\beta$ ist.*

Beweis. Mit den Lemmata 23.2.2a) und 23.2.3. □

Lemma 23.2.4. *Aus $\gamma \in \mathrm{C}_n(\alpha)/\Omega$ folgt $\gamma < \vartheta\alpha$.*

Beweis durch Induktion nach n.
1. $\gamma \in \mathrm{C}_n(\alpha)/\Omega$ gelte nach (C1). Dann hat man $\gamma \le \alpha^* < \vartheta\alpha$.
2. $\gamma \in \mathrm{C}_n(\alpha)/\Omega$ gelte nach (C2). Dann folgt $\gamma < \vartheta\alpha$ aufgrund von $\vartheta\alpha \in \mathrm{E}$ aus der I.V.
3. $\gamma \in \mathrm{C}_n(\alpha)/\Omega$ gelte nach (C3). Dann hat man $\gamma = \vartheta\beta$ mit $\beta < \alpha$ und $\beta \in \mathrm{C}(\alpha)$. Nach Lemma 23.2.3 folgt $\beta^* < \vartheta\alpha$. Nach Lemma 23.2.2d) folgt $\gamma = \vartheta\beta < \vartheta\alpha$. □

Korollar 23.2.2. *$\gamma \in \mathrm{C}_n(\alpha)/\Omega$ gilt genau dann, wenn $\gamma < \vartheta\alpha$ ist.*

Beweis. Mit Lemma 23.2.4 und der Definition von $\vartheta\alpha < \Omega$. □

Lemma 23.2.5.
a) *Aus $\alpha < \Omega$ folgt $\alpha < \vartheta\alpha$.*
b) *Aus $\Omega \le \alpha$ folgt $\vartheta\alpha < \alpha$.*

Beweis. a) folgt aus $\alpha \in \mathrm{C}(\alpha)$ und Lemma 23.2.4.
b) folgt aus $\vartheta\alpha < \Omega$. □

Induktive Definition der Ordinalzahlenmenge OT(ϑ) und des *Grades* gr(γ) einer Ordinalzahl $\gamma \in \mathrm{OT}(\vartheta)$.

(T1) $0, \Omega \in \mathrm{OT}(\vartheta)$ mit $\mathrm{gr}(0) = \mathrm{gr}(\Omega) := 0$.
(T2) Gilt $\gamma =_{\mathrm{NF}} \omega^\alpha + \beta$ und $\alpha, \beta \in \mathrm{OT}(\vartheta)$, so sei $\gamma \in \mathrm{OT}(\vartheta)$ mit $\mathrm{gr}(\gamma) := \max\{\mathrm{gr}(\alpha), \mathrm{gr}(\beta)\} + 1$.
(T3) Ist $\alpha \in \mathrm{OT}(\vartheta)$, so sei $\vartheta\alpha \in \mathrm{OT}(\vartheta)$ mit $\mathrm{gr}(\vartheta\alpha) := \mathrm{gr}(\alpha)+1$.

Offenbar ist $\varepsilon_{\Omega+1}$ die kleinste Ordinalzahl, die größer als jede Ordinalzahl aus OT(ϑ) ist. Aus den Lemmata 23.1.1, 23.2.1 und 23.2.5 geht hervor, daß die Definitionsregeln (T1) bis (T3) ein eindeutiges Bezeichnungssystem für die Ordinalzahlen aus OT(ϑ) liefern. Daher ist der Grad gr(γ) jeder Ordinalzahl $\gamma \in \mathrm{OT}(\vartheta)$ eindeutig bestimmt.

Lemma 23.2.6. *Aus $\gamma \in \mathrm{C}_n(\alpha)$ mit $\alpha^* = 0$ folgt $\gamma \in \mathrm{OT}(d)$.*

Beweis. Durch Induktion nach *n*. □

Definitionen. $\mathrm{u}(0) := \Omega+1$, $\mathrm{u}(n+1) := \omega^{\mathrm{u}(n)}$,
$V := \sup\{\vartheta(\mathrm{u}(n)) : n < \omega\}$.

Lemma 23.2.7. $\mathrm{OT}(\vartheta) = \bigcup\{\mathrm{C}(\mathrm{u}(n)) : n < \omega\}$.

Beweis. Durch Induktion nach n ergibt sich $\mathrm{u}(n)^* = 0$ für alle $n < \omega$. Nach Lemma 23.2.6 folgt $\bigcup\{\mathrm{C}(\mathrm{u}(n)) : n < \omega\} \subseteq \mathrm{OT}(\vartheta)$. Durch Induktion nach $\mathrm{gr}(\gamma)$ ergibt sich auch $\gamma \in \bigcup\{\mathrm{C}(\mathrm{u}(n)) : n < \omega\}$ für alle $\gamma \in \mathrm{OT}(\vartheta)$. □

Lemma 23.2.8.
a) $V = \min\{\eta : \eta \notin \mathrm{OT}(\vartheta)\}$.
b) $\mathrm{OT}(\vartheta)/\Omega$ *ist gleich dem Abschnitt aller Ordinalzahlen* $< V$.

Beweis. Mit Lemma 23.2.7 und Korollar 23.2.2. □

Im folgenden bezeichnen $\alpha, \beta, \gamma, \delta$ (auch mit Indizes) stets Ordinalzahlen $< V$.

Lemma 23.2.9.
a) $(\Omega{\cdot}\gamma{+}\alpha)^* = \max\{\gamma^*, \alpha^*\}$.
b) *Für* $\alpha \neq 0$ *ist* $(\Omega^\gamma{\cdot}\alpha)^* = \max\{\gamma^*, \alpha^*\}$.
c) *Für* $\mu = \Omega^{\gamma_m}{\cdot}\alpha_m + \cdots + \Omega^{\gamma_0}{\cdot}\alpha_0$ *mit* $\gamma_m > \cdots > \gamma_0$ *und* $\alpha_i \neq 0$ *für alle* $i = 0, \ldots, m$ *gilt* $\mu^* = \max\{\gamma_m^*, \alpha_m^*, \ldots, \gamma_0^*, \alpha_0^*\}$ *und* $\gamma_m, \alpha_0, \ldots, \alpha_m < \vartheta\mu$.

Beweis. a) gilt für $\gamma = 0$ trivialerweise. Andernfalls hat man $\gamma = \omega^{\gamma_n} + \cdots + \omega^{\gamma_0}$ und $\Omega{\cdot}\gamma + \alpha = \omega^{\Omega+\gamma_n} + \cdots + \omega^{\Omega+\gamma_0} + \alpha$ mit $\gamma_n \geq \cdots \geq \gamma_0$. Dann erhält man $(\Omega{\cdot}\gamma + \alpha)^* = \max\{\gamma_n^*, \ldots, \gamma_0^*, \alpha^*\} = \max\{\gamma^*, \alpha^*\}$.
b) Man hat $\alpha = \omega^{\alpha_n} + \cdots + \omega^{\alpha_0}$ und $\Omega^\gamma{\cdot}\alpha = \omega^{\Omega\cdot\gamma+\alpha_n} + \cdots + \omega^{\Omega\cdot\gamma+a_0}$ mit $\alpha_n \geq \cdots \geq \alpha_0$. Nach a) folgt $(\Omega^\gamma{\cdot}\alpha)^* = \max\{\gamma^*, \alpha_n^*, \ldots, \alpha_0^*\} = \max\{\gamma^*, \alpha^*\}$.
c) Nach b) hat man $\mu^* = \max\{\gamma_m^*, \alpha_m^*, \ldots, \gamma_0^*, \alpha_0^*\}$.
Ist $\gamma_m = \gamma_0$, so hat man trivialerweise $\gamma_0, \alpha_0 < \vartheta\alpha_0 = \vartheta\mu$. Sei nun $\gamma_m > 0$. Dann hat man $\gamma_m, \alpha_0, \ldots, \alpha_m < \mu$ und $\gamma_m^*, \alpha_0^*, \ldots, \alpha_m^* \leq \mu^* < \vartheta\mu$. Nach Lemma 23.2.2d) folgt $\gamma_m, \alpha_0, \ldots, \alpha_m < \vartheta\mu$. □

Lemma 23.2.10. *Für* $\mu = \Omega^{\gamma_m}{\cdot}\alpha_m + \cdots + \Omega^{\gamma_0}{\cdot}\alpha_0$ *und* $\eta = \Omega^{\delta_n}{\cdot}\beta_n + \cdots + \Omega^{\delta_0}{\cdot}\beta_0$ *mit* $\mu < \eta$, $\gamma_m > \cdots > \gamma_0$, $\delta_n > \cdots > \delta_0$, $\beta_n > 0$ *und* $\alpha_i > 0$ *für alle* $i = 0, \ldots, m$ *gilt:*
a) *Sind alle* $\alpha_0, \ldots, \alpha_m < \vartheta\eta$, *so ist* $\vartheta\mu < \vartheta\eta$.
b) *Gibt es ein* $\alpha_i \geq \vartheta\eta$ $(i = 0, \ldots, m)$, *so ist* $\vartheta\eta < \vartheta\mu$.

Beweis. a) Aus $\mu < \eta$ folgt $\gamma_m \leq \delta_n$. Nach Lemma 23.2.9c) folgt $\gamma_m < \vartheta\eta$. Hiermit und mit $\alpha_0, \ldots, \alpha_m < \vartheta\eta$ erhält man $\gamma_i^* \leq \gamma_m < \vartheta\eta$ und $\alpha_i^* \leq \alpha_i < \vartheta\eta$ für alle $i = 0, \ldots, m$. Nach Lemma 23.2.9c) folgt $\mu^* < \vartheta\eta$. Mit $\mu < \eta$ folgt $\vartheta\mu < \vartheta\eta$.
b) folgt aus Lemma 23.2.9c). □

23.3 Der Ackermannsche Ordinalzahlenabschnitt

W. Ackermann hat in [1] einen Ordinalzahlenabschnitt charakterisiert, in dem die additiven Hauptzahlen durch Tripel (α,β,γ) gemäß folgender Anordnung symbolisiert sind.

Definition. $(\alpha_1,\beta_1,\gamma_1) < (\alpha_2,\beta_2,\gamma_2)$ gelte in genau folgenden fünf Fällen:
a) $\alpha_1 < \alpha_2$ und $\max\{\beta_1,\gamma_1\} < (\alpha_2,\beta_2,\gamma_2)$.
b) $\alpha_1 = \alpha_2$, $\beta_1 < \beta_2$ und $\gamma_1 < (\alpha_2,\beta_2,\gamma_2)$.
c) $\alpha_1 = \alpha_2$, $\beta_1 = \beta_2$ und $\gamma_1 < \gamma_2$.
d) $\alpha_1 = \alpha_2$, $\beta_2 < \beta_1$ und $(\alpha_1,\beta_1,\gamma_1) \leq \gamma_2$.
e) $\alpha_2 < \alpha_1$ und $(\alpha_1,\beta_1,\gamma_1) \leq \max\{\beta_2,\gamma_2\}$.

Die betreffenden Ordinalzahlen lassen sich folgendermaßen in OT(ϑ) ausdrücken.

Lemma 23.3.1. *Für $\alpha,\beta,\gamma < \vartheta(\Omega^3)$ gilt:*
a) $(0,0,\gamma) =_{\mathrm{NF}} \omega^{\gamma+1}{+}0$, *wenn* $\gamma = \delta{+}n$ *mit* $\omega^\delta = \delta$ *und* $n < \omega$ *ist.*
b) $(0,0,\gamma) =_{\mathrm{NF}} \omega^{\gamma}{+}0$, *in jedem anderen Fall.*
c) $(0,1{+}\beta,\gamma) = \vartheta(\Omega{\cdot}\beta{+}\gamma)$.
d) $(\alpha,\beta,\gamma) = \vartheta(\Omega^2{\cdot}\alpha{+}\Omega{\cdot}\beta{+}\gamma)$ *für* $\alpha > 0$.

Beweis. Mit Lemma 23.2.10 durch Induktion nach $\Omega^2{\cdot}\alpha{+}\Omega{\cdot}\beta{+}\gamma$. □

Lemma 23.3.2. $\vartheta(\Omega^3)$ *ist der Ordnungstyp des Ackermannschen Ordinalzahlenabschnitts.*

Beweis. Mit Lemma 23.3.1. □

23.4 Die Veblen-Hierarchie der ε-Zahlen

Mit A, B bezeichnen wir zweizeilige Matrizen $\begin{pmatrix}\alpha_0...\alpha_n\\ \gamma_0...\gamma_n\end{pmatrix}$ mit $\gamma_0 < \cdots < \gamma_n$.
$A = B$ gelte genau dann, wenn A mit B übereinstimmt oder wenn B durch Streichung oder (und) Hinzufügung von Spalten $\begin{matrix}0\\ \gamma_i\end{matrix}$ aus A hervorgeht.

Hiernach gibt es zu je zwei Matrizen A, B Matrizen $\begin{pmatrix}\alpha_0...\alpha_n\\ \gamma_0...\gamma_n\end{pmatrix} = A$ und $\begin{pmatrix}\beta_0...\beta_n\\ \gamma_0...\gamma_n\end{pmatrix} = B$ mit gleicher unterer Zeile.
Ist $A \neq B$, so gibt es einen größten Index m mit $\alpha_m \neq \beta_m$.
Wir setzen $A < B$ ($B < A$), wenn $\alpha_m < \beta_m$ ($\beta_m < \alpha_m$) ist.

Induktive Definition von φA (gemäß [4, (2.1)–(2.3)]).

1. $\varphi\begin{pmatrix}\alpha\\ 0\end{pmatrix} := \varepsilon_\alpha$.
2. $\varphi A = \varphi B$ für $A = B$.

3. Für $A = \begin{pmatrix} \alpha_0 \alpha_1 \dots \alpha_n \\ 0\ \gamma_1 \dots \gamma_n \end{pmatrix}$ mit $\alpha_1 \neq 0$ sei $\alpha_0 \mapsto \varphi A$ die Ordnungsfunktion der Menge aller Ordinalzahlen η mit $\varphi\begin{pmatrix} \eta\ \alpha_1^- \alpha_2 \dots \alpha_n \\ \gamma_1^-\ \gamma_1\ \gamma_2 \dots \gamma_n \end{pmatrix} = \eta$ für alle $\gamma_1^- < \gamma_1$ und alle $\alpha_1^- < \alpha_1$.

Diese Funktionen sind nach [4, (3.2) und (3.5)] Normalfunktionen.

Definition von $\overline{\varphi}A$.
Für $A = \begin{pmatrix} \alpha_0 \alpha_1 \dots \alpha_n \\ 0\ \gamma_1 \dots \gamma_n \end{pmatrix}$ sei $\overline{\varphi}A := \varphi\begin{pmatrix} \alpha_0{+}1 \alpha_1 \dots \alpha_n \\ 0\quad \gamma_1 \dots \gamma_n \end{pmatrix}$, wenn es η und $m < \omega$ mit $\alpha_0 = \eta + m$ gibt derart, daß $\varphi\begin{pmatrix} \eta \alpha_1 \dots \alpha_n \\ 0 \gamma_1 \dots \gamma_n \end{pmatrix}$ gleich η oder gleich einem α_i oder γ_i mit $1 \le i \le n$ ist.
In jedem anderen Fall sei $\overline{\varphi}A := \varphi A$.

Lemma 23.4.1. *Für* $A = \begin{pmatrix} \alpha_0 \dots \alpha_n \\ \gamma_0 \dots \gamma_n \end{pmatrix} < B$ *gilt:*

a) *Sind alle* $\alpha_0, \dots, \alpha_n < \overline{\varphi}B$, *so ist* $\overline{\varphi}A < \overline{\varphi}B$.
b) *Gibt es ein* $\alpha_i \geq \overline{\varphi}B$ $(i = 0, \dots, n)$, *so ist* $\overline{\varphi}B < \overline{\varphi}A$.

Beweis. Dies gilt nach [4, (8.3) und (8.4)]. □

Lemma 23.4.2. $\overline{\varphi}\begin{pmatrix} \alpha_0 \dots \alpha_n \\ \gamma_0 \dots \gamma_n \end{pmatrix} = \vartheta(\Omega^{\gamma_n} \cdot \alpha_n + \dots + \Omega^{\gamma_0} \cdot \alpha_0)$.

Beweis. Mit den Lemmata 23.2.10 und 23.4.1 durch Induktion nach $\Omega^{\gamma_n} \cdot \alpha_n + \dots + \Omega^{\gamma_0} \cdot \alpha_0$. □

Definition. Eine Ordinalzahl γ heißt *stark kritisch*, wenn $\overline{\varphi}\begin{pmatrix} \alpha \beta \\ 0 1 \end{pmatrix} < \gamma$ für alle $\alpha, \beta < \gamma$ gilt.

Lemma 23.4.3.

a) *Die Menge der stark kritischen Ordinalzahlen* $< V$ *besteht aus den Ordinalzahlen* $\vartheta\eta$ *mit* $\eta \geq \Omega^2$.
b) $\Gamma_0 = \vartheta(\Omega^2)$ *ist die kleinste stark kritische Ordinalzahl.*

Beweis. Mit den Lemmata 23.4.1 und 23.4.2. □

Definition. Eine Ordinalzahl δ heiße eine *schwache (starke) Veblenzahl*, wenn $\overline{\varphi}\begin{pmatrix} \alpha_0 \dots \alpha_n \\ \gamma_0 \dots \gamma_n \end{pmatrix} < \delta$ für alle $\alpha_0, \dots, \alpha_n < \delta$ und $\gamma_n < \omega$ $(\gamma_n < \delta)$ gilt.

Lemma 23.4.4.

a) *Die Menge der schwachen Veblenzahlen* $< V$ *besteht aus den Ordinalzahlen* $\vartheta\eta$ *mit* $\eta \geq \Omega^\omega$.
b) $\vartheta(\Omega^\omega)$ *ist die kleinste schwache Veblenzahl.*

c) *Die Menge der starken Veblenzahlen* $< V$ *besteht aus den Ordinalzahlen* $\vartheta\eta$ *mit* $\eta \geq \Omega^\Omega$.
d) $\vartheta(\Omega^\Omega)$ *ist die kleinste starke Veblenzahl.*
e) V *ist eine starke Veblenzahl.*

Beweis. Mit den Lemmata 23.4.1 und 23.4.2. □

Anmerkungen.

1. δ ist genau dann eine starke Veblenzahl, wenn $\delta = \varphi\binom{1}{\delta}$ ist.
2. Aus Lemma 23.4.4b) geht hervor, daß $\vartheta(\Omega^\omega)$ der Ordnungstyp der in [5, §11] definierten Wohlordnung ist.

Beweis von Lemma 23.4.2 für $\alpha_i, \gamma_i < \mathrm{d}(\Omega^\Omega)$.[1]

Abk.: $\mu =_{\mathrm{NF}} \Omega^{\gamma_n}\alpha_n + \cdots + \Omega^{\gamma_0}\alpha_0 \;:\Leftrightarrow\; \begin{cases} \mu = \Omega^{\gamma_n}\alpha_n + \cdots + \Omega^{\gamma_0}\alpha_0 \wedge \\ 0 < \alpha_0, \ldots, \alpha_n < \Omega \wedge \\ \gamma_0 < \cdots < \gamma_n < \Omega. \end{cases}$

Definition von γ' für $\gamma < \mathrm{d}(\Omega^\Omega)$.
1. $0' := 0$.
2. Für $\gamma =_{\mathrm{NF}} \omega^\alpha + \beta$ sei $\gamma' := \omega^{\alpha'} + \beta'$.
3. Für $\gamma \in \mathrm{E}$ folgt mit Korollar 23.2.2 und Lemmata 23.2.2e), 23.2.9c), die Existenz von (eindeutig bestimmten) $\mu < \Omega^\Omega$ und $\alpha_0, \ldots, \alpha_n, \gamma_0, \ldots, \gamma_n < \gamma$, so daß $\gamma = \mathrm{d}\mu$ und $\mu =_{\mathrm{NF}} \Omega^{\gamma_n}\alpha_n + \cdots + \Omega^{\gamma_0}\alpha_0$. — Wir setzen $\gamma' := \overline{\varphi}\binom{\alpha'_0 \ldots \alpha'_n}{\gamma'_0 \ldots \gamma'_n}$.

(1) $$\gamma < \delta < \mathrm{d}(\Omega^\Omega) \Rightarrow \gamma' < \delta' < \eta_0$$

Beweis. Durch Induktion nach $\gamma\#\delta$ zeigen wir:

$$\gamma \neq \delta \;\Rightarrow\; (\gamma < \delta \wedge \gamma' < \delta') \vee (\delta < \gamma \wedge \delta' < \gamma').$$

Wir behandeln dabei nur den Fall $\gamma, \delta \in \mathrm{E}$. Dann gibt es
$\mu =_{\mathrm{NF}} \Omega^{\gamma_m}\alpha_m + \cdots + \Omega^{\gamma_0}\alpha_0$ mit $\alpha_0, \ldots, \alpha_m, \gamma_m < \gamma = \mathrm{d}\mu$, und
$\eta =_{\mathrm{NF}} \Omega^{\delta_n}\beta_n + \cdots + \Omega^{\delta_0}\beta_0$ mit $\beta_0, \ldots, \beta_n, \delta_n < \delta = \mathrm{d}\eta$.
Ohne Einschränkung können wir $\mu < \eta$ annehmen.
Nach I.V. gilt dann auch $\mu'' := \binom{\alpha'_0 \ldots \alpha'_m}{\gamma'_0 \ldots \gamma'_m} < \binom{\beta'_0 \ldots \beta'_n}{\delta'_0 \ldots \delta'_n} =: \eta''$.
Fall 1: $\alpha_0, \ldots, \alpha_m < \delta$. Mit $\mu < \eta$ und Lemma 23.2.10a) folgt $\gamma = \mathrm{d}\mu < \mathrm{d}\eta = \delta$.
Mit I.V. folgt $\alpha'_0, \ldots, \alpha'_m < \delta' = \overline{\varphi}\eta''$ und weiter, mit $\mu'' < \eta''$ und Lemma 23.4.1a), $\gamma' = \overline{\varphi}\mu'' < \overline{\varphi}\eta'' = \delta'$.

[1] Von Wilfried Buchholz.

Fall 2: $\delta \leq \alpha_i$ für ein $i \in \{0,\ldots,m\}$. Mit $\mu < \eta$ und Lemma 23.2.10b) folgt $\delta = \mathrm{d}\eta < \mathrm{d}\mu = \gamma$. Mit I.V. folgt $\overline{\varphi}\eta'' = \delta' \leq \alpha_i'$ und weiter, mit $\mu'' < \eta''$ und Lemma 23.4.1b), $\delta' = \overline{\varphi}\eta'' < \overline{\varphi}\mu'' = \gamma'$. □

Definition. $\eta_0 := \min\{\eta : \varphi\binom{1}{\eta} = \eta\}$.

(2) $$\gamma < \mathrm{d}(\Omega^\Omega) \Rightarrow \gamma' < \eta_0.$$

Beweis. Durch Induktion nach γ mit Hilfe von [4, (5.3)]. □

(3) $$\delta < \eta_0 \Rightarrow \exists\gamma < \mathrm{d}(\Omega^\Omega)(\gamma' = \delta).$$

Beweis. Durch Induktion nach δ mit Hilfe von [4, (8.1)] und Lemmata 23.2.2d), 23.2.9c). □

Aus (1), (2), (3) folgt:

(∗) $$\forall\gamma < \mathrm{d}(\Omega^\Omega)(\gamma = \gamma').$$

Aus (∗) erhält man Lemma 23.4.2 für $\alpha_i, \gamma_i < \mathrm{d}(\Omega^\Omega)$ wie folgt:

$$\mu =_{\mathrm{NF}} \Omega^{\gamma_n}\alpha_n + \cdots + \Omega^{\gamma_0}\alpha_0 \overset{\text{L.23.2.2d),L.23.2.9c)}}{\Longrightarrow} \mathrm{d}\mu < \mathrm{d}(\Omega^\Omega) \Rightarrow$$

$$\mathrm{d}(\Omega^{\gamma_n}\alpha_n + \cdots + \Omega^{\gamma_0}\alpha_0) = \mathrm{d}\mu \overset{(*)}{=} (\mathrm{d}\mu)' = \overline{\varphi}\begin{pmatrix}\alpha_0'\ldots\alpha_n'\\ \gamma_0'\ldots\gamma_n'\end{pmatrix} \overset{(*)}{=} \overline{\varphi}\begin{pmatrix}\alpha_0\ldots\alpha_n\\ \gamma_0\ldots\gamma_n\end{pmatrix}.$$

Bemerkungen zum Ordinalzahlensystem OT(ϑ)

Ω sei die kleinste überabzählbare Ordinalzahl und V die kleinste nicht in OT(ϑ) enthaltene Ordinalzahl. Dann gilt nach [3]:

Satz 23.1. *Die Menge derjenigen Ordinalzahlen aus* OT(ϑ), *die* $< \Omega$ *sind, ist gleich dem Abschnitt aller Ordinalzahlen* $< V$.

Mit α, β, γ bezeichnen ich Ordinalzahlen $< V$ und mit η eine beliebige Ordinalzahl aus OT(ϑ).

Definition von $\overline{\varphi}(\alpha, \beta)$ für die 2-stellige Veblenfunktion φ.
$\overline{\varphi}(\alpha, b) := \varphi(\alpha, \beta{+}1)$ in folgenden zwei Fällen:
a) $\varphi(\alpha, 0) = \alpha$ und $\beta < \omega$.
b) $\beta = \gamma{+}n$ mit $\varphi(\alpha, \gamma) = \gamma$ und $n < \omega$.
In allen anderen Fällen sei $\overline{\varphi}(\alpha, \beta) := \varphi(\alpha, \beta)$.

Satz 23.2. $\overline{\varphi}(1{+}\alpha, \beta) = \vartheta(\Omega{\cdot}\alpha + \beta)$.

Satz 23.3. *Die Menge der stark kritischen Ordinalzahlen* $< V$ *besteht aus den Ordinal* $\vartheta\eta$ *mit* $\eta \geq \Omega^2$, *und es ist* $\Gamma_0 = \vartheta(\Omega^2)$.

Satz 23.4. $\vartheta(\Omega^3)$ *ist der Ordnungstyp der von W. Ackermann in* [1] *definierten Wohlordnung.*

Die *Veblen-Hierarchie* der ε-Zahlen besteht aus Ordinalzahlen $\varphi\binom{\alpha_0 \ldots \alpha_n}{\beta_0 \ldots \beta_n}$ mit $\beta_0 < \cdots < \beta_n$ die bezüglich $\varphi\binom{\alpha}{0} = \varepsilon_\alpha$ gemäß [4] definiert sind.

γ heiße eine *schwache (starke) Veblenzahl*, wenn $\varphi\binom{\alpha_0 \ldots \alpha_n}{\beta_0 \ldots \beta_n} < \gamma$ für alle $\alpha_0, \ldots, \alpha_n < \gamma$ und $\beta_n < \omega$ ($\beta_n < \gamma$) gilt.

Satz 23.5. $\vartheta(\Omega^\omega)$ *ist die kleinste schwache Veblenzahl. Dies ist der Ordnungstyp der in* [5, §11] *entwickelten Wohlordnung.*

Satz 23.6. *V ist eine starke Veblenzahl.*

Literatur

1. Wilhelm Ackermann. Konstruktiver Aufbau eines Abschnitts der zweiten Cantorschen Zahlenklasse. *Mathematische Zeitschrift*, 53:403–413, 1951.
2. Michael Rathjen. Beweistheorie: Ausgewählte Kapitel. Lectures on Proof theory (selected topics), Winter semester 1990, Universität Münster, 1990.
3. Michael Rathjen and Andreas Weiermann. Proof-theoretic investigations on Kruskal's theorem. *Annals of Pure and Applied Logic*, 60:49–88, 1993.
4. Kurt Schütte. Kennzeichnung von Ordnungszahlen durch rekursiv definierte Funktionen. *Mathematische Annalen*, 127:16–32, 1954.
5. Kurt Schütte. *Beweistheorie*. Springer, 1960.
6. Oswald Veblen. Continuous increasing functions of finite and transfinite ordinals. *Transactions of the American Mathematical Society*, 9:280–2nzb, 1908.

Kapitel 24
Zur Beweistheorie von KPM

Kurt Schütte

Abstract This is the last paper on ordinal analysis that Schütte published himself. It is concerned with the calibration of the set theory KPM which formalizes a recursively Mahlo universe of sets. In an unpublished precursor [6], Schütte by and large followed Rathjen's approach of [4] and [5], except for the Schütte calculus setting and some technical changes. The treatment in [5] is very involved as it simultaneously uses two majorizing relations to control infinitary deductions. In the meantime Buchholz [3] presented a simpler analysis based on his operator controlled approach [2], though working with an ordinal representation system different from [4]. Schütte then modified both approaches in that he uses Rathjen's ordinal representation system [4] with two kinds of collapsing functions but employs Buchholz' operator controlled approach to carry out the cut elimination, albeit proceeding in two neatly separated steps pertaining to the two types of collapsing functions and working in his own proof system rather than in a sequent calculus. The paper was first published as [7]. We include it in this book since it is emblematic of Schütte's Spätwerk and, of course, to make it more widely known.

Einleitung

Ein Teilsystem KPM der Mengenlehre, das die Existenz eines rekursiven Mahlo-Universums postuliert, wurde zuerst von M. Rathjen [5] mittels einer sehr komplizierten Erweiterung eines von W. Buchholz [1] eingeführten Majorisierungsverfahrens beweistheoretisch analysiert mit dem Ergebnis, daß die Widerspruchsfreiheit von KPM durch transfinite Induktion bis zur Ordinalzahl $\psi\Omega_1(\psi(\mathrm{Z}\varepsilon_{\mathrm{M}+1})0)$ eines von M. Rathjen eingeführten Ordinalzahlsystems $\mathfrak{T}(\mathrm{M})$ beweisbar ist. Hierbei bezeichnet M die kleinste schwache Mahlozahl und Z eine nichtmonotone injektive

K. Schütte (†)
Mathematisches Institut, LMU München, Germany

R. Kahle and M. Rathjen (eds.), *The Legacy of Kurt Schütte*,
https://doi.org/10.1007/978-3-030-49424-7_24

Abbildung von der Menge aller Ordinalzahlen $< \Gamma_{M+1}$ in die Menge der unerreichbaren Ordinalzahlen < M, während die $\psi\sigma$ für überabzählbare reguläre Ordinalzahlen $\sigma <$ M schwach monotone Kollabierungsfunktionen sind. Die Ordinalzahlenanalyse von M. Rathjen beruht im wesentlichen auf einer ersten imprädikativen Schnitt-Elimination, zu der die Funktion Z gebraucht wird, und einer zweiten imprädikativen Schnitt-Elimination, zu der die Funktionen $\psi\sigma$ gebraucht werden.
Schließlich entwickelte W. Buchholz [2] ein neuartiges Verfahren von H-kontrollierten Herleitungen, mit dem er in [3] die beweistheoretische Analyse von KPM bedeutend einfacher durchführen konnte. Er benutze hierbei ein von $\mathfrak{T}$(M) verschiedenes Ordinalzahlensystem, in dem die beiden verschiedenartigen imprädikativen Schnitt-Eliminationen in komplexer Weise zusammengefaßt werden.
In der vorliegenden Note soll nun gezeigt werden, daß sich das Buchholzsche Verfahren der H-kontrollierten Herleitungen in einer sehr einfach modifizierten Weise vom Buchholzschen Ordinalzahlensystem unmittelbar auf das formal einfachere und natürlichere Rathjensche Ordinalzahlensystem übertragen läßt. Wir beschreiben zunächst im 1. Abschnitt das formale System KPM und im 2. Abschnitt ein dem Rathjenschen Ordinalzahlensystem $\mathfrak{T}$(M) entsprechendes Ordinalzahlensystem OT(M). Im 3. Abschnitt wird dann unter Bezugnahme auf [2], [3] und [5] die beweistheoretische Analyse vom KPM im Ordinalzahlensystem OT(M) skizziert.

24.1 Das mengentheoretische formale System KPM

Grundzeichen des Systems KPM:
1. Je abzählbar unendlich viele freie und gebundene Variablen.
2. $\bot$, $\in$, Ad, $\rightarrow$, $\exists$ und runde Klammern.

Unter einer n-stelligen *Nennform* ($n \geq 1$) verstehen wir eine nichtleere endliche Zeichenreihe, die außer Grundzeichen des formalen Systems höchstens die *Nennzeichen* $*_1, \ldots, *_n$ enthält. Ist $\mathcal{F}$ eine n-stellige Nennform und sind $e_1, \ldots, e_n$ nichtleere endliche Zeichenreihen, so bezeichnet $\mathcal{F}[e_i, \ldots, e_n]$ denjenigen Ausdruck, der aus der Nennform $\mathcal{F}$ dadurch entsteht, daß jedes Nennzeichen $*_i$ überall, wo es in $\mathcal{F}$ auftritt, durch e_i ersetzt wird. Eckige Klammern werden im folgenden nur in dieser Bedeutung im Zusammenhang mit Nennformen verwendet.

Induktive Definition der *Formeln* des Systems KPM.
1. $\bot$ (falsum) ist eine Formel.
2. Sind a und b freie Variablen, so sind $(a \in b)$ und Ad(a) Formeln.
3. Sind A und B Formeln, so ist auch $(A \rightarrow B)$ eine Formel.
4. Ist a eine freie Variable, $\mathcal{A}$ eine 1-stellige Nennform derart, daß $\mathcal{A}[a]$ eine Formel ist, und x eine gebundene Variable, die in $\mathcal{A}$ nicht auftritt, so ist $\exists x.\mathcal{A}[x]$ eine Formel mit einem *unbeschränkten Quantor* $\exists x$ und $(\exists x \in a).\mathcal{A}[x]$ eine Formel mit einem *beschränkten Quantor* $(\exists x \in a)$.
   Formeln, die keinen unbeschränkten Quantor enthalten, heißen Δ_0-*Formeln.*

Induktive Definition der P-*Formen* und N-*Formen*.
1. $*_1$ ist eine P-Form.
2. Ist $\mathcal{P}$ eine P-Form und A eine Formel, so ist $\mathcal{P}[(A \to *_1)]$ eine P-Form und $\mathcal{P}[(*_1 \to A)]$ eine N-Form.
3. Ist $\mathcal{N}$ eine N-Form, so ist $\mathcal{N}[(*_1 \to \bot)]$ eine P-Form.

Nach dieser Definition sind die P-Formen und die N-Formen 1-stellige Nennformen, die das Nennzeichen $*_1$ an genau einer Stelle enthalten und bei Ersetzung dieses Nennzeichens durch eine Formel in Formeln übergehen.
Eine Formel C heißt ein *Positivteil* (*Negativteil*) einer Formel F, wenn es eine P-Form $\mathcal{P}$ (N-Form $\mathcal{N}$) gibt derart, daß $\mathcal{P}[C]$ ($\mathcal{N}[C]$) die Formel F ist.
Ein Positivteil einer Formel heißt *minimal*, wenn er nicht die Gestalt $(A \to B)$ hat.
Ein Negativteil einer Formel heißt *minimal*, wenn er nicht die Gestalt $(A \to \bot)$ hat.
Für Formeln F und G bedeute $F \overset{S}{\models} G$ (aus F folgt strukturell G), daß jeder minimale Positivteil von F, der nicht die Formel $\bot$ ist, auch als Positivteil von G auftritt und jeder minimale Negativteil von F auch als Negativteil von G auftritt.

Als *Mitteilungszeichen* verwenden wir
1. a, b, c für freie Variablen,
2. v, w, x, y, z für gebundene Variablen,
3. A, B, C, F, G für Formeln,
4. $\mathcal{A}$ für 1-stellige Nennformen, so daß $\mathcal{A}[a]$ eine Formel ist,
5. $\mathcal{B}$ für 2-stellige Nennformeln, so daß $\mathcal{B}[a, b]$ eine Formel ist,
6. C für 3-stellige Nennformen, so daß $C[a, b, c]$ eine Formel ist,
7. $\mathcal{P}$ für P-Formen und
8. $\mathcal{N}$ für N-Formen.

Diese Mitteilungszeichen werden auch mit Indizes verwendet.

Definitionen.
1. $\neg A := (A \to \bot)$,
2. $(A \vee B) := (\neg A \to B)$,
3. $(A \wedge B) := \neg(A \to \neg B)$,
4. $\forall x \mathcal{A}[x] := \neg\exists x \neg\mathcal{A}[x]$,
5. $(\forall x \in a)\mathcal{A}[x] := \neg(\exists x \in a)\neg\mathcal{A}[x]$,
6. $(a = b) := ((\forall x \in a)(x \in b) \wedge (\forall x \in b)(x \in a))$,
7. $\mathrm{Tran}[a] := (\forall x \in a)(\forall y \in x)(y \in a)$.

Zur Abkürzung lassen wir teilweise runde Klammern in Formeln fort, wenn es nicht mißverständlich ist.

Axiome des Systems KPM:
(Taut) $F \to F$,
(Ext) $(a = b) \to (\mathcal{A}[a] \to \mathcal{A}[b])$,
(Fund) $\forall x((\forall y \in x)\mathcal{A}[y] \to \mathcal{A}[x]) \to \forall x \mathcal{A}[x]$,

(Ad 1) $(\mathrm{Ad}(a) \wedge \mathrm{Ad}(b)) \to ((a \in b) \vee (a = b) \vee (b \in a))$,
(Ad 2) $\mathrm{Ad}(a) \to \mathrm{Tran}(a)$,
(Ad 3) $\mathrm{Ad}(a) \to (\forall x \in a)(\forall y \in a)(\exists z \in a)((x \in z) \wedge (y \in z) \wedge \mathrm{Tran}[z])$,
(Ad 4) $\mathrm{Ad}(a) \to (\exists x \in a)((\exists y \in x)(\bot \to \bot) \wedge (\forall y \in x)(\exists z \in x)(y \in z))$
(auf Ad beschränktes Unendlichkeitsaxiom),
(Ad 5) $\mathrm{Ad}(a) \to (\forall v \in a)(\forall w \in a)(\forall x \in a)(\exists y \in a)((\forall z \in y)(C[v,w,z] \wedge (z \in x)) \wedge (\forall z \in x)(C[v,w,z] \to (z \in y)))$, wenn $C[a,b,c]$ eine Δ_0-Formel ist und C keine freie Variable enthält
(auf Ad beschränkte Δ_0-Separation),
(Ad 6) $\mathrm{Ad}(a) \to (\forall v \in a)(\forall x \in a)((\forall y \in x)(\exists z \in a)C[v,y,z] \to (\exists w \in a)(\forall y \in x)(\exists z \in w)C[v,y,z])$,
wenn $C[a,b,c]$ eine Δ_0-Formel ist und C keine freie Variable enthält
(auf Ad beschränkte Δ_0-Kollektion).

Schlußregeln des Systems KPM:
(Str) $F \vdash G$, wenn $F \stackrel{S}{\models} G$ gilt.
(S1) $\mathcal{N}[\neg A], \mathcal{N}[B] \vdash \mathcal{N}[(A \to B)]$,
wenn B nicht die Formel $\bot$ ist.
(S2a) $\mathcal{P}[\mathcal{A}[a]] \vdash \mathcal{P}[\exists x \mathcal{A}[x]]$.
(S2b) $\mathcal{P}[((a \in b) \wedge \mathcal{A}[a])] \vdash \mathcal{P}[(\exists x \in b)\mathcal{A}[x]]$.
(S3a) $\mathcal{N}[\mathcal{A}[a]] \vdash \mathcal{N}[\exists x \mathcal{A}[x]]$,
wenn a in der Konklusion nicht auftritt.
(S3b) $\mathcal{N}[((a \in b) \wedge \mathcal{A}[a])] \vdash \mathcal{N}[(\exists x \in b)\mathcal{A}[x]]$,
wenn a in der Konklusion nicht auftritt.
(S4) $\mathcal{P}[\forall x \exists y \mathcal{B}[x,y]] \vdash \mathcal{P}[\exists z(\mathrm{Ad}(z) \wedge (\forall x \in z)(\exists y \in z)\mathcal{B}[x,y])]$,
wenn $\mathcal{B}[a,b]$ eine Δ_0-Formel ist.
(Schnitt) $C \vee F, C \to F \vdash F$.

24.2 Das Ordinalzahlensystem OT(M)

Wir gehen aus von einer Mengentheorie mit Auswahlaxiom nach Zermelo-Frankel, in der zusätzlich die Existenz einer schwachen Mahlozahl vorausgesetzt wird, und bezeichnen mit $\alpha, \beta, \gamma, \delta, \eta, \mu$ zunächst beliebige Ordinalzahlen, mit i, k, m, n Ordinalzahlen $< \omega$ und mit M die kleinste schwache Mahlozahl.
Für eine Ordinalzahlenklasse X sei $X \cap \alpha := \{\eta \in X : \eta < \alpha\}$.
$X < \alpha$ ($X \leq \alpha$) bedeute, daß $\eta < \alpha$ ($\eta \leq \alpha$) für alle $\eta \in X$ gilt.

Induktive Definition der *α-kritischen* Ordinalzahlen und der Funktionen φ_α.
1. Die 0-kritischen Ordinalzahlen seien die additiven Hauptzahlen.
2. φ_α sei die Ordnungsfunktion der α-kritischen Ordinalzahlen.
3. Für $\alpha > 0$ seien die α-kritischen Ordinalzahlen die gemeinsamen Fixpunkte aller Funktionen φ_η mit $\eta < \alpha$.

Wir schreiben $\varphi\alpha\beta$ anstatt $\varphi_\alpha(\beta)$.

Lemma 24.2.1.

a) $\varphi 0\beta = \omega^\beta$ *und* $\varphi 1\beta = \varepsilon_\beta$.

b) $\varphi\alpha\beta$ *ist streng monoton und stetig in* β.

c) *Für* $\gamma < \alpha$ *ist* $\varphi\gamma(\varphi\alpha\beta) = \varphi\alpha\beta$.

d) $\varphi\alpha\beta$ *ist eine additive Hauptzahl, und zu jeder additiven Hauptzahl* γ *gibt es eindeutig* α, β *mit* $\beta < \gamma = \varphi\alpha\beta$.

e) $a \leq \varphi\alpha 0$.

Definitionen.

1. Eine Ordinalzahl α heißt *stark kritisch*, wenn $\varphi\alpha 0 = \alpha$ ist.
2. SC sei die Klasse der stark kritischen Ordinalzahlen.
3. $\mathrm{M}^\Gamma := \min\{\eta \in \mathrm{SC} : \mathrm{M} < \eta\} = \Gamma_{\mathrm{M}+1}$.
4. $\delta =_{\mathrm{NF}} \varphi\alpha\beta + \gamma$ (δ hat die Normalform $\varphi\alpha\beta + \gamma$) bedeute, daß $\delta = \varphi\alpha\beta + \gamma$ ist mit $\alpha < \delta$, $\beta < \varphi\alpha\beta$ und $\gamma < \delta$.

Lemma 24.2.2.

a) *Gilt* $\delta =_{NF} \varphi\alpha\beta + \gamma$, *so sind* α, β, γ *eindeutig durch* δ *bestimmt.*

b) *Zu* δ *gibt es* α, β, γ *mit* $\delta =_{NF} \varphi\alpha\beta + \gamma$ *genau dann, wenn* $0 < \delta \notin \mathrm{SC}$ *ist.*

Induktive Definition von $\delta^\star$ für $\delta < \mathrm{M}^\Gamma$.

1. $0^\star = \mathrm{M}^\star = 0$.
2. Für $\delta =_{\mathrm{NF}} \varphi\alpha\beta + \gamma < \mathrm{M}^\Gamma$ sei $\delta^\star := \max\{\alpha^\star, \beta^\star, \gamma^\star\}$.
3. Für $\delta \in \mathrm{SC} \cap \mathrm{M}$ sei $\delta^\star := \delta$.

Nach dieser Definition ist $\delta^\star \in \{0\} \cup \mathrm{SC} \cap \mathrm{M}$ für alle $\delta < \mathrm{M}^\Gamma$.

Definitionen.

1. $\Omega_0 := 0$ und $\Omega_\alpha := \aleph_\alpha$ für $\alpha > 0$.
2. K sei die Klasse aller Ordinalzahlen Ω_α.
3. R sei die Klasse der überabzählbaren regulären Ordinalzahlen.

Offenbar ist R eine echte Teilklasse von K und M unerreichbar, d.h. $\Omega_\mathrm{M} = \mathrm{M} \in \mathrm{R}$.

Induktive Definition der Ordinalzahlenmengen $\mathrm{B}_n(\alpha, \beta)$, $\mathrm{B}(\alpha, \beta)$ und Ordinalzahlen $\mathrm{Z}\alpha$ für $\alpha < \mathrm{M}^\Gamma$.

(B1) $0, \mathrm{M} \in \mathrm{B}_n(\alpha, \beta)$ und $\eta \in \mathrm{B}_n(\alpha, \beta)$ für alle $\eta < \beta$.

(B2) Gilt $\delta =_{\mathrm{NF}} \varphi\delta_1\delta_2 + \delta_3$ und $\delta_1, \delta_2, \delta_3 \in \mathrm{B}_n(\alpha, \beta)$, so sei $\delta \in \mathrm{B}_{n+1}(\alpha, \beta)$.

(B3) Ist $\mu < \Omega_\mu < \mathrm{M}$ und $\mu \in \mathrm{B}_n(\alpha, \beta)$, so sei $\Omega_\mu \in \mathrm{B}_{n+1}(\alpha, \beta)$.

(B4) Ist $\gamma < \delta \in \mathrm{R} \cap \mathrm{M}$ und $\delta \in \mathrm{B}_n(\alpha, \beta)$, so sei $\gamma \in \mathrm{B}_{n+1}(\alpha, \beta)$.

(B5) Ist $\gamma < \alpha$ und $\gamma \in \mathrm{B}_n(\alpha, \beta)$, so sei $\mathrm{Z}\gamma \in \mathrm{B}_{n+1}(\alpha, \beta)$.

(B6) $\mathrm{B}(\alpha, \beta) := \bigcup\{\mathrm{B}_n(\alpha, \beta) : n < \omega\}$.

(B7) $\mathrm{Z}\alpha := \min\{\eta \in \mathrm{R} : \alpha \in \mathrm{B}(\alpha, \eta)$ und $\mathrm{B}(\alpha, \eta) \cap \mathrm{M} < \eta\}$.

Lemma 24.2.3. *Für* $\alpha, \beta < \mathrm{M}^\Gamma$ *gilt:*

a) $\alpha \in \mathrm{B}(\alpha, \mathrm{Z}\alpha)$ *und* $\mathrm{B}(\alpha, \mathrm{Z}\alpha) \cap \mathrm{M} < \mathrm{Z}\alpha$.

b) $\Omega_{\mathrm{Z}\alpha} = \mathrm{Z}\alpha \in \mathrm{R} \cap \mathrm{M}$.

c) *Aus* $\alpha < \beta$ *und* $\alpha \in \mathrm{B}(\beta, \mathrm{Z}\beta)$ *folgt* $\mathrm{Z}\alpha < \mathrm{Z}\beta$.
d) *Aus* $\mathrm{Z}\alpha = \mathrm{Z}\beta$ *folgt* $\alpha = \beta$.
e) $\mathrm{Z}\alpha \neq \alpha$.
f) $\alpha \in \mathrm{B}(\beta, \mathrm{Z}\beta)$ *gilt genau dann, wenn* $\alpha\star < \mathrm{Z}\beta$ *ist.*

Beweis nach der Definition von $\mathrm{Z}\alpha$ und nach [4] Lemma 3.6.

Induktive Definition der Ordinalzahlen $\mathrm{C}_n(\alpha,\beta), \mathrm{C}(\alpha,\beta)$ und Ordinalzahlen $\psi\Omega_{\mu+1}\alpha$, $\psi(\mathrm{Z}\gamma)\alpha$ für $\mu < \mathrm{M}$ und $\gamma < \mathrm{M}^\Gamma$.
(C1) $0, \mathrm{M} \in \mathrm{C}_n(\alpha,\beta)$ und $\eta \in \mathrm{C}_n(\alpha,\beta)$ für alle $\eta < \beta$.
(C2) Gilt $\delta =_{\mathrm{NF}} \varphi\delta_1\delta_2 + \delta_3$ und $\delta_1, \delta_2, \delta_3 \in \mathrm{C}_n(\alpha,\beta)$, so sei $\delta \in \mathrm{C}_{n+1}(\alpha,\beta)$.
(C3) Ist $\eta < \Omega_\eta < \mathrm{M}$ und $\eta \in \mathrm{C}_n(\alpha,\beta)$, so sei $\Omega_\eta \in \mathrm{C}_{n+1}(\alpha,\beta)$.
(C4) Ist $\eta < \mathrm{M}^\Gamma$ und $\eta \in \mathrm{C}_n(\alpha,\beta)$, so sei $\mathrm{Z}\eta \in \mathrm{C}_{n+1}(\alpha,\beta)$.
(C5) Ist $\eta < \mathrm{M}$ und sind $\Omega_{\eta+1}, \delta \in \mathrm{C}_n(\alpha,\beta)$ mit $\delta < \alpha$ und $\delta \in \mathrm{C}(\delta, \psi\Omega_{\eta+1}\delta)$, so sei $\psi\Omega_{\eta+1}\delta \in \mathrm{C}_{n+1}(\alpha,\beta)$.
(C6) Ist $\eta < \mathrm{M}^\Gamma$ und sind $\mathrm{Z}\eta, \delta \in \mathrm{C}_n(\alpha,\beta)$ mit $\delta < \alpha$ und $\delta \in \mathrm{C}(\delta, \psi(\mathrm{Z}\eta)\delta)$, so sei $\psi(\mathrm{Z}\eta)\delta \in \mathrm{C}_{n+1}(\alpha,\beta)$.
(C7) $\mathrm{C}(\alpha,\beta) := \bigcup\{\mathrm{C}_n(\alpha,\beta) : n < \omega\}$.
(C8) $\psi\Omega_{\mu+1}\alpha := \min\{\eta : \eta \notin \mathrm{C}(\alpha, \Omega_\mu + 1)\}$.
(C9) $\psi(\mathrm{Z}\gamma)\alpha := \min\{\eta : \mathrm{Z}\gamma \in \mathrm{C}(\alpha,\eta)$ und $\mathrm{C}(\alpha,\eta) \cap \mathrm{Z}\gamma < \eta\}$.

Lemma 24.2.4. *Für* $\mu < \mathrm{M}$ *und* $g < \mathrm{M}^\Gamma$ *gilt:*
a) $\Omega_\mu < \psi\Omega_{\mu+1}\alpha < \Omega_{\mu+1}$ *mit* $\psi\Omega_{\mu+1}\alpha \in \mathrm{SC}$.
b) $\gamma^\star < \psi(\mathrm{Z}\gamma)\alpha < \mathrm{Z}\gamma$ *mit* $\Omega_{\psi(\mathrm{Z}\gamma)\alpha} = \psi(\mathrm{Z}\gamma)\alpha \notin \mathrm{R}$.

Beweis nach der Definition von $\psi\Omega_{\mu+1}\alpha$ und [3] Theorem 1.3.

Bezeichnungen. Mit π, σ bezeichnen wir die erreichbaren regulären Ordinalzahlen $\Omega_{\mu+1} < \mathrm{M}$ sowie die unerreichbaren Ordinalzahlen $\mathrm{Z}\gamma$ mit $\gamma < \mathrm{M}^\Gamma$. Für $\sigma = \Omega_{\mu+1}$ sei $\sigma^- := \Omega_\mu$, und für $\sigma = \mathrm{Z}\gamma$ sei $\sigma^- := \gamma^\star$.

In bekannter Weise folgt für alle π und σ:

Lemma 24.2.5.
a) $\sigma^- < \psi\sigma\alpha < \sigma < \mathrm{M}$.
b) $\mathrm{C}(\alpha, \psi\sigma\alpha) \cap \sigma < \psi\sigma\alpha$.
c) *Aus* $\pi^- < \psi\sigma\alpha$ *folgt* $\pi \in \mathrm{C}(\alpha, \psi\sigma\alpha)$.
d) $\sigma \in \mathrm{C}(\alpha, \psi\sigma\alpha)$.
e) *Aus* $\psi\sigma\alpha \le \pi \le \sigma$ *folgt* $\psi\sigma\alpha \le \pi^-$.
f) *Aus* $\psi\sigma\alpha = \psi\pi\beta$ *folgt* $\pi = \sigma$.
g) *Aus* $\alpha \in \mathrm{C}(\alpha, \psi\sigma\alpha)$ *und* $\alpha < \beta$ *folgt* $\psi\sigma\alpha < \psi\sigma\beta$.
h) *Aus* $\psi\sigma\alpha = \psi\sigma\beta$ *mit* $\alpha \in \mathrm{C}(\alpha, \psi\sigma\alpha)$ *und* $\beta \in \mathrm{C}(\beta, \psi\sigma\beta)$ *folgt* $\alpha = \beta$.

Induktive Definition der Ordinalzahlenmenge $\mathsf{OT}(\mathrm{M})$.
(T1) $0, \mathrm{M} \in \mathsf{OT}(\mathrm{M})$.
(T2) Gilt $\delta =_{\mathrm{NF}} \varphi\alpha\beta + \gamma < \varepsilon_{\mathrm{M}+1}$ und $\alpha, \beta, \gamma \in \mathsf{OT}(\mathrm{M})$, so sei $\delta \in \mathsf{OT}(\mathrm{M})$.

(T3) Ist $\mu < \Omega_\mu <$ M und $\mu \in$ OT(M), so sei $\Omega_\mu \in$ OT(M).
(T4) Ist $\gamma < \varepsilon_{M+1}$ und $\gamma \in$ OT(M), seo sei $Z\gamma \in$ OT(M).
(T5) Sind $\sigma, \alpha \in$ OT(M) mit $\alpha <$ M und $\alpha \in C(\alpha, \psi\sigma\alpha)$, so sei $\psi\sigma\alpha \in$ OT(M).

Aus den Lemmata 24.2.2 bis 24.2.5 geht hervor, daß jede Ordinalzahl aus OT(M) auf genau eine Weise durch die Definitionsregeln (T1) bis (T5) bestimmt ist. Daher liefern diese Definitionsregeln ein eindeutiges Bezeichnungssystem für die Ordinalzahlen der Menge OT(M).

Lemma 24.2.6. *Die Ordinalzahlenmenge* OT(M) $\cap\, \Omega_1$ *besteht aus dem Abschnitt aller Ordinalzahlen* $< \psi\Omega_1(\psi(Z\varepsilon_{M+1})0)$.

Beweis nach [4] Theorem 6.3.

Im folgenden bezeichnen $\alpha, \beta, \gamma, \delta, \eta, \mu, \nu, \xi, \pi, \varrho, \sigma$ stehts Ordinalzahlen aus OT(M), wobei $\pi, \sigma \in$ R $\cap$ M sind.

24.3 Das geschichtete halbformale System RS(M)

Grundzeichen des Systems RS(M):
1. Abzählbar unendlich viele gebundene Variablen.
2. Bezeichnungen der Ordinalzahlen aus OT(M).
3. $\bot, \in$, Ad, $\rightarrow, \exists$, L, Doppelpunkt, runde und geschweifte Klammern.

Nennformen werden in RS(M) entsprechend wie in KPM verwendet. Dabei setzen wir von jeder Nennform voraus, daß kein Nennzeichen zwischen geschweiften Klammern auftritt.

Induktive Definition der *Terme* des Systems RS(M), der *Schicht* St und *Komponentenmenge* k(t) eines Terms t, der *Formeln* des Systems RS(M) und der *Komponentenmenge* k(F) der Formel F.
1. Für $\alpha \leq$ M ist Lα ein Term der Schicht α mit der Komponentenmenge $\{\alpha\}$.
2. Ist $0 < \alpha <$ M, $\alpha \notin$ R, $\mathcal{A}$ eine 1-stellige Nennform derart, daß $\mathcal{A}$[L0] eine Formel mit k($\mathcal{A}[0]$) $< \alpha$ ist, und x eine gebundene Variable, die in $\mathcal{A}$ nicht auftritt, so ist $\{x \in \mathrm{L}\alpha : \mathcal{A}[x]\}$ ein Term der Schicht α mit der Komponentenmenge k($\mathcal{A}$[L0]) $\cup \{\alpha\}$.
3. $\bot$ ist eine Formel mit leerer Komponentenmenge.
4. Sind s und t Terme von Schichten $<$ M, so ist $(s \in t)$ eine Formel mit der Komponentenmenge k(s) $\cup$ k(t).
5. Ist t ein Term von einer Schicht $<$ M, so ist Ad(t) eine Formel mit der Komponentenmenge k(t).
6. Sind A und B Formeln, so ist $(A \rightarrow B)$ eine Formel mit der Komponentenmenge k(A) $\cup$ k(B).

7. Ist t ein Term, $\mathcal{A}$ eine 1-stellige Nennform derart, daß $\mathcal{A}[\mathrm{L}0]$ eine Formel ist, und x eine gebundene Variable, die weder in t noch in $\mathcal{A}$ auftritt, so ist $(\exists x \in t)\mathcal{A}[x]$ eine Formel mit der Komponentenmenge $\mathrm{k}(t) \cup \mathrm{k}(\mathcal{A}[\mathrm{L}0])$.

Eine Formel heiße *einfach*, wenn sie nicht die Gestalt $(A \to B)$ hat.

Als *Mitteilungszeichen* für Terme verwenden wir s, t, u. Im übrigen verwenden wir in RS(M) die entsprechenden Mitteilungszeichen wie in KPM. P-Formen, N-Formen, $F \overset{S}{\models} G$, $\neg A$, $(A \vee B)$, $(A \wedge B)$, $(\forall x \in t)\mathcal{A}[x]$, $(s = t)$ und $\mathrm{Tran}[t]$ werden in RS(M) entsprechend wie in KPM definiert.

Induktive Definition der $\Sigma(\alpha)$-*Formeln* und $\Pi(\alpha)$-*Formeln* für $\alpha \in \mathrm{R}$.
1. Eine Formel F, die weder die Gestalt $(A \to B)$ noch $(\exists x \in t)\mathcal{A}[x]$ hat, ist genau dann eine $\Sigma(\alpha)$-Formel ($\Pi(\alpha)$-Formel), wenn $\mathrm{k}(F) < \alpha$ ist.
2. Eine Formel $(A \to B)$ ist genau dann eine $\Sigma(\alpha)$-Formel ($\Pi(\alpha)$-Formel), wenn A eine $\Pi(\alpha)$-Formel ($\Sigma(\alpha)$-Formel) und B eine $\Sigma(\alpha)$-Formel ($\Pi(\alpha)$-Formel) ist.
3. Eine Formel $(\exists x \in t)\mathcal{A}[x]$ ist genau dann eine $\Sigma(\alpha)$-Formel ($\Pi(\alpha)$-Formel), wenn $\mathrm{S}t \leq \alpha (\mathrm{S}t < \alpha)$ und $\mathcal{A}[\mathrm{L}0]$ eine $\Sigma(\alpha)$-Formel ($\Pi(\alpha)$-Formel) ist.

Eine Formel F heißt eine $\Delta(\alpha)$-*Formel* (für $\alpha \in \mathrm{R}$), wenn $\mathrm{k}(F) < \alpha$ ist.

Definition der Formel $t(u)$ für $\mathrm{S}u < \mathrm{S}t$.
1. $t(u) := \neg(u \in \mathrm{L}0)$, wenn t ein Term $\mathrm{L}\alpha$ ist.
2. $t(u) := \neg(u \in \mathrm{L}0) \wedge \mathcal{A}[u]$, wenn t ein Term $\{x \in \mathrm{L}\alpha : \mathcal{A}[x]\}$ ist.

Induktive Definition des *Ranges* $\mathrm{rn}(F)$ einer Formel F.
1. $\mathrm{rn}(\bot) := 0$.
2. $\mathrm{rn}(s \in t) = \max\{\omega \cdot \mathrm{S}s + 6, \omega \cdot \mathrm{S}t + 1\}$ für $\mathrm{S}s, \mathrm{S}t < \mathrm{M}$.
3. $\mathrm{rn}(\mathrm{Ad}(t)) := \omega \cdot \mathrm{S}t + 5$ für $\mathrm{S}t < \mathrm{M}$.
4. $\mathrm{rn}(\neg A) := \mathrm{rn}(A)$.
5. $\mathrm{rn}(A \to B) := \max\{\mathrm{rn}(A), \mathrm{rn}(B)\} + 1$, wenn B nicht die Formel $\bot$ ist.
6. $\mathrm{rn}((\exists x \in t)\mathcal{A}[x]) := \max\{\omega \cdot \mathrm{S}t, \mathrm{rn}(\mathcal{A}[\mathrm{L}0]) + 2\}$.

Lemma 24.3.1. *Für* $\mathrm{S}u < \mathrm{S}t$ *ist* $\mathrm{rn}(t(u)) < \omega \cdot \mathrm{S}t$.

Lemma 24.3.2.
a) $\mathrm{rn}(t(u) \wedge (s = u)) < \mathrm{rn}(s \in t)$ *für* $\mathrm{S}s < \mathrm{M}$ *und* $\mathrm{S}u < \mathrm{S}t < \mathrm{M}$.
b) $\mathrm{rn}(\mathrm{L}\sigma = t) < \mathrm{rn}(\mathrm{Ad}(t))$ *für* $\sigma \leq \mathrm{S}t < \mathrm{M}$.
c) $\mathrm{rn}(t(u) \wedge \mathcal{A}[u]) < \mathrm{rn}((\exists x \in t)\mathcal{A}[x])$ *für* $\mathrm{S}u < \mathrm{S}t$.

Lemma 24.3.3. *Jede einfache Formel vom Rang* $\alpha \in \mathrm{R}$ *ist eine Formel* $(\exists x \in \mathrm{L}\alpha)$ $\mathcal{A}[x]$ *mit einer* $\Delta(\alpha)$-*Formel* $\mathcal{A}[\mathrm{L}0]$.

Axiome des Systems RS(M):
(Ax.$\bot$) $\mathcal{N}[\bot]$.
(Ax. $\in$) $\mathcal{N}[(s \in \mathrm{L}0)]$ für $\mathrm{S}s < \mathrm{M}$.

(Ax.Ad) $\mathcal{N}[\mathrm{Ad}(t)]$ für $\mathrm{S}t < \Omega_1$.
(Ax.∃) $\mathcal{N}[(\exists x \in \mathrm{L0})\mathcal{A}[x]]$.

Hauptschlüsse des Systems RS(M):

($\to$) $\neg A \to \mathcal{N}[(A \to B)], B \to \mathcal{N}[(A \to B)] \vdash \mathcal{N}[(A \to B)]$,
wenn B nicht die Formel $\perp$ ist.
Dieser Hauptschluß hat die *Indexmenge* $\{\mathrm{L0}, \mathrm{L1}\}$.

(P. ∈) $(t(u) \wedge (s = u)) \vee \mathcal{P}[(s \in t)] \vdash \mathcal{P}[(s \in t)]$,
wenn $\mathrm{S}s < \mathrm{M}$ und $\mathrm{S}u < \mathrm{S}t < \mathrm{M}$ ist.
Dieser Hauptschluß hat die *Indexmenge* $\{u\}$.

(P.Ad) $(\mathrm{L}\sigma = t) \vee \mathcal{P}[\mathrm{Ad}(t)] \vdash \mathcal{P}[\mathrm{Ad}(t)]$,
wenn $\sigma \leq \mathrm{S}t < \mathrm{M}$ ist.
Dieser Hauptschluß hat die *Indexmenge* $\{\mathrm{L0}\}$.

(P.∃) $(t(u) \wedge \mathcal{A}[u]) \vee \mathcal{P}[(\exists x \in t)\mathcal{A}[x]] \vdash \mathcal{P}[(\exists x \in t)\mathcal{A}[x]]$,
wenn $\mathrm{S}u < \mathrm{S}t$ ist.
Dieser Hauptschluß hat die *Indexmenge* $\{u\}$.

(Cl_σ) $(\forall x \in t)(\exists y \in \mathrm{L}\sigma)\mathcal{B}[x, y] \vee F \vdash F$, wenn $\mathrm{S}t < \sigma$, $\mathcal{B}[\mathrm{L0}, \mathrm{L0}]$ eine $\Delta(\sigma)$-Formel und F eine Formel $\mathcal{P}[(\exists z \in \mathrm{L}\sigma)(\forall x \in t)(\exists y \in z)\mathcal{B}[x, y]]$ ist.
Dieser Hauptschluß hat die *Indexmenge* $\{\mathrm{L}\sigma\}$.

(Cl_M) $(\forall x \in \mathrm{LM})(\exists y \in \mathrm{LM})\mathcal{B}[x, y] \vee F \vdash F$,
wenn $\mathcal{B}[\mathrm{L0}, \mathrm{L0}]$ eine $\Delta(\mathrm{M})$-Formel und F eine Formel $\mathcal{P}[(\exists z \in \mathrm{LM})(\mathrm{Ad}(z) \wedge (\forall x \in z)(\exists y \in z)\mathcal{B}[x, y])]$ ist.
Dieser Hauptschluß hat die *Indexmenge* $\{\mathrm{L0}\}$.

(N. ∈) $(t(u) \wedge (s = u)) \to \mathcal{N}[(s \in t)]$ für alle u mit $\mathrm{S}u < \mathrm{S}t \vdash \mathcal{N}[(s \in t)]$,
wenn $\mathrm{S}s < \mathrm{M}$ und $0 < \mathrm{S}t < \mathrm{M}$ ist.
Dieser Hautpschluß hat die *Indexmenge* $\{u : \mathrm{S}u < \mathrm{S}t\}$.

(N.Ad) $(\mathrm{L}\sigma = t) \to \mathcal{N}[\mathrm{Ad}(t)]$ für alle $\sigma \leq \mathrm{S}t \vdash \mathcal{N}[\mathrm{Ad}(t)]$,
wenn $\Omega_1 \leq \mathrm{S}t < \mathrm{M}$ ist.
Dieser Hauptschluß hat die *Indexmenge* $\{\mathrm{L}\sigma : \sigma \leq \mathrm{S}t\}$.

(N.∃) $(t(u) \wedge \mathcal{A}[u]) \to \mathcal{N}[(\exists x \in t)\mathcal{A}[x]]$ für alle u mit $\mathrm{S}u < \mathrm{S}t$
$\vdash \mathcal{N}[(\exists x \in t)\mathcal{A}[x]]$,
wenn $\mathrm{S}t > 0$ ist.
Dieser Hauptschluß hat die *Indexmenge* $\{u : \mathrm{S}u < \mathrm{S}t\}$.

Der bezeichnete minimale Positiv- oder Negativteil der Konklusion eines Hauptschlusses heißt der *Hauptteil* des betreffendes Hauptschlusses. Die Prämissen eines Hauptschlusses mit der Indexmenge J haben die *Indizes* $u \in J$.

Schnitte des Systems RS(M):

$C \vee F, C \to F \vdash F$.

Die mit C bezeichnete Formel in den Prämissen eines Schnittes heißt die *Schnittformel* des betreffenden Schnittes. Der *Rang* eines Schnittes ist der Rang $\mathrm{rn}(C)$ seiner Schnittformel C.

Im folgenden bezeichnen X, Y endliche Teilmengen von OT(M).

Definition.

$$\mathrm{H}_\gamma^\beta(X) := \begin{cases} \bigcap\{\mathrm{B}(\xi,\eta) : X \subset \mathrm{B}(\xi,\eta) \text{ und } \beta < \xi\} \cap \\ \bigcap\{\mathrm{C}(\xi,\eta) : X \subset \mathrm{C}(\xi,\eta) \text{ und } \gamma < \xi\} \cap \mathsf{OT}(\mathrm{M}) \end{cases}$$

Lemma 24.3.4.
a) *Für* $\beta \leq \mu$, $\gamma \leq \nu$ *und* $X \subseteq Y$ *gilt* $\mathrm{H}_\gamma^\beta(X) \subseteq \mathrm{H}_\nu^\mu(Y)$.
b) $X \subset \mathrm{H}_\gamma^\beta(X)$.
c) *Aus* $Y \subset \mathrm{H}_\gamma^\beta(X)$ *folgt* $\mathrm{H}_\gamma^\beta(X \cup Y) = \mathrm{H}_\gamma^\beta(X)$.
d) $0, \mathrm{M} \in \mathrm{H}_\gamma^\beta(X)$.
e) $\mu =_{NF} \varphi\alpha\delta + \eta \in \mathrm{H}_\gamma^\beta(X)$ *gilt genau dann, wenn* $\alpha, \delta, \eta \in \mathrm{H}_\gamma^\beta(X)$ *sind.*
f) $\Omega_\mu \in \mathrm{H}_\gamma^\beta(X) \cap \mathrm{M}$ *gilt genau dann, wenn* $\mu \in \mathrm{H}_\gamma^\beta(X) \cap \mathrm{M}$ *ist.*
g) *Aus* $\alpha \in \mathrm{H}_\gamma^\beta(X)$ *und* $\alpha \leq \beta$ *folgt* $\mathrm{Z}\alpha \in \mathrm{H}_\gamma^\beta(X)$.
h) *Aus* $\sigma, \alpha \in \mathrm{H}_\gamma^\beta(X) \cap \mathrm{M}$, $\alpha \in \mathrm{C}(\alpha, \psi\sigma\alpha)$ *und* $\alpha \leq \gamma$ *folgt* $\psi\sigma\alpha \in \mathrm{H}_\gamma^\beta(X)$.

Induktive Definition von $\mathrm{H}_\gamma^\beta(X) \overset{\alpha}{\underset{\varrho}{\vdash}} F$ (durch Induktion nach α). $\mathrm{H}_\gamma^\beta(X) \overset{\alpha}{\underset{\varrho}{\vdash}}$ gelte genau dann, wenn $\{a\} \cup \mathrm{k}(F) \subset \mathrm{H}_\gamma^\beta(X)$ gilt und einer der folgenden sechs Fälle vorliegt.
a) F ist ein Axiom des Systems $\mathsf{RS}(\mathrm{M})$.
b) F ist die Konklusion eines Hauptschlusses $(\rightarrow)$, und es gibt $\mathrm{f}(i) < a$ mit $\mathrm{H}_\gamma^\beta(X) \overset{\mathrm{f}(i)}{\underset{\varrho}{\vdash}} F_i$ für die Prämsissen $F_i (i = 0, 1)$.
c) F ist die Konklusion eines Hauptschlusses (P. $\in$), (P.Ad), (P.$\exists$) oder (Cl_σ) mit der Indexmenge $\{u\}$, und es gibt δ mit $\mathrm{S}u \leq \delta \leq \alpha$ und $\mathrm{H}_\gamma^\beta(X) \overset{\delta}{\underset{\varrho}{\vdash}} F_u$ für die Prämisse F_u.
d) F ist die Konklusion eines Hauptschlusses $(\mathrm{CL_M})$, und es gibt δ mit $\delta + \mathrm{M}+ < \alpha$ und $\mathrm{H}_\gamma^\beta(X) \overset{\delta}{\underset{\varrho}{\vdash}} F_0$ für die Prämisse F_0.
e) F ist die Konklusion eines Hauptschlusses (N. $\in$), (N.Ad) oder (N.$\exists$) mit der Indexmenge J, und es gibt $\mathrm{f}(u) < \alpha$ mit $\mathrm{H}_\gamma^\beta(X \cup \mathrm{k}(u)) \overset{\mathrm{f}(u)}{\underset{\varrho}{\vdash}} F_u$ für jede Prämisse F_u vom Index $u \in J$.
f) F ist die Konklusion eines Schnittes mit der Schnittformel C von einem Rang $< \varrho$, und es gibt δ mit $\mathrm{k}(C) \leq \delta < \alpha$ und $\mathrm{H}_\gamma^\beta(X) \overset{\delta}{\underset{\varrho}{\vdash}} F_i$ für die Prämissen $F_i (i = 0, 1)$.

Lemma 24.3.5.
a) *Aus* $\mathrm{H}_\gamma^\beta(X) \overset{\alpha}{\underset{\varrho}{\vdash}} F$, $\beta \leq \mu$, $\gamma \leq \nu$ *und* $X \subseteq Y$ *folgt* $\mathrm{H}_\gamma^\beta(Y) \overset{\alpha}{\underset{\varrho}{\vdash}} F$.
b) *Aus* $\mathrm{H}_\gamma^\beta(X) \overset{\alpha}{\underset{\varrho}{\vdash}} F$, $\alpha < \delta$ *und* $\delta \in \mathrm{H}_\gamma^\beta(X)$ *folgt* $\mathrm{H}_\gamma^\beta(X) \overset{\delta}{\underset{\varrho}{\vdash}} F$.
c) *Aus* $\mathrm{H}_\gamma^\beta(X) \overset{\alpha}{\underset{\varrho}{\vdash}} F$ *und* $\varrho < \mu$ *folgt* $\mathrm{H}_\gamma^\beta(X) \overset{\alpha}{\underset{\mu}{\vdash}} F$.

Lemma 24.3.6 (Strukturschlußlemma).
Aus $\mathrm{H}_\gamma^\beta(X) \overset{\alpha}{\underset{\varrho}{\vdash}} F \overset{S}{\vdash} G$ *und* $\mathrm{k}(G) \subset \mathrm{H}_\gamma^\beta(X)$ *folgt* $\mathrm{H}_\gamma^\beta(X) \overset{\alpha}{\underset{\varrho}{\vdash}} G$.

Beweis durch Induktion nach α.

Lemma 24.3.7.

a) *Aus* $\mathrm{H}_\gamma^\beta(X) \stackrel{\mathrm{f}(i)}{\underset{\varrho}{\vdash}} \mathcal{P}[A_i]$ *mit* $\mathrm{f}(i) < \alpha$ $(i = 0, 1)$ *und* $\alpha \in \mathrm{H}_\gamma^\beta(X)$ *folgt*
$\mathrm{H}_\gamma^\beta(X) \stackrel{\alpha}{\underset{\varrho}{\vdash}} \mathcal{P}[(A_0 \wedge A_1)]$.

b) *Ist* $\mathrm{S}u < \mathrm{S}t$ *und gilt* $\mathrm{H}_\gamma^\beta(X) \stackrel{\delta}{\underset{\varrho}{\vdash}} \mathcal{N}[(t(u) \rightarrow \mathcal{A}[u])]$ *mit* $\mathrm{S}u \leq \delta < \alpha$ *und* $\alpha, \mathrm{S}t \in \mathrm{H}_\gamma^\beta(X)$*, so folgt*
$\mathrm{H}_\gamma^\beta(X) \stackrel{\alpha}{\underset{\varrho}{\vdash}} \mathcal{N}[(\forall x \in t)\mathcal{A}[x]]$.

c) *Ist* $\mathrm{S}t > 0$ *und gilt* $\mathrm{H}_\gamma^\beta(X \cup \mathrm{k}(u)) \stackrel{\mathrm{f}(u)}{\underset{\varrho}{\vdash}} \mathcal{P}[(t(u) \rightarrow \mathcal{A}[u])]$ *mit* $\mathrm{f}(u) < \alpha$ *für alle* u *mit* $\mathrm{S}u < \mathrm{S}t$ *sowie* $\alpha, \mathrm{S}t \in \mathrm{H}_\gamma^\beta(X)$*, so folgt*
$\mathrm{H}_\gamma^\beta(X) \stackrel{\alpha}{\underset{\varrho}{\vdash}} \mathcal{P}[(\forall x \in t)\mathcal{A}[x]]$.

Beweis aufgrund der Definitionen von $\wedge$ und $\forall$ mit Lemma 24.3.6.

Lemma 24.3.8 (Inversionslemma).

a) *Aus* $\mathrm{H}_\gamma^\beta(X) \stackrel{\alpha}{\underset{\varrho}{\vdash}} \mathcal{N}[(A \rightarrow B)]$ *folgt*
$\mathrm{H}_\gamma^\beta(X) \stackrel{\alpha}{\underset{\varrho}{\vdash}} \mathcal{N}[\neg A]$ *und* $\mathrm{H}_\gamma^\beta(X) \stackrel{\alpha}{\underset{\varrho}{\vdash}} \mathcal{N}[B]$.

b) *Aus* $\mathrm{H}_\gamma^\beta(X) \stackrel{\alpha}{\underset{\varrho}{\vdash}} \mathcal{N}[(s \in t)]$ *mit* $\mathrm{S}s < \mathrm{M}$ *und* $\mathrm{S}u < \mathrm{S}t < \mathrm{M}$ *folgt*
$\mathrm{H}_\gamma^\beta(X \cup \mathrm{k}(u)) \stackrel{\alpha}{\underset{\varrho}{\vdash}} \mathcal{N}[(t(u) \wedge (s = u))]$.

c) *Aus* $\mathrm{H}_\gamma^\beta(X) \stackrel{\alpha}{\underset{\varrho}{\vdash}} \mathcal{N}[\mathrm{Ad}(t)]$ *mit* $\sigma \leq \mathrm{S}t < \mathrm{M}$ *folgt*
$\mathrm{H}_\gamma^\beta(X \cup \{\sigma\}) \stackrel{\alpha}{\underset{\varrho}{\vdash}} \mathcal{N}[(\mathrm{L}\sigma = t)]$.

d) *Aus* $\mathrm{H}_\gamma^\beta(X) \stackrel{\alpha}{\underset{\varrho}{\vdash}} \mathcal{N}[(\exists x \in t)\mathcal{A}[x]]$ *mit* $\mathrm{S}u < \mathrm{S}t$ *folgt*
$\mathrm{H}_\gamma^\beta(X \cup \mathrm{k}(u)) \stackrel{\alpha}{\underset{\varrho}{\vdash}} \mathcal{N}[(t(u) \wedge \mathcal{A}[u])]$.

e) *Aus* $\mathrm{H}_\gamma^\beta(X) \stackrel{\alpha}{\underset{\varrho}{\vdash}} \mathcal{P}[(\forall x \in t)\mathcal{A}[x]]$ *mit* $\mathrm{S}u < \mathrm{S}t$ *folgt*
$\mathrm{H}_\gamma^\beta(X \cup \mathrm{k}(u)) \stackrel{\alpha}{\underset{\varrho}{\vdash}} \mathcal{P}[(t(u) \rightarrow A[u])]$.

Beweis durch Induktion nach α mit Lemma 24.3.6.

Lemma 24.3.9 (Beschränkungslemma).

a) *Aus* $\mathrm{H}_\gamma^\beta(X) \stackrel{\alpha}{\underset{\varrho}{\vdash}} \mathcal{P}[(\exists x \in \mathrm{L}\delta)\mathcal{A}[x]]$ *mit* $\alpha \leq \mu < \delta \leq \mathrm{M}$ *und* $\mu \in \mathrm{H}_\gamma^\beta(X)$ *folgt*
$\mathrm{H}_\gamma^\beta(X) \stackrel{\alpha}{\underset{\varrho}{\vdash}} \mathcal{P}[(\exists x \in \mathrm{L}\mu)\mathcal{A}[x]]$.

b) *Aus* $\mathrm{H}_\gamma^\beta(X) \stackrel{\alpha}{\underset{\varrho}{\vdash}} \mathcal{P}[(\forall x \in t)(\exists y \in \mathrm{L}\delta)\mathcal{B}[x, y]]$ *mit* $\alpha \leq \mu < \delta \leq \mathrm{M}$ *und* $\mu \in \mathrm{H}_\gamma^\beta(X)$ *folgt*
$\mathrm{H}_\gamma^\beta(X) \stackrel{\alpha}{\underset{\varrho}{\vdash}} \mathcal{P}[(\forall x \in t)(\exists y \in \mathrm{L}\mu)\mathcal{B}[x, y]]$.

c) *Aus* $\mathrm{H}_\gamma^\beta(X) \stackrel{\alpha}{\underset{\varrho}{\vdash}} \mathcal{N}[(\exists x \in \mathrm{L}\delta)\mathcal{A}[x]]$ *mit* $\mu < \delta \leq \mathrm{M}$ *und* $\mu \in \mathrm{H}_\gamma^\beta(X)$ *folgt*
$\mathrm{H}_\gamma^\beta(X) \stackrel{\alpha}{\underset{\varrho}{\vdash}} \mathcal{N}[(\exists x \in \mathrm{L}\mu)\mathcal{A}[x]]$.

Beweis durch Induktion nach α.

Definition. Unter einer RS(M)-*Interpretation* einer Formel F des Systems KPM verstehen wir eine Formel des Systems RS(M), die aus F dadurch entsteht, daß für jede in F auftretende freie Variable ein Term einer Schicht $< \mathrm{M}$ eingesetzt wird und jeder in F auftretende unbeschränkte Quantor $\exists x$ durch $(\exists x \in \mathrm{LM})$ ersetzt wird.

Satz 24.3.1 (Einbettungssatz).
Zu jeder in KPM *herleitbaren Formel F gibt es* $k, m < \omega$ *derart, daß*

$$\mathrm{H}_0^0(\mathrm{k}(F^{\mathrm{M}})) \mathrel{\vdash^{\mathrm{M}\cdot k}_{\mathrm{M}+m}} F^{\mathrm{M}}$$

für jede RS(M)-*Interpreation* F^{M} *von F gilt.*

Beweis entsprechend wie in [3] oder [5].

Lemma 24.3.10 (Prädikative Schitt-Elimination).

a) *Ist C eine einfache Formel vom Rang* $\varrho \notin \mathrm{R}$ *und gilt* $\mathrm{H}_\gamma^\beta(X) \mathrel{\vdash^{\alpha}_{\varrho}} C \to F$ *und* $\mathrm{H}_\gamma^\beta(X) \mathrel{\vdash^{\delta}_{\varrho}} C \vee G$ *mit* $\mathrm{k}(C) \le \alpha$, *so folgt* $\mathrm{H}_\gamma^\beta(X) \mathrel{\vdash^{\alpha+\delta}_{\varrho}} F \vee G$.
b) *Aus* $\mathrm{H}_\gamma^\beta(X) \mathrel{\vdash^{\alpha}_{\varrho+1}} F$ *mit* $\varrho \notin \mathrm{R}$ *folgt* $\mathrm{H}_\gamma^\beta(X) \mathrel{\vdash^{\varphi 0\alpha}_{\varrho}} F$.
c) *Gilt* $\mathrm{H}_\gamma^\beta(X) \mathrel{\vdash^{\delta}_{\mu+\varphi 0\alpha}} F$ *mit* $\alpha, \delta < \mathrm{M}$ *und* $\alpha \in \mathrm{H}_\gamma^\beta(X)$, *wobei es kein* $\zeta \in \mathrm{R}$ *mit* $\mu \le \zeta < \mu + \varphi 0\alpha$ *gibt, so folgt* $\mathrm{H}_\gamma^\beta(X) \mathrel{\vdash^{\varphi\alpha\delta}_{\mu}} F$.

Beweis von a) ergibt sich durch Induktion nach δ.
b) ergibt sich mit a) durch Induktion nach α.
c) ergibt sich mit b) durch Hauptinduktion nach α und Nebeninduktion nach δ.

Definitionen.

a) $\mathrm{NF}(\alpha, \beta)$ bedeute, daß entweder $\alpha = 0$ ist oder $\eta + \beta < \alpha + \beta$ für alle $\eta < \alpha$ gilt.
b) Für $\mu \in \mathrm{K} \cap \mathrm{M}$ sei $\mu' := \mu + 1$, wenn $\mu \in \mathrm{R}$ ist, sonst $\mu' := \mu$.

Lemma 24.3.11 (Imprädikative Schnitt-Elimination mit Kollabierung).

a) *Gilt* $\mathrm{H}_\gamma^\beta(X) \mathrel{\vdash^{\alpha}_{\mathrm{M}+1}} F$ *für eine* $\Sigma(\mathrm{M})$-*Formel* F *mit* $\beta \in \mathrm{H}_\gamma^\beta(X)$, $X \subset \mathrm{B}(\beta + 1, \mathrm{Z}(\beta + 1))$ *und* $\mathrm{NF}(\beta, \varphi 0\alpha)$, *so folgt* $\mathrm{H}_\gamma^{\beta+\varphi 0\alpha}(X) \mathrel{\vdash^{\mathrm{Z}(\beta+\varphi 0\alpha)}_{\mathrm{Z}(\beta+\varphi 0\alpha)}} F$.
b) *Gilt* $\mathrm{H}_\gamma^\beta(X) \mathrel{\vdash^{\alpha}_{\mu'}} F$ *für eine* $\Sigma(\sigma)$-*Formel* F *mit* $\mu \in \mathrm{K} \cap \mathrm{M}$, $\alpha < \mathrm{M}$, $\gamma, \mu, \sigma \in \mathrm{H}_\gamma^\beta(X)$ *und* $X \subset \mathrm{C}(\gamma + 1, \psi\pi(\gamma + 1))$ *für alle* $\pi \ge \sigma$, *so folgt* $\mathrm{H}_{\gamma+\varphi 0(\mu+\alpha)}^\beta(X) \mathrel{\vdash^{\psi\sigma(\gamma+\varphi 0(\mu+\alpha))}_{\psi\sigma(\gamma+\varphi 0(\mu+\alpha))}} F$.

Beweis entsprechend wie in [3] mit dem Unterschied, daß Hauptschlüsse ($\mathrm{Cl_M}$) nur im Beweis von a) und Hauptschlüsse (CL_σ) nur im Beweis von b) auf Hauptschlüsse ($\mathrm{P}.\exists$) zurückzuführen sind.

Lemma 24.3.12. *Ist eine* RS(M)*-Interpretation* F^{M} *einer in* KPM *herleitbaren* Δ_0*-Formel* F *eine* $\Sigma(\Omega_1)$*-Formel mit* $\mathrm{k}(F^{\mathrm{M}}) \in \mathrm{H}_0^0(\emptyset)$*, so gibt es* $\alpha < \Omega_1$, β *und* $\gamma < \mathrm{M}$ *mit* $\mathrm{H}_\gamma^\beta(\emptyset) \mid\frac{\alpha}{0}\; F^{\mathrm{M}}$.

Beweis. Nach Satz 24.3.1 hat man $k, m < \omega$ mit $\mathrm{H}_0^0(\emptyset) \mid\frac{\mathrm{M}\cdot k}{\mathrm{M}+m}\; F^{\mathrm{M}}$.
Mit Lemma 24.3.10 b) folgt, daß es δ mit $\mathrm{H}_0^0(\emptyset) \mid\frac{\delta}{\mathrm{M}+1}\; F^{\mathrm{M}}$ gibt.
Für $\beta := \varphi 0\delta$ folgt nach Lemma 24.3.11 a) $\mathrm{H}_0^\beta(\emptyset) \mid\frac{\mathrm{Z}\beta}{\mathrm{Z}\beta}\; F^{\mathrm{M}}$.
Für $\gamma := \varphi 0(\mathrm{Z}\beta + \mathrm{Z}\beta) < \mathrm{M}$ folgt nach Lemma 24.3.11 b) $\mathrm{H}_\gamma^\beta(\emptyset) \mid\frac{\psi\Omega_1\gamma}{\psi\Omega_1\gamma}\; F^{\mathrm{M}}$.
Für $\alpha := \varphi(\psi\Omega_1\gamma)(\psi\Omega_1\gamma)$ folgt nach Lemma 24.3.10 c) $\mathrm{H}_\gamma^\beta(\emptyset) \mid\frac{\alpha}{0}\; F^{\mathrm{M}}$ mit $\alpha < \Omega_1$.

Satz 24.3.2 (M. Rathjen). *Die beweistheoretische Ordinalzahl des formalen Systems* KPM *ist* $\leq \psi\Omega_1(\psi(\mathrm{Z}\varepsilon_{\mathrm{M}+1})0)$.

Beweis. Dies folgt aus den Lemmata 24.3.12 und 24.2.6.

Literatur

1. W. Buchholz. Ein Mengensystem ohne Kollabierungsfunktionen. Seminar München 1984.
2. W. Buchholz. A simplified version of local predicativity. In S. Wainer P. Aczel, J. Simmons, editor, *Proof Theory*, pages 115–147. Cambridge University Press, Cambridge, 1992.
3. W. Buchholz. A note on the ordinal analysis of KPM. In J. Väänänen J. Oikkonen, editor, *Logic Colloquium '90*, pages 1–9. Springer Verlag, Berlin, 1993.
4. M. Rathjen. Ordinal notations based on a weakly Mahlo cardinal. *Arch. Math. Logic*, 29:249–263, 1990.
5. M. Rathjen. Proof-theoretic analysis of KPM. *Arch. Math. Logic*, 30:377–403, 1991.
6. K. Schütte. Beweistheoretische Analyse von KPM. Type-written manuscript, München, 30.10.1991, 25 pages, 1991.
7. K. Schütte. Zur Beweistheorie von KPM. *Bayerische Akademie der Wissenschaften, Mathematisch-Naturwissenschaftliche Klasse*, Sitzungsberichte 93:19–34, 1994.

Kapitel 25
Zur Beweistheorie von KP+ Π_3-Ref

Kurt Schütte

Abstract This article together with the foregoing on KPM belongs to the last two papers Schütte wrote on ordinal analysis. It is from 1992 and deals with the strongest system for which an ordinal analysis was available at the time: Kripke-Platek set theory with Π_3-reflection. A crucial tool is the stationary collapsing technique. Schütte sketches a modified treatment of the analysis given in Rathjen's Habilitationschrift [1], later published as [2]. Schütte furnishes a very elegant treatment in his calculus of positive and negative forms, making direct use of structural inferences that are harder to engineer in a sequent calculus. Moreover, he achieves the operator control on derivations in a very economical way.

Skizzierung einer modifizierten Fassung der beweistheoretischen Analyse von KP+ Π_3-Ref nach M. Rathjen. Die Beweise der hier zusammengestellten Lemmata sind, soweit sie nicht ausgeführt sind, entweder wohlbekannt oder der Habilitationsschrift [1] von M. Rathjen zu entnehmen.

25.1 Das formale System KP+ Π_3-Ref

Grundzeichen des Systems KP+ Π_3-Ref:
1. Je abzählbar unendliche viele freie und gebundene Variablen.
2. $\bot, \in, \rightarrow, \exists$ und runde Klammern.

Unter einer n-stelligen *Nennform* ($n \geq 1$) verstehen wir eine nichtleere endliche Zeichenreihe, die außer Grundzeichen des formalen Systems höchstens die *Nennzeichen* $*_1, \ldots, *_n$ enthält. Ist $\mathfrak{F}$ eine n-stellige Nennform und sind $e_1, \ldots, e_n$ nichtleere endliche Zeichenreihen, so bezeichnet $\mathfrak{F}[e_1, \ldots, e_n]$ denjenigen Ausdruck, der aus

K. Schütte (†)
Mathematisches Institut, LMU München, Germany

R. Kahle and M. Rathjen (eds.), *The Legacy of Kurt Schütte*,
https://doi.org/10.1007/978-3-030-49424-7_25

der Nennform $\mathfrak{F}$ dadurch entsteht, daß jedes Nennzeichen $*_i$ überall, wo es in $\mathfrak{F}$ auftritt, durch e_i ersetzt wird. Eckige Klammern werden im folgenden nur mit dieser Bedeutung im Zusammenhang mit Nennformen verwendet.

Induktive Definition der *Formeln* des Systems KP+ Π_3-Ref.
1. $\bot$ (falsum) ist eine Formel.
2. Sind a und b freie Variablen, so ist $(a \in b)$ eine Formel.
3. Sind A und B Formeln, so ist auch $(A \rightarrow B)$ eine Formel.
4. Ist a eine freie Variable, $\mathfrak{A}$ eine 1-stellige Nennform derart, daß $\mathfrak{A}[a]$ eine Formel ist, und x eine gebundene Variable, die in $\mathfrak{A}$ nicht auftritt, so ist $\exists x \mathfrak{A}[x]$ eine Formel mit einem *unbeschränkten Quantor* $\exists x$ und $(\exists x \in a)\mathfrak{A}[x]$ eine Formel mit einem *beschränkten Quantor* $(\exists x \in a)$.
   Formeln, die keinen unbeschränkten Quantor enthalten, heißen Δ_0-*Formeln*.

Induktive Definition der P-*Formen* und N-*Formen*.
1. $*_1$ ist eine P-Form.
2. Ist $\mathfrak{P}$ eine P-Form und A eine Formel, so ist $\mathfrak{P}[(A \rightarrow *_1)]$ eine P-Form und $\mathfrak{P}[(*_1 \rightarrow A)]$ eine N-Form.
3. Ist $\mathfrak{N}$ eine N-Form, so ist $\mathfrak{N}[(*_1 \rightarrow \bot)]$ eine P-Form.

Nach dieser Definition sind die P-Formen und N-Formen 1-stellige Nennformen, die das Nennzeichen $*_1$ an genau einer Stelle enthalten und bei Ersetzung dieses Nennzeichens durch eine Formel in Formeln übergehen.
Eine Formel C heißt ein *Positivteil* (*Negativteil*) einer Formel F, wenn es eine P-Form $\mathfrak{P}$ (N-Form $\mathfrak{N}$) gibt derart, daß F die Formel $\mathfrak{P}[C]$($\mathfrak{N}[C]$) ist. Ein Positivteil einer Formel heißt *minimal*, wenn er nicht die Gestalt $(A \rightarrow B)$ hat. Ein Negativteil einer Formel heißt *minimal*, wenn er nicht die Gestalt $(A \rightarrow \bot)$ hat.
Für Formeln F und G bedeute $F \overset{S}{\models} G$ (aus F folgt strukturell G), daß jeder minimale Positivteil von F, der nicht die Formel $\bot$ ist, auch als Positivteil von G auftritt und jeder minimale Negativteil von F auch als Negativteil von G auftritt.

Als *Mitteilungszeichen* verwenden wir
1. a, b, c für freie Variablen,
2. w, x, y, z für gebundene Variablen,
3. A, B, C, F, G für Formeln,
4. $\mathfrak{A}$ für 1-stellige Nennformen derart, daß $\mathfrak{A}[a]$ eine Formel ist,
5. $\mathfrak{B}$ für 2-stellige Nennformen derart, daß $\mathfrak{B}[a, b]$ eine Formel ist,
6. $\mathfrak{C}$ für 3-stellige Nennformen, so daß $\mathfrak{C}[a, b, c]$ eine Formel ist,
7. $\mathfrak{P}$ für P-Formen und
8. $\mathfrak{N}$ für N-Formen.

Diese Mitteilungszeichen werden auch mit Indizes verwendet.

Definitionen.
1. $(A \vee B) := (\neg A \rightarrow B)$,
2. $(A \wedge B) := \neg(A \rightarrow \neg B)$,

3. $\forall x\mathfrak{A}[x] := \neg\exists x\neg\mathfrak{A}[x]$,
4. $(\forall x \in a)\mathfrak{A}[x] := \neg(\exists x \in a)\neg\mathfrak{A}[x]$,
5. $(a = b) := ((\forall x \in a)(x \in b) \wedge (\forall x \in b)(x \in a))$,
6. $\mathrm{Tran}[a] := (\forall x \in a)(\forall y \in x)(y \in a)$.

Zur Abkürzung lassen wir teilweise runde Klammern in Formeln fort, wenn es nicht mißverständlich ist.

Axiome des Systems KP+ Π_3-Ref:

(Ax.1) $F \rightarrow F$,

(Ax.2) $(a = b) \rightarrow (\mathfrak{A}[a] \rightarrow \mathfrak{A}[b])$
(Extensionalität),

(Ax.3) $\forall x((\forall y \in x)\mathfrak{A}[y] \rightarrow \mathfrak{A}[x]) \rightarrow \forall x\mathfrak{A}[x]$
(Fundierung),

(Ax.4) $\exists x((a \in x) \wedge (b \in x) \wedge (\forall y \in x)((y = a) \vee (y = b)))$
(Paarmengenaxiom),

(Ax.5) $\exists x((\forall y \in a)(\forall z \in y)(z \in x) \wedge (\forall z \in x)(\exists y \in a)(z \in y))$
(Vereinigungsmengenaxiom),

(Ax.6) $\exists x((\exists y \in x)(y \in x) \wedge (\forall y \in x)(\exists z \in x)(y \in z))$
(Unendlichkeitsaxiom),

(Ax.7) $\exists x((\forall y \in x)((y \in a) \wedge \mathfrak{B}[y,b]) \wedge (\forall y \in a)(\mathfrak{B}[y,b] \rightarrow (y \in x))$,
wenn $\mathfrak{B}[a,b]$ eine Δ_0-Formel ist und $\mathfrak{B}$ keine freie Variable enthält
(Δ_0-Separation),

(Ax.8) $(\forall x \in a)\exists y \mathrm{C}[x,y,c] \rightarrow \exists z(\forall x \in a)(\exists y \in z)\mathrm{C}[x,y,c]$,
wenn $\mathfrak{C}[a,b,c]$ eine Δ_0-Formel ist und $\mathfrak{C}$ keine freie Variable enthält
(Δ_0-Kollektion).

Schlußregeln des Systems KP+ Π_3-Ref:

(Str) $F \vdash G$, wenn $F \overset{S}{\models} G$ gilt.

(S1) $\mathfrak{N}[\neg\mathfrak{A}], \mathfrak{N}[B] \vdash \mathfrak{N}[(\mathfrak{A} \rightarrow B)]$,
wenn B nicht die Formel $\bot$ ist.

(S2a) $\mathfrak{P}[A[a]] \vdash \mathfrak{P}[\exists x\mathfrak{A}[x]]$.

(S2b) $\mathfrak{P}[((a \in b) \wedge \mathfrak{A}[a])] \vdash \mathfrak{P}[(\exists x \in b)\mathfrak{A}[x]]$.

(S3a) $\mathfrak{N}[\mathfrak{A}[a]] \vdash \mathfrak{N}[\exists x\mathfrak{A}[x]]$,
wenn a in der Konklusion nicht auftritt.

(S3b) $\mathfrak{N}[((a \in b) \wedge \mathfrak{A}[a])] \vdash \mathfrak{N}[(\exists x \in b)\mathfrak{A}[x]]$,
wenn a in der Konklusion nicht auftritt.

(S4) $\mathfrak{P}[\forall x\exists y\forall z\mathfrak{C}[x,y,z]] \vdash \mathfrak{P}[\exists w(\mathrm{Tran}[w] \wedge (\exists x \in w)(x \in w) \wedge (\forall x \in w)(\exists y \in w)(\forall z \in w)\mathfrak{C}[x,y,z])]$,
wenn $\mathfrak{C}[a,b,c]$ eine Δ_0-Formel ist
(Π_3-Reflexion).

(Schnitt) $C \vee F, C \rightarrow F \vdash F$.

25.2 Ordinalzahlentheoretische Grundbegriffe

Wir gehen aus von einer Mengentheorie mit Auswahlaxiom nach Zermelo-Fraenkel, in der zusätzlich die Existenz einer schwach kompakten Kardinalzahl vorausgesetzt wird, und bezeichnen mit $\alpha, \beta, \gamma, \delta, \eta$ zunächst beliebige Ordinalzahlen und mit i, m, n Ordinalzahlen $< \omega$.

Für eine Ordinalzahlenklasse X sei $X \cap \alpha := \{\eta \in X : \eta < \alpha\}$.
$X < \alpha$ bedeute, daß $\eta \leq \alpha$ für alle $\eta \in X$ gilt.

Induktive Definition der *α-kritischen* Ordinalzahlen und der Funktionen φ_α.
1. Die 0-kritischen Ordinalzahlen seien die additiven Hauptzahlen.
2. φ_α sei die Ordnungsfunktion der α-kritischen Ordinalzahlen.
3. Für $\alpha > 0$ seien die α-kritischen Ordinalzahlen die gemeinsamen Fixpunkte aller Funktionen φ_η mit $\eta < \alpha$.

Wir schreiben $\varphi\alpha\beta$ anstatt $\varphi_\alpha(\beta)$.

Lemma 25.2.1.
a) $\varphi 0\beta = \omega^\beta$ *und* $\varphi 1\beta = \varepsilon_\beta$.
b) $\varphi\alpha\beta$ *ist streng monoton und stetig in* β.
c) *Für* $\eta < \alpha$ *ist* $\varphi\eta(\varphi\alpha\beta) = \varphi\alpha\beta$.
d) $\varphi\alpha\beta$ *ist eine additive Hauptzahl, und zu jeder additiven Hauptzahl* γ *gibt es eindeutig* α *und* $\beta < \gamma$ *mit* $\varphi\alpha\beta = \gamma$.
e) $\alpha \leq \varphi\alpha 0$.

Definitionen.
1. Eine Ordinalzahl α heißt *stark kritisch*, wenn $\varphi\alpha 0 = \alpha$ ist.
2. SC sei die Klasse der stark kritischen Ordinalzahlen.
3. $\delta =_{\text{NF}} \varphi\alpha\beta + \gamma$ (δ hat die Normalform $\varphi\alpha\beta + \gamma$) bedeute, daß $\delta = \varphi\alpha\beta + \gamma$ ist mit $\alpha < \delta, \beta < \varphi\alpha\beta$ und $\gamma < \delta$.

Lemma 25.2.2.
a) *Gilt* $\delta =_{NF} \varphi\alpha\beta + \gamma$, *so sind* α, β, γ *eindeutig durch* δ *bestimmt.*
b) *Zu* δ *gibt es* α, β, γ *mit* $\delta =_{NF} \varphi\alpha\beta + \gamma$ *genau dann, wenn* $0 < \delta \notin \text{SC}$ *ist.*

Definitionen.
1. $\Omega_0 := 0$ und $\Omega_\alpha := \aleph_\alpha$ für $\alpha > 0$.
2. R sei die Klasse der überabzählbaren regulären Ordinalzahlen.
3. Eine Ordinalzahl $\alpha \in \text{R}$ heißt *unerreichbar*, wenn $\Omega_\alpha = \alpha$ ist. Andernfalls heißt sie *erreichbar*.

Lemma 25.2.3.
a) *Für* $\alpha > 0$ *ist* $\Omega_\alpha \in \text{SC}$.
b) *Eine Ordinalzahl* $\alpha \in \text{R}$ *ist genau dann erreichbar, wenn es* η *mit* $\alpha = \Omega_{\eta+1}$ *gibt.*

Definitionen.

1. Eine Ordinalzahlenmenge X heißt *unbeschränkt* in einer Limeszahl β, wenn es zu jeder Ordinalzahl $\alpha < \beta$ ein $\eta \in X$ mit $\alpha \leq \eta \leq \beta$ gibt.
2. Ene Ordinalzahlenmenge X heißt *abgeschlossen* in $\gamma \in \mathrm{R}$, wenn für jede Limeszahl $\beta < \gamma$ gilt: Ist X unbeschränkt in β, so ist $\beta \in X$.
3. Eine Ordinalzahlenmenge X heißt *stationär* in $\gamma \in \mathrm{R}$, wenn jede Ordinalzahlenmenge, die unbeschränkt und abgeschlossen in γ ist, mindestens ein Element von X enthält.

Lemma 25.2.4. *Ist eine Ordinalzahlenmenge X stationär in $\gamma \in \mathrm{R}$ und eine Ordinalzahlenmenge Y unbeschränkt und abgeschlossen in γ, so ist der Durchschnitt $X \cap Y$ unbeschränkt in γ.*

Bezeichnungen.

1. K sei die kleinste schwach kompakte Kardinalzahl.
2. K^Γ sei die kleinste stark kritische Ordinalzahl $> \mathrm{K}$.

Induktive Definition von $\delta^\star$ für $\delta < \mathrm{K}^\Gamma$.

1. $0^\star = \mathrm{K}^\star := 0$.
2. Für $\delta =_{\mathrm{NF}} \varphi\alpha\beta + \gamma < \mathrm{K}^\Gamma$ sei $\delta^\star := \max\{\alpha^\star, \beta^\star, \gamma^\star\}$.
3. Für $\delta \in \mathrm{SC} \cap \mathrm{K}$ sei $\delta^\star := \delta$.

Lemma 25.2.5. *Für $\delta > \mathrm{K}^\Gamma$ ist $\delta^\star < \mathrm{K}$.*

Im folgenden bezeichnen $\alpha, \beta, \gamma, \delta, \eta, \mu, \nu$ stets Ordinalzahlen $< \mathrm{K}^\Gamma$ und π, σ Ordinalzahlen aus $\mathrm{R} \cap \mathrm{K}$.

25.3 Die Ordinalzahlenmengen M^α und Ordinalzahlen $\Xi(\alpha)$

Induktive Definition der Ordinalzahlenmengen $\mathrm{B}_n(\alpha,\beta), \mathrm{B}(\alpha,\beta), \mathrm{M}^\alpha$ und Ordinalzahlen $\Xi(\alpha)$.

(B1) $0, \mathrm{K} \in \mathrm{B}_n(\alpha,\beta)$ und $\eta \in \mathrm{B}_n(\alpha,\beta)$ für alle $\eta < \beta$.
(B2) Gilt $\delta =_{\mathrm{NF}} \varphi\delta_1\delta_2 + \delta_3$ mit $\delta_1, \delta_2, \delta_3 \in \mathrm{B}_n(\alpha,\beta)$, so sei $\delta \in \mathrm{B}_{n+1}(\alpha,\beta)$.
(B3) Ist $\delta < \Omega_\delta < \mathrm{K}$ und $\delta \in \mathrm{B}_n(\alpha,\beta)$, so sei $\Omega_\delta \in \mathrm{B}_{n+1}(\alpha,\beta)$.
(B4) Ist $\delta < \pi$ und $\pi \in \mathrm{B}_n(\alpha,\beta)$, so sei $\delta \in \mathrm{B}_{n+1}(\alpha,\beta)$.
(B5) Ist $0 < \delta < \alpha$ und $\delta \in \mathrm{B}_n(\alpha,\beta)$, so sei $\Xi(\delta) \in \mathrm{B}_{n+1}(\alpha,\beta)$.
(B6) $B(\alpha,\beta) := \bigcup\{\mathrm{B}_n(\alpha,\beta) : n < \omega\}$.
(B7) M^0 sei die Menge der Limeszahlen $< \mathrm{K}$. Für $\alpha > 0$ sei M^α die Menge der $\pi \in \mathrm{R} \cap \mathrm{K}$ mit $\alpha \in \mathrm{B}(\alpha,\pi)$, $\mathrm{B}(\alpha,\pi) \cap \mathrm{K} < \pi$ und M^μ stationär in π für alle $\mu \in \mathrm{B}(\alpha,\pi) \cap \alpha$.
(B8) $\Xi(\alpha) := \min(\mathrm{M}^\alpha \cup \{\mathrm{K}\})$.

Lemma 25.3.1.
a) *Für* $\alpha \leq \gamma$ *und* $\beta \leq \delta$ *gilt* $\mathrm{B}(\alpha,\beta) \subseteq \mathrm{B}(\gamma,\delta)$.
b) *Für jede Limeszahl* α *ist* $\mathrm{B}(\alpha,\beta) = \bigcup\{\mathrm{B}(\eta,\beta) : \pi < \alpha\}$.
c) *Für jede Limeszahl* β *ist* $\mathrm{B}(\alpha,\beta) = \bigcup\{\mathrm{B}(\alpha,\eta) : \eta < \beta\}$.
d) *Für* $\beta < \mathrm{K}$ *hat* $\mathrm{B}(\alpha,\beta)$ *eine Kardinalität* $< \mathrm{K}$.

Lemma 25.3.2.
a) M^α *ist stationär in* K.
b) *Für* $\alpha > 0$ *ist* $\Xi(\alpha)$ *eine unerreichbare Ordinalzahl aus* M^α.
c) $\alpha \in \mathrm{B}(\alpha,\Xi(\alpha))$ *und* $\mathrm{B}(\alpha,\Xi(\alpha)) \cap \mathrm{K} < \Xi(\alpha) < \mathrm{K}$.
d) *Für* $\alpha < \mathrm{K}$ *ist* $\alpha < \Xi(\alpha)$, *und für* $\mathrm{K} \leq \alpha$ *ist* $\Xi(\alpha) < \alpha$.
e) *Aus* $\alpha < \beta$ *und* $\alpha \in \mathrm{B}(\beta,\Xi(\beta))$ *folgt* $\Xi(\alpha) < \Xi(\beta)$.
f) *Aus* $\Xi(\alpha) = \Xi(\beta)$ *folgt* $\alpha = \beta$.
g) *Aus* $\pi \in \mathrm{M}^\alpha$ *und* $\mu \in \mathrm{B}(\alpha,\pi) \cap \alpha$ *folgt* $\pi \in \mathrm{M}^\mu$.

Lemma 25.3.3.
a) *Für* $\delta =_{NF} \varphi\delta_1\delta_2\delta_3$ *gilt* $\delta \in \mathrm{B}(\alpha,\beta)$ *genau dann, wenn* $\delta_1,\delta_2,\delta_3 \in \mathrm{B}(\alpha,\beta)$ *sind.*
b) *Für* $\delta \in \Omega_\delta < \mathrm{K}$ *gilt* $\Omega_\delta \in \mathrm{B}(\alpha,\beta)$ *genau dann, wenn* $\delta \in \mathrm{B}(\alpha,\beta)$ *ist.*
c) $\delta \in \mathrm{B}(\alpha,\Xi(\alpha))$ *gilt genau dann, wenn* $\delta^\star \in \Xi(\alpha)$ *ist.*
d) $\alpha^\star \in \Xi(\alpha)$.

Definition der *Höhe* $\mathrm{h}(\pi)$ von $\pi \in \mathrm{R} \cap \mathrm{K}$. $\mathrm{h}(\pi) := \sup\{\eta < \mathrm{K}^\Gamma : \pi \in \mathrm{M}^\eta\}$.

Lemma 25.3.4. *Für alle* $\pi \in \mathrm{R} \cap \mathrm{K}$ *gilt:*
a) $\mathrm{h}(\pi) \leq \mathrm{K}^\Gamma$.
b) $\mathrm{h}(\pi) > 0$ *genau dann, wenn* π *unerreichbar ist.*

Lemma 25.3.5.
a) *Für* $\mu \in \mathrm{B}(\mathrm{h}(\pi),\pi) \cap \mathrm{h}(\pi)$ *ist* $\pi \in \mathrm{M}^\mu$ *und* M^μ *stationär in* π.
b) *Ist* $\pi \in \mathrm{M}^\alpha$ *und* M^α *nicht staionär in* π, *so ist* $\mathrm{h}(\pi) = \alpha$.
c) *Für* $\alpha > 0$ *ist* $\mathrm{h}(\Xi(\alpha)) = \alpha$.

Beweis. a) Ist $\mathrm{h}(\pi) < \mathrm{K}^\Gamma$ und $\pi \in \mathrm{M}^{\mathrm{h}(\pi)}$, so gilt die Behauptung.
Sonst ist $\mathrm{h}(\pi)$ eine Limeszahl und man hat $\eta < \mathrm{h}(\pi)$ mit $\pi \in \mathrm{M}^\eta$ und $\mu \in \mathrm{B}(\eta,\pi) \cap \eta$, womit sich ebenfalls die Behauptung ergibt.
b) Da M^α nicht stationär in π ist, hat man $\alpha > 0$. Dann ist $\alpha \in \mathrm{B}(\alpha,\pi)$. Für alle $\beta > \alpha$ folgt $\alpha \in \mathrm{B}(\beta,\pi) \cap \beta$. Wäre $\pi \in \mathrm{M}^\beta$, so würde folgen, daß M^α stationär in π ist im Widerspruch zur Voraussetzung. Somit ist in diesem Fall $\mathrm{h}(\pi) = \alpha$.
c) folgt aus b). □

25.4 Die Ordinalzahlen $\psi_\pi^\mu(\alpha)$ und $\psi\Omega_{\gamma+1}\alpha$

Induktive Definition der Ordinalzahlenmengen $\mathrm{C}_n(\alpha,\beta)$, $\mathrm{C}(\alpha,\beta)$ und Ordinalzahlen $\psi_\pi^\mu(\alpha)$, $\psi_{\gamma+1}^\Omega\alpha$ für $\mu \in \mathrm{B}(\mathrm{h}(\pi),\pi) \cap \mathrm{h}(\pi), \gamma < \mathrm{K}$.

(C1) $0, \mathrm{K} \in \mathrm{C}_n(\alpha,\beta)$ und $\eta \in \mathrm{C}_n(\alpha,\beta)$ für alle $\eta < \beta$.

(C2) Gilt $\delta =_{\mathrm{NF}} \varphi\delta_1\delta_2 + \delta_3$ mit $\delta_1,\delta_2,\delta_3 \in \mathrm{C}_n(\alpha,\beta)$, so sei $\delta \in \mathrm{C}_{n+1}(\alpha,\beta)$.

(C3) Ist $\delta < \Omega_\delta < \mathrm{K}$ und $\delta \in \mathrm{C}_n(\alpha,\beta)$, so sei $\Omega_\delta \in \mathrm{C}_{n+1}(\alpha,\beta)$.

(C4) Ist $0 < \delta \in \mathrm{C}_n(\alpha,\beta)$, so sei $\Xi(\delta) \in \mathrm{C}_{n+1}(\alpha,\beta)$.

(C5) Sind $\delta,\nu,\sigma \in \mathrm{C}_n(\alpha,\beta)$ mit $\delta < \alpha, \nu \in \mathrm{B}(\mathrm{h}(\sigma),\sigma) \cap \mathrm{h}(\sigma)$ und $\delta,\sigma \in \mathrm{C}(\delta,\sigma)$, so sei $\psi_\sigma^\nu(\delta) \in \mathrm{C}_{n+1}(\alpha,\beta)$.

(C6) Sind $\delta,\eta \in \mathrm{C}_n(\alpha,\beta)$ mit $\delta < \alpha, \eta < \mathrm{K}$ und $\delta \in \mathrm{C}(\delta,\Omega_{\eta+1})$, so sei $\psi\Omega_{\eta+1}\delta \in \mathrm{C}_{n+1}(\alpha,\beta)$.

(C7) $\mathrm{C}(\alpha,\beta) := \bigcup\{\mathrm{C}_n(\alpha,\beta) : n < \omega\}$.

(C8) $\psi_\pi^\mu(\alpha) := \min(\{\eta \in \mathrm{M}^\mu : \alpha,\pi \in \mathrm{C}(\alpha,\eta), \mathrm{C}(\alpha,\eta) \cap \pi < \eta\} \cup \{\pi\})$.

(C9) $\psi\Omega_{\gamma+1}\alpha := \min(\{\eta : \eta \notin \mathrm{C}(\alpha,\Omega_{\gamma+1})\})$.

Lemma 25.4.1.

a) *Für* $\alpha \leq \gamma$ *und* $\beta \leq \delta$ *gilt* $\mathrm{C}(\alpha,\beta) \subseteq \mathrm{C}(\gamma,\delta)$.

b) *Für jede Limeszahl* α *ist* $\mathrm{C}(\alpha,\beta) = \bigcup\{\mathrm{C}(\eta,\beta) : \eta < \alpha\}$.

c) *Für jede Limeszahl* β *ist* $\mathrm{C}(\alpha,\beta) = \bigcup\{\mathrm{C}(\alpha,\eta) : \eta < \beta\}$.

d) *Für* $\beta < \pi$ *hat* $\mathrm{C}(\alpha,\beta)$ *eine Kardinalität* $< \pi$.

Lemma 25.4.2. *Für* $\gamma = \psi_\pi^\mu(\alpha)$ *mit* $\mu \in \mathrm{B}(\mathrm{h}(\pi),\pi) \cap \mathrm{h}(\pi)$ *und* $\alpha,\pi \in \mathrm{C}(\alpha,\pi)$ *gilt:*

a) $\alpha,\pi \in \mathrm{C}(\alpha,\gamma)$ *und* $\mathrm{C}(\alpha,\gamma) \cap \pi < \gamma < \pi$.

b) *Im Fall* $\mu > 0$ *ist* $\gamma \in \mathrm{M}^\mu$ *und* M^μ *nicht stationär in* γ, *also* $\gamma \in \mathrm{R} \cap \mathrm{K}$ *mit* $\mathrm{h}(\gamma) = \mu$.

c) *Im Fall* $\mu = 0$ *ist* $\Omega_\gamma = \gamma \notin \mathrm{R}$.

d) $\gamma \neq \Xi(\eta)$ *und* $\gamma \notin \{\alpha,\mu,\pi\}$.

Lemma 25.4.3. *Aus* $\psi_\pi^\mu(\alpha) = \psi_\sigma^\nu(\beta)$ *mit* $\mu \in \mathrm{B}(\mathrm{h}(\pi),\pi) \cap \mathrm{h}(\pi), \alpha,\pi \in \mathrm{C}(\alpha,\pi)$, $\nu \in \mathrm{B}(\mathrm{h}(\sigma),\sigma) \cap \mathrm{h}(\sigma)$ *und* $\beta,\sigma \in \mathrm{C}(\beta,\sigma)$ *folgt* $\alpha = \beta, \mu = \nu$ *und* $\pi = \sigma$.

Lemma 25.4.4. *Für* $\gamma,\delta < \mathrm{K}$ *gilt:*

a) $\Omega_\gamma < \psi\Omega_{\gamma+1}\alpha < \Omega_{\gamma+1}$ *mit* $\psi\Omega_{\gamma+1}\alpha \in \mathrm{SC}$.

b) *Aus* $\psi\Omega_{\gamma+1}\alpha = \psi\Omega_{\delta+1}\beta$ *folgt* $\gamma = \delta$.

c) *Aus* $\alpha \in \mathrm{C}(\alpha,\Omega_\gamma + 1)$ *folgt* $\mathrm{C}(\alpha,\psi\Omega_{\gamma+1}\alpha) \cap \Omega_{\gamma+1} < \psi\Omega_{\gamma+1}\alpha \neq \alpha$.

d) *Aus* $\alpha \in \mathrm{C}(\alpha,\Omega_\gamma + 1)$ *und* $\alpha < \beta$ *folgt* $\psi\Omega_{\gamma+1}\alpha < \psi\Omega_{\gamma+1}\beta$.

e) *Aus* $\psi\Omega_{\gamma+1}\alpha = \psi\Omega_{\gamma+1}\beta$ *mit* $\alpha \in \mathrm{C}(\alpha,\Omega_\gamma + 1)$ *und* $\beta \in \mathrm{C}(\beta,\Omega_\gamma + 1)$ *folgt* $\alpha = \beta$.

Lemma 25.4.5.

a) *Für* $\delta =_{NF} \varphi\delta_1\delta_2 + \delta_3$ *gilt* $\delta \in \mathrm{C}(\alpha,\beta)$ *genau dann, wenn* $\delta_1,\delta_2,\delta_3 \in \mathrm{C}(\alpha,\beta)$ *sind.*

b) *Für* $\delta < \Omega_\delta < \mathrm{K}$ *gilt* $\Omega_\delta \in \mathrm{C}(\alpha,\beta)$ *genau dann, wenn* $\delta \in \mathrm{C}(\alpha,\beta)$ *ist.*

c) *Aus* $\delta^\star < \beta$ *folgt* $\delta \in \mathrm{C}(\alpha,\beta)$.

25.5 Die Ordinalzahlenmenge T(K)

Induktive Definition der Ordinalzahlenmenge T(K) und des *Grades* $\mathrm{gr}(\delta)$ einer Ordinalzahl $\delta \in \mathrm{T(K)}$.

(T1) $0, \mathrm{K} \in \mathrm{T(K)}$ mit $\mathrm{gr}(0) = \mathrm{gr(K)} := 0$.

(T2) Gilt $\delta =_{\mathrm{NF}} \varphi\alpha\beta + \gamma < \varepsilon_{\mathrm{K}+1}$ mit $\alpha, \beta, \gamma \in \mathrm{T(K)}$, so sei $\delta \in \mathrm{T(K)}$ mit $\mathrm{gr}(\delta) := \max\{\mathrm{gr}(\alpha), \mathrm{gr}(\beta), \mathrm{gr}(\gamma)\} + 1$.

(T3) Ist $\alpha < \Omega_\alpha < \mathrm{K}$ und $\alpha \in \mathrm{T(K)}$, so sei $\Omega_\alpha \in \mathrm{T(K)}$ mit $\mathrm{gr}(\Omega_\alpha) := \mathrm{gr}(\alpha) + 1$.

(T4) Ist $0 < \alpha \in \mathrm{T(K)}$, so sei $\Xi(\alpha) \in \mathrm{T(K)}$ mit $\mathrm{gr}(\Xi(\alpha)) := \mathrm{gr}(\alpha) + 1$.

(T5) Sind $\alpha, \mu, \pi \in \mathrm{T(K)}$ mit $\mu \in \mathrm{B}(\mathrm{h}(\pi), \pi) \cap \mathrm{h}(\pi)$ und $\alpha, \pi \in \mathrm{C}(\alpha, \pi) \cap \mathrm{K}$, so sei $\psi_\pi^\mu(\alpha) \in \mathrm{T(K)}$ mit $\mathrm{gr}(\psi_\pi^\mu(\alpha)) := \max\{\mathrm{gr}(\alpha), \mathrm{gr}(\mu), \mathrm{gr}(\pi)\} + 1$.

(T6) Sind $\alpha, \Omega_{\gamma+1} \in \mathrm{T(K)} \cap \mathrm{K}$ mit $\alpha \in \mathrm{C}(\alpha, \Omega_\gamma + 1)$, so sei $\psi\Omega_{\gamma+1}\alpha \in \mathrm{T(K)}$ mit $\mathrm{gr}(\psi\Omega_{\gamma+1}\alpha) := \max\{\mathrm{gr}(\alpha), \mathrm{gr}(\Omega_{\gamma+1})\} + 1$.

Aus den Lemmata der Abschnitte 25.2 bis 25.4 geht hervor, daß jede Ordinalzahl aus T(K) auf genau eine Weise durch (T1) bis (T6) bestimmt ist. Daher ist der Grad $\mathrm{gr}(\delta)$ jeder Ordinalzahl $\delta \in \mathrm{T(K)}$ eindeutig bestimmt.

Lemma 25.5.1. $\varepsilon_{\mathrm{K}+1}$ *ist die kleinste Ordinalzahl, die größer als jede Ordinalzahl aus* T(K) *ist.*

Lemma 25.5.2. *Für* $\sigma \in \mathrm{T(K)} \cap \mathrm{R} \cap \mathrm{K}$ *kommen genau folgende drei Fälle in Betracht.*

a) $\sigma = \Xi(\alpha)$ *mit* $0 < \alpha \in \mathrm{T(K)}$ *und* $\mathrm{h}(\sigma) = \alpha$.

b) $\sigma = \psi_\pi^\mu(\alpha)$ *mit* $\alpha, \mu, \pi \in \mathrm{T(K)}$, $0 < \mu \in \mathrm{B}(\mathrm{h}(\pi), \pi) \cap \mathrm{h}(\pi)$, $\alpha, \pi \in \mathrm{C}(\alpha, \pi) \cap \mathrm{K}$ *und* $\mathrm{h}(\sigma) = \mu$.

c) $\sigma = \Omega_{\gamma+1}$ *mit* $\gamma \in \mathrm{T(K)} \cap \mathrm{K}$ *und* $\mathrm{h}(\sigma) = 0$.

Lemma 25.5.3. *Für alle* $\delta \in \mathrm{T(K)} \cap \mathrm{K}$ *ist* $\delta < \Xi(\varepsilon_{\mathrm{K}+1})$.

Beweis durch Induktion nach $\mathrm{gr}(\delta)$. Gilt $\delta \in \mathrm{T(K)} \cap \mathrm{K}$ nach (T4), so hat man $\delta = \Xi(\alpha)$ und nach I.V. $\alpha^\star < \Xi(\varepsilon_{\mathrm{K}+1})$. Es folgt $\alpha \in \mathrm{B}(\varepsilon_{\mathrm{K}+1}, \Xi(\varepsilon_{\mathrm{K}+1}))$. Mit $\alpha < \varepsilon_{\mathrm{K}+1}$ folgt $\delta = \Xi(\alpha) < \Xi(\varepsilon_{\mathrm{K}+1})$. In allen anderen Fällen gilt die Behauptung trivialerweise oder folgt unmittelbar aus der I.V. □

Lemma 25.5.4. *Für alle* $\delta \in \mathrm{T(K)}$ *gilt* $\delta^\star < \psi^0_{\Xi(\varepsilon_{\mathrm{K}+1})}(0)$.

Beweis durch Induktion nach $\mathrm{gr}(\delta)$.
Gilt $\delta \in \mathrm{T(K)}$ nach (T4), so hat man $\delta = \Xi(\alpha)$ und nach I.V. $\alpha^\star < \psi^0_{\Xi(\varepsilon_{\mathrm{K}+1})}(0)$. Es folgt $\alpha \in \mathrm{C}(0, \psi^0_{\Xi(\varepsilon_{\mathrm{K}+1})}(0))$ und $\delta = \Xi(\alpha) \in \mathrm{C}(0, \psi^0_{\Xi(\varepsilon_{\mathrm{K}+1})}(0))$. Mit Lemma 25.5.3 folgt $\delta^\star = \delta \in \mathrm{C}(0, \psi^0_{\Xi(\varepsilon_{\mathrm{K}+1})}(0)) \cap \Xi(\varepsilon_{\mathrm{K}+1}) < \psi^0_{\Xi(\varepsilon_{\mathrm{K}+1})}(0)$. In allen anderen Fällen gilt die Behauptung trivialerweise oder folgt unmittelbar aus der I.V. □

Lemma 25.5.5. *Für alle* $\delta \in \mathrm{T(K)}$ *gilt* $\delta \in \mathrm{C}(\psi^0_{\Xi(\varepsilon_{\mathrm{K}+1})}(0), 1)$.

Beweis durch Induktion nach $\mathrm{gr}(\delta)$ mit Lemma 5.4. □

Lemma 25.5.6. *Aus* $\delta \in \mathrm{C}_n(\psi^0_\Xi(\varepsilon_{\mathrm{K}+1})(0), 1)$ *mit* $\delta^\star < \psi^0_{\Xi(\varepsilon_{\mathrm{K}+1})}(0)$ *folgt* $\delta \in \mathrm{T(K)}$.

Beweis durch Indukton nach n. □

Lemma 25.5.7. $\mathrm{T(K)} \cap \Omega_1$ *ist gleich dem Abschnitt aller Ordinalzahlen* $< \psi\Omega_1(\psi^0_{\Xi(\varepsilon_{\mathrm{K}+1})}(0))$.

Beweis mit den Lemmata 25.5.4, 25.5.5 und 25.5.6. □

Im folgenden bezeichnen $\alpha, \beta, \gamma, \delta\eta, \mu, \nu, \xi, \pi, \rho, \sigma$ stets Ordinalzahlen aus T(K), wobei $\pi, \sigma \in \mathrm{R} \cap \mathrm{K}$ sind.

25.6 Das geschichtete halbformale System RS(K)

Grundzeichen des Systems RS(K):
1. Abzählbar undendlich viele gebundene Variablen.
2. Bezeichnungen der Ordinalzahlen aus T(K).
3. $\bot, \in, \mathrm{Ad}, \rightarrow, \exists, \mathrm{L}$, Doppelpunkt, runde und geschweite Klammern.

Nennformen seien in RS(K) entsprechend wie in KP+ Π_3-Ref definiert, wobei jedoch in keiner Nennform Nennzeichen zwischen geschweiften Klammern auftreten sollen.

Induktive Definition der *Terme* des Systems RS(K), der *Schicht* St und *Komponentenmenge* k(t) eines Terms t, der *Formeln* des Systems RS(K) und der *Komponentenmenge* k(F) einer Formel F.
1. Für $\alpha \leq \mathrm{K}$ ist $\mathrm{L}\alpha$ ein Term der Schicht α mit der Komponentenmenge $\{\alpha\}$.
2. Ist $0 < \alpha < \mathrm{K}$, $\alpha \notin \mathrm{R}$, $\mathfrak{A}$ eine 1-stellige Nennform derart, daß $\mathfrak{A}[\mathrm{L}0]$ eine Formel mit $\mathrm{k}(\mathfrak{A}[\mathrm{L}0]) < \alpha$ ist, und x eine gebundene Variable, die in $\mathfrak{A}$ nicht auftritt, so ist $\{x \in \mathrm{L}\alpha : \mathfrak{A}[x]\}$ ein Term der Schicht α mit der Komponentenmenge $\mathrm{k}(\mathfrak{A}[\mathrm{L}0]) \cup \{\alpha\}$.
3. $\bot$ ist eine Formel mit leerer Komponentenmenge.
4. Sind s und t Terme von Schichten $< \mathrm{K}$, so ist $(s \in t)$ eine Formel mit der Komponentenmenge $\mathrm{k}(s) \cup \mathrm{k}(t)$.
5. Ist t ein Term von einer Schicht $< \mathrm{K}$, so ist $\mathrm{Ad}^\mu(t)$ eine Formel mit der Komponentenmenge $\mathrm{k}(t)$.
6. Sind A und B Formeln, so ist $(A \rightarrow B)$ eine Formel mit der Komponentenmenge $\mathrm{k}(A) \cup \mathrm{k}(B)$.
7. Ist t ein Term, $\mathfrak{A}$ eine 1-stellige Nennform derart, daß $\mathfrak{A}[\mathrm{L}0]$ eine Formel ist, und x eine gebundene Variable, die weder in t noch in $\mathfrak{A}$ auftritt, so ist $(\exists x \in t)\mathfrak{A}[x]$ eine Formel mit der Komponentenmenge $\mathrm{k}(t) \cup \mathrm{k}(\mathfrak{A}[\mathrm{L}0])$.

Eine Formel heißt *einfach*, wenn sie nicht die Gestalt $(A \rightarrow B)$ hat.

Als *Mitteilungszeichen* für Terme verwenden wir s, t, u (auch mit Indizes). Im übrigen verwenden wir in RS(K) die entsprechenden Mitteilungszeichen wie in KP+ Π_3-Ref. P-Formen, N-Formen, $F \overset{S}{\vdash} G$ $\neg A$, $(A \vee B)$, $(A \wedge B)$, $(\forall x \in t)\mathfrak{A}[x]$, $(s = t)$ für $\mathrm{S}s, \mathrm{S}t < \mathrm{K}$ und $\mathrm{Tran}[t]$ für $\mathrm{S}t < \mathrm{K}$ seien in RS(K) entsprechend wie in KP+ Π_3-Ref definiert.

Induktive Definition des *Ranges* $\mathrm{rn}(F)$ einer Formel F.
1. $\mathrm{rn}(\bot) := 0$.
2. $\mathrm{rn}(s \in t) := \max\{\omega \cdot \mathrm{S}s + 6, \omega \cdot \mathrm{S}t + 1\}$ für $\mathrm{S}s, \mathrm{S}t < \mathrm{K}$.
3. $\mathrm{rn}(\mathrm{Ad}^{\mu}(t)) := \omega \cdot \mathrm{S}t + 5$ für $\mathrm{S}t < \mathrm{K}$.
4. $\mathrm{rn}(\neg A) := \mathrm{rn}(A)$.
5. $\mathrm{rn}(A \rightarrow B) := \max\{\mathrm{rn}(A), \mathrm{rn}(B)\} + 1$, wenn B nicht die Formel $\bot$ ist.
6. $\mathrm{rn}((\exists x \in t)\mathfrak{A}[x]) := \max\{\omega \cdot \mathrm{S}t, \mathrm{rn}(\mathfrak{A}[\mathrm{L}0]) + 2\}$.

Lemma 25.6.1. *Jede einfache Formel vom Rang* $\alpha \in \mathrm{R}$ *ist eine Formel* $(\exists x \in \mathrm{L}\alpha)\mathfrak{A}[x]$ *mit* $\mathrm{k}(\mathfrak{A}[\mathrm{L}0]) < \alpha$.

Definition der Formel $\mathrm{t}(u)$ für $\mathrm{S}u < \mathrm{S}t$.
1. $t(u) := \neg(u \in \mathrm{L}0)$, wenn t ein Term L_{α} ist.
2. $t(u) := (\neg(u \in \mathrm{L}_0) \wedge \mathfrak{A}[u])$, wenn t ein Term $\{x \in \mathrm{L}\alpha : \mathfrak{A}[x]\}$ ist.

Lemma 25.6.2.
a) $\mathrm{rn}(t(u) \wedge (s = u)) < \mathrm{rn}(s \in t)$ *für* $\mathrm{S}s < \mathrm{K}$ *und* $\mathrm{S}u < \mathrm{S}t < \mathrm{K}$.
b) $\mathrm{rn}(\mathrm{L}\eta = t) < \mathrm{rn}(\mathrm{Ad}^{\mu}(t))$ *für* $\eta \leq \mathrm{S}t < \mathrm{K}$.
c) $\mathrm{rn}(t(u) \wedge \mathfrak{A}[u]) < \mathrm{rn}((\exists x \in t)\mathfrak{A}[x])$ *für* $\mathrm{S}u < \mathrm{S}t$.

Axiome des Systems RS(K):

(Ax.$\bot$) $\mathfrak{N}[\bot]$.
(Ax.$\in$) $\mathfrak{N}[(s \in \mathrm{L}_0)]$ für $\mathrm{S}s < \mathrm{K}$.
(Ax.Ad^{μ}) $\mathfrak{N}[\mathrm{Ad}^{\mu}(t)]$ für $\mathrm{S}t < \Xi(\mu)$.
(Ax.$\exists$) $\mathfrak{N}[(\exists x \in \mathrm{L}_0)\mathfrak{A}[x]]$.

Hauptschlüsse des Systems RS(K):

($\rightarrow$) $A \vee \mathfrak{N}[(A \rightarrow B)], B \rightarrow \mathfrak{N}[(A \rightarrow B)] \vdash \mathfrak{N}[(A \rightarrow B)]$,
wenn B nicht die Formel $\bot$ ist.
Dieser Hauptschluß hat die *Indexmenge* $\{\mathrm{L}0, \mathrm{L}1\}$.

(P. $\in$) $(t(u) \wedge (s = u)) \vee \mathfrak{P}[(s \in t)] \vdash \mathfrak{P}[(s \in t)]$,
wenn $\mathrm{S}s < K$ und $\mathrm{S}u < \mathrm{S}t < \mathrm{K}$ ist.
Dieser Hauptschluß hat die *Indexmenge* $\{u\}$.

(P.Ad^{μ}) $(\mathrm{L}\eta = t) \vee \mathfrak{P}[\mathrm{Ad}^{\mu}(t)] \vdash \mathfrak{P}[\mathrm{Ad}^{\mu}(t)]$,
wenn $\eta \leq \mathrm{S}t < \mathrm{K}$ mit $\eta \in \mathrm{M}^{\mu}$ ist.
Dieser Hauptschluß hat die *Indexmenge* $\{\mathrm{L}\eta\}$.

(P.$\exists$) $(t(u) \wedge \mathfrak{A}[u]) \vee \mathfrak{P}[(\exists x \in t)\mathfrak{A}[x]] \vdash \mathfrak{P}[(\exists x \in t)\mathfrak{A}[x]]$,
wenn $\mathrm{S}u < \mathrm{S}t$ ist.
Dieser Hauptschluß hat die *Indexmenge* $\{u\}$.

(Ref_{K}) $(\forall x \in \text{LK})(\exists y \in \text{LK})(\forall z \in \text{LK})\mathfrak{C}[x,y,z] \vee \mathfrak{P}[C] \vdash \mathfrak{P}[C]$,
wenn $\text{k}(\mathfrak{C}[\text{L}0,\text{L}0,\text{L}0]) < \text{K}$ und C die Formel
$(\exists w \in \text{LK})(\text{Tran}[w] \wedge (\exists x \in w)(x \in w) \wedge$
$(\forall x \in w)(\exists y \in w)(\forall z \in w)\mathfrak{C}[x,y,z])$ ist.
Dieser Hauptschluß hat die *Indexmenge* $\{\text{LK}\}$.

(Ref_π^μ) $(\forall y \in \text{L}x)(\exists z \in \text{L}\pi)\mathfrak{C}[t,y,z] \vee \mathfrak{P}[C] \vdash \mathfrak{P}[C]$,
wenn $\text{k}(\mathfrak{C}[t,\text{L}0,\text{L}0]) < \mu$ und C eine Formel
$(\exists w \in \text{L}\pi)(\text{Ad}^\mu(w)) \wedge (\exists w \in w)(\forall y \in w)(\exists z \in w)\mathfrak{C}[x,y,z])$ mit
$\mu \in \text{B}(\text{h}(\pi),\pi) \cap \text{h}(\pi)$ oder $\mu = \text{h}(\pi) = 0$ ist.
Dieser Hauptschluß hat die *Indexmenge* $\{\text{L}\pi\}$.

(N. $\in$) $(t(u) \wedge (s = u)) \to \mathfrak{N}[(s \in t)]$ für alle u mit $\text{S}u < \text{S}t$
$\vdash \mathfrak{N}[(s \in t)]$, wenn $\text{S}s < \text{K}$ und $0 < \text{S}t < \text{K}$ ist.
Dieser Hauptschluß hat die *Indexmenge* $\{u : \text{S}u < \text{S}t\}$.

(N.Ad^μ) $(\text{L}\eta = t) \to \mathfrak{N}[\text{Ad}^\mu(t)]$ für alle $\eta \leq \text{S}t$ mit $\eta \in \text{M}^\mu$
$\vdash \mathfrak{N}[\text{Ad}^\mu(t)]$, wenn $\Xi(\mu) \leq \text{S}t < \text{K}$ ist.
Dieser Hauptschluß hat die *Indexmenge* $\{\text{L}\eta : \text{S}t \geq \eta \in \text{M}^\mu\}$.

(N.$\exists$) $(t(u) \wedge \mathfrak{A}[u]) \to \mathfrak{N}[(\exists x \in t)\mathfrak{A}[x]]$ für alle u mit $\text{S}u < \text{S}t$
$\vdash \mathfrak{N}[(\exists x \in t)\mathfrak{A}[x]]$, wenn $\text{S}t > 0$ ist.
Dieser Hauptschluß hat die *Indexmenge* $\{u : \text{S}u < \text{S}t\}$.

Der bezeichnete minimale Positiv- oder Negativteil der Konklusion eines Hauptschlusses heißt der *Hauptteil* des betreffenden Hauptschlusses. Die Prämissen eines Hauptschlusses mit der Indexmenge J haben die *Indizes* $u \in J$.

Schnitte des Systems RS(K):
$C \vee F, C \to F \vdash F$.

Die mit C bezeichnete Formel in den Prämissen eines Schnittes heißt die *Schnittformel* des betreffenden Schnittes. Der *Rang* eines Schnittes ist der Rang $\text{rn}(c)$ seiner Schnittformel C.

25.7 H-kontrollierte Herleitungen

Im folgenden bezeichnen X und Y endliche Teilmengen von T(K).

Definition.

$$\text{H}_\gamma^\beta(X) := \begin{cases} \bigcap\{\text{B}(\xi,\eta) : & X \subset \text{B}(\xi,\eta) \text{ und } \beta < \xi\} \cap \\ \bigcap\{\text{C}(\xi,\eta) : & X \subset \text{C}(\xi,\eta) \text{ und } \gamma < \xi\} \cap \text{T(K)}. \end{cases}$$

Lemma 25.7.1.
a) *Für* $\beta \leq \mu, \gamma \leq \nu$ *und* $X \subseteq Y$ *gilt* $\text{H}_\gamma^\beta(X) \subseteq \text{H}_\nu^\mu(Y)$.
b) $X \subset \text{H}_\gamma^\beta(X)$.

c) *Aus* $Y \subset \mathrm{H}^{\beta}_{\gamma}(X)$ *folgt* $\mathrm{H}^{\beta}_{\gamma}(X \cup Y) = \mathrm{H}^{\beta}_{\gamma}(X)$.
d) $0, \mathrm{K} \in \mathrm{H}^{\beta}_{\gamma}(X)$.
e) *Für* $\delta =_{NF} \varphi\delta_1\delta_2\delta_3$ *gilt* $\delta \in \mathrm{H}^{\beta}_{\gamma}(X)$ *genau dann, wenn* $\delta_1, \delta_2, \delta_3 \in \mathrm{H}^{\beta}_{\gamma}(X)$ *sind.*
f) *Für* $\delta < \Omega_\delta < \mathrm{K}$ *gilt* $\Omega_\delta \in \mathrm{H}^{\beta}_{\gamma}(X)$ *genau dann, wenn* $\delta \in \mathrm{H}^{\beta}_{\gamma}(X)$ *ist.*

Lemma 25.7.2.
a) *Aus* $\alpha \in \mathrm{H}^{\beta}_{\gamma}(X)$ *mit* $\alpha \leq \beta$ *folgt* $\Xi(\alpha) \in \mathrm{H}^{\beta}_{\gamma}(X)$.
b) *Aus* $\alpha, \mu, \pi \in \mathrm{H}^{\beta}_{\gamma}(X)$ *mit* $\mu \in \mathrm{B}(\mathrm{h}(\pi), \pi) \cap \mathrm{h}(\pi), \alpha, \pi \in \mathrm{C}(\alpha, \pi) \cap \mathrm{K}$ *und* $\alpha \leq \gamma$ *folgt* $\psi^{\mu}_{\pi}(\alpha) \in \mathrm{H}^{\beta}_{\gamma}(X)$.
c) *Aus* $\alpha, \delta \in \mathrm{H}^{\beta}_{\gamma}(X) \cap \mathrm{K}$ *mit* $\alpha \in \mathrm{C}(\alpha, \Omega_\delta + 1)$ *und* $\alpha \leq \gamma$ *folgt* $\psi\Omega_{\delta+1}\alpha \in \mathrm{H}^{\beta}_{\gamma}(X)$.

Induktive Definition von $\mathrm{H}^{\beta}_{\gamma}(X) \left|\frac{\alpha}{\rho}\right. F$ (durch Induktion nach α).
$\mathrm{H}^{\beta}_{\gamma}(X) \left|\frac{\alpha}{\rho}\right. F$ gelte genau dann, wenn $\{\alpha\} \cup \mathrm{k}(F) \subset \mathrm{H}^{\beta}_{\gamma}(X)$ gilt und einer der folgenden sechs Fälle vorliegt.

1. F ist ein Axiom des Systems RS(K).
2. F ist die Konklusion eines Hauptschlusses $(\rightarrow)$, und es gibt $f(i) < \alpha$ mit $\mathrm{H}^{\beta}_{\gamma}(X) \left|\frac{f(i)}{\rho}\right. F_i$ für die Prämissen $F_i (i = 0, 1)$.
3. F ist die Konklusion eines Hauptschlusses $(\mathrm{P.} \in), (\mathrm{P.Ad}^{\mu}), (\mathrm{P.}\exists)$ oder $(\mathrm{Ref}_{\mathrm{K}})$ mit der Indexmenge $\{u\}$, und es gibt δ mit $\mathrm{S}u \leq \delta < \alpha$ und $\mathrm{H}^{\beta}_{\gamma}(X) \left|\frac{\delta}{\rho}\right. F_u$ für die Prämisse F_u.
4. F ist die Konklusion eines Hauptschlusses $(\mathrm{Ref}^{\mu}_{\pi})$ mit $\mu \in \mathrm{H}^{\beta}_{\gamma}(X)$, und es gibt δ mit $\pi < \delta + 1 < \alpha$ und $\mathrm{H}^{\beta}_{\gamma}(X) \left|\frac{\delta}{\rho}\right. F_0$ für die Prämisse F_0.
5. F ist die Konklusion eines Hauptschlusses $(\mathrm{N.}\exists), (\mathrm{N.Ad}^{\mu})$ oder $(\mathrm{N.}\exists)$ mit der Indexmenge J, und es gibt $f(u)$ mit $\mathrm{S}u \leq f(u) < \alpha$ und $\mathrm{H}^{\beta}_{\gamma}(X \cup k(u)) \left|\frac{f(u)}{\rho}\right. F_u$ für jede Prämisse F_u vom Index $u \in J$.
6. F ist die Konklusion eines Schnittes mit der Schnittformel C von einem Rang $< \rho$, und es gibt $\delta < \alpha$ mit $\mathrm{k}(C) < \delta + 1$ und $\mathrm{H}^{\beta}_{\gamma}(X) \left|\frac{\delta}{\rho}\right. F_i$ für die Prämissen $F_i (i = 0, 1)$.

Lemma 25.7.3.
a) *Aus* $\mathrm{H}^{\beta}_{\gamma}(X) \left|\frac{\alpha}{\rho}\right. F, \beta \leq \mu, \gamma \leq \nu$ *und* $X \subseteq Y$ *folgt* $\mathrm{H}^{\mu}_{\nu}(Y) \left|\frac{\alpha}{\rho}\right. F$.
b) *Aus* $\mathrm{H}^{\beta}_{\gamma}(X) \left|\frac{\alpha}{\rho}\right. F, \alpha < \delta$ *und* $\delta \in \mathrm{H}^{\beta}_{\gamma}(X)$ *folgt* $\mathrm{H}^{\beta}_{\gamma}(X) \left|\frac{\delta}{\rho}\right. F$.
c) *Aus* $\mathrm{H}^{\beta}_{\gamma}(X) \left|\frac{\alpha}{\rho}\right. F$ *und* $\rho < \mu$ *folgt* $\mathrm{H}^{\beta}_{\gamma}(X) \left|\frac{\alpha}{\mu}\right. F$.

Lemma 25.7.4 (Strukturschlußlemma).
Aus $\mathrm{H}^{\beta}_{\gamma}(X) \left|\frac{\alpha}{\rho}\right. F, F \left|\frac{\mathrm{S}}{}\right. G$ *und* $\mathrm{k}(G) \subset \mathrm{H}^{\beta}_{\gamma}(X)$ *folgt* $\mathrm{H}^{\beta}_{\gamma}(X) \left|\frac{\alpha}{\rho}\right. G$.

Lemma 25.7.5.
a) *Aus* $\mathrm{H}^{\beta}_{\gamma}(X) \left|\frac{f(i)}{\rho}\right. \mathfrak{P}[A_i]$ *mit* $f(i) < \alpha (i = 0, 1)$ *und* $\alpha \in \mathrm{H}^{\beta}_{\gamma}(X)$ *folgt* $\mathrm{H}^{\beta}_{\gamma}(X) \left|\frac{\alpha}{\rho}\right. \mathfrak{P}[(A_0 \wedge A_1)]$.
b) *Ist* $\mathrm{S}u < \mathrm{S}t$ *und gilt* $\mathrm{H}^{\beta}_{\gamma}(X) \left|\frac{\delta}{\rho}\right. \mathfrak{N}[(t(u) \rightarrow \mathfrak{A}[u])]$ *mit* $\mathrm{S}u \leq \delta \leq \alpha$ *und* $\alpha, \mathrm{S}t \in \mathrm{H}^{\beta}_{\gamma}(X)$, *so folgt* $\mathrm{H}^{\beta}_{\gamma}(X) \left|\frac{\alpha}{\rho}\right. \mathfrak{N}[(\forall x \in t)\mathfrak{A}[x]]$.

c) *Ist* $\mathrm{S}t > 0$ *und gilt* $\mathrm{H}^\beta_\gamma(X \cup \mathrm{k}(u)) \mathrel{\vphantom{\vdash}\smash{\overset{f(u)}{\underset{\rho}{\vdash}}}} \mathfrak{P}[(t(u) \to \mathfrak{A}[u])]$ *mit* $\mathrm{S}u \leq f(u) < \alpha$ *für alle* u *mit* $\mathrm{S}u < \mathrm{S}t$ *sowie* $\alpha, \mathrm{S}t \in \mathrm{H}^\beta_\gamma(X)$, *so folgt* $\mathrm{H}^\beta_\gamma(X) \overset{\alpha}{\underset{\rho}{\vdash}} \mathfrak{P}[(\forall x \in t)\mathfrak{A}[x]]$.

Lemma 25.7.6 (Inversionslemma).
a) *Aus* $\mathrm{H}^\beta_\gamma(X) \overset{\alpha}{\underset{\rho}{\vdash}} \mathfrak{N}[(A \to B)]$ *folgt* $\mathrm{H}^\beta_\gamma(X) \overset{\alpha}{\underset{\rho}{\vdash}} \mathfrak{N}[\neg A]$ *und* $\mathrm{H}^\beta_\gamma(X) \overset{\alpha}{\underset{\rho}{\vdash}} \mathfrak{N}[B]$.
b) *Aus* $\mathrm{H}^\beta_\gamma(X) \overset{\alpha}{\underset{\rho}{\vdash}} \mathfrak{N}[(s \in t)]$ *mit* $\mathrm{S}s < \mathrm{K}$ *und* $\mathrm{S}u < \mathrm{S}t < \mathrm{K}$ *folgt* $\mathrm{H}^\beta_\gamma(X \cup \mathrm{k}(u)) \overset{\alpha}{\underset{\rho}{\vdash}} \mathfrak{N}[(t(u) \wedge (s = u))]$.
c) *Aus* $\mathrm{H}^\beta_\gamma(X) \overset{\alpha}{\underset{\rho}{\vdash}} \mathfrak{N}[\mathrm{Ad}^\mu(t)]$ *mit* $\eta \leq \mathrm{S}t < \mathrm{K}$ *und* $\eta \in \mathrm{M}^\mu$ *folgt* $\mathrm{H}^\beta_\gamma(X \cup \{\eta\}) \overset{\alpha}{\underset{\rho}{\vdash}} \mathfrak{N}[(\mathrm{L}\eta = t)]$.
d) *Aus* $\mathrm{H}^\beta_\gamma(X) \overset{\alpha}{\underset{\gamma}{\vdash}} \mathfrak{N}[(\exists x \in t)\mathfrak{A}[x]]$ *mit* $\mathrm{S}u < \mathrm{S}t$ *folgt* $\mathrm{H}^\beta_\gamma(X \cup \mathrm{k}(u)) \overset{\alpha}{\underset{\rho}{\vdash}} \mathfrak{N}[(t(u) \wedge \mathfrak{A}[u])]$.

Lemma 25.7.7 (Beschränkungslemma).
a) *Aus* $\mathrm{H}^\beta_\gamma(X) \overset{\alpha}{\underset{\gamma}{\vdash}} \mathfrak{P}[(\exists x \in \mathrm{L}\delta)\mathfrak{A}[x]]$ *mit* $\alpha \leq \mu < \delta$ *und* $\mu \in \mathrm{H}^\beta_\gamma(X)$ *folgt* $\mathrm{H}^\beta_\gamma(X) \overset{\alpha}{\underset{\rho}{\vdash}} \mathfrak{P}[(\exists x \in \mathrm{L}u)\mathfrak{A}[x]]$.
b) *Aus* $\mathrm{H}^\beta_\gamma(X) \overset{\alpha}{\underset{\gamma}{\vdash}} \mathfrak{N}[(\exists x \in \mathrm{L}\delta)\mathfrak{A}[x]]$ *mit* $\mu < \delta$ *und* $\mu \in \mathrm{H}^\beta_\gamma(X)$ *folgt* $\mathrm{H}^\beta_\gamma(X) \overset{\alpha}{\underset{\gamma}{\vdash}} \mathfrak{N}[(\exists x \in \mathrm{L}\mu)\mathfrak{A}[x]]$.

25.8 Einbettung von KP+ Π_3-Ref in RS(K)

Induktive Definition der *Norm* $\mathrm{no}(F)$ einer RS(K)-Formel F.
1. Für jede einfache Formel A sei $\mathrm{no}(A) := \omega^{\mathrm{rn}(A)}$.
2. $\mathrm{no}(\neg A) := \mathrm{no}(A)$.
3. $\mathrm{no}(A \to B) := \mathrm{no}(A) \,\#\, \mathrm{no}(B)$, wenn B nicht die Formel $\bot$ ist, wobei # für die natürliche Summe steht.

Bezeichnung. $\Vdash F$ bedeute, daß $\mathrm{H}^0_0(\mathrm{k}(F)) \overset{\mathrm{no}(F)}{\underset{0}{\vdash}} F$ gilt.

Lemma 25.8.1.
a) $\Vdash F \to F$.
b) $\Vdash t = t$ *für* $\mathrm{S}t < \mathrm{K}$.
c) $\Vdash t \in \mathrm{L}\alpha$ *für* $\mathrm{S}t < \alpha < \mathrm{K}$.
d) $\Vdash \mathrm{Tran}[\mathrm{L}\alpha]$ *für* $\alpha < \mathrm{K}$.
e) $\Vdash \mathrm{Ad}^\mu(\mathrm{L}\eta)$ *für* $\eta \in \mathrm{H}^\mu$.
f) $\Vdash \neg(t \in t)$ *für* $\mathrm{S}t < \mathrm{K}$.

Lemma 25.8.2 (Extensionalität).
Für alle Terme $s_1, \ldots, s_n, t_1, \ldots, t_n$ *von Schichten* $< \mathrm{K}$ *gilt*
$\Vdash (s_1 = t_1 \wedge \ldots \wedge s_n = t_n) \to (\mathfrak{F}[s_1, \ldots, s_n] \to \mathfrak{F}[t_1, \ldots t_n])$.

Lemma 25.8.3 (Fundierung).
$\Vdash (\forall x \in \mathrm{LK})((\forall y \in x)\mathfrak{A}[y] \to \mathfrak{A}[x]) \to (\forall x \in \mathrm{LK})\mathfrak{A}[x]$.

Definition. Unter einer RS(K)-*Interpretation* einer Formel A des Systems KP + Π_3-Ref verstehen wir eine Formel des Systems RS(K), die aus der Formel A dadurch entsteht, daß für jede in A auftretende freie Variable ein Term von einer Schicht $<$ K eingesetzt wird und jeder in A auftretende unbeschränkte Quantor $\exists x$ durch $(\exists x \in \mathrm{LK})$ ersetzt wird.

Lemma 25.8.4 (Einbettungslemma).
Zu jeder in KP + Π_3-Ref3 *herleitbaren Formel A gibt es* $m, n < \omega$ *derart, daß* $\mathrm{H}_0^0(\mathrm{k}(F)) \mathrel{\Big|\!\!\frac{\mathrm{K}\cdot\varphi 0m}{K+n}} F$ *für jede* RS(K)-*Interpretation F von A gilt.*

25.9 Herleitungsreduktionen in RS(K)

Lemma 25.9.1 (Prädikative Schnitt-Elimination).

a) *Ist C eine einfache Formel vom Rang* $\rho \notin \mathrm{R}$ *und gilt* $\mathrm{H}_\gamma^\beta(\mathrm{K}) \mathrel{\Big|\!\!\frac{\alpha}{\rho}} C \to F$ *und* $\mathrm{H}_\gamma^\beta(X) \mathrel{\Big|\!\!\frac{\delta}{\rho}} C \vee G$ *mit* $\mathrm{k}(C) < \alpha + 1$, *so folgt* $\mathrm{H}_\gamma^\beta(X) \mathrel{\Big|\!\!\frac{\alpha+0}{\rho}} F \vee G$.
b) *Aus* $\mathrm{H}_\gamma^\beta(X) \mathrel{\Big|\!\!\frac{\alpha}{\rho+1}} F$ *mit* $\rho \notin \mathrm{R}$ *folgt* $\mathrm{H}_\gamma^\beta(X) \mathrel{\Big|\!\!\frac{\varphi 0\alpha}{\rho}} F$.
c) *Gilt* $\mathrm{H}_\gamma^\beta(X) \mathrel{\Big|\!\!\frac{\delta}{\mu+\varphi 0\alpha}} F$ *mit* $\alpha, \delta < \mathrm{K}$ *und* $\alpha \in \mathrm{H}_\gamma^\beta(X)$, *wobei es kein* $\zeta \in \mathrm{R}$ *mit* $\mu \le \zeta < \mu + \varphi 0\alpha$ *gibt, so folgt* $\mathrm{H}_\gamma^\beta(X) \mathrel{\Big|\!\!\frac{\varphi\alpha\delta}{\mu}} F$.

Beweis von a) durch Induktion nach δ.
b) ergibt sich mit a) durch Induktion nach α.
c) ergibt sich mit b) durch Hauptinduktion nach α und Nebeninduktion nach δ. □

Induktive Definition der einfachen Formeln, die in einer Formel des Systems RS(K) *positiv* oder *negativ* auftreten.

1. In einer einfachen Formel F tritt nur F positiv und keine einfache Formel negativ auf.
2. In einer Formel $(A \to B)$ tritt eine einfache Formel genau dann positiv (negativ) auf, wenn sie in A negativ (positiv) oder in B positiv (negativ) auftritt.

Definitionen für $\alpha \in \mathrm{R}$.

1. Eine RS(K)-Formel F heiße eine $\Delta_0(\alpha)$-*Formel*, wenn $\mathrm{k}(F) < \alpha$ ist.
2. Eine RS(K)-Formel F heißt eine $\Sigma_3(\alpha)$-Formel ($\Pi_3(\alpha)$-Formel), wenn jede einfache Formel, die in F positiv (negativ) auftritt, eine Formel $(\exists x \in \mathrm{L}\alpha)(\forall y \in \mathrm{L}\alpha)(\exists z \in \mathrm{L}\alpha)\mathfrak{C}[x, y, z]$ oder $(\exists z \in \mathrm{L}\alpha)\mathfrak{A}[z]$ oder eine $\Delta_0(\alpha)$-Formel ist und jede einfache Formel, die in F negativ (positiv) auftritt, eine Formel $(\exists y \in \mathrm{L}\alpha)\neg(\exists z \in \mathrm{L}\alpha)\mathfrak{B}[y, z]$ oder $(\exists z \in \mathrm{L}\alpha)\mathfrak{A}[z]$ oder eine $\Delta_0(\alpha)$-Formel ist, wobei $\mathfrak{A}[\mathrm{L}0]$, $\mathfrak{B}[\mathrm{L}0, \mathrm{L}0]$ und $\mathfrak{C}[\mathrm{L}0, \mathrm{L}0, \mathrm{L}0]$ $\Delta_0(\alpha)$-Formeln sind.

3. Für eine RS(K)-Formel F sei $F^{(t,\alpha)}$ diejenige Formel, die aus der Formel F dadurch entsteht, daß jeder darin auftretende Quantor $(\exists x \in \mathrm{L}\alpha)$ durch $(\exists x \in t)$ ersetzt wird.

Lemma 25.9.2. *Ist F eine $\Sigma_3(\pi)$-Formel und gilt* $\mathrm{H}_\gamma^\beta(X) \stackrel{\alpha}{\underset{\rho}{\vdash}} F \vee G$ *mit* $\pi < \alpha = \varphi 0\alpha$, $\pi + \omega < \rho$, $0 < \mu \in \mathrm{B}(\mathrm{h}(\pi), \pi) \cap \mathrm{h}(\pi)$ *und* $\mu \in \mathrm{H}_\gamma^\beta(X)$, *so folgt* $\mathrm{H}_\gamma^\beta(X) \stackrel{\alpha+2}{\underset{\rho}{\vdash}} (\exists w \in \mathrm{L}\pi)(\mathrm{Ad}^\mu(w) \wedge F^{(w,\pi)}) \vee G$.

Beweis. Die $\Sigma_3(\pi)$-Formel F hat eine pränexe Normalform A der Gestalt $(\exists x \in \mathrm{L}\pi)$ $(\forall y \in \mathrm{L}\pi)(\exists z \in \mathrm{L}\pi)\mathfrak{C}[x, y, z]$ mit $\mathrm{k}(\mathfrak{C}[\mathrm{L}0, \mathrm{L}0, \mathrm{L}0]) < \pi$. Mit den Voraussetzungen
(1) $\mathrm{H}_\gamma^\beta(X) \stackrel{\alpha}{\underset{\rho}{\vdash}} F \vee G$, $\pi < \alpha = \varphi 0\alpha$ und $\mu \in \mathrm{B}(\mathrm{h}(\pi), \pi) \cap \mathrm{h}(\pi)$ erhält man
(2) $\mathrm{H}_\gamma^\beta(X) \stackrel{\alpha}{\underset{\rho}{\vdash}} F \rightarrow A$ und
(3) $\mathrm{H}_\gamma^\beta(X) \stackrel{\alpha}{\underset{\rho}{\vdash}} (\exists w \in \mathrm{L}\pi)(\mathrm{Ad}^\mu(w) \wedge \mathrm{Ad}^{(w,\pi)}) \rightarrow (\exists w \in \mathrm{L}\pi)(\mathrm{Ad}^\mu(w) \wedge F^{(w,\pi)})$.
Mit Lemma 25.8.1 a), (Ref_π^μ) und $(\mathrm{N}.\exists)$ erhält man auch
(4) $\mathrm{H}_\gamma^\beta(X) \stackrel{\alpha}{\underset{\rho}{\vdash}} A \rightarrow (\exists w \in \mathrm{L}\pi)(\mathrm{Ad}^\mu(w) \wedge A^{(w,\pi)})$.
Aufgrund von $\pi + \omega < \rho$ folgt die Behauptung aus (1) und (4) mit dem Strukturschlußlemma und Schnitten. □

Definitionen.
1. $\mathrm{NF}(\alpha, \beta)$ bedeute, daß entweder $\alpha = 0$ ist oder $\eta + \beta < \alpha + \beta$ für alle $\eta < \alpha$ gilt.
2. $\underline{\mathrm{B}}(X, \beta)$ bedeute, daß $\beta \in \mathrm{H}_0^\beta(X)$ und $x \subset \mathrm{B}(\beta + 1, \Xi(\beta + 1))$ gilt.

Lemma 25.9.3. *Unter den Voraussetzungen* $\underline{\mathrm{B}}(X, \beta)$, $\alpha \in \mathrm{H}_0^\beta(X)$, $\mathrm{NF}(\beta, \mathrm{K}^\alpha)$ *und* $\mu \in \mathrm{M}^{\beta+\mathrm{K}^\alpha}$ *gilt folgendes:*
a) $\mathrm{H}_0^\beta(X) \cap \mathrm{K} < \Xi(\beta + 1)$.
b) $\Xi(\beta + \mathrm{K}^\alpha + \pi) \in \mathrm{H}_0^{\beta+\mathrm{K}^\alpha+\pi}(X \cup \{\pi\})$.
c) *Für* $\delta < \alpha$ *mit* $\delta \in \mathrm{H}_0^\beta(X)$ *ist* $\Xi(\beta + \mathrm{K}^\delta + \pi) < \Xi(\beta + \mathrm{K}^\alpha + \pi)$.
d) *Für* $\mathrm{S}u \leq f(u) < \alpha, f(u) \in \mathrm{H}_0^\beta(X \cup \mathrm{k}(u))$ *und* $g(u) = \beta + \omega^{\mathrm{K}\cdot f(u)+\mathrm{S}u}$ *gilt* $\underline{\mathrm{B}}(X \cup \mathrm{K}(u), g(u))$ *und* $g(u) + \mathrm{K}^{f(u)} \in \mathrm{H}_0^{g(u)}(X \cup \mathrm{k}(u))$.
Für $\mathrm{S}u < \pi$ *gilt auch* $\Xi(g(u) + \mathrm{K}^{f(u)} + \pi) < \Xi(\beta + \mathrm{K}^\alpha + \pi)$ *und* $\pi \in \mathrm{M}^{g(u)+\mathrm{K}^{f(u)}}$.

Lemma 25.9.4 (Erste imprädikative Schnitt-Elimination mit Kollabierung). *Gilt* $\mathrm{H}_0^\beta(X) \stackrel{\alpha}{\underset{\mathrm{K}+1}{\vdash}} F$ *für eine* $\Pi_3(\mathrm{K})$*-Formel* F *mit* $\mathrm{NF}(\beta, \mathrm{K}^\alpha)$ *und* $\underline{\mathrm{B}}(X, \beta)$, *so folgt* $\mathrm{H}_0^{\beta+\mathrm{K}^\alpha+\pi}(X \cup \{\pi\}) \stackrel{\Xi(\beta+\mathrm{K}^\alpha+\pi)}{\underset{\Xi(\beta+\mathrm{K}^\alpha+\pi)}{\vdash}} F^{\mathrm{L}\pi,\mathrm{K}}$ *für alle* $\pi \in \mathrm{M}^{\beta+\mathrm{K}^\alpha}$.

Beweis durch Induktion nach α mit Lemma 25.9.3. Wir betrachten nur den Fall, daß $\mathrm{H}_0^\beta(X) \stackrel{\alpha}{\underset{\mathrm{K}+1}{\vdash}} F$ durch einen Hauptschluß $(\mathrm{Ref}_\mathrm{K})$ erschlossen ist. Dann ist F eine Formel $\mathfrak{P}[C]$, und man hat $\mathrm{K} \leq \delta < \alpha$ mit $\mathrm{H}_0^\beta(X) \stackrel{\delta}{\underset{\mathrm{K}+1}{\vdash}} B \vee F$, wobei B eine Formel $(\forall x \in \mathrm{LK})(\exists y \in \mathrm{LK})(\forall z \in \mathrm{LK})\mathfrak{C}[x, y, z]$ mit $\mathrm{k}(\mathfrak{C}[\mathrm{L}0, \mathrm{L}0, \mathrm{L}0]) < \mathrm{K}$ und C die Formel $(\exists w \in \mathrm{LK})(\mathrm{Tran}[w] \wedge (\exists x \in w)(x \in w) \wedge \mathrm{B}^{(w,\mathrm{K})}$ ist. Nach I.V. folgt

$$\mathrm{H}_0^{\beta+\mathrm{K}^\delta+\sigma}(X \cup \{\sigma\}) \stackrel{\Xi(\beta+\mathrm{K}^\delta+\sigma)}{\underset{\Xi(\beta+\mathrm{K}^\delta+\sigma)}{\vdash}} B^{(\mathrm{L}\sigma,\mathrm{K})} \vee F^{(L_\sigma,\mathrm{K})} \text{ für alle } \sigma \in \mathrm{M}^{\beta+\mathrm{K}^\delta} \cap \pi.$$

Mit $\Vdash \mathrm{Tran}[\mathrm{L}\sigma]$, $\Vdash (\exists x \in \mathrm{L}\sigma)(x \in \mathrm{L}\sigma)$, Lemma 25.7.5 a) und (P.$\exists$) folgt

$$\mathrm{H}_0^{\beta+\mathrm{K}^\alpha+\pi}(X \cup \{\pi,\sigma\}) \mathrel{\Big|\!\frac{\Xi(\beta+\mathrm{K}^\delta+\sigma)+\omega}{\Xi(\beta+\mathrm{K}^\alpha+\pi)}} C^{\mathrm{L}\pi,\mathrm{K}} \vee F^{(\mathrm{L}\sigma,\mathrm{K})}.$$

Nach dem Strukturschlußlemma folgt

$$\mathrm{H}_0^{\beta+\mathrm{K}^\alpha+\pi}(X \cup \{\pi,\sigma\}) \mathrel{\Big|\!\frac{\Xi(\beta+\mathrm{K}^\delta+\sigma)+\omega}{\Xi(\beta+\mathrm{K}^\alpha+\pi)}} F^{\mathrm{L}\sigma,\mathrm{K}} \vee F^{(\mathrm{L}\pi,\mathrm{K})}.$$

Für $\sigma \leq \mathrm{S}u < \pi$ folgt mit Lemma 25.8.2 und einem Schnitt

$$\mathrm{H}_0^{\beta+\mathrm{K}^\alpha+\pi}(X \cup \{\pi,\sigma\} \cup \mathrm{k}(u)) \mathrel{\Big|\!\frac{\Xi(\beta+\mathrm{K}^\delta+\mathrm{S}u)+\omega+1}{\Xi(\beta+\mathrm{K}^\alpha+\pi)}} ((\mathrm{L}\sigma = u) \to F^{(u,\mathrm{K})}) \vee F^{(\mathrm{L}\pi,\mathrm{K})}.$$

Für $\mu := \beta + \mathrm{K}^\delta$ folgt mit (Ax.Ad$^\mu$) und (N.Ad$^\mu$)

$$\mathrm{H}_0^{\beta+\mathrm{K}^\alpha+\pi}(X \cup \{\pi\} \cup \mathrm{k}(u)) \mathrel{\Big|\!\frac{\Xi(\beta+\mathrm{K}^\delta+\beta u)+\omega+2}{\Xi(\beta+\mathrm{K}^\alpha+\pi)}} (\mathrm{Ad}^\mu(u) \wedge \neg F^{(u,\mathrm{K})}) \to F^{(\mathrm{L}\pi,\mathrm{K})}.$$

Mit (N.$\exists$) folgt

$$\mathrm{H}_0^{\beta+\mathrm{K}^\alpha+\pi}(X \cup \{\pi\}) \mathrel{\Big|\!\frac{\Xi(\beta+\mathrm{K}^\delta+\pi)}{\Xi(\beta+\mathrm{K}^\alpha+\pi)}} (\exists w \in \mathrm{L}\pi)(\mathrm{Ad}^\mu(w) \wedge \neg F^{(w,\mathrm{K})}) \to F^{(\mathrm{L}\pi,\mathrm{K})}.$$

$\neg F^{(\mathrm{L}\pi,\mathrm{K})}$ ist eine $\Sigma_3(\pi)$-Formel. Aus $\Vdash \neg F^{\mathrm{L}\pi,\mathrm{K}} \vee F^{(\mathrm{L}\pi,\mathrm{K})}$ folgt daher mit Lemma 25.9.2

$$\mathrm{H}_0^{\beta+\mathrm{K}^\alpha+\pi}(X \cup \{\pi\}) \mathrel{\Big|\!\frac{\Xi(\beta+\mathrm{K}^\delta+\pi)}{\Xi(\beta+\mathrm{K}^\alpha+\pi)}} (\exists w \in \mathrm{L}\pi)(\mathrm{Ad}^\mu(w) \wedge \neg F^{(w,\mathrm{K})}) \vee F^{(\mathrm{L}\pi,\mathrm{K})}.$$

Mit einem Schnitt folgt die Behauptung. □

Definitionen.
1. Kard sei die Menge der Ordinalzahlen Ω_η mit $0 < \eta < \mathrm{K}$.
2. Für $\kappa \in$ Kard sei $\kappa' := \kappa + 1$, wenn $\kappa \in \mathrm{R}$ ist, sonst $\kappa' := \kappa$.
3. Für $\mathrm{h}(\pi) = 0$ und $\alpha < \mathrm{K}$ sei $\psi_\pi^0\alpha := \psi\pi\alpha$.
4. $\underline{\mathfrak{A}}(X,\beta,\gamma,\pi,\mu,\kappa)$ bedeute, daß folgendes gilt:
   $\beta,\gamma,\pi,\mu,\kappa \in \mathrm{H}_\gamma^\beta(X)$, $\kappa \in$ Kard, $\pi \leq \kappa$, $X \subset \mathrm{C}(\gamma+1,\psi_\sigma^0(\gamma+1))$ für alle $\sigma \geq \pi$ und $\mu \in \mathrm{B}(\mathrm{h}(\pi),\pi) \cap \mathrm{h}(\pi)$ oder $\mu = \mathrm{h}(\pi) = 0$.

Lemma 25.9.5. *Unter den Voraussetzungen* $\underline{\mathfrak{A}}(X,\beta,\gamma,\pi,\mu,\kappa)$ *und* $\alpha \in \mathrm{H}_\gamma^\beta(X) \cap \mathrm{K}$ *gilt folgendes:*

a) $\psi_\pi^\mu(\gamma+\kappa^\alpha) \in \mathrm{H}_{\gamma+\kappa}^\beta\alpha(X)$ *und* $\psi_\pi^\mu(\gamma+\kappa^\alpha) \in \mathrm{M}^\mu \cap \pi$.
b) *Für alle* $\sigma \geq \pi$ *gilt* $\mathrm{H}_\gamma^\beta(X) \subseteq \mathrm{C}(\gamma+1,\psi_\sigma^0(\gamma+1))$.
c) *Für* $\delta < \alpha$ *mit* $\delta \in \mathrm{H}_\gamma^\beta(X)$ *ist* $\psi_\pi^\mu(\gamma+\kappa^\delta) < \psi_\pi^\mu(\gamma+\kappa^\alpha)$.
d) *Für* $\mathrm{S}u < \pi$ *und* $g(u) = \gamma + \omega^{\kappa\cdot\alpha+\mathrm{S}u}$ *gilt* $\underline{\mathfrak{A}}(X \cup \mathrm{k}(u),\beta,g(u),\pi,\mu,\kappa)$.

Definition. Eine RS(K)-Formel F heiße eine $\Sigma_1(\pi)$-*Formel*, wenn jede einfache Formel, die in F positiv auftritt, eine Formel $(\exists x \in \mathrm{L}\pi)\mathfrak{A}[x]$ mit $\mathrm{k}(\mathfrak{A}[\mathrm{L}0]) < \pi$ oder eine $\Delta_0(\pi)$-Formel ist und jede einfache Formel, die in F negativ auftritt, eine $\Delta_0(\pi)$-Formel ist.

Lemma 25.9.6 (Zweite imprädikative Schnitt-Elimination mit Kollabierung)**.** *Gilt* $\underline{\mathfrak{A}}(X, \beta, \gamma, \pi, \mu, \kappa)$, $\alpha < \kappa$ *und* $\mathrm{H}^\beta_\gamma(X) \,\frac{\alpha}{\kappa'}\, F$ *für eine* $\Sigma_1(\pi)$-*Formel* F, *so folgt* $\mathrm{H}^\beta_{\gamma+\kappa^\alpha}(X) \,\frac{\psi^\mu_\pi(\gamma+\kappa^\alpha)}{\psi^\mu_\pi(\gamma+\kappa^\alpha)}\, F$.

Beweis durch Hauptinduktion nach κ und Neeninduktion nach α mit Lemma 25.9.5 Wir betrachten nur den Fall, daß $\mathrm{H}^\beta_\gamma(X) \,\frac{\alpha}{\kappa'}\, F$ durch einen Hauptschluß (Ref^μ_π) erschlossen ist. Dann ist F eine Formel $\mathfrak{P}[(\exists w \in \mathrm{L}\pi)(\mathrm{Ad}^\mu(w) \wedge B^{(w,\pi)})]$, wobei B eine Formel $(\exists x \in \mathrm{L}\pi)(\forall y \in \mathrm{L}\pi)(\exists z \in \mathrm{L}\pi)\mathrm{C}[x, y, z]$ mit $\mathrm{k}(\mathrm{C}[\mathrm{L}0, \mathrm{L}0, \mathrm{L}0]) < \kappa$ ist, und man hat

(1) $$\mathrm{H}^\beta_\gamma(X) \,\frac{\delta}{\kappa'}\, (\forall y \in \mathrm{L}\pi)(\exists z \in \mathrm{L}\pi)\mathrm{C}[t, y, z] \vee F$$

mit $\mathrm{S}t < \pi$ und $\pi < \delta + 1 < \alpha$. Tritt t in (1) nicht auf, so können wir $t = \mathrm{L}0$ annehmen. Dann hat man in jedem Fall

(2) $$\mathrm{S}t \in \mathrm{H}^\beta_\gamma(X) \cap \pi \subseteq \mathrm{C}(\gamma + 1, \psi^0_\pi(\gamma + 1)) \cap \pi < \psi^0_\pi(\gamma + 1).$$

Für alle u mit $\mathrm{S}u < \pi$ folgt aus (1) durch Inversion

$$\mathrm{H}^\beta_\gamma(X \cup \mathrm{k}(u)) \,\frac{\delta}{\kappa'}\, (\neg(u \in \mathrm{L}0) \to (\exists z \in \mathrm{L}\pi)\mathrm{C}[t, u, z]) \vee F.$$

Für $g(u) := \gamma + \omega^{\kappa\cdot\delta+\mathrm{S}u}$ und $f(u) := \psi^\mu_\pi(g(u) + \kappa^\delta)$ folgt nach der N.I.V.

$$\mathrm{H}^\beta_{g(u)+\kappa^\delta}(X \cup \mathrm{k}(u)) \,\frac{f(u)}{f(u)}\, (\neg(u \in \mathrm{L}0) \to (\exists z \in \mathrm{L}\pi)\mathrm{C}[t, u, z]) \vee F.$$

Für $\eta := \psi^\mu_\pi(\gamma + \kappa^{\delta+1})$ folgt nach Lemma 25.7.7 a)

$$\mathrm{H}^\beta_{\gamma+\kappa^\alpha}(X \cup \mathrm{k}(u)) \,\frac{f(u)}{f(u)}\, (\neg(u \in \mathrm{L}0) \to (\exists z \in \mathrm{L}\eta)\mathrm{C}[t, u, z]) \vee F.$$

Nach Lemma 25.7.5 c) folgt

$$\mathrm{H}^\beta_{\gamma+\kappa^\alpha}(X) \,\frac{\eta}{\eta}\, (\forall y \in \mathrm{L}\eta)(\exists z \in \mathrm{L}\eta)\mathrm{C}[t, y, z] \vee F.$$

Aus (2) folgt $\mathrm{S}t < \eta$. Mit $\Vdash \neg(t \in \mathrm{L}0)$, Lemma 25.7.5 a) und (P.$\exists$) folgt

$$\mathrm{H}^\beta_{\gamma+\kappa^\alpha}(X) \,\frac{\eta+2}{\eta}\, \mathrm{B}^{(\mathrm{L}\eta,\pi)} \vee F.$$

Mit $\Vdash \mathrm{Ad}^\mu(\mathrm{L}\eta)$, $\Vdash \neg(\mathrm{L}\eta \in \mathrm{L}0)$, Lemma 25.7.5 a) und (P.$\exists$) folgt die Behauptung. □

Lemma 25.9.7. *Ist eine* RS(K)*-Interpretation F einer in* KP+ Π_3-Ref *herleitbaren* Δ_0*-Formel eine* $\Sigma_1(\Omega_1)$*-Formel mit* $\mathrm{k}(F) \subset \mathrm{H}_0^0(\emptyset)$*, so gibt es* $\alpha < \Omega_1$*,* β *und* $\gamma < \mathrm{K}$ *mit* $\mathrm{H}_\gamma^\beta(\emptyset) \,\frac{\alpha}{0}\, F$.

Beweis. Nach Lemma 25.8.4 gibt es $m, n < \omega$ mit $\mathrm{H}_0^0(\emptyset) \,\frac{\mathrm{K}\cdot\varphi 0m}{\mathrm{K}+n}\, F$.
Dann gibt es nach Lemma 25.9.1 b) ein δ mit $\mathrm{H}_0^0(\emptyset) \,\frac{\delta}{\mathrm{K}+1}\, F$.
Für $\beta := \mathrm{K}^\delta + \Xi(\mathrm{K}^\delta)$ folgt nach Lemma 25.9.4 $\mathrm{H}_0^\beta(\emptyset) \,\frac{\Xi(\beta)}{\Xi(\beta)}\, F$.
Für $\gamma := \Xi(\beta)^{\Xi(\beta)} < \mathrm{K}$ folgt nach Lemma 25.9.6 $\mathrm{H}_\gamma^\beta(\emptyset) \,\frac{\psi\Omega_1\gamma}{\psi\Omega_1\gamma}\, F$.
Für $\alpha := \varphi(\psi\Omega_1\gamma)(\psi\Omega_1\gamma) < \Omega_1$ folgt nach Lemma 25.9.1 c) $\mathrm{H}_\gamma^\beta(\emptyset) \,\frac{\alpha}{0}\, F$. □

Theorem. *Die beweistheoretische Ordinalzahl des formalen Systems* KP+ Π_3-Ref *ist* $\leq \psi\Omega_1(\psi^0_{\Xi(\varepsilon_{\mathrm{K}+1})}(0))$.

Beweis mit den Lemmata 25.5.7 und 25.9.7. □

Literatur

1. M. Rathjen. *Eine Ordinalzahlanalyse der* Π_3*-Reflexion*. Habilitationsschrift, Mathematik, Universität Münster, 1992.
2. M. Rathjen. Proof theory of reflection. *Annals of Pure and Applied Logic 68*, pages 181–224, 1994.

The manufacturer's authorised representative in the EU is Springer Nature Customer Service Centre GmbH, Europaplatz 3, 69115 Heidelberg, Germany. If you have any concerns regarding our products, please contact ProductSafety@springernature.com

Printed and bound by CPI Group (UK) Ltd, Croydon, CR0 4YY
30/09/2026
02225650-0019